Springer Praxis Books

Astronautical Engineering

More information about this series at http://www.springer.com/series/5495

Alex Ellery

Planetary Rovers

Robotic Exploration of the Solar System

 Springer

Alex Ellery
Department of Mechanical and Aerospace
  Engineering
Carleton University
Ottawa, Ontario
Canada

Published in association with Praxis Publishing, Chichester, UK

ISSN 2365-9599   ISSN 2365-9602   (electronic)
Springer Praxis Books
ISBN 978-3-642-03258-5   ISBN 978-3-642-03259-2   (eBook)
DOI 10.1007/978-3-642-03259-2

Library of Congress Control Number: 2015959579

Printed on acid-free paper

This Springer imprint is published by SpringerNature
The registered company is Springer-Verlag GmbH Berlin Heidelberg

Contents

Preface

As space exploration missions evolve from planetary flybys and orbiters towards in situ surface missions, in particular taking advantage of the mobility offered by rovers, so it seems appropriate to provide a textbook to bring together the many disparate aspects of planetary rover technology into a single source. Of course, I cannot claim any pretext for comprehension as such a text would be impossible in a volume of this size. However, I have attempted to provide a wide coverage to entail most of the major robotics aspects associated with rovers per se without sacrificing depth. Furthermore, although I have attempted to review the literature thoroughly, there will inevitably be unintentional omissions—for these I apologize. Please feel free to provide me with omitted information that may be incorporated in future editions.

The target audience for this book is anyone who requires an intimate and detailed technical knowledge of planetary rovers and how they are likely to evolve in their capabilities in the near future. For the most part that would mean primarily graduate students, final year undergraduates, and industry engineers who are involved with planetary rover development. This text has grown from a course I teach at Carleton University on planetary rovers based on my experiences through the years ranging from general spacecraft development at several institutions both industrial and academic, laboratory and computer modeling of mobile robotics projects with the Surrey Space Centre (U.K.), the ExoMars Rover program in the U.K. to the design and development of the major robotics systems of the Kapvik microrover prototype for the Canadian Space Agency. This latter vehicle was designed as an analog platform from the ground up for planetary exploration (as opposed to retro-fitted) with a clear path to flight qualification. Although administered by MPB Communications of Montreal as prime contractor, my team's role at Carleton University was as the primary technical authority in providing the design and the core mobile robotics aspects—mobility system, motor control avionics, autonomous navigation, vision-processing, camera control, and other motorized units. Indeed, the design had been based on the Vanguard Mars rover micromission developed by the author years before in the U.K. However, these platforms are used as illustrations and the text is applicable to any planetary rover project.

Acknowledgments

I would like to thank my former and current colleagues but in particular those at Surrey Space Centre which had a profound effect on my education as a spacecraft engineer showing that even an experienced engineer can learn a great deal. I would like to thank my former and current graduate students and postdoctoral researchers who have supplied many of the images presented including Dr. Yang Gao, Dr. Ala Qadi, Dr. Nildeep Patel, Dr. Elie Allouis, Dr. Brian Lynch, Adam Mack, Tim Hopkins, Tim Setterfield, Cameron Frazier, Marc Gallant, Rob Hewitt, Chris Nicol, Matt Cross, Helia Sharif, Mark Swarz, Jesse Hiemstra, Adam Brierley, and others. I apologize to anyone I have inadvertently omitted.

Alex Ellery
Space Exploration Engineering Group (SEEG)
Carleton University, Ottawa

Figures

Tables

Acronyms and abbreviations

A-DREAMS	Advanced DREAMS
ACWP	Actual Cost of Work Performed
ADC	Acoustic Digital Current
ADC	Aide-De-Camp
ADC	Analog–Digital Converter 152
ADC	Analog-to-Digital Conversion 161
ALU	Arithmetic Logic Unit
AMCW	Amplitude Modulated Continuous Wave
APS	Active Pixel Sensor
APXS	Alpha Particle X-ray Spectrometer 6
APXS	Alpha Proton X-ray Spectrometer 421
ARM	Anthropomorphic Robotic Manipulator 423
ARM	Articulated Robotic Manipulator 134
ARMA	AutoRegressive Moving Average
ASIC	Application-Specific Integrated Circuit
ASM	Augmented Finke Statement Machine
ASPEN	Automated Scheduling and Planning ENvironment
ATHLETE	All-Terrain Hex-Legged Extra-Terrestrial Explorer
ATP	Adenosine triphosphate
AUDOS	Aberdeen University Deep Ocean Submersible
AuRA	Autonomous Robot Architecture
AUV	Autonomous Underwater Vehicle
AVATAR	Autonomous Vehicle Aerial Tracking and Reconnaissance
AWG	American Wire Gauge
AXS	Alpha-X-ray Spectrometer
BACON	General scientific law
BCWP	Budgeted Cost for Work Performed
BCWS	Budgeted Cost for Work Scheduled

BD   Bayesian Dirichlet
BEES   Bio-inspired Engineering of Exploration Systems
BF   Boundary Following
BISMARC   Biologically Inspired System for Map-based Autonomous
                Rover Control
BN   Bayesian Network
BRISK   Binary Robust Invariant Scalable Keypoint
CAMPOUT   Control Architecture for Multirobot Planetary OUTpost
CAN   Controller Area Network
CANX   CANadian Advanced Nanospace eXperiment
CASPER   Continuous Activity Scheduling Planning and Replanning 325
CASPER   Continuous Activity Scheduling Planning Execution and
                Replanning 413
CCD   Charge Coupled Device
CCIR   Consultative Committee on International Radio
CDH   Command and Data Handling
CEBOT   CEllular roBOT
CELSS   Closed-loop Environmental Life Support System
CER   Cost–Estimation Relationship
CESAR   Common controller for European Space Automation and
                Robotics
CFRP   Carbon Fiber Reinforced Polymer
CG   Center of Gravity
CHIMRA   Collection and Handling for In situ Martian Rock Analysis
CHIRPS   Cryo-Hydro Integrated Robotic Penetrator System
CI   Cone Index
CID   Charge Injection Device
CKF   Cubature Kalman Filter
CLL   488
CMDL   Conditional Minimal Description Length
CMOS   Complementary Metal Oxide Semiconductor
CNES   Centre National d'Etudes Spatiales
CogAff   Cognition and Affect
COTS   Commercial Off The Shelf
CPM   Critical Path Method
CPT   Conditional Probability Table
CPU   Central Processing Unit
CSA   Canadian Space Agency
CTD   Conductivity Temperature and Depth
CV   Cost Variance
DAC   Digital-to-Analog Conversion
DAMN   Distributed Architecture for Mobile Navigation
DARE   Directed Aerial Robot Explorer
DARPA   Defense Advanced Research Projects Agency
DC   Direct Current

DeeDri	Deep Driller
DEM	Digital Elevation Map 204, 305, 331
DEM	Digital Environment Mapping 267
DENDRAL	Mass spectrometer data
DLR	German Aerospace Center
DNA	Deoxyribonucleic acid
DOF	Degree Of Freedom
DOG	Difference-of-Gaussian (function)
DP	Drawbar Pull
DPSK	Differential Phase Shift Keying
DRAM	Dynamic RAM
DREAMS	Dust characterization, Risk assessment, and Environment Analyzer on the Martian Surface
DS2	Deep Space 2
DSP	Digital Signal Processor
DTM	Digital Terrain Map 305
DTM	Digital Terrain Model 331
EDAC	Error Detection and Coding
EDL	Entry Descent and Landing
EDLS	Entry Descent and Landing System
EEPROM	Electrically Erasable Programmable Read Only Memory
EIA	Electronic Industries Alliance
EKF	Enhanced Kalman Filter 282
EKF	Extended Kalman Filter 184
ELF	Extremely Low Frequency
ELMS	Elastic Loop Mobility System
EMF	ElectroMotive Force
EMI	ElectroMagnetic Interference
EPFL	École Polytechnique Fédérale de Lausanne
ESA	European Space Agency
ESOC	European Space Operations Centre
EVA	ExtraVehicular Activity
FAMOUS	Flexible Automation Monitoring & Operation Users Station
FET	Field Effect Transistor
FIDO	Field Integrated Design and Operations
FIR	Far InfraRed
FIRAS	Force Involving Artificial Repulsion from the Surface
FLOPS	FLoating-point Operations Per Second
FMCW	Frequency Modulated Continuous Wave
FOC	Focus Of Contraction
FOE	Focus Of Expansion
FORMID	FORmal robotic Mission Inspection and Debugging (tool)
FOV	Field Of View
FPGA	Field Programmable Gate Array
FREAK	Fast REtinA Keypoint

FSK	Frequency Shift Keying
FSMs	Finite State Machine of registers and timers with output wires
GAP	Gas Analysis Package
GCMS	Gas Chromatography Mass Spectrometer
GCOM	Gray level Co Occurrence Matrix
GESTALT	Grid-based Estimation of Surface Traversibility Applied to Local Terrain
GLCOM	Gray-Level Co-Occurrence Matrix
GNRON	Goal Not Reachable with Obstacles Nearby
GPS	Global Positioning System
GVF	Gradient Vector Flow field
H&S	Health & Safety
HARVE	Unmanned ultra High Altitude Reconnaissance VEhicle
HERRO	Human Exploration using Real-time Robotic Operations
HMM	Hidden Markov Model
HO	Human Operator
HSV	Hue, Saturation, and Value
HUF	HangUp Failure
IARES	Illustrateur Autonome de Robotique mobile pour l'Exploration Spatiale
ICD	Interface Control Document
ICE	Integrated Cryobot Experimental
ICER	Image compressor
IDD	Instrument Deployment Device
IDDS	Inchworm Deep Drilling System
IIR/FIR	Infinite Impulse Response/Finite Impulse Response
IMU	Inertial Measurement Unit
IP	Internet Protocol
IQM	Image Quality Measurement
IRR	Internal Return on Revenue
ISRU	In Situ Resource Utilization
ISS	International Space Station
JPEG	Joint Photographic Experts Group
JPL	Jet Propulsion Laboratory
KREEP	Potassium, Rare Earth Elements, Phosphorus
LAAS	Local Area Augmentation System
LAN	Local Area Network
LAVA	Lunar Advanced Volatile Analysis
LCROSS	Lunar Crater Remote Observation and Sensing Satellite
LED	Light Emitting Diode
LIBS	Laser Induced Breakdown Spectroscopy
LIDAR	LIght Detection And Ranging
LITA	Life In The Atacama
LocSyn	Locomotion Synthesis
LOG	Laplacian of Gaussian (filter)

LRM	Lunar Rover Mockup
LRO	Lunar Reconnaissance Orbiter
LRV	Lunar Rover Vehicle
LUMOT	Locomotion Concepts Analysis for Moon Exploration
LVDS	Low-Voltage Differential Signal
LWF	Line Weight Function
MANTA	Mars Nano/Micro-Technology Aircraft
MAP	Maximum A Posteriori
MARTE	Mars Analog Research and Technology Experiment 460
MARTE	Mars Astrobiology Research and Technology Experiment 447
MASE	Miniature Autonomous Submersible Explorer
MASSIVA	MArs Surface Sampling and Imaging VTOL Aircraft
MAV	Micro-Air Vehicle
MBO	Management By Objectives
MDA	MacDonald, Dettwiler & Associates Ltd.
MECA	Microscopy Electrochemistry and Conductivity Analyzer
MEMS	MicroElectroMechanical System
MER	Mars Exploration Rover
MFP	Mean Free Path
MI	Mobility Index
MICE	Mars Ice Cap Explorer
MIDD	Mobile Instrument Deployment Device
MINERVA	MIcro Nano-Experimental Robot Vehicle for Asteroids
mini-TES	miniature Thermal Emission Spectrometer
MISUS	Multi-rover Integrated Science Understanding System
MITEE	MIniature reacTor EnginE
MLE	Maximum Likelihood Estimation
MLI	MultiLayer Insulation
MMP	Mean Maximum Pressure 97
MMP	Mean Metric Pressure 97
MOLA	Mars Orbiter Laser Altimeter
MPF	Mars PathFinder
MRF	Markov Random Field
MSL	Mars Science Laboratory
MSSL	Mullard Space Science Laboratory
MtG	Motion-to-Goal
MUROCO	Mathematical specification and verification tool
MWD	Measurement While Drilling
NASA	National Aeronautics and Space Administration
NASREM	NASA-NBS Standard REference Model
NBV	Neutral Buoyancy Vehicle
NEAR	Near Earth Asteroid Rendezvous
NEMO	Nuclear Europa Mobile Ocean
NEO	Near-Earth Object
NGP	Nominal Ground Pressure

NIF	Nose-In Failure
NIR	Near InfraRed
NP	Nondeterministic Polynomial
NP-hard	Nondeterministic Polynomial time hard
NPV	Net Present Value
NRMM	NATO Reference Mobility Model
NTSC	National Television System Committee
OASIS	Onboard Autonomous Science Investigation System 469
OASIS	Operations And Science Instrument System 322
OBC	OnBoard Computer
OBS	Organization Breakdown Structure
OKR	OptoKinetic Response
OVEN	Oxygen and Volatile Extraction Node
PAW	Position-Actuated Workbench
PCA	Principal Components Analysis
PCB	Printed Circuit Board
PCI	Peripheral Component Interconnect
PD	Proportional Derivative
PDC	Polycrystalline Diamond Compact
PDF	Probability Density Function
PDM	Precedence Diagram Method
PERT	Program Evaluation and Review Technique
PET	Polyethylene terephthalate
PI	Proportional Integral
PIC	Programmable Interface Controller
PID	Proportional Integral Derivative 525
PID	Proportional, Integral, Derivative 144
PL	Pathfinder Lander
PLUTO	PLanetary Underground TOol
PoE	Power over Ethernet
PPTA	Poly-p-phenyleneterephthalimide
PR	Penetration Rate
PrOP-F	Pribori Otchenki Prokhodimosti-Fobos
PROSPECTOR	Mineral prospecting data
PSK	Phase Shift Keying
PTFE	Polytetrafluoroethylene
PWM	Pulse Width Modulation
PZT	Lead zirconate titanate
QA	Quality Assurance
QAM	Quadrature Amplitude Modulation
RAM	Random Access Memory
RANSAC	RANdom SAmple Consensus
RASSOR	Regolith Advanced Surface Systems Operations Robot
RAT	Rock Abrasion Tool
RCET	Rover Chassis Evaluation Tool

SLIM	Source LIkelihood Map
SMC	Sliding Mode Control
SNC	Shergottite, Nakhlite, and Chassigny
SOLID	Signs Of LIfe Detector
SOTF	Science On The Fly
SPADE	Sample Processing and Distribution Experiment
SPARCO	SPAce Robot COntroller
SPDHS	Sample Processing, Distribution, and Handling System
SPDS	Sample Processing and Distribution System
SRAM	Static RAM
SRR	Sample Return Rover
SRTMR	Self-Reconfigurable Tracked Mobile Robot
SSD	Sum of Squared Differences
SSX	Mars Subsurface Explorer
SURF	Speed-Up Robust Feature
SVD	Singular Value Decomposition
TAN	Tree-Augmented Naive
TANDEM	Titan AND Enceladus Mission
TAPAS	TAsk-Planning tool
TCP	Transmission Control Protocol
TDI	Time Delay and Integration
TDLAS	Tuned Laser Absorption Diode Spectrometer
TEGA	Thermal Evolved Gas Analyzer
THEMIS	THermal EMission Imaging System
TOF	Time Of Flight
TVTC	Temporal VTC
UCS	Unconfined Compressive Strength
UHF	Ultra High Frequency
UKF	Unscented Kalman Filter
USB	Universal Serial Bus
USDC	Ultrasonic/Sonic Driller/Corer
UUV	Unmanned Underwater Vehicle
UV	UltraViolet
VCI	Vehicle Cone Index
VL1	Viking Lander site 1
VL2	Viking Lander site 2 103
VL2	Viking Landing site 2 104
VLSI	Very Large Scale Integration
VMC	Visual Monitoring Camera
VME	Virtual Machine Environment
VOR	Vestibular Ocular Reflex
VR	Virtual Reality
VTC	Visual Threat Cue
VTOL	Vertical TakeOff and Landing
VXI	VME eXtensions for Instrumentation

WAC	Wide Angle Camera 216
WAC	Wide-Angle stereo Camera 210
WBS	Work Breakdown Structure
WEB	Warm Electronics Box
WES	U.S. Army Engineers' Waterways Experiment Station
WIMPS	Windows Icons Menus Pointers

About the Author

Professor Alex Ellery is a Canada Research Chair in Space Robotics & Space Technology in the Mechanical & Aerospace Engineering Department at Carleton University, Ottawa, Canada. He holds a BSc (Hons) in Physics, MSc in Astronomy and PhD in Astronautics & Space Engineering. He is also an alumnus of the International Space University. He is a Chartered Engineer and holds Fellowships at the Institution of Engineering & Technology, Institution of Mechanical Engineers, and the Royal Aeronautical Society. He was formerly at the Surrey Space Centre at the University of Surrey where he was awarded the George Stephenson medal by the Institution of Mechanical Engineers (2005). He was formerly Chair of the Astrobiology Society of Britain (2005/2006).

1

Introduction

Robots are generally deployed for dull, dirty, or dangerous tasks. They have the advantage that they do not tire or become inattentive due to boredom. Human attentiveness is often the weakest link in any system. However, this is not the rationale for the use of robots in space—robots in the space environment project human influence where humans cannot yet go. Planetary rovers are uniquely useful for almost all types of planetary missions on planets with solid surfaces ranging from small bodies such as asteroids and comets to the moons of gas giants and our own Earth to terrestrial-type planets such as Mars. As exploration probe flyby missions (such as Voyager) give way to more focused planetary orbiter and landed in situ missions, the emphasis on planetary rovers is increasing as the key to detailed planetary exploration with the capability to move to different locations for wider area exploration. Operationally, they offer adaptability and flexibility allowing judicious selection of in situ targets for scientific analysis. Indeed, for this reason, rovers are preferred over static landers [2]. Astrobiological investigation has recently become a unifying theme in planetary science which imposes even greater demands on planetary rover platforms. The first rovers to be placed onto the surface of another planetary body were Lunokhods 1 and 2 on the Moon which were wheeled teleoperated lunar rover vehicle laboratories controlled from Earth by five-man teams. The U.S. Apollo lunar rover, although manually operated by astronauts, had the capability of being operated remotely by ground control for unmanned exploration similar to an unmanned rover. After a long hiatus in planetary surface exploration following the Viking program, the U.S. has led the field in planetary rovers with Mars Pathfinder's (MPF) Sojourner and the remarkably successful Mars Exploration Rovers (MER), Spirit and Opportunity, which have operated far in excess of their design lifetimes. Most recently, the Mars Science Laboratory (MSL) rover Curiosity has begun its exploration of the

© Springer-Verlag Berlin Heidelberg 2016
A. Ellery, *Planetary Rovers*, Springer Praxis Books,
DOI 10.1007/978-3-642-03259-2_1

Martian surface. During the proofing stage of this book, the Chinese Chang'e-3 lunar mission successfully emplaced a 120 kg six-wheeled lunar rover vehicle *Yutu* (Jade Rabbit) onto the lunar surface at Mare Imbrium (Sea of Rains) towards the end of 2013 for a three-month traverse. The term rover was coined during the Apollo missions for their manned lunar ground vehicles but the term had previously been used for a class of automobile built in Britain. Apparently, there was no objection from the British automobile firm because it was considered by the powers that be at the time that the U.S. lunar rover vehicle was irrelevant to their terrestrial market. So began a tradition of naming robot vehicles after dogs— rover, fido, rhex, etc. Why should the dog be so maligned or exalted? Perhaps because the dog was the first animal assistant recruited to humanity's service. However, genetic evidence indicates that dogs originated from wolves over 100,000 years ago, far earlier than the archeological record indications of 14,000 years ago, the latter coinciding with the human transition from nomad hunter-gatherer to sedentary agricultural lifestyles [3]. No matter the reason, the term "rover" has stuck to differentiate the rover for planetary exploration from the mobile robot of more terrestrial origin. But this dichotomy is illusory. Space robotics has been considered an esoteric sideline of robotics in general. I disagree. Space application of robotics—particularly for planetary exploration—is, and will, drive mobile robotics forward beyond its current apparently marginalized condition. Robotics (like its sister discipline artificial intelligence) has not achieved its promises (or worries) in developing pervasive robots useful to society in general (the Roomba vacuum cleaner does not suffice). It is my contention that necessity is the mother of invention and that planetary robotics imposes such demanding requirements that this will be the fountainhead of truly useful and ubiquitous robots for society in the near future.

Future missions—regardless of the details and the waxing and waning of specific space exploration missions—will involve further exploration of the Moon and Mars in particular, and rovers will play a significant part in this. At the time of writing, the U.S. Mars rover Curiosity has just successfully landed on Mars— expectations are high. Mars will always be an explicit or implicit goal for human missions and it is of course of intense interest as a near planetary neighbor and for its astrobiological potential. Human missions will always require significant support from robotic elements. Beyond Mars, interest is also focused on small bodies of the solar system and several moons of the gas giants for similar reasons. Rovers in one form or other will likely comprise a significant contribution to the exploration of these bodies. It is timely therefore for a textbook on planetary rovers to consolidate current knowledge and point the way towards future developments to advance the capabilities of rovers. The $2.5B Curiosity is likely to be the last of the large flagship missions to Mars for at least a decade, but more modest robotic devices will pick up the mantle in the future, ExoMars notwithstanding. Cheap, fast exploration missions using large numbers of low-mass ~1–2 kg robotic explorer devices have been proposed representing the other end of the scale [4, 5]. Regardless of size, rovers will be instrumental in the future exploration of our solar system.

1.1   WHY ROVERS?

A planetary rover is a spacecraft in its own right with many of the same functions and subsystems limited by the same constraints plus those imposed by the planetary environment that it is designed to explore [6]: (i) traction, mobility, and suspension system (propulsion); (ii) vehicle control and autonomous navigation system; (iii) structural system; (iv) command and data-handling system; (v) communications system; (vi) power generation and storage system; (vii) thermal control system; (viii) scientific payload. All planetary rovers possess a locomotion system for traction and maneuvrability on rugged terrain [7]. The most important of these constraints are the severe limits on mass budget, power budget, computational resources, communications bandwidth, and line-of-sight windows. Extraterrestrial deployment implies exposure to a high-radiation environment during the cruise through space but UV exposure on Mars is of negligible import compared with ionizing radiation on the Moon or other atmosphereless bodies. Such constraints rarely afflict terrestrial mobile robots. Terrestrial mobile robotics research generally considers only subsystem (ii) with little regard to the other subsystems. Almost all planetary rover designs to date with a few exceptions are six-wheeled vehicles. Furthermore, planetary rovers have scientific tasks to perform—they are robotic scientific instrument deployment devices [8]. The scientific instrument suite defines their payload—they are thus task oriented. Scientific instrument deployment typically includes sample acquisition and sample handling or processing. This requires an imaging system to support teleoperation or autonomous navigation (or mixed mode) between scientific sample sites in addition to the scientific instrments. This same facility provides the basis for scientific site surveying prior to the deployment of scientific instruments. All planetary rovers undergo a series of design phases: (i) requirement definition of performance; (ii) candidate rover concepts, principally chassis designs; (iii) characterization of terrain performance of each chassis concept; (iv) tradeoff of rover concepts based on predicted performance; (v) detailed chassis and overall vehicle design; (vi) development testing of breadboard hardware; (vii) numerical simulation of complete vehicle over reference terrain. A typical rover mass distribution comprises 30% chassis, 20% scientific instruments, and 50% subsystems. Space and planetary robot systems must be designed to survive the highly adverse environmental conditions of space and the environment in which they are to operate. Furthermore, the remote distances of operation from Earth imply the need for autonomous operation. The basic performance requirements for a planetary rover typically include both forward and reverse motion capability, on-the-spot turning capability, and a well-defined obstacle negotiation capability.

There are four types of planetary rover and all but the last have flown in previous missions: (i) robotic "autonomous" vehicles (e.g., Sojourner, MERs and Curiosity); (ii) teleoperated vehicles (e.g., Lunokhods); (iii) human-operated vehicles (unpressurized) (e.g., Apollo Lunar Rover Vehicle, LRV); (iv) human-operated vehicles (pressurized). The advantages of manned over robotic rovers are evident from previous exploration missions: the manned Apollo lunar rover vehicle and the Mars exploration rover Opportunity traversed similar distances (35 km) but over very different timescales; the manned Apollo lunar missions returned 382 kg of lunar rock including drill cores to depths of 2–3 m compared with the Russian Luna missions that returned 0.32 kg from the surface and near-surface (though the former involved twice as many missions). Human missions require the deployment of significant infrastructure which may be exploited for scientific investigations far beyond those achievable through robotic missions. There is little doubt that human exploration affords greater capabilities than robotic ones but the expense of human missions cannot be justified on the basis of scientific gains, only political or economic ones. Pressurized rovers offer astronauts radiation protection—they are built around a cylindrical pressure vessel of metal alloy and/or carbon fiber composite and generally require several kilowatts of power for crewed missions [9]. A pressurized manned lunar rover requires a self-contained life support system to support unsuited astronauts. Ideally, it should be at least partially regenerative to provide a near closed-loop environmental life support system (CELSS). The pressurized rover serves as a mobile lunar base to extend the exploration range of the lunar infrastructure to ~100 km radius. It should support sortie durations of 14 days (one lunar night) per sortie (assuming a traverse speed of 5 km/h). Equipping the rover with in situ resource utilization (ISRU) facilities would extend its capacity and range further. An airlock must provide for astronaut EVAs (extravehicular activities) in such a fashion to minimize dust contamination. All EVA operations must involve two astronauts with at least one astronaut remaining within the vehicle. Unlike for robotic rovers, ride comfort is an important consideration. The Chariot manned lunar rover has six independent sets of motorized double wheels which enable the vehicle to turn on the spot and move sideways (Figure 1.1b). It is a pressurized vehicle of 8,000 kg (5,314 kg pressurized cabins and 2,618 kg chassis). A two-person suit lock acts as the airlock. The rear payload section includes tooling—winch, cable reel, crane, and backhoe. The suitport virtually eliminates dust contamination within the pressurized cabin. ATHLETE (All-Terrain Hex-Legged Extra-Terrestrial Explorer) is a cargo-handling lunar rover based on six three-jointed wheeled legs (wheel-on-leg) which can raise and lower the payload cabin for high mobility [10] (Figure 1.1a). A hexagonal chassis frame provides attachment points for the leg assemblies. The wheels may be emplaced as feet in walking mode or be rolled in different kinematic leg configurations. It can walk and roll, stepping over obstacles. Each leg may behave as a manipulator through a tool interface to allow different tools (such as a gripper) to be attached.

Rovers may be deployed in multiple roles in rough terrain including scouting for astronauts and larger rover platforms for risk reduction. In this role, they can survey hazardous or challenging locations without risking astronauts and their large manned rover platforms thereby acting as remotely deployable eyes and hands to recover samples to be examined at leisure by more advanced instrumentation in laboratory environments.

Planetary rovers have traditionally functioned as mobile scientific instrument deployment devices whereby in situ measurements are performed on selected

Figure 1.1. Pressurized lunar rover prototypes: (a) ATHLETE; (b) Chariot [credit NASA].

targets. Their range opens up the choices available to target selection thereby enriching the scientific data return. Planetary rovers are thus an enabling technology for planetary science—MERs have been characterized as "robotic field geologists" operating as remote geological tools. Their primary scientific instrument is the camera which provides for surveying and mapping of the local environment. Additional scientific tools may include sample acquisition and sample-processing devices and further analytical instrumentation. Rovers are unique in offering the mobility to traverse to different locations unlike a lander which is limited in reach no more than $\sim$1–2 m. Rovers greatly increase the range of scientific exploration and provide the flexibility to select targets from an area of the surface that increases with the square of the range—even a small increase in range from the lander dramatically increases the number of sites available for investigation. Indeed, during the Viking mission to Mars (1976), the science team were frustrated in not being able to traverse to locations of interest that they could see from the downloaded images. This led to the emphasis on planetary rover missions that have been conducted to Mars ever since—Pathfinder Sojourner, the Mars Exploration Rovers, and now Curiosity.

This, however, has not led to the demise of the fixed lander (such as the Mars Scout lander Phoenix). The lander is a stable platform that allows implementation of a greater instrument complement than can be achieved on a rover. The lander can be larger and deploy larger solar arrays to support a more varied instrument package. Rover power resources are limited to the areal availability on the upper deck. Landers are suited to static measurements such as meteorological and geophysical sensors. These are sensors that require a stable platform and can be deployed in arrays on multiple landers for more global measurements (e.g., NetLander, now canceled). Rovers by virtue of their mobility have restrictions in this regard and tend to be limited in their instrument complement. Their mobility places restrictions on their size, mass, and power. Furthermore, they are ideally suited to the deployment of regional measurements which vary over short distances (e.g., mineralogical and astrobiological instrumentation). It is these and other physical characteristics that vary regionally and exhibit high diversity on planetary surfaces such as Mars. This is not a rigid classification, however, as landers can also carry regional instruments (though they are of course restricted to directly accessible local environs). Landers including those carrying rovers are generally emplaced on locations that impose the lowest risk to the landing but they do not usually constitute locations of high scientific merit—mobility in this case is a necessity to traverse from the landing site to regions of scientific interest. It is certainly conceivable to deploy the lander and rover in combination (e.g., Pathfinder) but this generally implies a local rover capability rather than a highly mobile platform. This is the realm of the microrover and/or nanorover [11]. In this case, the lander would carry the heavier in situ analysis instrumentation while the rover provides access to multiple local sites for samples which are fetched back for analysis on the lander. The rover may be free ranging (such as Sojourner) or tethered (such as the European Nanokhod originally proposed for the BepiColombo Mercury mission), the latter being an extension of the robot arm (such as that which was to be deployed on the failed Beagle 2 lander).

Although rovers (such as Sojourner and the Mars Exploration Rovers) have deployed long-integration time instruments (such as the Alpha Particle X-ray Spectrometer, APXS and Mössbauer spectrometer), these are not ideal instruments to deploy on a rover as they reduce the duration of traverse due to their long integration time of several hours. Of particular relevance in future missions is the central role to be played by rovers in planetary sample return missions, Mars sample return in particular. Rovers are the key scientific multiplier in providing access to the geographical range and geological diversity of samples within their geological context allowing careful selection of returned samples through in situ measurements to maximize the scientific return of such missions. The retention of geological context to each sample is of paramount importance (unlike grab bag locations in which geological context is lost). This is of enormous scientific value.

Planetary rovers may be classified according to mass: macrorovers of mass over 100 kg (a value somewhat arbitrarily set). All current and planned future macrorovers have significantly higher mass: the Mars Exploration Rovers massed 180 kg while the original ExoMars rover was designed to a mass budget of 240 kg); a general trend to the escalation of mass is indicated by the Mars Science Laboratory rover, Curiosity, which masses a tonne. The ExoMars rover was required to traverse a maximum of 25° slopes covered with sandy soils both cross-hill and up/downhill with an average slope-climbing ability of 15° slopes over 1 m [12]. The current ExoMars mission envisages an orbiter mission launched in 2016 followed by a rover mission launched in 2018 (Figure 1.2). The 2016 Trace Gas Orbiter will act as a relay for the 2018 rover and analyze the atmosphere for methane. The second phase involves delivering the ExoMars rover and lander to the Martian surface in 2018 [13]. A Mars Sample Return Mission is planned at some subsequent but as yet undetermined date. This will require both a lander with an ascent vehicle and a 60 kg sample fetch rover with a good range. This rover may be sent as part of the 2018 mission jointly with the ExoMars rover. This rover will require extensive autonomous capabilities for both traverse and scientific selection. It will seal samples on the Martian surface for later recovery during the sample return phase. Given the need to minimize exploration time, a premium will be placed on speed—it is expected that the sample fetch rover will traverse at an average speed of 200 m/h (6 cm/s).

Minirovers of mass range 50–100 kg (again, somewhat arbitrarily as there exist few current prototypes in this range except Marsokhod which masses around 70–100 kg depending on configuration). Extremely small picorover-class devices have been developed—a 7 mm long miniature vehicle comprising a wheeled chassis, shell body, and electromagnetic motor has been constructed capable of a maximum speed of 100 mm/s [14]. The core of the microvehicle device was a 1 mm diameter electromagnetic four-pole stepper motor. However, such picorovers have yet to offer any useful function in planetary exploration. Nanorovers of mass 5–10 kg and dimensions $35 \times 25 \times 20$ cm offer reasonable exploration ranges of <10 m. Their dimensions restrict their obstacle negotiation capability to 10 cm height. They have restricted scientific payload capacity ~1–2 kg. The severe limits on power consumption <2 W (average) and 3 W (peak) limit the rover traverse speed to 5 m/h and maximum grade negotiation capability to 15–20° (i.e., fairly flat terrain). The basic architecture relies on supplying power and data through a deployable tether connected to the lander for operations in the vicinity of the

Figure 1.2. European ExoMars rover [credit ESA].

lander. Such nanorovers act as "extended robotic arms" to around 10 m range due to the length of the tether (limited by DC voltage transmission losses). The payload of a rover typically comprises around 5–15% of the rover's mass, but nanorovers increase the payload ratio by offloading communications and power generation to the lander. A lander-based pan–tilt stereoscopic imaging head enables the use of a large baseline of 0.5 m and height of 1.5 m from the ground for a 20 m navigation range which could not be accommodated on board the rover. An example is the European Nanokhod rover prototype, a tracked vehicle with a payload of 1.1 kg for scientific instruments (nominally an APXS, Mössbauer spectrometer, and close-up stereo imager) while its own mass is just 2.45 kg connected to a lander with a 20–30 m tether supplying power and data

Figure 1.3. Nanorover [credit NASA JPL].

transfer. The Nanokhod stereo cameras had limited fields of view ($23 \times 23°$) with a 60% overlap so the environment was segmented into $30 \times 12°$ azimuth arcs and three elevation arcs focusing at 1.5, 3.9, and 20 m ranges [15]. Motorized levers on the actuated cab allowed adjustment of ground clearance and "kneeling" to point scientific instruments. Steering was accomplished by differential driving of opposite wheels. Nanorovers are limited in their capabilities both as platforms for instrument deployment and in their mobilities. The U.S. Nanorover technology demonstrator was a free-ranging nanorover of around 1.3 kg in mass with low power consumption (2.5 W) (Figure 1.3) [16]. It was a four-wheeled 20 cm long vehicle which mounted its wheels on levers to alter its wheel footprint and steered through skid steering. The 6.5 cm diameter wheels were independently driven on movable struts which allowed body pitching. The main cab comprised two structural kevlar printed circuit boards the outside of which were covered with solar cells. It was designed for extreme temperature ranges ($-170–125°C$) [17].

Nanorover was to fly with the Japanese MUSES-C Hayabusa near-Earth asteroid mission mounting an imaging camera with filter wheel, near-infrared spectrometer, temperature sensors, and a miniaturized alpha-X-ray spectrometer (AXS). It was to be deployed for slow traverse $\sim$1 mm/s and ballistic hops using the levers but it exceeded its budget and was dropped from the payload manifest of the spacecraft.

Microrovers with a mass range of 10–50 kg or so offer the greatest applicability for regional surface exploration in the near term [18]—they offer reasonable scientific payload capacities, modest mass overheads, modest volumetric requirements, and reasonable ranges ~1 km. The Sojourner rover is of this class but Mars rovers have since been putting on weight. The successes and limitations of the Sojourner microrover inspired the U.S. to follow a path of increasing rover size to enhance their range and speed (and their scientific instrument capacity much more modestly). This has emulated the path followed in general spacecraft design whereby smaller, limited functionality spacecraft gave way to enhancing their payload capacities and functionality—the ultimate example of such multi-instrumented, massive, and complex spacecraft is Cassini–Huygens. In spacecraft, of course, this trend has more recently been reversed with an emphasis on increasing flight opportunities. The approach today is to fly less ambitious exploration missions with more focused goals and more frequent flight opportunities. Planetary rovers have yet to follow this trend so the microrover is a neglected but promising class given recent advances in miniaturized technology. Sojourner provides a baseline for the performance of microrovers [19–22]. It was designed for a primary mission lifetime of 7 days but exceeded this in functioning for 30 days during which it traversed just over 100 m. It carried its own power and its own (limited) onboard computer. It was limited to a range of 10 m from the lander, the Carl Sagan Memorial Station, not due to any deficiencies in its tractive capabilities but due to the fact that its navigation cameras resided on the lander and that the lander–rover communications link was limited by the incidence of holes in the radiation pattern beyond this range. There is thus no inherent limitation in ranging capability if its carries its own NavCam and provided the rover–lander/orbiter communications link can be maintained. Furthermore, Sojourner was driven highly conservatively (it was the first time that a rover had been deployed by the U.S.)—the Mars Exploration Rovers were driven similarly initially but their traverses became more ambitious as the missions progressed. Furthermore, Sojourner was deployed at a grab bag site with high rock coverage whereas this was not the case for the Mars Exploration Rovers which experienced more benign environments. Sojourner represents an old technology—its chief limitation was that its onboard computational resources were extremely limited. The limited autonomous capability of both Sojourner and the Mars Exploration Rovers impacts their scientific return—the MERs are controlled through a four-day cycle of command–execute–telemetry–deploy which the scientific community is keen to reduce. This requires more extensive autonomous capabilities to make decisions and recover from fault conditions without ground intervention. An example of a more modern state-of-the-art microrover is the Canadian microrover prototype Kapvik of 30 kg mass with a demonstrated traverse capability of several kilometers and an average speed of 80 m/h (close to the MER top speed of 100 m/h). This microrover developed by a Canadian team (Carleton University, MPB Montreal, Ryerson University, University of Toronto, MDA, Xiphos and the University of Winnipeg) for the Canadian Space Agency (CSA) will figure prominently in this text [23, 24] (Figure 1.4).

The Kapvik microrover was designed by the author and his team (Space Exploration Engineering Group) at Carleton University. The design philosophy followed the Surrey University approach which pioneered the development of low-cost miniaturized spacecraft. This involves the use of COTS components especially to exploit the superior performance of off-the-shelf computers while implementing EDAC (error detection and correction) techniques to enhance their reliability in radiation environments. This allows emplacement of engineered complexity into software while minimizing hardware complexity. To exploit the latest advances and trends, FPGA (field programmable gate array) technology was adopted as a core capacity in the Kapvik rover to ensure high degrees of autonomous capability. The mechanical design was modular at the subsystem level to maximize its reconfigurability—however, the avionics were integrated through a common backplane to minimize connectors, brackets, and cabling [25]. The Kapvik proto-type was derived from the Vanguard Mars mission proposal for an astrobiology-focused microrover for post-Beagle 2 Mars exploration. The Vanguard Mars mission objectives were [26–29]:

(i) to characterize the geological microhabitats that might have supported life on the surface of Mars indicated by the evidence of water, carbonate deposits, and dimishment of the oxidizing layer below the Martian surface;
(ii) to detect, identify, and characterize any signs of organic and/or biomolecules indicative of extinct life below the surface of Mars;
(iii) to provide subsurface geophysical and meteorological data on Mars.

To achieve these objectives, it was considered that both surface and subsurface mobility were essential. There were two derivative robotics requirements:

(i) multiple sample datasets from different surface locations—this implies the need for surface mobility;
(ii) subsurface data sample sets from a depth of 2–3 m—this implies the need for subsurface penetration.

The Vanguard mission concept was based on a triad of surface elements [30]—a small base station lander, a microrover and a set of three ground-penetrating moles (Figure 1.5). The surface assets of a 30 kg microrover supported by a 35 kg base lander amounts to 70 kg including the moles. The Vanguard microrover itself was designed to mount a number of scientific instruments that required little or no sample processing while Kapvik was designed with a manipulator to acquire soil samples. The entry descent and landing system (EDLS) was perceived to be similar to that for Beagle 2 based on an atmospheric entry ablation shield, parachute, and airbag impact absorption providing a landing error ellipsoid of 50×100 km. This implied a Mars entry mass of 140 kg—very modest in comparison with the ever increasing sizes of the Mars Exploration Rovers, ExoMars, and Curiosity.

Figure 1.4. Kapvik microrover prototype [courtesy of Brian Lynch, Carleton University].

Although Kapvik does not currently mount a subsurface capability such as a drill assembly, it does mount a winch to enable it to abseil down a cliff from two secure anchors. Such climbing robots have been proposed for the exploration of Martian cliffs which would provide access to geological layering (e.g., Vallis Marineris). One such proposal is Cliffbot—the inspiration for Kapvik's abseiling capability—which can abseil down cliffs using a tether supported by two anchor robots at the top of the cliff paying out cable (Figures 1.6 and 1.7) [31].

Dante II was an example of an eight-legged abseiling robot which adopted a pantographic frame-walking system to eliminate energy losses during vertical body

Figure 1.5. Kapvik microrover [courtesy of Ian Sinclair, MPB Communications].

movement. Robot descent was controlled by a tensioned tether which snagged and broke on its maiden voyage into a volcanic crater in Alaska [32, 33]. This illustrates the chief problem that tethers impose single-point failure modes with the potential for entanglement and breakage.

There are several roles to which the microrover can be applied. They have more limited range than larger rovers but if targeted to specific high-risk locations they can take advantage of their small size. There is the question of performance versus cost and the mass–cost multiplier for the launch and cruise is even more acute. I estimate that a microrover gives 1/5 the performance at 1/30 the cost

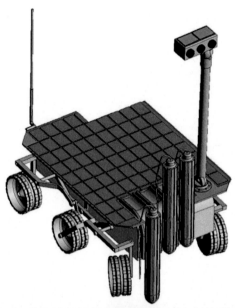

Figure 1.6. Two versions of the Vanguard Mars microrover concepts with three vertically mounted ground-penetrating moles: (a) wheeled chassis; (b) tracked chasses [credit Elie Allouis, Surrey Space Centre].

based on mass. An advantage of microrovers is that they can be deployed in multiple units for rapid exploration enabled by their small size. This provides high redundancy. This enables either multiple rovers each with their own specific sensors or alternatively identical rovers according to the mission requirements. Furthermore, multiple rovers offer more rapid and accurate self-localization and mapping by providing distributed sensing. One microrover deployment scenario is the Vanguard Mars micromission concept which adopts the microrover as its core element (similar scenarios are applicable to other planetary surfaces such as the Moon). The major roles envisaged for the microrover include (but not limited to):

(i) An astronaut aide-de-camp (ADC) and scout reduces the risk to astronauts and their manned vehicle. Astronauts located at a base or a large manned rover may deploy microrovers to investigate hazardous regions such as craters. The microrovers may be deployed by teleoperation, telerobotics, or supervised autonomy. They survey the hazardous region using their scientific instruments perhaps recovering a sample for return to the mother rover or base. The astronauts are relieved of potentially dangerous EVA or from driving the manned rover into a dangerous situation. Larger manned rovers may be driven along a transect with microrovers deployed to survey the region until recovered on large-rover return—this massively increases the geographical coverage of the large rover.

(ii) Sample return for a large rover is a variant of the above application. However, microrovers are uniquely suited to robotic sample return missions. The high cost of robotic sample return mitigates against the notion of a grab-and-go sample return to Mars for instance. Scientific samples recovered will require a scientific context in order to interpret their complex history—this is the primary scientific reason for returned samples over meteorites which lack such context. A microrover with a scientific instrument suite offers the ability to select the samples returned and provide data on their context.

(iii) Scientific exploration surface mobility (e.g., Mars) uses the microrover as a scientific instrument deployment device (IDD) for in situ measurement (this would be required by the robotic sample return mission defined above). Microrovers may be deployed as part of a lander mission to provide a much greater areal coverage for exploration. Although limited to miniaturized instruments with low power requirements, this category of scientific instrument is becoming increasingly large as instruments shrink in size. The Vanguard mission concept is of this nature.

1.2 SPACE PROJECT MANAGEMENT

All space missions—rovers included—are managed as projects. Rover projects in particular tend to involve extensive requirements for validation and experimental testing that are both costly and time consuming. The timely delivery of such projects is the concern of the project manager and his/her team. Management

Figure 1.7. (a) Cliffbot rapelling down a cliff [credit NASA]; (b) and (c) Kapvik abseiling up a test slope [credit Ian Sinclair, MPB Communications].

requires economy, efficiency, and effectiveness—economy involves acquisition of the appropriate resources at the lowest cost; effectiveness ensures that the output from economic activity achieves the desired objectives; and efficiency provides the ability to maximize useful output at minimal cost in resources. Effectiveness is the essential key to the business of the project as economy and efficiency are of no value without effectiveness. In particular, if fewer resources are available, the temptation is to improve efficiency over the effectiveness of the level of delivery which may be put at risk by such measures. A program shows the major steps required to reach objectives, the organizational units responsible for each step, and the order and timing of each step. Projects are the smaller separate portions of a program. They are defined by a set of goals, objectives, requirements, a life-cycle cost, a beginning, and an end. The definition must have clear definitions of performance success criteria. They involve safety, cost, schedule, technical and managerial requirements. The task of the project manager is to identify which techniques in any given situation will best realize attainment of project goals. The project must meet customer requirements and this is usually defined in the bid. Bidding for projects through a proposal is the commonest form of research grant application. ESA adopts a certain standard grading of proposals according to:

- Background and experience of the proposers
- Understanding of the requirements and objectives
- Quality and suitability of the proposed program of work
- Adequacy of the management, costing, and planning
- Compliance with the administrative tender conditions

An ESA proposal comprises a Technical Proposal and a Financial, Management, and Administrative Proposal. The Technical Proposal demonstrates understanding of the requirements and objectives, a plan of work, indication of conformance of the work plan to the requirements, and a full development/test plan. The Financial, Management, and Administrative Proposal should include the background experience, organizational and management structure, management procedures, facilities to be used, curriculum vitae of key personnel, work breakdown structure, work package descriptions, costings, schedule and milestones, travel/subsistence plan, and a list of deliverable items. Most ESA calls for proposals define firm fixed price contracts which impose a maximum price for the work defined in the Statement of Work. Fixed price contracts have essentially superseded cost-plus contracts in order to control costs. The primary marketing strategy is based on unique technical knowledge, effective research and development, effective manufacturing capabilities, and patent protection.

1.2.1   Project planning

It is essential that the first step is to understand and meet customer requirements prior to the planning process. The process of management is essentially the process of planning and controlling the activities of the members of the project and of utilizing all other organizational resources to achieve the project goals. Planning involves selection of goals to determine its objectives and programs of action to achieve those objectives. Decision making is the selection of a course of action to solve a given problem. There may be alternative courses of action for reaching the desired goals which require making a choice for the most suitable alternative. In human flight operations, decision making is structured through GRADE: (i) Gather all information available from all sources to diagnose the problem; (ii) Review the information, resolve ambiguities, and discard irrelevancies; (iii) Analyse options, discard irrelevancies and search for a route to the objectives; (iv) Decide what to do and act upon it by delegating activities; (v) Evaluate the outcome by analyzing feedback to ensure that the solution has worked correctly. Planning has four steps: (i) establish a set of goals; (ii) define the present situation; (iii) identify aids and barriers to the goals; (iv) develop a set of actions for reaching those goals. Planning assumes methods of devising courses of action to achieve goals, organizing implies coordination of resources to effect those plans, leading implies influencing subordinates to work effectively, and controlling implies monitoring and steersmanship of those plans. This requires balancing competing objectives through compromise and analytically breaking down problems into their component parts to find feasible solutions. This assumes both problem solving and decision making. All managers require organized sets of behaviors for their subordinates, peers, and superiors. In the interpersonal role, the manager is a figure of authority, a leader for employees, and a liaison to peers. In the informational role, the manager's responsibility is the communication of information within the organization (be it a large corporation, academia, or a small enterprise) which will not necessarily be conducive to the project management role.

As a useful lesson for project management, there are eight attributes of innovative and successful companies: (i) bias for action with informal task forces; (ii) good customer-oriented products/services; (iii) internal autonomy and risk-taking entrepreneurship; (iv) productivity through value of human resources; (v) emphasis on shared values; (vi) emphasis on internal diversification rather than conglomeration; (vii) simple divisional structure with lean staffing; (viii) autonomy pushed down the hierarchy. Project management requires monitoring of business activities, distribution of relevant information to employees, and the transmission of information outside the unit of responsibility with superiors. The process of management provides an integral medium of communication between the organization structure and its employees. This stresses the information role of the manager in acquiring information, prioritizing it and making decisions on it. Such decision processes can be autocratic, consultative or group-based. The concept of synergy suggests that the whole is greater than the sum of the organization's parts, emphasizing the need for cooperation between departments in the same organization. In the decision-making role, the manager is responsible and accountable for his subordinates' performance in their duties, mediate between disputes, deal with difficulties, allocate resources, and negotiate contracts. They must make difficult decisions in compromising between different organizational goals. Of all management functions, human relations is most important in dealing with subordinates.

Structured project management was first introduced to the U.S. space program and is now widely adopted in almost all theaters of productive activity. The statement of work is a contractual definition of the project's end product with start and finish dates and a well-defined budget. It is the basis of project management planning. The project is a sequence of logically related tasks leading to the attainment of a defined objective within limited resource constraints (including time). Project management involves organization, planning, directing, monitoring, and control of all elements of the project. The entire contract must be planned in terms of work, resources, time, and budget. The primary role of project management is to define the process of achieving the project goals using the available (usually scarce) resources. The project is characterized by time, cost, and performance. The project is first defined by well-defined start and finish points—project completion should be delivered on time. Project durations, however, can

slip which may be countered by increasing costs to apply additional manpower.
Indeed, this is usually preferable to delays in delivery which themselves often incur
financial penalties. The project requires sufficient manpower—of sufficient quality,
expertise, skills, and experience—to complete the tasks within the specified timescale. The project is planned and then tracked throughout its duration by a
number of measurement tools such as the schedule and budget. The basic project
management functions are: (i) manage the project scope be defining the goals and
work to be performed; (ii) manage the human resources; (iii) manage communications to keep the project on track; (iv) manage time through planning and keeping
to the schedule; (v) manage the quality of work to ensure satisfactory results;
(vi) manage costs within budget. The laws of project management are (somewhat
tongue-in-cheek): (i) it is well known that work expands to fill the available
schedule and that no major project is installed on time within budget or with the
same staff throughout; (ii) projects progress quickly until they become 80% complete and then the remaining 20% take forever to complete; (iii) when things are
going well, something will go wrong; when things cannot get worse, they do; when
things appear to be improving, something has been overlooked; (iv) if the project
is allowed to change in content, the rate of change will exceed the rate of progress;
(v) no system is ever completely debugged; (vi) a carelessly planned project takes
three times longer to complete than expected while a carefully planned project
takes only twice as long; (vii) project teams detest progress reporting as it demonstrates their lack of progress; (ix) based on the premise that administrators
multiply their subordinates, Parkinson's Law states that the average rate of
increase in administrative staff without any change in work volume is around
5–7% per year [34]. Space mission objectives must be achievable, clear, and un-
ambiguous. The number of objectives should be as few as possible (preferably
only one though this is rare!). Project goals should be specific, measurable, agreed,
realistic, and time bounded (the latter may be imposed by launch windows). The
project is defined by a set of manageable and well-defined tasks—these are the
work packages and must be allocated the appropriate resources and staffed with
the appropriate teams reflecting the required technical skills. The role of the
project manager is to motivate the staff to achieve the project goals. This requires
leadership which requires intelligence, energy, and resourcefulness. The project
manager should have appropriate skills including the ability to generate financial
plans, manage the contract, manage creative thinkers, and inspire creative problem
solving. Problem solving is begun by problem definition, outlining alternative solutions, comparative analysis of each alternative, selection of the solution, and
execution of that plan. Decision making serves to reduce uncertainty. It is essential
to be able to adapt as no project proceeds exactly as planned. The encouragement
of informal structures and lines of communication is also productive. The basis of
project management is the project plan (project terms of reference). The most
likely problems to frustrate good project progress involve:

● poor communication with the project team;
● poorly defined project goals and requirements;

- micromanagement rather than individual responsibility;
- disagreements with customer and/or subcontractors;
- lack of compliance to standards and regulations;
- personality conflicts within the team;
- lack of environmental support (e.g., poor company relations, etc.).

There are several strategies to reduce the communications overhead and increase the capacity to process information: (i) creation of slack resources by increasing planning targets either by extended completion dates, budget targets, and/or resource inventories—these approaches are cost limited; (ii) creation of multiple self-contained tasks and the necessary resources to achieve self-containment based on output such as the product line to reduce conflicts of resource allocation, but highly specialized labor functions must be centralized to retain the division of specialized labor and economies of scale (e.g., finance and legal services); (iii) creation of lateralized relationships through liaisons to bypass lengthy communication lines, multidepartment task forces to remove problems from higher levels of the hierarchy, project management of integrated teams. The deliverables are important as they are tangible and either are or are not produced. They also provide a means for tracking progress.

1.2.2 Work breakdown structure

Plans are implemented through detailed actions designed to realize set objectives. This introduces the control function of management to ensure that actions do conform to plans. Control and planning are intimately linked. If planned and actual events diverge, adjustment must be made to the actions to ensure that they match the plan more closely, or to the plan itself to be replanned, or to the controls to ensure that they are monitoring the plan effectively. An important issue is traceability such that project control can identify problems causing deviations. This is achieved primarily through the organization breakdown structure (OBS) and the work breakdown structure (WBS). The OBS identifies the members of functional departments and sections responsible for the performance of project work from the management level down to the skill level. It establishes relationships between sections and so defines the reporting structure. A WBS work package is assigned as the responsibility of the human element of the OBS. The accomplishment of work must be measured—this requires work to be easily monitored. Reporting structures must be defined to ensure that all resources have all the information required. This enables forecasting the impact of changes on planned dates and costs. Reporting structures defining communication channels are of three types—regular written reports, regular meetings, and presentations during reviews. Schedule reporting should include monthly status/progress reports. They provide an opportunity to review progress and include: (i) overall accomplishments to date; (ii) accomplishments this month and programmatic status; (iii) expenditure to date according to budget; (iv) activities to be achieved; (v) deliverables accepted or to be delivered to the client; (vi) current problems and their proposed solutions; (vii) anticipated problems and their proposed solutions; (viii) status of previous actions. The project work requirement is defined through the WBS of the project. Project resource allocation is defined through the organizational breakdown structure. The WBS defines the different activities and specific tasks required to complete the project in a hierarchical manner. It relates all work elements to each other logically without duplication or redundancy. Tasks in the WBS are assigned to one functional element of the OBS (i.e., one individual). The WBS is derived directly from functional requirements and comprises a functional decomposition tree. This functional tree is then broken down into a product tree which defines the system's elements—the tasks to be performed. For a planetary rover, this will comprise a breakdown of all the onboard subsystems at a minimum of three levels of resolution, the first being the rover vehicle and the rover's scientific instrument suite.

McKinsey's 7-S framework describes seven interdependent aspects of organizing work—structure, strategy, systems (and procedures), style of management, skills, staff, and shared values. The dominant theme here is work organized by specific tasks. The tasks required to produce each element define the WBS which identifies the scope of the project. A task is assigned to a manager responsible for that task. Each work package so defined is assigned a responsible person to lead it who interfaces with the overall management. The designation of detailed work to team leaders is defined in the responsibility assignment matrix. Each work package is characterized by its resource requirements and constraints. Each work package must be measurable with clearly identified inputs and outputs defining the scope of the work, the duration between start and finish dates (typically, a month or more), and a budget. It should utilize a single element of cost, either labor or material but not both. There is always a separately defined project management WBS. The WBS comprises the basis for developing the schedule which gives a timeline for each activity against which progress is monitored. A typical WBS breaks the project into its constituent work packages to the level at which each work package is the responsibility of an individual or principal investigator for its delivery. An example set of generalized work packages is given by Table 1.1:

1. Program management, financial accounting, and administrative support.
2. Reliability and QA.
3. Structure (*mass budget*).
4. Power subsystem (*power budget*).
5. Thermal control (*thermal budget*).
6. Communications subsystem (*link budget*).
7. CDH subsystem (*data budget*).
8. RF/Electromagnetic compatibility.
9. Orbit/attitude control subsystem.
10. Propulsion system (Δv *budget*).
11. Payload instruments.
12. Integration and test campaign (including wiring harness).
13. Ground support equipment (mechanical/electrical).

Table 1.1. Work package description.

Project		WP No.
WP Title		
Contractor		
Start event		Planned start time
End event		Planned end time
WP manager		Issue No.
		Review date
Inputs required to start WP		
Tasks included in WP		
Tasks explicitly excluded in WP		
WP outputs		

14. Launch campaign.
15. Mission operations.

The technical budgets—mass budget, power budget, etc.—are maintained by the systems engineer who also controls interfaces defined through interface control documents (ICDs).

1.2.3 Project schedule

A detailed time-based schedule is developed to meet contractual requirements. Project scheduling involves planning of timetables and dates for the allocation of resources, activities, and deliverables. Scheduling requires estimates of the duration of activities, the precedential relationships between them within the constraints of resources, budget, and delivery times. It is essential to differentiate between critical tasks and those which can be delayed. Scheduling the project proceeds from the WBS where the work packages are scheduled over time. Gantt (bar) charts are commonly used for project planning and tracking into parallel and serial task paths while reporting progress against work. Each work activity is listed on a calendar. The Gantt chart is a labeled bar chart which shows the time relationships between events. Each WBS package must be represented on the vertical axis with its duration on the horizontal axis. Generally, the maximum number of activities on a Gantt chart is around 40. The bars represent the duration of an activity over time. A hollow bar designates planned activity while a filled bar denotes progress for that activity. Activities are represented as arrows with milestones as nodes. Milestones may be defined as major occurrences or key events in the project lifecycle which act as checkpoints along the plan to monitor work as it progresses—they should be defined for all major phases of the project and should be verifiable. The top level of the plan is the summary milestone schedule that identifies milestones in the contract. The master project schedule indicates milestones that mark events that represent major accomplishments of the project relevant to functional work. Examples include: (i) project kickoff; (ii) completion of requirements analysis; (iii) preliminary design review; (iv) critical design review; (v) integration and test complete; (vi) quality assurance review and delivery; (vii) customer acceptance test complete. Intermediate schedules show integration points between the WBS and the network schedule, and so make references to functional managerial responsibility and individual functional plans. Gantt charts can also represent the dependency relationships between different tasks according to the start and completion times. A precedence relationship indicates that activity A must be completed before the onset of activity B, thereby generating a sequence of activities and activities that can be achieved in parallel. A project comprises a set of interdependent activities. A rule of thumb is that each activity is 0.5–2% of the project length. Generally, it is better to reduce the time risk of late delivery by increasing the engineering budget. The Gantt chart is defined by four times associated with each event—early start, early finish ($= ES + L$), late start ($= LF - L$) and late finish times where $L =$ activity duration. Total slack time is defined as: $TS = LS - ES = LF - EF$. Slack time is thus defined as the latest allowable start time less the earliest expected finish time in chains of activities. Activities that have no slack are critical activities and a sequence of activities connecting the start and finish of the project defines the critical path as any delay in any constituent activity delays the project as a whole—the critical path defines the highest sensitivity of the project to delay. Project control should focus on the critical path. The best approach to increasing slack time on the critical path is to parallelize activities and redistribute manpower. A budget status may be included that shows expenditure. The budget comprises several components: discrete work effort (labor), material costs, and other costs. Most projects are resource limited. When resources exceed availability, slack in non-critical activities may be used up to a point. The PERT (program evaluation and review technique) and CPM (critical path method) graphically represent the relationship between tasks and milestones. The CPM assumes deterministic activity times while the PERT assumes activity times are random with optimistic/pessimistic/most likely durations. Project management must set up an efficient and transparent means of progress reporting usually through monthly status reports and three-monthly progress meetings (or higher frequency during crisis management).

1.2.4 Project budget

The most important role of project management is to optimize the deployment of financial and other resources—this implies the need for prioritization. Budgets are statements of financial resources for each activity over a given period of time.

They include income and expenditure and provide targets. Budget development is a key planning process for the coordination of projects. Budgeting is the commonest link between planning and controlling. It guides decisions about resource allocation to achieving the objectives. An overrunning budget is an early symptom of activities not running to plan. The project budget derives from the WBS and project schedule. Planning involves network scheduling at different levels of detail, and an authorized budget is assigned to these tasks over a period of time. The Precedence Diagram Method (PDM) is the network method of choice which allocates activities as functions that consume time and resources and their relationships which define the sequence of activities. Once each cost account has been planned, the total budget of all cost accounts becomes the performance measurement baseline or project schedule. The basis of monitoring and control is given by budgeted cost for work performed (BCWP) (i.e., planned budget for work accomplished). It is the planned cost for work accomplished. The actual cost of work performed (ACWP) gives the actual cost incurred by the work accomplished. This enables calculation of the cost variance, $CV = BCWP - ACWP$. The budgeted cost for work scheduled (BCWS) allows calculation of the scheduled variance, $SV = BCWP - BCWS$. These are the measurements that provide inputs to performance measurement reports. Costs are directly against work performed (e.g., labor and material costs). Cost account management involves planning, scheduling, budgeting, and controlling all budgetary expenditure and labor support. Cost accounts are identified by the matrix of responsibility constructed by plotting the WBS against the OBS. Performance is enabled by tracking work done against the network schedule and forecasting future needs while monitoring the cost of the work to be done. The performance measurement system is a formalized process to establish cost and schedule control of the project. It clarifies accountability in the project workforce and ensures the time reference for the project program. Tracking of earned value allows the cost of scheduled work to be determined based on resource use and financial procurement. The budget is effectively a projected profit/loss statement and balance sheet. Timesheets which record the amount of time spent on different activities provide the basis for project effort tracking. Typical time robbers include incomplete or poor work by subordinates, inadequate delegation or having to monitor delegated work, day-to-day administration with lack of clerical support, and excessive crisis management. Project costs include: (i) staff (development and operations); (ii) materials; (iii) travel; (iv) consumables/equipment including special equipment (e.g., calibration equipment, jigs, test facilities, handling equipment); (v) consultancy; (vi) administration; (vii) licenses; (viii) overheads (energy, IT, library, H&S, estates); (ix) special items (import/export duties, transport containment, insurance). Direct costs are those directly related to design, development, build, and operation of the system. These include non-recurring costs such as R&D, design, facilities, tooling, and testing, and recurring costs which occur throughout the life of the program such as manufacture, planning, preflight checkout, flight operations, maintenance, consumables, and training. Indirect costs are those related to the business side including management, profit, non-operational facilities, and overheads. There are a number of cost estimation methods available. The rule of thumb/analogy approach requires expert judgment in similar previous projects (typically used as an initial estimate). This adopts the 6/10th rule:

$$C_i = C_{i-1} \left(\frac{S_{i-1}}{S_i} \right)^{\gamma}$$

where $C_{i,i-1}$ = cost of project i based on project $i-1$, $S_{i,i-1}$ = size of project i based on project $i-1$, and $\gamma = 0.6$. The bottom–up approach uses a detailed WBS at the work package level (ESA requirement). The parametric approach correlates cost with manpower and product and yields cost–estimation relationships (CERs) based on historical data (non-applicable to new technology). Cost estimation ratios have the form:

$$C(2005\$M) = a(m_i(kg))^b$$

where a = similarity multiplier × regressive best fit and $0.3 < b < 1.0$. For example, one attitude and orbit control system CER based on subsystem mass is given by:

$$C(2000\$K) = 464(m_i(kg))^{0.867}$$

The learning curve can be defined as the effort (time, cost, etc.) to achieve the task, which decreases with repetition (Table 1.2). The Crawford formulation states that doubling the production run yields a fractional reduction in effort such that the second unit costs 80% of the first, the fourth unit is 80% of the second, etc.:

$$C_n = C_1 n^p \quad \text{where} \quad p = \frac{\log(C_2/C_1)}{\log(2)}$$

Average cost is given by:

$$\bar{C}_n \approx C_1 \frac{n^p}{1+p}$$

The business plan is based on NPV (net present value) analysis of a project. Cost is discounted in the future based on the opportunity cost of money—a fixed sum of money is worth progressively less in the future which is discounted analogously to the compound annual interest rate. NPV is defined as:

$$C_i = \frac{C_{i+n}}{(1+r)^n}$$

Table 1.2. Example costs of production.

Spacecraft type	Non-recurring cost, a	Non-recurring cost, b	First unit production, a	First unit production, b
Planetary spacecraft	13.89	0.55	1.007	0.662
Earth orbital spacecraft	3.93	0.55	0.4464	0.662
Scientific instrument	2.102	0.50	0.2974	0.70

where $r =$ discount rate (typically 10–15%). NPV assuming constant annual payments of R:

$$C_i = R\frac{1 - (1 + r)^{-n}}{r}$$

Investors generally require specific minimum values of IRR (internal return on revenue)—venture capitalists generally require 70–100% IRR within 18 months payback illustrating the difficulty in obtaining long-term investments for techno-logical products. Costs can be spread by varying costs annually with low startup costs. Costs are usually estimated using a beta function which defines the fraction of total program cost $(0 \leq C \leq 1)$:

$$C(\tau) = 10\tau^2(1 - \tau)^2(A + B\tau) + \tau^4(5 - 4\tau)$$

where $\tau =$ fraction of total program time $(0 \leq \tau \leq 1)$ and $A, B =$ shape parameters $(0 \leq A + B \leq 1)$. Shape parameters are defined by the location of maximum C and the width of peak P:

$$\text{If } C < 0.5, \quad A = \frac{(1 - P)(C - 0.1875)}{0.625} \quad \text{and} \quad B = P\frac{C - 0.1875}{0.3125}$$

$$\text{If } C \geq 0.5, \quad A = \frac{P(C - 0.8125) + (C - 0.1875)}{0.625} \quad \text{and} \quad B = P\frac{0.8125 - C}{0.3125}$$

Cost overruns are endemic in space projects due to lack of effective control and unreasonable expectations on the part of agencies. The use of COTS (commercial off-the-shelf) components offers one means to minimize development costs of space-rated components that are expensive. Sometimes, this can be achieved by enhancing component reliability in space by prototyping—this may involve repackaging or adopting software approaches (particularly for electronic components).

1.2.5 Development plan

The development plan comprises the major document in a development project. It includes the following components: (i) an introduction which summarizes the scope of the development plan, a list of applicable reference documents, and a brief description of all the major subsystems; (ii) deliverables based on the work breakdown structure, identifying the work packages and a description of the deliverables which must be related to the product tree, and reliance on suppliers and subcontractors; (iii) constraints describing requirements regarding their docu-mentation, main delivery dates, and responsibilities of interacting groups; (iv) a development strategy defining the phases of the project and sequence of activities, identifying where work will be performed, and the dates of review meetings; (v) a verification plan describing how it will be achieved; (vi) work breakdown breaks the project into individual work packages and defines the scope of work, the activities to be performed, duration and resources required, and the team leader of each work package; (vii) organizational breakdown of the project with allocated responsibilities and authority; (viii) a risk plan identifying key risks, their effects, and how they may be mitigated; (ix) a schedule identifying tasks defining the work packages and milestones to monitor progress; (x) cost breakdown to identify staff resources including their costs.

1.2.6   Personnel management

At high-management levels, technical skills are less important than conceptual skills of strategic coordination and integration when more time is spent on ∼planning rather than supervision. However, all project managers require the fundamental skills of working with other people—this is their most important skill. Leaders who have good relationships with subordinates will have more influence than those who do not. The project manager must use persuasion, compulsion, and example to manage the team. It is important that managers have clear direction, communicate clearly, support their subordinates, and resolve conflicts effectively. At a personal level, they should inspire trust, respect, and common values. An important consideration often ignored or glossed over is the necessity of managing oneself [35]. Furthermore, the project environment is a changing environment and this change must be managed effectively through innovation and creativity to respond to opportunities. The basis for this is the management of information and knowledge to support effective decision making. Leading by example is the basis for action-centered leadership. Effective leadership is characterized by the assumption that work is a source of satisfaction, that team members exercise self-direction and commitment to the objectives based on rewards, that delegation of responsibility is desirable and requires the sharing of information through good communications. Subordinates should be accorded respect and value. Good communication is essential and should be characterized by clarity. This encourages an employee-centered form of supervision rather than job-centered supervision, allowing employees maximum participation in decision making.

To reduce micromanagement, delegation is essential. Tasks are assigned to subordinates which define their responsibilities and imposes both accountability on the subordinate and the authority to achieve those responsibilities. Hence, the delegation process assigns that task and the authority to achieve it. This provides the basis for a consultative decision-making process which generally improves productivity. It is important that tasks to be achieved and team roles are meaningful, implement useful training, and provide self-development opportunities (team roles should be balanced within a competent team with shared leadership, mutual accountability, and collective work). This is the basis of job enrichment, variety, and staff commitment in human resource management which promotes wider utilization of talents, capabilities, and greater personal motivation and satisfaction of subordinates. Those who flourish best in such an environment exhibit leadership, initiative, judgment, decision-making ability, self-discipline, and diligence, all qualities required by a leader. Lack of team member commitment is one of the most difficult problems for the project manager. There are a number of

Table 1.3. Job terms of reference.

Job title	
Line manager	
Scope of job	
Overall purpose of job	
Detailed task requirements	

Completed/Ongoing key task list	Performance standards	Supporting data
New key task list	Performance requirements	Target dates

tasks which increase the perception of job boredom—form-filling administration, meaningless tasks, repetition, and lack of task completion. Although some administration is necessary, excessive administration, a characteristic of all these common complaints, is unfortunately a characteristic of bureaucratic organizations. It is such mindless, impersonal bureaucracies, typically implemented by the lackluster managers and administrators fostered within such environments, that stifle creativity and innovation (government departments are curiously fond of such inane bureaucracies). Innovation is based on knowledge and the knowledge base of an organization is its primary asset. Lack of senior management support inevitably affects resource availability and commitments to projects and is manifest as a lack of senior management support for projects and their project managers. Each staff member should have staff terms of reference (objectives) defining the work to be achieved in hierarchical fashion, thereby devolving responsibility to the level of the individual worker (Table 1.3).

These should be defined in terms of weekly objectives. This is management by objectives (MBO) which seeks to integrate organizational goals with the needs of employees. It focuses on results rather than processes and is characterized by regular performance appraisal. Management by objectives is based on the establishment of common goals by managers with employees in consultation. Each person is responsible for the achievement of the objectives. Effective planning is determined by well-defined specific objectives for each manager to provide a focus for the business. Managers must be actively involved in the objective-setting process. The relationship of each individual's objectives to the common goal is of central importance in the efficiency of attaining company objectives. The individual should have a wide range of discretion and autonomy in determining and selecting the means for achieving his/her objectives in order to utilize initiative and creativity. Such objectives should be set by consultation between individual and supervisor—the greater the mutual participation especially subordinate participation, the more likely goals will be achieved. It is difficult, however, to coordinate the overarching objectives of the organization with multiple individual

personal objectives. Fundamentally, it involves manager/subordinate agreeing and
reviewing job responsibilities, performance objectives, and appraisal with a degree
of self-assessment. Individual goals clearly define job responsibilities and objectives
in relation to the organization's goals. The end result is improved performance in
planning, coordination, control, and communication. The MBO system has several
features. Commitment to the program is necessary at all levels of the organization
to achieve maximum functionality and to ensure compatible personal and organizational objectives. Management by objectives also has value in performance
assessment. A good appraisal system provides the opportunity for advancement as
well as a mechanism for coping with unsatisfactory performance. Periodic reviews
of performance in progress to achieving objectives are required to resolve
problems. Such performance reviews must be regular to provide feedback on performance. Individuals who set or participate in their own goals tend to improve
on past performance generally. Constructive feedback on performance to employees also generally leads to better performance, especially when specific. MBO
has many strengths—individuals are involved and understand organizational
goals, which in turn tend to be more realistic due to improved communication.
Realistic and clear objectives are absolutely fundamental to success. There are
eight different business areas for objectives—market standing, innovation, productivity, physical and financial resources, profitability, manager performance and
development, worker performance and attitude, and public responsibility. The
chief problem of MBO involves appraisal by objective measurements which are
often difficult to achieve. Furthermore, for success, MBO requires management
style to be adaptable tempered with a high level of skill in interpersonal relations.
This mutual involvement in decision making generally fosters reduced antagonism.
However, a poor or abused appraisal process is worse than none at all. The best
appraisal process is truly participatory with a combination of self-appraisal,
superior appraisal, and subordinate appraisal. The project team is organized to fit
the project to its available resources including staffing. The team is structured
according to the complexity of the project and the size of the project team. It will
evolve as the project proceeds. Organizational structure is based on everyone's
capacity to process information. The project manager should delegate to a few
subordinate managers. Good project managers should maintain a psychological
distance between themselves and their team. Psychological distance is enhanced by
being task oriented, minimizing gossip by maintaining effective communication,
maintaining fairness, and not fraternizing with team members. All team members
must be made aware of their duties—what is to be done over what timescale using
what available resources with what authority. Team members need to know the
division of responsibilities and know the formal reporting procedures. Space
mission project teams are populated by highly motivated individuals. All space
missions are performed by people who are the most important resource. Development work characteristic of space missions is people intensive. Maslow's hierarchy
of human needs defines people's motivation for work:

(i) physiological needs—hunger, thirst, sleep, etc.;

 (ii) safety—security from danger;
(iii) social needs—social inclusion and relationships of family and social groups;
(iv) self-esteem—self-respect, status, and recognition;
 (v) self-actualization—personal development, growth, and accomplishment.

People's needs form a hierarchy, the lowest levels of which need to be satisfied before higher levels. At the lowest level are physical needs which in modern society tend to be satisfied generally. Each basic need must be at least partially satisfied before needs at higher levels can be satisfied. Individuals will be motivated to fulfill whatever need is most powerful at the time depending on the current situation and recent experiences. Basic physiological needs must be satisfied by a wage sufficient to feed, shelter, and protect each employee and their families, and employers must provide a safe working environment. Beyond this, incentives may provide employees with esteem, social context of close associations with colleagues, and opportunities for advancement. Security needs imply job security, freedom from coercion, and clear regulations. Generally, physiological and security needs are generally satisfied. The work environment can provide a good social environment. The desire for achievement and status may be achieved by challenging work assignments and due recognition. Beyond this, there is the need for self-actualization to provide advancement and meaning in work. The need for achievement may be defined as the desire to excel in competitive and challenging situations. This generally implies willingness to take responsibility, the tendency to set challenging personal goals, and a great need for feedback on performance. Employees with high achievement needs thrive on challenging and stimulating work. They welcome autonomy and variety. Job satisfaction is related to the nature of the job and the rewards gained from performance (i.e., achievement, responsibility, recognition, and advancement). Job dissatisfaction derives from the individual's job and its relationship to the organisation (i.e., salary, interpersonal relations, working conditions, and, especially, company policy). Motivation is concerned with the satisfaction of needs which may vary over time or in a given situation. These needs drive goal-directed behavior in an attempt to satisfy them. Goals depend on expectancy of reward as a result of a given type of behavior. For space mission project personnel, the last three are the primary drivers. Hersberg defined two scales for measuring job satisfaction and productivity: maintenance factors (company policy, administration, working conditions, supervision, interpersonal relations, job security, pay) and motivational factors (achievement, recognition, advancement and growth, responsibility, technical aspects of the job). If an organization meets the maintenance factors only, the employee is not satisfied but only not dissatisfied.

To improve productivity, an organization must meet motivational factors. Motivation may be defined as that which "causes" people's behavior—people are self-motivated to a large degree and desire autonomy to pursue their jobs in their own way. Traditional schools of management utilized wage incentives to motivate workers to perform boring, repetitive labor in the most efficient way. Human relations schools sought to reduce boredom and repetitiveness which themselves demotivate workers by using informal work groups to provide a social dimension to work. Human resources schools suggest that motivation is multifactorial—wage reward, job satisfaction, meaningful work and that good performance provides satisfaction, particularly if given responsibility. Personnel policy includes issues such as employee benefits and methods of reward for individual employees such as salary advances by promotion. Organizational culture such as the degree of autonomy and decision making is also important in influencing motivation. The immediate work environment of peers and supervisors, in particular, strongly influences motivation and performance as they are the most important trans- mitters of organizational culture. These extrinsic outcomes are provided by the employer. Intrinsic outcomes are experienced directly by the individual as a result of successful task performance in terms of accomplishment, self-esteem, and mastery of new skills. The expectancy approach to motivation attempts to account for differences between individuals and situations. It assumes that behavior is determined by a combination of forces originating from both the individual as expectations and the environment as situations. Individuals each with different goals and needs decide their behaviors based on their expectations that a certain behavior will have certain desired effects. The value of the expected reward to the individual combines with the perception of effort involved in attaining the reward and the probability of achieving the reward given the level of effort. This effort combines with the individual's abilities and the way the task is done to yield a specific performance level which generates the rewards. The rewards valued by each subordinate must be determined. The desired performance must also be determined. The reward should be limited to the degree of performance but should be adequate according to the individual's evaluation of the fairness of the reward by comparison with rewards to others for similar performance. Managers wishing to influence subordinate behavior must ensure that the appropriate con- sequences are entailed by individual subordinate behavior. Generally, it is more effective to reward desired behavior than to punish undesirable behavior. Positive reinforcement encourages repetition of a desired behavior and often results in major gains in efficiency, cost reduction, and productivity. Punishment provides negative reinforcement to improper behavior but can cause resentment or humilia- tion which is counterproductive—an alternative variation is extinction where reinforcement is absent following moderately undesired behavior and this can be effective. Continuous reinforcement rewards desired behavior immediately when it arises which leads to rapid transient effects. Partial reinforcement provides rewards intermittently which generate more permanent, long-lasting effects. Fairness is essential for reward allocation.

In any application of learning theory to the work environment, managers should identify undesired behavior, measure its frequency over time, determine the causes of undesired behavior, intervene to alter the consequences of such behavior to discourage it, and evaluate progress. The entire system of forces operating on any employee must be considered in order to understand an employee's motiva- tion and behavior. These variables affect organizational motivation: individual characteristics such as interests, attitudes and needs; job characteristics are attributes of employee tasks including degree of responsibility; work situation characteristics are factors in the work environment including colleague encouragement and management attitude. Meaningful and challenging jobs are characterized by five core job requirements for motivation: skill variety, task meaningfulness, task significance, autonomy (responsibility), and feedback of job effectiveness. It is important to note, however, that there are two types of workers—high-growth-need individuals who need challenges and low-growth-need individuals who need directive supervision. Theory X holds that people are generally lazy, dislike work and tend to avoid it, and evade responsibility, so management must coerce people to work. Theory Y holds that work is natural and that people will commit themselves to gain rewards through achievement. They will exercise self-control, take responsibility, and utilize their capacity for ingenuity and creativity. In reality, the potential of the average human being will only be partially realized, and satisfaction and reward are the key to commitment to achievement. However, the welfare of the individual worker is enhanced if they can achieve their full potential as human beings. In this case, every worker would perform his job, preparing for the next highest job and train his successor simultaneously. Only when both individual and organizational goals and needs are satisfied can an organization function effectively. Ethical considerations that underlie political democracy and economic freedoms argue for participatory management as a form of workplace democracy over more autocratic forms of management [36]. This is the only consistent model with the assumption of the value of human beings and the most appropriate way to administer human social organizations (i.e., an ethical foundation in which justice prevails). This is well established in some European and more recently Japanese forms of organization while many American organizations are still dictatorial and feudal in spirit, even to the extent of denial of basic civil liberties in corporate cultures.

The 9,9 team management style, which is built on trust, stresses the career development of individuals. This suggests participative group management with economic rewards, high-performance goals, good working practices, and accurate communications. In summary, managers must consider the importance of motivation in the achievement of both personal and organizational goals:

  (i) managers must actively motivate their subordinates;
 (ii) managers must recognize that employees have different motives and abilities;
(iii) rewards should be related to performance rather than non-merit–based considerations;
 (iv) jobs should be challenging and have variety;
  (v) management should be founded on a performance-based organizational culture;
 (vi) managers should be in close touch with employees and their problems;
(vii) employees should play an active part in organizational achievement.

Employees require their roles and status to be clearly defined in a mutually acceptable way. The main criterion about the importance of a job is its length of time between evaluated decisions. Fayoll's 14 principles of management are: (i) division of labor allows specialization and efficiency; (ii) authority requires relevant experience to achieve personal authority; (iii) discipline requires both penalties and rewards; (iv) unity of command implies that any one person is answerable to only one line manager; (v) unity of direction implies that operations with the same objective should be under one manager; (vi) subordination of individual interests to the common interests of the organization; (vii) remuneration for work should be fair; (viii) centralization implies that managers should have final responsibility but delegate sufficient authority to enable jobs to be done effectively; (ix) line of authority in an organization should be hierarchical; (x) resources should be in the right place at the right time; (xi) managers should be equitable and fair to subordinates; (xii) employee turnover rate should be low; (xiii) subordinates should be given sufficient freedom to use their initiative; (xiv) team spirit should be fostered. Well-educated employees are less accepting of formal authority so human relations have an impact on the organizational structure. Leadership in management should come from managers' greater knowledge and expertise rather than formal authority. Financial incentives alone are not enough to motivate productivity improvements. Sympathetic management towards employee welfare tends to increase motivation to work productively (Hawthorne effect). Work group dynamics and pride influence productivity greatly providing social motivation of rewarding work relationships. The social environment in the workplace is an important dimension to productivity.

The project management team includes a number of significant personnel: (i) program manager responsible for project schedule, task assignment, status monitoring, and financial planning (the project manager should have previous experience of the project sector, have strategic leadership expertise, be technically competent in project decision making, possess interpersonal skills, and have proven managerial ability); (ii) product assurance engineer is responsible for reliability and quality standards; (iii) systems engineer responsible for overall systems engineering, interfaces, and systems integration (ICD); (iv) principal scientist (PI) responsible for payload system. According to Belbin, there are nine roles within a team for it to achieve a good balance (though most team failures are due to intellectual inadequacy of team members):

(i) plant—creative and unorthodox problem solver;
(ii) resource investigator—extrovert and communicative networker and project fixer;
(iii) chairman—confident, goal-oriented decision maker;
(iv) shaper—dynamic driver thriving on surmounting obstacles;
(v) monitor/evaluator—strategic judge in assessing options;
(vi) teamworker—cooperative diplomat who eases tensions;
(vii) implementer—reliable and conservative pragmatist;
(viii) completer/finisher—conscientious attendance to detail;
(ix) specialist—self-starter focusing on specific problems.

The team should have a good mix of roles, the emphasis of these roles shifting as the project proceeds. Curiously, Belbin found that most managers were weak completer/finishers. The distribution of manpower over time required to perform a project follows a Puttman–Norden–Rayleigh curve:

$$y(t) = 2K \cdot a \cdot t \cdot e^{-at}$$

where K = manpower effort in man-months and a = initial slope of the curve. There is a tradeoff between project duration and manpower:

$$K = C/t^4$$

where C = constant. Manpower can be used to shorten development time but is expensive.

2

Planetary environments

The primary goal for a planetary rover is to navigate and traverse an unknown, hostile terrain, recognize and negotiate obstacles, deploy scientific instrumentation, and acquire samples from scientific targets. The range of planetary environments is vast ranging from the relatively benign and flat to extremely rocky and hostile. Most terrestrial mobile robotics platforms are operated in relatively benign environments such as office corridors and the like despite recent emphasis on "embodied" or "situated" robotics paradigms which emphasize the necessity for dealing with realistic (and so uncompromising) environments [37, 38]. Terrestrial mobile robotics research tends to employ primitive mobility systems such as differential drives and tricycle configurations involving two powered wheels and a castor wheel [39]. Planetary robotics does not have that luxury—planetary environments are rugged, hostile, and a priori unknown (e.g., the Martian terrain is both rocky and sandy, Figure 2.1). Planetary rovers have a number of additional critical constraints that are generally absent from traditional terrestrial mobile robots [40]:

(i) adverse terrain characterized by rocks, cliffs, crevasses, etc. with few features for self-localization;
(ii) lack of a priori data on specific features of this environment to be explored;
(iii) extensive signal time-of-flight and limited communication windows to Earth implying a need for high degrees of autonomy;
(iv) hostile ambient conditions including thermal extremes and dust environments;
(v) limited power availability;
(vi) high-reliability requirements which limit mechanical complexity.

It is these issues, which impose more stringent constraints on planetary rovers than are normally traditionally associated with terrestrial mobile robots, that have significant impact on the design and methodologies employed in planetary rovers. The chassis typically comprises around 30% of the rover's total mass.

© Springer-Verlag Berlin Heidelberg 2016
A. Ellery, *Planetary Rovers*, Springer Praxis Books,
DOI 10.1007/978-3-642-03259-2_2

Figure 2.1. Terrain on Mars: (a) rocky and (b) sandy [credit NASA JPL].

The NATO Reference Mobility Model (NRMM) emphasizes performance characteristics based on [41]: (i) maximum speed and turning radius; (ii) traction for overcoming resistive forces to motion; (iii) vehicle maneuvrability for obstacle avoidance; (iv) ride comfort. Performance parameter (iv) is not considered further as suspension is not generally regarded as high priority for robotic rovers and traversal speeds of planetary rovers are low ~10–20 cm/s—low speeds under 3 m/s imply that quasistatic analysis is sufficient to characterize their dynamics. Performance parameter (i) is determined by parameters (ii) and (iii). Maximum speed will be determined by: (i) motor torques; (ii) slopes; (iii) incidence of obstacles (which determines mean free path); (iv) surface traction on soil. Turning radius will depend on the geometry of the vehicle and the nature of the turning mode and strongly influences parameter (iii)—skid steering which is adopted in tracked vehicles and small microrover vehicles offers the highest turning maneuverability at the expense of power consumption. Most vehicles with forward and aft motion capability can turn through skid steering. Ackermann steering is commonly implemented on planetary rovers at or above the microrover class.

All terrestrial surfaces are covered in fragmented grains resultant from ~physical fracturing of surface rock by impact flux, thermal cycling, and chemical or other types of weathering—regolith which is similar in constitution to terrestrial soil but without soil's moisture and organic constituents. Lunar soil is more finely fragmented with fewer boulders than Mars due to the lack of an atmosphere to filter out microtektite bombardment. We consider the most popular venues for rover missions—the Moon and Mars—but rover missions are not exclusive to these planets (e.g., Venus, gas giant moons, small bodies of the Solar System, etc.). There are certain features common to most terrestrial planets, the Moon and Mars especially. The properties of the lunar regolith are known from both in situ measurements and returned samples. The properties of Martian soil are known to a much lesser degree through in situ measurements. In general, extraterrestrial environments are characteristically dry which has implications for the physical properties of the regolith. This is not the case for gas giant moons of interest— Europa (Figure 2.2), Ganymede, Callisto, and Enceladus—which are essentially covered in icy surfaces with evidence of liquid water flows.

The nature of planetary terrain will have the greatest influence on the selection of the rover chassis design which in turn imposes implications for rover mass and power design budgets. For this reason, planetary rover design begins with a reference terrain. Similarly, like all spacecraft that have landed on planetary surfaces, rovers are designed for hostile environmental conditions that are rare or absent on Earth. The planetary day/night cycle imposes thermal cycling and deep thermal loads: for instance, a Martian day (sol) is slightly longer than Earth's and Mars is farther from the Sun than Earth but the ambient temperature is mediated by the Martian atmosphere; the Moon has a 2-week day/2-week night and resides at essentially the same distance from the Sun as Earth but there is direct exposure to the Sun or the cold of deep space. The distance to the Earth impacts the rover communications architecture limiting teleoperation to the Moon and necessitating

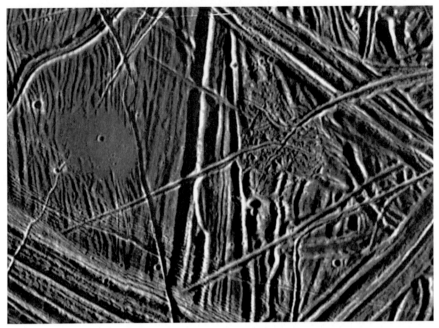

Figure 2.2. Surface of Europa [credit NASA JPL].

autonomous behavior for Mars operations. General spacecraft design issues relevant to planetary rover design are covered in other texts [42].

2.1 THE MOON

The Moon is a high priority for robotic exploration prior to human missions [43]. Although U.S. intentions towards the Moon are unclear at present, China's ambitions are focused on the Moon. This began with its first manned launch in 2003, followed by robotic missions—Chang'e-1 and Chang'e-2 orbiters and Chang'e-3 the lander/rover—to the Moon. These are due to be followed by the Chang'e-4 carrying another lander and rover to the lunar surface, and finally Chang'e-5 to return samples from the Moon. China is also committed to its manned space station laboratory Tiangong-1. The Moon has a surface gravity of $1.62\,\text{m/s}^2$ and a temperature variation of 120–380 K at the equator and as low as 40 K at the poles. It has no atmosphere. Its crust has an average thickness of 70 km being thinner on the nearside than on the farside. Below the crust is its mantle which overlies a small solid core of ~340 km radius. The Moon therefore has no global magnetic field but some of its surface rocks exhibit remanent magnetism. Its center of mass is offset from its geometric center by 2 km towards

Earth. The Moon is tectonically dead and is tidally locked into a resonant orbital revolution around the Earth which equates to the Moon's axial spin (synodic day) of around 29 days. The sub-solar point remains within $1.59°$ of the equator throughout the lunar cycle, so high equatorial regions exhibit high solar inclination angles. The nearside equatorial sites visited by the Apollo missions were geochemically unusual with elevated concentrations of KREEP (potassium, rare earth elements, and phosphorus) elements resultant from impact ejecta from the formation of the 1200 km diameter Imbrium impact crater $\sim$3.85–3.90 Gyr ago. The Moon's surface is characterized by rugged highlands characterized by anorthosite rich in Ca and Al and younger smooth "maria" plains of basalt rich in Fe and Mg, the highlands representing the more ancient crust. The smooth maria are concentrated on the nearside facing the Earth. Due to constant meteoritic flux over the eons, the lunar surface is heavily scarred by impact basins and craters. In particular, the lack of atmosphere allows microtektitic bombardment which has pulverized the surface layer into a fine-grained regolith.

Lunar rocks are constructed from minerals and glasses that formed under negligible oxygen partial pressure and in the absence of water. Minerals are solid crystalline solutions. Only four minerals—plagioclase feldspar, pyroxene, olivine, and ilmenite—comprise over 98% of the lunar crust. The remaining <2% is primarily potassium feldspar, metal oxide minerals (such as rutile TiO_2, and chromite $FeCr_2O_4$ but little evidence of hematite or magnetite), metal iron grains, troilite FeS, zircon $ZrSiO_4$, and calcium phosphate. Feldspar is an aluminosilicate while plagioclase is a type of feldspar being a Ca-Na-aluminosilicate. Most lunar feldspare are of the plagioclase type which varies between albite ($NaAlSi_3O_8$) and anorthite ($CaAl_2Si_2O_8$). Anorthite is a Ca-rich form of plagioclase feldspar which is dominant on the Moon. Almost all lunar Al exists in plagioclase form. Pyroxene is a Mg-Fe-Ca silicate (e.g., enstatite $MgSiO_3$, wollastonite $CaSiO_3$, and ferrosilite $FeSiO_3$ end members), clinopyroxene ($Ca(Mg,Fe)Si_2O_6$) is a pyroxene containing Ca (associated with mare basalts), while orthopyroxene (($Mg,Fe)SiO_3$) is a pyroxene containing negligible Ca (associated with highland rock). Olivine is a Mg-Fe silicate whose composition varies between forsterite Mg_2SiO_4 and fayalite Fe_2SiO_4. Most mare basalt olivine minerals are 80% forsterite and 20% fayalite. Almost all Fe and Mg are incorporated in pyroxenes, olivine, and ilmenite. Lunar rocks have a near constant ratio of Fe/Mn of 70. Lunar highlands are anorthositic due to the high incidence of the plagioclase feldspar mineral anorthite and small amounts of pyroxene and olivine. Lunar basalts of the maria are comprised mostly of silicates (olivines, plagioclase feldspars, clinopyroxenes). Lunar silicates are the most abundant minerals which include pyroxene ($Ca,Fe,Mg)_2Si_2O_6$, plagioclase feldspar ($Ca,Na)(Al,Si)_4O_8$, olivine ($Mg,Fe)_2SiO_4$. Lunar oxides are the next most abundant minerals, particularly in the mare basalts which include ilmenite ($Fe,Mg)TiO_3$, spinel $MgAl_2O_4$, chromite $FeCr_2O_4$, ulvospinel Fe_2TiO_4, and hercynite $FeAl_2O_4$. In the northwest nearside of the Moon lies the Procellarum KREEP Terrane which is enriched in pyroxene rather than plagioclase with enrichment in K and rare earth elements. Silica minerals SiO_2, such as quartz, and most sulfide minerals are rare on the Moon. The paucity of silicate minerals due to the lack of hydrothermal precipitation of silica yields little granite (lunar granite does not contain mica or amphibole). However, the silica minerals concentrate in association with KREEP elements. Since sulfides are rare, consequently elements Cu, Zn, As, Se, Hg, Ag, and Pb are in very low abundance. Water-containing minerals such as clays, micas and amphiboles are absent—most ore-forming processes on Earth are based on water so ore concentrations do not exist on the Moon.

Lunar water has been detected in the hydrous mineral apatite, a constituent of mare basalts and highland anorthosite [44]. Hydrogen isotope analysis indicates that much of this water may originate from cometary delivery. The Moon is expected to have water ice at its poles in permanently shadowed craters where temperatures are $\sim$100 K acting as cold traps. Volatiles including water would have been delivered to the Moon by cometary impacts over the eons. Characterization of any water ice deposits is one major rationale for robotic rover missions in order to support such human exploration. The U.S. lunar orbiter missions Clementine (1994) using S-band bistatic radar and Lunar Prospector (1998) using neutron spectroscopy mapped the lunar farside and polar regions and their instruments suggested that $\sim$10^9 tonnes of water ice may reside in permanently shadowed craters at both poles—these data were indirect in the form of radar reflectivity and hydrogen signatures, respectively, and the form of the "water ice" is still unknown until further measurements are made (other possibilities include methane ice or hydrogen impregnation). The Indian Chandrayaan-1 spacecraft's mini-synthetic aperture radar (SAR) suggest the existence of water ice at the lunar North Pole. The Lunar Reconnaissance Orbiter (LRO) mission (2009) performed orbital surveying of the lunar South Pole and worked in tandem with LCROSS (Lunar Crater Remote Observation and Sensing Satellite). LCROSS impacted the lunar South Pole region ejecting a plume of debris and volatiles which was analyzed spectrally by the LRO. The measurements suggest the existence of molecular hydrogen (amongst other things), possibly derived from water ice estimated to comprise 5.6% of the floor of the shadowed crater Cabeus that was targeted. It is not clear if Cabeus is unusually enriched as might be expected from a former cometary impact site or representative of permanently shadowed craters. Nevertheless, the existence of water ice has yet to be confirmed with confidence. Confirmation of lunar water deposits, locating mineral supplies and the creation of topographic maps are high priorities using high-resolution cameras, laser altimeters, IR spectrometers, X-ray/gamma-ray spectrometers for future human mission scenarios. The existence of water resources on the Moon will enable support of human lunar missions at much lower cost particularly for colonization.

The solar wind and cosmic rays have implanted volatiles into the lunar regolith. Solar wind implantation of gases is enhanced within the polar regions of the Moon by virtue of its low temperature <100 K reducing their desorption rate which more than compensates for lower solar wind flux at high latitudes [45]. This suggests that the polar regions are the most promising sites for extraction of both hydrogen (which may be molecular gases or water ice) and He-3 with a much reduced concentration of $[He\text{-}3] = 3(0.025)(4 \times 10^{-4})H_{max}$. The Moon is thus the

nearest source of He-3 (rare on Earth) for clean nuclear fusion. Such solar wind–
implanted gases may be released by heating to 700°C which may be possible with
lunar infrastructure development. In situ resource utilization (ISRU) at the lunar
South Pole has been proposed to exploit water ice deposits. Water deposits offer
the prospect of life support and the manufacture of propellant, oxidizer, and fuel
cell reagents. Water is the most critical commodity for life support as well as
providing a source of oxygen on electrolysis. Such water deposits are expected to
lie at shallow depths ∼1 m [46]. Although deep drilling requires human intervention, sample collection and scientific instrument deployment can be performed
robotically. The hydrogen distribution as measured by Clementine and Lunar
Prospector was limited to ∼0.5–10 km/pixel, insufficient for targeting sites for
landing (but the U.S. Lunar Reconnaissance Orbiter launched in 2009 is designed
to investigate potential target sites for landing). Lunar mineralogy also offers a
promising in situ resource of oxygen, silicon, iron and titanium in high-yield
concentrations through carbothermic reduction. Such in situ resource capabilities
offer the possibility of long-term commercial potential.

There are a number of science objectives that make the Moon an attractive
target [47]. The Moon has scientific interest on its own merits with a rich geology.
Regions such as impact crater interiors exhibiting ∼30 K temperatures will trap
and retain water ice and other volatiles derived from cometary and asteroid
impacts over the eons. Given the high priority of characterizing the water deposits
at the polar regions, the South Pole Aitken basin is favored for robotic rover
missions. The South Pole Aitken crater (180°W, 57°S) on the farside is 2,250 km
in diameter. This makes it the largest impact crater in the Solar System. It is the
oldest and deepest impact basin on the Moon, formed ∼3.9 Gyr ago during the
terminal cataclysm phase of early solar system accretion. It has an altitude which
varies from 6 km above the local horizontal (e.g., Malapert Mountain ∼120 km
from the South Pole) to −5 km depth below the local horizontal (i,e.. a topographic profile of 13 km range). Its depth suggests that the impact penetrated the
crust and excavated lunar mantle material, which has not so far been sampled.
The mass of the impactor that formed the South Pole Aitken crater is estimated
at 1.5×10^{19} kg. The surface is covered with regolith of pulverized rock and dust.
Areas of the basin floor are in permanent shadow and so may have trapped
volatiles. The following analysis of potential landing sites and rover missions has
been developed by Philip Stooke at the University of Western Ontario. The 39 km
diameter Shoemaker–Faustini crater (87.3°S, 77.0°E) is a promising target with
[H] = 0.17% = 1,700 ppm and [He-3] = 10^{-7} ∼ 100 ppb where the temperature is
estimated at ∼100 K. This was the primary landing site for the Hopper sample
return mission concept. However, there are other possibilities.

A cross section through the South Pole Aitken basin exhibits a stair-like
topography (Figures 2.3 and 2.4). Three concentric rings may be defined through
the Aitken basin. Ring 1 has a diameter of 2,400 km and the elevation drops
sharply by 3–4 km. Ring 2 has a diameter of 2,100 km and the elevation drops by
2–3 km. Ring 3 has a diameter of 1,500 km enclosing the deepest parts of the basin
with a sharp drop of 2–3 km depth on transition to rings 2 and 3. Now, a good

Figure 2.3. South Pole Aitken (SPA) basin [credit NASA].

landing site requires: (i) good illumination for solar power generation; (ii) good view to Earth for communication; (iii) feasible rover routes to permanently dark crater regions; (iv) accessibility for orbital imaging; (v) reasonably rock-free and flat terrain. As the Moon's axis is tilted $5°$ from the Earth–Moon ecliptic plane, communications from the Moon to Earth require a minimum elevation of $5°$. Polar landing sites offer a line of sight to Earth with a maximum azimuth $7°$ above the horizon during midsummer with an azimuth variation of $±5.5°$ over a lunar day. These margins are too severe to guarantee continuous communications (accounting for topographic obstruction) suggesting the need for a communications relay satellite in lunar orbit. The Malapert Mountain peak ($86.0°$S, $2.7°$E) has a continuous line of sight to Earth but $<74\%$ illumination. Older craters at the South Pole exhibit rounded edges with thick ejecta blankets. However,

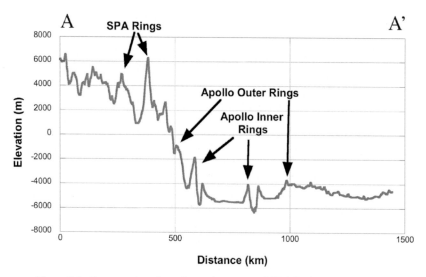

Figure 2.4. Cross section through northern part of SPA basin [credit NASA].

boulders ~1 m diameter may be present at and nearby the rim but may reach as much as 70 m in diameter (this will impact rover operations). Younger craters with steeper sides ~20–25° will present greater challenges to rover traction suggesting that the traditional rocker-bogie six-wheeled chassis design will provide insufficient traction. This favors a tracked solution such as a larger version of Nanokhod or the elastic loop mobility system (ELMS). Near the lunar poles, local topography will obscure solar incidence to certain locations. There will be topographically elevated regions that will be illuminated by the Sun near-continuously—these may lie in close proximity to the permanently shadowed regions. There are several promising regions where high illumination (86% of the year) occurs in close proximity to permanently shaded craters [48]: (i) Point A on the rim of the Shackleton crater (89.68°S, 166.0°W); (ii) Point B on a ridge close to the Shackleton crater (89.44°S, 141.8°W); (iii) Point C on the rim of the De Gerlache crater (88.71°S, 68.7°W); (iv) Point D (99.79°S, 124.5°E) on a ridge from the Shackleton crater rim along the 120°E longitude line. The most illuminated regions occur on the western rim of the Shackleton crater (Figure 2.5).

A, B, and C are the most illuminated regions with a minimum of >70% illumination—A (on the rim of Shackleton crater) and B are 10 km apart and both are collectively illuminated for 98% of the time but no single spot is continuously illuminated (peaks of eternal light). Point D offers illumination >86% of the year, and during the summer it is illuminated continuously for 5 months while individual eclipses during the 7-month eclipse season are short. However, nearby access to permanently shadowed regions to detect volatiles is necessary (within 10 km preferably). The proposed Hopper lander which hops between three sites

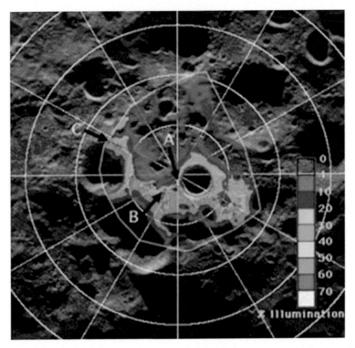

Figure 2.5. The 20 km diameter by 3 km deep Shackleton crater indicating % solar illumination [credit NASA].

50 m apart using thrusters would be insufficient to locate permanently frozen water ice deposits given the separation of peaks of eternal light and permanent shadow. A lunar rover must have access to nearby, permanently shadowed craters. The most appropriate landing sites are selected on the ejecta blankets of the Shackleton and de Gerlache craters.

Sites 1 and 2 on the Shackleton rim provide potential access to the interior of the 20 km diameter and 3 km deep Shackleton crater. However, Shackleton is relatively young, implying steep walls and reduced volatile inventory. The exterior slope of the Shackleton crater is less severe than its internal slope. Sites 3, 5, and 6 offer short drives into nearby small craters but small craters are likely to have limited volatile inventory. Site 4 requires a 20 km traverse into an older crater with both shallow interior slopes and a likely significant volatile fraction. Communication may be maintained from the lander at site 4. As the rover traverses, it will lose line-of-sight communication with the lander implying the need for relay deployment (mass expensive) or through an orbiter satellite. Site 7 offers several routes to the de Gerlache crater. The baseline landing site is site 4 implying the need for significant mobility for outward and return trips from the lander (Figure 2.6).

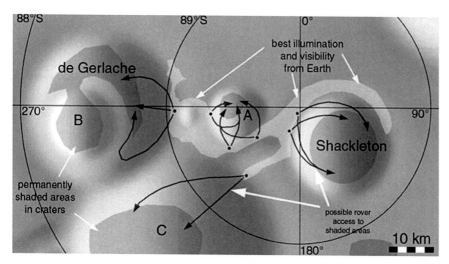

Figure 2.6. Potential rover routes from peaks of eternal light [credit Phil Stooke, University of Western Ontario].

The lander must land at an eternally sunlit region for constant solar power adjacent to a permanently shaded region within the Shackleton crater with a high-mobility rover on board. The baseline rover traverse will be highly challenging—20 km each way over extremely rugged terrain within the Shackleton crater with high vertical variation. There will be limited and variable solar illumination for the rover. The traverse is assumed to be restricted by the 14-day daytime window, implying a minimum speed of under 3 km/24 h day. If operational around the clock, this imposes a minimum speed of 120 m/h. This does not include the scientific investigation at the target site. This will be highly challenging.

As an example, the proposed lunar resource prospector/RESOLVE (regolith and environment science and oxygen and lunar volatile extraction) mission for 2018 is an ISRU mission to locate near-subsurface volatiles, such as water, extract and analyze samples of such, and demonstrate chemical processing of lunar regolith at a shadowed region near the lunar South Pole. The selected site requires periods of sunlight to allow the use of 250 We solar panels for energy (supported by 3.5 kWh rechargeable batteries) and continuous line-of-sight visibility to Earth (Antarctica stations) yet provide access to shadowed regions with high hydrogen concentration and high ice stability. The lunar resource prospector is based on a 243 kg robotic lunar rover (including payload) for a surface mission of 1–3 km over 5–7 days to the Cabeus crater (85.75°S, 45°W) [49]. The landed mass is 1,285 kg while the launch mass on the Atlas V is 3,476 kg. The 72 kg integrated RESOLVE payload comprises four subsystems: (i) a sample site selection system consisting of a neutron spectrometer and near-infrared spectrometer to locate hydrogen/water sources; (ii) a sample acquisition and transfer subsystem comprising a 1 m drill/auger and core transfer mechanism to perform three to five 1 m drill cores; (iii) the oxygen and volatile extraction node (OVEN) for sample chemical processing consisting of a sample heating reactor oven to 900°C; and (iv) the lunar advanced volatile analysis (LAVA) system for evolved volatile characterization comprising a gas chromatograph–mass spectrometer (GCMS) and a water capture device involving a condensation chamber. A 1 m core is divided into eight segments, which are to be processed. They are to be heated to 150°C for volatile extraction and to 900°C with hydrogen for hydrogen reduction to extract oxygen. The evolved volatiles are to be analyzed by the GCMS.

Lunar dust is ubiquitous and adheres to everything due to its frictional and electrostatic properties. The fine-grained dust is levitated by solar ultraviolet radiation during the day and by solar wind flux during the night [50]. Lunar soils have increasing abundances of agglutinate glasses and nanophase metallic iron content with decreasing size (Gaussian average size is 3.0 μm) [51]. There are a significant number of particles <2.5 μm which are regarded as toxic to the respiratory system. Lunar soil has an average aspect ratio of 0.7 and exhibits angular jagged shapes making it highly adhesive. This presents a major problem for extended lunar operations and will require aggressive mitigation techniques [52]. Dust mitigation to prevent settling of dust particles and their adhesion by van der Waal forces onto solar arrays, thermal radiators, and optical surfaces may be achieved through several mechanisms—vibromechanical, (electro)magnetic, and electrostatic removal. Although magnetic approaches are simple [53], electrostatic ones are favored as the most effective [54–56]. Lunar soil of all particle sizes is also a significant problem for mechanical components such as motors and gears used on rovers, manipulators, and drills. For robotic applications, rotary seals must be used as the primary or secondary mechanism for dust mitigation. Rotor/stator airgap, motor alignment, brush/commutator wear, gearing friction and meshing, sheave seating, connector coupling integrity, bearings, and motor temperature are highly susceptible to dust contamination. Rotary seals are used for sealing between moving parts. While static seals rely on solid–solid contact, rotary seals require fluid lubrication, usually hydrocarbon oil of low viscosity. Moving parts require lubrication to minimize friction and wear. Most wet lubricants operate on the basis of their viscosity and thin-film adhesion. Wet lubricants are made of petroleum-based and synthetic chemicals. For space use, PTFE seals are most appropriate as a dry lubricant which is resistant to temperature excursions and has high chemical stability. Teflon O-ring seals have been tested and found to be inadequate for high-dust environments (modeled with angular silica dust) [57]. Magnetic fluid rotary seals can be used in vacuum and low pressures offering zero leakage for fluid sealing [58]. Two opposing magnets create a magnetic circuit and contain a ferrofluid. Ferrofluids are colloidal suspensions of paramagnetic particles of 10 nm diameter within a fluid which imparts paramagnetic behavior to the fluid as a whole. The application of magnetic fields allows control of such fluids which can be used to seal two regions. This makes them ideal for rotary seals in motors.

2.2 MARS

The Martian terrain comprises five major divisions: (i) highland rock; (ii) lowland
rock; (iii) volcanic regions; (iv) channel systems; (v) polar regions (Figure 2.7). The
northern hemisphere is flat with an average elevation of -1 to -3 km compared
with the mountainous southern hemisphere with an average elevation of $+4$ km.
The northern hemisphere has more moderate seasons than the southern hemisphere—the mean daily temperature ranges from 215 K at the equator to 150 K at
the poles. At the equator, night-time temperatures drop to 150 K and reach 290 K
at noon, but Martian soil remains close to the average of 215 K. The most obvious
features are water-cut channels and flood basins etched across the Martian
surface. The heavily cratered ancient ($\sim$3.5–4 Gyr old) highlands of the southern
hemisphere are characterized by river valley networks contrasting with the
smoother, younger low-lying plains of the northern hemisphere with ancient shorelines separated by an equatorial region of canyons and volcanoes of the Tharsis
plateau. The Martian crust is 40 km thick in the northern hemisphere and 70 km
thick in the southern hemisphere. The crust is thickest under the Tharsis volcanoes
and thinnest at the giant Hellas impact basin in the southern hemisphere. The
main volcanic regions of Mars are Tharsis and Elysium. Tharsis at a mean altitude
of 10 km at the equator represents the region where tectonic faults and volcanoes
are generated. The Tharsis bulge was created 3.7 Gyr ago but the caldera floors of
the Tharsis mountains including Olympus Mons were formed only 100–150 Myr
ago indicating recent volcanic activity.

   Martian history is divided into three major epochs—Noachian (southern
highlands), Hesperian (northern lowlands), and Amazonian (volcanic regions).

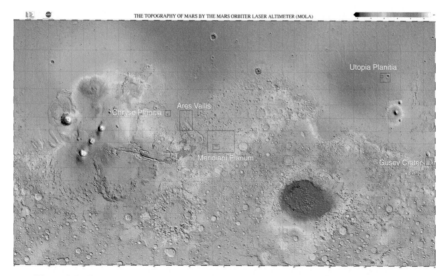

Figure 2.7. Topographic map of Mars generated from MOLA data [credit NASA].

The boundaries are calibrated from the lunar cratering record—Noachian/Hesperian (3.5–3.7 Gyr), Hesperian/Amazonian (2.9–3.3 Gyr). Of its 20 large volcanoes, Mars sports the largest shield volcano in the solar system—Olympus Mons east of Tharsis—with a height of 22 km, caldera diameter of 60×90 km, and depth 3 km. The Martian surface has been substantially altered by the action of liquid water in two phases—an early epoch when the valleys were formed and a later epoch when outflow channels were formed, especially at the interface between northern lowland and southern highlands where there is an elevation discontinuity of around 5 km [59–61]. There is evidence of substantial quantities of water ice 1–2 m below the surface regolith of Mars at latitudes exceeding $50°$ sufficient to flood the planet to a depth of 500 m. Further near-surface water is indicated at much lower latitudes though its form in not known (e.g., chemically or physically bound in minerals). Adsorbed water has much stronger bonds due to van der Waal forces. The Viking experiments heated soil samples up to temperatures of $500°C$ to release water of hydration indicating an integrated average of 0.2% water for all regolith. The discovery of both goethite and jarosite by the MERs indicate chemically bound water in minerals on the Martian surface. The outflow channels and valley networks on Mars have long indicated the presence of liquid water on the Martian surface during different epochs. The outflow channels are large fluvial features formed by catastrophic flooding in the northwest and Chryse regions while the older valley networks were formed by surface runoff fed by groundwater seepage [62]. Valley networks in the southern highlands may have formed near the end of the heavy bombardment phase $\sim$3.5 Gyr ago, which would have maintained clement conditions due to hot ejecta [63]. Groundwater sapping is caused by geothermal melting of near-surface ice. There would be a net migration of groundwater ice from warm equatorial regions where water is unstable to sublimation to the polar regions where it is more stable as subsurface permafrost at shallow depths $<$1 m [64]. Although this subsurface ice is most stable at shallow depths poleward of $40°$ latitude, its incidence is more widespread. The largest outflow channel is Kasei Vallis. Outflow channels were dominant during the Hesperian but some may have originated during the Amazonian.

The valley networks throughout the ancient heavily cratered terrains which formed after the late heavy bombardment suggest that the early Martian atmosphere was capable of supporting a hydrological cycle in a warm, wet climate (1–5 bar CO_2) during the Noachian epoch [65]. A thick CO_2 greenhouse atmosphere of $\sim$1 bar would have been insufficient by itself to maintain warm enough conditions for liquid water. However, the formation of CO_2 ice clouds may have trapped thermal radiation in the surface [66]. Alternatively, greenhouse gases H_2O and SO_2 (the latter being a minor constituent supplied by periodic volcanic emissions but which inhibits the formation of CO_2 cloud condensation) ensured stable bodies of liquid water on the surface [67–69]. CO_2 has a strong vibration absorption band at 15 μm and several weaker less significant bands near 9 and 10 μm. Water vapor has rotational absorption bands longward of 20 μm and a weak continuum absorption band between 8 and 12 μm well above 273 K. SO_2 has vibrational absorption bands near 7, 9, and 19 μm and rotational transitions longward of 40 μm. NH$_3$ has vibration absorption bands at 10 μm and a weaker line at 16 μm. CH$_4$ has an absorption peak at 7.7 μm. Reduced gases CH$_4$ and NH$_3$ are unstable to solar UV action, however. For CO$_2$ and CH$_4$, absorption longward of 20 μm is only important at higher pressures. There exist younger (<2.5 Gyr old) valleys in the Tharsis region (Alba Patera) which were formed by local surface runoff while the Martian climate was similar to the present day [70]. There is evidence that the northern lowlands indicate ancient equipotential shorelines of a large ocean below which the topography is unusually smooth and flat with an average depth of 3.8 km below the geoid and a capacity of 1.4×10^7 km^3 [71]. This ancient ocean—Oceanus Borealis—may have persisted up to the late Hesperian. Outflow channels empty into the northern lowlands from Chryse and Amazonis. The northern regions include the Elysium plains. The tectonics of the Tharsis region has not changed appreciably since the late Noachian though it may have been the source of episodic flooding on the northern plains.

The Valles Marineris rift canyon has a length of 4,000 km, a width of 65 km, and a depth of 8–10 km formed with the creation of Tharsis (Figure 2.8). Its layered deposits suggest large amounts of hydrated sulfate salts indicating that catastrophic outflows have occurred triggered by volcanic heat sources. During the late Noachian and early Hesperian, Valles Marineris underwent radical tectonic activity generating outflow channels with the Valles Marineris structures forming during the late Hesperian and early Amazonian. However, volcanic and tectonic activity have continued until the present day (within the last 10 Myr) indi-

Figure 2.8. Valles Marineris [credit NASA].

cated by features in the Tharsis Montes region and evidence of recent water flows—gullies, floods, polar layered northern ice cap [72]. There is evidence of intermittent volcanism within the last 100 Myr at Tharsis and Elysium—Cerberus Fossae to the southeast of Elysium Mons indicates magma flows within the last 3 Myr—this is within the approximate cool-down timescale for magma chambers. Furthermore, this is believed to be the source of the lowland flooding near the north side of the equator indicated by the fragmented plates believed to be ice rafts overlain by volcanic ash and dust erupted from Cerberus Fossae. Exposed Martian gulleys on pole-facing steep crater walls between 30–70° latitudes in both hemispheres indicate recent erosion by liquid water flows within the last ~1–5 Myr [73]. This indicates that there are multiple layers of ice deposits from 0.2 to 1.0 km depth. This recent activity is primarily due to a recent obliquity change to high obliquity exceeding 35° (within the last 5 Myr) for ~10^5 yr. This would have generated sublimation of carbon dioxide and water to give atmospheric pressures >25 mbar generating the precipitation of snow which can flow as glaciers.

Subsurface ice is indicated by the existence of landforms resulting from viscous deformation such as fluidized ejecta craters which requires a minimum of 28% ice at 1 km depth in the permafrost (with 40% ice near the surface) [74]. This implies a global water inventory to a depth of 1 km. Most of the planet, apart from the equatorial regions, has ground ice to high depths. The cryosphere is stable at depths of 3–5 km at the equator to 8–13 km at the poles below which liquid water may exist. Ground ice is unstable at shallow depths at latitudes below 40° where ice sublimes. The freezing point of water may be depressed by salts increasing the latitude range of water melting. Dilute aqueous solutions of H_2SO_4 can depress the freezing point of water by 70°C. The seasonal change in polar deposits on Mars is primarily due to carbon dioxide ice. However, the Martian North Pole is comprised mostly of water ice with variable amounts of dust while the southern cap is also primarily water ice with little dust. Polar ice and permafrost may potentially yield ancient bacteria (a few million years old or so) and gases such as methane, ammonia, hydrogen sulfide, sulfur oxides, and nitrogen oxides. The Martian poles comprise ice/dust layers over 10^6 km² with a thickness of 3–4 km which varies seasonally but which has existed for 10^5–10^8 yr. They possess extensive ground ice in the upper layers of the soil, probably overlying deeper older ice. In particular, the South Pole represents heavily cratered ancient terrain. The large-obliquity cycles of Mars varying by 45° with periods of 10^5–10^6 yr suggest that the poles receive more sunlight at high obliquity. The current Martian obliquity is 25.2° from the plane of the Sun but varies substantially over a few million years. The obliquity of Mars' spin axis has varied chaotically from 0–60° while Mercury and Venus are tidally stabilized by the Sun and the Earth is tidally stabilized by the Moon [75, 76]. For Mars, spin axis precession is driven by solar torques while orbit precession is caused by perturbations by other planets. The Martian obliquity oscillations due to spin–orbit resonance over ~10^6–10^7 years drive the Martian climate [77]. In the past 20 Myr, Mars' obliquity was closer to 45° which would have warmed higher latitudes and vaporized ground ice near the surface. It can lean as far as 60° generating a thick atmosphere due to polar ice evaporation. Periodic catastrophic outflows indicate enormous flood releases $\sim 10^6$–10^8 m^3/s. These short episodes of regional flooding may have persisted for $\sim 10^3$ yr rather than $\sim 10^6$ yr. At greater planetary inclinations of 40–60°, greater sunlight evaporates the northern polar water ice which migrates to lower latitudes, increasing atmospheric pressure ~ 30 mbar, greenhouse effect, and surface temperature [78]. This has a direct effect on the Martian climate including generating precipitation from a water-saturated atmosphere at low latitudes. This may cause permafrost to warm to a thermal damping depth of

$$d^2 = \frac{KT}{\pi}$$

where $K =$ thermal diffusivity and $T =$ oscillation period. This suggests that drilling to a depth of 1 km is required to access undisturbed permafrost. Such ancient permafrost is reckoned to be the most likely candidate for preservation of fossilized remains of biological organisms on Mars [79]—this permafrost is around 3.5 Gyr old compared with Siberian permafrost (5 Myr old) and Antarctic permafrost (8–20 Myr old). Martian permafrost temperatures are much lower ($< -90°$C) possibly allowing longer preservation of biological viability. However, the radiation flux from U, Th, and K decay of 0.2 rad/yr makes such long-term survival unlikely. There is substantial evidence of glacial forms over a wide range of longitudes, latitudes, and altitudes including equatorial regions—this suggests that the Martian climate has changed relatively recently (within the last 5 Myr). There are glacial features indicating recent surface ice at equatorial regions as well as polar features resembling eskers and karsts [80]. This provides support for obliquity-driven climate in which the stability of surface water varies with time. The near-surface soil across much of the Martian subsurface may be 50% water ice.

Heat loss from the Martian interior caused a decline in volcanic activity while water evaporated as the Martian atmosphere thinned. The Martian climate has undergone significant evolution. It may have been lost during the heavy bombardment phase due to impact erosion [81]. Alternatively, it occurred subsequently by a combination of solar wind–induced sputtering, Jeans escape of H, and photochemical loss of N and O [82]. The history of Martian volatiles is indicated by measurements of D/H, ^{18}O/^{16}O, and ^{38}Ar/^{36}Ar indicating that there was a massive early loss of Martian atmosphere [83]. There is a large deuterium enrichment D/H in the atmospheric water vapor indicating major loss of H equivalent to a global ocean thickness of 100 m. Martian atmospheric density is low at 0.02 kg/m^3 having been sputtered over the eons by the solar wind after the freezing of the Martian core (which also provides an explanation for the lack of detected carbonate deposits). The atmospheric pressure on Mars is around 6–7 mbar on average but varies seasonally by 25% due to the condensation/sublimation of carbon dioxide at the poles. The average wind velocity at the surface is 2–9 m/s but can gust up to 60 m/s. The largest dust storms begin in southern spring as Mars heats up, sometimes encircling the whole planet. A typical dust storm is 200–300 m in diameter and tends to occur at 2–3 PM local time. During summer the surface winds are low (~ 2–7 m/s) but increase to 5–10 m/s during autumn and rise to

20–30 m/s during dust storms. The NASA Ames Mars general circulation model indicates a 6-day average oscillation in atmospheric heating rates generating localized dust storms in the northern spring in lowland regions [84]. These dust devils are generated by local surface heating generating plumes of warm air in the surrounding cooler air which then circulate in convection cells.

In situ analysis of Martian rock indicates basaltic and andesitic rock enriched in oxides and sulfides. The dominant silicate minerals are olivine, plagioclase feldspars, pyroxenes (clinopyroxene and orthopyroxene in Hesperian and Noachian deposits, respectively), and quartz mixed with Ti–Fe oxides. Most Martian meteorites are basalts with pyroxene–olivine secondary minerals. Basalt dominates the southern hemisphere while andesite is common in the northern hemisphere. This follows a well-defined pattern of igneous rock melting separating out the silica content. Mantle material is primarily olivine which, if melted, forms basalt. Basalt if melted yields andesite, melted andesite yields dacite, and dacite when melted yields granite. Only miniscule amounts of granite exist on Mars. It appears that Mars has been host to transient lakes which have undergone repeated cycles of drying and wetting generating acidic salt flats built from layers. The Meridiani plains also exhibit distinctive arc wave patterns indicative of shallow-water lakes. Hematite and goethite are common iron (oxy)hydroxides found on Mars. In Meridiani Planum, Opportunity discovered "blueberries" of localized gray hematite (α-Fe_2O_3) deposits formed in water associated with evaporitic deposits. Mineral grain size distribution is consistent with water deposition rather than wind. Goethite (α-$FeOOH$) is a hydrated iron mineral commonly associated with hematite suggesting aqueous formation. Ferrihydrite ($5Fe_2O_3 \cdot 9H_2O$) is the anhydrous precursor to hematite and goethite—goethite is formed at low temperature and low pH while hematite is formed at higher temperature and higher pH. The discovery of the mineral jarosite suggests that water was acid and briny. Martian mineralogy is reviewed in Chevrier and Mathe (2007) which stresses the importance of the formation of sulfate salts by acidic solutions due to water evaporation and hydrothermal processes indicated by SNC meteorites [85]. Phyllosilicates—clays, serpentine, and talc—are the common derivaives of water action on silicates. Nakhlite meteorites indicate smectite–illite clay mineral veins and other evaporite minerals such as gypsum and salts within olivine clinopyroxenites [86]. Clay minerals have been detected in the oldest regions of Mars indicative of aqueous processes. Endeavour crater contains clay minerals that require near-neutral pH to form. Acidic solutions destroy carbonates and inhibit the formation of clays. Rudimentary dating of K-40/Ar-40 ratios by the Curiosity rover's sample analysis at Mars (SAM) instrument suggests that clay deposits were laid down as recently as 80 Myr ago. Maghemite is believed to be a constituent of Martian rock varnish resulting from weathering under Martian conditions. Magnetization in Martian rocks is dominated by magnetite (γ-Fe_3O_4), titanomagnetite ($Fe_{3-x}Ti_xO_4$), and maghemite (γ-Fe_2O_3). The soil has high Fe and S content but low Si and Al content compared with terrestrial references. Fe is generally in the Fe_3O_4 oxide and oxyhydroxide form. The high Cl and S content suggests the presence of soluble sulfate salts. Kieserite ($MgSO_4 \cdot H_2O$), gypsum ($CaSO_4 \cdot 2H_2O$), and epsomite ($MgSO_4 \cdot 7H_2O$) have been identified, which typically comprise over 50% bound water. High sulfate abundances are derived from sulfide minerals (such as troilite FeS) in basalts by acid aqueous conditions. However, the existence of adjacent olivine deposits suggests that the aqueous conditions were short lived. Jarosite (hydrated iron sulfate mineral $(K,Na,H)Fe_3(SO_4)_2(OH)_6$) has been identified on the Martian surface and is indicative of aqueous deposition. Hydrothermal processes are generally implied—Río Tinto in Spain may be representative in which jarosite and iron oxides are deposited in acidic conditions under which acid-tolerant organisms thrive. Martian soil appears to be saturated in the salt perchlorate, such as $Mg(ClO_4)_2$ which are found only in arid environments on Earth. Perchlorate salts can significantly reduce the freezing point of water providing one possible explanation for groundwater runoff. Perchlorates may be used as an energy source when coupled to a reduced reagent in the presence of water (ammonia perchlorate is used as an oxidizer in heated rocket fuel). One explanation for the Viking results was that when heating soil to release organics, perchlorates become powerful oxidizers. It was originally assumed that chlorinated organic molecules were the result of cleaning agents used to clean the Viking spacecraft. However, perchlorates may be induced through UV photochemistry at the surface of the soil which diffuses into the subsurface. This diffusion in combination with the UV radiation into the near surface would destroy any organic materials and any biomarkers. The predicted large-scale precipitation of carbonates such as siderite ($FeCO_3$) derived from an earlier thicker atmosphere have yet to be identified (detectable with signatures near 2.5, 4, and 7 μm). Low concentrations <5% of carbonate—hydrated magnesium carbonates—have been detected in the soil but their paucity requires explanation. One possibility is that their low solubility means that they are overlain by more soluble minerals such as sulfates. Alternatively, the early Martian atmosphere included volcanically outgassed greenhouse gases SO_2 and H_2S. Together with ferric iron in solution, this would have created an acidic (H_2SO_4) oceanic environment and inhibited siderite formation, the first carbonate mineral to precipitate from solution [87, 88]. Soluble sulfur-bearing minerals result in acidic solutions which destroy carbonates. There is evidence in support of weathering under low pH (acidic) conditions [89].

The primary geological objective is to find sedimentary deposits which have been found across large depression areas of the Martian surface—craters, canyon floors, central plains, etc. Layered terrain which is widespread on Mars provides evidence of stratigraphic sedimentary deposition and of intermittent and ancient water flows which created this stratigraphic layering (Figure 2.9). In many impact craters (in particular, those with well-defined rims) there is evidence of standing bodies of water (e.g., the 1,000 km diameter Argyre basin appears to have been filled with water).

The search for evidence of water on the Martian surface is the primary scientific goal because of its implications for Martian life early in its history [90]. During the wet Hesperian period, conditions may have been suitable for the origin and evolution of microbial life. Archean life arose rapidly on Earth, indicated by their (disputed) geological record, suggesting that archea may have arisen under

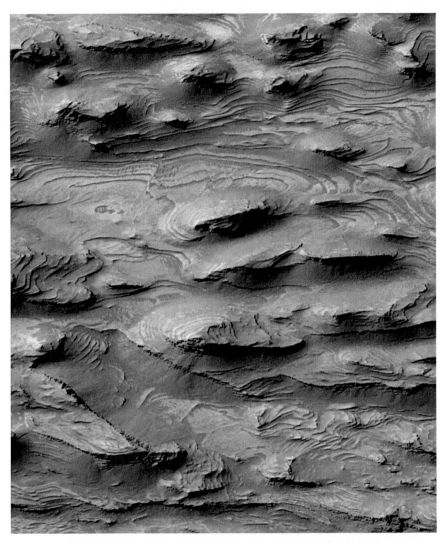

Figure 2.9. Layered sedimentary deposits on Mars [credit NASA].

similar conditions on Mars at this time. Assuming Earth-like evolution with similar starting ingredients and conditions, early surface pioneers of Mars may have resembled the cyanobacteria found in the 3.5 Gyr old record of Apex Chert on Earth which currently dominate in Antarctic deserts. Liquid water lakes could have persisted under protective ice cover for significant periods even as the climate gets progressively colder resembling the ice-covered lakes of the Antarctic Dry Valleys. These lakes remain liquid by geothermal energy input but are overlain by

3–5 m of ice (e.g., Lake Hoare). Hydrothermal systems may result from volcanic hot spots for several million years and cause the flow of mineral-rich fluids through rifts and faults though there is no evidence of shallow extant volcanic activity [91]. The most likely site for such is the Tharsis region which is the youngest volcanic region on Mars. It is plausible that Antarctic conditions are representative—endolithic organisms exhibit metabolic activity with little water and there may exist similar or dormant reservoirs of organisms on Mars. Photosynthetic organisms in polar soil biofilms, in the stromatolites of the beds of ice-covered lakes, and in the endolithic communities of translucent sandstone of the Trans-Antarctic Mountains represent potentially diverse habitats. Some microbial compounds (e.g., hopanoids, porphyrins, and isoprenoids) are recalcitrant and are found in the fossil record up to 2.5 Gyr ago. These represent potential analogs to possible extinct Martian biota. There is little doubt that Martian lacustrine deposits represent high-priority targets for astrobiological exploration. Although there exists little evidence of Noachian lacustrine activity due to erosion, lacustrine activity indicated by paleolake deltas was maximum during the Upper Hesperian period (2.5–2.1 Gyr ago) which decreased during the Amazonian epoch [92]. There are deepwater paleolakes indicating wave-cut terraces near their rims and flat floors with deltas at the mouths of inflowing channels and rarer dry palaeolakes which have high albedo [93]. The former are likely to contain increasingly fine sediment towards the central regions of the lake—gravel, sand, silt, and clays. As the water supply evaporated, evaporates were precipitated from the briny solutions. Carbonates are likely to be deposited centrally from both chemical and biological processes. Dry lakes may experience only periodic flooding and comprise evaporite salts such as carbonates, sulfates, etc. precipitated during the evaporation of water. Evaporites provide a signature for the former presence of water under hot conditions. Large evaporate deposits on Mars have yet to be identified but evaporitic conditions are expected to have existed on Mars (e.g., Wegener crater at 64.3°S, 4°W). Indeed, remnants of life may have been preserved within saline brine pockets within permafrost during Mars' dehydration and cooling which subsequently dried to form evaporites—predominantly halite (NaCl), gypsum ($CaSO_4 2H_2O$) and anhydrite ($CaSO_4$) on Earth. It is conceivable that viable archeal and bacterial halophiles may survive in the frozen state, though the duration of such survival has yet to be established [94]. Caves would be a primary site for investigation as they represent the sites of underground water channel conduits. One of the most promising future sites for exploration is the Elysium Planitia adjacent to the 160 km diameter Gusev crater which terminates in a large branch of the Ma'adim Vallis valley network. They indicate evidence of a subsurface lake of ice beneath the soil.

The problem of dust also afflicts the Martian surface though its behavior is different. Mars is an eolian environment with seasonal dust storms and ever-present dust devils. The problem of dust sealing of mechanisms on the Moon and Mars are similar. On Sojourner, dust sealing was accomplished through the use of Delrin balls as output bearings in anodized Al races without lubrication at movable joints which provided tolerance to dust contamination.

3

Survey of past rover missions

All planetary rover missions to date have adopted wheeled chassis designs across a range of rover sizes for mechanical simplicity and high reliability. This trend looks set to continue for the foreseeable future. One exception was the Prop-M nano-rover on the Russian Mars 3 lander (1971). The 4.5 kg Prop-M used a pair of skis mounted onto legs. Regrettably, the mission failed after apparently landing successfully.

3.1 LUNOKHOD

The first planetary rovers were the two Russian eight-wheeled Lunokhods which landed on the Moon in 1970 and 1973 (Table 3.1) [95].

Table 3.1. Comparison of Lunokhod rovers.

	Lunokhod 1	Lunokhod 2
Operational duration (da)	302	125
Distance traversed (km)	10.5	37.0
Average speed (km/h)	0.14	0.34
Maximum slope (°)	22–27	22–27
Specific energy expenditure (Wh/m)	0.2–0.22	0.2–0.22
Motion resistance coefficient	0.15–0.25	0.15–0.25
Average slippage (%)	5–7	5–7

© Springer-Verlag Berlin Heidelberg 2016
A. Ellery, *Planetary Rovers*, Springer Praxis Books,
DOI 10.1007/978-3-642-03259-2_3

Figure 3.1. Lunokhod rover [credit NASA].

The Lunokhod design was the basis of the Luna probes (less wheels). These vehicles were teleoperated in near-real time from Earth and were based on an eight-wheeled chassis (Figure 3.1). Eight wheels were selected to minimize the wheel size while ensuring high-tractive ability. A deployable rear ninth unpowered trailing wheel measured actual traversed distance without slip. The bathtub design was based around a centralized pressure vessel within which electronics, a radio-isotope heat source and a battery were maintained in air at pressure. A deployable lid with solar cells for power exposed a thermal radiator for daytime operation— at night, the lid was closed over the radiator in survival mode. Lunokhod 1 (Luna 17) had a mass of 756 kg and explored Mare Imbrium traversing 11 km for 11 months (average 0.14 km/h speed) through teleoperation. The 840 kg Lunokhod 2 (Luna 21) traversed 37 km over the more rugged terrain of Mare Serenitatis over 8 weeks averaging 0.34 km/h due to its improved control system and improved TV camera configuration. The duration of uninterrupted motion increased from 50 s for Lunokhod 1 to 7 min for Lunokhod 2. They both employed eight rigid spoked wheels of 51 cm diameter and 20 cm width with a torsion bar–based chassis. The rims of each wheel were constructed from steel wire mesh reinforced by three helical ribbon spring hoops of titanium corrugated with transverse titanium cleats. Both Lunokhods had two gear speeds of 0.8 km/h and 2 km/h and were capable of overcoming a step height of 0.4 m. Steering was accomplished through skid steering. They were controlled from Earth by two five-man teams of teleoperators (commander, pilot, navigator, engineer, and radio operator) on two-hour alternating shifts. Feedback to the pilot was through still image frames every 20 s with a 2.5 s signal delay for Lunokhod 1 which increased to a frame rate of one frame every 3 s for Lunokhod 2. Lunokhod 2 overheated when loose soil fell onto the radiator while the craft climbed out of a crater.

3.2 LUNAR ROVING VEHICLE (LRV)

The Apollo 15, 16, and 17 lunar rover vehicles (LRV) driven by suited astronauts on the Moon in 1971–1972 were manned, four-wheeled vehicles but could be teleoperated from ground if necessary if the two-astronaut crew were incapacitated [96, 97]. The open "buggy-style" lunar rover had a mass of 218 kg with a payload capacity of 490 kg (fully laden mass 708 kg) (Figure 3.2). It was designed for a range of 4 × 30 km traverses within its 78 h operational lifetime at a continuously variable speed of 0–16 km/h. It was 3.1 m in length with a wheelbase of 2.29 m constructed from Al alloy tubing. The Apollo lunar rovers employed wheels mounted onto its chassis by suspension arms attached to a torsion bar. They used flexible, interwoven Zn-coated metallic wire mesh–based wheels. The wheels each comprised an Al hub and had 81.8 cm diameter × 23.0 cm wide tyres. Ti chevrons covering 50% of the tyre surface acted as grousers to improve traction. Each wheel was independently driven by a 190 W DC motor with an 80:1 harmonic drive housed within the wheel hub. The torque generated at the drive shaft of each wheel was 45 Nm. Unlike the Lunokhods, it employed double Ackermann steering. Wheel fenders were used for dust protection of the astronaut crew and payload. The Apollo lunar rover was limited to slopes of 19–23°. Apollo lunar rovers generated a drawbar pull of 239 N per wheel fully loaded on a level surface with an average wheel slip of 2–3% [98]. The fully loaded LRV averaged a mean maximum pressure of 4.2 kN/m^2 and required an average energy consumption of 35–56 Wh/km.

Comparisons between the Lunokhods and the LRV are shown in Table 3.2.

3.3 MARS PATHFINDER (MPF) SOJOURNER

The Mars Pathfinder lander landed at Ares Vallis (19.4°N, 33.1°W) carrying the Sojourner microrover in 1997 [99–102]. The Mars Pathfinder microrover Sojourner had a mass of 11.2 kg (Mars weight of 38.6 N) and dimensions of 63 cm (length) × 28 cm (height) × 48 cm (width) (Figure 3.3; Table 3.3). It used a six-wheeled rocker-bogie lever assembly as its suspension system which has become almost a standard in U.S. robotic rover chassis design. Each wheel was powered

Figure 3.2. Apollo lunar rover vehicle (LRV) [credit NASA].

by a tractive motor plus four additional motors on the outer wheels for steering. It was capable of moving over the Martian terrain at a maximum speed of 0.6 cm/s (0.4 cm/s nominally) using its six 12.5 cm diameter × 7.9 cm wide wheels. It had a ground clearance of 13 cm with its center of mass close to the centroid of the vehicle such that it could withstand a 45° tip angle.

Sojourner carried three cameras—a forward-pointing monochrome stereo pair and a rear color camera for instrument pointing. However, its main navigation stereo panoramic camera pair resided on the Pathfinder lander on a telescopic mast. Sojourner had 16 × 0.127 mm thick steel cleats per wheel which protruded 1 cm on each wheel. The vehicle could turn on the spot with a 37 cm turning radius and a top steering speed of 7°/s; steering angle feedback was provided by potentiometers. Sojourner traveled at speeds of 15 cm/s and stopped for hazard detection every 6.5 cm (one wheel radius). Sojourner drew 4 W to drive the wheels, 1 W for the microcontroller, and 1 W for onboard navigation. Hazard detection

Table 3.2. Comparison of LRV and Lunokhod.

Parameter	Lunokhod 1	LRV
Mass (kg)	756	708 (loaded)
Chassis mass (kg)	105	
Load/wheel (N)	153.4	287.3
Wheel diameter (cm)	51	81
Wheel width (cm)	20	22
Speed (km/h)	0.8/2.0	13 (max)
Base (m)	1.7	2.29 (2.06 platform only)
Track (m)	1.6	1.83
Clearance (m)	0.38	0.36
Turning radius (m)	2.7	3
Obstacle negotiation (m)	0.35	0.3 (height), 0.7 (crevasse), 25° (slope)

was performed using two hazard cameras and five laser diode–based laser stripe emitters. Sojourner traversed a total of 106 m at 0.036 km/h over its month-long mission duration [103]. However, it never ventured farther than a 10 m range from the Pathfinder lander—this range limitation was imposed by the rover–lander communications link radiation pattern integrity and the line-of-sight resolution of the Pathfinder lander camera which was used for rover navigation.

Sojourner steered autonomously (dead reckoning) to avoid obstacles using its wheel odometry, potentiometers, gyroscopes, and accelerometers to generate steering requirements to reach commanded goal locations. Sojourner hazard detection was based on proximity sensors including a frontal stereo camera pair, five laser striping projectors, and frontal contact sensors.

Sojournor traversed a total of 106 m (but within 10 m of the lander) via 114 commanded movements. This range limitation was imposed by the rover–lander communications link and the line-of-sight resolution of the Pathfinder lander camera which was used for rover navigation.

3.4 MARS EXPLORATION ROVERS

Much of the robotics research for planetary rovers has been developed on the NASA/JPL FIDO (Field Integrated Design and Operations) rover testbed [104, 105]. It has a mass of 70 kg with dimensions of 85 (width) × 105 (length) × 55 (height) cm and a ground clearance of 23 cm. As a development from Sojourner

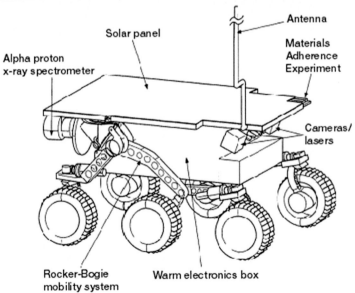

Figure 3.3. Pathfinder Sojourner microrover: (a) actual and (b) schematic [credit JPL/NASA].

Table 3.3. Sojourner microrover properties [adapted from Moore et al., 1997].

Sojourner rover parameter	Value
Deployed length	62 cm
Deployed width	47 cm
Deployed height	32 cm
Mass	10.5 kg
Weight on Mars	39.1 N
Mass/wheel axle (front)	27%
Mass/wheel axle (middle)	36%
Mass/wheel axle (rear)	37%
Nominal speed	0.4 m/min
Rough terrain speed	0.25 m/min
Turning speed	7°/s
Wheel diameter	13 cm
Wheel width	7 cm
Drive motor no-load current	9–13 mA
Drive motor torque factor	3.9–4.4 Nm/mA
Nominal motor voltage	15.5 V DC
Motor stall torque	3.0–4.5 Nm
Stall current at 0°C	196 mA

and other rover testbeds such as the Rocky series, it is a six-wheeled rocker-bogie chassis–based rover (though all six wheels are steerable). It has an average speed of 200 m/h over favorable terrain or 100 m/day over rough terrain with a nominal range of 10 km. Each wheel is driven and steered independently with 35 Nm torque per wheel, each wheel having a diameter of 20 cm to provide <9 cm/s vehicle speed. It carries a 4 DOF (degree of freedom) mast with a mast-mounted stereo panoramic camera, an infrared point spectrometer, Raman and Mossbauer spectrometers, minirock corer and a 4 DOF robot arm with an arm-mounted microcamera and gripper. It possesses an inertial navigation unit incorporating three-axis gyroscopes and accelerometers, sun sensor, and wheel odometry to support multisensory fusion and navigation.

The dual-mission Mars Exploration Rovers (MER)—Spirit at the Gusev crater and Opportunity at the Meridiani Planum were identical—first began operations in 2004 and have far exceeded their original 90-day design lifetimes. Their mission was to characterize the geology of their local landing sites as "robotic geologists" in search of clues for aqueous processes contextual to Mars' astrobiology potential. Both rovers had a mass of 174 kg with a total vehicle length of 1.6 m and wheel baseline width 1.22 m and length 1.41 m [106, 107] (Figure 3.4). The chassis was a six-wheeled rocker-bogie springless suspension design inherited from the Sojourner rover. The rover structure was based on composite panels with titanium alloy fittings while the rocker suspension was constructed from titanium alloy mounting six aluminum alloy wheels. Each wheel had a diameter of 25 cm diameter with the six-wheel configuration defining a 1.4 m length × 1.2 m wide footprint. Each wheel was independently driven, the four corner wheels being steerable for on-the-spot turning (with turn radius of 1.9 m). However, skid steering could reduce the turning circle to 0.9 m if necessary. Each wheel was cleated for increased traction. The design average traverse speed was 100 m/day including stops constrained by both energy consumption limitations and the risks inherent in target designation beyond 100 m. A 1.4 m tall pan–tilt PanCam mast assembly mounted both navigation and science stereo camera platforms and thermal emission sensors. Hazard camera pairs were mounted onto the front and back of the rover. With the PanCam mast assembly deployed, each MER was 1.54 m high with a ground clearance of 0.3 m while each wheel was 0.25 m in diameter defining the maximum negotiable obstacle height. The mast mounted both the scientific stereoscopic PanCam and the traverse-supporting stereoscopic NavCams. The maximum speed of the rover on flat ground was 4 cm/s but hazard avoidance would reduce this to 1 cm/s. The center of mass of the rover resided close to the rocker-bogie pivot giving it 45° lateral stability though software fault protection flagged any tilt exceeding 30° with an alarm condition. Each rover carried a Litton LN-200 inertial measurement unit incorporating three-axis tilt and rate data. MOLA data were used to localize the rover by triangulation from orbit initially. UHF Doppler tracking from orbit provides coarse navigation to within 100 m accuracy supplemented by additional in situ techniques. Onboard self-localization error using onboard sensors was 10% and cumulative but the adoption of visual odometry reduced this error to 1%. The rover moves 30 cm traverse segments at 5 cm/s at a time separated by stops of 20 s for navigation functions with daily traverses usually limited to around 10 m (though this constraint was relaxed later in the mission). Hazard detection was enabled through the hazard cameras (HazCam) the images from which were processed while static. As the terrain is imaged, they are collected incrementally into world model maps of 10 × 10 m. All other onboard data—wheel encoders, motor currents, steering headings, lever angles, tilt, etc.—are sampled at 8 Hz.

The MERs employed a highly limited scientific sensor suite comprising: (i) multispectral scientific PanCam stereo pair with two 1024 × 1024 CCDs each with a 14-band filter wheel sensitive to 400–1,100 nm with 0.28 mrad/pixel angular resolution; (ii) miniature thermal emission spectrometer based on the Michelson

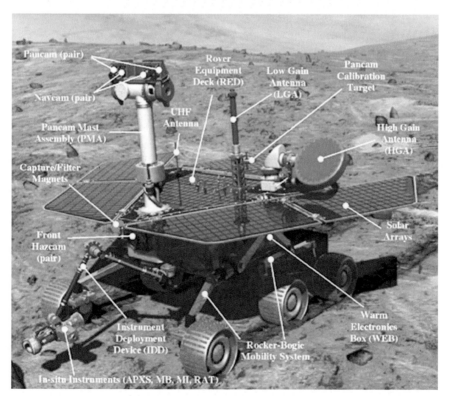

Figure 3.4. Mars Exploration Rover (MER): (a) actual and (b) schematic [credit NASA JPL].

Figure 3.5. Sojourner, Mars Exploration Rover, and Curiosity [credit NASA JPL].

interferometer with a spectral resolution of 10 cm over the wavelength range 5–29 μm for mineral infrared absorption spectra; (iii) a 5 DOF 0.7 m long robotic instrument deployment arm for pointing science instruments; (iv) rock abrasion tool with mechanical grinding heads at the end of the arm to remove 5 mm deep weathered rind from rocks; (v) close-up microscopic imager with a resolution of 30 μ per pixel; (vi) alpha-particle proton X-ray spectrometer for rock and soil elemental analysis; (vii) Mössbauer spectrometer to determine rock/soil oxidation state and Fe-bearing mineral identification by measuring resonant absorption of γ-rays from a ^{57}Co source. The MERs have been an unprecedented success. In 2009, Spirit's wheels sank into a sandtrap while climbing a slope forcing it to become a fixed station after traversing 7.73 km (6 years, 3 months). Opportunity had covered 33.2 km by August 2011 (7 years, 6 months). The trend of increasing rover size (and by implication capability) from Sojourner to MER has continued with Curiosity (Figure 3.5) .

3.5   MARS SCIENCE LABORATORY

The Mars Science Laboratory (MSL) Curiosity rover is a development from MER which was launched and successfully landed on the Martian surface in 2012 (its

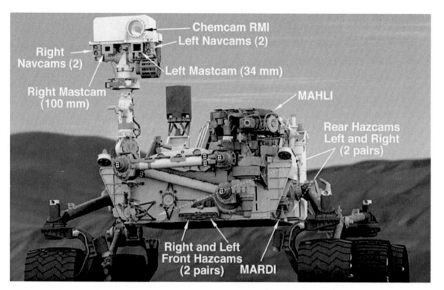

Figure 3.6. Curiosity rover [credit NASA].

mission had just begun at the time of writing) (Figure 3.6). It is significantly larger at 900 kg with 51 cm diameter wheels in a rocker-bogie configuration. Curiosity's rocker-bogie chassis is based on 0.5 m diameter wheels with a similar design to Spirit and Opportunity. Its top speed is 4 cm/s (0.1 mph) but it is nominally operated at half-speed. It is capable of traversing at 90 m/h and climbing obstacles up to 75 cm in height. Its primary mission in the 154 km diameter Gale crater is to cover 5 km possibly as much as 20 km from its landing site over two years enabled by a radioisotope thermal generator power source. Apart from its scientific instrumentation, its basic design and capabilities are derived from the MERs.

It is currently planned to launch a clone of the Curiosity rover to Mars in 2020.

4

Rover mobility and locomotion

It is essential that rover navigation take into account the capabilities of the rover chassis. There must be a strong interaction between the navigation system and the terrain adaptation system. The dynamic systems approach to implementing adaptive cognitive behavior suggests that rover–environment interaction is the key to intelligent systems capabilities [108]. There have been a number of reviews of the many robotic mobility concepts proposed for planetary exploration [109–113]. There are five classes of locomotion system in mobile robots which are applicable to planetary exploration rovers: (i) wheels rolling (e.g., automobile locomotion); (ii) tracks rolling (e.g., armored vehicle locomotion, and rotary drill or screwing); (iii) legs walking (e.g., animal locomotion); (iv) body articulation crawling/sliding (e.g., snake undulation); (v) non-contact locomotion (hopping/flying). Small-legged rovers are capable of traversing rough terrain, climbing obstacles, and are tolerant of falls. The BEES (bio-inspired engineering of exploration of systems) concept is an extension of this approach in which mobility, control, and sensing are biomimetic [114]. One example is the Scorpion robot with eight legs each with 3 DOF which may be deployed as a scout rover supporting a larger wheeled vehicle [115]. A microcontroller uses a central pattern generator in conjunction with local reflexes for leg control. Legged rovers are often eliminated from consideration for flight by virtue of their high mechanical and control complexity. It has been proposed that rigid frame walkers which decouple vertical and horizontal movement may offer a solution for planetary deployment [116]. However, the problem is not just one of complexity of control rather than the high consumption of power inherent in driving so many degrees of freedom. There are a number of candidate mobility systems that have been proposed specifically for Mars rovers [117]. A tetrahedral rover concept has been proposed which acts as a reconfigurable skeleton with a strut-based configuration connecting nodes [118]. Such a configuration can tumble like tumbleweed or walk (with 26 struts and 9 nodes forming 12 interlinked tetrahedral). A genetically evolved neural network may be used to generate appropriate control behaviors. Most planetary rovers are wheeled vehicles due to their simplicity, maturity, reliability, and ease of control [119]. There are a number of options and variations that may be adopted for the chassis design [120]. The use of kinematic linkages can greatly improve their obstacle negotiation capabilities beyond that of simple wheels. Six wheels give an advantage over four wheels in requiring lower motor torques to provide a given locomotion performance and superior obstacle negotiation capability. They offer the best compromise between the number of wheels and performance. Eight wheels do not offer significant increases in mobility for the additional complexity.

The basic performance requirements for a planetary rover typically include both forward and reverse motion capability, turning capability, and a well-defined obstacle negotiation capability. Rover locomotion performance requirements define the chassis selection process. For the ExoMars rover, the locomotion performance requirements were thus: (i) step-climbing ability of 0.3 m height; (ii) gradability up to 18–25° depending on soil type; (iii) static stability of 40°; (iv) locomotion speed of 125 m/sol (average equating to 20 m/h over a 6.25 h day) including localization, trajectory planning, obstacle negotiation, etc. including localization, trajectory planning, obstacle negotiation, etc, and 100 m/h (maximum for 20 min) on reference terrain; (v) point-turning capability. The Phase A ExoMars rover chassis selection process is outlined in Ellery et al. (2005) [121]. There were five candidate suspension concepts selected from an initial 19 reviewed to ensure 0.3 m step-climbing ability with slope-climbing ability up to 25° on loose sandy soil. The selection criteria (based on a multiattribute decision-making scheme) which was weighted (1 = insufficient; 2 = below average; 3 = average; 4 = above average; 5 = very good) in decreasing priority included (i) payload capabilities; (ii) locomotion capabilities; (iii) resource efficiency; (iv) control requirements; (v) inherent design characteristics. The baseline wheel diameter was 0.3 m with 5 mm grousers. Rover design and performance simulation is a difficult and specialized task for which there exist few adequate simulation environments particularly for modeling vehicle–terrain interaction through soils. Dynamic analysis allows determination of the forces and moments to be determined and their effects on vehicle behavior. Given the low speed of planetary rovers, quasi-static assumptions are valid and analysis of configuration kinematics is sufficient for control system design. Kinematic modeling allows relative motion of multiple-jointed bodies to be analyzed. Obstacle negotiation capability may be analyzed using multibody dynamics software (most software packages do not provide traction analysis).

Traction analysis using terramechanics models allows vehicle behavior in soil to be assessed but there is a paucity of software tools designed for traction analysis [122]. One such environment is ROAMS which is specialized for rocker-bogie–type chassis designs [123]. The soil–wheel interaction was modeled as a spring damper system in contact with a surface with a coulomb friction law [124]. This is not an effective model of the complex interaction between rovers and their terrain. RCAST is an integrated suite of software tools. It includes a 3D multibody dynamics simulator based on the SimMechanics toolbox and incorporating an

Figure 4.1. Kapvik microrover chassis undergoing field trials near Ottawa.

AESCO wheel–soil interaction mechanics module AS^2TM [125]. RCAST can simulate wheel walking which may be employed to mount slopes where wheel rolling would yield 100% slippage. Patel et al. (2004) adopted a traction-modeling system RMPET (rover mobility prediction and evaluation tool) which enabled wheel–soil parameters to be defined for input to a Cosmos-Motion 3D multibody dynamics engine [126]. It also included a computational facility similar to LocSyn (locomotion synthesis) for the configuration design for wheeled rovers such as Nomad based on the mechanics of vehicle–terrain interactions modeled by Bekker's equations [127]. RMPET was part of a larger suite of software tools developed for the European Space Agency—RCET (rover chassis evaluation tool) [128]. The primary means of validation, verification, and test is through physical field trials for which there is no substitute, but prior theoretical analysis saves significant time, resources, and cost (Figure 4.1).

4.1 WHEELED VEHICLES

The complexity of the kinematic configuration of the chassis is a tradeoff between mechanical complexity and obstacle negotiation capability. The U.S. six-wheeled rocker-bogie chassis adopts a single cab mounted onto kinematic lever linkages

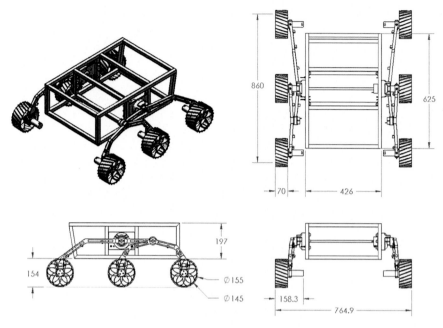

Figure 4.2. Kapvik rocker-bogie chassis [credit Tim Setterfield, Carleton University].

and has become the U.S. standard planetary mobility system [129–134]. The rocker-bogie wheeled concept has been flight-tested on the U.S. Sojourner, Mars Exploration Rovers, and Curiosity but was developed through the Rocky series of prototype microrovers [135, 136]. Suspension systems involving springs were avoided because in wheel climbing the spring is deflected producing an upward reaction which reduces the traction of the other wheels that assist in climbing. The rocker-bogie is a springless suspension system with the property of maintaining equal pressure on all six wheels while negotiating obstacles through passive compliance.

The Kapvik microrover prototype chassis developed by Carleton University is a rocker-bogie mechanism 0.782 m in width by 0.50 m in length (Figure 4.2) [137, 139]. The rocker-bogie chassis comprises two pairs of rocker-bogie arm assemblies, one each side of the vehicle. Each rocker has a rear wheel connected at one end and a secondary rocker (bogie) connected to the other. Each rocker pivots at its midpoint where it is attached to the main rover body using a differential jointed torsion bar which maintains an average pitch angle of the body between the two rockers (Figure 4.3).

The Kapvik wheels had a radius of 75 mm and width of 70 mm. This includes the 5 mm high grousers arranged helically at an angle of 19.4° to give favorable forward force while driving. The two rockers are constrained to ±16° by hard

Figure 4.3. Rocker-bogie differential mechanism (top–bottom) [credit Tim Setterfield, Carleton University].

Figure 4.4. Kapvik rocker-bogie assembly [credit Tim Setterfield, Carleton University].

stops. The forward ends of each rocker are connected directly to the front wheels. The aft end of the rocker is connected to the rotary bogie link. The bogie joint was limited to rotations of $\pm 30°$ by hard stops. At each end of the bogie are attached the mid and rear wheel. A differential torsion bar ensures that the joint angles of the two rockers are equal and opposite with respect to the rover body—this minimizes pitching of the rover body. The MER differential across the rover body tilts up to $45°$ (software limits this to $30°$). All six wheels are driven independently and attached to the articulated frame and four corner wheels were independently steered to provide double Ackermann steering. This mechanism can negotiate step heights of 1.5 times the wheel diameter and cross crevasses up to 40% of the vehicle length [140]. The rocker-bogie kinematic linkage may be modeled as a planar system of rover body–wheel dynamics (Figure 4.4) [141].

The use of force sensors above each wheel station integrated into the rocker-bogie mechanism of Kapvik was so designed to provide a means for traction control and, more importantly, online estimation of vehicle terramechanics. Sherborne Sensors SS4000M single-axis load cells with a 200 N range were installed above each wheel station (Figure 4.5).

These force measurements together with forward velocity measurements allow

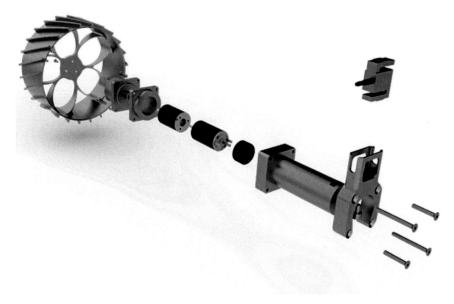

Figure 4.5. Integration of load force sensor to wheel assembly [credit Tim Setterfield, Carleton University].

estimation of drawbar pull and resistive torque as functions of normal load and slippage given soil properties through empirically determined fourth-order polynomial fits [142]. This was achieved through the use of two simultaneous unscented Kalman filters one of which estimates resistive torques and slippage while the other estimates normal loads and drawbar pull. It has been suggested that rocker-bogie performance might be enhanced if some of the linkages incorporated compliance. Furthermore, onboard manipulators may be used to improve a rover's tipover stability by adjusting the location of its center of mass, but this is an unlikely strategy to be adopted in a planetary rover [143]. There are a number of rocker-bogie–type variants that attempt to overcome the limitations of wheeled vehicles. The Japanese lunar microrover Micro5 has five wheels of which four lie at the corners of the vehicle in a conventional fashion but a fifth wheel attached to front axle lies beneath the centerpoint of the vehicle to give additional obstacle-climbing robustness but adds mass to the chassis [144]. It incorporated an averaging bridge frame between the two sides of the rover which rotated relative to each other.

A variation on the rocker-bogie principle of front-wheel obstacle negotiation is EPFL's (École Polytechnique Fédérale de Lausanne) Solero rover prototype (and its predecessor Shrimp) which accomplishes much the same approach as the rocker-bogie but with six independently actuated wheels in a four-bar rhombic

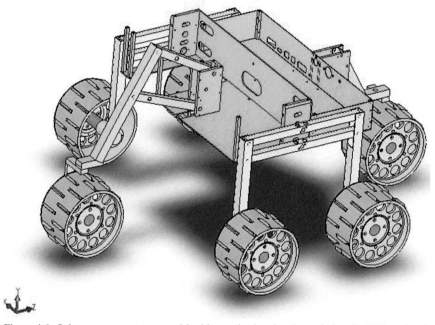

Figure 4.6. Solero rover prototype and its kinematic chassis schematic [credit Nildeep Patel, Surrey Space Centre].

linkage configuration [145, 146]. The rhombus configuration has the front wheel mounted on a spring-loaded fork, a wheel in the rear attached directly to the rover chassis and two bogie wheels each side (Figure 4.6). Two center-wheel bogie assemblies are mounted onto a support that can rotate freely on a central pivot. The front steering wheel is attached to the rover chassis by a parallelogram linkage-articulated fork which has spring suspension to maintain ground contact of all wheels and to elevate the front wheel over any obstacles. The virtual center of bogie rotation lies at the height of the wheel axes. In traversing a vertical step, the front fork's suspension spring compresses as it climbs the step. This stored energy enables the front bogie wheel to traverse the step [147, 148]. Solero can overcome obstacles twice its wheel diameter and enables slope negotiation of 40° while maintaining good stability. The 88 cm long × 40 cm wide × 38 cm high Solero prototype is a 12 kg mass rover with six 14 cm diameter × 9 cm wide wheels, each with 12 grousers.

Further variants of the rocker-bogie mechanism include the RCL series of six-wheeled chassis designs proposed for the ExoMars rover [149]. All RCL concepts were based on passive rigid suspensions and all four corner wheels have individual steering drives for on-the-spot turning [150]. RCL concepts C and D add an additional lever to make kinematic structure longitudinally symmetric double rocker-bogie designs [151]). RCL-C (Figure 4.7) is simpler than RCL-D (Figure 4.8) which utilizes double rather than single levers and so imposes greater mass overhead. RCL-C suffers a suspension lever hangup problem when climbing obstacles—this is caused by the center wheels lifting and drawing the outer wheels together. RCL-E (Figure 4.9) represents a return to the longitudinally asymmetric design in which the rear wheels are linked laterally to the longitudinal axis by orthogonal levers. The front and middle wheels are also linked longitudinally through levers. Unlike the rocker-bogie system, the rear wheels are not connected through levers to middle/front wheel assemblies. Its independent transverse bogie eliminates the need for a differential torsion bar. The RCL-E design ensures near-vertical wheel movement without relative wheel horizontal displacement which was the primary problem with RCL-C and RCL-D. This chassis is designed to improve obstacle negotiation with a lower mass overhead than RCL-C and RCL-D designs. Unfortunately, RCL-E suffers from poor static stability on high slopes. One way to improve this was by kinking the linkages into inverted V-shapes increasing the reaction force on the downslope wheels [152]. However, the rocker-bogie mechanism retains superior terrain adherence and obstacle negotiation capability.

The suspension configuration must allow compaction for stowage in the lander and subsequently for deployment—the motors used for deploying the rover chassis can be used for wheel walking. The Sojourner rover needed to be stowed flat on the inside of one of the lander side petals which limited the stowage envelope to 200 mm height (compared with the rover deployed height of 300 mm). The rocker-bogie mechanism had to be deconstructed for stowage by adding a passive joint into each rocker arm (Figure 4.10). Rover deployment occurred by driving the rear wheels forward while locking the other wheels to raise the rover body. This caused a relative rotation of the two half-rockers until a spring latch locked the two pieces of the rocker together [153]. In addition to being collapsed for stowage, the rover was tied to the lander mount through holddowns: it was mounted in a kinematic arrangement of two bipods (X/Z-axes), one monopod (Z-axis), and one shear pin (Y-axis). The bipods and monopod were each pre-loaded to the rover body via tensioned stainless steel cables which were cut pyrotechnically after landing to free the vehicle.

Not all wheeled chassis designs involve rocker-bogie–type linkages. For instance, SpaceCat was a tri-wheeled "wheel-walking" nanorover comprising six independently driven wheels mounted onto two triangular wheel frames on either side of the rover forming stepped triple wheels—the frames on each side could rotate independently with respect to the main body so the wheels can be lifted onto obstacles [154, 155]. The 3.1 kg MIDD (mobile instrument deployment device) nanorover used front wheels mounted on folding levers allowing the payload cab to touch the soil for instrument pointing [156, 157]. These folding levers also provided a rerighting capability by shifting the vehicle's center of mass. MIDD was originally slated as a fetch rover for the Beagle 2 Mars mission but was dropped in the face of mass budget constraints [158].

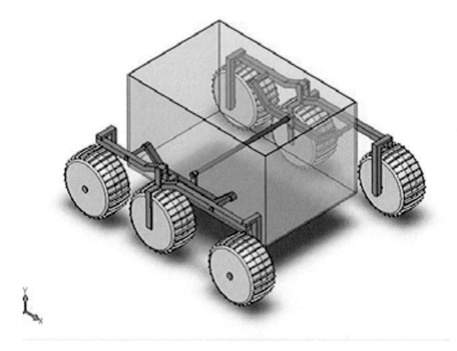

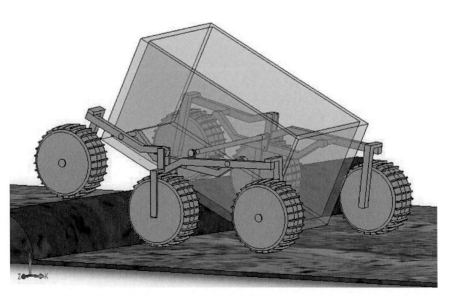

Figure 4.7. RCL-C six-wheeled chassis [credit Nildeep Patel, Surrey Space Centre].

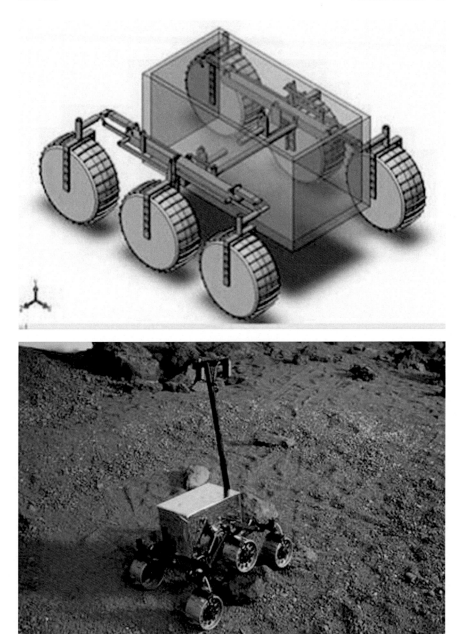

Figure 4.8. (a) RCL-D six-wheeled chassis [credit Nildeep Patel, Surrey Space Centre] and (b) its implementation on ESA's Exomader chassis demonstrator [credit ESA].

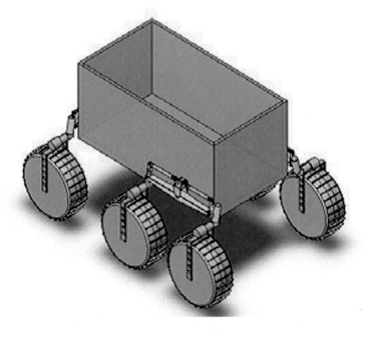

Figure 4.9. RCL-E six-wheeled chassis [credit Nildeep Patel, Surrey Space Centre] and its implementation on ExoMars [credit ESA].

Figure 4.10. Sojourner in stowed configuration [credit NASA JPL].

4.2   ARTICULATED CHASSIS

Articulated chassis have been developed for planetary rover prototypes but not yet flown—the Marsokhod rover in particular has been used extensively for planetary rover development. The novelty of the six-wheeled Marsokhod chassis lies in the use of three articulated units and the wheel design. The use of articulation by means of compliant linkages between rover cab segments provides robust terrain-ability but at the cost of increased mechanical complexity [159]. The Russian Marsokhod rover represents this type of approach to articulated chassis (Figure 4.11) [160, 161]. In particular, articulated chassis provide the basis for wheel walking to climb inclines close to the angle of repose of the soil. Marsokhod has a complex articulated chassis and undercarriage design including a torsional spine which comprises a significant fraction of its total mass [162, 163]. This approach using multiple vehicle cabs and a number of large drum-like wheels provides optimal traction. Marsokhod possesses three cabs/axles connected through two actuated rotational joints to provide the benefits of a flexible spine between axles. The Marsokhod rover chassis is 1.5 m long × 1.0 m wide within a mass of 35 kg generating a total vehicle mass of 100 kg. Each conically tapered cylindrical wheel is 35 cm in diameter and 70 cm in width and attached to a 3 DOF articulated frame for lateral and longitudinal movement for good terrain adaptability. Each wheel possesses volume for the integration of support hardware such as wiring harness and batteries to lower the center of gravity of the vehicle. Each wheel can move relative to each other and the front and rear axles have longitudinal freedom of motion. Marsokhod can negotiate obstacles up to twice the wheel diameter with a slope-climbing capability of 45°. It has a maximum velocity of 0.5 km/h (wheels

Figure 4.11. Marsokhod rover configurations [credit NASA].

only) or 0.02 km/h (wheel walking). The wheels occupy almost all the under-carriage volume to prevent the chassis from getting stuck between underlying obstacles. Its power consumption is high at 25–30 Wh/km. Aft and rear wheels are attached to levers to allow wheel walking to climb high-angled slopes up to the angle of repose of the soil—this involves a peristaltic sequence of alternately locking, stepping, and driving the wheels. Wheel walking is of questionable value in improving slope-climbing ability as backsliding can occur during the step cycle. There are many variants on the Marsokhod chassis including the IARES-L which adopts six-wheeled multiple cabs with rotational degrees of freedom between them but sporting more conventional wheels [164]. The IARES-L rover has a chassis mass of 80 kg which takes up a significant fraction of its all-out mass of 150 kg indicating the disadvantage of this system. The CNES LAMA is a six-wheeled articulated Marsokhod chassis demonstrator with a mass of 160 kg (of which 70 kg is payload) and has a maximum speed of 0.2 m/s (650 m/h) [165, 166].

A 70 kg small Marsokhod rover was selected for the canceled Russian Mars 2001 mission to carry a 15 kg scientific payload inherited from the canceled Russian Mars-98 mission [167] (Table 4.1). Integration of support and payload electronics in the Marsokhod chassis design represents a considerable challenge [168]. Their chief difficulty lies in payload integration into small cabs and the hollow wheels. Body articulation may be employed in varying degrees from a single articulation joint between two body segments [169], or multiple articulation joints across many body segments, the most extreme form of which is the snake. Articulated robots provide greater rough terrain adaptability by introducing separate sections connected by articulated joints—this enables each section to rotate independently about the longitudinal axis. Such articulation allows maximum contact between the wheels and the surface by adapting to the surface

Table 4.1. Different Marsokhod size scales [adapted from Kermidjan et al., 1992].

Parameter	Micro Marsokhod (conceptual)	Small Marsokhod (Mars-96)	Large Marsokhod
Mass (kg)	4–10	70	200+
Payload (kg)	1–5	15	100
Speed (km/h)	10–20	0.5	1.6
Wheel diameter (m)		0.35	0.51
Wheelbase (m)		0.7 (min)	1.4 (min)
Extension (m)		0.3	1.1
Obstacle height (m)		0.5–0.75	1.0
Max. slope/° (wheel walk)		20 (30–35)	20 (30–35)

Figure 4.12. Nomad Antarctic rover with its transforming chassis [credit NASA].

contours to minimize the need for avoidance maneuvers. Indeed, animal obstacle negotiation capability results primarily from the flexibility of the backbone rather than the mobility of the legs. It is this articulation that allows animal legs to follow ground contours. Body articulation essentially provides a peristaltic capability [170]. Rovers other than Marsokhod have adopted articulation mechanisms. The 725 kg four-wheeled Nomad rover in service in Antarctica deployed in the search for meteorites used a transforming chassis (Figure 4.12). It exploits a transforming chassis to vary its ground footprint by 60% from 1.8 to 2.4 m^2 [171–173]. The transforming chassis expands and contracts driven by two pairs of four-bar linkages connected to each wheel by two electric motors which also provide steering. The transforming chassis allows shifting of the center of gravity for enhanced obstacle negotiation capability. This virtually eliminates the possibility of getting stuck in obstacle fields. Nomad could use this chassis articulation for explicit steering which alters the direction of the wheels—such explicit steering provides better odometry with lower power consumption than skid steering at the expense of a higher number of actuators, parts count, and volume sweep. Nomad required half the power for explicit steering than for skid steering.

Figure 4.13. Sample return rover (SRR) prototype with variable footprint [credit NASA JPL].

Similarly, the SRR (sample return rover) prototype is a four-wheeled vehicle with each of its four wheels mounted onto struts to provide a variable footprint (Figure 4.13). This is a 9 kg "fetch" rover prototype designed to collect samples to be delivered to an ascent vehicle for return to Earth.

4.3   TRACKED VEHICLES

Tracks provide terrain contour following which makes them favorable for general off-road applications by virtue of their high ground contact area and high traction on soft ground [174]. Tracked vehicles have similar properties to wheeled vehicles and are generally favored for their higher tractive effort over more rugged terrain as they spread load over a much wider area—tracked vehicle will almost invariably offer higher drawbar pull than four, six, or eight-wheeled vehicles of the same mass due to its lower ground pressure. However, tracked vehicles are mechanically less efficient than wheeled vehicles imposing higher power requirements. Tracked vehicles generally comprise tread, driver sprocket, idler wheel (or alternatively it may be powered), and optionally number of supporting bogie wheels. Unpowered bogie wheels in tracked vehicles between the end wheels provide more continuous support for the track and reduce peak ground pressure by spreading the load uniformly. Tracked vehicles do not suffer from HUF (hangup failure where the bottom of the vehicle clashes with the obstacle during negotiation) which can occur with four-wheeled vehicles and are generally designed to ensure that NIF (nose-in failure when the front end of the vehicle interferes with the obstacle during negotiation) cannot occur under most conditions. Their high power loss is due to track/ground friction and bogie wheel/ track friction so the power requirement for tracked vehicles is much higher than for wheeled vehicles. The vehicle suspension system comprises a torsion bar, road-arm, inclined spring-based shock absorber and bump stops. Military tracked vehicles employ hydraulic damper suspension units. In military vehicles, tracks comprise interconnected steel links pinned together to give them flexibility but pin breakage due to debris causes catastrophic loss of the vehicle. There are two types of track drive systems. A friction drive relies on friction between the drive wheel and inner surface of the track—this requires high track tension (up to 80% vehicle weight). A positive track drive relies on sprocket teeth engaging either holes or complementary drive teeth on the inner surface of the track. The powered sprocket wheels tend to increase track tension. This allows the interwheel track to take up the vehicle load—loose tracks equate to a multiwheeled vehicle. The shape of the track between road wheels in contact with the terrain is given by [175]:

$$T\frac{d^2z}{dx^2} - p\left(1 + \left(\frac{dz}{dx}\right)^2\right)^{3/2} = 0$$

where T = track tension and p = normal pressure. Detracking requires that there is sufficient excess slack to override the sprocket and a sufficiently large sideways force on the vehicle. Tracks employ skid steering by differentially driving tracks for turning. This relies on track slippage so vehicle odometry on such vehicles is lost, though this is not a major disadvantage as wheel odometry is very inaccurate and must be augmented with external sensing modalities. The 20 kg Urban II robotic vehicle is a tracked urban rescue robot which includes smaller tracked "arms" which can rotate to enable it to climb steps (Figure 4.14) [177]. NASA's

Figure 4.14. Urban II rescue robot [credit NASA JPL].

tracked RASSOR robot bears some similarity to Urban II but has bucket-excavator wheels mounted at the end of levers for excavating soil.

Another tracked urban robot which can climb steps is the self-reconfigurable tracked mobile robot (SRTMR) [178]. The 3.2 kg Nanokhod prototype is a microrover package $250 \times 160 \times 65$ mm which used a continuous foil track wound around two wheels mounted into sealed rigid locomotion units [179–181] (Figure 4.15). Between the track units a $\pm 180°$ tiltable payload cab is mounted on cab levers attached to the two locomotion units. Similarly, the whole track can be rotated 90° to act as a wheel and allows the chassis to be maintained in a horizontal direction. Each locomotion unit comprises a frame, a track with cleats, a single powered wheel, a second unpowered wheel, and two gaskets. The tracks are constructed from thin stainless steel foil with metal cleats attached. The DC brush track motors and planet/worm gearing were sealed into track enclosures by means of plastic brush sealing for the tracks for dust mitigation and PTFE dry lubrication for the bearings. The sealing was plastic but did not prevent small particles entering the track body but additional sealing of motors, bearing, etc. could readily be achieved [Bertrand, R., private communication]. The two locomotion units contain the DC brush motors, planet/worm gears, and rover support subsystems for power distribution, communications, and control mounted onto the inner wall of the locomotion units. Each track is driven by the powered front wheel with a single idler wheel at the aft end of the track, both residing in the locomotion

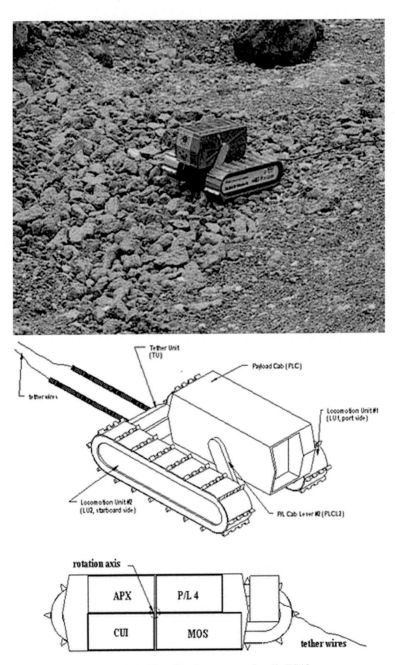

Figure 4.15. Nanokhod nanorover [credit ESA].

unit. The single-drive motor of each track is implemented with a planet and worm gear assembly. The locomotion units are connected by a mechanical bridge at the rear of the vehicle which contains the two tether spools for deploying them. Each spool comprises 30 copper subwires each for power and communication with the lander. It was capable of supporting currents for 100 V DC without plasma discharge effects. The low specific mass (0.4 g/m) was enabled by minimal tensioning as the tether pays out the rover traverses without respooling (unlike the MIDD tether capable of supporting <1 N but with a higher specific mass of 22.2 g/m). Nanokhod had a maximum speed of 5 m/h and was capable of overcoming obstacles up to 10 cm in height. Although Nanokhod eliminated the need for bogie wheels and hydraulic dampers, the maximum ground contact area cannot be maintained over rough terrain profile due to the rigid locomotion units.

The elastic loop mobility system (ELMS) overcomes this problem by offering high-mobility capabilities (Figure 4.16) [182]. It eliminates many of the disadvantages inherent in traditional tracks such as high internal friction, mechanical complexity, and link breakage problems. The elastic loop comprises a single continuous band around two end wheels, one being the drive wheel, the other the idler wheel (though there is no reason both wheels may not be powered). There are no bogie wheels and so there are reduced internal friction losses. Loop elasticity is due to preformed longitudinal and transverse curvature. The highly elastic metal band curls over its width so that sections between end wheels are flat and taut due to preformed transverse curvature. ELMS was traditionally manufactured from Ti(III) β alloy (Ti with 11.5% Mo, 6.5% Zr, and 4.6% Sn). The track stiffens longitudinally along straight ground due to preformed transverse curvature which uniformly distributes the load over a large footprint:

$$f_{\text{load}} = \frac{\pi b}{r^2} \frac{Et^3}{12(1 - \nu^2)}$$

where $E = 83$ GPa for NiTi in austenitic phase or 300 MPa for NiTi in superelastic phase. It eliminates multiple interlocking segments connected by locking pins which can break. The forward and aft wheels were raised so that the track acted as a spring suspension to elevate the vehicle and wheels to ensure that rocks and soil fall from the track before they could jam the wheels. The elastic loop due to its high rigidity ensures that it requires no bogie wheels. This substantially reduces the power requirement of ELMS in comparison with traditional tracks. The tight fit between two end roller wheels and track is limited to the upper-third portion of the loop which reduces the possibility of jamming by soil and rock particles. Wheel/track contact is maintained by suspension arms under variable load aided by the one-way damping action of shock absorbers. Loop elasticity provides spring suspension while shock absorbers provide damping. Obstacle climbing involves driving the chassis into the obstacle which pitches the nose up, deflecting the rear part of the loop. The front wheel climbs the obstacle along the vertical section of the loop independent of the obstacle's friction coefficient. Step obstacles as high as the loop length can be negotiated. Trials with a prototype vehicle

Elastic Loop Mobility System

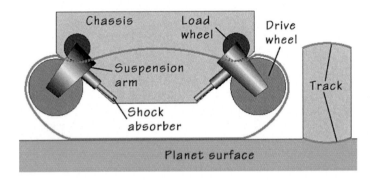

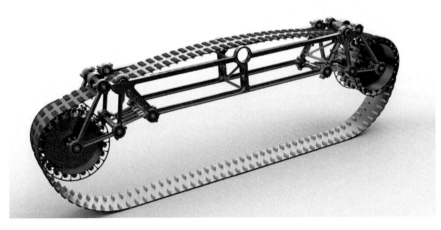

Figure 4.16. (a) Elastic loop mobility system [credit NASA] and (b) ELMS design for the Kapvik microrover [credit Cameron Frazier, Carleton University].

indicate twice-the-step-height negotiation of traditional tracked systems with the same geometry [183].

The ELMS system has been proposed for both lunar and Martian applications [184]—indeed, the canceled Viking 3 lander employed elastic loop mobility system legs for mobility but never flew. The Kapvik microrover was designed as a reconfigurable system to accommodate different chassis designs, the ELMS being its preferred track solution (Figure 4.17). Configurability was based on a modular system where simple interfaces allow chassis changeout. Chassis changeout on Kapvik takes less than five minutes to accomplish including electrical connections. ELMS offers high potential for robust mobility making it

Figure 4.17. Kapvik's modular exchangeable chassis design allows implementation of a track-based chassis [credit Cameron Frazier, Carleton University].

an excellent candidate for planetary application [185, 186]. The chief difficulties for the ELMS concept will be in dust/rock mitigation but reduced-scale prototypes indicate that this may not be as problematic as envisaged. Currently, this approach has undergone only sporadic development.

4.4 LOCOMOTION METRICS

Quick-look metrics of tractive performance may be used to provide rapid go/no-go assessment of vehicle design viability. In particular, the number and sizing of wheels may be selected on the basis of these metrics or further analysis to determine their tractive and obstacle negotiation capabilities. Pressure-based estimates must not exceed the Terzaghi bearing strength of a soil which is given by:

$$W_c = 2\gamma l b^2 N_\gamma + 2 l b \sigma N_q + \tfrac{4}{3} l b C N_c$$

where C = soil cohesion, N = bearing capacity factor dependent on friction angle, $\gamma = \rho g$ = specific gravity, b = wheel/track width, l = wheel/track length, $N_c = N_q - 1/\tan\phi$ = cohesion multiplier, $N_q = e^{2(\frac{3}{4}\pi - \frac{\varphi}{2})\tan\phi}/2\cos^2(45 + \phi/2)$ = overburden multiplier, $N_\gamma = (2(N_q + 1)\tan\phi)/(1 + 0.4\sin 4\phi)$ = wedge weight multiplier, and φ = soil friction angle.

Often, the middle parameter is excluded as the surcharge σ acting on the soil surface is assumed to be null. To increase the maximum vehicle load in frictional soil without soil failure, wheel or track width should be increased over the track length or wheel diameter: $W = 2b^2 l\gamma N_\gamma$. The critical load for cohesive soil is dependent on the contact area of the vehicle: $W = \tfrac{4}{3}blCN_c$. Typical bearing values for terrestrial soils are given in Table 4.2.

For lunar soil, the ultimate bearing capacity for a 1 m diameter footprint is 6,000 kPa. There are a significant number of different metrics of ground pressure. Cone index (CI) is a measure of soil strength defined as force/unit cone base area required to force a 30° cone-shaped probe into the top 15 cm of soil at a steady rate. It reflects the effects of both the shear and cohesive strength of the terrain and vehicle–terrain friction/adhesion. The performance of a single wheel based on WES (U.S. Army Engineers' Waterways Experiment Station) soil may be quantified through several measures [187]:

$$N_c = \frac{Cbd}{W}\left(\frac{\delta}{h}\right)^{1/2}\frac{1}{1 + (b/2d)}$$

for purely cohesive clays (clay numeric), where C = cone index and h = wheel height less sinkage depth

$$N_\phi = \frac{G(bd)^{3/2}}{W}\left(\frac{\delta}{h}\right)$$

for purely frictional sand (sand numeric), where G = sand penetration resistancegradient = 0.9–5.4 MPa/m for desert Yuma sand

$$N_{c\phi} = \frac{Cbd}{W}$$

Table 4.2. Common soil-bearing strengths.

Soil	Bearing values (kN/m^2)
Dense gravel	>600
Medium dense gravel/sand mixture	200–600
Loose gravel/sand mixture	<200
Dense sand	>300
Medium dense sand	100–300
Loose sand	<100
Hard clay	300–600
Stiff clay	150–300
Firm clay	75–150
Soft clay/silt	<75

for both cohesive and frictional soils (Wismer–Luth relation). High values are favored for off-road vehicles. There are equivalent measures for tracked vehicles [188]. The Freitag–Turnage mobility number [189] is an additional metric but rarely used. The mobility index (MI) is an alternative mobility metric; for a wheeled vehicle, the MI is given by:

$$\mathrm{MI_{wheel}} = \left(\frac{CF}{TF} + WL - \frac{C}{10}\right)$$

where $CF = W/nrb$ = contact pressure factor, $TF = (10 + b)/100$ = width factor, $WL = W/n$ = wheel load factor, and C = ground clearance. The MI for a tracked vehicle is given by:

$$\mathrm{MI_{track}} = \left(\frac{CF}{TF} + B - \frac{C}{10}\right)$$

where $CF = W/A$ = contact pressure factor, A = vehicle–soil contact patch area, $TF = b/100$ = width factor, and $B = W/10nA$ = weight factor. The vehicle cone index (VCI) is used in the NATO Reference Mobility Model (NRMM) based on the use of the cone penetrometer for characterizing soil strength from which the MI can be determined. It represents the minimum soil strength required for a vehicle to make x numbers of passes (Table 4.3). The VCI may be computed from the MI and can also accommodate multipass conditions. For a wheeled vehicle,

$$\mathrm{VCI_1} = 11.48 + 0.2\mathrm{MI} - \left(\frac{39.2}{\mathrm{MI} + 3.74}\right) \quad \text{and} \quad \mathrm{VCI_{50}} = 28.23 + 0.43\mathrm{MI} - \left(\frac{92.67}{\mathrm{MI} + 3.67}\right)$$

Table 4.3. Recommended VCI ranges.

	Multipass	Satisfactory	Single pass
Wet fine-grained sand	90	140	240
Snow	10	25–30	40

For a tracked vehicle:

$$\text{VCI}_1 = 7.0 + 0.2\text{MI} - \left(\frac{39.2}{\text{MI} + 5.6}\right) \quad \text{and} \quad \text{VCI}_{50} = 19.3 + 0.43\text{MI} - \left(\frac{125.8}{\text{MI} - 7.1}\right)$$

Alternatively, the VCI for wheeled and tracked vehicles, respectively, for one pass without reliance on MI values may be expressed as the vehicle-limiting cone index [190, 191]:

$$\text{VCI}_{\text{wheel}} = \frac{1.85W}{2nb^{0.8}d^{0.8}(\delta/d)^{0.4}} \quad \text{and} \quad \text{VCI}_{\text{track}} = \frac{1.56W}{2nba(pd)^{0.5}}$$

where W = vehicle weight, n = number of axles, d = wheel diameter, b = wheel width, δ/d = tyre radial deflection fraction, p = track link ground pitch (length), a = track link area as a proportion of total track area = A/pb (= 1 for continuous track), and A = track link area. The VCI correlates well in cohesive soils such as clays but it is often inaccurate in non-cohesive soils.

Rating cone index (RCI) is a measure of soil strength under repeated traffic while the VCI is a mobility measure defining the minimum soil strength for a given vehicle over a specified number (1 or 50) of passes. The RCI is not used often. There are a number of ground pressure metrics in use for quantifying wheeled vehicle performance [192]. These measure the vertical pressure exerted by the vehicle on soil. Nominal ground pressure (NGP) is defined as:

$$\text{NGP}_{\text{wheel}} = \frac{W}{nrb}$$

where W = vehicle weight (mg), m = vehicle mass, n = number of wheels, r = wheel radius (assuming maximum sinkage), and b = wheel width

$$\text{NGP}_{\text{track}} = \frac{W}{2lb}$$

where l = track ground contact length and b = track width. Increasing the width of a wheel or track increases the bearing capacity of the vehicle more than increasing the length of contact, a factor that is not incorporated in NGP. The NRMM used the VCI as a performance metric but this has been essentially superseded by mean metric pressure (MMP). MMP is the most commonly used metric for determining the pressure exerted by a vehicle on the soil. MMP models the effects of a number of vehicle parameters. MMP differentiates between variations in wheel width and wheel diameter and for wheel deflection [193]. Soil thrust must exceed the MMP of the vehicle and the maximum MMP must be <10 kPa (a highly conservative limit based on the load-carrying capacities of 25–55 kPa for lunar soils but it will be lower at crater rims and steep slopes). Maximum permissible MMP for terrestrial snow is 40 kPa (preferably <30 kPa). Some soil parameters have been measured for the Venusian surface by the Russian Venera 13 and 14 landers— rock strength and electrical resistance were measured by a dynamic penetrometer, rock strength from landing impact loads, mechanical properties of rock through drilling, and visual imaging of the landing sites. Bearing strength was estimated at $(2.6–10) \times 10^4$ kg/m^2 (Venera 13) representing a sedimentary source and $(6.5–25) \times 10^5$ kg/m^2 (Venera 14) representing a volcanic source. Mean maximum pressure (MMP) quantifies the mean peak pressure under each roadwheel station of a vehicle and is regarded as preferable over NGP. The MMP metric for wheeled and tracked vehicles is defined through Rowland's formula [194]:

$$\text{MMP}_{\text{wheel}} = \frac{KW}{2nb^{0.85}d^{1.15}(\delta/d)^{0.5}} \quad \text{and} \quad \text{MMP}_{\text{track}} = \frac{1.26W}{2nba(pd)^{0.5}}$$

where n = number of wheel axles, d = wheel or track diameter, b = wheel or track width, p = track pitch, a = fraction of link to total track area = A/bp, δ/d = fractional radial tyre deflection <0.8 generally, and K = a parameter defined by the proportion of axles driven. K is defined by the proportion of axles driven (as given in Table 4.4).

Some NGP and MMP values for a number of planetary rovers are given in Table 4.5. Both NGP and VCI are expected to correlate more closely to actual ground pressure than MMP but this is difficult to measure in practice as it assumes measurement from the soil surface without sinkage. These empirical values are of use only in comparative analysis and suffer from limited range of applicability. They often give inconsistent results but MMP has become the dominant metric. If ground pressure is under 7 kPa, vehicle traction will not be problematic on the Moon.

Table 4.4. Axle parameter.

No. axles/fraction of driven axles	1	$\frac{3}{4}$	$\frac{2}{3}$	$\frac{1}{2}$	$\frac{1}{3}$	$\frac{1}{4}$
2	3.65			4.4		
3	3.9		4.35		5.25	
4	4.1	4.4		4.95		6.05

Table 4.5. Planetary rover NGP and MMP measures.

	Mass (kg)	W (N)	No. of wheels n	Wheel load (N)	Wheel diameter d (cm)	Wheel width b (cm)	Deflection δ (cm)	K	MMP (kPa)	NGP (kPa)
Lunokhod	756.0	1227.2	8	153.4	51.0	20.0	—	2.05	8.7	1.5
Apollo Lunar Rover Vehicle	708.0 (loaded)	1149.3	4	287.3	81.0	22.0	7.5	1.83	8.0	1.6
Sojourner	10.5	38.8	6	4.8 (front) 7.2 (center) 7.4 (rear)	13.0	8.0	—	1.95	4.9	0.7
Mars Exploration Rover	180.0	669.6	6	101.0 (front) 120.5 (center) 112.6 (rear)	25.0	16.0	—	1.95	10.3	8.0

4.5 TERRAIN ROUGHNESS AND VEHICLE SUSPENSION

The Martian environment is predominantly flat, rarely exceeding a few degrees in slope, though there are localities which may exceed this such as craters and, of course, rocks. More than 20% of the Martian northern hemisphere lies within 200 m of the mean elevation as measured by the Mars Orbiter Laser Altimeter (MOLA) with low elevation correlating with low surface roughness [195]. Again, MOLA data are limited in horizontal resolution for the purposes of terrain modeling. However, the topography is rough at finer resolution requiring the rover to have high mobility across such terrains. Terrain roughness is responsible for vehicle vibration which may be modeled as the power-spectral density of terrain surfaces. The power spectrum of terrain input to the vehicle is defined as the square of the height of the terrain above arbitrary zero as a function of frequency (i.e., power spectral density). This provides the basis for analysis of terrain roughness modeled through generalized harmonic analysis. Terrain roughness may be modeled as ground surface spectral density of the form [196]:

$$S(k) = Q|k|^{-w}$$

where Q = terrain profile spectral density, w = dimensionless constant = 2 for poor terrain, and k = wavenumber of terrain profile spectra (cycles/m) = $1/\lambda$. This may be converted into ground height profile $h(x)$ at regular intervals k:

$$h_j(k) = \sum_{j=0}^{n-1} \sqrt{S_j}\, e^{i\left(\theta_j + \frac{2\pi/r}{k}\right)}$$

where $S_j = (2\pi/nr)S(k)$, r = distance interval of ground profile, θ_j = set of random phase angles uniformly distributed between 0 and 2π. Terrain elevation is assumed to occur above an arbitrary zero level which is essentially random along the path of locomotion. Terrain-spectral density may be modeled by the terrain wave equation Fourier series:

$$h(k) = \sum_{j=0}^{n} 2a_j \cos\left(\frac{k}{l_j} + \phi_j\right)$$

where $l_j = w_j/v_0$, w_j = ground wave velocity, and v_0 = vehicle speed. Mean square roughness may be determined:

$$\bar{h}^2(k) = \int_0^\infty S(k) \cdot dk$$

The power-spectral density $S(k)$ of the ground profile may be approximated by an exponential equation with a straight line fit on a log–log plot: $S(k) = Ck^{-w}$ where k = spatial frequency and $C \sim 10^{-3}$ for rough terrain. Models of the Martian topography suggest three scales modeled by similar power laws of the form $S(k) \propto k^{-w}$ but with different values of w; however, the data are currently too coarse to be of much use for the terrain analysis required for rover operations [197].

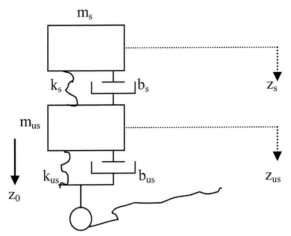

Figure 4.18. Vehicle–terrain model.

Although robotic rovers do not travel fast enough for dynamic analysis to be necessary, this will not be the case for manned rovers where ride comfort will be an important consideration. The vehicle suspension system is responsible for filtering the soil–vehicle interaction forces as the terrain contours vary by providing elasticity and damping. The vehicle is modeled as a 6 DOF system with two masses—sprung mass and unsprung mass. The elastic spring is a wishbone spring that isolates the vehicle (sprung mass) from the terrain with a shock-absorbing damper to dissipate wheel (unsprung mass) oscillations. We consider the $1/n$-vehicle model comprising a sprung mass (vehicle) and an unsprung mass (wheel) (Figure 4.18).

For the sprung mass:

$$m_s \ddot{z}_s + b_s(\dot{z}_s - \dot{z}_{us}) + k_s(z_s - z_{us}) = 0$$

For the unsprung mass:

$$m_{us} \ddot{z}_{us} + b_s(\dot{z}_{us} - \dot{z}_s) + k_s(z_{us} - z_s) + b_{us}\dot{z}_{us} + k_{us}z_{us} = F(t) = b_{us}\dot{z}_0 + k_{us}z_0$$

where $F(t) =$ wheel excitation induced by surface irregularities, $m_s =$ sprung mass $\sim 10m_{us}$, $m_{us} =$ unsprung mass, $b_s =$ shock absorber damping coefficient ~ 0.2–0.4, $b_{us} =$ tyre-damping coefficient, $k_s =$ suspension stiffness, $k_{us} =$ tyre spring stiffness $\sim 10k_s$, and $z_0 =$ surface elevation. Assuming free vibrations with $b = 0$ such that $F(t) = 0$, the solutions have the form:

$$z_s = z_s \cos w_n t$$

$$z_{us} = z_{us} \cos w_n t$$

where w_n = undamped natural frequency. Substitute:

$$(-m_s w_n^2 + k_s) z_s - k_s z_{us} = 0$$

$$-k_s z_s + (-m_{us} w_n^2 + k_s + k_{us}) z_{us} = 0$$

$$\begin{vmatrix} (-m_s w_n^2 + k_s) & -k_s \\ -k_s & (-m_s w_n^2 + k_s + k_{us}) \end{vmatrix} = 0$$

Expanding the determinant yields the characteristic equation:

$$w_n^4 (m_s m_{us}) + w_n (-m_s k_s - m_s k_s - m_{us} k_s) + k_s k_{us} = 0$$

Hence,

$$w_n^2 = \frac{B_1 \pm \sqrt{B_1^2 - 4A_1 C_1}}{2A_1}$$

where $A_1 = m_s m_{us}$, $B_1 = m_s k_s + m_s k_{us} + m_{us} k_s$, and $C_1 = k_s k_{us}$. This gives an approximation:

$$w_{n(s)} = \sqrt{\frac{k_s k_{us}/(k_s + k_{us})}{m_s}}$$

$$w_{n(us)} = \sqrt{\frac{k_s + k_{us}}{m_{us}}}$$

The vibration amplitude of the sprung mass is given by:

$$\frac{z_s}{z_0} = \frac{\sqrt{A_2}}{\sqrt{B_2 + C_2}}$$

where $A_2 = (k_s k_{us})^2 + (b_s k_{us} w)^2$, $B_2 = [(k_s - m_s w^2)(k_{us} - m_{us} w^2) - m_s k_s w^2]^2$, $C_2 = (b_s w)^2 (m_s w^2 + m_{us} w^2 - k_{us})^2$, $w = 2\pi v/l_w$ = excitation frequency, v = vehicle speed, and l_w = ground profile wavelength. The vibration amplitude of the unsprung mass is given by:

$$\frac{z_{us}}{z_0} = \frac{\sqrt{A_3}}{\sqrt{B_2 + C_2}}$$

where $A_3 = [k_{us}(k_s - m_s w^2)]^2 + (b_s k_{us} w)^2$. As

$$\frac{w_{n(ur)}}{w_n(s)} \gg 1$$

the sprung mass contribution is very low indicating vibration isolation. The dynamics of the $(1/n)$-vehicle are given by [198]:

$$\ddot{z}_{rel} = \sqrt{\frac{1}{T} \int_0^T \left(\frac{\ddot{z}}{g}\right)^2 \cdot dt} \quad \text{and} \quad F_{rel} = \sqrt{\frac{1}{T} \int_0^T \left(\frac{F_{z,dyn}}{F_{z,stat}}\right)^2 \cdot dt}$$

Spring-based suspension is appropriate when the vehicle weight is much greater than the wheel and the spring-restoring force. The commonest are coil springs placed around a damper housing shock absorber filled with oil. Leaf springs which comprise arched sections of carbon steel or composite fiber are the most versatile in that they do not require suspension control devices. Anti-roll bars are lateral springs which give resistance to lateral roll but this is unlikely to be necessary for a manned rover. Torsion bars are constructed from alloy spring steel and are commonly used for front suspension with coil/leaf springs for rear suspension. The commonest suspension axle systems are designed to keep wheels on the ground during camber changes. The simplest is the I-beam axle in which a long–short arm suspension axle connects the wheels with a longer lower-control arm and a shorter upper-control arm. The MacPherson strut suspension has the wheel spindle attached to the bottom of the vertical/damper strut which controls camber through the angle of the strut from the vertical.

4.6   ROVER MEAN FREE PATH

The rover must be able to accommodate obstacles such as rock fields on the planet's surface. Rock obstacles are defined in terms of their height above the ground. For a rocker-bogie mechanism, the maximum traversable height is nominally given by the wheel diameter with a 50% margin. In addition, vehicle ground clearance also influences maximum height negotiable with a margin (typically 20%). Surface rock distribution impacts on ground clearance and straight-line trajectory segments. Mean free path (MFP) may be defined as the expected straight line path distance that a vehicle can traverse over a given terrain before a heading change is required due to incidence of an insurmountable obstacle (i.e., it is related to maneuvrability). High mean free path reduces the requirement for steering changes due to obstacles. One of the most important aspects of MFP is that it defines the artificial intelligence requirements of the robot. The capabilities of the chassis in negotiating rock fields define what may be regarded as obstacles. First, the rock size–frequency distribution needs to be ascertained. Martian terrain is varied at the different landing sites—Viking Lander 1, Viking Lander 2, Pathfinder, Spirit, and Opportunity. Although our focus is Mars, rock distribution models will be relevant on other terrestrial-type bodies such as the Moon. More surprisingly, it appears from the descent of NEAR Shoemaker prior to impact on the asteroid Eros that there are rocks embedded in water-deposited but now dry sandy regolith which may pose obstacle hazards for asteroid rover missions.

Spatial density of obstacles and their separations may be given by a Poisson distribution of the form:

$$p(n) = \frac{(\rho a)^n}{n!} e^{-\rho a}$$

where a = areal coverage, n = number of objects, and ρ = spatial density. If n individual obstacles are placed randomly over area a, the probability $p(n)$ of n individual obstacles in area a is $p = n/a$. Similarly, the probability of no obstacles in area a with $n = 0$ is given by $P(0) = e^{-\rho a}$. If $a = \pi r^2$, then $p(r) = 2\pi r \rho e^{-\pi r^2 \rho}$.

However, we are interested in rock distribution according to rock diameter. This resembles the power laws for determining the distribution of craters on the Martian surface according to diameter D:

$$N(D) = CD^{-b}$$

where $b = 3.32$ for small craters $<1.54\,\text{km}$ diameter and $C = 2.5 \times 10^{12}$. Such crater models can indicate the likelihood of having to negotiate craters, which generally means steep slopes as obstacles to motion. Fresh impact craters on regional scales potentially impose up to $40°$ interior slopes but slopes exceeding $20\text{–}30°$ are rare on Mars. The MER Opportunity rover experienced great difficulty in escalating a steep slippery sandy slope of the crater in which it landed in Meridiani Planum. Of more immediate importance, however, is the rock (obstacle) distribution. Similar to crater density distribution, the cumulative number of rocks/m^3 with diameter $D > D(m)$ is given by:

$$N(D) = \kappa D^{-2.66}$$

where $\kappa = 0.013$. The corresponding cumulative fractional area covered by rocks with diameter D is given by:

$$\rho(D) = cD^{-0.66}$$

where $c = 0.0408$. These power laws of rock distribution are adequate for rock diameters exceeding $20\,\text{cm}$ at Viking Lander 2 site (VL2). However, it over-estimates the number and fractional coverage of rocks for small and large rocks [199–201]. A better approximation for rock size–frequency distribution based on an exponential function was found to predict the Mars rock distribution at the Viking lander sites with a 98–99% correlation (Table 4.6) [202–205]. The cumulative number of rocks/m^3 with diameter $D > D(m)$ may be given by:

$$N(D) = Le^{-sD}$$

where $L = $ cumulative number of rocks of all sizes/m^2, $L_{\text{VL1}} = 5.61$, $L_{\text{VL2}} = 6.84$, $L_{\text{MPF}} = 20.00$, $L_{\text{MER-A}} = 40.00$, $s_{\text{VL1}} = 12.05$, $s_{\text{VL2}} = 8.30$, $s_{\text{MPF}} = 11.00$, and $s_{\text{MER-A}} = 24.00$. The frequency distribution (or probability density) of rocks for Viking Lander sites 1 and 2 of diameter D are given by:

$$\rho(D) = Ke^{-q(K)D}$$

Table 4.6. Martian rock distribution parameters [courtesy of Lutz Richter].

	Viking 1	Viking 2	MPF	MER-A (landing point)
k	0.069	0.176	0.187	0.070
L	5.61	6.84	20.00	40.00
s	12.05	8.30	11.00	24.00

where $K =$ cumulative fractional area covered by rocks of all sizes, $q(K) =$ exponentiation coefficient $= 1.79 + 0.152/K$, $K_{VL1} = 0.069$, $K_{VL2} = 0.176$, $K_{MPF} = 0.187$, $K_{MER-A} = 0.07$, $q_{VL1} = 4.08$, and $q_{VL2} = 2.73$. These are related by the expression

$$F(D) = \frac{\pi}{4} L e^{-sD}(D^2 + 2D/s + 2/s^2)$$

As $D \to 0$, $F(D) \to k$ for total rock coverage. Viking Lander site 1 (VL1) and Viking Landing site 2 (VL2) had terrain similar to that at the Surveyor 7 site on the Moon with a similar rock–frequency distribution and a roughness similar to that of the more rugged lunar maria. VL1 and VL2 were unusually rocky as expected from an outflow channel. Large rocks at VL2 were twice as common as those at VL1. Pathfinder site predictions extrapolated from Viking data were consistent—the largest rocks were 20–50 cm in diameter with rocks greater than 0.5 m high covering less than 1% of the region and rocks exceeding 1.0 m high covering much less than 1% of the surface. The Gusev crater environment was highly rock strewn similar to that at the Viking and Pathfinder locations. Indeed, most landing sites of scientific interest are likely to be populated by large numbers of rocks as these are appropriate grab-bag sites. The long MER traverses have been across relatively benign environments. For assessment purposes, VL2 rock distribution is usually selected as the worst-case rock field scenario. These statistical models may be used to generate random instantiations of representative Martian rock fields (Figure 4.19).

The rock distribution impacts the scale of a robotic rover through the mean free path (MFP) of a vehicle over that terrain. The MFP is determined by rock vertical height rather than rock diameter due to embedding in soil. Rocks of diameter D are assumed to be distributed randomly according to a Poisson distribution with an expectation value in proportion to area. The average rock height may be determined by limiting rock diameter D_0:

$$H = 0.365D + 0.008 \quad \text{assuming that rock height for VL1 is } \tfrac{3}{8} \text{ rock diameter}$$

$$H = 0.506D + 0.008 \quad \text{assuming that rock height for VL2 is } \tfrac{1}{2} \text{ rock diameter}$$

Hence, $H = (0.25 + 1.4k)D$ for both VL1 and VL2 with 0.1% of the surface covered with the highest rocks of 0.2 m at VL1, and 1% of the surface covered with the highest rocks of 0.5 m at VL2. The JPL Mars Yard uses granitic sand and Viking Lander rock distribution for their rover traverse testing. The maximum negotiable rock height is dependent on the chassis design: for the rocker-bogie design, it is 1.5 times the wheel diameter (though anything over one wheel diameter would be flagged as an alarm condition). Note that most formulations for obstacle negotiation capability limit the ratio of obstacle height to wheel diameter to unity (e.g., Jindra, 1966) [206]. For a vehicle to move distance m, it must sweep out a rectangle of length $(m + D/2)$ by width $(r + D)$ for rocks of diameter D in order to avoid a rock's position defined as a point particle. To determine mean distance between nearest neighbor rocks, the first moment of the

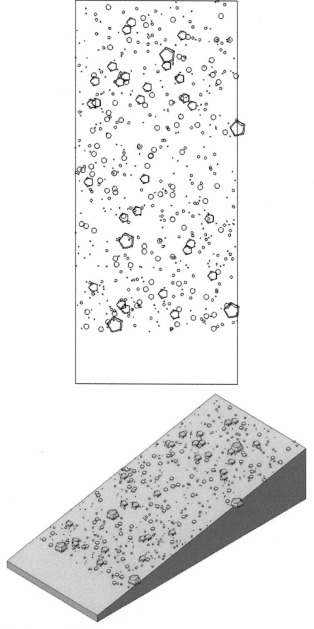

Figure 4.19. Random statistically generated rock distribution representative of VL2 site [credit Elie Allouis, Surrey Space Centre].

probability density function is required. For rocks larger than D_0 [207, 208]:

$$\int_{D_0}^{\infty} (m + D/2)(r + D)\rho(D) \cdot dD = 1$$

where $\rho(D) = Ke^{-qD}$ = probability density of rock centers/m^2 for rocks between D and $D + \delta D$ and r = minimum turning radius of the vehicle. The solution for MFP is given by [209]:

$$m = \frac{1 - (r/2)\int_{D_0}^{\infty} D\rho(D)\, dD - (1/2)\int_{D_0}^{\infty} D^2\rho(D)\, dD}{r\int_{D_0}^{\infty} \rho(d)\, dD + \int_{d_0}^{\infty} D\rho(D)\, dD}$$

The proper integrals may be evaluated thus:

$$\int_{D}^{\infty} Ke^{-qD} \cdot dD = \frac{K}{q} e^{-qD_0}$$

$$\int_{D}^{\infty} DKe^{-qD} \cdot dD = K\left(\frac{D_0 e^{-qD_0}}{q} + \frac{e^{-qD_0}}{q^2}\right)$$

$$\int_{D}^{\infty} D^2 Ke^{qD} \cdot dD = K\left(\frac{D_0^2 e^{-qD_0}}{q} + \frac{2De^{-qD_0}}{q^2} + 2e^{\frac{-qD_0}{q^3}}\right)$$

For on-the-spot turning, the largest diagonal of the vehicle must be defined $r = \sqrt{l^2 + w^2}$, where l = rover length and w = rover width; for skid steering the turning radius may be larger. Sojourner could turn on the spot with a 0.37 m turning radius. The problem of reliance on such statistical models was exemplified by Sojourner's difficulties in traversing the rock garden which had an areal coverage of 24.6% of rocks over 3 cm diameter compared with 16% on average at that location. However, these models do serve to enable mean free path performance metrics to be determined according to chassis obstacle negotiation capability and rover sizing (Figure 4.20). From Table 4.7, it can be seen that increased obstacle negotiation capability increases the mean free path assuming comparable vehicle dimensions. The MFP computed for the Kapvik microrover on a terrain similar to VL2 site was 15 m.

4.7   VEHICLE–SOIL INTERACTION—BEKKER THEORY

For planetary rovers, the robot–environment interaction must be considered, which takes the form of traction analysis [210]. Planetary terrains are extremely challenging with regard to tractive effort—typically, most planetary bodies are covered in a layer of loose sand-like regolith interspersed by distributions of rocks. Bekker analysis provides the basis for determining the theoretical performance (quantified by the drawbar pull metric) of a vehicle traversing natural terrain. It is semiempirical and is based on pressure–sinkage tests using a metal plate

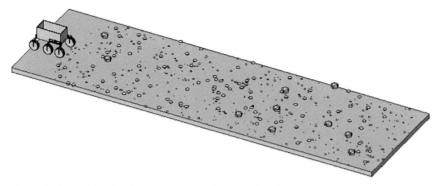

Figure 4.20. RCL-E chassis rover on Martian terrain with VL2 rock distribution [credit Nildeep Patel, Surrey Space Centre].

Table 4.7. Mean free path estimates of several microrovers.

	Obstacle height limit (m)	MFP VL1 (m)	MFP VL2 (m)
Sojourner (nominal)	0.13	66.3	14.9
Sojourner (maximum)	0.20	72.0	15.9
Nanokhod	0.25	76.3	16.7
ELMS	0.325	83.4	18.04

(bevameter) simulating the contact area of the vehicle on the soil. The basis for Bekker traction analysis is consideration of the wheel–soil contact area, similar in many respects to pressure metrics but far more elaborate [211–215]. In this case, mobility performance is quantified through vehicle drawbar pull (DP—defined as the difference between soil thrust and motion resistance). Locomotion requires traction to provide forward thrust on the ground in excess of resistance to motion which is dominated by sinkage resistance. An important aspect of designing planetary rovers is consideration of soil mechanics properties—shear strength, compressive strength, bearing capacity, penetration resistance, etc. Almost all of these parameters are interrelated (e.g., bearing capacity is dependent on penetration resistance). Most important soil parameters are related to shear strength, which is quantified by the Mohr–Coulomb relation. From effective soil shear stress (which is determined by soil cohesion and the internal angle of friction), rover trafficability and other parameters can be determined. A comprehensive review of Bekker theory and its variants is provided by Plackett (1985) [216]. Traction analysis requires estimation of regolith mechanical properties but this will not necessarily be available. Soil shear strength is characterized by soil cohesion C and internal friction angle ϕ quantified by the Mohr–Coulomb law such that shear stress parallel to the plane of failure:

$$\tau = C + \sigma \tan \phi$$

where τ = soil shear strength, C = cohesive strength of soil (related to cementation), μ = soil coefficient of friction = $\tan \phi$, ϕ = soil internal angle of friction, and σ = normal stress = pressure = W/A. When a wheel deforms the soil, the pressure on the contact patch is highest under the center of the wheel and lowest at the edges of the contact patch. Note that the Mohr–Coulomb relation substitutes for soil friction coefficient (nominally 0.3–0.5 for sliding friction but >0.5 for dry soils reaching 0.8 for rocks) in homogeneous media—adhesion between the wheel and soil is given by a similar relation $\tau = c_a + \sigma \tan \delta$ where δ = wheel roughness angle and c_a = adhesion = c = soil cohesion to a first approximation. Soil cohesion arises due to weak bonds between soil particles (e.g., clays have high cohesion). The friction angle defines the slope of the soil which retards interparticle sliding (e.g., dry sand has a high friction angle). The cohesion of sand varies from 0 (dry) to $10 \, \text{kN/m}^2$ (wet) while that of clay ranges across 50–200 kN/ m^2. The friction angles of sand can reach up to 50° while those of clay range from 0 to 20°. Hence, dry sand is a frictional soil with $C_0 \rightarrow 0$ while clay is a cohesive soil with $\phi \rightarrow 0$.

Martian regolith is fine grained with some sandy silicate material ~0.1–1.0 mm in size with up to 40% pore space filled with frozen volatiles and cemented with sulfate and chloride salts. It has an unusually high concentration of iron, almost 20% Fe_2O_3. The dominant regolith grain size is between 100–1,000 μm which accounts for 55% of the regolith, the rest following a linear relation from 0.1 to 100 μm. Soil bulk density was computed to be 1.2–2.0 g/cm^3 (average 1.52 g/cm^3) similar to clayey silt with embedded sand and pebbles. Martian soils are generally of dry sandy type derived from basalt cemented by hydrated sulfate salts formed under aqueous conditions. The classification of granular material follows this convention: 5 mm–2 mm (granules); 2 mm–62.5 μm (sand—mean grain size of Martian drift and cloddy soil was 20–40 μm); 62.5 μm–12 μm (silt); and 12 μm–0 (clay). Generally, the differences in Martian soils were mechanical rather than mineralogical. Sand particles are transported through saltation which occurs on Mars and are generally limited to angles of repose of <30°. Clay-type soils are unlikely on planetary surfaces though clay minerals (smectites), a particular type of phyllosilicate formed under aqueous conditions, have been detected at several locations on Mars—the largest abundance of clay outcrops occur at Mawrth Vallis, an outflow channel. Clay minerals are hydrated minerals such as montmorillonite or kaolinite. Much Martian soil, particularly near-horizontal sediments, is similar to sandy loam but it generally comprises a number of different types of soil. The Viking landers encountered three types of surface material—homogeneous drift material with easy diggability (VL1); crusty to cloddy material layered with a crust above a cloddy material that broke into 4 cm sized chunks which was difficult to dig (VL2); and a blocky material that was difficult to pulverize and dig (VL1). In general, the fine-grained ferric oxide–rich dust overlies all deposits. Many of the Martian rocks at the Pathfinder landing site resembled basalt–andesite with feldspar, orthopyroxene, and quartz mixed with Ti–Fe oxides probably derived from magma originating from the Martian mantle (indicated by the high SiO_2 content) during Mars' early geological history. Martian soil comprises a windblown eolian fraction of very fine dust (drift material) of mean grain size 20–40 μm (drift material) overlying a duricrust and a denser and cohesive "blocky" or "cloddy" soil which covers the greater part of Martian lander sites. Drift material (possibly clayey) had a density of $1,300\,kg/m^3$. Cloddy soil was a poorly sorted mixture similar to terrestrial clayey silts impregnated by sand and pebbles with cohesion due to cementation by salts, probably sulfates. Cohesions were generally inversely correlated with the friction angle [217]. Drift material ($C = 0.01$–$0.30\,N/cm^2$, $\varphi = 18°$) was the weakest soil with homogeneous texture. Crusty-to-cloddy soil ($C = 0.02$–$0.58\,N/cm^2$, $\sigma = 30$–$35°$) was intermediate in strength and layered with a crust which broke into thin slabs and an underlying material which broke into thicker chunks. Blocky material ($C = 0.15$–$0.90\,N/cm^2$) was the strongest soil which was difficult to pulverize. Soil density was estimated to be $1,200$–$2,000\,kg/m^3$. VL1 was dominated by drift and blocky soils while VL2 was dominated by crusty–cloddy material. The Viking landers showed slightly higher cohesive strength measurements than those experienced by Sojourner with $C = 0.02$–$0.58\,N/cm$ and $\phi = 30$–$35°$. Friction angles varied from 32–41° (average 36.6°), repose angles varied between 30 and 38° (average 34.2°), and cohesion varied between 0.12 and 0.356 kPa (average 0.238 kPa) [218]. There were regions of weak, porous deposits with $\phi = 26°$ and $C = 0.53\,kPa$. Most of the deposits were crusty to cloddy similar to that at VL2 with $\phi = 34.5° \pm 4.7°$ and $C = 4.4 \pm 0.8\,kPa$. VL1 was mostly crusty–blocky with $C = 5.5 \pm 2.7\,kPa$. Sojourner carried a wheel abrasion experiment to assess wheel wear—an atomically thin metal film on a black anodized Al strip attached to one of the rover wheels. A photocell monitored changes in the film's reflectivity. The Sojourner rover locked all but one of its six wheels so that the free wheel could be powered to dig a trench into the soil while monitoring the motor current (as a measure of wheel torque). Wheel–soil shear strength can thus be estimated. One Pathfinder Martian soil has the following properties: $\rho =$ bulk density $= 1,550\,kg/m^3$, $C = 220\,Pa$, $\phi = 33.1°$. Pathfinder estimates of soil cohesion are lower than for Viking probably due to Pathfinder estimates being limited to 4.3 cm (by wheel sinkage) compared with Viking's 10 cm sampling depth (by trenching)—Viking's estimates include increased cohesion due to compaction with depth. Viking data are regarded as more reliable. Average values of $C = 0.24\,kPa$ and $\phi = 35°$ with an average density of $1.52\,g/cm^3$ are representative baseline values for the Martian surface (Table 4.8). Martian drift soil represents the worst-case environment while the VL2 crusty–cloddy soil is the most representative soil environment for Mars. MER soil analysis is consistent with the Viking/Pathfinder results [219]. It is recommended that three soils be defined for the minimum, maximum, and median parameters, respectively.

Table 4.8. Summary of important Martian soil physical properties [courtesy of Lutz Richter].

Soil	Bulk density (kg/m^3)	Soil cohesion (Pa)	Friction angle (°)	K_c (N/m^{n+1})	K_ϕ (N/m^{n+2})	Deformation coefficient n
DLR soil simulant A	1,140	188	24.8	2,370	60,300	0.63
DLR soil simulant B	1,140	441	17.8	18,773	763,600	1.10
DLR soil simulant C	1,140	41	25.6	1,342	265,114	0.86
DLR soil stimulant D	—	13	13.4	19,152	667,500	1.80
VL1 drift	1,153	1,600 ± 1,200	18 ± 2.4	1,400	820,000	1.0
VL1 blocky	1,605	5,500 ± 2,700	30.8 ± 2.4	1,400	820,000	1.0
VL2 crusty–cloddy	1,403	1,100 ± 800	34.5 ± 4.7	1,400	820,000	1.0
VL rocks	2,600	1,000–10,000	40-60	—	—	—
MPF drift	1,172	380 ± 200	23.1 ± 8.0	1,400	820,000	1.0
MPF cloddy	1,532	170 ± 180	37.0 ± 2.6	1,400	820,000	1.0
Meridiani Planum		4,800	20.0	28,000	7,600,000	1.0
Dry sand	1,524	1,040	28	990	1,528,000	1.1
Sandy loam	1,524	1,720	29	5,270	1,515,000	0.7
Clayey soil	1,524	4,140	13	13,190	692,200	0.5
Snow	<500	150	20.7	10,550	66,080	1.44
MER-A "loose"	1,231	1,000	20	TBD	TBD	TBD
MER-A "dense"	1,484	15,000	25	TBD	TBD	TBD
MER-B "slope soil"	1,185	500	20	6,800	210,000	0.8
MER-B "sandy loam"	1,333	5000	20	28,000	7,600,000	1.0

TBD = to be determined.

Regarding the Moon, penetrometers were used by the Apollo astronauts and deployed on the Lunokhod 1 and 2 rovers on the Moon to measure lunar soil parameters. The PROP payload of the Lunokhods comprised an odometer wheel and a cone/vane shear penetrometer mounted at the rear of the Lunokhods. For the Moon, the recommended lunar soil parameters are $C = 0.017$ (variation 0–0.035) N/cm^2, $\varphi = 35°$ (variation ±4°), $K = 1.78$ (variation ±0.76) cm, $n = 1$,

Table 4.9. DLR lunar soil simulant properties [credit Lutz Richter].

	Region A	*Region B*
n	0.8	1.0
k_ϕ (N/m^{n+2})	3.63×10^5	6.02×10^5
k_c (N/m^{n+1})	2.41×10^3	7.77×10^3

$k_c = 0.14$ (range 0–0.28) N/cm^2, and $k_\varphi = 0.82$ N/cm^3, which are partly replicated by crushed basalt [220]. DLR lunar soil simulants exhibit the properties given in Table 4.9 emulating two different regions.

Solid rock has an average density of around 2.50–2.70 g/cm^3. For lunar soil, the top 15 cm has a density of 1.5 ± 0.05 g/cm^3. In the top 5 cm, the lunar soil had a relative density of 65% increasing to 70–75% at 5 cm depth and 75–85% at 10 cm depth. Soil density increases according to depth from 20 to 60 cm averaging 1.66 ± 0.05 g/cm^3. On reaching 30 cm depth, the relative density of lunar soil reached its maximum at 90%. Ice such as that found on cometary bodies and polar ice caps has an average cohesion of 0.08–0.28 MPa, an internal friction angle of 40–60° and a bulk density of 0.05–1.5 g/cm^3. Ice impregnation of regolith is likely to alter soil parameters significantly The hyperbolic equation relates density to depth by the relation:

$$\rho = 1.92 \left(\frac{z + 12.2}{z + 18} \right)$$

giving a density of 1.30 g/cm^3 at $z = 0$ and 1.76 g/cm^3 at $z = 50$ cm for lunar soil. For the top 30 cm of lunar soil, average porosity is 49%. Other parameters vary with depth (see Table 4.10).

Generally, the frictional forces between soil particles may be considered to be viscous forces. Soil has both elastic and plastic behaviors which may be modeled

Table 4.10. Lunar soil property variation with depth.

Depth (cm)	*Cohesion (average)* (kPa)	*Cohesion (range)* (kPa)	*Friction angle (average)* (°)	*Friction angle (range)* (°)
0	0.45	0.35–0.70	36	30–50
0–15	0.52	0.44–0.62	42	41–43
0–30	0.90	0.74–1.10	46	44–47
30–60	3.0	2.40–3.80	54	52–55
0–60	1.6	1.30–1.90	49	48–51

through stiffness/viscosity and compression/yield, respectively. Ground deformation under the wheels acts like a nonlinear spring and acts to reduce vibration in the vehicle generated by ground profile roughness. With increasing weight, the soil has elastic and plastic regimes such that soil deformation varies as:

$$W = \begin{cases} k_e z & \text{for } z < z_e \text{ where } k_e = \text{elastic coefficient} \\ k_e z + k_p(z - z_e) & \text{for } z > z_e \text{ where } k_p = \text{plastic coefficient} \\ & z_e = \text{elastic limit depth} \end{cases}$$

The soil–vehicle interaction may also be characterized through a stiffness parameter determined by means of plate load tests:

$$K = \frac{4Gr}{1-v} = \frac{2rE}{1-v^2}$$

where $r =$ plate radius, $G = E/(2(1+v)) =$ modulus of rigidity $= 0.2\,\text{N/mm}^2$ for sandy soil, $E =$ Young's modulus $= 0.6\,\text{N/mm}^2$ for sandy soil, and $v =$ Poisson's ratio $= 0.35$ for soil. The U.S. Federal Highway Administration adopts an experimental version of this formulation:

$$K = \frac{1.77rE}{1-v^2}$$

For example, the stiffness of DLR soil simulant B is 84.72 N/mm. Soil also possesses damping properties: $F_d = -C_d \dot{z}$. Clays tend to exhibit higher damping than sandy soils. Soil damping is given by:

$$C_d = \frac{3.4r^2}{1-v} \sqrt{G \rho_s}$$

where $\rho =$ soil density $= 1,140\,\text{kg/m}^3$. DLR soil simulant B has a damping coefficient of 12.24 N/(mm/s). The equilibrium of forces between the elastic and plastic components may be used to determine soil pressure [221, 222]:

$$p = c_f(z_{\text{el}} - z_{\text{pl}}) + \eta(\dot{z}_{\text{el}} - \dot{z}_{\text{pl}}) = \mu z_{\text{pl}} + u_y$$

where $c_f =$ soil stiffness, $\eta =$ soil (Newtonian) viscosity, $\mu =$ soil compression constant, $u_y =$ yield constant, and $\dot{z} =$ sinkage velocity. Differentiation and rearrangement gives:

$$\dot{p} + \frac{c_f + \mu}{\eta} p = \mu \dot{z}_{\text{el}} + \frac{c_f \mu}{\eta} z_{\text{el}} + \frac{c_f u_y}{\eta}$$

Assuming that the elastic component of sinkage corresponds to Bekker sinkage, this gives a formulation of ground pressure to sinkage of the form:

$$p = \left(\left(\frac{\mu}{c_f + \mu} \right)^2 \eta \dot{z} + \left(\frac{c_f u_y}{c_f + \mu} \right) \left(1 - e^{-\frac{c_f + \mu}{\eta} z} \right) \right) + \frac{c_f \mu}{c_f + \mu} z$$

This may be compared with the much simpler Bernstein–Goriatchkin formulation introduced later in this section. The major problem is these soil parameters are generally not known and are difficult to measure. Bekker theory is a more com-

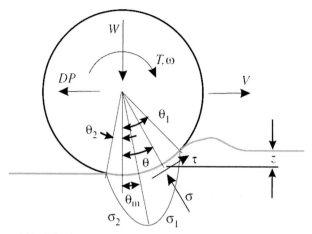

Figure 4.21. Wheel geometry [credit Tim Setterfield, Carleton University].

monly utilized approach for traction modeling of planetary rovers. The force balance of the wheel is given by drawbar pull DP horizontally and wheel loading W vertically with a torque τ applied to the wheel. The wheel has angular velocity w giving a translational velocity of v (Figure 4.21). The angle from vertical at which the wheel makes contact with the terrain is θ_1 while the angle at which the wheel loses contact with the surface is θ_2, which define the limits of the wheel–soil contact angle. Stress is created at wheel–soil interface σ which can be decomposed into normal and parallel components. Maximum stress occurs at wheel–soil angle θ defined to lie halfway between the extrema.

Forces on a wheel are defined by the normal (weight) and tangential (drawbar pull) forces:

$$F_n = W = rb\left(\int_{\theta_1}^{\theta_2} \sigma(\theta) \cos\theta \, d\theta + \int_{\theta_1}^{\theta_2} \tau(\theta) \sin\theta \, d\theta\right)$$

$$F_t = DP = rb\left(\int_{\theta_1}^{\theta_2} \tau(\theta) \cos\theta \, d\theta - \int_{\theta_1}^{\theta_2} \sigma(\theta) \sin\theta \, d\theta\right)$$

where σ = normal stress at the wheel–soil interface and τ = tangential shear stress at the wheel–soil interface. Increased vehicle weight on sandy soil increases drawbar pull. Wheel torques are given by:

$$T = r^2 b \int_{\theta_1}^{\theta_2} \tau(\theta) \cdot d\theta$$

where r = wheel radius and b = wheel width. For the ExoMars rover, peak wheel torques are estimated to be around 15 Nm. Only powered wheels produce traction and this traction must exceed resistance forces for forward movement—this is drawbar pull (DP), which is the difference between soil thrust H and motion resistance R:

$$DP = H - \sum R$$

where H = soil thrust as a result of soil–wheel interaction, $\sum R$ = soil resistance to motion = $R_c + R_b + R_g$, R_c = compaction resistance due to sinkage, R_b = bulldozing resistance, and R_g = grade resistance due to slope. Drawbar pull defines trafficability. Rolling resistance is sometimes used to refer to combined compaction and bulldozing resistance but the term is to be discouraged. For the Apollo LRV rolling resistance accounted for 15% of total energy losses with the rest being mechanical. Compaction resistance due to wheel sinkage is by far the most dominant form of soil resistance on flat terrain—on slopes, slippage increases rapidly becoming dominant at high inclines. Tractive soil thrust H is based on the Mohr–Coulomb relation for soil shear strength on soil failure for vehicle contact area A and is given by:

$$H = AC + W\mu = AC + W \tan \phi$$

where A = wheel/track ground contact area = $(\pi/4)bl$, l = ground contact length = $(d/2) \cos^{-1}(1 - (2z/d))$, b = wheel/track width, W = vehicle weight, d = wheel diameter, and z = soil sinkage. This is the Bernstein–Bekker equation which assumes that vehicle ground pressure is uniform if applied to the vehicle as a whole. For N wheeled/tracked vehicle, soil thrust is given by:

$$H = N(blC + W \tan \phi)$$

where N = number of wheels and l = ground track length. The wheel width-to-diameter ratio commonly selected is 1:2.5, though the Sojourner wheel width-to-diameter ratio was closer to 1:2.2 (i.e., favoring increased width). The wheels are generally assumed to have a flat profile but MER wheels were curved because the motor drive was inclined to the wheel hub (Figure 4.22).

Soil thrust is modified by slippage which affects both cohesion and friction (but can be modeled independently [223]). Bekker theory is highly sensitive to model parameters, particularly slippage, though good experimental agreement occurs at 20% slip. Gee-Clough (1976) [224] incorporates slip into rolling resistance (compaction and bulldozing) rather than against soil thrust leading to more complex formulations. If slippage occurs, this must be modified to account for soil deformation using the Janosi–Hanamoto relation describing the shear stress–strain of the soil for a wheeled vehicle. Normal displacement is given by:

$$N(\theta) = \int_{\theta_e}^{\theta} r_r \cos \theta \cdot d\theta = r_r(\sin \theta - \sin \theta_e)$$

where θ_e = wheel/soil contact angle and r = wheel radius. This gives normal stress:

$$\sigma_n = kN(\theta)^n = k(r_r(\sin \theta - \sin \theta_e))^n$$

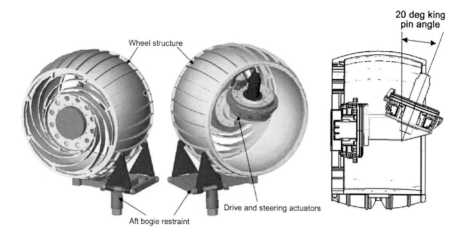

Figure 4.22. MER wheel configuration [credit NASA JPL].

Tangential displacement is given by:

$$T(\theta) = \int_{\theta_e}^{\theta} (r - r_r \sin\theta) \cdot d\theta = r(\theta - \theta_e) + r_r(\cos\theta - \cos\theta_e)$$

This gives tangential stress:

$$\sigma_t = (C + \sigma_n \tan\phi)(1 - e^{-T(\theta)/K})$$

$$= (C + \sigma_n \tan\phi)(1 - e^{-(r(\theta-\theta_e)-r_r(\cos\theta-\cos\theta_e))/K})$$

This is the Janosi–Hanamoto equation relating shear stress to shear displacement. This is effective for certain types of sand, saturated clay, snow, and peat. Soil shear stress displacement has a maximum value in brittle soils but not in plastic soils [225]. Maximum radial stress does not act directly beneath the wheel axle but shifts forward at increasing slip. The stress distribution under a point load at different angles is given by the Fröhlich equation of which the Boussinesq equation is a special case:

$$\sigma_r = \frac{mW}{2\pi R^2} \cos^m \theta$$

where $m = 3$–6 ($=3$ for Boussinesq equation) and $R =$ radius of the contact patch. Its derivation is complex. At different soil depths, vertical stress under the wheel is given by:

$$\sigma_z = \sigma_0 \left(1 - \frac{z^3}{(z^2 + r_0^2)^{3/2}}\right)$$

Slip velocity,

$$v_j = rw(1 - (1 - s)\cos\theta)$$

where $\theta =$ the angle between the track/wheel and the horizontal. Shear displacement along the wheel–soil interface is given by:

$$j = \int_0^t v_j \, dt = \int_0^t rw(1 - (1 - s)\cos\theta)dt = \int_0^{\theta_0} rw(1 - (1 - s)\cos\theta)\frac{d\theta}{w}$$

$$= r[(\theta_0 - \theta) - (1 - s)(\sin\theta_0 - \sin\theta)]$$

where $\theta_0 =$ entry angle from horizontal soil contact from wheel vertical, $\theta =$ angle between wheel vertical to perpendicular tangential contact of wheel with soil, and where

$$s = \begin{cases} \dfrac{rw - v}{rw} & \cdots \quad rw > v \\ \dfrac{rw - v}{v} & \cdots \quad rw < v \end{cases} = \text{slip during driving/braking such that } -1 \leq s \leq 1$$

Slip in the lateral direction may be modeled by means of the slip angle [226]:

$$\beta = \tan^{-1}\left(\frac{v_y}{v_c}\right)$$

Hence, shear stress is given by [227, 228]:

$$\tau = (C + \sigma \tan \phi)(1 - e^{-j/K})$$

$$\tau(\theta) = (c + \sigma(\theta) \tan \phi)(1 - e^{-(r/K)(\theta_0 - \theta - (1-s)(\sin \theta_0 - \sin \theta))})$$

for wheeled vehicles where j = the soil deformation rate constant = $\int_0^t v \cdot dt = r[\theta_1 - \theta - (1-s)(\sin \theta_1 - \sin \theta)]$, K = soil slip shear deformation coefficient at maximum soil stress which depends on the wheel–soil contact area = 0.01–0.05 m [229] (1.6 cm nominally for lunar soil), $\sigma = (k_c + k_\phi b)(z/b)^n$ = normal stress, θ = wheel angular location, θ_1 = angle from the vertical at which the wheel first makes contact with the terrain, and $z(\theta) = r(\cos \theta - \cos \theta_1)$ = soil sinkage. Shear deformation modulus K (dependent on the wheel–soil contact area) defines the fraction of maximum soil shear strength mobilized due to wheel slip (i.e., the coefficient of soil slip in centimeters). It indicates how quickly the maximum shear strength of a soil is exploited upon shearing [Richter, L., private communication]:

$$\frac{K_2}{K_1} = \sqrt{\frac{A_2}{A_1}}$$

where

$$K = 1.78 \text{ cm for LRV on lunar soil (variation } \pm 0.76 \text{ cm)}$$

$$= 0.8 \text{ cm for RCL-E wheel on Martian terrain}$$

$$= 0.2 \text{ cm for Solero wheel}$$

$$= 3.0 \text{ cm for MER wheel}$$

For a track vehicle, shear displacement is given by [230–232]:

$$j = \int_0^t v_j \, dt = \int_0^t rw(1 - (1-s)\cos \theta) \, dt = \int_0^l rw(1 - 1 - s)\cos) \frac{dl}{rw}$$

$$= \int_0^l \left(1 - (1-s)\frac{dx}{dt}\right) dl = l - (1-s)x$$

where l = distance between track from initial shearing and x = horizontal distance on soil from initial shear. Hence, shear stress is

$$\tau(x) = (C + \sigma(x) \tan \phi)(1 - e^{-(l-(1-s)x/K)})$$

for tracked vehicles. Hence, soil thrust is given by:

$$F = b\int_0^l \tau \, dx = b\int_0^l (c + \sigma \tan \phi)(1 - e^{-j/K}) \, dx = b\int_0^l \left(c + \frac{W}{bl}\tan \phi\right)(1 - e^{-ix/K}) \, dx$$

$$= (Ac + W \tan \phi)\left(1 - \frac{K}{il}(1 - e^{-il/K})\right)$$

where l = track contact length. For wheeled terrestrial vehicles, slip is generally around 2–3% (as for the Apollo LRV) but the Lunokhods exhibited slippage in the 5–7% range and the MERs experienced slip as high as 10–20% on flat ground. The slip for tracked vehicles is an order of magnitude less than that for wheeled vehicles (rarely beyond 1–12%). Slippage is one of the most difficult processes to model yet is the most critical issue for planetary rovers, particularly on slopes— soil possesses a self-organized criticality phenomenon typical of granular media at far from equilibrium conditions. Soil is composed of agglomerated particles with complex behaviors intermediate between solid and fluid depending on its stress state. It exhibits both solid-like and liquid-like behaviors separated by phase transition [233]. It is the cohesion and friction between grains that allows maintenance of solid-like behavior, but this is unstable beyond a certain stress threshold. Experiments with sand piles suggest that catastrophic dune avalanches occur once a certain configuration is reached according to the power law dependence between avalanche frequency and dune size and results in a soil configuration that is barely stable. Above a certain critical slope angle, soil avalanches behave like a fluid when subject to small perturbations. This represents a phase transition similar to the $1/f$ noise phenomenon at low frequencies with its self-similar structure at all scales. Self-organized criticality is exhibited by piles of sand such that the pile grows with increasing slope as sand accumulates until it reaches its angle of repose; thereafter, further addition of sand causes the sand to collapse back to the angle of repose [234]. Self-organized criticality involves critical states which, when perturbed, reorganize the system into a new critical state. The distribution of avalanche events s for different sized sandpiles follows a power law behavior such that $p(s) \sim s^{-\tau} G(s/L^D)$ where $\tau = 1$ and $D = 2.23 \pm 0.03$ [235]. Growing sandpiles will steepen until the slope reaches a critical value beyond which sand grains will cascade into a new stable configuration [236]. At 100% slip, the wheel does not move forward and the soil acts like a fluid. This is accentuated on slopes. The angle of repose is defined as the maximum angle at which unconsolidated material is stable. It is the angle between the normal and resulting forces when soil failure occurs in response to shear stress. Most soils on Earth have angles of repose varying between 23 and 40° (e.g., the angle of repose for sand is 23° where slip becomes 100%). Wheel slip has the general form:

$$\frac{F}{W} = \frac{\tau}{rw} = a(1 - e^{-bs})$$

The JPL/Caltech (Lindemann–Voorhees) team employed a slip graph generated from empirical experiments using their MER model on sand dunes at Californian beaches. A simple polynomial function may be represented as: $s = ab^\theta$, where $s = $ slip (%), $a = $ constant $= $ slip on a flat plane (when $\theta = 0$), $b = $ constant $= $ 1.0893. A cubic spline interpolation for minimum slip of 2% on a flat surface and maximum slip of 100% on a 23° slope was developed which matched the empirical data well except at extreme values:

$$S = a\theta^3 + b\theta^2 + c\theta + d$$

where $\theta = $ slope angle and $a, b, c, d = $ coefficients of interpolation. A more sophisticated polynomial model successfully equates wheel slip as a function of slope angle in the x-direction where it is sextic and in the y-direction where it is quartic with respect to slope angle assuming minimum slip of 5% (due to MER data) on a flat plane ($\theta = 0°$) and maximum slip of 100% at the angle of repose ($\theta = 23°$) for the slope [237]. The slip–slope relation may also be well represented by a sigmoid function assuming minimum slip of 5% on a flat surface ($\theta = 0°$) and a maximum slip of 100% at the angle of repose:

$$S = \frac{a}{1 + be^{(-c\theta)}}$$

where $a = 100$, $b = 67$, and $c = 0.32$. This equation follows the interpolation curve very closely and predicts the result within <10% of the accuracy for soils/sands having an angle of repose of 23°. The reason this model appears to work is not currently understood.

Generally, if wheel elasticity is incorporated, the wheel contact patch area is altered [238]. For a flexible wheel, its deflected diameter is given by:

$$d = \frac{2}{\sqrt{F/\pi bB}}$$

where F = load, b = wheel width, $B = Et^3/(12(1 - v^2))$ = bending rigidity, E = Young's modulus = 71 GPa for Al alloy, t = tyre thickness, and v = Poisson's ratio = 0.3. Flexible wheels increase the ground contact area, enhancing the soil thrust contribution to drawbar pull. Pneumatic tyres are unsuitable for planetary rovers as they are susceptible to puncture and polymer degradation under space conditions. The Tweed non-pneumatic tyre used on ATHLETE uses flexible spokes attached to an outer rim and an inner rim. The Nepean wheeled vehicle performance model has been used to assess two candidate flexible wheels for a lunar rover—a wire mesh wheel and hoop spring wheel [239]. The approach taken by the DLR is to use metal leaf springs as the principal elastic elements with an outermost ring providing additional elasticity. These were baselined for the ExoMars rover (Figure 4.23). Although a wheel stiffness of ~10 kPa appears to be optimal for traction on multiple soils, a slightly lower stiffness of 7 kPa gives 13% radial deflection with advantages of greater traction. The ExoMars flexi-wheel was 0.3 m in diameter and 0.1 m wide with an inner hub 0.16 m in diameter and a bump stop disk 0.22 m in diameter giving a maximum deflection of 0.04 m. There are 2×6 outer circular spokes of dimension 70 mm diameter and 24 mm wide and six inner circular spokes of the same diameter but 48 mm wide. Radial stiffness of the wheel was designed to be 5.9 N/mm giving 2.54 cm deflection with a 150 N wheel load (assuming a 240 kg rover with equal load on all six wheels under Martian gravity conditions of 0.38g). Using Ti6Al4V alloy with a Young's modulus of 110 GPa, the rim, spoke, and hub thicknesses are 0.5 mm, 0.5 mm, and 4.0 mm giving respective masses of 0.208 kg, 0.280 kg, and 0.889 kg (total wheel mass of 1.377 kg). Bump stop deflection requires a load of 350 N.

Soil thrust requires modification to include grousers of height h which substantially increase soil thrust. Wheels require an optimum number of grousers such that intergrouser spacings do not clog up with soil—typically ~10–12

Figure 4.23. Flexible wheel for ExoMars [credit ESA–DLR].

grousers per wheel. The 4 kg DLR wheeled MIDD nanorover used 4 × 160 mm
diameter × 22 mm wide wheels studded with 26 × 5 mm high grousers [240–242].

The Kapvik microrover's wheels were 145 mm in diameter and included 24
5 mm long grousers (Figure 4.24). Grouser shape is also important—sharp blades
increase the friction of the wheel and thereby assist climbing over rocks. Chevron-
shaped grousers may have two orientations—forward-pointing chevrons provide
self-cleaning of intergrouser spacings in cohesive soil, while rearward-pointing
chevrons increase grouser thrust in sandy soils. The Apollo lunar rovers used 54
alloy chevron grousers of 1.6 mm height over 50% of the wheel surface (separated
by 2.4 cm). While Sojourner used spike-based grousers, the MERs adopted 0.25 m
diameter wheels shaped as truncated spheres with flat grousers while Curiosity has
evolved a further design variation (Figure 4.25). This was imposed by the fact that
the wheel drive shaft was angled so a spherical sectioned wheel was necessary for
steering. The ExoMars rover phase A design adopted 0.3 m diameter wheels with
12 × 5 mm high grousers. Dynamic analysis suggests that doubling the number of
grousers increases drawbar pull by 30% [243, 244]. Discrete element analysis

Figure 4.24. Kapvik rover wheel with cleats.

Figure 4.25. Sojourner, MER, and Curiosity rover wheel and grouser configurations [credit NASA JPL].

suggests that a greater number of smaller cleats generate better traction than fewer larger cleats [245]. Hence, the addition of grousers improves drawbar pull with minimal impact on mass overhead. For a vertical grouser, passive pressure distribution is given by:

$$\sigma_p = \gamma z N_\phi + q N_\phi + 2c\sqrt{N_\phi}$$

where $\gamma = \rho g =$ weight density of soil, $z =$ depth, $q =$ surcharge pressure on soil surface, and $N_\phi = \tan^2(45 + \phi/2)$. The resultant tractive force due to grouser use is given by:

$$F_p = \int_0^{h_b} \sigma_p \, dz = \tfrac{1}{2}\gamma h_b^2 N_\phi + q h_b N_\phi + 2c h_b \sqrt{N_\phi}$$

where $h_b =$ grouser penetration depth. When part of a wheel, grousers shear the soil along an internal failure surface overcoming resistive friction between the soil and blade—the fundamental equation of earth-moving describes this effect:

$$\tau = F(r + \tfrac{2}{3}h_b)$$

where $F = b(\tfrac{1}{2}\gamma h_b^2 N_\phi + \sigma h_b N_\phi + 2C h_b \sqrt{N_\phi}) =$ shear force, $\gamma =$ soil specific density $= \rho g = 1.52$ g/cm^3 for average Martian soil, $\sigma =$ normal stress under the wheel at bottom center, and $N_\phi = \tan^2(45 + \phi/2) =$ soil-bearing coefficient. Overall wheel axle torque for a grousered wheel is given by:

$$\tau = r^2 b \int_0^{\theta_0} \tau(\theta) \, d\theta + F(r + \tfrac{2}{3}h_b)$$

Corresponding tractive force/wheel in the drive direction is given by:

$$H = rb \int_0^{\theta_0} \tau(\theta) \cos\theta \, d\theta + F$$

For a grousered N-wheeled vehicle, with n grousers in contact with the soil, soil thrust is given by:

$$H = N(T_0 + W \tan\phi)$$

where $T_0 = n\tau_0(hb/(\cos\beta\sin\beta) + h^2/(\cos\beta))$, $\beta = 45 + (\phi/2) =$ soil rupture angle, $\tau_0 =$ soil shear stress, and $h =$ grouser height. For an N-tracked vehicle with n grousers per track, the Bekker formulation yields (for $h/b < 0.2$):

$$H = N\left(nblC\left(1 + \frac{2h}{b}\right) + W\tan\phi\left(1 + 0.64\left(\frac{h}{b}\cot^{-1}\left(\frac{h}{b}\right)\right)\right)\right)$$

Some general observations regarding rover performance in relation to traction theory have been summarized (e.g., grouser size and distribution are the primary determinants for enhancing drawbar pull) [246].

Acting in opposition to soil thrust are resistances to motion. To obtain drawbar pull, resistive forces must be subtracted from soil thrust. There are a number of different sources of resistance to forward motion which are dominated by soil compaction with other resistance sources such as soil bulldozing being negligible in comparison. Soil stiffness is given by equating ground pressure p to sinkage depth z:

$$p = k_{eq}z$$

where $k_{eq} = ((6.94)/b + 505.8)\,\text{kN/m}^3$ for dry sand and $k_{eq} = ((74.6)/b + 2,080)\,\text{kN/m}^3$ for sandy loam. In fact, this is not linear:

$$p = kz^n$$

where $k =$ modulus of soil deformation due to sinkage (soil consistency) (N/m^{n+2})

$\quad n =$ soil deformation exponent where $0 < n < 1.2$ (note, for $n > 3, z \rightarrow \infty$)

$\quad = 0.5$ for Bernstein soil

$\quad = 1$ for Gerstner soil (such as sand)

Compaction resistance may be computed from the Bernstein–Goriatchkin pressure–sinkage relation as a result of the work involved in forming a rut:

$$\sigma(z) = (k_c + k_\phi b)\left(\frac{z}{b}\right)^n = \left(\frac{k_c}{b} + k_\phi\right)z^n = kz^n$$

where $\sigma(z) =$ ground pressure load, $k = k_c + bk_\phi$, $k_c =$ modulus of cohesion of soil deformation, and $k_\phi =$ modulus of friction of soil deformation. This describes the sinkage of a flat plate in soil as a function of normal pressure and soil deformation. Unfortunately, parameters k_c, k_ϕ, and n are unknown for Mars and are unlikely to be measurable on the fly—either lunar values [247] or Mars soil stimulant values must be used:

$\quad k_c =$ modulus of cohesion of soil deformation

$\quad\quad = 0.14$ (variation 0-0.28) N/cm^2 for the Moon

$\quad k_\phi =$ modulus of friction of soil deformation

$\quad\quad = 0.82\,\text{N/cm}^3$ for the Moon

$\quad n =$ soil deformation exponent

$\quad\quad = 1.0$ for the Moon

However, estimates from Viking lander footpad pressure and its radius give for the Viking landing site: $k = 1{,}481.0\,\text{kPa}$, $n = 1.8$, $k_c = 306.8\,\text{kPa}$, and $k_\phi = 20.0$ kPa/m. It must be noted that these are estimated values with high uncertainty. Soil parameters can make a significant difference to sinkage and compaction—a low exponent n reduces soil sinkage and compaction resistance significantly. The most important factor is k_ϕ which must be high to keep sinkage and compaction resistance low. An alternative to the Bernstein–Goriatchkin relation is the Reece load–sinkage equation:

$$\sigma(z) = (Ck_c + \gamma bk_\phi)\left(\frac{z}{b}\right)^n$$

where $\gamma = \rho g$. This formulation requires dimensionless moduli of sinkage k_i which eliminates the dependence of deformation moduli on n but will not in general be available. It gives essentially the same results as the Bernstein–Goriatchkin formulation. A further variation on the Bernstein–Goriatchkin relation is to add a damping coefficient:

$$\sigma = kz^n - c\dot{z}$$

where $c = \eta k = $ damping coefficient proportional to stiffness coefficient. For snow, the pressure–sinkage relation becomes:

$$\sigma = \sigma_0\left(-\ln\left(1 - \frac{z}{z_0}\right)\right)$$

where $z = z_0(1 - e^{-p/p_0})$. Sinkage into the soil depends on properties of both the soil and the vehicle. Wheel sinkage may be determined:

$$W = R_c l = \int_0^{\theta_0} b\sigma r \cos\theta \cdot d\theta$$

Soil compaction resistance due to each wheel is given by [248]:

$$R_c = b\int_0^z \left(\frac{k_c}{b} + k_\phi\right)z^n \cdot dz = bk\frac{z^{n+1}}{n+1} = \frac{z^{n+1}}{n+1}(k_c + bk_\phi)$$

Hence,

$$z = \left(\frac{p}{k/b}\right)^{1/n} = \left(\frac{W}{A(k/b)}\right)^{1/n} = \left(\frac{3W}{(3-n)(k_c + bk_\phi)\sqrt{d}}\right)^{2/(2n+1)}$$

for a single wheel or track. This is valid if $n < 1.3$ (i.e., wheel sizes greater than 500 mm diameter and vertical wheel loads exceeding 45 N). Hence, soil compaction resistance can be determined by:

$$R_c = \int_0^{\theta_0} b\sigma r \sin\theta \cdot d\theta$$

$$= \left(\frac{1}{(3-n)^{(2n+2)/(2n+1)}(n+1)b^{1/(2n+1)}(k_c/b + k_\phi)^{1/(2n+1)}}\right)\left(\frac{3W}{\sqrt{d}}\right)^{(2n+2)/(2n+1)}$$

$$= \left(\frac{(k_c/b + k_\phi)b}{n+1}\right)z^{n+1} = \left(\frac{bk}{n+1}\right)\left(\frac{3W}{(3-n)bkd^{1/2}}\right)^{2n+2/2n+1}$$

For a tracked vehicle, the work done in compacting terrain to a rut depth z determines motion resistance:

$$W = R_c l = bl\int_0^{z_0} p \cdot dz = bl\left(\frac{k_c}{b} + k_\phi\right)\left(\frac{z_0^{n+1}}{n+1}\right) = \frac{bl}{(n+1)(k_c/b + k_\phi)^{1/n}}\left(\frac{W}{bl}\right)^{(n+1)/n}$$

Hence,

$$R_c = \frac{1}{(n+1)(k_c/b + k_\phi)^{1/n} b^{1/n}} \left(\frac{W}{l}\right)^{(n+1)/n}$$

Sinkage and compaction resistance decreases with increasing wheel width but it is more effective to increase wheel diameter rather than wheel width. The adoption of tracks significantly reduces sinkage by more than an order of magnitude—indeed, track width affects the modulus of soil deformation k counteracting the advantage of ground areal increase of increased wheel width. Small diameter-wheeled vehicles have excessively low predicted performance according to Bekker theory which makes assumptions concerning minimum wheel sizes [249]. Wheel sinkage should generally be <4 cm, preferably up to a maximum of $0.3d$. These expressions are valid only for $n < 1.3$ and for moderate sinkages—for wheels under 50 cm in diameter, sinkage predictions are less accurate in dry, sandy soils.

For the front wheels of a vehicle, weight is related to sinkage:

$$W = bk \int_0^{z_0} \frac{z^n \sqrt{d}}{2\sqrt{z_0 - z}} dz$$

Additional second wheel axle sinkage is given by $z_0 + z_2$ [Richter, L., private communication]:

$$W = bk\sqrt{dz_2}\left((z_0 + z_2)^n - \frac{n}{3}(z_0 + z_2)^{n-1} z_2\right)$$

The third wheel axle generates further sinkage of z_3 where:

$$W = bk\sqrt{dz_3}\left((z_0 + z_2 + z_2)^n - \frac{n}{3}(z_0 + z_2 + z_3)^{n-1} z_3\right)$$

The effect of grousers on sinkage is given by [Richter, L., private communication]:

$$z_s = 2h \cdot s$$

where $h = 1.2h_b$ and $h_b =$ standoff height of grousers. However, Richter et al. (2006) suggest a nonlinear relationship [250]:

$$z_s = 2h(s - \tfrac{1}{2}s^2)$$

Hence, total sinkage is given by:

$$z = z_0 + z_s$$

Soil sinkage may be increased by slip-induced slippage of the form:

$$z_s = z(1 + s)^{n/(2n+1)}$$

If sinkage due to slippage is included, soil radial stress normal to the wheel is given by:

$$\sigma_{av} = \frac{1}{s_0} \int_0^{s_0} kz^n (1+s)^n \cdot ds = \frac{kz^n(1+s)^n}{n+1}$$

Although compaction resistance dominates soil resistance, bulldozing resistance occurs due to soil accumulating in front of the advancing wheels. Bulldozing resistance due to horizontal soil displacement is given by Hegedus' bulldozing resistance equation:

$$R_b = \frac{\cot\beta + \tan(\beta+\phi)}{1 - \tan\alpha \cdot \tan(\beta+\phi)} b\left(z \cdot C + \tfrac{1}{2}\rho z^2\left((\cot\beta - \tan\alpha) + \frac{(\cot\beta - \tan\alpha)^2}{\tan\alpha + \cot\phi}\right)\right)$$

where $\alpha = \cos^{-1}(1 - (2z/d)) = $ approach angle and $\beta = 45 - \phi/2 = $ angle between soil level and soil shear plane. Bekker gives a slightly different formulation whereby only the first term is relevant for tracked vehicles while both terms apply to wheeled vehicles:

$$R_b = N\left(\frac{b\sin(\alpha+\phi)}{2\sin\alpha\cos\phi}(2zCk_c + \gamma z^2 k_\gamma) + \left(\frac{\pi\gamma l^3(90-\phi)}{540} + \frac{\pi C l^2}{180} + C l^2 \tan(45+\phi/2)\right)_w\right)$$

where $k_c = (N_c - \tan\phi'')\cos^2\phi'' = $ soil-bearing capacity factor, $k_\gamma = (2N_\gamma/(\tan\phi'') + 1)\cos^2\phi'' = $ soil-bearing capacity factor, and $N_{c,\gamma} = $ Terzaghi coefficient of passive Earth pressure such that

$$\sigma = 2C\sqrt{N_\phi} + \gamma z N_\phi$$

$$N_c = \frac{W}{2lbC}$$

$$N_\gamma = \frac{W}{2b\gamma}$$

$$N_\varphi = \tan^2\left(45 + \frac{\phi}{2}\right)$$

$$l = z\tan^2(45 - \phi/2) = \text{distance of rupture}$$

$$\phi'' = \tan^{-1}(\tfrac{2}{3}\tan\phi)$$

The latter equation appears to be the most reliable. Bulldozing resistance is affected by changes in both wheel/track diameter and wheel/track width—narrow wheels reduce bulldozing resistance significantly while a decrease in wheel width increases bulldozing resistance by a much lower degree. Narrow wheels are favored to reduce bulldozing as more soil is pushed to the sides of the wheel. Bulldozing occurs primarily for front wheels as the rearward wheels follow in the ruts created by the forward wheels. Bulldozing resistance becomes a problem when $z/d > 0.06$. Bulldozing is particularly important during steering when there is significant lateral movement in the soil quantified by the lateral slip angle $\psi = \tan^{-1}(v_\gamma/v_x)$ [251]. In general, bulldozing has a minor effect on motion resistance in comparison with compaction resistance due to sinkage. Evidently, tracks have superior tractive performance than wheeled vehicles (Table 4.11) [252].

The rover performance evaluation tool (RPET) comprises three main modules—rover mobility performance evaluation tool (RMPET), mobility synthesis (MobSyn), and rover generator for SolidWorks (RoverGen) (Figure 4.26).

Table 4.11. Comparison of drawbar pull metric for wheeled versus tracked vehicles.

	Soil thrust (N)	Compaction resistance (N)	Drawbar pull including slip (N)
Sojourner	36.8	23.1	6.9
Nanokhod	950.3	0.03	943.5
ELMS	2,085.4	0.03	2,078.6

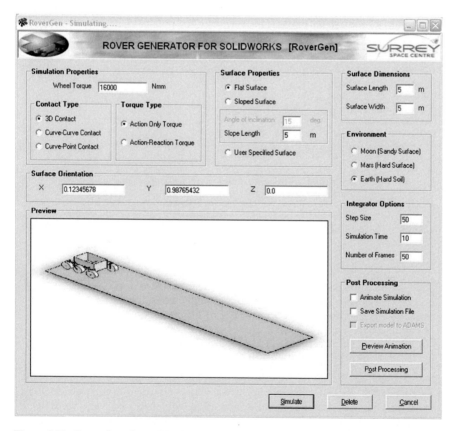

Figure 4.26. Screenshot from RMPET software tool [credit Nildeep Patel, Surrey Space Centre].

RMPET computes rover performance based on three different metrics—nominal ground pressure/mean maximum pressure, drawbar pull, and mean free path. MobSyn reverses the flow logic of RMPET and is similar to LocSyn in this respect. RoverGen creates a multibody kinematic/dynamic assembly of the rover based on SolidWork/ADAMS software.

For the six-wheeled Solero rover, the weight was 12 kg, drawbar pull was 24 N at a velocity of 30 mm/s and 39 N at a velocity of 50 mm/s, and typical wheel torques were 0.3 Nm on average at the latter speed. To climb a 15° slope, the torques required were 0.52 Nm for the back wheels, 0.3 Nm for the mid-wheels, and 0.28 Nm for the front wheels. Its maximum slope-climbing ability was 33° at 50 mm/s (Figures 4.27 and 4.28).

Based on Bekker analysis, the six 15 cm diameter × 7 cm wide wheels with 12 × 5 mm cleats of the Kapvik microrover generates a theoretical drawbar pull of 363 N in reference lunar soil and a theoretical slope capability of 50°. Traverse requires <25 W of power at a nominal speed of 40 m/h assuming 10% slip. For an early version of the 240 kg ExoMars rover, drawbar pull estimates were calculated for different representative soils (Table 4.12).

There are a few minor additional factors that may be taken into account in computing drawbar pull. The Bekker–Semonin empirical relation gives motion resistance due to wheel flexure:

$$R_f = \frac{3.58bd^2(W/A)\varepsilon(0.035\theta_a - \sin 2\theta_a)}{\theta_a(d - 2\delta)}$$

where $\theta_a = \cos^{-1}((d - 2\delta)/d) =$ wheel–soil contact angle, $\epsilon = 1 - e^{-k\delta/h}$, $h =$ wheel height, and where

$$k = \text{empirical wheel parameter} = 15 \text{ for bias ply tyres}$$

$$= 7 \text{ for radial ply tyres}$$

The tracked vehicle turning resistance moment due to lateral track sliding is given by [253]:

$$M = \tfrac{1}{4}\mu WL$$

where $L =$ track length and $W =$ vehicle weight. Increased track width improves vehicle mobility especially on cohesive soils. Finally, consideration should be given to resistance offered by climbing an obstacle of height h:

$$R_o = \frac{W(l_f + d/2)(\mu - \mu_r)(\mu_r \sin \theta + \cos \theta)}{(\cos \theta + \mu_r \sin \theta - \mu_r \sin \theta)(h(\mu - \mu_r) + l_f + l_r + d/2}$$

where $\theta = \sin^{-1}((d - 2h)/d) =$ contact angle and $\mu =$ adhesion coefficient for rock–metal $= 0.5$. In order to maximize drawbar pull when traversing different types of terrain, the position of the center of gravity needs to vary by shifting the vehicle's main body center of mass (e.g., in traversing obstacles, maximum drawbar pull is achieved by shifting the center of mass towards the downhill wheels). The maximum slopes that must be negotiated on Mars are the inner

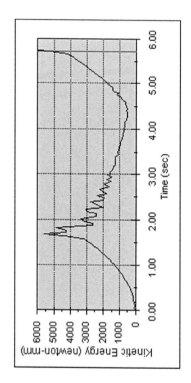

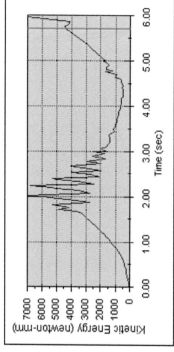

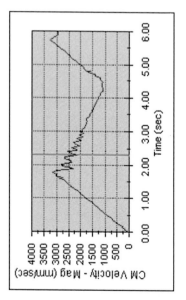

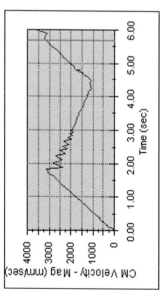

Figure 4.27. Solero rover simulation at constant wheel torque of 300 Nm for front and rear wheels [credit Nildeep Patel, Surrey Space Centre].

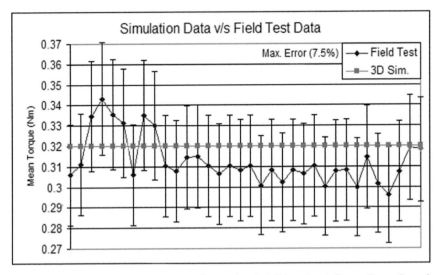

Figure 4.28. Mean torque of Solero on 15° slope [credit Nildeep Patel, Surrey Space Centre].

slopes of recent craters which may reach up to 30–40°, but wheeled vehicles are generally limited to gradients <25°. Drawbar pull coefficient DP/W dictates the maximum acceleration rate of the vehicle on flat ground or the maximum grade of negotiable slope. Gradability is approximately equal to the drawbar-pull-to-wheel-loading ratio so is a direct function of mobility efficiency:

$$\theta_{max} = \tan^{-1}\left(\frac{DP}{W_{wheel}}\right)$$

Traction considerations alone suggest that tracks offer far superior tractive performance over multiaxled wheeled vehicles due to the high soil contact area—similar considerations suggest that elastic wheels increase the soil contact area and so tractive performance, but these effects are less pronounced on frictional sandy soil than on cohesive clayey soils [254]. Although Bekker theory has a long history, it is reliant on calibration, accurate soil measurements, and direct applicability to vehicle size (small vehicles require modification to Bekker theory [255]).

4.8   VEHICLE–SOIL INTERACTION—THE RUSSIAN APPROACH

Russian planetary rover wheel theory defines so-called "generalized interaction parameters", which are related to parameters measurable in soil channels made by single wheels and which completely characterize wheel tractive behavior (LUMOT) [256] (i.e., slip s and wheel-steering angle ψ normalized to wheel load W).

Table 4.12. ExoMars rover drawbar pull estimates for different soils.

Soil	Specific gravity (ρg)	Cohesion (Pa)	Friction angle (°)	k_c (N/m^{n+1})	K_ϕ (N/m^{n+1})	$k = k_c + b k_\phi$	n	Drawbar pull
DLR soil sim A	4.24	188	24.8	2,370	60,300	8,400	0.63	112.7
DLR soil sim B	4.24	441	17.8	18,773	763,600	95,133	1.1	155.0
VL1 drift	4.29	1,600	18	1,400	820,000	83,400	1.0	151.28
VL1 blocky	5.97	5,500	30.8	1,400	820,000	83,400	1.0	319.5
VL2 crusty–cloddy	5.22	1,100	34.5	1,400	820,000	83,400	1.0	378.8
PL drift	4.36	380	23.1	1,400	820,000	83,400	1.0	215.2
PL cloddy	5.70	170	37	1,400	820,000	83,400	1.0	421.5
Dry sand	5.67	1,040	28	990	1,528,000	153,790	1.1	293.2
Sandy loam	5.67	1,720	29	5,270	1,515,000	156,770	0.7	298.8
Clayey soil	5.67	4,140	13	13,190	692,200	82,410	0.5	79.2
MER-B sandy loam	4.24	4,800	20.0	28,000	7,600,000	788,000	1.0	202.7
MER-B sloping soil	4.24	500	20.0	6,800	210,000	27,800	0.8	137.2

The gross pull coefficient (coupling coefficient) is given by:

$$\varphi(s, \psi) = \frac{H}{P_z} = \frac{H}{W}$$

The tractive resistance coefficient is given by:

$$\rho(s, \psi) = \frac{R}{P_z} = \frac{R}{W}$$

The specific traction coefficient (defining turning capability) is given by:

$$T(s, \psi) = \frac{\tau/r}{P_z} = \frac{\tau/r}{W}$$

The drawbar pull coefficient is given by:

$$K(s, \psi) = \frac{P_x}{P_z} = \frac{\mathrm{DP}}{W}$$

The lateral resistance coefficient is given by:

$$\mu(s, \psi) = \frac{P_y}{P_z} = \frac{P_y}{W}$$

The slippage coefficient is given by:

$$S = \frac{s_{\text{theor}} - s}{s_{\text{theor}}}$$

Now, $R = H - DP$ and $\tau \approx \varphi$:

$$\rho(s, \psi) = \frac{R}{W} = \frac{H - DP}{W} = \phi(s, \psi) - K(s, \psi) \approx T(s, \psi) - \mu(s, \psi)$$

The Russian approach is highly experiment based and is little used in the West currently.

5

Rover sensorimotor control systems

There are four major hardware components to a rover: the chassis, main computer, motors and motor controller, and sensors. The software binds these systems together through the control system.

5.1 DRIVE MOTORS

Wheels are driven by permanent magnet DC motors with position and velocity sensors to provide closed-loop speeds. Motors for space application must have low rotor inertia (implying rare earth flat disk rotors), low-mass but high-torque and high-positioning accuracy. Stepper motors lack torque at high speed limiting their slew rate—this favors servomotors which offer high torque at speed with smoother action (though microstepping offsets this problem in stepper motors). DC motors are generally used in space systems for mechanisms including robotic systems such as rovers and manipulators [257]. The motors of choice for spacecraft applications are electronically commutated brushless DC motors with integrated electronics because of their long life, high torque, light weight, high efficiency, and low heat dissipation. Most small mobile DC motors are rated at 12, 24, or 48 V. The Nomad rover prototype used brushless DC wheel motors with a motor torque constant of 0.56 Nm/A in conjunction with harmonic gear drives with a gear reduction of 218:1 to drive at a speed of 0.15 m/s. A brushless DC motor is comprised of a series of stator coils with a permanent magnet (ferrite or rare earth) rotor without the need for a commutator. Brushless DC motors use electronic commutation to control current through the windings and no power loss occurs through brushes. The commutation logic and switching electronics provide sensory feedback integrated into the motor. They offer high speeds up to 100,000 rpm, double the output torque over brush DC motors and greater reliability. The motors are integrated with their electronic controllers and

© Springer-Verlag Berlin Heidelberg 2016
A. Ellery, *Planetary Rovers*, Springer Praxis Books,
DOI 10.1007/978-3-642-03259-2_5

commutators. The brushless motor adopts fixed soft-iron stator poles wound with coils and a permanent magnet rotor similar to the synchronous AC motor. The absence of brushes eliminates coulomb friction and power is transmitted through Be-Cu roller contacts rather than slip rings. Brushless motors use solid state switching rather than a commutator and brushes. The permanent magnet in the rotor of a brushless DC motor generates an armature field independent of current. They need shaft position-sensing feedback and complex electronic control but are popular and efficient for space applications. The choice of standard, disk, bell, or toroidal motors depends on the application—they are generally built as thin rotors (i.e., disk or pancake motors) which have the space pedigree. Be may be used for mechanical parts by virtue of its high thermal expansion coefficient (similar to steel), low density, good thermal conductivity, and very high modulus of elasticity. Several innovative approaches offer the promise of low-mass motors (e.g., replacement of copper with super enameled aluminum for winding wires, replacement of wire coils with etched film coil disk patterns, elimination of iron core from armature coils). One of the chief difficulties with electric motors is the incidence of static friction (stiction) which generally increases over the motor life.

Brush DC motors use commutators and carbon brushes to apply current through the windings connected to the commutator ring. Spring-loaded brushes carry current through the commutator. Carbon brushes in sliding contact with a revolving commutator are subject to friction and wear so brushless DC motors are preferred. Commutation electronics typically include a Hall effect sensor, signal conditioning, and amplifier circuitry. Commutation limits the motor velocity beyond which sparking can occur—the motor performance levels are bounded by the limits ~0.1–10,000 rpm. Brush motors should not be used for long-duration missions as carbon brushes wear excessively in vacuum. However, brush DC motors were used on Sojourner and MER wheels and for the MER due to their limited mission life designs. The Maxon RE series has flight heritage having supplied 11 motors to Sojourner and 39 motors to the MERs. They are efficient ~90% with high reliability and are customizable. Beagle 2 adopted Maxon motors with a 100:1 gear ratio for its articulated robotic manipulator (ARM). Sojourner's wheels were driven by Maxon RE016 motors with a wheel torque of 1 Nm in motion (12 Nm stall torque) through 2,000:1 gear transmission for the wheels to run at 0.9 rpm (or equivalently 0.4 m/min). The front wheel axle generated 1 Nm torque (30.8 N tractive force for 0.065 m radius wheels), the rear wheel axle generated 0.5 Nm torque (15.4 N traction), while the mid-wheels operated at zero load (i.e., 46.2 N traction in total). Each motor required 116 mA at 15.5 V to produce 1 Nm of torque at −80°C. The motor assemblies developed for Sojourner were customised from commercial motors for −80°C operation comprising Maxon DC brush motors, planetary gearboxes, and lubricant. All structural components of the motor, gearing, and bearings were constructed from Ti with carbon fiber end fittings to minimize thermal expansion. Capacitors were incorporated to reduce the high breakaway torque required to start the motor at low temperatures to limit brush arcing—the Martian atmosphere has a breakdown voltage of 100 V. The motors were vacuum-impregnated with Braycote 814 oil, chosen for its very low pour point. Maxon RE016 motors are 16 mm in diameter and 41 mm in length and have a mass of 38 g with precious metal brushes and neodynium magnets which can operate at a nominal voltage of 16 V and a power rating of 2.5 W for 100,000 rpm. Maxon brush motors were selected for their superior torque-to-mass ratio while drawing less than 1.5 W. The smallest Maxon motors have a mass of 2.8 g and include Hall effect sensors. They are 6 mm in diameter yet offer 80,000 rpm while drawing only 1.2 W at a nominal voltage of 9 V and maximum current of 500 mA. Their maximum continuous torque is 0.26 mNm but use with 6 mm planetary gears provide a higher torque at lower speeds. Maxon motors have also become virtually standard in space robotic mechanisms. Each of the four wheels of the sample return rover (SRR) is driven by a DC brush Maxon RE025 motor to drive its 20 cm diameter wheels. The MER used 39 Maxon motors—8 × RE25 for lander deployment, 2 × RE20 for chassis suspension, 5 × RE20 for solar panel deployment, 6 × RE25 for wheel drives, 4 × RE25 for wheel steering, 2 × RE20 for antenna drive, 5 × RE20 for robotic arm, 1 × RE25 and 2 × RE20 for rock abrasion tool, and 1 × RE25 and 3 × RE20 for camera pointing.

On Sojourner, a 157 Ω heater provided 1.2 W at 14 V for each motor. Each motor required 116 mA at 15.5 V to produce 1 Nm of torque at −80°C (or 30.8 N tractive force for 0.065 m radius wheels). A motor at 1 Nm torque output applies a force of 15.4 N at 0.065 m (wheel radius). Stall torque at −80°C was 12 Nm drawing 196 mA at 15.5 V. The two front wheels produced 9 in.-lb (1 Nm, i.e., 30.8 N force), the two rear wheels produced 4.5 in.-lb (0.5 Nm, i.e., 15.4 N force), and the two mid-wheels operated at no load (total 46.2 N force while the rover weight was 38.6 N). Hence, 8.5 W were required to drive the vehicle. In addition, 1 W was required for the microcontroller and 1 W for onboard navigation computations. The Kapvik microrover was heavier and faster than Sojourner necessitating the use of more powerful motors. We adopted commercial off-the-shelf 130 g Maxon RE25 motors with graphite brushes (Figure 5.1) with a Maxon two-stage planetary gearhead GP26B (14:1), harmonic drive to a GP26B inter-mediate plate, and harmonic drive gear CSF-11-2XH-F (100:1) giving a total gear ratio of 1,400:1. The RE25 motor provided a nominal 29.1 mNm torque at 3,820 rpm consuming 10 W at 12 V. The gearing was integrated into a protective enclosure at the wheel base. Wheel odometry was provided by motor-integrated incremental encoders. The RE25 motor uses magnetoresistive encoders (500 counts per turn). In addition, single-axis load sensors were mounted on top of each wheel station to measure vertical forces. Each wheel ws driven by a Maxon EPOS 24/1 motor controller to which all sensors were connected and sampled at 1 kHz. Additional motors could be readily added to the CAN bus. The power allocated to the six motors was 24 W (87% efficiency) with a maximum nominal speed of 80 m/h. Maxon motors offer good torque-to-mass performance and could be retrofitted for space such as the addition of capacitors for improved power efficiency. Maxon motors have become the de facto standard onboard robotic platforms for spaceflight by virtue of their compactness, high efficiency and per-formance specifications, high reliability, low electromagnetic interference, and

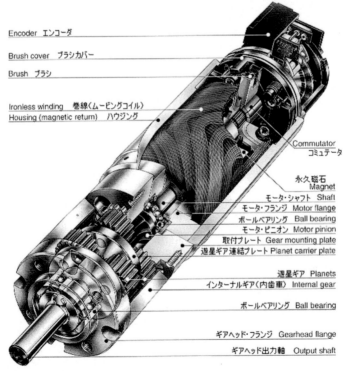

Encoder エンコーダ

Brush cover ブラシカバー

Brush ブラシ

Ironless winding 巻線（ムービングコイル）
Housing (magnetic return) ハウジング

Commutator
コミュテータ

永久磁石
Magnet
モータ・シャフト Shaft
モータ・フランジ Motor flange
ボールベアリング Ball bearing
モータ・ピニオン Motor pinion
取付プレート Gear mounting plate
遊星ギア連結プレート Planet carrier plate

遊星ギア Planets
インターナルギア（内歯車） Internal gear

ボールベアリング Ball bearing

ギアヘッド・フランジ Gearhead flange
ギアヘッド出力軸 Output shaft

Figure 5.1. Maxon RE025 motor and its schematic [reproduced with permission from Maxon Motors].

adaptability to multiple combinations of gears, feedback, and control electronics. The highest performance is provided by their RE series based on NdFeB magnets while lowest cost is offered by their S series based on AlNiCo magnets.

Geared motors are complex dynamic systems which are characterized by sliding teeth losses, teeth-meshing losses, and lubricant winding losses. Gearing is typically characterized by high transmission ratios from first-stage planetary gears and final-stage worm gears. Epicyclic gears in planetary gearboxes are commonly used in robotic mechanisms such as the Shuttle Remote Manipulator System (RMS). Planetary gears are commonly used in planetary rovers (e.g., Lunokhods which adopted two drive speeds with gear ratios ~80–200). Double planetary gears provide a high reduction ratio of 1:600 for torque amplification. Gear wheels are commonly made from anodized Al alloy 6802 (for light loads) or hardened stainless austenitic steel (440C corrosion resistant), though high loads may require the use of plasma-nitrided, precipitation-hardened steel (e.g., Shuttle RMS epicyclic gearboxes). Shafts and housings are often made from Ti alloy which has similar coefficients of thermal expansion as bearing steel. It is preferable that motor assemblies suffer from low friction and backlash—this favors the harmonic drive gearing system which eliminates backlash via its wave generator teeth arrangement with circular splines (Figure 5.2). The circular gear has two more teeth than the flexspline. Harmonic drive gearing offers 80:1 speed reduction nominally with 75% mechanical efficiency. Harmonic drive gearing was adopted on the Apollo LRV with an 80:1 step-down ratio. The elimination of gearing substantially reduces mechanical complexity and the requirements for lubricant. However, harmonic drives are limited to 1:100 gear ratios so they must be augmented by planetary gears. Direct drive motors do not utilize mechanical transmission between actuator and link offering the advantages of compactness and low mass. However, direct drive motors are limited in torque capability and tend to be heavier than geared motors. In Kapvik's wheel motors, the harmonic drive is connected to a two-stage planetary gearhead giving a total gear ratio of 1,400:1. Kapvik had a total of 24 W of drive power available.

All space-based gearing systems include a brake to lock the motors under emergency or hold conditions (e.g., maintaining a rover on a slope). Brakes are almost exclusively friction disk brakes with electromagnetic drives installed on the engine shaft.

Motor selection is a critical facet of rover design and is based on the load requirement—this must be converted to the equivalent load at the motor shaft through the gear ratio $n = w_m/w_i$ where $w_{m,l}$ = angular speed of the motor and load, respectively. The peak motor torque is dominated by the inertial torque. The motor torque available at maximum speed is less than stall torque. Generally, maximum load speed is converted into half the maximum motor speed. The torque–speed curve of the motor provides the basis of motor performance—it involves plotting the no-load armature current versus stall torque and the no-load motor speed versus stall torque. The line from zero torque/zero current to stall torque/stall current indicates motor current as a function of motor torque. A plot of mechanical output power versus stall torque shows the power requirement given

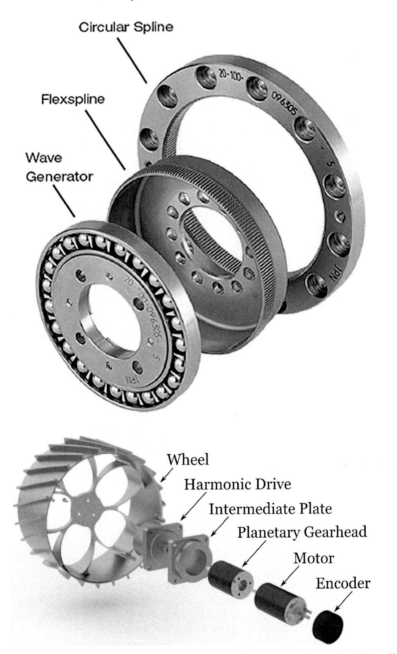

Figure 5.2. Harmonic drive [reproduced with permission from Maxon Motors] and Kapvik's motor-gear assembly [credit Tim Setterfield, Carleton University].

by $P = \tau w$. Power supply is typically DC power to a servo amplifier. A current flowing through the armature results in a power loss of $I^2 R$. Maximum armature winding temperature is $120°C$ which may impose a requirement for a maximum current and/or the implementation of a heat sink. Motor selection involves a number of specific steps:

(i) Load torque which resists motor rotation—this comprises friction torque (required to overcome frictional resistance; it is approximately constant but rises with speed) and acceleration torque (required to accelerate/decelerate the payload; it increases proportionally with acceleration). Power delivered to load is a function of output torque and motor shaft speed: $T = (f(K_t w))$. Motor friction torque must be added to the output load torque.

(ii) Winding resistance—this is the back EMF which generates opposing voltage resistance across windings (it increases with speed until back EMF equals the applied voltage when a no-load speed condition occurs with zero torque output). Maximum torque output is stall torque. This component dominates at low motor speeds but should typically be a few watts in order to make thermal dissipation tractable.

(iii) Compute the motor torque constant K_m which relates power loss in the windings to motor torque output T: $K_m = T/\sqrt{P_w}$ where P_w = power loss in windings. Add a 15% margin to K_m to cover winding variation.

(iv) Peak torque from the motor should be two to three times maximum design load torque.

(v) Select the winding wire gauge which affects resistance and the torque-winding constant, $K_t = T/I$ where I = winding current (A).

(vi) Solve:

$$K_m = \frac{K_t}{\sqrt{R}} \quad \text{and} \quad T = \frac{K_t}{R}(V_i - f(K_t w))$$

where V_i = input voltage.

(vii) Calculate the back EMF constant K_b.

(vii) Find the stall torque and no-load speed for a given voltage:

$$T_{stall} = \frac{V_i K_t}{R} \quad \text{and} \quad w_{no\text{-}load} = \frac{K V_i}{K_b}$$

where K = proportionality constant.

(ix) Motor performance will vary with temperature which may be quantified by the quality factor,

$$Q = \frac{1}{1 - P\left(\dfrac{\delta T}{T}\right)}$$

where T = ambient temperature and δT = motor temperature increase.

(x) Torque/speed characteristics may be used to determine the voltage required for a given motor speed and load torque.

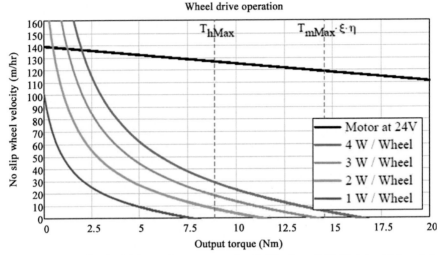

Figure 5.3. Kapvik wheel drive performance tests [credit Tim Setterfield, Carleton University].

(xi) The gear ratio requirement may be estimated from $n = \sqrt{I_{load}/I_{motor}}$ where $I_{l,m}$ = moment of inertia of load/motor—a high gear ratio reduces the payload inertia reflected onto the motor shaft.

There are also certain safety features required in actuator design: (i) emergency dynamic braking in addition to power cutout; (ii) software limits backed up by hardware limits in range of movement; (iii) overload power protection switches across all axes of movement; (iv) temperature and voltage output measurement of motors to detect power overload. For the Kapvik microrover, Maxon motor assemblies were subjected to performance tests (Figure 5.3).

From torque requirements power consumption can be determined—it may be assumed that motor efficiency is 0.7 nominally, gearbox efficiency is 0.6 nominally for planetary gears, and drive train efficiency is 0.7 nominally. Engine power is related to motor torque such that maximum torque to the wheels must be greater than the moment of all resistance forces about the center of the wheel plus the energy to supply the tractive effort:

$$P = R\sum_{i=1}^{m} I_i^2 = \frac{Rn^2}{K_t^2}\sum_{i=1}^{m} \tau_i^2 = \frac{Rn^2r^2}{K_t^2}\sum_{i=1}^{m} T_i^2$$

where R = motor resistance, $I_i = r_m/k_t$ = motor current, K_t = motor torque constant, n = motor gear ratio, $\tau = \sum(R + H) \times r$ = motor torque, $\tau_i = \tau_m/n$ = output motor torque, T_i = tractive effort, and r = wheel radius. A motor torque should include grouser effects which are given by:

$$\tau \approx F(r + \tfrac{2}{3}h_b)$$

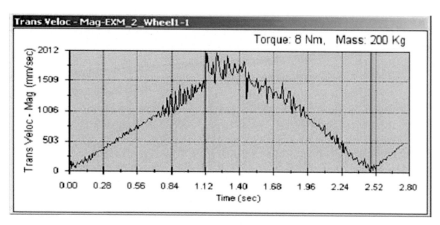

Figure 5.4. Simulated ExoMars wheel velocity (no slip) for a given wheel torque over a step obstacle [credit Nildeep Patel, Surrey Space Centre].

where h_b = grouser height and

$$F = b(\tfrac{1}{2}\gamma_s h_b^2 N_\phi + \sigma h_b N_\phi + 2Ch_b\sqrt{N_\phi})$$

Total wheel torque is thus given by:

$$\tau = r^2 b \int_0^{\theta_a} \tau(\theta) \cdot d\theta + F(r + \tfrac{2}{3}h_b)$$

Wheel torques or wheel speed will vary with the terrain—simulations of a 240 kg ExoMars rover employing the six-wheeled RCL-C chassis over a step obstacle (no slip) for a single wheel allow rover velocity to be determined at any given wheel torque (Figure 5.4).

Principal power losses in electric motors must be accounted for:

(i) I^2R joule losses in the windings of the stator and rotor (copper losses):

$$P_{Cu} = \sum_{i=1}^{n} I^2 A\rho l_{turn}$$

where n = number of rotor/stator slots, I = current density, A = slot area, ρ = specific resistance of copper, and l_{turn} = length of winding turn.

(ii) Hysteresis and eddy current losses in the armature core (iron losses):

$$P_{Fe} = k_e B^2 f_{rot}^2 + k_e B^2 f_{stat}^2 m_{stat} + k_h B^2 f_{stat} m_{stat}$$

where k_e = eddy current constant of material = 50 Hz, k_h = hysteresis constant of material = 50 Hz, B = maximum magnetic flux density, f = frequency, and m = mass.

(iii) Mechanical losses primarily due to bearing and air friction. Motors typically suffer from frictional losses, viscous friction being readily modeled. However, Coulomb friction (characterized by constant amplitude acting in opposition to the direction of motion) and stiction (characterized by high threshold and acting only under static conditions) are less easily modeled.

The power requirement for tracked vehicles is much higher than for wheeled vehicles at high masses—for low mass vehicles, this is not necessarily the case as power/unit weight ratings become comparable. The mass of the small Marsokhod of Mars 96 was dominated by the undercarriage (31 kg) and the control apparatus (35 kg). Its power consumption was high at 25–30 Wh/km. The power system for a rover is a critical part of its operational capabilities. Sojourner adopted certain specific features [258]:

(i) current limiter circuits to ensure that the power bus did not drop below 13.5 V to protect the CPU if the drive motors became excessively loaded;
(ii) battery bypass switches allowed the batteries to power the motors in the event of solar panel failure;
(iii) a three-battery string switch relay closed only after surface landing so that the batteries were offline during the cruise;
(iv) input latching relay allowed power to scientific instruments in the event of CPU failure;
(v) power monitor circuit to allow graceful shutdown at the end of each day;
(vi) load shedder circuit to monitor power bus voltage and switch off all devices if bus voltage dropped below a specified value to protect the CPU from power surges;
(vii) lander-controlled power switch was magnetically coupled to the lander to allow the lander to wake the rover up during the cruise for health checks;
(viii) alarm clock–controlled power switch controlled by an independent alarm clock to wake the rover as backup.

A power-scheduling algorithm would greatly extend the operational duration of the microrover and provide it with a graceful degradation profile. As part of this algorithm, a safe mode should be invoked under critical conditions in which all non-critical functions are switched off. Sojourner implemented a simple power-checking system whereby battery drainage was monitored prior to load switch off and the execution of driving telecommands. Power management will require the ability to measure the remaining capacity of a battery. One approach is to incorporate a sensor to count the coulombic charge/discharge rate by measuring the voltage across a resistor at one of the battery terminals [259]. An ADC samples the voltage periodically and a voltage-to-frequency converter time-integrates this to yield the quantum charge count. Measurement of the battery current to the load over time provides an estimate of the available Peukert capacity.

5.2 MOTOR CONTROLLERS

The purpose of the motor controller is to drive the motor in either direction in a controlled fashion. Two supply voltage inputs to the motor provide independent logic and power channels, the latter commonly through a voltage regulator. A limiting resistor protects against high currents during motor stall, and provides a means of measuring motor current. For the Kapvik microrover, the PIC18F4431 microcontroller (successor to the PIC18F8680) was considered for the motor controller given its compatibility with CAN-bus 2.0, but the Maxon solution was selected. The Kapvik microrover used EPOS 24/1 positioning control units for the control of its Maxon DC motors compatible with its 24 V DC power bus and its CAN bus for data transmission (Figure 5.5). A CAN bus can link all the motor controllers in order to implement the distribution of commands from a central controller. It implements the RS-232 serial protocol with an asynchronous serial data transfer packet comprising 1 start bit, 8-data-bit opcode, and 1 stop bit.

Each EPOS 24/1 motor controller has power, data, and communications connections. It provides adjustable current, speed, or position control for a single DC motor. It accepts reference commands and activates the motor by means of a

Figure 5.5. Maxon motor EPOS 24/1 motor controller [credit Matt Cross, Carleton University].

closed-loop PID control at a sample rate of 1 kHz or 10 kHz for speed/position and current control, respectively. EPOS 24/1 offers adjustable P, PI, and PID gains with CAN bus connectivity and 9–24 V DC bus compatability.

Up to 16 EPOS 24/1 nodes can be hard-configured on a single CAN bus (offering data rates of up to 1 Mbps), which is compatible with Kapvik requirements for wheel drive motors, an onboard manipulator and several pan–tilt platforms (Figure 5.6). The EPOS 24/1 motor controllers—rated to −10°C—were subjected to environment tests to ensure their suitability for subzero temperatures with storage at −40 to +50°C and operations at −20 to +40°C in northern Canada and equatorial Mars [260]. They can be utilized to −30°C without modification though lubrication viscosity increased below −20°C. The Kapvik wheel drive system comprised six motor controllers connected through the CAN bus. Rover translation velocity varied from 3–4 cm/s (108–144 m/h) for MER rovers but could increase to 5–10 cm/s in more advanced rovers. The maximum MER traverse velocity was 5 cm/s but motor heating limited traverses to 3.75 cm/s. Fuke et al. (1994) compared the advantages/disadvantages of centralized and decentralized drive schemes (Table 5.1) [261].

It is clear that the use of individual drives for each wheel simplifies the drive train and eliminates single-point failure modes—this has been the approach with planetary rovers. Current sensors are required at the motors to measure torque, batteries, and the power regulation system. The power regulation system also requires bus voltage measurement. The motor drive/controller board controls motor actuation with its own dedicated processor, most commonly a microcontroller. Pulse width modulation (PWM) is the most common method of motor speed control. PWM amplifiers are relatively noisy at high frequencies but linear amplifiers do not suffer from deadbands. Motor drivers control wheel-driving torque using PWM whereby the torque applied is in proportion to the duty cycle. Sojourner was driven through bang-bang on–off control of the drive motors but the Rocky test platforms used PWM for the generation of motor currents while sensor data were fed back in a PID control scheme—this would be the most flexible option. The Rocky III/IV prototypes to Sojourner controlled each wheel's speed and direction by a microcontroller via PWM through a L293D motor drive controller IC per wheel. With microprocessor control, PWM is commonly used. Each motor is controlled by a PWM signal generated by H-bridges which regulate motor voltage. All planetary rover platforms use health monitoring, diagnostics, and fault protection. Traditionally, functionally different boards are mounted separately, though to save mass different functions may be placed on the same board to minimize duplication. In this case, the motherboard may implement motor driver, power management, and sensor driver interfaces. The power system and voltage DC/DC regulator are mounted on a separate PCB, however. Implementation as double-sided chip-on-board electronics reduces board area and volume. It is important to note that there are several caveats to applying terrestrial technology to space-qualifiable platforms. A common approach is to adopt the PC/104 processor architecture running a Linux operating system which communicates with external devices via a wireless local area network (LAN) such as

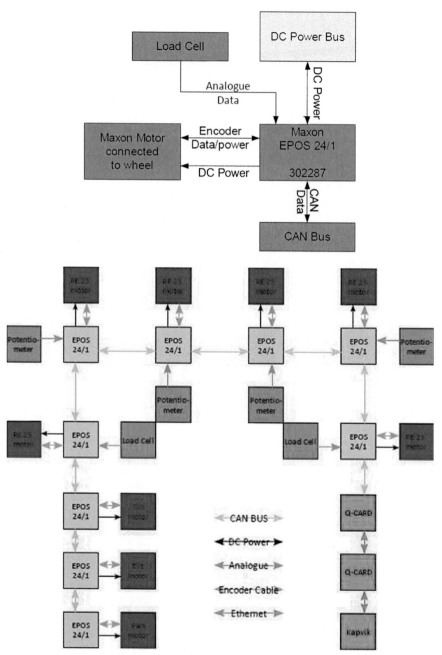

Figure 5.6. Kapvik motor control architecture [credit Ala Qadi, Carleton University].

Table 5.1. Centralized versus decentralized motor drives [adapted from Fuke et al., 1994].

Function	Centralized drive (two actuators)	Individual drive (six actuators)
Thermal protection	Simpler as actuators are in rover body	Harder as actuators are in wheels
Effect of transmission failure	Critical	Tolerant
Electric cabling	Cabling inside body	Cable routing between body and wheels
Drive train	Complex due to high friction	Simple with low friction
Power	Inefficient power coupling	Efficient power coupling
Torque distribution	Differential/viscous coupling	Slip or torque control

Ethernet. PC/104+ offers a small form factor which is advantageous for rovers. ESA's Lunar Rover Mockup (LRM) processor is a 700 MHz Pentium III with 256 MB RAM and 20 GB hard disk plus I/O interfaces. The FIDO autonomous navigation system was implemented on a 266 MHz Pentium CPU running the VxWorks 5.4 operating system. Similarly, RWI Pioneer—a research platform workhorse—uses Intel Pentium Pro as its primary processor mounting the Linux operating system and Ethernet TCP/IP network interfaces to maximize development environment options. TCP/IP is not appropriate for planetary operations.

Real time control requires the use of real time operating systems to provide reliable responses to triggering events. Real time operating systems schedule tasks, run tasks in parallel, process messages and respond to interrupts in real time. The core of such an operating system includes a central kernel which implements basic functions such as Push, Pop, Stop. Several real time operating systems are available including RTEMS, real time Linux, or VxWorks. VxWorks is the most popular multitasking operating system for embedded systems especially robotic systems such as rovers. The Marsokhod rover prototype Lama contains four CPUs running VxWorks for data acquisition (except imaging), locomotion, environment modeling, and planning. Similarly, the planetary rover testbed FIDO employs VxWorks. RTEMS is more robust but Linux is more flexible. Linux uses open-source code with a UNIX-like kernel which can be freely distributed without licensing fees. It is a multitasking, multiuser, multiprocessor operating system supported by a wide range of hardware platforms and its software is supported by a large number of developers. The Linux operating system offers extensions to its real time kernel for time-critical and event-driven interrupts and has a C compiler and C libraries to support application programs. The operational Nomad Antarctic meteorite–hunting mobile robot used three onboard computers—one VxWorks Pop CH62 68060 processor was mounted on a VME backplane for real time control, one 133 MHz Pentium processor on a PCI bus using RedHat Linux for autonomous navigation, and one 200 MHz Pentium Pro processor on a PCI bus using RedHat Linux to run autonomous science analysis of the instrument data. The Mars Pathfinder lander implemented the VxWorks operating system to perform fault protection, closed-loop thermal control, high-gain antenna pointing, attitude determination and control, and all aspects of EDL. Multiple processor boards with different functions offer greater flexibility and may be integrated into a VME backplane card cage with processors and DAC/ADC boards—the MERs implemented VME serial interfaces to other devices.

Dedicated microcontrollers are normally recommended for motor control to ensure that motion control is accomplished with sufficient bandwidth for the application [262]. Microcontrollers are microprocessors with memory and input–output functions on the same chip making them compact. They are used for real time embedded controllers making them suited to motor control as they often have useful built-in hardware such as timers, interrupts, ADC channels, ROM/RAM, significant number of I/O lines, and low power consumption. Micro-controllers are commonly built in to motion control applications such as electric motors. PD digital microcontrollers typically require low-pass filters to remove gear jitter which can lead to instability in the control loop. The processor must run at a sufficient update rate in order for the control loop to operate in real time (>20 Hz minimum). Most embedded microcontrollers are based on the 32-bit Intel 8051/8052/x86 series or Motorola 68K series. Embedded processors provide for independent distributed control. A distributed system was adopted on board the Urban II robot for urban surveying applications: its motors were controlled by a low-level input/output Motorola 68332 processor which read proximity sensors and sent commands to the motors. It communicated over an RS-232 serial inter-face to a PC-104 bus to two high-level host processors: a 166 MHz Pentium II navigation processor and a 232 MHz Pentium II vision processor. The micro-controller needs sufficient input channels for the sensors and output channels for the motors. Microcontroller units are required to deliver commands to the motors. The microcontroller output channels interface to the motor driver circuits for each motor—this is necessary as the output is insufficient to drive a motor which generally requires 1–2 A current. This is the purpose of the motor driver. A high-voltage output is required to drive inductive loads such as motors. MER avionics included a 9U VME bus with closed-loop motor microcontrollers and H-bridges, parallel camera interfaces with digitizing framegrabbers, and RS-232 serial inter-faces. MER traverse performance was based on straight line driving at 3.75 cm/s and turning at 2.1°/s—the PID motor drive controller was driven at a rate of 1 kHz.

Microcontrollers need to interface to a suite of sensors through the required number of ADC channels. One input channel is dedicated to each sensor—motor shaft encoders for odometry, inertial measurements, proximity detectors, range sensors, etc. in the example of robotics. Microcontrollers usually implement PWM output for motor control. Microcontrollers can generate n independent PWM signals for each of the motors at the required frequency. The Nanokhod nano-rover required only a limited 40 MHz 8051 microcontroller within its payload cab as all image-processing and navigation functions were performed remotely. Its PIC-based motor drivers provided PWM-based motor control of the left/right track motors, the payload cab articulation motor, and lever articulation motors each employing angular feedback encoders. A 4-bit microcontroller for distributed networks of small spacecraft has also been developed [263]. The Motorola MC68HC11 controller has spawned many variants and developments—indeed, the Khepera mobile robot uses the MC68311 processor variant. The M68HC11 microcontroller has on-chip timer and counter circuits that can directly generate square-wave PWM signals for DC electric motor control. The MC68HC11 provides the basis for the Handy Board robot controller which is programmable in interactive C [264]. The Rocky III/IV microrovers prototypes to Sojourner used an Intel 8085 processor and variants of the MC68HC11 microcontroller [265] (compatible with C programs) with 40 kB memory for the main computer (of which only 20 kB were used). The Rocky III/IV prototypes to Sojourner used variants of the Motorola MC68HC11 processor with 40 kB memory (of which 20 kB were used). Three Intel 8751 microcontrollers were connected as slave processors to the 6811 through a synchronous serial bus to increase the number of I/O pins. Sojourner itself used the radiation-hardened Intel 80C85 processor. An 8-bit radiation-tolerant microcontroller based on the 8 MHz Motorola 68HC11 is being developed for Mars surface applications [266]. The 40 MHz ARM7 computer with 1 MB internal flash memory, 64 kB internal SRAM, 256 MB external flash memory, and three 2 MB external SRAM chips (with triple-voted EDAC) is a standard University of Toronto Space Flight Lab computer with flight heritage which consumes less than 500 mW of power. The Freescale (formerly Motorola) 68HC11 is being developed in a radiation-hard variant. Pulse modulation can be implemented using a MC68HC11 microcontroller or even a PIC—the PIC family of microcontrollers from Microchip has been proposed for the Surrey Space Centre Palmsat onboard computer (PIC18F8680) which has since been updated with the 8-bit PIC18F4431 offering more programmable memory, RAM, and higher speed than its predecessor but retaining the low power with high clock frequency. They are also compatible with the CAN 2.0B bus interface. The Aeroflex UT69RH051 microcontroller is a radiation-hard microcontroller for space applications. The ESA ADV 80S52 microcontroller is a space application microcontroller fully compliant with the Intel 8052 architecture including standard peripherals [267]. More sophisticated microcontrollers exist today. It must offer PWM support with n channels to support n motors—10 drive motors $(6 + 4)$/18 drive motors $(6 \times 3\,\mathrm{DOF}$ legs). It must interface to n MOSFET H-bridge circuits to control motor speed/direction. Motors must incorporate shaft encoders for position/velocity feedback. The architecture should be CAN bus compatible to support space heritage—microcontrollers typically include CAN bus and serial peripheral interfaces. The Surrey Space Centre Palmsat OBC is based on the PIC family of microprocessors—the 20 MHz PIC18F8680 microcontroller replacing the original PIC16F877 selection because it supports a full CAN 2.0B interface. The Microchip 8-bit 40 MHz PIC18F4431 with flash ROM was selected for the Kapvik motor controller.

For the Kapvik microrover, housekeeping and rover control computers were implemented separately. Its onboard computer, power distribution/control, and communications systems were derived from CANX microsatellite flight heritage. The subsystem computer for housekeeping was based on the 150 g 32-bit RISC ARM7 processor with triple-vote EDAC memory controller for external SRAM. It is supported by 64 kB internal SRAM/1 MB internal FLASH, 3×2 MB external SRAM, and 256 MB external FLASH memory. At its maximum switching speed of 40 MHz, it consumes 325 mW at 4.5 V. Internal SRAM/FLASH stores the bootloader while the application program is stored in internal FLASH/external SRAM. It proves RS-422 interfaces.

Image capture takes around 125 ms independent of frame rate. IEEE 194 interfaces are high-speed interfaces that offer high frame rates with uncompressed video. The framegrabber is used to act as a DAC board to digitize images and provide an image buffer. There are several common digital communications protocols—I2C, IEEE 1394 (FireWire), MIL-STD-1553 (spacecraft standard), PC/104+, PCI (peripheral component interconnect), compact PCI, RS232/RS422 serial, USB, and VME/VXI are bus architecture/interface protocols. RS-232 and RS-422 serial interfaces are being superseded by SpaceWire. The most favored architectures are VME, PC/104+, and PCI which are terrestrial standards: the VME system bus has flight heritage but PCI has superior performance. The VME offers a 32-bit and 64-bit databus at 40–500 MBps through a 21-slot backplane. The VME backplane is more rugged than PC-104 bus connectors. PCI offers 32 and 64-bit databuses with 133–533 MBps throughput and a 8–16-slot backplane (this was a candidate for JPL X2000 architecture). PC 104+ is a compact form with a 5–132 MBps self-stacking bus without a backplane. The latter includes 32-bit PCI bus connectivity. An independent clock is required in addition to on-chip timers. SpaceWire is a rapidly developing protocol for space use.

Planetary rovers require significant computing resources. The main onboard computer provides most of the onboard signal processing, interfaces, and communication functions. Traditionally, it would be a 32-bit radiation-hard flight-qualified processor—25 krad cumulative dose is sufficient for most mission scenarios [268]. The processor speed must support event-based real time control and navigation ($\sim$100+ MHz). Non-volatile memory is necessary to store scientific and other data prior to the communications downlink. The implementation of rad-hard memory may require off-chip allocation. An additional digital signal processor (DSP) may be required for signal/image processing. The limits to processing capability are a major constraint but recent space missions have been adopting off-the-shelf processors for higher performance. For example, the Clementine lunar probe adopted a traditional rad-hard 1.7 MIPS MIL-STD 1750A processor for general functions supported by an 18 MIPS 32-bit RISC processor for image processing. High-performance coprocessors are ideal for dedicated applications in embedded real time systems that can be built with parallel processing architectures based on high-speed serial links; for example, T805 transputers were used on the CNES IARES-L mobile robot for locomotion control and another for manipulation of the mounted manipulator.

The Sojourner power, navigation, and computer electronics were all integrated to prevent duplication—the CPU and power boards were interconnected to each other within the WEB (warm electronics box). Sojourner's main computer was a 0.5 kg flight-qualified 2 MHz Intel 8-bit 80C85 microprocessor offering 100 KIPS, 16 kB PROM boot memory, 176 kB EEPROM program storage, and 512 kB temporary RAM storage in 16 kB page-swapping mode. It implemented a multiplexed 8-bit address/data bus to support input/output to 90 sensor channels while drawing only 1.5 W (compared with the Rocky 7 terrestrial testbed which carried a 68060 computer offering 100 MIPS). Rad-hard RAM held operational programs loaded from non-volatile RAM. The 48 kB where the operating system resided was radiation-hardened to protect it from latchups. Sojourner's behavior control software used 90 kB of EEPROM for autonomous navigation between waypoints selected by the ground station operator. To overcome its severe processing limitations, it used laser stripers to aid its lander stereo cameras mounted 1.5 m on a mast above the lander but most of the image processing was performed on the ground. The total board area was 360 cm^2, height 1.9 cm in a mass of 995 g. The Athena Software Development Model (SDM) used a highly distributed approach in adopting a 12 MHz/10 MIPS R3000 CPU (drawing 2–3 W of power) supported by FPGAs for low-level motor control [269]. It included 32 MB DRAM, 4 MB EEPROM, 64 MB flash memory, and 128 kB boot PROM. The CPU board was built on a 32-bit Synova Mongoose-V processor with a R3000 RISC architecture supported by 2 kB on-chip data cache and 4 kB on-chip instruction cache. The board has a I2C serial interface capable of reading/writing at 400 kbps. This was greater than the 100 KIPS 8085 processor used on the Sojourner microrover. The I/O board implements image capture and ADC functions. The remote engineering unit comprises an FPGA-based PID motor controller and ADC for motor encoder sensors. It implements a profiler to ensure a trapezoidal velocity profile for smooth motor operation. The inertial measurement unit includes a three-axis accelerometer and rate gyroscopes.

The Pathfinder lander used a 20 MHz RISC RAD 6000 radiation-hardened processor (based on an early PowerPC design) with a 32-bit architecture plus 128 kB of DRAM memory, 11 MB EEPROM, and 256 MB flash memory. It was capable of a 22 MIPS processing speed with 128 MB mass memory. The MER onboard computer was also a 20 MHz RAD6000 computer. In contrast, the FIDO test rover implemented a 266 MHz Intel Pentium processor (consuming 15 W) with four RS-422 serial I/O ports, two USB ports, and one parallel port supported by a 384 MB solid state flash disk. Additional boards included a 100 Mbps Ethernet board, two color framegrabber boards, digital I/O board, 12-bit ADC board, 8-bit DAC board, two 16-bit encoder boards, five low-pass filter boards, and PWM power amplifiers and motor drivers for its six 15 V motors. On FIDO, a closed-loop PID motor control operated at 50 Hz while velocity synchronization was performed at 10 Hz. All measurement devices were interfaced through an RS232 serial link and measurement data collected at a sample rate of 200 Hz. A PowerPC processor is used by Robonaut (see later) for real time control across a VME backplane running the VxWorks operating system.

The Mars Science Laboratory employs a space-qualified PowerPC 750 with 100 MHz processing speed that is still far behind those of terrestrial processors. MER avionics implemented closed-loop motor controllers and H-bridges for the motors, two parallel camera interfaces for simultaneous stereo image capture, instrument interfaces, telecommunications interfaces, and a 1553 interface bus to the lander during the cruise. An inertial measurement unit measured rover attitude. Four additional wide-field HazCam cameras were mounted as two stereo pairs at the front and back of the WEB to implement autonomous hazard detection. The onboard software was coded in the C software language. Within the VME cage, different cards were dedicated to auxiliary functions such as image acquisition, motor drives, and scientific instrument processing in parallel. Sensory data were reported at a rate of 8 Hz.

Motor control has become increasingly sophisticated with three layers of control [270]—inner loop for current control, outer loop for position control, and mid loop for speed control. The inner loop has around five times the response time as the outer loop. DC motor controllers based on application-specific integrated circuits (ASICs) offer certain advantages such as high processing capabilities for real time control within compact electronics [271]. The ASIC combines the fast interrupt response of microcontrollers with the optimized repetitive computations of digital signal processors (DSPs) within a reduced instruction set computing (RISC) platform employing the Harvard architecture. DSPs can be used for high-frequency motor control implementing filter algorithms efficiently though general control is better implemented using a master central processing unit [272]. The integration and differentiation operators required for PID control are implemented through difference equations using trapezoidal approximations, and multiply/divide operators are implemented using the add/subtract and shift functions. Image processing for embedded systems may be enabled through DSPs which offer superior performance to general purpose computing architectures and FPGAs (though their development time is much longer) [273, 274]. The 32-bit Analog Devices ADSP2120 digital signal processor offers 15 MIPS/45 MFLOPS at 15 MHz clock speeds for specialized functions. The TSC21020E is a rad-hard version of the 20 MHz ADSP 21020 DSP which has an enhanced Harvard Architecture with simultaneous memory read/write capability. Digital signal processing used for vision systems and embedded processing may be implemented on the TSC21020 DSP (the space-qualified version of the ADSP21020 DSP board) providing a computation capacity of 20 MIPS and 60 MFLOPS [275]. More sophisticated approaches to motor drive control based on artificial intelligence methods have been proposed without sensors to provide observer-based speed prediction based on feedforward neural networks, neuro-fuzzy networks, and the self-organizing Takagi–Sugeno rule base, the latter being superior [276]. The chief criticism of ASICs is their high development time, cost, and risk. FPGAs are superseding the use of DSPs for signal-processing functions [277]. FPGA reprogrammability gives a dramatic increase in functionality over ASICs. Thus they are used instead of ASICs as their development time is much shorter. FPGAs have been implemented in a limited capacity in several rovers including the MERs. For example, ESA's Lunar Rover Mockup's (LRM) low-level control software runs on a microcontroller that receives velocity and motor angle data through a serial line. A PID controller generates commands that are sent to the H-bridges for steering and to an FPGA where the PID controller is implemented.

The MUSES-C nanorover exhibits use of more advanced computational platforms such as ASICs and FPGAs. Its electronic control system was centered on a 32-bit Synova Mongoose V processor with 1.5 MB SRAM, 1 MB EEPROM, an ASIC for glue logic, and interfacing supported by an FGGA for motor control. The electronics were implemented on chip-on-board assemblies including a computer board, two memory boards, an analog signal–processing and ADC board, a motor driver board, and a power-switching regulator board. The onboard electronics included a 64-channel analog signal–processing chain with 10-bit analog–digital converter (ADC), 10 brushless DC motors (4 for wheels, 4 for struts, and 2 for optics) with Hall sensor feedback and PWM motor driver control, a bidirectional 9,600-baud radio interface, 20×0.3–4.4 V switching power regulators, and a 2.5 W solar panel. The electronics also interfaced to a 256×256 CMOS active pixel sensor, a 256-channel IR spectrometer, an APXS, 14 temperature channels, 6 solar sensors (one for each body side).

Future computational resources such as multicore architectures and FPGAs will offer enhanced complexity navigation algorithms. The use of FPGA technology for digital image processing will enhance the navigation process. JPL is developing a rover avionics module based on the Xilinx Virtex-II Pro FPGA which includes two embedded PowerPC 405 processor cores with a processing speed of 300 MHz with the rest of the FPGA processing at 100 MHz. The PPC405 cores include the Linux operating system with Ethernet, PCI bus, Compact Flash, and serial I/O and capable of running all rover applications software. The adoption of FPGAs is exemplified in the design of the Kapvik microrover is by implementing FPGA electronics to process images faster for autonomous navigation. More rapid autonomous navigation enables higher speeds of traverse. On Kapvik, the space-rated Xiphos Q5 card was configured with a PowerPC architecture with the Linux operating system (not real time as yet) (Figure 5.7). The Q5 comprises 250,000 to 1,500,000 gates of reprogrammable logic with a 300 MHz PowerPC 405 processor core supported by 32/64 MB SDRAM and 16 MB flash memory. Code is split between the hardware CPU and the reprogrammable logic. It has a temperature tolerance of $-65°C$ to $+150°C$ (storage) and $-40°C$ to $+85°C$ (operating).

A daughterboard bridges the Q5 to other avionics through Ethernet/RS-232/RS-422 protocols. The Xiphos Q6 card was improved with the MicroBlaze architecture incorporating a Xilinx Spartan 6 application (with LEON-3 option) and Actel ProASIC3 FPGAs with the Linux 2.6 operating system with some features of real time Linux. The Q6 card offers 4 GB mass storage with reduced power consumption of 2 W. The motor controllers were connected to a Xiphos Q6 computer card via a controller area network (CAN) bus. The controllers operated in three primary control modes: (i) constant velocity maintained all wheels at the

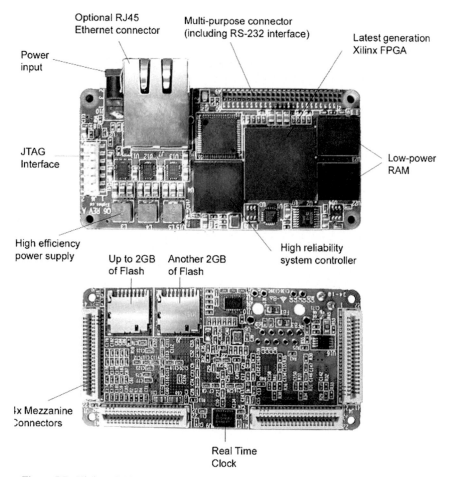

Figure 5.7. Xiphos Q6 FPGA processor card [reproduced with permission from Xiphos].

same rotational speed through PID control; (ii) fine position with constant wheel velocity constrained by initial and final positions using PID control; (iii) fixed current mode which was not used on Kapvik. The Q6 card was stacked with an Ethernet/CAN power card and a USB/FireWire/RS-422 card. The CAN Bus is highly reliable and has flight heritage with CAN Bus 2.0 offering speeds up to 1 MHz. Player/Stage was implemented as the development environment because of its wide compatibility and driver availability. Both Pathfinder and the MERs have implemented the VxWorks operating system. A real time operating system such as Lynx RTOS is also suitable for space robotics but RT Linux is also a common choice (Kapvik uses the Linux operating system). The approach favoring real time Linux was selected for Kapvik as a compromise between a real time operating system (such as RTEMS or VxWorks) and shortened development time (Linux/ Player) considering that Kapvik requires only soft real time operation due to its low speed (though Player has now been superseded by ROS).

The overall avionics architecture for Kapvik is shown in Figure 5.8. Kapvik's avionics are based on a modular backplane with conventional point-to-point design mounted into the rear segment of the rover (Figure 5.9). The avionics enclosure includes a slide-in access panel, and the onboard battery is mounted above the electronics boards.

The MER control system comprised a number of layers: motor control, direct drive maneuvers (turn-in-place, etc.), and goal-based driving (goto waypoint) [278]. Commands could be executed immediately on receipt of the command, or more typically, stored for subsequent execution serially. The rover onboard software provided three major functions: terrain assessment, path selection, and visual pose update. Terrain assessment built an environmental map from the rover cameras. Path selection performed searches composed of candidate arcs forming paths to the goal. Rover state estimation based on wheel odometry and IMU was updated at a rate of 8 Hz while driving. The onboard software constantly monitored for off-nominal conditions such as rover body tilt, rocker-bogie configuration, motor drive temperature, etc. Whilst static, visual odometry was used to refine these estimates. Virtual pose update implemented visual odometry to self-localize the rover from the camera imagery. The MER processor measured an average of 15,000–50,000 xyz points in each image pair. Each image pair required 3 min to acquire and process. Although sufficient for static image processing, there was thus insufficient processor speed for image processing while traversing.

5.3   LOCOMOTION CONTROL

Locomotion control is the lowest level of autonomous navigation and implements motor commands. It requires closed-loop trajectory control to limit error propagation. This involves two control loops, one to control speed and steering angle (including traction control and force distribution), and a higher level control to ensure tracking of the reference trajectory through the required waypoints with minimal energy consumption. The autonomous rover control architecture must support real time sensing, control, and planning. It must implement obstacle avoidance and accommodate simple non-nominal deviations. Relying on "move_straight" commands for long traverses will generate deviations as the left and right motors will not be exactly equal in behavior. A continuous control strategy with a bang-bang controller with a deadband centered on the rover's heading will apply corrections with minimal deviation. Straight line motion control and steering may be implemented using a PID (lead–lag compensator) control law. If the rover encounters unexpected situations, the autonomous control system halts the rover and requests recovery interdiction from the ground control station. A two-wheeled differential drive robot is the most basic vehicle

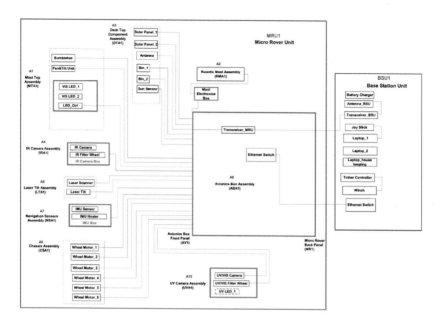

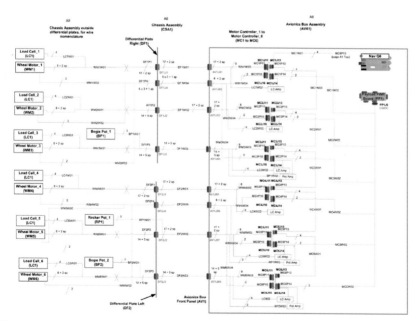

Figure 5.8. Kapvik microrover avionics architecture and chassis electrical diagram [reproduced with permission from Ian Sinclair, MPB Communications].

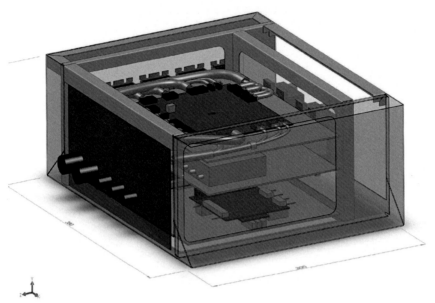

Figure 5.9. Kapvik removable avionics box with backplane configuration [credit Adam Brierley, Carleton University].

model. Vehicle position and velocity state is defined with reference to an external global reference frame by:

$$p_k = \begin{pmatrix} x_k \\ y_k \\ \theta_k \end{pmatrix} = \begin{pmatrix} x_{k-1} + r\cos\theta_k \\ y_{k-1} + r\sin\theta_k \\ \theta_{k-1} + \Delta\theta \end{pmatrix}$$

where (x, y) = robot centroid position, θ = steering angle, r = wheel radius, and x, y, θ define the Cartesian position coordinates and steering angle of the vehicle at time k. Average steering angle is given by:

$$\theta = \tan^{-1}\left(\frac{2d}{\sqrt{\rho^2 - d^2}} \right)$$

where d = half-distance between steering axles, ρ = distance from center of motion to vehicle centroid, and $\Delta\theta$ = angle between current direction and desired direction. Assuming a horizontal flat surface with pure rolling (no slip) the kinematics is simple to derive with the velocity of a robot about its two sides being given by $(v, w)^T$:

$$\begin{pmatrix} v \\ w \end{pmatrix} = \frac{1}{2} \begin{pmatrix} \frac{1}{2}(v_r + v_i) \\ (v_r - v_l)/l \end{pmatrix} = \begin{pmatrix} \frac{1}{2} & \frac{1}{2} \\ -1/l & 1/l \end{pmatrix} \begin{pmatrix} v_r \\ v_l \end{pmatrix} = \begin{pmatrix} \frac{1}{2}r(w_r + w_l) \\ \frac{r}{b}(w_r - w_l) \end{pmatrix}$$

where l = distance between two wheels. Directional velocity can be controlled by varying left and right-wheel tangential velocities:

$$v_l = v - \frac{i}{2}w$$

$$v_r = v + \frac{i}{2}w$$

Turning radius is given by:

$$R = \frac{l}{2}\left(\frac{v_l - v_r}{v_r - v_l}\right) = \frac{l}{2}\left(\frac{w_l + w_r}{w_r - w_l}\right)$$

Now,

$$\dot{P} = \begin{pmatrix} \dot{x} \\ \dot{y} \\ \dot{\theta} \end{pmatrix} = \begin{pmatrix} \cos\theta & 0 \\ \sin\theta & 0 \\ 0 & 1 \end{pmatrix} \begin{pmatrix} v \\ w \end{pmatrix}$$

such that

$$\tan\theta = \frac{\dot{y}}{\dot{x}}$$

where

$$J = \begin{pmatrix} \cos\theta & 0 \\ \sin\theta & 0 \\ 0 & 1 \end{pmatrix} = \text{Jacobian matrix of the rover}$$

The motion of the vehicle is given by:

$$\dot{x} = v\sin\theta \quad \text{and} \quad \dot{y} = v\cos\theta$$

Now, (x, y, θ) is the location of the midpoint of the rear axle, and the motion of the rover is subject to the non-holonomic constraint:

$$-\dot{x}\sin\theta + \dot{y}\cos\theta = 0 \quad \text{and} \quad \dot{\theta} = \frac{v}{L}\tan\phi$$

where φ = steering angle. If $\tan\theta$ is constant, the equation is integrable. The resultant force is divided into two components, one in the direction of the vehicle heading to control the speed, and the other perpendicular to it to control the steering angle.

$$\begin{pmatrix} x_k \\ y_k \\ \theta_k \end{pmatrix} = \begin{pmatrix} x_{k-1} \\ y_{k-1} \\ \theta_{k-1} \end{pmatrix} + \begin{pmatrix} (L/\tan\varphi)(\sin\theta_k - \sin\theta_{k-1}) \\ (L/\tan\varphi)(\cos\theta_k - \cos\theta_{k-1}) \\ v\Delta t(\tan\phi)/L \end{pmatrix}$$

where $L = v/w$ = radial coordinates. The rover must stop within a distance from an obstacle such that $d = v^2/2a$ where a = deceleration. Control will need to be maintained to minimize deviation from the selected path.

Ackermann steering mechanically coordinates the angle of the wheels that steer so that during a turn, the wheels roll without skidding and follow a curved path with a common origin (Figure 5.10 for double Ackermann geometry). Thus,

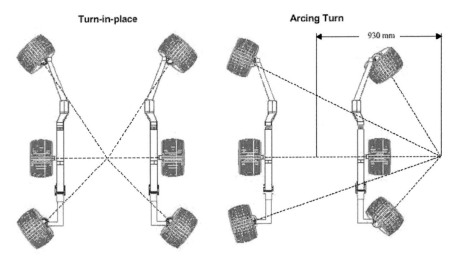

Figure 5.10. Double Ackermann steering for MER [credit NASA JPL].

a common centerpoint for the turning arc of the inside and outside wheels ensures that the inside wheel is rotated at a slightly sharper angle than the outside wheel to eliminate wheel slippage. This may be implemented on the front wheels (with a specified turning circle), end wheels (for point turning), or all wheels (for sideways driving). From geometry:

$$\cot \theta_i - \cot \theta_o = \frac{B}{L}$$

where θ_i = inner wheel steer angle, θ_o = outer wheel steer angle, B = wheelbase, and L = longitudinal wheel separation. Vehicle steering angle is given by:

$$\tan \theta = \frac{L}{0.5B}$$

So,

$$\cot \theta = \cot \theta_i + \frac{B}{2L} \quad \text{and} \quad \cot \theta = \cot \theta_o - \frac{B}{2L}$$

Ackerman steering geometry turns the front wheels differentially so that the vehicle trajectory curvature is given by [279]:

$$k(t) = \frac{1}{r(t)} = \frac{d\theta(t)}{ds} = \frac{\tan \phi}{B}$$

where ϕ = steering angle, r = radius of curvature, and θ = vehicle orientation. Now,

$$\dot{\theta} = \frac{d\theta}{ds}\frac{ds}{dt} = k(t)v(t) = \frac{\tan \phi}{B}v(t)$$

Similarly,

$$\dot{x} = v(t) \cos \theta(t) \quad \text{and} \quad \dot{y} = v(t) \sin \theta(t)$$

The outer wheels generally exhibit higher torques than the inner wheels (up to 80% of total drive torque). Terramechanics issues associated with skid steering are addressed in [280]. Skid steering generally consumes twice the power of Ackerman steering: For a four-wheeled vehicle with equal weight distribution on each wheel, the steering moment is given by:

$$\left(\frac{B}{2}\right)(F_l - F_r) = \frac{\mu WL}{2}$$

The power required for a turn is given by:

$$P = P_{str} + \frac{\mu WL}{4}\dot{\phi}$$

where $\dot{\phi}$ = turn velocity and μ = coefficient lateral resistance = 0.6 for steel on hard ground. Typically, $P_{skid} = 2.5 P_{str}$ though the length-to-width ratio should be minimized to reduce the power overhead of skid steering. A more sophisticated steering law from potential field vectors may be applied to drive the wheels of planetary rovers:

$$v_{\text{left}} = v - K_p \Delta\theta - K_v \dot{\theta} \quad \text{and} \quad v_{\text{right}} = v + K_p \Delta\theta + K_v \dot{\theta}$$

where v = speed command, $\Delta\theta$ = steering command, $\theta = (v_r/B) \tan \phi$, B = wheelbase between left and right wheels, K_p = proportional steering gain, and K_v = damping steering gain. Kinematic analysis is illustrated based on a simplified Marsokhod chassis modeled by two rods joining the three axles [281]. Quasistatic force balance analysis determines whether the rover wheels slip or the rover topples. More complex quasistatic rover models have been developed to incorporate the linkage relations of the rocker-bogie mechanism [282]. Toppling is an ever present possibility in an unstructured environment which can yield complex situations that are impossible to predict. The kinematic configuration of the rover should allow for 45° static stability on slopes in all directions without tipping— this means that the center of mass of the vehicle does not lie outside the closed-wheel footprint. For a vehicle to topple, its center of gravity must be applied outside the vehicle's area of contact with the ground in one or more vertical plane. This can occur due to the slope of the ground or obstacle, an altered distribution of the mass of the vehicle, or the external forces applied to the vehicle. Tilt sensors can provide prior warning of this condition. The ExoMars rover's static stability must accommodate a tilt of 40° in any direction. For static stability on slopes, we require:

$$\theta_{\max} = \tan^{-1}\left(\frac{Y_{CG}}{Z_{CG}}\right) = \tan^{-1}\left(\frac{L/2}{H_{CG}}\right) = 59°$$

for up/downhill stability with CG located at wheel diameter height

$$\theta_{max} = \tan^{-1}\left(\frac{X_{CG}}{Y_{CG}}\right) = \tan^{-1}\left(\frac{B/2}{H_{CG}}\right) = 76°$$

for cross-hill stability with CG located at wheel diameter height. Keeping the CG low to maximize stability will be difficult as the linkage bar between the left and right sides of the chassis must pass through the CG. If static stability cannot be maintained, consideration of self-righting behaviors must be considered [283].

5.4   ROVER LOCOMOTION SENSORS

An extensive suite of onboard sensors is required for rover obstacle avoidance, navigation, power management, and fault protection. Sensors are characterized by a number of important properties that define their performance [284]: (i) input–output sensitivity; (ii) linearity; (iii) measurement range; (iv) response time (time constant); (v) frequency response (including hysteresis); (vi) accuracy; (vii) stability to drift and repeatability (calibration); (viii) resolution; (ix) bandwidth; (x) reliability. Sensors provide the basis for robot navigation which requires estimation of position [285]. Robotic sensors are of two major types: internal sensors that provide proprioceptive information about the current state of the robot, and external sensors that provide exteroceptive information about the current state of the environment in which the robot is embedded. External sensors include contact sensors that directly sense immediately adjacent space and ambient sensors that sense larger spatial ranges. Ambient sensors provide ambiguous signals as they integrate environmental information from regions of space and this process is non-invertible. There are a number of sources of uncertainty in sensor outputs including random noise and aliasing which must be accounted for. These may be due to limited resolution, misalignments, dimension errors, etc. For instance, wheel odometry suffers from cumulative drift. Similarly, there are uncertainties in effector outputs—these may be errors in motor commanding, motor execution, motor alignments, etc. As sensors vary in terms of range, directionality, sensitivity, resolution, specificity, and accuracy, a suite of sensors is usually required.

Path integration (dead reckoning) relies on proprioceptive sensors such as wheel encoders and/or gyroscopes, but it is accurate only for short distances due to cumulative errors such as slippage. Reference coordinates are necessary to overcome the limitations of dead reckoning which typically yields ~5–10% error. Sojourner's trajectory had a standard deviation of 25 cm per 10 m traverse distance. There are three main types of sensors required to support navigation [286]—odometers (e.g., wheel encoders, potentiometers, tachometers, inclinometers), inertial navigation sensors (e.g., gyroscopes and accelerometers), and visual localization (e.g., pan–tilt cameras, laser stripers, LIDAR) though there are additional sensors that may be included such as proximity sensors (e.g., whiskers are contact force sensors that detect collisions with obstacles). Typical odometry uses optical encoders attached to the motor shafts or wheel axles to measure wheel rotation. The accuracy of odometry is a direct function of the kinematic design of the vehicle. Wheels usually adopt optical encoders while potentiometers are used to determine chassis kinematic configuration. Odometry measurement is most widely used to measure motor rotations (such as within a wheel). The steering direction may be measured through a gyroscopic compass or angular potentiometers/optical encoders at the steering wheels. Steering angle encoders (nominally four) are mounted at the corner wheels with on-the-spot turning. Odometry provides an estimate of the distance traveled which may be fused with yaw gyroscope measurements to provide a heading estimate. Kinematic levers are mounted with chassis angle encoder sensors to measure bogie and differential positions (i.e., chassis configuration). Hence, odometry includes measurement of both wheel revolutions and chassis articulations which must be integrated for measurement of rover position. Motor currents may be measured to generate motor torque estimates ($I \propto \tau$). Most COTS motors are integrated with Hall effect sensors to measure the magnetic flux of the motor and thereby rotor position [287]. Hall effect sensors may thus be used to measure displacement but optical shaft encoders are more accurate: they measure the rotation of motorized shafts to determine vehicle movement through path integration. All wheel and steering motors are mounted with encoders to feed back wheel position and velocity. Rather than employing a separate tachometer, it is best to employ a single-motor encoder for both position and velocity feedback with the velocity controller running faster than the position controller. A tachometer measures rotational velocity for velocity feedback and infers relative position—they are often built into DC motors. Optical encoders are the most common shaft displacement sensors for electric motors—their digital nature ensures that no ADC (analog-to-digital conversion) is required. An optical encoder uses a beam of light aimed at a photodetector which is interrupted by a coded pattern on a rotary disk attached to the motor's output shaft. There are two types of optical encoder—amplitude disk and interferometric. Holosense encoders are based on digital diffractive optics that can implement many different encoder patterns on the same disk. Optical encoders offer high accuracies and are tolerant of high EMI environments. Sojourner and the MERs carried wheel optical encoders for each wheel. Odometry on Sojourner was averaged every 6.5 cm of distance traveled (one wheel radius) to update the estimated vehicle position. FIDO adopted optical wheel encoders with an accuracy of 0.027° equating to a 0.0485 mm linear traverse for 20 cm diameter wheels. The Kapvik wheel encoders were three-channel Maxon MR encoders (type ML) giving 500 counts per revolution geared down by 1,400:1 (i.e., $5 \times 10^{-4\circ}$) resolution at the output shaft. The rocker-bogie angular sensors were Inscale GL series hollow-shaft potentiometers (two GL60s for the bogie and one GL100 for the rocker) both offering 340° range (Figure 5.11). Hard stops limit the rocker and bogie angle ranges (Figure 5.12).

The chief difficulty in utilizing motor rotation measurements is that they suffer from systematic errors (due to gearing) and non-systematic errors (due to slippage). The use of low-pass filters on odometry and inertial navigation sensors filter out most noise. Navigation also requires measurement of body attitude

Figure 5.11. Potentiometer used in Kapvik's chassis.

angles using gyroscopes to provide data on terrain tilt. Wheel odometry and inertial measurement unit (IMU) measurements were used to estimate relative rover pose between traverse segments on the MERs. The IMU involves the measurement of attitude to measure the rover's traverse in the third dimension. The IMU comprises accelerometers and gyroscopes in a single integrated package which measures attitude to determine rover tilt (especially pitch) and measure heading (yaw). Sojourner carried a three-axis rate gyroscope and three linear accelerometers for inertial navigation. There are three broad approaches to micro-accelerometers: (i) a silicon proof mass suspended by silicon cantilever beams with ion-implanted piezoresistive strain gauges; (ii) a silicon proof mass suspended between microcapacitor plates; (iii) detection of resonant frequency shift of a silicon proof mass. Accelerometers tend to suffer from a poor signal-to-noise ratio at low accelerations and suffer from extensive drift especially while traversing rough ground. For this reason, accelerometers are rarely used in mobile robotics. However, Sojourner used three accelerometers aligned to each axis of the vehicle to act as inclinometers to ensure that no slope over $30°$ was negotiated. Gyroscopes can be absolute sensors (rate gyroscope) or relative sensors (rate-integrating gyroscope). A rate gyro measures inertial angular velocity and is based on a spinning rotor or vibrating crystal which exploits the Coriolis effect. The optical ring laser rate gyro uses a fiber optic cable and measures the differential time of travel (Sagnac effect). Micromachined gyroscopes generally employ the principle of a vibrating proof mass to detect rotation-eschewing rotating parts. They are based on the transfer of energy between vibration modes generated by Coriolis forces given by $a = 2v \times w$. They comprise self-tuned resonators combined with displacement measurement such as a capacitance sensor. Tuning fork gyro-

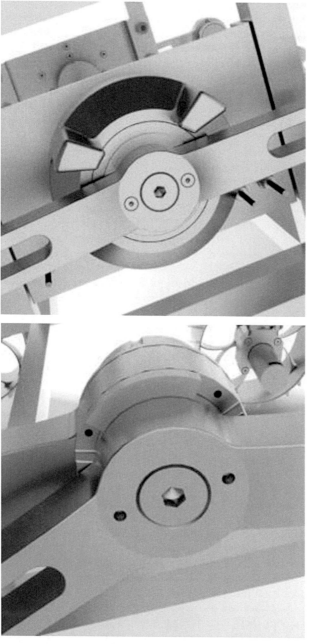

Figure 5.12. Rocker-bogie hard stops [credit Tim Setterfield, Carleton University].

scopes comprise a pair of proof masses connected to a junction bar that resonate in equal but opposite directions. Rotation of the proof masses induces an orthogonal vibration (bending or torsional) in proportion to the angular rate. Vibrating wheel gyroscopes vibrate about the wheel axis of symmetry. Rotation induces a detectable wheel tilt. The piezoelectric plate gyroscope uses a vibrating PZT plate as its base to measure coriolis forces. These transducers are mounted on three sides of a triangle and are excited at their resonant frequency ~8 kHz. Coriolis forces during rotation cause a difference in intensity of vibration. Micromachined inertial sensors are dependent on temperature which influences sensor bias due to thermal drift.

MER attitude was measured by a 0.75 kg Litton LN-200S Inertial Measurement Unit (IMU) comprising a three-axis accelerometer and a three-axis angular rate gyroscope. The accelerometers had a range of 80 g and resolution of 2.4 mg with a data rate of 400 Hz but was sampled at 8 Hz. Such sampling reduced the effective noise to 300 μg. MER onboard position and attitude estimates were performed at 8 Hz during movement. FIDO sported a differential GPS receiver (for reference navigation) and a Crossbow DMU-6X inertial navigation system of three accelerometers and three gyroscopes. The rate gyroscopes were vibrating ceramic plates that measured the Coriolis force to output angular rate information with a range of ±50°/s with a sensitivity of 25°/s. The accelerometers were micromachined Si capacitors that sensed linear acceleration across a range of ±2 g with a sensitivity of 1 g.

The Crossbow IMU 400 series (with a power requirement of 3 W) was adopted on both the Shrimp and Lama mobile robots. Surrey's GSTB-V2/A spacecraft used the MEMS Systron Donner QRS-11 gyro with a mass of 240 g, a power consumption of 1.2 W, a range of 10°/s, and average drift of <20°/h. The 34 g Tethers Unlimited Inc. nanosat MEMS-based IMU has dimensions of 38.1 × 50.8 × 15.2 mm including 6 DOF motion sensing including three-axis MEMS rate gyroscopes (range ±300°/s), dual (±1.2 and ±10 g) range three-axis MEMS accelerometers and three-axis magnetometers (±1,000 μT range). It has a power draw of only 40 mW at 5 V with shock survival to 1,000 g and temperature range of −40°C to 85°C. The chief problem with MEMS inertial sensors is that the small mass of the inertial component is subject to thermal noise. Kapvik adopted the Memsense H3-IMU HP02-0300 IMU which includes three mutually orthogonal accelerometers (range ±2 g) and three orthogonal gyroscopes to measure rover pitch, yaw, and roll (Figure 5.13). Wheel motors can act as odometric sensors as well as actuators—the DC electric current to drive a motor varies in proportion to the torque generated by wheel–soil friction. Wheel torque sensors measure these motor currents, which through the reduction gear measures wheel torque indirectly (and inaccurately). It is thus possible to estimate interaction forces from motor current feedback although this information will be corrupted. The measurement of small currents under 100 mA is subject to noise due to brushes, amplifiers, and torque ripple. One way to overcome this is to design a low-pass filter stage for motor current measurement [288]. Tactile sensors are contact sensors that provide for the detection of collisions with obstacles. The

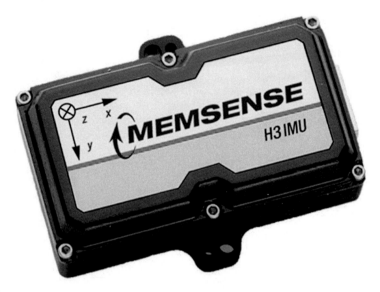

Figure 5.13. Memsense H3 HP02-0300 IMU [credit Rob Hewitt, Carleton University].

simplest are contact closure switches which may be configured as whiskers, antennas, or bumpers. All operate on the same principle such that an obstacle in direct contact deflects the sensor/switch resulting in detection. Tactile sensors provide last resort detection of obstacles if range sensors fail.

The simplest range detectors to measure distances to environmental objects by non-contact means include proximity, time-of-flight, and phase range sensors. Proximity sensors detect only the presence or absence of obstacles and do not measure distances. Commonly used proximity sensors include NIR proximity sensors operating at ~880 nm typically comprising transmitter/receiver units to detect objects which are either opaque or reflective of NIR radiation. Sojourner used bump detector switches as proximity sensors but non-contact proximity sensors such as capacitative sensors may be a more appropriate substitute for bump sensors. Proximity sensors are rarely used on planetary rovers (LIDAR serves this purpose).

These locomotion-monitoring sensors involve high-frequency control loops ~10–100 Hz. MER hazards were checked reactively with real-time interrupts operating at a sampling rate of 8 Hz—these included pitch–roll, rocker-bogie lever suspension angles, motor stalls, limit cycles (no forward motion), and resource conflicts. Self-localization was performed at 8 Hz using rover wheel odometry and inertial measurements from its three-axis accelerometers and three-axis angular rate gyroscopes. These fast control loops would be supplemented by visual localization data with much longer control loop periods (varying between 0.5 and 2.0 m depending on the terrain and therefore speed). Low-bandwidth navigation algorithms should run on a separate onboard computer from the motion controller. They are employed to generate the necessary reference commands for microcontrollers. Proximity sensing of nearby objects at close quarters has not been adopted on planetary robots except through vision (hazard cameras). Visual localization—tracking of pixels in stereovision is essential to recover position changes with respect to external landmarks. A polarization-sensitive Sun sensor is also essential to provide an external global reference frame under a wide variety of meteorological conditions. A set of CCD cameras are required for both proximity hazard detection and long-range navigation imaging. They also serve scientific imaging.

5.5 SUN SENSING

A Sun sensor is considered essential for self-localization of rovers on planetary surfaces. It provides an external reference with respect to the Sun to overcome the problems incurred in rover odometry inaccuracies. The MER PanCam was used for imaging the Sun to determine the Sun vector for rover steering angle measurement and rover attitude measurement without using a dedicated Sun sensor [289]. Onboard accelerometers determine the vector from gravity while PanCam images determine the solar vector a couple of times separated by 10 min equating to a solar transit of around 2.5° to provide inertial rover attitude about all three axes. The narrow FOV of the PanCam requires a gimbaled search to acquire the solar image at 880 and 440 nm wavelengths. In order not to saturate the PanCam, neutral density filters of ~5 are also used. The solar disk projects onto ~22 pixels of the PanCam. Dark current is subtracted from the solar image and the solar image centroid computed. The Sun sensor of the FIDO rover comprised a CCD imager with a wide field of view. The centroid of the Sun is determined using a thresholding and centroid-finding algorithm with outlier rejection. The Sun is located in the 2D image plane from which a ray vector is computed from the calibrated camera parameters modeled as a fish-eye lens. The ray vector is converted into a gravity-down reference frame from the rover inertial measurement system. The Sun position in the sky is correlated with thetime of day and the rover position on the planetary surface (latitude and longitude). The NASA/JPL Sun sensor provides for estimates of planetary rover heading measurements with minimal computational overhead [290]. The Sun is bright and its angular radius is small enough to be considered a point source (0.3° at 1 AU). The Sun sensor is a monochrome CCD camera with a wide-angle lens (the JPL Sun sensor has a FOV of 120 × 84°) and solar filters attached to the lens to filter out non-solar low-intensity incident light. The Sun sensor uses a dedicated framegrabber to minimize computational overhead. The Sun sensor integrates light input across its field of view and outputs a signal defining the centroid of the Sun image. The centroid of the Sun image is extracted by fixed thresholding, artifact removal, and center of mass and circularity determination from first and second-order shape moment computations. Sun vector is given by:

$$\begin{pmatrix} s_x \\ s_y \\ s_z \end{pmatrix} = \begin{pmatrix} \cos\alpha\sin\zeta \\ \cos\zeta \\ \sin\alpha\sin\zeta \end{pmatrix}$$

where $\alpha =$ Sun azimuth with respect to sensor and $\zeta =$ Sun elevation with respect to sensor. Whereas the MERs used PanCam images of the Sun to extract the Sun vector, a dedicated Sun sensor is useful for rover navigation. The Sinclair Interplanetary Sun sensor was adopted for the Kapvik microrover (Figure 5.14).

The sun sensor with a 140° conical field of view on Kapvik provides absolute measurement of yaw. The Sinclair Interplanetary SS-411 Sun sensor is a low-power device with a mass of 30 g and an accuracy of 0.1° with a wide 140° field of view ideal for rover deployment [292]. In combination with a gravity vector measurement by an inclinometer or inertial measurement unit, the rover rotation matrix may be estimated (assuming that yaw is zero) as it traverses the terrain [293]:

$$T = \begin{pmatrix} \cos P & 0 & -\sin P \\ \sin R \sin P & \cos R & \sin R \cos P \\ \cos R \sin P & -\sin R & \cos R \cos P \end{pmatrix}$$

where $R =$ roll and $P =$ pitch. Rover acceleration is given by:

$$\begin{pmatrix} a_x \\ a_y \\ a_z \end{pmatrix} = \begin{pmatrix} -\sin P \\ \sin R \cos P \\ \cos R \cos P \end{pmatrix}$$

Solution is given by:

$$P = \tan^{-1}\left(\frac{-a_x}{\sqrt{1-a_x^2}}\right)$$

$$R = \tan^{-1}\left(\frac{a_y}{a_z}\right)$$

Hence,

$$T = \begin{pmatrix} \sqrt{1-a_x^2} & 0 & a_x \\ -\dfrac{a_y a_x}{\sqrt{1-a_x^2}} & \dfrac{a}{\sqrt{1-a_x^2}} & a_y \\ -\dfrac{a_z a_x}{\sqrt{1-a_x^2}} & -\dfrac{a_y}{\sqrt{1-a_x^2}} & a_z \end{pmatrix}$$

For a given universal time (UT), the astronomical position of the Sun is determined from ephemeris data in a stored astronomical almanac using Kepler's equations. From this, the rover heading can be determined. During autonomous navigation, position error shows error growth while steering (yaw) error is constrained by the Sun sensor (Figure 5.15). Rover localization can be achieved using wheel odometry, inertial navigation measurements, and Sun sensor measurement

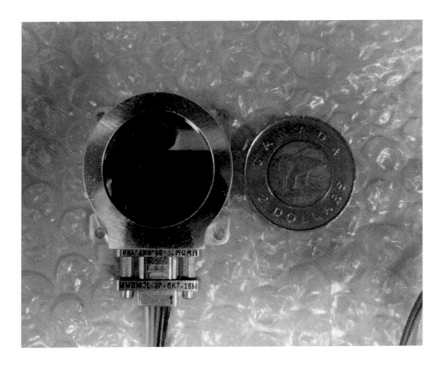

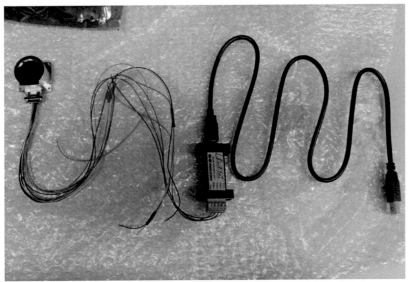

Figure 5.14. Sinclair Interplanetary SS-411 Sun sensor [credit Jesse Hiemstra, Carleton University].

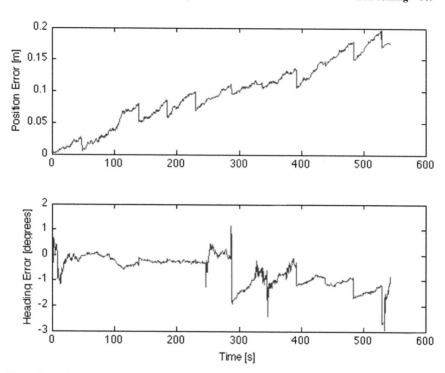

Figure 5.15. Sun sensor constraint on steering error [credit Rob Hewitt, Carleton University].

fused to provide an integrated state estimate of position and orientation based on an extended Kalman filter, similar to that used in traditional spacecraft attitude control systems.

Use of the Sun for navigation requires compensation for movement of the Sun through the day. On Earth, the rate of change of the Sun's azimuth varies with the time of day (7.7°/h at dawn/dusk to 56.1°/h at noon), date, and latitude—desert ants extrapolate from the most frequently observed rate of movement for accurate navigation rather than assume a constant solar rate [294]. Polarization sensitivity would increase the robustness of Sun sensing under less-than-ideal conditions due to dust-induced opacity and cloud cover (applicable only to planets with atmospheres). Skylight polarization occurs due to sunlight scattering in the atmosphere which partially plane-polarizes skylight with the degree of polarization being greatest at 90° from the solar direction. Polarization patterns result from Rayleigh scattering through the atmosphere. The direction of polarization across the sky forms a regular pattern with mirror symmetry about the solar and anti-solar meridian plane. The degree of polarization is highest for light scattered at 90° to the sunlight. On Mars, dust deposition and elevation by dust devils and global dust storms generates scattering from particles ~10–100 μm in size. Rayleigh scattering in the Martian atmosphere occurs at UV wavelengths below 450 nm with a maximum 5% effect at 200 nm. On Earth, the direction of linear polarization (e-vectors) forms a regular pattern over the sky with mirror symmetry, the line of symmetry formed by the Sun and zenith where the e-vectors are oriented orthogonally to the solar azimuth. The dominant e-vector orientation is orthogonal to the line of sight to the Sun. The electric vector polarization rotates about the zenith as the Sun changes its elevation due to its westward movement by 15°/h. This provides the basis for compass orientation in insects. Some insects use visual navigation based on landmarks and polarization patterns on the sky. Insect compound eyes are comprised of many eyelets (omnatidia). Specialized polarization-sensitive omnatidia (compound eye pixels) in some insects such as bees reside at the dorsal rim of the eye and point upward with a 60° field of view [295, 296]. Each omnatidium contains two sets of polarization-sensitive photoreceptors that are tuned orthogonally to each other in a cross-analyzer arrangement. Many insects navigate according to the orientation of the plane of polarization of linearly polarized light from the Sun. Ants and bees orient themselves with respect to the Sun as a visual compass reference by measuring the polarization (e-vector) of sunlight in the sky. In desert ants, information from both the polarization compass and odometry are integrated to estimate self-navigation [297]. Small arrays of polarization-sensitive photodiodes with a linear polarizer on planetary rovers can act in a similar way to provide a robust solar compass. They would be directed to the zenith with one sensor being aligned with its polarization axis 90° to the other to act as a cross-analyzer. Three such pairs may be aligned to 0, 60, and 120° to the rover forward direction. An analog polarized light sensor with a 60° field of view has been developed [298]. It comprises a pair of photodiodes with linear polarizers that feed into a log ratio amplifier. Each pair is fed into a log ratio amplifier: $V_o = K(\log I_1 - \log I_2)$. This compresses the photoresponse thereby enlarging the intensity range. There are three possible mounting arrangements [299]—scanning by rotating the entire body in the vertical axis to scan the sky using only one opponent pair of sensors, more robust extended scanning that uses three pairs for fine-tuning, and simultaneous scanning using three sensor pairs aligned at three orientations (e.g., 0, 60, 120). The latter can be operated continuously to provide errorless navigation. The view of the sky through a rotating linearly polarizing filter yields an incident intensity that depends sinusoidally on the orientation of the filter. The intensity, polarization angle, and degree of polarization vary with the position of observation defining a pattern of partial polarization on the sky that depends on the relative location of the sun. The sensitivity of a polarization-sensitive photoreceptor to partially polarized light is given by:

$$S(\phi, D) = KI\left(1 + \frac{d(S_p - 1)}{(S_p + 1)}\cos(2\phi - 2\phi_{\max})\right)$$

where $S_p = s_{\max}/s_{\min} = $ polarization sensitivity, $\phi = \frac{1}{2}\tan^{-1}((s_1(\phi) + 2s_2(\phi) - \frac{3}{2})/(\sqrt{3}(s_1(\phi) - \frac{1}{2})) = $ polarization angle with respect to the solar meridian, $\phi_{\max} = $ maximum polarization angle with maximum e-vector that maximizes output $S(\phi)$, $d = (I_{max} - I_{min})/(I_{max} + I_{min}) = $ degree of polarization, $K = $ constant, $I = I_{max} + I_{min} = $ total intensity, $I_{max} = $ maximum intensity, and $I_{min} = $ minimum intensity. The amplitude of the output from the log ratio amplifier varies in proportion to the degree of polarization, which changes through the day. The use of such a polarization-sensitive Sun sensor would provide an external global reference for navigation. Alternatively, rover self-localization to within 20 km may be achieved at night by using Phobos or Deimos as external references against the background of two stars.

5.6   LASER RANGEFINDING

Infrared sensors are the simplest type of non-contact sensor—infrared light is emitted and reflections from close objects in the environment are detected. The intensity is very low so they are extremely limited in range ($\sim$50 cm or so). Active sensors emit a signal that is scattered back and detected by a receiver. The simplest active sensing method is laser striping—laser stripers either side of frontal cameras project vertical stripes onto objects in the visual field (structured light). On Sojourner, five laser diode–based stripers were used as hazard proximity sensors when halted. When cameras take images the project stripe forms a signature dependent on the object shape. Similarly, by controlling the toe-in of each laser striper, the distance to the object can be determined. Structured light can be used to resolve ambiguities and reduce computational complexities in images. The striped laser projects a strip or grid of light onto the scene which shows displacement proportional to depth, kinks indicating a change in plane geometry, and discontinuities indicating gaps, that is captured by the camera. Depth is defined as:

$$z = \frac{b}{x/f + \tan \alpha}$$

where $b = $ baseline between laser and camera optics, $f = $ camera lens focal plane, $x = $ image location where laser stripe is detected, and $\alpha = $ laser projection angle with respect to z-axis. Such structured light approaches provide more robust estimates of depth [300]. A variation on this is the time-to-range laser that continuously sweeps the scene and computes the range when a pixel sees the laser stripe. Scannerless approaches eliminate scanning mechanisms. Such approaches use a laser pulse to cover the whole scene, with detection by a CCD detector array that computes the time of flight at each pixel. Depth resolution is dependent on the length of the baseline, focal length, and image resolution. Generally, depth resolution degrades to <1 m at 5 m distance, $\sim$3 m at 10 m distance, $\sim$7 m at 15 m distance, and $\sim$12 m at 20 m distance. Sojourner's laser diode–based stripers projected up to 30 cm in front of the vehicle and 13 cm to the left and right of it. The stripes undergo distortion if there is an elevation change which may be measured to determine object curvature—no ambient light is required. Unfortunately, Sojourner's laser stripers were susceptible to false positives (crevasses, slopes, and obstacles) on the Martian terrain. More sophisticated external range sensors are

based on either time-of-flight measurement of an emitted pulse and its echo or
phase shift measurement of a continuous wave emission. Pulsed methods may be
ultrasonic, radiofrequency, or optical. Ultrasonic range sensors offer range detection by generating and emitting an ultrasonic chirp. Its reflection from an object
and detection by a receiver allows measurement of the time of flight and so
distance. Specular reflections corrupt ultrasonic range data, and ultrasonic arrays
can generate crosstalk interference. They are inaccurate in terms of direction resolution from their large beamwidth, suffer from temperature dependence, significant
specular reflection, and crosstalk due to multiple reflections from objects. For
microwave radar, the radar equation shows that electromagnetic attenuation
occurs as $1/r^2$ but if a collimated beam such as a laser is used, there is little or
no spreading loss on the outward path, giving $1/r$ path attenuation. Laser range-
finders use optical beams rather than ultrasonic signals to measure time-of-flight
or phase difference between emitted and reflected signals [301]. Optical (laser)
radars have considerably greater accuracy than microwave (maser) radars. Laser
diodes with a typical power output of 30–100 mW emit a collimated light beam at
a single frequency in conjunction with a photodiode detector. Phase shift measurements typically involve heterodyning the reference and received signals with an
intermediate frequency. Relative phase shift is given by:

$$\phi = \frac{4\pi d}{\lambda} \rightarrow d = \frac{\phi c}{4\pi f}$$

LIDAR (light detection and ranging) is a time-of-flight optical range detection
scheme between an optical transmitter and receiver telescope that does not require
ambient lighting [302]. LIDAR systems employ narrow beams with little beam
spread and the wavelength of light is less than object surface undulation
dimensions. Hence, beam bounce from false inclines and specular reflection will
not occur. Some of the reflected rays strike the receiver and time-of-flight measurement gives the distance to the target. Laser scans provide information on the
angles and positions of objects within its field of view generating depth and height
profiles of objects in the scene. Doppler signal processing can also generate motion
data. LIDAR was adopted on the U.S. MUSES-C nanorover as its primary navigation and obstacle avoidance sensor with a range of 10 m. The MUSES-C
nanorover's laser rangefinder was also to be used to demonstrate direct optical
communication to Earth at 100 bps. Nomad similarly carried a laser rangefinder
as its primary navigation sensor. Pulsed laser light time of flight between emitted
and reflected beams is measured when detected by a photomultiplier. Alternatively, a continuous wave light beam undergoes phase shift between transmitted
and received signals which is measured. A typical scanning laser radar scans a
laser beam across the scene and a detector measures the reflected light [303].
Scanning laser range finders are of three basic types which differ in how they
measure range. AMCW lasers measure the phase difference between emitted and
received signals. Time-of-flight (TOF) lasers measure the travel time of a pulse.
FMCW lasers measure the frequency shift of a frequency-modulated laser.
AMCW lasers are faster and perform best at close to medium range (up to 50 m).

TOF scanners perform best at long range and in outdoor environments but their acquisition rates are lower. FMCW lasers are more accurate but are complex and bulky. Detector response times of ~50 ns are required for accuracies in range-finding of a few meters or so. Millimeter radar (77 or 94 GHz) is not subject to the opacity limitations inherent in LIDAR. They offer reduced resolution but greater range and can generate a terrain image but terrain topography introduces shadowing and occlusion errors as with LIDAR. A 3D scanner creates a point cloud of geometric distance to the scanned object from which the surface is extrapolated (reconstruction). Multiple scanning requires registration to a common reference in order to create the 3D model. The locus method of generating range/direction/height map $(\rho\phi z)$ at points (x,y) from a range sensor and terrain feature matching transformations between two images to generate a digital elevation map are described by Kweon and Kanade (1992) [304]. This was the method adopted for the terrain mapper of Ambler, the large six-legged walking rover prototype that used a camera image [305]. It used a laser rangefinder to plan its footfalls over rough terrain based on the terrain map [306]. Typically, an artificial grid is imposed that holds height, slope, roughness and obstacle occupancy data (with confidence estimates) [307]. The Sick LM series time-of-flight laser with 20 Hz 2D scanning offers 20 cm accuracy to a distance of 50 m. The Hokuyo URG-04LX 2D scanning laser rangefinder is similar to the Sick LMS 200 laser rangefinder which has become standard in mobile robotics but without its large size, weight (1.1 kg), and power requirements (12 W) [308]. The Sick LMS 111 has a longer range of 20 m and an angular resolution of 0.5°. The Hokuyo laser rangefinder was mounted onto a pan–tilt unit on the Kapvik microrover and was selected for its good performance with low power consumption (2.5 W) and mass (0.16 kg) (Figure 5.16). This enabled it to take 3D scans by tilting its sweep. An

Figure 5.16. Hokuyo URG-04LX laser scanner [credit Rob Hewitt, Carleton University].

infrared laser beam is scanned by a rotating mirror and the phases of emitted and returned light are compared to extract distance. The mirror sweeps out a 240° field of view with an angular resolution of 0.36° at a scan rate of 10 Hz. It can transfer up to 750 kbps through the RS-232C serial interface and consumes 4 W maximum power. Its usable range is up to 4 m with a range error of only 0.3%.

3D laser imaging may also be applied to rock face geometry determination through strike and dip measurements. However, LIDAR was not adopted on the MERs and not baselined for the ExoMars rover as stereovision in both cases is used for traverse planning and obstacle avoidance. It is considered that Martian terrain has sufficient texture for stereovision to be adequate in terms of feature density. Nevertheless, combining LIDAR with vision-based methods eliminates the computational overhead of disparity map generation inherent in stereovision.

5.7   AUTOMATED TRACTION CONTROL

Navigation over rough terrain requires consideration of the physical interaction with the terrain. One aspect concerns the negotiation of obstacles. It has been suggested that the following algorithm can be used to minimize impact shocks of a rocker-bogie chassis on obstacles at high speed by lifting the front wheels off the ground [309]:

  (i) run all wheels at the same nominal speed;
 (ii) when the obstacle approaches the front wheels within one wheel diameter
      —speed up the middle wheel by 30%,
      —slow the rear wheel by 30%;
(iii) when the obstacle is directly under the front wheel
      —return all the wheels to the same nominal speed;
 (iv) when the obstacle approaches the middle wheel within one wheel diameter
      —slow down the front wheel by 30%,
      —speed up the rear wheel by 30%;
  (v) when the obstacle is under the middle wheel
      —return all the wheels to the same nominal speed.

More generally, during the traverse, wheel slip must be minimized to reduce both energy consumption and odometric errors as well as wheel force distribution to ensure that individual wheels drive the rover cooperatively. The integration of wheel encoder data to infer the rover position is inaccurate because of rugged terrain giving variable slopes and wheel slippages. Even implementation of an inertial sensing unit does not recover vertical data as it too is subject to cumulative drift. Furthermore, odometry is strongly dependent on the kinematic structure of the chassis which introduces inaccuracies. The chief problem is slippage. Opportunity was required to climb a 15° to exit its 22 m crater in which it landed and experienced slippage of 45–50%. Spirit experienced a 100% slip driving up a 16° slope attempting to reach "Larry's Lookout" on sol 399, yet it

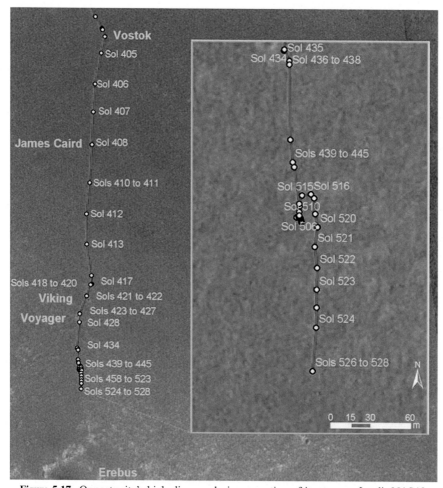

Figure 5.17. Opportunity's high slippage during a portion of its traverse [credit NASA].

experienced only 20% slip on a 19° slope a few meters away. On sol 454, Spirit halted when it detected slippage of over 90% and submerged its wheels almost a full wheel radius in the soil. Similarly, in 2005, Opportunity experienced immobilization due to sinkage up to wheel axle level in the drift soil of a sand dune (Figure 5.17).

Slippage invokes high-power consumption without tractive progress so should be minimized. Traction control reduces this power consumption by attempting to reduce slippage. The basis of all automated traction control schemes is rover slip estimation in order to minimize slippage [310]. Wheel slip occurs at the wheel–soil interface whereby wheel angular velocity does not match equivalent ground velocity. Odometry based on proprioception (i.e., odometry) is very inaccurate so exteroception is essential (i.e., a form of external reference sensing to measure vehicle progress). In order to correct for slip, some form of external reference measurement is required to monitor ground traverse directly. Slip may be measured from odometry compared with an external reference sensor:

$$s = 1 - \left(\frac{v}{rw}\right)$$

where $rw > v$ during acceleration, $rw < v$ for deceleration, $v =$ translation velocity of the vehicle, $r =$ wheel radius, and $w =$ wheel rotational velocity. This may be differentiated to yield:

$$\dot{s} = \frac{1}{wr}\left(-\dot{v} + \frac{v}{wr}\dot{w}r\right)$$

to describe slip evolution. Wheel odometry provides data on the distance traveled but this is corrupted by wheel slippage in the soil. Accurate wheel odometry requires a free-wheeling wheel as used on the Lunokhods but this still suffers from inaccuracy during turning ($\sim$10% error) at the cost of mass. Visiodometry has been proposed to measure ground translation displacement whereby two downward-facing cameras are mounted either side of the rover to measure normalized phase correlation between camera images [311, 312]. It is based on normalized phase correlation between two images as measured by pixel shift between two images:

$$I_1(x + \Delta x, y + \Delta y) = I_2(x, y)$$

This technique operates on featureless surfaces. An optical mouse requires placement near the surface within a few millimeters over which the rover traverses in order to track its movement with respect to the surface. The separation distance between the sensor and the surface determines the projection area on the imaging sensor—this makes its use over rough terrain difficult where the separation will vary. Furthermore, operation close to the surface restricts its speed range to $\sim$30 cm/s. Adaptation to higher altitudes for robotics applications may be achieved through the addition of an amplifying lens and an illumination source [313, 314]. Optic flow microsensors have been developed with a mass of 5–20 g in conjunction with a 10 MHz PIC 18C252 microcontroller and integrated vision chip [315]. A pulse is generated whenever intensity at a pixel increases rapidly, with visual motion detected through sequences of pulses from adjacent pixels. Laser-based devices are superior to LED-based devices. In fact, two-axis illumination improves 2D position measurement accuracy over a single illumination source [316]. Alternatively, a pair of optical mice may be used [317]. A differential optical sensor which uses two optical mice at different heights and a beamsplitter compensates for height variations and may be used for rugged terrains [318]. Alternatively, optic flow measurements from a single downward-pointing camera may be adopted in conjunction with a Kalman filter, a Newton–Raphson estimator [319], or a sliding mode observer [320]. However, these approaches have some mass impact. Alternatively, a laser speckle velocimeter, which uses optical feedback from a semiconductor diode laser that acts as both transmitter and receiver in a single optical axis, may be employed [321–323]. The speckle effect results from interference of coherent beams of radiation from many secondary point sources from a rough surface. Typical size of a speckle is defined as:

$$d = 0.61 \frac{\lambda}{NA} = 1.22 \lambda f_\#$$

where NA = numerical aperture and $f_\#$ = f-number. Contrast (visibility) in the speckle image is given by:

$$C = \frac{I_{\max} - I_{\min}}{I_{\max} + I_{\min}} = \frac{\Delta I}{T_T}$$

The laser projects a laser source onto a rough surface which generates interference (speckles) due to the different pathlengths traveled by light beams. The detected image from the ground surface back-couples into the laser cavity and interferes with transmitted light which causes random power output variations of the spectrum measured by a rear photodiode. If translated over a rough surface whose RMS height is greater than the wavelength of laser light, its in-plane motion generates random speckle effects due to the interference effect of different pathlengths between transmitted and received light. Autocorrelation is used to detect the self-mixed signal that is generated in which the velocity of the generated random speckle pattern is extracted as a linear function of the inverse correlation time of random amplitude fluctuations in the speckle pattern. Prefiltering the images using a binary threshold or Canny edge detection filter improves tracking capability. Normalized cross-correlation between speckle images enables tracking of features between image frames with errors of 0.75% at speeds of 50 mm/s compared with 10% for wheel odometry. Once slip has been measured, slip control may be implemented. As the rover traverses rough terrain, each wheel experiences a different loading due to terrain topography. This can cause wheels to "fight" each other as each wheel is independently driven closed loop to maintain its commanded velocity setpoint. The maximum tractive efficiency $\eta = 1 - R/H$ for a four-wheeled vehicle occurs when front and rear wheel speeds and slippages are equal [324, 325]. Baumgartner et al. (2001) proposed a relatively simple velocity synchronization algorithm [326]. Deviation of wheel velocity from the nominal wheel velocity profile is given by:

$$\delta_i = \sum_{i=1, i \neq m}^{n} |\theta_m - \theta_i|$$

where θ_m = distance traveled by mth wheel, $m = 1, \ldots, n$, θ_i = distance traveled by ith wheel, and $i \neq m$. Average deviation is given by:

$$\delta_{av} = \frac{1}{n} \sum_{i=1}^{n} \delta_m$$

A voting scheme was used to generate a composite score of average small deviations and exclude outliers:

$$d = \sum_{m=1}^{n} \begin{cases} \theta_m & \cdots & \delta_m \leq \delta_{av} \\ 0 & \cdots & \delta_m > \delta_{av} \end{cases}$$

with $c = c + 1$ acting as a counter if $\delta_m \leq \delta_{av}$. Total distance is computed as $D = d/c$. The required change in commanded velocity setpoint relative to nominal velocity setpoint for the mth wheel:

$$e_m = D - \theta_m$$

$$v = v_{ref} + \Delta v$$

where

$$\Delta v = \begin{cases} e_m v_{ref} & \cdots & e_m v_{ref} \leq \Delta v_{th} \\ \Delta v_{th} & \cdots & e_m v_{ref} > \Delta v_{th} \end{cases}$$

and where $\Delta v_{th} = $ maximum velocity change. More sophisticated approaches are possible, however. It is recommended that model-based feedforward control (based on estimated wheel forces computed from Bekker analysis) be combined with sensor-based feedback control (based on slip estimation through motor current measurements) for rover locomotion [327]. The use of forward and inverse kinematics models of rocker-bogie linkage systems has been explored to provide a reference for estimating slip supported by visual odometry and Kalman filtering avoiding the use of traction modeling [328]. Unfortunately, such an approach is difficult to implement online while on the move due to the latency of the vision system. The slip ratio can be expressed as a curvature regardless of drawbar force and sinkage (i.e., slip ratio can be estimated from the ratio between T/W). This requires measurement of vertical wheel load and drive torque along with data on soil type. Online terrain parameter estimation algorithms based on Bekker theory have been developed using simplified force balance equations for drawbar pull, weight, and wheel torque [329]. Most traction control algorithms involve consideration of the wheel–terrain interaction determined through Bekker theory and, in particular, the Janosi–Hamamoto relation [330]. This type of approach is based on computation of the load–traction factor defined as the ratio between drawbar pull (tangential force) and vertical loading (nominally weight over each wheel) (DP/W) which is maximized to reduce slip [331]. Wheel torque is determined from the Mohr–Coulomb relation in conjunction with the Janosi–Hamamoto relation and is given by:

$$\tau = r^2 b \int_{\theta_a}^{\theta_d} (c + \sigma \tan \phi)(1 - e^{(r/K)(\theta_a - \theta - (1-s)(\sin \theta_a - \sin \theta))})$$

where $r = $ wheel radius, $b = $ wheel width, $c = $ soil cohesion, $\varphi = $ soil angle of friction, $\sigma = (k_c + bk_\phi)(z/b)^n = $ normal stress (Bernstein–Goriatchkin load–sinkage relation), $k_c = $ modulus of cohesion of soil deformation, $k_\phi = $ modulus of friction of soil deformation, $s = $ slip, $j = r[(\theta_a - \theta) - (1 - s)(\sin \theta_a - \sin \theta)] = $ shear displacement along wheel/soil interface, K = empirically determined shear constant, θ_a = entry angle from vertical at which wheel makes contact with terrain, θ = wheel angular location, z = sinkage, and n = soil deformation exponent (nominally unity defining a Gerstner soil). A similar relation can be determined for tracked vehicles [332]:

$$\tau = b \int_0^l (c + \sigma \tan \phi)(1 - e^{-(l - (1-s)x/K)})$$

where l = ground track length and x = track displacement. Slip compensation requires several component capabilities—visual odometry and IMU data to provide input to a Kalman filter estimator to determine rover motion; output is compared with a kinematic model of the vehicle using motor angles to estimate slippage; and a path generation system compensates for slippage [333]. A PI-based slip control law may be implemented with error $\Delta s = s^d - s$ to reduce slippage to zero $\Delta s \to 0$ from current measured slip determined from an empirical model such that:

$$\delta = k_p e + k_i \int e \cdot dt$$

This formulation requires knowledge of soil parameters which will in general be unknown. Iagnemma et al. (2002) [334–337] have proposed using traction modeling estimation with some simplifications facilitating symbolic computations using closed-form relationships. Their method provides the basis for estimating the terrain parameters of soil cohesion and soil internal friction angle online through linear least squares estimation which also has scientific value as well as contributing to the efficiency of traverse:

$$\begin{pmatrix} c \\ \tan \phi \end{pmatrix} = (K_2^T K_2)^{-1} K_2^T K_1$$

where

$$K_1 = \frac{4\tau \cos \theta_a - 8\tau(\theta_a/2) + 4\tau}{1 - A}$$

$$K_2 = \begin{pmatrix} 2r^2 w \theta_a (\cos \theta_a - 2 \cos(\theta_a/2) + 1) \\ -(4\tau \sin \theta_a + W \theta_a^2 r - 8\tau \sin(\theta_a/2)) \end{pmatrix}$$

$$A = e^{-\frac{r}{K}\left(\frac{\theta_a}{2} + (1-s)(-\sin \theta_a + \sin(\theta_a/2))\right)}$$

$$\tau = \tfrac{1}{2} r^2 b \tau_m \theta_1$$

$$\tau_m = \text{motor torque}$$

The moduli of soil deformation will not generally be known. However, measurement of normal load over each wheel using strain gauges mounted on the shafts of each wheel will eliminate this requirement and have a minimal impact on mass. Motor torques τ can be estimated from measurements of current input and knowledge of the current–torque relation for each motor (though there will be nonlinearities through the gearbox). Wheel position and motion may be measured by in-wheel encoders/tachometers. Wheel sinkage and the wheel–soil contact angle will be difficult to measure online without external vision sensors. One proposal is to introduce a tactile wheel to measure the wheel–soil contact angle [Siegwart, R. et al., private communication]: various methods have been proposed including a laser source and detector based on intensity, triangulation, phase shift measurement, ultrasonic range sensor (though not feasible for Mars due to thin Martian atmosphere), and eddy current sensor to measure distance and thereby wheel deformation. Also proposed was a strain gauge sensor (based on a SAW sensor) to measure wheel deformation—this last would be useful in determining the deformation of the wheel. A tactile (haptic) wheel will have a severe mass and complexity impact on a planetary rover. Iagnemma et al. [338–340] proposed employing an extended Kalman filter that uses only vehicle inclinometers and wheel tachometers as a measure of terrain topography to estimate the wheel–soil contact angle. However, their formulation assumed point contact on a rigid terrain and did not include sinkage. Both wheel–soil contact angle and sinkage are essentially unobservable using simple sensors. A simple method to determine sinkage may be determined from its geometric relation to sinkage (for rigid wheels only):

$$\theta_a = \cos^{-1}\left(\frac{r-z}{r}\right)$$

Sinkage may be determined from the normal loading on each wheel (Figure 5.18):

$$z = \left(\frac{3W}{b(3-n)(k_c/b+k_\phi)\sqrt{d}}\right)^{2/(2n+1)}$$

for each wheel or track. An alternative, more complex formulation is given by Richter and Hamacher (1999) to account for additional sinkage of trailing wheels in the ruts made by the front wheels in the form:

$$W = -bk\int_0^{z_0} z^n \frac{2(z_0-z)-d}{2\sqrt{d(z_0-z)-(z_0-z)^2}}\,dz$$

for front wheels (sinkage z_0)

$$W = bk\sqrt{dz_2}\left((z_0+z_2)^n - \frac{n}{3}(z_0+z_2)^{n-1}z_2\right)$$

for middle wheels (sinkage $z_0 + z_2$)

$$W = bk\sqrt{dz_3}\left((z_0+z_2+z_3)^n - \frac{n}{3}(z_0+z_2+z_3)^{n-1}z_3\right)$$

for rear wheels (sinkage $z_0 + z_2 + z_3$). Wheel force measurement of the wheel loading may be measured by a single or multi-axis strain gauge force sensors mounted on each wheel shaft. Wheel force measurements by load sensors incorporated above each wheel station provide the basis for converting wheel loads into sinkage estimates for extracting soil parameters (Figure 5.19).

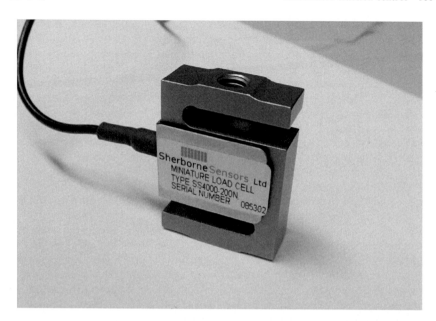

Figure 5.18. (a) Sherborne SS4000M-200 N load sensor (with amplifier) (b) mounted above each Kapvik wheel station to measure wheel loads [credit Tim Setterfield, Carleton University].

Two multilayer perceptrons were used to learn a Bekker–Wong model of traction in order to extract a condinuous data set of both soil cohesion and friction (Figure 5.19). The soil extraction algorithm is strongly dependent on slip estimation implying that slip measurements will be essential (Figure 5.20). Slip measurement may be accomplished using a small 512×512 pixel camera with a frame rate of 0.5 Hz (1 image/20 cm travel) at a speed of 10 cm/s. Assuming that slip will be constant over one vehicle length of 1 m this imaging rate can be reduced to once every 10 s. This approach has minimal mass impact on the rover. Traction control offers significantly lower power consumption and slippage than (constant) velocity control. The rover "feels" the terrain through inputs from wheel torque, load, and sinkage. The chief limitation of these approaches is that they use simplified models of drawbar pull and do not account for flexible wheels or grousers. Automated traction control systems are still currently under development and have yet to be implemented in planetary rovers.

Hazard detection is an important part of obstacle avoidance and the navigation function during traverses. Hazards include obstacles, excessive slopes, challenging terrain roughness, and low-traction loose soils. Geometric analysis of slope and roughness through the extraction of statistical properties allows classification of terrain as sandy, rough, rocky, untraversable, or unknown. For instance, roughness may be computed from standard deviation:

$$\sigma = \sqrt{\frac{\sum_{i,j}(I_{\text{av}} - I_{i,j})^2}{nm}} \quad \text{where } I_{\text{av}} = \frac{\sum_{i,j} I_{i,j}}{nm}$$

The image parameter I may be estimated from terrain elevation z, which represents challenging terrain, and/or terrain slope θ, which increases wheel slippage. Rougher terrains have higher standard deviation. Visual analysis of terrain is the key to this process. Different terrains require different image-processing techniques (e.g., ice is highly reflective while sand is variable over time). Slip prediction requires the use of visual information to assess the terrain (e.g., sandy slopes yield high degrees of slippage which are best avoided). Classification of terrain may be through rule-based, artificial neural network, fuzzy logic–based or Bayesian-based approaches. A fuzzy traversibility index may be defined that encapsulates four traversability measures (roughness, slope, discontinuity, and hardness) [341]. The fuzzy logic–based terrain classifier gave superior performance with higher computational cost than the worst performing but rapidly computed rule-based terrain classifier [342]. Essentially, the complexity of the classifier determines its performance [343]. Terrain is range dependent so this must be accounted for. It is possible to learn correlations between terrain texture features and the incidence of slippage [344]. Neural networks are highly suited to learning such associations through experience. Images of terrains may be classified based on terrain features such as soil texture using neural networks (assuming shadowing is absent) [345]. Once learned, slip can then be predicted from visual features of the terrain. Visual odometry serves to measure slip and online terrain parameter estimation is used for characterizing some aspects of

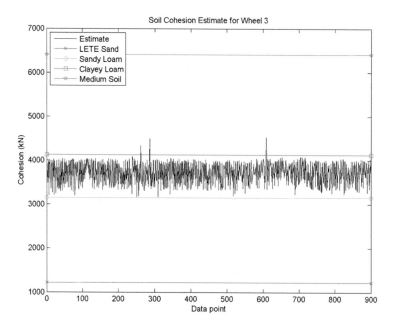

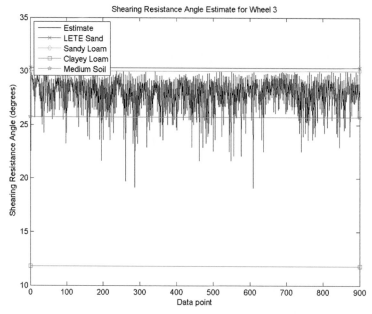

Figure 5.19. Soil cohesion and friction angle indicating sandy loam [credit Matt Cross, Carleton University].

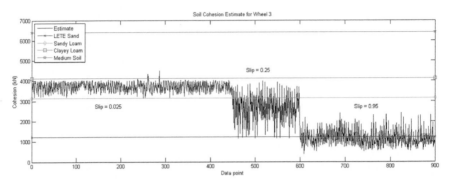

Figure 5.20. Effect of slip on soil cohesion estimates [credit Matt Cross, Carleton University].

terrain. Visual imagery based on visual textures (textons) of terrain patches provides the basis for recognition of terrain classes. Slip is a nonlinear function of terrain slope which may be modeled through a neural network that may be used by an extended Kalman filter to estimate rover pose [346, 347]. Unusually, this work also used the EKF to estimate neural network weights, learning rules that successfully compensated for slip. Terrain maps may incorporate slip prediction [348].

It is desirable to identify non-geometric hazards such as loose sand in which the rover can become trapped. Vibration signatures from acceleration data generated by a vehicle during its traverse may characterize different underlying terrains. Vibration signatures encode terrain data in the frequency domain. The power spectral density may be analyzed and subjected to principal component analysis to allow Fisher discriminant analysis to be applied. The Mahalanobis distance may then be used to classify new terrains. These methods are limited as they are restricted to the specific speeds at which training occurs for effective terrain classification. Different velocities produce different vibration responses on the same terrain. The same applies to motor current monitoring correlating to terrain [349]. These are local methods. One approach relies on knowing the vibration transfer function of the rover which defines the vibration frequency response of the rover for terrain inputs [350]. It is preferable to learn terrain classification from long-range visual data to be incorporated in a map-based representation [351]. The commonest map representation is the grid-based elevation map [352]. It appears that simple color analysis such as HSV (hue, saturation, and value) may not be sufficient for this purpose. Visual feature analysis may be based on HSV color, or texton-based texture analysis may be used to define terrain classes. This requires the analysis of texture such as that performed by Gabor filters, but these are costly. An alternative is to use Haar wavelet decomposition. Similar performance can be achieved using 2D normalized color histograms based on rg color ($r = R/(R + G + B)$ and $g = G/(R + G + B)$) which is less illumination sensitive at reduced computational cost. Classification may be through a naive Bayes classifier to populate the cells of a Cartesian map. The Bayesian classifier provides for sensor fusion of both tactile and visual features extracted using statistical methods [353, 354]. This potentially provides slip prediction remotely prior to traverse. This may be combined with a previously trained vibration-based terrain classification system that serves as input to a visual classifier [355]. Singular value decomposition may be used for such a classification [356]. A learning procedure has been explored which encapsulates predictions of the tractive properties of soil from its visual texture (classified through Haar wavelets) based on previously traversed visually classified terrain associated with vibration signatures [357]. Estimates of drawbar pull and slippage of the Kapvik rocker-bogie system have been computed using an unscented Kalman filter based on polynomial soil–traction models and measurements of wheel torque and normal wheel loads [359]. Attempts to extract wheel-based sensor data to measure changes in soil composition have been made in the context of water ice detection for in situ resource utilization [360]. Proposed wheel sensors include an impedance spectrometer to measure soil electric permittivity, a conductivity sensor to measure electric conductivity, an electrometer to measure electrostatic properties, and a magnetometer to measure the magnetic properties of ferrous minerals. The efficacy of such measurements from in-wheel sensors was unclear and the complexity overhead is unwarranted. A neural network–based method has also been developed which successfully extracts soil parameters (cohesion and friction angle) during traverses based on measured wheel torques, normal wheel loads, and slippage [362]. The wheel loading provides the key measurements that allow estimation of c and ϕ assuming constraints on other soil parameters n, k_c, and k_ϕ. This continuous measurement of the cohesion and friction angle of soil as a rover traverses may enable detection of water ice without dedicated in-wheel sensors.

5.8 SENSOR FUSION

Multisensor fusion involves merging information from multiple sensory modalities. It captures different characteristics of the environment allowing combinations of features for more accurate and robust classification [363]. For complex environments, it is desirable to integrate such sensory data simultaneously from diverse multiple sources. A distributed and diverse sensor suite enables the rover system to cope with failures gracefully due to functional redundancy. Sensor suites offer redundant information with different fidelities or offer complementary information of different modalities. Each sensor output is degraded by noise, so if multiple sensors are used to determine the same or a similar property, uncertainty is reduced. Multisensor data fusion improves the overall performance of interpretation and classification by providing more reliable and robust estimation of the environment than any sensor considered alone. It reduces sensory uncertainty by exploiting the redundancy in different sensory modalities. Sensory integration in humans follows a Bayesian process based on subjective probabilities (or fuzzy logic) which specifies how different pieces of evidence are combined [364]. Human multisensory fusion provides the basis for ventriloquism where the ventriloquist's voice appears to originate from the dummy to coincide with the movement of its lips (i.e., visual information is dominant over auditory information as visual information is more accurate and reliable). Several problems are immediately apparent in sensor fusion: (i) different sensors provide different physical transduction mechanisms, different locations, with different sensitivities to environments to which they are exposed, have different bandwidths, different spectral detectivity profiles, and differing data-processing rates; (ii) the data points of one sensor corresponding to data points on another sensor need to be registered; (iii) merging data from widely differing forms of representation into a unified representation medium. Sensor fusion can occur at different levels from the integration of raw data to the integration of the object properties of a world model [365]. A common central environment for integration of sensor data may be provided by a distributed blackboard architecture to facilitate high-level sensor fusion with each sensor module passing messages to the blackboard [366]. Blackboards are a problem-solving approach in which multiple knowledge bases utilize a global blackboard data structure for processing. However, it is apparent that human cognition integrates sensory data at a much lower hierarchical level [367]. Sensor fusion of vision and haptic perceptions in the human brain is performed through maximum likelihood integration which minimizes variance in the combined estimate [368]:

$$\hat{s} = \sum_i w_i \hat{s}_i \quad \text{where } w_i = \frac{1/\sigma_i^2}{\sum_i 1/\sigma_j^2}$$

$\hat{s}_i = f_i(s) =$ estimated variable, $f =$ neural transduction operator, and $\sigma_i =$ variance. Variance of the final estimate is less than either visual or haptic variances individually:

$$\sigma_{vh}^2 = \frac{\sigma_v^2 \sigma_h^2}{\sigma_v^2 + \sigma_h^2}$$

This is consistent with Bayesian approaches. Multisensor fusion can adopt a number of different methods—weighted average, Kalman filter, Bayesian estimation (Markov), Dempster–Shafer probability, fuzzy logic, or neural networks [369–373]. As well as numerical methods for dealing with uncertainty, there are also symbolic methods such as non-monotonic default logics [374]. If redundant data are in conflict, a mechanism for assigning confidence to each data value is needed. Confidence may be assigned to provide weighted averages with weightings given to each sensor value. The probability-based approach is founded on Bayes theorem to find likelihood ratios whose weights are determined by confidence values. The simplest fusion strategy involves combining raw sensor measurements of the same property obtained from multiple sensors (direct fusion). Before fusion can occur, consistency between the diverse data sources must be determined to check whether they represent the same physical entity. The Mahalanobis distance of the sample from the class mean can be determined from standard deviation. The Mahalanobis distance, which should be small for the same representative object, can determine this [375]:

$$d = \tfrac{1}{2}(x_i - x_j)^T C_k^{-1}(x_i - x_j)$$

where $x_k =$ sensor output and $C_k =$ diagonal variance–covariance matrix. The Mahalanobis distance will follow a χ^2 distribution. For two sensors with scalar outputs this reduces to:

$$d = \frac{(x_1 - x_2)^2}{\sqrt{\sigma_1^2 + \sigma_2^2}}$$

where $\sigma_i =$ standard deviation of sensor measurement. The simplest way to fuse data is to use the Bayes maximum likelihood ratio to model sensor uncertainty. Bayesian-based perception represents noisy sensory data as a conditional probability density function rather than a single value from each sensor allowing optimal estimation to occur during multisensory fusion [376]. Data from each sensor are represented as the degree of belief in an event [377]. Now, the probability of both events H and X occurring is given by:

$$p(H, X) = p(H)p(X|H) = p(X)p(H|X)$$

Bayes theorem assigns a probability measurement of the degree of belief on the basis of evidence. Bayes rule defines the posterior probability of hypothesis H occurring given event or evidence X:

$$p(H|X) = \frac{p(X|H)p(H)}{p(X)}$$

where $X =$ sensor output (evidence), $H =$ object property (hypothesis), $p(X|H) =$ conditional probability of output being X given property H is true, $p(H|X) =$ conditional probability of property being H given output X, $p(H) =$ prior probability of H being true, and $p(X) = \sum_{i=1}^{n} p(X|H)p(H) =$ probability of evidence X being true. Now, $p(X|H)$ is determined from the sensor model:

$$p(H|X) = \frac{p(X|H)}{\sum_{i=1}^{k} p(X|H)}$$

The best estimate of object property H using k sensor readings is given by the likelihood estimate with H chosen so that it is maximized:

$$p\left(\sum_{i=1}^{k} X_i | H\right) = \sum_{i=1}^{k} p(X_i|H)$$

As logarithms are easier to manipulate:

$$L(H) = \sum_{i=1}^{k} \log[p(X_i|H)] \quad \text{where} \quad p(X_i|H) = \frac{1}{\sqrt{(2\pi)^n C_i}} e^{-\frac{1}{2}(X_i - H)^T C_i^{-1}(X_i - H)}$$

Hence,

$$L(H) = \sum_{i=1}^{k} (-\tfrac{1}{2}\log[(2\pi)^n C_i] - \tfrac{1}{2}(X_i - H)^T C_i^{-1}(X_i - H))$$

The best estimate $\hat{H}$ is found by differentiating $L(H)$ with respect to H and equating to zero:

$$\hat{H} = \frac{\sum_{i=1}^{k} C_i^{-1} X_i}{\sum_{i=1}^{k} C_i^{-1}}$$

For two sensors with scalar output, this reduces to:

$$\hat{H} = \frac{\sigma_i^2 x_i + \sigma_j^2 x_j}{\sigma_i^2 + \sigma_j^2}$$

Bayes theorem may be used to assign confidence factors in expert systems by quantifying degrees of truth through probabilistic induction [378]. Mycin and Prospector were early expert systems that used certainty factors defining a numerical scale in the range $[0, 1]$. However, certainty factors were not probabilities as they did not satisfy the probability summation to unity (i.e., $p(H) + p(\sim H) = 1$) due to the inapplicability of this axiom with regard to evidence. In this case, belief B and disbelief D where $D + B \neq 1$ are defined by:

$$B = \frac{p(h|e) - p(h)}{1 - p(h)} \quad \text{and} \quad D = \frac{p(h) - p(h|e)}{p(h)}$$

The certainty factor was defined as:

$$CF = \frac{B - D}{1 - \min(B, D)}$$

The certainty factor is a measure of the association between premises and actions. The primary criticism of Bayes theorem stems from the fact that it cannot represent ignorance—if evidence is only partially in favor of an hypothesis, it must also partially support the negation of the hypothesis such that $p(H|E) + p(\sim H|E) = 1$. This is clearly counter-intuitive and contradictory. Any probability not assigned to evidence in favor of the truth of the proposition H is assigned to evidence in favor of $\sim H$.

An alternative approach is the Dempster–Shafer theory of evidence which addresses some of the problems inherent in Bayes theorem in sensor fusion [379–382]. It differentiates between evidence and plausibility and disbelief and unlike

Bayes theorem can represent ignorance. Sensor observations are treated as evidence to infer hypotheses about the world. The Dempster–Shafer theory considers uncertain evidence as representative of subjective ignorance. It allows a distinction between disbelief and ignorance and distinguishes between degrees of belief (based on evidence for a hypothesis) and plausibility (based on the measure of evidence that fails to refute the hypothesis). Evidence does not give a probability distribution over a set of hypotheses but is represented as a belief function. The belief function is based on a basic probability assignment. A frame of discernment h is a finite set of mutually exclusive hypotheses $H = \{H_i\}$. Parts of the belief may be distributed to any subset of hypotheses $H = \{H_1, \ldots, H_n\}$. The effect of evidence is defined by a basic probability assignment function m which assigns evidential weight to each subset of the frame of discernment:

$$m: 2^h \rightarrow [0, 1]$$

where h = set of exclusive atomic hypotheses (frame of discernment), 2^h = set of all subsets of h (power set of h), $m(\phi) = 0$ and $m(H_i) \geq 0$ for $H_i \subset H$, $H = \sum_{H \subseteq h} m(H) = 1$, and $m(H_i) = \Sigma p(H_i)/(1 - \Sigma p(H_i))$. This is the Dempster rule. The basic probability assignment (belief mass) m defined over the range $[0, 1]$ represents the strength of the evidence—$m = 0$ for the empty set and $m = 1$ when evidence is complete. This measures the belief that H exactly is true. The rest of the belief does not have to be assigned to the negation of the hypothesis, separating ignorance and disbelief (i.e., there is no probability distribution over the set of hypotheses). This distribution can be over any subset of H without assigning contradictory evidence to members of $\sim H$. Belief is the degree of support for a hypothesis. A belief of zero indicates no evidence supporting a hypothesis (not disbelief) while a belief of unity indicates absolute belief in the hypothesis. We define the uncertainty of H over the interval $[\text{Bel}(H), \text{Plaus}(H)]$. Belief represents the lower probability representing reasons to believe proposition H while plausibility represents an upper probability of the hypothesis H representing belief if all unknown facts supported H'. $m(H)$ represents the belief in hypothesis H such that $m(H) + m(\sim H) \leq 1$ where $m(\sim H)$ is the degree of ignorance. The belief function is defined on all proper subsets of A:

$$\text{Bel}: 2^h \rightarrow [0, 1] \quad \text{such that Bel}(H) = \sum_{H_i \subseteq H} m(H_i)$$

There is one-to-one correspondence between the belief function and the basic probability function. For an atomic hypothesis H, $\text{Bel}(H) = m(H)$. Plausibility describes the extent to which evidence does not support negation of the hypothesis. The plausibility function is defined as:

$$\text{Pl}(H) = \sum_{H_i \cap H = \phi} m(H_i) = 1 - \text{Bel}(\sim H)$$

where $\text{Bel}(\sim H) = \text{Dou}(H)$ and $\text{Bel}(H) \leq \text{Pl}(H)$. This is the measure of doubt in H—it indicates how much evidence there is for the hypothesis not to be true. True belief in H lies in the interval $[\text{Bel}, \text{Pl}]$. A belief of zero and plausibility of zero indicates there is no evidence available. The interval between $\text{Bel}(H)$ and $\text{Pl}(H)$ denotes the uncertainty about proposition H and true belief will lie in this interval. Should this uncertainty be zero, the Dempster–Shafer theory defaults to Bayes theorem. The commitment of a belief to a subset does not require the remaining belief to be committed to its negation (i.e., $\text{Bel}(H) + \text{Bel}(\sim H) < 1$). The amount of belief committed to a set cannot be subdivided into further subsets. Evidence is combined using Dempster's rule of combination which combines beliefs as a normalized sum of products of masses rather than Bayes rule.

$$\text{Bel}(H|E) = \frac{\text{Bel}(H \cup \bar{E})\text{Bel}(\bar{E})}{1 - \text{Bel}(\bar{E})} \quad \text{and} \quad \text{Pl}(H|E) = \frac{\text{Pl}(H \cap E)}{\text{Pl}(E)}$$

This allows revision of beliefs H given evidence E. The basic probability assignment is updated with new evidence to yield a new basic probability assignment. Any two belief functions A and B may be combined into a new belief function $C = A \cap B$ using Dempster's rule of combination which specifies the combined mass:

$$m(C_k) = (m_1 \oplus m_2) = \frac{\displaystyle\sum_{A_i \cap B_j = C_k \neq \phi} m(A_i)m(B_j)}{1 - \displaystyle\sum_{A_i \cap B_j = C_k = \phi}} \quad \text{and} \quad (m_1 \oplus m_2)\phi = 0$$

by definition. For two sensor outputs, uncertainty may be characterized as belief functions which may be combined using Dempster's rule of combination:

$$\text{Bel}(A_i B_j) = \frac{\text{Bel}(A_j)\text{Bel}(B_i)}{1 - \displaystyle\sum_u \sum_v \text{Bel}(A_u)\text{Bel}(B_v)}$$

with $A_u B_v = \Phi$. This represents the orthogonal sum of two belief functions. It is possible to measure the weight of conflict as a measure of disagreement between belief functions:

$$\text{Con}(\text{Bel}_1, \text{Bel}_2) = \log\left(\frac{1}{1 - \kappa}\right) \quad \text{where} \quad \sum_{A_i \cap B_j = \phi} m(A_i)m(B_j)$$

This gives an indication that one or more sensors have malfunctioned or the environment has changed. The Dempster–Shafer belief theory has a high computational complexity of $O(n^{2^m})$ where $2^m = $ number of basic probability assignments and $n = $ number of specific evidence parameters. This is higher than Bayes theorem computational complexity—for m hypotheses, there are 2^m possible combinations of hypotheses to be tested (i.e., computational complexity of $O(2^m)$). Furthermore, it is unclear whether Dempster–Shafer can detect certain inconsistencies.

The classical approach to classification defines crisp categories containing elements with common properties—this is modeled through set theory. The universe of discourse comprises two sets: those members of the set and those elements that lie outside the set—membership is all or nothing in the set $\{0, 1\}$ (i.e., $x \in A$ or $x \notin A$). However, human thinking is fuzzy—classes of objects in the real world have the property that the transition between classes is gradual rather than well defined (i.e., vague) [383–385]. Fuzzy set theory provides the basis of possibility theory (as opposed to probability theory) which attempts to model vagueness in human reasoning, though it has been criticized [386]. Fuzzy set theory is an approach to possibilistic logic based on linguistic variables for approximate reasoning. Fuzzy logic is a generalization of traditional logic which expresses natural language with its vagueness and non-statistical imprecision, whereas conventional sets possess precise properties for membership based on discrete truth values (1 or 0). Fuzzy truth values are multivalued in the real interval $[0, 1]$ represented by linguistic quantifiers but, rather than being expressed as a specific value, they may be expressed as a possibility distribution over the unit interval—a membership function $\mu(x)$ defining the truth value for the element in the fuzzy set. The membership function encapsulates graduations of truth. Fuzzy membership represents similarities between ill-defined properties while probabilities represent statistical relative frequencies. Fuzzy sets allow formal set-theoretic concepts with partial degrees of truth and abide by the same rules as crisp sets:

(i) Equality: $A = B \leftrightarrow \mu_A(x) = \mu_B(x)$
(ii) Containment: $A \subset B \leftrightarrow \mu_A(x) \leq \mu_B(x)$
(iii) Complement (not-A): $\mu_{\sim A}(x) = 1 - \mu_A(x)$

Fuzzy sets are not considered mutually exclusive and therefore resemble linguistic measures. The intersection (conjunction) of fuzzy sets is represented by $A \cap B = \min(A, B)$:

$$\mu_{A \cap B}(x) = \min(\mu_A(x), \mu_B(x))$$

this may be generalized to the T-operator. This intersection is equivalent to an AND connective such that $x \in A$ and $x \in B$. Similarly, the union of fuzzy sets has membership function defined by:

$$\mu_{A \cup B}(x) = \min(\mu_A(x), \mu_B(x))$$

this may be generalized to the S-operator. This union is equivalent to an OR connective such that $x \in A$ or $x \in B$. The union of a fuzzy set and its component does not necessarily equal unity—this disallows the use of a yes/no dichotomy unlike the case in subjective probabilities. The law of the undistributed middle that every statement is either true or false is disallowed. However, the laws of distribution and association apply. For two fuzzy sets A and B, the following axiom holds:

$$\mu(A \cup B) = \mu(A) + \mu(B) - \mu(A \cap B)$$

Fuzzy sets reflect our vague categorizations used in linguistic labels: "tall", "slow", "cold", etc. These linguistic variables are used relative to their context and have undefined boundaries. Fuzzy sets describe uncertainty as fuzzy linguistic variables—very low, low, medium, high, very high, etc.—rather than as probabilities. As a many-valued logic, a fuzzy set assumes subjective degrees of truth—very true, true, rather true, more or less, not very true, slightly true, etc.: very true $=$ true^2, not-true $=$ true$'$, false $p =$ true-p, more or less true $= \sqrt{\text{true}}$. The use of "quite" true is ambiguous as it can mean "fairly" or "absolutely" dependent on its context. A fuzzy proposition contains such fuzzy adjectives and adverbs and defines them as fuzzy sets. There may be additional fuzzy qualifiers such as "very", "extremely", "quite", etc. and fuzzy quantifiers such as "most", "few", "normally", etc. Fuzzy quantifiers enable information to be communicated in a generalized way and are closely allied to linguistic variables. Fuzzy sets provide for degrees of truth through the degree of membership μ that represents uncertainty in the interval $[0, 1]$: $\bar{\mu}_{i+1} = \lambda \bar{\mu}_i + (1 - \lambda)\mu_i$ where $0 \le \lambda \le 1$ and $\mu_A(x) = 1$ if $x \in A$ or 0 otherwise. Fuzzy measures incorporate belief, plausibility, possibility, probability, etc. In this way, it can accommodate possible worlds similar to modal and default logics. For a set of possible worlds W, there is a possibility distribution on W with a possibility function $\pi: W \rightarrow [0, 1]$. The membership function represents the partial membership of a set to characterize vague concepts. If $\mu(x) = 1$ then $x \in A$ and if $\mu(x) = 0$ then $x \notin A$ with a gradual range of membership values existing between these extremes. In traditional classical sets, the membership of an element is either unity (x is a member of the set) or nil (x is not a member of the set). Classical sets can partition classifications into mutually exclusive classes (e.g., for human height, "short" is defined by <1.62 m, "average" by 1.62–1.78 m, and "tall" by >1.78 m). Each class is separated by a step function representing a discontinuity. In fuzzy sets, the membership function allows intermediate degrees of truth (or beliefs). The degree of membership is determined by adverbial hedges (modifiers) (e.g., tiny, small, medium, large, huge, etc. when describing size). The degree of membership may be based on the degree of prototypicality of a given category [387]. This allows a more blended classification in which classes overlap with each other. Fuzzy membership functions may have many different shapes—triangular, trapezoidal, sinusoidal, Gaussian, generalized bell, or sigmoidal. The membership function is commonly triangular with overlap between classes. The triangular function has the form:

$$\mu(x) = \max\left(\min\left(\frac{x - a}{b - a}, \frac{c - x}{c - b}\right), 0\right)$$

The triangular-shaped function provides the basis for fuzzification and defuzzification of fuzzy variables with real numbers. The fuzzy number 2 may be specified (Figure 5.21): nearly $2 = 2 \pm 5\%$, about $2 = 2 \pm 10\%$, roughly $2 = 2 \pm 25\%$, or crudely $2 = 2 \pm 50\%$.

Other common fuzzy set shapes are trapezoidal, piecewise linear, Gaussian curve, S-curve, etc. The trapezoidal function is given by:

$$\mu(x) = \max\left(\min\left(\frac{x - a}{b - a}, 1, \frac{d - x}{d - c}\right), 0\right)$$

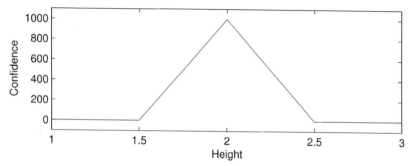

Figure 5.21. Fuzzy number 2.

The bell function has the form

$$\mu(x) = \frac{1}{1 + \left[\left(\dfrac{x - \bar{x}}{a}\right)^2\right]^b}$$

where $\bar{x}$ = membership function centerpoint, a = halfwidth, b = gradient at crossover point. A normalized Gaussian membership function has the form

$$\mu(x) = a \exp\left(-\frac{x - \bar{x}}{2\sigma^2}\right)$$

This correlates with Shepard's universal generalization function which is concave and has an exponential drop-off with distance [388, 389]. The sigmoidal/logistic function is given by:

$$\mu(x) = \left(1 + e^{-m(x-c)}\right)^{-1}$$

where c = crossover point and m = slope at c.

Fuzzy sets have been applied to control systems [390–392], but are particularly suited to representing inadequate knowledge of expert systems [396–398]. Fuzzy rules of inference provide for deductions similar to standard logical inferencing ($A \rightarrow B$ or $\sim A \cup B$) but use linguistic variables defined by the membership function. This may be illustrated by the concept of the fuzzy relation (composition) between two sets A and B which link them—a relation is a subset of a continuous product:

$$\pi R(x, y) = A(x) \times B(y)$$

this is the Cartesian product such that

$$\pi_B(y) = \max[\min(\pi R(x, y), A(x))]$$

where (x, y) form an ordered pair of elements. Fuzzy if–then rules have an antecedent–consequent form: "if A is X then B is Y" where membership functions quantify the propositions. For example, if velocity is high then force is very high where velocity and force are linguistic variables and (very) high are linguistic values determined by the membership function. Propositions in fuzzy rules can be compounded in the form: if A_1 is X_1 and A_2 is X_2, etc. then B is Y; if A_1 is X_1 or A_2 is X_2, etc. then B is Y. Multiple sets of fuzzy if–then rules constitute a fuzzy rule base. A fuzzy model comprises a set of if–then rules, each mapping a specific input–output behavior. A_i techniques may be incorporated into controllers commonly in the form of if–then production rules in which matching sensory input invokes actions dictated by the rules [399]. The fuzzy controller comprises a set of fuzzy rules determining the control action to be taken under different process states. The fuzzy rules encapsulate expert knowledge and have a fuzzy inferencing form thus: "if X_1 is A_1 and X_2 is A_2 and … then Y is B" where A_i, B are linguistic variables comprising part of the fuzzy sets. Sets of such fuzzy rules constitute a fuzzy rule base in which the fuzzy rules are equated with certainty factors. This illustrates the linguistic interpretation of fuzzy sets. Control involves the use of linguistic rules such as "if error is large and positive, then response is large and negative" and vice versa. These rules act as input–output functions. Fuzzy logic controllers have been used to model control rules which are defuzzified to define numerical values. In robotic navigation, each cell of the rectangular grid may be defined by a Gaussian membership function of the form:

$$\mu_i = e^{-|x-p_{x,y}|^2/\sigma_{x,y}^2}$$

where $p_{x,y}$ = center of membership function and $\sigma_{x,y}$ = width of membership function. Typically, fuzzification involves partitioning the variable into a number of fuzzy sets (e.g., far left, medium left, close left, ahead, close right, medium right, and far right constituting the inputs for an obstacle in the visual field). This allows the creation of fuzzy rule sets such as "if close left or ahead then turn right quickly"; "if medium right then turn left slowly", etc. The outputs of fuzzy if–then rules are fuzzy actuator actions. As crisp control signals are required, defuzzification of fuzzy if–then rules into crisp values is necessary. The inputs and outputs to a fuzzy controller are crisp values and there are two general approaches to fuzzy control [400, 401]. In the Mamdani inferencing system, fuzzy rules have the form:

$$\text{If } x_1 \text{ is } A_1 \text{ and } … \text{ and } x_n \text{ is } A_n \text{ then } y \text{ is } B$$

The membership function for the implication (AND) is defined by:

$$\mu_{A \to B} = \mu_A \cdot \mu_B = \min[\mu_A, \mu_B]$$

where $A = A_1 \times \cdots \times A_n$. All active fuzzy rules are summed (OR/maximum):

$$\mu_C = \sum_{j=1}^{m} \mu_{A \to B}(j)$$

Hence, the Mamdani inference applies min and max operators for fuzzy AND (intersection) and OR (union) operators. Mamdani fuzzy inference is based on linguistic variables (rough–smooth) and the max–min inference operator while Takagi–Sugeno fuzzy inference uses (non)-linear mathematical functions which may be considered to be universal approximators. In Takagi–Sugeno inferencing systems, fuzzy rules have fuzzy antecedents and a crisp output:

$$\text{If } x_1 \text{ is } A_1 \text{ and } \ldots \text{ and } x_n \text{ is } A_n \text{ then } y = f(x_i)$$

A defuzzifier maps fuzzy sets into real crisp points. Deffuzification usually proceeds by computing the area under the membership function and then the fuzzy centroid of the area. The weighted strength of each output membership function is multiplied by their respective output membership function center points and summed. This area is divided by the sum of weighted membership function strengths to give a crisp output. Defuzzification is computed as a weighted average:

$$x_{i+1} = \frac{w_1 x_1 + \cdots + w_n x_n}{w_1 + \cdots + w_n}$$

where $w_1 = \mu_1 = $ weights and $x_i = $ center of triangular membership function. The commonest approach to defuzzification is through the center of gravity (centroid—ratio of moment to area) method due to its similarity to expectation values in probability distribuions given by:

$$C = \frac{\sum\limits_{i=1}^{n} x_i \int \mu_i}{\sum\limits_{i=1}^{n} \int \mu_i}$$

where $\int \mu_i = $ area of membership function. Alternatively, a combined mean of the maximum and center of area defuzzification can be used:

$$W = \frac{\sum\limits_{j=1}^{n} \mu_j \left(\sum\limits_{i=1}^{k} w_i / k \right)}{\sum\limits_{i=1}^{n} \mu_j}$$

where $\mu_i = $ maximum membership function of each set and $w_i = $ weight value of each set. The use of the Takagi–Sugeno model significantly reduces the number of fuzzy rules required to represent the fuzzy knowledge base. The Sugeno method converts linguistic membership values into a single real value. It has been shown that Takagi–Sugeno fuzzy models are equivalent to autoregressive models and IIR/FIR filters, in particular [402]. Takagi–Sugeno fuzzy systems generally model the plant as a weighted sum of linear state equations of the form:

$$\text{IF } x_1 \text{ is } a_1, \text{ and } \ldots, \text{ and } x_n \text{ is } a_n, \text{ etc.}$$

$$\text{THEN } \dot{x} = A_i x + B_i u_i$$

where $x^T = (x_1, \ldots, x_n)$, $u^T = (u_1, \ldots, u_n)$, and $a_i =$ fuzzy sets. For open-loop systems, the antecedent becomes

$$\dot{x} = a_{0i}x_0 + \cdots + a_{ni}x_n$$

Takagi–Sugeno fuzzy rules involve fuzzy sets only in the premise part of the rule (e.g., if velocity is high then force $= k^*v^2$). The zero-order Takagi–Sugeno model where $f(x) =$ constant is a special case of the Mamdani fuzzy model. The membership function takes the mean of all central values and weights them by the height of the membership function:

$$\mu = \frac{\sum\limits_{i=1}^{n} c_i w_i f_i}{\sum\limits_{i=1}^{n} w_i f_i}$$

where $f_i =$ height, $c_i =$ peak value of domain element, and $w_i =$ weighting factor. The final output for a pair (x, u) is given by:

$$\dot{x} = \frac{\sum w_i (A_i x + B_i u)}{\sum w_i}$$

where $w_i = \prod \mu_j(x)$. The chief advantage of the Takagi–Sugeno formulation is that it merges universal approximation with symbolic if–then rules making it suitable for integrated planning and control in mobile robots [403–405]. However, Takagi–Sugeno rules are difficult to construct for a large number of design parameters. The use of fuzzy logic in artificial intelligence systems is also well established. In this case, fuzzy logic representation is characterized by fuzzy predicates, fuzzy quantifiers, and approximate reasoning processes with numerical membership functions. Fuzzy reasoning provides the basis of fuzzy inferencing through fuzzy "if–then" rules [406, 407]. Fuzzy logic can be utilized directly through fuzzy modus ponens which is central to fuzzy reasoning. The classical logic version of modus ponens has the form:

$$\begin{array}{ccc} A & \to & B \\ \underline{A} & & \\ & & B \end{array}$$

Fuzzy modus ponens is a generalization of classical modus ponens. Fuzzy reasoning attempts to emulate human reasoning of the antecedent–consequent form (the fuzzy ponens rule being common):

  If x is A then y is B where $A, B =$ fuzzy sets

    Or $A \xrightarrow{m} B$ where $m =$ logical quantifier that suggests A implies B
        to some extent

We shall not consider this further. A neuro-fuzzy vehicle controller whose fuzzy if–then control rules (concerning fuzzy robot motor outputs based on fuzzy distance measurements) and their membership functions were evolved according to a standard fitness function of the form

$$F = v(1 - \sqrt{\Delta v})(1 - s)$$

where v = wheel rotation speed, Δv = left/right wheel speed difference, and s = value of maximum proximity sensor [408]. This demonstrates the use of fuzzy logic in robotic vehicle controllers.

Fuzzy sets deal with possibilities generated from imprecise past or current events while probability deals with the random nature of future events [409]. More specifically, probability theory deals with precise sets to describe imprecise events, while possibility theory deals with imprecise sets to describe precise events. Hence, the degree of possibility may not necessarily correlate with the same degree of probability. Hence, if a variable x can take values $u_1,\ldots,u_n$ with respective possibilities $\pi = (\pi_1,\ldots,\pi_n)$ and probabilities $p = (p_1,\ldots,p_n)$, the degree of consistency of probability and possibility distributions may be given by $\gamma = \pi_1 p_1 + \cdots + \pi_n p_n$. However, it is not clear that the membership function cannot be described as a probability [410]. Fuzzy quantifiers such as "certain", "very likely", and "always" correlate to probabilities in the range 0.95–1.0 in most interpretations. The probability of fuzzy event A may be defined by the Lebesgue–Stieltjes integral:

$$P(A) = \int \mu_A(x) \cdot dP$$

However, determination of the appropriate membership function from the probability function is non-trivial. The degree of membership permits $\mu(A) + \mu(\sim A) \leq 1$. This represents indecision and such indecision may be compensated for through Dempster–Shafer plausibility which satisfies: $\text{Plaus}(\sim A) = I - \mu(A)$ and $\text{Plaus}(A) = I - \mu(\sim A)$ such that $\text{Plaus}(A) \geq \mu(A)$, $\text{Plaus}(\sim A) \geq \mu(\sim A)$, and $\text{Plaus}(A) + \text{Plaus}(\sim A) \geq 1$. Plausibility is greater than belief for the same evidence so is weaker than belief. There is the problem that assigning a possibility of 1 to an event does not have clear meaning—whether this event is certain or still only possible [411].

More complex fusion strategies involve such techniques as weighted least squares fit and the recursive Kalman filter. The Kalman filter naturally fuses together measurements from different instruments at different rates including sensor dropouts. Bayesian methods are a generalization of the Kalman filter which is a highly efficient type of Bayesian estimate. Measurement is usually modeled as Gaussian probability distribution requiring estimates of the mean and the variance–covariance matrix. The recursive Kalman filter estimation algorithm can be used to merge data from multiple images into a composite by assuming Gaussian errors [412, 413].

6

Rover vision—fundamentals

No single modality sensor can provide sufficient data to extract all the relevant features of the environment but vision is the most information-rich modality. It is the primary sensory modality for planetary rovers in providing distance observation for navigation and obstacle avoidance. Furthermore, it is the most information-rich form of sensory data—in autonomous rovers it is the primary means of generating maps of the locality representing distance information about the external world. Stereovision is required for the recognition of objects, the determination of their positions and orientations in space, and for visual servoing. Vision also provides the basis for scientific analysis by the science team and as the first step towards autonomous science. Natural environments such as those on planetary surfaces are unstructured making image processing more difficult. Furthermore, sensor data are always corrupted by noise and characterized by limited observability.

6.1 GENERAL CONSIDERATIONS FOR ROVER CAMERA SYSTEMS

Rover cameras serve three primary functions—to support navigation during traverse, to support the deployment of sample acquisition devices and scientific instruments, and general scientific observations. Indeed, it is the primary instrument on board the rover. Such cameras are often mounted on telescopic masts with pan–tilt motion capabilities which are a type of manipulator (from a control perspective). Viking carried a stereoscopic pair of cameras mounted 1 m apart each on a separate mast 1.3 m above the ground with pan–tilt rotation of 360° and −40° to 60°, respectively. All planetary rovers carry a stereoscopic panoramic multispectral camera system (PanCam), typically mounted on a mast of 1.5–2 m in height ideally: a 1 m height gives visibility to the local horizon of 1.8 km while a 2 m high mast provides a horizon at 3.5 km distance (for Mars). PanCam fulfills

© Springer-Verlag Berlin Heidelberg 2016
A. Ellery, *Planetary Rovers*, Springer Praxis Books,
DOI 10.1007/978-3-642-03259-2_6

the first mission goal in generating a 360° panoramic survey view of the landing site. This immediately implies the need for a pan–tilt mechanism. Color imaging is essential as color provides a good approximation to reflectance and allows differentiation between obstacles and shadows. Detecting albedo changes provides the basis of the retinex theory of color vision whereby color is estimated under a wide variety of intensities by using surface reflectance at different wavelength distributions. Color imaging is better for accurate terrain perception. Multispectral scanners estimate surface color by recovering albedo independently in three primary color spectral bands. The standard photovisual filter is usually RGB which ranges across visible/near-UV regimes offering tripeak (red, green, and blue) maximum responses at 3,650, 4,400, and 5,500 A. For ultraviolet wavelengths (900–3,000 A), LiF coatings are used. For nanorovers, PanCam is typically lander-mounted which severely limits the range of the rover to <10 m typically. The panoramic survey mosaic constructed from a combination of different views (Figure 6.1) lends itself to appearance-based matching for "homing" behaviors but this has yet to be implemented on planetary rovers [414]. Although not useful for robotic exploration robots, this may have utility in robotic vehicles that undertake construction tasks where homing periodically to the same location is required. In this approach, the environment is memorized as a sequence of small local view snapshots defined around maximum intensity values along the horizontal axis. The set of all these local views defines a place in the environment. In homing back to the home location, the robot subsequently compares the currently observed image with stored snapshot images and matched using a correlation procedure. Once the best match image is selected representing the "home" place, a displacement between the template image and the current image is computed.

The rover camera system should have low spatial and spectral distortion. There is conflict between cameras tasked with scientific observations and those tasked with navigation functions. The latter require a high field of view (FOV). Distortion occurs for a large FOV (e.g., a 65° FOV gives a distortion of 0.1–1.0%). The MER HazCam had a very high distortion due to its fish-eye lens configuration of the objective (Figure 6.2). Scientific observations have a much reduced FOV to minimize distortion to geological images. A narrower FOV generally implies higher resolution. Cameras construct views through mosaics of individual images such as in the creation of panoramic images (Figure 6.3). Wide-angle images reduce the number of images required to construct a 360° panoramic mosaic.

Most rover vision systems have been based on solid state CCD (charge coupled device) or CID (charge injection device) technology though CMOS (complementary metal oxide semiconductor)-based APS (active pixel sensor) technology is becoming available. CCD imagers are arrays of analog shift registers in which the electronic charge accumulates in proportion to incident light intensity. They are integrating detectors where electrons are generated from the flux of photons. These electrons accumulate in a CMOS pixel and their chief advantage is the direct readout electronics generally in the form of a FET (field effect transistor). The electronic charge is trapped in potential wells (acting as capacitors) produced

Figure 6.1. Panoramic surveys of (i) Pathfinder, (ii) Spirit, (iii) Spirit, (iv) Opportunity, and (v) Curiosity [credit NASA JPL].

by electrodes insulated from each other by highly doped material. CCD arrays are fast and robust with high geometric fidelity and good resolution. They do not suffer appreciably from distortion, blooming, or electronic drift. However, if a CCD potential well becomes saturated, then additional light flux causes charge leakage into neighboring pixels—this is blooming. They can be constructed with low-mass, small-volume, and low-power consumption offering high signal-to-noise ratios with efficiencies close to 100%. Planetary landers have traditionally used line-by-line raster scanning for image construction but the mobility introduced by planetary rovers dictates against this. Most CCD cameras comprise $1,024 \times 1,024$ pixels. The most demanding real time vision tasks require video image frame rates of 25–30 Hz (25 Hz for European CCIR standard and 30 Hz for American EIA standard) but this requires high-performance dedicated framegrabber hardware which is not currently feasible. This can be relaxed considerably for planetary

Figure 6.2. Linearization radiometric correction and coloring of distortion on MER HazCam image [credit NASA JPL].

Figure 6.3. Opportunity's mosaic imaging of El Capitán indicating aqueous processes [credit NASA JPL].

rover imaging systems due to the slow speed of planetary rovers and limitations on the computational processing resources on board.

APSs are also based on CMOS technology but with on-chip circuitry for real time signal conditioning such as low-level image processing, control logic, calibration, and interfacing—smart camera systems of photodiodes similar to the human retina [415]. CMOS technology introduces pixel-specific transistors to each pixel to measure and amplify pixel signals. This provides for parallel operation circumventing the tardiness of serial shift operations in CCDs. They offer low-voltage operation, low-power consumption, compatibility with integrated on-chip electronics, random access of image data. Such systems have reached 0.5–0.35 μm resolution technology. Their performance is not quite that of CCDs with higher noise but their lower power is attractive for planetary rover applications offering high frame and readout rates. This is enabled by the integration of pixel readout, ADC, and dual-port SRAM onto a single chip. Readout of each row can be performed while ADC is performed on the next row. A $1,024 \times 1,024$ $8\,\mu m^2$ pixel CMOS APS can achieve throughputs of 60 frames/s with low power consumption of 250 mW at a clock speed of 74 MHz [416]. They are currently in advanced stages of development.

For the purpose of rover navigation, there are a number of specific vision requirements recommended [417]:

(i) stereo cameras should be wide angle and have a minimal resolution of 256×256 pixels coded in at least 8 bits per pixel—currently, PanCams have a standard resolution of $1,024 \times 1,024$ pixels with 16 bits per pixel;

(ii) stereo cameras should be located on top of a mast assembly at a height h above the ground where $h > 1.5\,m$ (preferably up to 2 m) to give the maximum view behind obstacles and a range to 10 m—a 1 m height gives a local horizon of 1.8 km on Mars not accounting for topographical features;

(iii) the stereo camera baseline should be around 0.5 m to give a 10 mm position accuracy at 10 m range;

(iv) stereo camera orientation should be adjustable along pan and tilt axes (180° to −180° pan and 0–90° tilt);

(v) stereo cameras should be located as close to the front of the rover along its longitudinal axis to minimize obscuring the front of the rover;

(vi) (optional) hazard avoidance cameras should be placed on the front side of the body of the rover to emergency-stop the rover in case of obstruction;

(vii) (optional) hazard avoidance cameras should have a minimal resolution of 128 × 128 pixels coded in at least 8 bits due to their short range;

(viii) the localization system should include three-axis attitude angles of the rover with an accuracy of 1° to control camera pan–tilt;

(ix) volatile (RAM) memory allocation for rover navigation software is >3 MB but 4–5 MB would allow additional software enhancement;

(x) non-volatile memory for executable rover navigation code is >100 kB.

Direct sunlight saturates camera pixels while reflected sunlight from metallic surfaces causes glare which must be avoided. Furthermore, the sunlight vector generates sharp shadowing which can make the edge extraction of features and physical properties such as centroids difficult. For the Kapvik microrover, it was considered that the autonomous avigation system should have a 120° visual field with a look-ahead radius of 4–5 m at a spatial resolution of 2 cm.

6.2   EXAMPLE CAMERA SYSTEMS

The failed Beagle 2 lander carried a color stereo camera pair similar to PanCam on the Pathfinder lander (Figure 6.4)—each was a 1,024 × 1,024-pixel CCD camera with integrated electronics including 8 MB local memory storage [418, 419]. The effective aperture was $f/18$ and focal length was 20 mm with a resolution of 1 mrad/pixel. Each camera had a 48 × 48° FOV and was mounted with a 209 mm baseline and 3.7° angle toe-in for a focus at 1.2 m at the end of the robotic arm. The pan–tilt mast had 360° pan and 68° tilt capability. A wide-angle mirror allowed viewing of the Martian surface before the robotic arm was deployed. An LED flashlight was mounted on each camera (total mass of 2.47 g) for general illumination. All optical components comprised BK7 silica glass. Each camera had an integrated stepper motor–driven filter wheel with 24 × 10 mm diameter filters— 18 for geological imaging in the red/infrared spectrum and 4 for RGB color (equivalent to human photometric response) across the spectral range 440–1,000 nm. Digital elevation map (DEM) construction with 670 nm filtering has an accuracy proportional to the stereo baseline and inversely proportional to a quadratic function of range. In addition, a stereo microcamera pair was mounted on the robot arm with a 195 mm baseline and a 20 mm focal length. The total camera system had a mass of 360 g with each camera contributing 52.5 g plus 1.4 g of optics, 89 g of electronics with drive mechanism, wiper blades, wiring, and

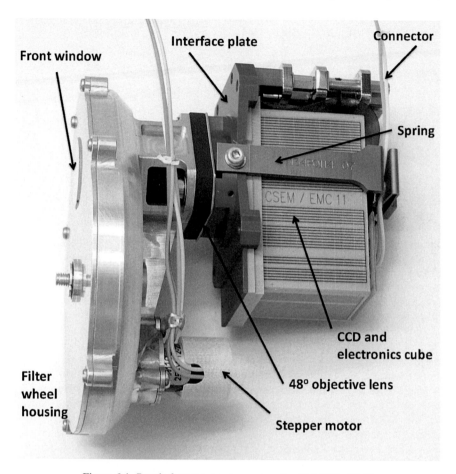

Figure 6.4. Beagle 2 stereo camera system [credit MSSL–ESA].

mounting comprising 19.5 g. The camera was rated to 100°C and had an average
power consumption of 1.8 W. A 10 Mbps RS-422 bus connected each camera head
to the common electronics in the lander base. Wavelet encoding was tasked with
reducing a 10 Mb image by 32:1 times prior to transmission.

Sojourner carried three cameras—a forward-pointing hazard camera pair and
a rear instrument camera—but its main navigation panoramic stereo camera pair
resided on the Pathfinder lander on a telescopic mast. The 0.5 kg Mars Pathfinder
camera pair mounted atop a 1.4 m pan–tilt imaging mast combined rover navigation and scientific imaging roles. It used 12 filters to provide a spectral range of
0.45–1.0 μ for geological observations. Sojourner itself used two forward-pointing
black-and-white cameras to provide stereoscopic imaging for navigation while the

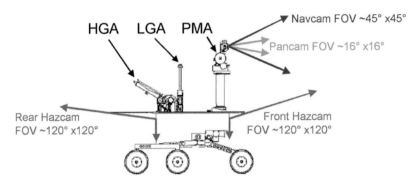

Figure 6.5. MER camera configuration [credit NASA JPL]. HGA = high-gain antenna; LGA = low-gain antenna; and PMA = panoramic mast assembly.

RGB (red-green-blue) color camera at the rear of the vehicle provided spectral information on surface geology. Each of Sojourner's CCD cameras was a 768×484-pixel imager with raster scanning acquisition due to the limitations of its onboard 80C85 processor. Each camera had a lens focal length of 4 mm, a stereoscopic baseline of 12.56 cm, and an angular resolution of 0.001 rad/pixel. The FOV for the forward cameras was $127.6 \times 94.5°$ and for the aft cameras it was $94.5 \times 127.5°$. It was limited to operation within a 10 m range of the lander and gave navigation support to the Sojourner microrover, illustrating the limitations of lander-based cameras. As well as line-of-sight restrictions, the differing viewing angle with range impose significant gaps in the image field.

The MERs carried nine cameras, of which eight were operated in stereo pairs which were used to generate range and 3D multiresolution terrain maps (Figure 6.5). The MERs each possessed two mast-mounted black-and-white NavCams for long-range imaging and four body-mounted black-and-white HazCams (two forward and two rear) for imaging within 3 m of the rover. For the mast-mounted color panoramic science camera pair, each had a mass of 270 g and 12 filters. The MERs were capable of monochromatic microscopic imaging. Also mounted on the mast was a 2.1 kg thermal emission spectrometer to survey rocks at a distance with the aim of selecting rocks for more detailed analysis. It could identify water and water-lain minerals such as carbonates. It could also perform atmospheric analysis of water vapor and generate temperature profiles up to 10 km altitude. All nine imaging cameras on board the MERs plus the descent camera used the same design and electronics but different optics—they were almost identical $1,024 \times 1,024$ CCD detector arrays of 12 μm^2 pixels with 12-bit grayscale images. Each camera had a mass <300 g and power consumption <3 W [420, 421]. Each camera consisted of a detector head (comprising a CCD array with supporting optics and an electronics box comprising CCD driver electronics), 12-bit ADC, rover interface electronics, and resistance heater to maintain a minimum temperature of $-55°$C. An Actel RT 1280 field-programmable gate array (FPGA) communicated with the rover electronics through a high-speed serial low-voltage differential signal (LVDS) interface. CCDs were operated in frame transfer mode between the photosensitive array from which the image was shifted into an image buffer storage region with a frame transfer rate of 5.1 ms with readout noise less than 50 electrons. Readout from the storage region took 5.4 s due to digitizing by a 12-bit ADC at a rate of 200,000 pixels/s.

The MERs sported a 1.54 m mast with 4 DOF for its PanCam and navigation cameras (NavCam) [422] (Figure 6.6). The pan–tilt mast was capable of 370° azimuth (pan) and 194° elevation (tilt) motion with a pointing accuracy of 0.1° for generating panoramic images out to 3.2 km horizon (assuming a spherical Mars) comprised of 10 stereo pairs spaced by 36° (Figure 6.6). All actuators for the mast were Maxon RE020 brush motors equipped with heaters for warming under cold conditions. The angular resolution of 0.28 mrad/pixel was equivalent to human 20/20 visual acuity thereby defining its role in supporting MER's role as a robotic geologist. This enabled obstacles of 20–25 cm to be detected over 5–8 pixels out to the maximum 100 m/sol range.

The MER PanCam was a color stereo camera pair with a 12 cm baseline, an 18° FOV, and an angular resolution of 0.28 mrad used for solar imaging and for science exploration while static [423]. It employed a filter wheel for multispectral imaging including RGB filters to emulate human photometric response. PanCam was used as a Sun sensor to determine the rover heading and was run prior and posterior to the navigation process. A high-resolution camera with a FOV of 3.5° was used for zoom functions. PanCam had a long focal length of 38 mm compared with the NavCam of 14.67 mm making it more suitable for more distant objects.

The MER NavCam was a black-and-white stereo pair with a 20 cm baseline and 45° FOV with lower angular resolution of 0.82 mrad/pixel mounted on the pan–tilt assembly. It was an $f/12$ camera pair with a focal length of 14.67 mm and a depth of field ranging from 0.5 m to infinity. The MERs also carried an under-belly camera with a $110 \times 90°$ FOV for imaging the deployment of the mini-corer instrument into Martian soil and two pairs of body-mounted hazard cameras (HazCam) with a 124° FOV, angular resolution of 2.1 mrad/pixel, and 10 cm base-line on the front and back for hazard avoidance. HazCams are $f/15$ cameras with a focal length of 5.58 mm and fish-eye lenses. The front HazCam boresight was pointed at an angle of 45° below the horizon while the rear HazCam boresight was pointed at an angle of 35° below the horizon. They were rigidly mounted 53 cm above the ground plane on the front and back sides. Their short focal length of 5.58 mm focused onto nearby objects. Images were acquired through a single Capture_Image command. A single image had an uncompressed size of $1,024 \times 1,024 \times 12 = 1.5$ MB. The flash memory of 256 MB capacity limited data storage to ∼130 images. Compression using JPL ICER wavelet image compression (similar to JPEG) gave compression of 48:1 on 12-bit images. The NavCams and HazCams had the same spectral sensitivity of 600–800 nm. The NavCam was designed to track terrain features to estimate rover motion while the 256×256-pixel HazCams were designed for local obstacle avoidance with a 3 m range. HazCams could detect ranges to accuracies better than 50 cm at a distance of 10 m while NavCams had an accuracy of 10 cm at 10 m distance by virtue of their

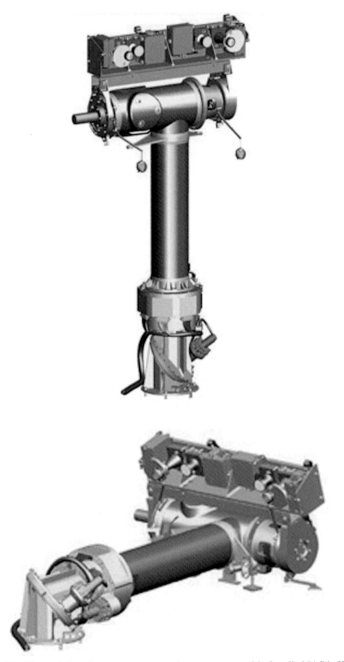

Figure 6.6. Mars exploration rover panoramic camera assembly [credit NASA JPL].

higher angular resolution. HazCams were used as the primary navigation cameras on Spirit at the Gusev crater as this was highly rock strewn. Images were acquired every 0.5–1.5 m look-ahead distance. They were also used on both rovers for planning the deployment of the 5 DOF instrument deployment device (robotic arm) for emplacement of the scientific instruments. However, Opportunity at Meridiani Planum, which had very few obstacles, used its more limited FOV NavCams as the primary navigation cameras. As more confidence in MER traverses was gained, short-range traverse planning was superseded by long-range traverse planning up to 100 m ahead by ground personnel identifying obstacle-free paths beyond which autonomous obstacle avoidance was implemented extending traverses up to 300–400 m per sol.

The ExoMars rover PanCam was baselined on the Beagle 2 stereo camera pair to perform both science camera and navigation camera functions [424]. Its main scientific objectives are to:

(i) provide self-localization of the landing site and rover position during rover traverse;
(ii) provide multispectral geological maps of the locality;
(iii) provide atmospheric data;
(iv) provide contextual support of other Pasteur scientific instruments.

The ExoMars PanCam is to be mast-mounted to a height of up to 2 m (Figure 6.7). Each 182 g wide-angle stereo camera (WAC) is based on CCD technology with 1,024 × 1,024 pixels, each encoded in 10 bits. Each camera has an FOV of 69 × 55° (1.1 mrad/pixel)—this is much larger than the FOV of the MER PanCam whose FOV was 18° and even larger than the FOV of the MER NavCam of 45°. The 65° (average) FOV allows panoramas with six image positions with 60° azimuth segments with a 5° image overlap. A 65° FOV gives a 25 mm resolution at 20 m. This wide FOV requirement was driven by the requirement to minimize image data volume to be returned to Earth. However, this large FOV makes it much more suitable for navigation functions as it is specifically designed for both digital terrain mapping and multispectral geological imaging. Its focal length of

Table 6.1. MER camera family.

MER camera pair	Field of view (°)	Resolution	Angular resolution (mrad/pixel)	Focal length (mm)	Baseline (cm)	Stereo range (m)	Altitude (cm)
NavCam	45	1,024 × 1,024	0.77	14.67	20	2–20	152
PanCam	18	1,024 × 1,024	0.27	43	28	4–70	152
HazCam	120	256 × 256	2.2		10	0.5–5	53

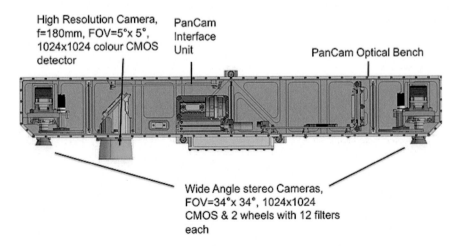

High Resolution Camera,   PanCam
f=180mm, FOV=5°x 5°,   Interface
1024x1024 colour CMOS   Unit
detector   PanCam Optical Bench

Wide Angle stereo Cameras,
FOV=34°x 34°, 1024x1024
CMOS & 2 wheels with 12 filters
each

Figure 6.7. ExoMars PanCam prototype and the ExoMars mounting bar [credit ESA].

10 mm gives a resolution of 1.1 mrad/pixel, much lower than the 0.28 mrad/pixel resolution of the MER PanCam, though a central high-resolution narrow-angle camera will provide a zoom function. They will all be mounted on a common pan–tilt assembly with a fixed stereo baseline of 50 cm—this is significantly larger than the 28 cm baseline of the MER mast-mounted PanCam suggesting greater accuracy and range of stereoscopic imaging. The pan–tilt assembly offers 360° azimuth and ±90° elevation driven by two stepper motors with 1 arcmin accuracy, the same as that for the MERs. WAC generates images of 10 Mb.

There are several low-power, low-mass, low-volume, and low-cost imaging cameras suitable for HazCam applications. This includes the radiation-hard (to 20 krad) VMC range [425]. The FUGAI5 black-and-white model with 512×512-pixel imaging offers logarithmic light intensity. The IRIS model has a color filter with 640×480-pixel imaging. They both consume 3 W with a 28 V DC supply and can support a 1 Mbps TTC-B-01 interface. Both have dimensions of $6 \times 6 \times 10$ cm and a mass of 430 g. Both have been flight-qualified and IRIS-1 flew on Mars Express to monitor the Beagle 2 separation. The CSEM $1{,}024 \times 1{,}024$-pixel CCD microcamera of mass 100 g offers 10 Mbps on an RS-422 interface with a tolerance down to $-100°$C. The embedded digital signal processing used in vision systems is typically based on the TSC21020 digital signal processor (the space-qualified version of the ADSP21020 DSP board). In conjunction with a 256 kword SRAM memory, this provides a computation capacity of 20 MIPS and 60 MFLOPS.

Kapvik carries a single stereo camera pair for both navigation and scientific data acquisition (Figure 6.8). The Bumblebee XB3 stereovision camera system offered a wide baseline of 240 mm and a UV spectrometer mounted at the midpoint (covering the mid camera). Each camera offered $1{,}280 \times 960$ pixels, each $3.75\,\mu m^2$ in size. Its scientific capabilities are limited as it does not use science filters.

The 505 g package plus spectrometer was mounted on a dedicated purpose-built 2 DOF pan–tilt unit designed and constructed at Carleton University (Figure 6.9). It was mounted on a 4 DOF robotic arm that doubled as a mast. The large pointing range provides flexibility of operations.

6.3   VISION-PROCESSING REQUIREMENTS

MER flight software was upgraded periodically by means of patches which included autonomous image-processing and navigation algorithms [426]. A description of MER image processing is outlined in Alexander et al. (2006) [427]. Typically, image acquisition time is the most significant temporal latency in the control of robotic devices. Vision processing typically requires high computational resources in terms of both processing and storage. Compression (nominally lossy wavelet with 30:1 compression ratios) is required for radio transmission but image processing, calibration, and map building for navigation requires raw images. This implies that PanCam images of $1{,}024 \times 1{,}024$ with 16-bit pixel encoding during navigation will yield a high requirement for data processing and

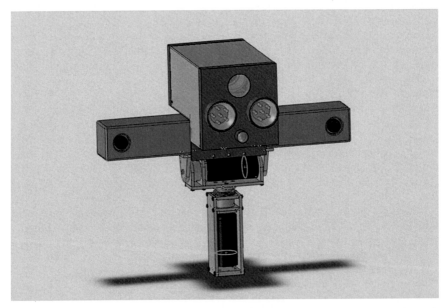

Figure 6.8. Kapvik's Bumblebee XB3 stereovision system and the UV imager mounted onto it [credit Brian Lynch, Carleton University].

storage. Distributed parallel processing offers an alternative more flexible solution to single centralized processor solutions, particularly with multiple processors sharing a single memory with asynchronous communication. This requires considerable design effort. The most simple multiprocessing solution is to use coprocessors such as ASICs (application-specific integrated circuits), DSPs (digital signal processors), or FPGAs (field-programmable gate arrays) for regular low-level image-processing functions.

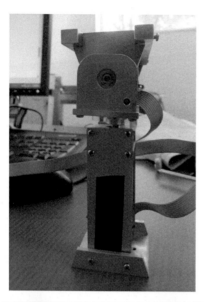

Figure 6.9. Kapvik pan–tilt unit [credit Brian Lynch, Carleton University].

Both the camera and image-processing requirements must first be determined and then checked to ensure that they fulfill the requirements. Constraints on the design of the vision system are well presented in Wagner et al. (2002) [428]. The mast should be mounted forward to minimize the size of the blind zone immediately in front of the rover. Field of view (FOV) relates to the angular size of the projection to be imaged and the required angular sampling of the scene. FOV is constrained by:

$$\text{FOV} \geq \left(\frac{w}{2\sqrt{h^2 + (R + y_{\text{sens}})^2}} \right)$$

where $w =$ ground width of projected camera FOV, $R =$ look-ahead distance (range), $y_{\text{sens}} =$ camera position from vehicle front, and $h =$ stereo camera height above ground. Once projected onto the ground, the FOV forms a wedge-shaped pattern. Hence, field of view (FOV) is determined by look-ahead distance, camera height, camera position from the front of the vehicle, and projected ground width of the FOV. The mutual intersection distance between stereo cameras must be equal to or greater than the rover width at the maximum range. Minimal resolution is given by:

$$h_{\text{min}} = d^2 \left(\frac{2\tan(\text{FOV}/2)}{bN_{\text{col}}} \right)$$

where $d =$ distance from camera to object, $b =$ camera baseline, and $N_{\text{col}} =$ number of pixel columns per image. Maximum range is given by:

$$R = \sqrt{\frac{h_{\text{obs}} b N_{\text{col}}}{2\tan(\text{FOV}/2)}}$$

where $h_{\text{obs}} =$ minimum obstacle height. The look-ahead range is generally limited to $<10\,\text{m}$ ($15\,\text{m}$ max) with stereovision images having a $0.5\,\text{m}$ baseline. Rough terrain imposes requirements on vertical FOV defined as $4\theta_{\text{max}}$ where $\theta_{\text{max}} =$ maximum body pitch angle. Sensor throughput is defined for a vision sensor as [429, 430]:

$$f_{\text{pixel}} = \frac{\psi}{(\text{PFOV})^2}$$

where PFOV $=$ sensor angular resolution, $\psi =$ HFOV $\times$ VFOV $\times f_{\text{image}} =$ sensor flux, HFOV $=$ horizontal FOV, VFOV $=$ vertical FOV, and $f_{\text{image}} =$ image frame rate. Vision processing generates a time delay as image processing is considerably more time consuming than proprioceptive servo control: each has a different sampling rate of kT and $k\tau$, respectively, such that $k\tau \ll kT$. The time delay in image processing is not a pure time delay as there are inputs to the system and delayed outputs depend on inputs. Thus, the image frame rate imposes a maximum speed on the rover. Processor load is defined as

$$f_{\text{cpu}} = f_{\text{pixel}} F$$

where $F = $ FLOPS/pixel. Hence,

$$f_{cpu} = F\frac{\psi}{(PFOV)^2} = F\frac{HFOV \times VFOV \times f_{image}}{(PFOV)^2}$$

Guaranteed detection is defined by the minimum acuity:

$$PFOV = \frac{1}{2}\frac{Lh}{R^2}$$

where $L = $ resolvable length, $R = $ range, and $h = $ camera height. This gives:

$$f_{cpu} = Ff_{pixel} = F\left(\frac{4R^4}{(Lh)^2}\right)\psi = F\left(\frac{4(t_{react}v(1+b))^4}{(Lh)^2}\right)\psi$$

There is thus a fourth-order polynomial increase in computational requirements with increased speed. Computing processing power requirements may be defined as:

$$P_{comp} = \frac{\gamma}{T_{clock}}$$

where $\gamma = $ scale factor. Scale factor is often not constant for a particular platform. The effect of computing time may be quantified as stopping distance:

$$d_{resp} = v_{robot}t_{resp}(1 + b)$$

where $t_{resp} = $ response time and $b = $ braking distance. The myopic problem occurs when sensor look-ahead distance is too small. The latency problem occurs when response time t is too long. Response time is dependent on image-processing times. If a distributed computation approach is adopted:

$$t_{resp} = \frac{t_{opt}}{N_{cpu}} + \frac{t_{net}(N_{cpu} - 1)}{N_{cpu}}$$

For a single processor, response time is determined only by image-processing computations that are dependent on CPU clock speed latency:

$$t_{opt} = \alpha t_{clock} + \beta$$

where $\alpha, \beta = $ constants dependent on software size and image size. Hence,

$$d_{resp} = v_{robot}\frac{\alpha t_{clock} + \beta + t_{net}(N_{cpu} - 1)}{N_{cpu}} = v_{robot}(\alpha t_{clock} + \beta)$$

Imaging time is determined by:

$$T_{tot} = T_{int} + T_{ftr} + T_{ro} + T_{str} + T_{tx}$$

where $T_{int} = $ image integration time, $T_{ftr} = $ frame transfer time, $T_{ro} = $ CCD readout time, $T_{str} = $ image storage time before the WAC is commanded to return an image, and $T_{tx} = $ image transmission to rover mass memory t. As well as processing, images require storage. Images dominate in determining computation memory storage requirements. Each image requires the following storage capacity:

$$N = \frac{(I + BWHf)t}{C}$$

where I = image header size (bytes), B = number of bytes/pixel = 1 for 256-color images, W = image width (number of pixels), H = image height (number of pixels), f = frame rate (Hz), t = duration of storage (s), and C = compression ratio. Lossless data compression can offer up-to-a-factor-of-10 compression rates though wavelet compression is reckoned to yield 20–30 factor compression rates.

6.4   CAMERA CALIBRATION

The optical bench for cameras must be calibrated in order to recover the geometric parameters required for image formation (e.g., focal length, pixel size, optic axis intersection with the image plane, distortion, baseline, etc.). Although this will normally be performed before launch, recalibration is necessary on landing following perturbations from launch loads, inflight cruise (particularly extreme temperature excursions), entry interface impulses, descent deceleration, buffeting, and extreme landing impulses. Vision calibration on landing is essential to ensure the correct mapping between world and image coordinates and to calibrate odometry [431]. Coordinates of a point on a real object (x_0, y_0, z_0) project onto an image (x_i, y_i) with $z_i = 1$:

$$x_i - x_c = -\frac{x_0 - x_c}{z_0} \quad \text{and} \quad y_i - y_c = -\frac{y_0 - y_c}{z_0}$$

A focal lens is usually placed at (x_c, y_c). Camera calibration is complex and is required to establish the 3D position and orientation of the camera with respect to a reference coordinate system. The distance from the camera to an object is given by:

$$d = \sqrt{\Delta x^2 + \Delta y^2 + \Delta z^2}$$

The apparent angular size of a feature such as a rock is given by:

$$\theta = 2 \tan^{-1}\left(\frac{r}{R}\right)$$

where r = radius and R = range. Measurement of object position and orientation requires consideration of coordinate frames defined by 4×4 homogeneous coordinate transformations [432]. Transform $^A T_B$ transforms coordinates in reference frame A into equivalent coordinates in frame B. Inverse transform from B to A is $^B T_A = {}^A T_B^{-1}$. Generally, this has the form:

$$^{\text{object}} T_{\text{world}} = {}^{\text{world}} T_{\text{image}} \, {}^{\text{object}} T_{\text{image}}$$

A pinhole lens model has a pinhole located at the focal center of the lens. An image point (u,v) is a projection of a point in space (x,y) and may be described by an affine transformation of the form:

$$\begin{pmatrix} u \\ v \end{pmatrix} = \begin{pmatrix} m_1 & m_2 \\ m_3 & m_4 \end{pmatrix} \begin{pmatrix} x \\ y \end{pmatrix} + \begin{pmatrix} t_x \\ t_y \end{pmatrix}$$

where (t_x, t_y) = translation and m_i = affine rotation, scale, and stretch parameters. This may be rewritten as:

$$Ax = b$$

where

$$A = \begin{pmatrix} x & y & 0 & 0 & 1 & 0 \\ 0 & 0 & x & y & 0 & 1 \\ & & \vdots & \vdots & & \end{pmatrix}$$

$$x = \begin{pmatrix} m_1 \\ m_2 \\ m_3 \\ m_4 \\ t_x \\ t_y \end{pmatrix}$$

$$b = \begin{pmatrix} u \\ v \\ \vdots \end{pmatrix}$$

A least sum of squares solution for x is given by:

$$x = (A^T A)^{-1} A^T b$$

The pinhole camera model projects point P at (x,y,z) in world coordinates relative to a camera frame of reference to point p at (u,v) in image coordinates:

$$k \begin{pmatrix} x \\ y \\ z \end{pmatrix} = \begin{pmatrix} u \\ v \\ \lambda \end{pmatrix} \rightarrow \begin{matrix} kx = u \\ ky = v \\ kz = \lambda \end{matrix}$$

Hence,

$$k = \frac{\lambda}{z}, \quad u = \lambda \frac{x}{z}, \quad \text{and} \quad v = \lambda \frac{y}{z}$$

where z = depth. These are the perspective equations derived from similar triangle geometry (law of sines) in which $\lambda = f/s$ (focal length in pixels) where s = pixel size and f = focal length (Figure 6.10). The collinearity equations generalize this

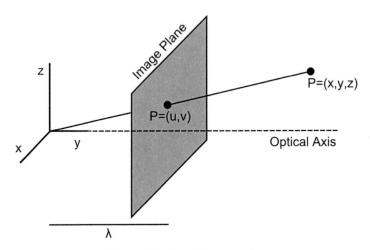

Figure 6.10. Pinhole lens model.

simple model to allow inclusion of camera orientation and translation:

$$u = -f\frac{r_{11}(x - x_0) + r_{21}(y - y_0) + r_{31}(z - z_0)}{r_{13}(x - x_0) + r_{23}(y - y_0) + r_{33}(z - z_0)} \rightarrow u_p = \frac{u}{s_x} + x_0 \quad \text{in pixel coordinates}$$

$$v = -f\frac{r_{12}(x - x_0) + r_{22}(y - y_0) + r_{32}(z - z_0)}{r_{13}(x - x_0) + r_{23}(y - y_0) + r_{33}(z - z_0)} \rightarrow v_p = \frac{v}{s_y} + y_0 \quad \text{in pixel coordinates}$$

where $(x_0, y_0, z_0)^T$ = camera translation. Now

$$R_{RPY} = \begin{pmatrix} r_{11} & r_{21} & r_{31} \\ r_{12} & r_{22} & r_{32} \\ r_{13} & r_{23} & r_{33} \end{pmatrix} = \text{camera } RPY \text{ rotation}$$

Tsai's calibration method [433] is the most widely used algorithm and decomposes the solution for 12 parameters (9 rotations and 3 translations) by introducing constraints. It is autonomous, accurate, efficient, and versatile. Knowledge of camera pose is necessary relative to a fixed reference frame for camera calibration. Beagle 2 used a fixed multiple color pattern designed by the artist Damien Hirst as its calibration target. Similarly, a radiometric calibration target for the ExoMars PanCam has been designed comprising a platform of different metal plates of different colors [434].

Step 1. Transformation from 3D world coordinates (x_w, y_w, z_w) to 3D camera coordinates (x_c, y_c, z_c):

$$\begin{pmatrix} x_c \\ y_c \\ z_c \\ 1 \end{pmatrix} = \begin{pmatrix} R & P \\ 0 & 1 \end{pmatrix} \begin{pmatrix} x_w \\ y_w \\ z_w \\ 1 \end{pmatrix}$$

where

$$R = \begin{pmatrix} r_1 & r_2 & r_3 \\ r_4 & r_5 & r_6 \\ r_7 & r_8 & r_9 \end{pmatrix} = \text{camera orientation}$$

$$P = \begin{pmatrix} p_x \\ p_y \\ p_z \end{pmatrix} = \text{camera translation position}$$

$$r_7 = r_2 r_6 - r_3 r_5$$
$$r_8 = r_3 r_4 - r_1 r_6$$
$$r_9 = r_1 r_5 - r_2 r_4$$

Camera roll–pitch–yaw coordinates can be converted readily into homogeneous 3×3 rotation coordinates (see Ellery, 2000 [435]).

Step 2. Transformation from 3D camera coordinates (x_c, y_c) to the 2D camera image plane (x_u, y_u) assuming a pinhole camera model with perspective projection of 3D features (x_u, y_u, z_u) onto the 2D image plane. Image formation relates object distance from the camera lens and the focal length of the lens. The camera model defines the perspective projection from the camera to image coordinates derived from similar triangle geometry

$$x_u = f \frac{x_c}{z_c}$$

$$y_u = f \frac{y_c}{z_c}$$

where f = effective focal length of camera z_c = range of optical axis of camera, and $(x_c, y_c, z_c)^T$ = object position with respect to camera coordinates.

Step 3. Account for radial distortion of the lens:

$$x_d(r_d^2 K + 1) = x_u \quad \text{and} \quad y_d(r_d^2 K + 1) = y_u$$

where (x_d, y_d) are actual distorted image coordinates, $r_d = \sqrt{x_d^2 + y_d^2}$, and K = first-order lens radial distortion coefficient.

Step 4. Translation of actual image plane coordinates into camera pixel frame coordinates (x_f, y_f):

$$x_f = C_x + \left(\frac{x_d}{d_x}\right)s_x \rightarrow x_d = (x_f - C_x)d_x$$

$$y_f = C_y + \left(\frac{y_d}{d_y}\right) \rightarrow y_d = (y_f - C_y)d_y$$

where $(C_x, C_y) = $ row/column computer fixed coordinates of center of computer frame memory, $d_x = $ center-to-center distance between pixels along the scanline, $d_y = $ center-to-center distance between pixels between scanlines, and $s_x = $ scaling factor. Camera height is given by:

$$h = r_{12}x_c + r_{22}y_c + r_{23}z_c$$

This algorithm is very straightforward but very versatile.

6.5   VISION PROCESSING

The primary purpose of vision is to detect features that serve to discriminate between different objects in the world. A feature is a distinctive character comprising an attribute and a value. This discrimination is based on a set of one or more features—these features may be hierarchically ordered at different scales. It is this capacity for discrimination that imparts the concept of semantic meaning to objects in the world. An excellent review of vision processing as it applies to planetary rovers on Mars has been compiled by Matthies et al. (2007) [436]. Vision processing is highly computationally demanding from the point of view of both processing and storage. Vision processing generally models objects as geometric constructions of cylinders with constraints that define the relative positions of features. Features provide the basis for automated target recognition while their topological relationship to other objects provides the basis for path planning. The fundamental problem of vision is that the number of possible scenes far exceeds the number of possible sensory measurements that can be made, so simplifying assumptions are made. Vision processing transforms sensor input comprising a 2D array of brightness levels representing the 2D projection of a 3D scene into an output that comprises a description of the 3D scene, the objects in it, and their relationships (i.e., vision involves the measurement of light intensity against background noise level). All 2D lines in an image correspond to the possible projection of an infinite number of 3D objects—this is the inverse optics problem. 3D information is lost in the projection to 2D and the single brightness level is also affected by the incident illumination, reflecting characteristics, and orientation of the viewed surface in relation to the viewer and the light source. Recovery of the properties of scenes is underconstrained (ill posed). Visual processing must extract invariants from objects in the environment—the invariants allow direct matching to stored templates of the objects in the image. The human visual system carves up the external world into unitary objects according to cohesiveness, boundedness, rigidity, and permanence. The first visual process is to determine edges in the scene through visual search—this is a categorization task to distinguish between relevant and irrelevant signals. The most basic image-processing methods involve edge detection, edge smoothing, and segmentation into spectrally uniform regions. Segmentation partitions the image into distinct homogeneous regions, the most fundamental being separation of foreground (objects) from the background. This assumes that objects are differentiated from the background by their edges which are characterized by high contrast from the background. Vision processing is thus the search for edges in images—an NP-hard problem. A 2D image is denoted $I(x, y)$ which represents the intensity (brightness level) at each pixel. The numbers of pixels in the horizontal and vertical directions define spatial resolution. Each pixel has four horizontal and vertical neighbors $(x + 1, y)$, $(x - 1, y)$, $(x, y + 1)$, $(x, y + 1)$ and four diagonal neighbors $(x + 1, y + 1)$, $(x + 1, y - 1)$, $(x - 1, y + 1)$, $(x - 1, y - 1)$ except edge pixels which are either ignored or padded with the same values. The Euclidean distance between pixels $p(x_1, y_1)$ and $q(x_2, y_2)$ is readily defined as

$$D(p, q) = \sqrt{(x_1 - x_2)^2 + (y_1 - y_2)^2}$$

A CCD camera automatically digitizes the intensity level at each pixel. Vision processing initially involves preprocessing to filter out noise, correct distortion, and smooth the image by applying "mask" operators which are equivalent to convolution integral filters. High gradient intensity transitions indicate high-frequency components so filtering to suppress noise tends to smooth sharp intensity transitions. Smoothing using low-pass filters is necessary since edge detection depends on differentiating the image. The basic philosophy behind image processing is to delineate objects in the world from the background based on finding their edges which have high contrast from the background. The image is thresholded to extract edges as lines of high contrast and then segmented into spectrally uniform regions. These uniform regions represent objects or parts of objects. From these spectrally uniform regions, we can extract invariant properties, shapes of features, and boundaries (perimeter, area, multiple radii vectors, center/ moment of area, etc.). Fortunately, in planetary exploration, these will essentially be a limited subset of geological features and objects. To reduce the computational overhead, pyramid-level reduction may be performed in which the average of neighboring pixels is computed for a reduced image. It is generally recommended that coarse to fine-resolution image processing be implemented for robust pattern recognition. Such multiresolution representations provide a hierarchical framework to extract different physical structures of the scene [437]. Multiscale transforms are useful for analyzing images through coarse to fine resolutions by successive refinement [438].

The first stage in vision processing is digitization of the input field into arrays of pixel values of image intensity $I(x, y)$ according to its gray level (brightness resolution)—this is the function of the framegrabber. A 2D array camera generates a 2D array of image intensity values, $I(x, y)$—typically resolved to 8 bits to give $2^8 = 256$ possible values. The light intensity at a point in the image is

determined by:
$$I = ER \cos i$$
where $E =$ incident illumination, $R =$ surface reflectance (albedo), and $i =$ angle of viewing incidence from the local surface normal. Recovery of E, R, and i from intensity array I requires additional constraints as well as distance and orientation. One such constraint is that the object surface is uniformly diffuse (Lambertian) with constant albedo while boundaries exhibit discontinuous albedo:

$$\frac{dE}{E} = \frac{dI}{I} + \frac{dR}{R} + \frac{d(\cos(i))}{\cos(i)} \rightarrow \frac{dE}{E} = \frac{dI}{I}$$

within regions and

$$\frac{dE}{E} = \frac{dR}{R}$$

at boundaries. Although color imaging by planetary rovers is essential, color may be reduced to a grayscale image eliminating hue while retaining luminescence in order to compress the image. RGB color may be represented as a 3D color cube with RGB at the corners of each axis, black at the origin, and white at the far vertex (Figure 6.11). For three 8-bit colors, red is at $(255, 0, 0)$ normalized to $(1, 0, 0)$. CCD pixels are often grouped into 2×2 sets of four with two pixels measuring green and one each measuring red and blue intensities. To convert RGB color to grayscale:

$$I = 0.299R + 0.587G + 0.114B$$

where $R, G, B =$ red, green, blue intensities of image. There are other color schemes such as (Y, U, V) which may be similarly converted:

$$\begin{pmatrix} Y \\ U \\ V \end{pmatrix} = \begin{pmatrix} 0.299 & 0.587 & 0.114 \\ -0.169 & -0.331 & 0.500 \\ 0.500 & -0.419 & -0.081 \end{pmatrix} \begin{pmatrix} R \\ G \\ B \end{pmatrix}$$

Color edge detection involves applying operators to each of the RGB color chan-

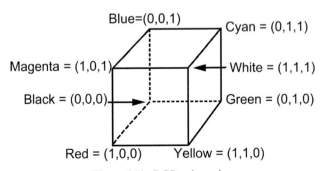

Figure 6.11. RGB color cube.

nels then combining the results additively from each. Color may be represented as a normalized variant of NTSC color space:

$$Y = 0.299R + 0.587G + 0.114B;$$
$$G = (0.596R - 0.274G - 0.322B)/Y;$$
$$Q = (0.221R - 0.523G + 0.312B)/Y$$

Hue, saturation, and value (HSV) of color offers greater robustness to illumination variations than RGB intensities. The definition of object boundaries is necessary prior to categorization and recognition. Visual features are usually corners and edges. Edges are represented by discontinuous changes in intensity, and an edge detection algorithm must detect and locate true edges with minimum false positives. Edge detection is thus the commonest image analysis function as edges define the outlines of objects separating them from the background. Once object outlines have been determined, physical properties such as area, perimeter, and shape can be determined. Vision processing applies a functional transformation of the raw input image such that the processed output image is defined as

$$I'(x, y) = f(I(x, y))$$

Preprocessing the image to correct distortions due to blurring and noise involves applying convolution masks (filters). A convolution mask is an $m \times m$ matrix kernel that is convolved over an $n \times n$ image matrix I to generate a filtered $n \times n$ image matrix I':

$$I'_{ij} = \sum_{p,q} k_{p,q} I_{i-p, j-q} \quad \text{where } 1 \le i, j \le n, 1 \le p, q \le m$$

The top left-hand corner of the mask is superimposed over each pixel in the image in turn and shifted across and down the image until all the pixels have been processed.

Noise can cause corruption of edges in images. The addition of noise suppression filtering of high-frequency components blurs the edge transitions. However, low-pass filters are widely used for smoothing at multiple spectral scales. As there are significant intensity changes within an image, filters must operate at multiple scales. A smoothing mask replaces each pixel intensity with the average of itself and its neighbors:

$$\begin{pmatrix} \frac{1}{9} & \frac{1}{9} & \frac{1}{9} \\ \frac{1}{9} & \frac{1}{9} & \frac{1}{9} \\ \frac{1}{9} & \frac{1}{9} & \frac{1}{9} \end{pmatrix}$$

By convolving this with the image

8	3	4	5
7	6	4	5
4	5	7	8
6	5	5	6

We get a smoothed image:

6.44	5.22	4.33	4.67
5.78	5.33	5.22	5.67
5.56	5.44	5.67	6.00
5.22	5.33	5.78	6.33

As small-scale texture contains a lot of intensity variations at high spatial frequencies, smoothing removes high spatial frequencies without distorting lower spatial frequencies. A two-dimensional mask is generated by multiplying the elements of two one-dimensional masks together. Hence, to reduce computation requirements, separable filters can be regarded as the convolution of two smaller filters; for example:

$$\begin{pmatrix} \frac{1}{9} & \frac{1}{9} & \frac{1}{9} \\ \frac{1}{9} & \frac{1}{9} & \frac{1}{9} \\ \frac{1}{9} & \frac{1}{9} & \frac{1}{9} \end{pmatrix} = \begin{pmatrix} 0 & \frac{1}{3} & 0 \\ 0 & \frac{1}{3} & 0 \\ 0 & \frac{1}{3} & 0 \end{pmatrix} \cdot \begin{pmatrix} 0 & 0 & 0 \\ \frac{1}{3} & \frac{1}{3} & \frac{1}{3} \\ 0 & 0 & 0 \end{pmatrix}$$

Ignoring the zeros, there are only six computations to perform rather than nine. In general, this reduces $(n+1)^2 m^2$ to $2(n+1)m^2$ computations. The smoothing (blurring) function is in fact a Jacobian matrix. The pseudoinverse is a means of solving overdetermined equations in a least-squares-error sense—for a function J, the pseudoinverse is given by:

$$J^+ = (J^T J)^{-1} J^T$$

This is used in vision processing such that if blurring is given by $g = Jf$, then restoration (deblurring) is achieved by:

$$\hat{f} = J^+ g$$

The blur matrix J may also be decomposed into eigenmatrices through singular value decomposition (SVD) as:

$$J = U\Lambda^{1/2} V^T \quad \text{so that} \quad J^+ = V\Lambda^{-1/2} U^T$$

There are a number of simple edge filters that may be adopted to suit different circumstances—this may necessitate using several different edge detectors [439]. Edge detection is highly sensitive to lighting conditions, ensembles of objects in the scene, and noise levels. To find edges, we need to find the highest gradients between intensity levels. Edges are detected as high-intensity changes (i.e., maximum of the first derivative or zeros of the second derivative). Anything above the threshold is set to 1 otherwise 0. The gradient of image $I(x, y)$ is defined as:

$$G(I(x, y)) = \begin{pmatrix} G_x \\ G_y \end{pmatrix} = \begin{pmatrix} \dfrac{\partial I}{\partial x} \\ \dfrac{\partial I}{\partial y} \end{pmatrix}$$

The gradients are computed as first-order differences:

$$G_x = \frac{\partial I}{\partial x} = I(x, y) - I(x - 1, y)$$

$$G_y = \frac{\partial I}{\partial y} = I(x, y) - I(x, y - 1)$$

Vector G points in the direction of the maximum rate of change of I at (x, y)— this is zero-crossing detection. The values for constant illumination regions will be small. From this a gradient edge map can be generated of the form:

$$|G(I(x, y)| = \sqrt{\left(\frac{\partial I}{\partial x}\right)^2 + \left(\frac{\partial I}{\partial y}\right)^2}$$

$$= \sqrt{(\Delta_x I)^2 + (\Delta_y I)^2} \quad \text{in the direction} \quad \theta = \tan^{-1}\left(\frac{\dfrac{\partial I}{\partial y}}{\dfrac{\partial I}{\partial x}}\right)$$

where

$$\Delta_x I = [I(x - 1, y - 1) + 2I(x, y - 1) + I(x + 1, y - 1)$$
$$- I(x - 1, y + 1) + 2I(x, y + 1) + I(x + 1, y + 1)]$$
$$\Delta_y I = [I(x + 1, y + 1) + 2I(x + 1, y) + I(x + 1, y - 1)$$
$$- I(x - 1, y + 1) + 2I(x - 1, y) + I(x - 1, y - 1)]$$

This is sensitive to spatial rates of change in pixel intensities to detect discontinuities (edges). This can be performed more effectively digitally using a convolution mask of the form:

Delta-x gradient: $\begin{pmatrix} -1 & 1 \\ 0 & 0 \end{pmatrix}$ Delta-y gradient: $\begin{pmatrix} -1 & 0 \\ 1 & 0 \end{pmatrix}$

To calculate the diagonal directions at 45° and 135° for diagonal neighbors only, the Roberts cross gradient operator can be used which may be applied through the following convolution masks:

$$\begin{pmatrix} 0 & 1 \\ -1 & 0 \end{pmatrix} \quad \text{and} \quad \begin{pmatrix} 1 & 0 \\ 0 & -1 \end{pmatrix}$$

These 2×2 edge detector masks can find edges of certain orientations without smoothing but larger masks can reduce noise through local averaging. The 3×3

spatial convolution mask is centered on each pixel in turn to compute the gradient of intensity change in neighboring pixels that replace the intensity value of the central pixel. The Sobel operator is a thresholded high-pass filter that enhances edges. It adopts two 3×3 masks:

$$|G| = \sqrt{S_1^2 + S_2^2}$$

where

$$\theta = \tan^{-1}\left(\frac{S_1}{S_2}\right), \quad S_1 = \begin{pmatrix} -1 & -2 & -1 \\ 0 & 0 & 0 \\ 1 & 2 & 1 \end{pmatrix}, \quad S_2 = \begin{pmatrix} -1 & 0 & 1 \\ -2 & 0 & 2 \\ -1 & 0 & 1 \end{pmatrix}$$

The equivalent 3×3 Sobel convolution mask is sensitive to edges of all different orientations:

$$\begin{pmatrix} -1 & 0 & 1 \\ -2 & 0 & 2 \\ -1 & 0 & 1 \end{pmatrix} \quad \text{and} \quad \begin{pmatrix} 1 & 2 & 1 \\ 0 & 0 & 0 \\ -1 & -2 & 1 \end{pmatrix}$$

Larger masks such as 5×5 convolution masks can be used to further suppress noise and smooth the image by local averaging. The Sobel convolution mask performs poorly on blurred edges.

Gaussian filters can implement low-pass, high-pass, and band-pass filters. The Gaussian convolution mask is an isotropic mask with a cross-section that yields a weight profile with a Gaussian normal distribution to smooth the image and remove noise optimally. The width of the Gaussian mask sets the frequency below which variation is removed, but above which variation is retained:

$$G(x, y) = \text{Gaussian weighting smoothing mask} = \frac{1}{2\pi\sigma^2} e^{-r^2/2\sigma^2}$$

$$r = \sqrt{x^2 + y^2} = \text{pixel distance from the center of the mask}$$

$$\sigma = \text{mask width}$$

Initially, images are filtered with a low-pass Gaussian filter to remove noise. Gaussian smoothing distributes noise through the larger image by blurring, thereby behaving as a low-pass filter removing high-frequency noise. The image is thus smoothed using a Gaussian to reduce noise errors prior to edge detection convolution. This may be accomplished using the kernel:

$$G = \begin{pmatrix} \frac{1}{16} & \frac{2}{16} & \frac{1}{16} \\ \frac{2}{16} & \frac{4}{16} & \frac{2}{16} \\ \frac{1}{16} & \frac{2}{16} & \frac{1}{16} \end{pmatrix}$$

This is essential to stabilize the derivatives of edge detectors. The Gaussian function is used to convolve an image [440]:

$$L(x,y) = G(x,y) * I(x,y) \quad \text{where} \quad G(x,y) = \frac{1}{2\pi\sigma^2} e^{-(x^2+y^2)/2\sigma^2}$$

with Fourier transform:

$$g(w_x, w_y) = \exp(-\tfrac{1}{2}\sigma^2(w_x^2 + w_y^2))$$

The difference-of-Gaussian function (DOG) at different scales convolved with the image is given by:

$$D(x,y) = [G(x,y,k\sigma) - G(x,y,\sigma)] * I(x,y)$$

that is

$$g(x,y) = k^2 e^{-(x^2+y^2)/2\sigma^2} - e^{-(x^2+y^2)/2(k\sigma)^2} \quad \text{where} \quad k = \frac{1}{\sqrt{2\pi}\sigma}$$

The Gaussian function gives a receptive field of the form:

$$w_i = e^{-\frac{1}{2}(x-c_i)^T D_i(x_i-c_i)}$$

where D_i = distance metric. To detect edges it is necessary to find zero-crossings of their second derivative. This is achieved through the orientation-independent Laplacian of a Gaussian filter. The Marr–Hildreth filter is a gradient-based operator that uses the Laplacian to compute the second derivative of the image. Any step difference in image intensity is represented as a zero-crossing in the second derivative. A 2D Laplacian involving second-order differentials may be applied to the image of the form:

$$\nabla^2 f = \frac{\partial^2 f}{\partial x^2} + \frac{\partial^2 f}{\partial y^2}$$

This is an edge detector that does not require a preset threshold and is computed as the second derivative of the pixel distribution. It emphasizes the central pixel at the expense of outer pixels which has a strong biological analog. This produces the most stable image features by virtue of its center–surround (Mexican hat) characteristics [441]. Biologically, the center–surround operation involves a neuron receiving excitatory input from photoreceptors near the receptive field center but inhibitory inputs from the surrounding region for on-center cells; off-center cells operate vice versa (i.e., lateral inhibition is the key to center surround receptive fields). The center surround mask has a biological analog that searches for edges in images. The center–surround configuration of photoreceptor density may be modeled by a log-polar mapping defined by:

$$q = (r, \phi) = \alpha(x,y) = \left(\log\sqrt{x^2 + y^2}, \tan^{-1}\frac{y}{x} \right)$$

where r = radial coordinate and ϕ = angular coordinate:

$$(x,y) = \alpha^{-1}(r,\phi) = (e^r \cos\phi, e^r \sin\phi)$$
$$I(r,\phi) = I(\alpha^{-1}(r,\phi)) \quad \text{and} \quad I(x,y) = I(\alpha(x,y))$$

To convert between log-polar and Cartesian coordinates, the Jacobian is defined:

$$J(x,y) = \begin{pmatrix} \dfrac{x}{x^2+y^2} & \dfrac{y}{x^2+y^2} \\ -\dfrac{y}{x^2+y^2} & \dfrac{x}{x^2+y^2} \end{pmatrix} \quad \text{and} \quad J^{-1}(r,\phi) = \begin{pmatrix} e^r\cos\phi & -e^r\sin\phi \\ e^r\sin\phi & e^r\sin\phi \end{pmatrix}$$

The DOG function (Mexican hat operator) is a close approximation of convolution with a Gaussian mask with a center–surround mask (Laplacian of Gaussian) of the form $\sigma^2\nabla^2 G$ [442]:

$$LG(x,y) = -\frac{1}{\pi\sigma^4}\left(1 - \frac{x^2+y^2}{2\sigma^2}\right)e^{-(x^2+y^2)/2\sigma^2}$$

This is inspired by the Marr–Hildreth theory of neurophysiological edge detection at multiple scales through lateral inhibition:

$$\nabla^2 G(x,y) = -k\left[\frac{(x^2+y^2)-2\sigma^2}{2\pi\sigma^6}\right]\exp\left[\frac{-(x^2+y^2)}{2\sigma^2}\right]$$

$$I'(x,y) = \nabla^2 G(x,y) * I(x,y)$$

The most widely used smoothing filter is the Laplacian of Gaussian (LOG) filter followed by the difference of Gaussian (DOG) filter which correlates with mammalian vision. The LOG filter suffers from susceptibility to noise. The DOG function replicates the receptive field properties of visual-processing neurons. The receptive field response has the general form of a convolution between an input image and a weighting function which determine their input responsivity. Two-dimensional Gaussian filters can be separated into simpler 1D Gaussian filters (horizontal and vertical):

$$G(x,y) = \frac{1}{\sigma\sqrt{2\pi}}e^{-(x^2+y^2)/2\sigma^2} = \frac{1}{\sigma\sqrt{2\pi}}e^{-x^2/2\sigma^2}e^{-y^2/2\sigma^2}$$

Edges are detected from local maxima/minima in the image derivatives or as zero-crossings in the second derivative. The output of the center–surround mask becomes positive and negative either side of zero-crossings (light intensity boundaries) (Figure 6.12).

The Laplacian is particularly suited to natural scenes [443]. The Laplacian of Gaussian is related to the heat diffusion equation given by:

$$\frac{\partial G}{\partial \sigma} = \sigma\nabla^2 G \approx \frac{G(x,y,k\sigma) - G(x,y,\sigma)}{k\sigma - \sigma}$$

Hence,

$$G(x,y,k\sigma) - G(x,y,\sigma) \approx (k-1)\sigma^2\nabla^2 G$$

where k = scale factor. Image-deblurrung methods assume that blurring is thus a

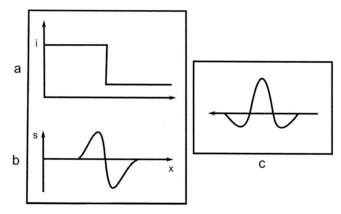

Figure 6.12. Difference of Gaussian filter (a) response on an edge giving (b) zero-crossing (first derivative) and (c) maximum response (second derivative).

diffusion satisfying

$$\frac{\partial^2 I}{\partial t^2} = k\left(\frac{\partial^2 I}{\partial x^2} + \frac{\partial^2 I}{\partial y^2}\right) = k\nabla^2 I$$

where $\nabla^2 I = L$ is the Laplacian operator. The blurred picture $\hat{I}$ is related to the sharp image by:

$$I = \hat{I} - K\nabla^2\hat{I}$$

where $0 < K < 1$. In the discrete case, the gradient measurements are given by:

$$\nabla^2 I(x,y) = \Delta_x I(x,y) - \Delta_x I(x-1,y) + \Delta_y I(x,y) - \Delta_y I(x,y-1)$$
$$= I(x+1,y) + I(x-1,y) - 2I(x,y) + I(x,y+1) + I(x,y-1) - 2I(x,y)$$
$$= I(x+1,y) + I(x-1,y) + I(x,y+1) + I(x,y-1) - 4I(x,y)$$

The Laplacian operators $I_{xx} = \partial^2 I/\partial x^2$ and $I_{yy} = \partial^2 I/\partial y^2$ can be approximated locally by the following 3×3 convolution mask:

$$\begin{vmatrix} 0 & -1 & 0 \\ -1 & 4 & -1 \\ 0 & -1 & 0 \end{vmatrix}$$

The Laplacian is zero in areas of constant intensity. A more complex Laplacian would include diagonal neighbors $I(x+1,y+1)$, $I(x-1,y-1)$, $I(x-1,y+1)$, and $I(x+1,y-1)$:

$$\begin{vmatrix} -1 & -2 & -1 \\ -2 & 12 & -2 \\ -1 & -2 & -1 \end{vmatrix}$$

As differentiation amplifies high-frequency noise, this approach requires smoothing the image prior to using a Gaussian filter. Filtering noise involves replacing points that differ from neighboring points by more than a threshold amount by the average of its neighbors. Thresholding separates edge points from non-edge points. The average of the pixel and its four neighbors $I(x \pm 1, y)$ and $I(x, y \pm 1)$ is given by:

$$\nabla^2 I(x, y) = 5[\tfrac{1}{5}(I(x, y) + I(x-1, y) + I(x+1, y) + I(x, y-1) + I(x, y+1) - I(x, y))]$$

$$= 5[\hat{I}_{av} - I]$$

If a scaling factor $K = \frac{1}{5}$ is used then $I = 2\hat{I} - \hat{I}_{av}$. This strengthens the high-frequency content of signal relative to low-frequency content generating a crisper image. The median filter replaces the pixel of interest with the median of its eight neighbors and itself removing outlier spikes. A normalized convolution mask for the Laplacian operator emphasizes the central pixel at the expense of the outer pixels which has a strong biological analog:

$$\begin{pmatrix} -\tfrac{1}{8} & -\tfrac{1}{8} & -\tfrac{1}{8} \\ -\tfrac{1}{8} & 1 & -\tfrac{1}{8} \\ -\tfrac{1}{8} & -\tfrac{1}{8} & -\tfrac{1}{8} \end{pmatrix}$$

It is isotropic as it utilizes all its nearest neighbors weighted equally in filtering noise. The Gaussian and Laplacian filters are the only smoothing filters that do not create zero-crossings as the scale is increased. The Gaussian filter gives the best tradeoff between localization in spatial and frequency domains simultaneously. Any arbitrary function can be approximated by a linear combination of Gaussian derivatives similar to Fourier series expansions in terms of sinusoidal basis functions. A circularly symmetric Gaussian filter is given by:

$$G(x, y) = e^{-(x^2 + y^2)}$$

The Gabor function and difference of Gaussian function emulate biological filtering. However, Gaussian derivatives are close to optimal for detecting edges. To detect arbitrary edge orientations, the Laplacian of Gaussian is preferred.

These filters can yield false positives and mixed positives in edge detection. By applying the filter at different scales defined by σ, a set of images with different levels of smoothing are obtained. The problem of multi-scaling suggests a number of approaches—fine-to-coarse and coarse-to-fine methods adopt iterative edge detection procedures with adaptive thresholding. A simpler approach is to define three scales such that:

$$\sigma_1 = \sigma_0 + \tfrac{1}{3}(\sigma_2 - \sigma_0)$$

where $\sigma_0 =$ finest detection scale at which edgels appear and $\sigma_2 =$ coarsest blurring scale at which edgels are present. Alternatively, a line weight function (LWF) for enhancing edges which combines the Gaussian function and its second derivative

(Laplacian):

$$f(x, y) = G''_{\sigma_1}(x) G_{\sigma_2}(y)$$

based on biological receptive fields in primates [444]—higher derivatives provide deblurring. Once filtered, the image can be convolved with the Sobel operator to detect edges by finding where the first derivative is maximum. LWF offers good localization of edges even with noise shifts.

The Laplacian of the Gaussian is orientation dependent and breaks down at corners and curves. Canny's edge detector combines Gaussian smoothing with horizontal and vertical difference operators to provide a smoothed gradient direction at each pixel. The Canny edge operator is a Gaussian edge operator that can extract occluding boundaries. It is well approximated by the first derivative of the Gaussian function which optimizes good edge detection and good localization [445]:

$$G'(x) \approx -\frac{r}{\sigma^2} e^{-(r^2/\sigma^2)}$$

in one dimension r. Canny's operator is similar to the Laplacian of the Gaussian operator for step edges but is more robust when used with adaptive thresholding (Figure 6.13). It is widely used and is the de facto standard with excellent performance outperforming many newer algorithms. It combines Gaussian smoothing with two perpendicular 1D Gaussians plus differentiation computations with gradients in both directions to extract edges. Magnitude and direction of the gradient is computed at each pixel. Edge thresholding is based on hysteresis using a high threshold and a low threshold. If a pixel value exceeds the high threshold, it is an edge; if it is above the low threshold and is adjacent to an edge pixel, it is an edge; if it is below the low threshold it is not an edge. Canny's filter is a recursive approximation to Gaussian convolutions that comprise 12 multiplies and 24 adds per pixel for a single 2D image. A simpler approach is to use a discrete kernel such as the 2×2 Roberts gradient-detecting operators in both directions:

$$|G| = \sqrt{R_1^2 + R_2^2}$$

where

$$R_1 = \begin{bmatrix} -1 & 0 \\ 0 & 1 \end{bmatrix} \quad \text{and} \quad R_2 = \begin{bmatrix} 0 & -1 \\ 1 & 0 \end{bmatrix}$$

The Prewitt operator uses two 3×3 masks:

$$|G| = \sqrt{P_2^2 + P_2^2}$$

where

$$\theta = \tan^{-1}\left(\frac{P_1}{P_2}\right), \quad P_1 = \begin{pmatrix} -1 & -1 & -1 \\ 0 & 0 & 0 \\ 1 & 1 & 1 \end{pmatrix}, \quad P_2 = \begin{pmatrix} -1 & 0 & 1 \\ -1 & 0 & 1 \\ -1 & 0 & 1 \end{pmatrix}$$

Gaussian edge filters are reviewed in Basu (2002) [446]. The Canny edge detector in conjunction with wavelet transforms is used for visual acquisition of line features in the terrain by the FIDO rover [447]. Derivatives of 2D Gaussian

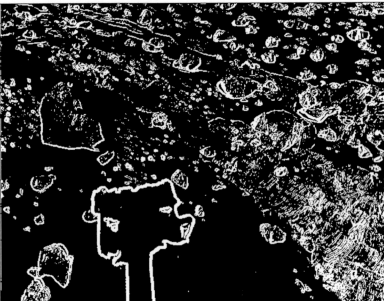

Figure 6.13. MER mast and scene subjected to Canny edge extraction [credit Adam Mack, Carleton University].

functions at different orientations may be used to analyze features at different orientations. A textel is a feature vector of filter responses from Gaussian derivative operators of the first three orders; an edgel feature uses an orthogonal pair of first-order derivatives. Spatial combinations of these primitives can describe a wide variety of shapes and textures.

Corner detectors attempt to detect the intersection of two lines (e.g., Harris and FAST, features from accelerated segment test), while blob detectors attempt to detect patches that differ from the background (e.g., SIFT, SURF, etc.). Blobs tend to be more uniquely identifiable but are more complex to compute. MERs use the Harris corner detector because of its simplicity, but as a corner detector it is inferior in performance to blob detectors. There are a number of blob detector algorithms for the matching of points between images: these include scale-invariant feature transforms (SIFT) which were superseded by speed-up robust features (SURF) based on Haar wavelets with similar performance but lower computational complexity, and more recently binary robust invariant scalable keypoints (BRISK) which replace Euclidean distance with Hamming distance measures. SIFT begin with the DOG filter at several scales to extract features that are scale invariant. The second stage is to filter out false features through a Taylor expansion of the DOG of the form:

$$D(x) = D + \frac{\partial D^T}{\partial x} x + \frac{1}{2} x^T \frac{\partial^2 D}{\partial x^2} x$$

low-contrast features where:

$$\frac{1}{2} x^T \frac{\partial^2 D}{\partial x^2} x < 0.003$$

typically are discarded. Low-stability edges with higher curvature across the edge than that along it are eliminated. The image is then Gaussian smoothed at different scales for each pixel around the center of each feature using an orientation-invariant mask. A set of orientation histograms is generated. SURF offer similar robustness as SIFT but with much faster computation time. A box filter replaces the second-order Taylor expansion which is faster to compute, and Haar wavelets are used to extract orientation-invariant features. Wavelets are a type of multiresolution function approximation that allow for hierarchical decomposition of a signal. At different scales, wavelets encode information about an image from a coarse to a fine scale. The Haar wavelet is the simplest wavelet. It may be applied in different orientations in the image at each scale (e.g., vertical, horizontal, and diagonal directions). The SURF feature extractor has become the de facto standard due to its greater efficiency than the SIFT (scale-invariant feature transform). However, the BRISK (binary robust invariant scalable keypoints) algorithm offers similar performance to SURF but is an order of magnitude faster (using features of the FAST algorithm) making it more suitable for planetary rover applications [447a]. Fast retina keypoint (FREAK) is based on human retinal feature extraction using a multiscale difference of Gaussian filter with exponentially varying spatial resolution from the fovea [448].

Information content provides a measure of the distinctiveness of visual features to define points of interest [449]. The information content of a message is inversely proportional to its probability: $I = -\log(p)$. Average information content per message of a set of messages is given by the entropy $S = -\sum_i p_i \log(p_i)$. Interest points include contours of high-curvature (corners) and high-contrast regions differentiable from the background due to contrast or differing texture, etc. Intensity differences may be quantified through the autocorrelation function. The Harris corner detector computes derivatives of the autocorrelation matrix of the image by convolving the image with a mask. It generates high entropy indicating its detection of distinctive points. It is stable making it useful for determining depth. It may be used to select points of interest but it is sensitive to changes in image scale. An improved version—PreciseHarris—offers superior properties. Corners, however, are ubiquitous in textured surfaces and are not easily applicable to natural scenes.

Contour detection algorithms can differentiate between intensity changes due to object contour edges and those due to texture [450]. This involves applying Bayesian denoise and surround inhibition following the edge detection stage at multiple resolutions. Surround inhibition takes into account the context of each pixel—it involves subtracting from the gradient magnitude of a pixel the integral of the gradient magnitude of the surrounding pixels. Of course, this approach is complex to compute. Edge detection may be regarded as data-driven statistical inference rather than model driven as in traditional edge detection approaches [452]. It gives superior performance to the Canny edge detector but requires modeling the statistics of the image background. The linear receptive field has been superseded by the nonlinear receptive field in describing the response of early neurons to visual inputs [453].

Bayesian estimation based on maximum a posteriori estimation of images may be used to compensate for image degradation [453]. This technique is similar to simulated annealing which adopts a global control parameter akin to temperature. A slowly descending simulated annealing schedule forms a Markov chain that relaxes to the lowest energy states of the image. This corresponds to the most probable states in an exponential Gibbs distribution of noise with the form $\pi(w) = \frac{1}{2} e^{-E(w)T}$ where $E(w)$ = energy function, $Z = \sum_w e^{-E(w)/T}$ = partition function (normalization constant). This Gibbs distribution characterizes the Markov random field giving a maximum a posteriori (MAP) estimate. The probability that a pixel is in a given state is determined by the probabilities for the states of neighboring cells.

Given the lack of color variance in lunar and Martian rocks, texture may be a more appropriate visual property to examine independent of color. Texture analysis involves several steps—textural feature extraction, texture segmentation and texture classification. Most effort is engaged in the first two as they are tightly related—texture segmentation involves partitioning an image into different textured regions which implies prior texture extraction. Markov random fields are suited as texture models but it does not model regular textures well [453]. The combination of Markov random field models to fuse discrete wavelet channels offers more efficient image texture segmentation than either technique alone [453]. Most textural feature extraction methods are based on statistical or transform approaches. An example rock library for the analysis of texture is given in Figure 6.14. Simple image statistics which form the basis of texture analysis provide some discriminatory capacity. Texture may be regarded as a measure of local spatial variations in visual intensity. Haralick parameters include contrast (inertia), angular second moment (energy), entropy, mean, variance, correlation, etc, the first three being most often used. The human visual system has maximum sensitivity to stimulus contrast ratio given by:

$$C = \frac{I_{\max} - I_{\min}}{I_{\max} + I_{\min}}$$

where $I_{\max,\min}$ = maximum, minimum illuminance of stimulus. Angular second moment is a measure of texture energy (homogeneity) which is a useful discriminant for texture. The first few Haralick parameters are given by [454, 455]:

$$\text{Energy,} \quad E = \sum_i \sum_j \Delta I_{ij}^2$$

$$\text{Variance,} \quad V = \sum_i \sum_j (i - \mu)^2 \Delta I_{ij}$$

$$\text{Entropy,} \quad S = -\sum_i \sum_j \Delta I_{ij} \log \Delta I_{ij}$$

$$\text{Contrast,} \quad C = \sum_i \sum_j (i - j)^2 \Delta I_{ij}$$

$$\text{Correlation,} \quad \chi = \frac{\sum_i \sum_j \Delta I_{ij} - \mu_x \mu_y}{\sigma_x \sigma_y}$$

where ΔI_{ij} = intensity difference between image pair pixels (i,j) a distance d pixels apart. These Haralick texture parameters are computed through a gray-level co-occurrence matrix (GCOM) which is orientation independent. They are stable in the range 2–6 m [456]. It appears that second-order statistics (based on pixel pairs) such as the Haralick co-occurrence matrix may offer superior textural discrimination than transform-based approaches such as wavelet transforms [457].

The commonest transform is the Fourier transform defined as:

$$F(w) = \int_{-\infty}^{\infty} f(x) \, e^{-jwt} \, . \, dx$$

The Fourier transform of an $n \times n$ image $f(x,y)$ is given by [458]:

$$F(w_x, w_y) = \int_{-\infty}^{\infty} \int_{-\infty}^{\infty} e^{-\sqrt{1}(w_x x + w_y y)} f(x,y) \, . \, dx \, . \, dy \approx \frac{1}{n^2} \sum_{i,j=0}^{n-1} f(i,j) \, e^{-\sqrt{-1}(iw_x + jw_y)}$$

The Fourier transform is a pure frequency representation across an infinite spatial domain. It cannot therefore represent localisation of information. Texture measurement requires both spatial and frequency information. To represent spatial

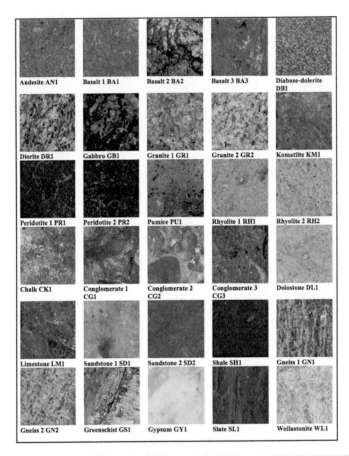

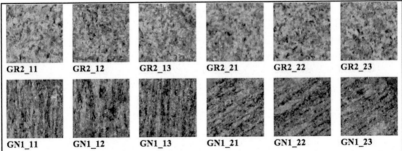

Figure 6.14. (top) Example image library of different rocks; (bottom) subset images of granite and gneiss; (opposite) intensity histograms of granite (mean = 141.37 and s.d. = 30.96) and gneiss (mean = 120.6 and s.d. = 32.91) indicating the discriminatory nature of simple statistical properties [credit Helia Sharif, Carleton University].

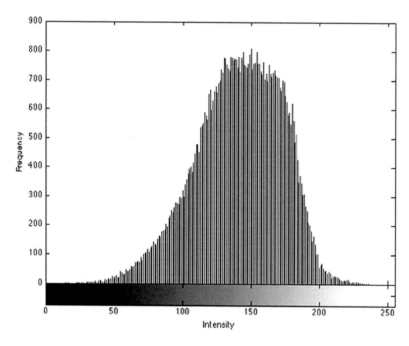

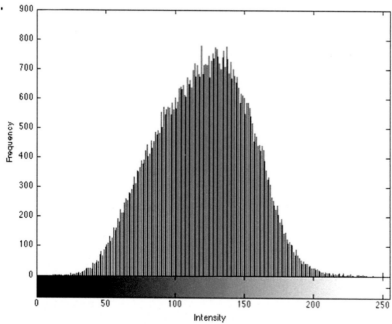

information the window Fourier transform is necessary. The window Fourier transform is subject to the uncertainty principle applied to the standard deviations of the window function $g(x)$ and its Fourier transform $G(w)$:

$$\sigma_x^2 \sigma_w^2 \geq \frac{\pi}{2}$$

where

$$\sigma_x^2 = \int_{-\infty}^{\infty} x^2 |g(x)|^2 \cdot dx \quad \text{and} \quad \sigma_w^2 = \int_{-\infty}^{\infty} w^2 |G(w)|^2 \cdot dx$$

The resolution in phase space is maximized when the window function is a Gaussian such as in the Gabor transform (i.e., the Gabor function is optimally localized in both space and frequency). A Gabor function is a sine wave modulated by a Gaussian envelope as the window function. The Gabor transform is a particular type of window Fourier transform with a receptive field response. The spatial window $g(x)$ in the Fourier integral is translated along the spatial axis to cover the entire signal. At position u for a frequency w, the window Fourier transform is given by:

$$Gf(w, u) = \int_{-\infty}^{\infty} e^{-jwt} g(x - u) f(x) \cdot dx$$

It measures the amplitude of the wave of frequency w in the region of position u. A Gabor filter comprises a complex sinusoid modulated by a 2D Gaussian that may be represented by:

$$G(x, y) = g(x, y) \cdot f(x, y) = g(x, y) \cdot e^{-j(w_x x + w_y y)} = \frac{1}{2\pi\lambda\sigma_x\sigma_y} e^{-\frac{1}{2}\left(\left(\frac{x/\lambda}{\sigma_x}\right)^2 + \left(\frac{y}{\sigma_y}\right)^2\right)} e^{-j(w_x x + w_y y)}$$

where $(w_x, w_y) = $ position of the filter in the frequency domain, $\theta = \tan^{-1}(w_x/w_y) = $ orientation of the filter, $w_0 = \sqrt{w_x^2 + w_y^2} = $ center frequency, $\lambda = $ aspect ratio, and $\sigma = $ scale factor.

Gabor functions represent Gaussian-modulated complex exponentials which may be modeled as radial basis functions to represent specific receptive fields (defining its label in a form of labeled line coding) [458]. Gabor filters are suited to the detection of objects in complex backgrounds and provide a means for texture-based segmentation [459]. They are considered to represent a biological analog with the center–surround characteristic for extracting oriented edges in the primary visual cortex, especially its orientation–selectivity. The human visual system has maximum sensitivity to signals at $0°$ and $90°$ orientation which decreases to a minimum of $45°$. They are band-pass filters that provide optimal resolution in frequency and space and conform to the receptive fields of cortical cells [460]. The Gabor filter is a product of an elliptical Gaussian function and a complex plane wave in log-polar coordinates given by:

$$G(x, y) = \frac{1}{2\pi\sigma_x\sigma_y} \exp\left(-\frac{1}{2}\left(\frac{x^2}{\sigma_x^2} + \frac{y^2}{\sigma_y^2}\right)\right) \cos(2\pi f_0(x \cos\theta + y \sin\theta) + \phi)$$

where f_0 = frequency of sinusoid, ϕ = phase of sinusoid, θ = filter wavelet orientation, σ_x/σ_y = aspect ratio of elliptical Gaussian envelope, $\sigma_x = \sigma_y = \sigma(\sqrt{\ln 2}(2^B + 1)/\sqrt{2}\pi f_0(2^B - 1)) \approx 0.5$ biologically, and B = frequency bandwidth (typically one octave). In more compact form:

$$G(x,y) = \exp\left[-\frac{1}{2}\left(\frac{x^2}{\sigma_x} + \frac{y^2}{\sigma_y}\right)\right] \exp[iw_0(x+y)]$$

The real (even) component has the form:

$$G(x,y) = \exp\left[-\frac{1}{2}\left(\frac{x^2}{\sigma_x} + \frac{y^2}{\sigma_y}\right)\right] \cos(w_0 x + \phi)$$

where w = filter radial frequency. The imaginary (odd) component has the form:

$$G(x,y) = \exp\left[-\frac{1}{2}\left(\frac{x^2}{\sigma_x} + \frac{y^2}{\sigma_y}\right)\right] \sin(w_0 x + \phi)$$

The Gabor filter is closely related to the Fourier transform which has an identical form:

$$F(x,y) = \exp\left[-\frac{1}{2}\left(\frac{x^2}{\tau_x} + \frac{y^2}{\tau_y}\right)\right] \exp[ix(w_x + w_y)]$$

(see [461]). Gabor filters may be designed to be sensitive to different scales and orientations of intensity variations—hence, Gabor-like filters can detect lines at different orientations. Any image must be convolved with a bank of Gabor filters to cover a range of different scales and orientations. Orientation-selective response may be implemented through Gabor wavelets. Thus, Gabor filters respond well to different shapes of different orientations (i.e., texture). Gabor wavelets are constructed with a rotationally symmetric Gaussian window multiplied by sinusoidal waves that propagate along different orientations with different phases. Early human visual processing has been characterized by a bank of Gabor filters across the spatial frequency domain [462]. A compromise filter bank is based on three size scales and four orientations between 0 and 180°. A Gabor filter comprises harmonic oscillations bounded within a Gaussian envelope. It represents a compromise between global Fourier analysis and localized line detectors. Wavelets based on Gaussian-modulated sinusoids (Gabor functions) are polar models of the visual cortex. Gabor functions can act as texton detectors [463]. Texture segmentation requires measurement of both spatial and spatial frequency domains. A Fourier transform of the Gabor filter (modulation transfer function) with $\phi = 0$ in the spatial frequency domain (u,v) is given by:

$$G(u,v) = 2\sqrt{\pi\sigma_x\sigma_y}\left(\exp\left(-\frac{1}{2}\left(\frac{(u-f_0)^2}{\sigma_u^2} + \frac{v^2}{\sigma_v^2}\right)\right) + \exp\left(-\frac{1}{2}\left(\frac{(u+f_0)^2}{\sigma_u^2} + \frac{v^2}{\sigma_v^2}\right)\right)\right)$$

The Gabor filter has optimal resolution in the spatial and spatial frequency domains. A bank of Gabor filters is implemented at discrete orientations (human orientation resolution is estimated at $5°$) and radial frequencies (human frequency resolution is estimated to have octave separations such as $(1\sqrt{2}, 2\sqrt{2}, 3\sqrt{2}, \ldots n/4\sqrt{2})$ where $n = $ whole number of pixels. Energy spectra of real world images have a power law form $1/f^\alpha$ where $\alpha \sim 1$–2 with the value of α being specific to each environment. Gabor filters for natural scenes should be computed at ± 30, ± 45, and $\pm 60°$ for computational efficiency (i.e., resolution of $15°$). The problem with Gabor filters is that they are complex to compute suggesting the use of specialized circuits to approximate the most time-consuming functions. Neural networks have also been deployed for 2D Gabor analysis of images yielding compact coding as Gabor coefficients from which the exact image can be reconstructed [465]. They also have a tendency to ring near edges due to their high-frequency response. For texture analysis, scale-invariant feature transform (SIFT) performs better than Gabor filters [466]. Nevertheless, Gabor filters offer tremendous potential for sophisticated image processing in the service of autonomous science. A visual discovery algorithm to extract different image intensities from the background based on a human visual cortical receptive field model using multiscale Gabor filters has been implemented. This has the advantage of detecting visually different objects from the background without a priori models. It is particularly suited to microscopic imaging. Gabor filters have been used in conjunction with center–surround and corner-sensitive filters. The center–surround filter identified bright or dark localized regions that differ from their surroundings. Recent preliminary work has been based on the use of Gabor filters for texture-based recognition [Mack, A., unpublished material]—this suggests that Gabor filters may be used to reliably extract stratification.

The window Fourier transform is ill suited to image processing—it decomposes a function into a set of frequency intervals of constant size. This limits the spatial resolution of the window transform in images but it is well suited to signal processing at the same scale. For natural textures, it is desirable to allow variable spatial resolution to represent and discriminate between different textures at multiple scales. The human visual system decomposes stimuli into a set of frequency channels with logarithmic bandwidths. The wavelet transform offers both frequency and spatial representation by dilating and translating a unique wavelet function $\psi(x)$ into a family of wavelet functions $\sqrt{s}\psi(s(x-u))$ to represent the signal [467]. The wavelet transform has the form:

$$Wf(s,u) = \int_{-\infty}^{\infty} f(x)\sqrt{s}\psi(s(x-u)) \cdot dx$$

where $s = $ dilation factor and $u = $ translation factor

When scale s is small, the resolution is coarse spatially and fine in frequency; when scale s is large, the resolution is fine spatially and coarse in frequency. If $s = 2^j$ and $u = 2^{-j}n$, an orthonormal wavelet results. The Gabor function is an admissible wavelet (though not orthonormal)—a transform constructed from a dilation of a Gabor function constitutes a Gabor wavelet not a Gabor transform. Gabor wavelets are constructed with a rotationally symmetric Gaussian window multiplied by sinusoidal waves that propagate along different orientations with different phases. A detailed analysis of Gabor wavelets is given by Lee (1996) [467]. Any function can be reconstructed from its decomposition into an orthonormal wavelet. An example of an orthonormal wavelet is the Haar wavelet of the form:

$$\Psi(x) = \begin{cases} 1 & \text{for } 0 \le x < 1 \\ -1 & \text{for } \frac{1}{2} \le x < 1 \\ 0 & \text{otherwise} \end{cases}$$

The wavelet decomposes an image into a series of frequency channels with a bandwidth that varies logarithmically. Image processing proceeds from coarse to fine resolution—this effectively implements multiresolution decomposition in which coarse information provides context for finer computations. For texture, energy measures computed from a set of wavelets is sufficient giving slightly superior performance of entropy measures [467]. Generalized Gabor functions combine position-based sampling with logarithmic frequency scaling characteristic of biological vision systems implementing a multiresolution pyramid with variable spectral width similar to the Laplacian pyramid [467]. If rotation–invariance is required, there are three terrain feature extraction techniques [467]: (i) rotation-invariant wavelet transforms; (ii) circularly symmetric Gabor filters; (iii) Gaussian–Markov random field with a circular set of neighbors; of which the wavelet and Gabor filters exhibit robustness to noise. Circular symmetry renders the Gabor filter rotation-invariant:

$$G(x,y) = \frac{1}{2\pi\sigma^2} e^{-(x^2+y^2)/2\sigma^2} \, e^{-jw_0\sqrt{x^2+y^2}}$$

where $\sigma = \sigma_x = \sigma_y$.

The filtered image may then be subjected to a nonlinear blob detector transformation with a sigmoidal-type form:

$$\psi(t) = \tanh(\alpha t) = \frac{1 - e^{-2\alpha t}}{1 + e^{-2\alpha t}}$$

where $\mu = \text{constant} = 0.25$ nominally. The textural features extracted are defined by an energy measure that may then be square error clustered into textural classes. Many other classifiers are suitable such as k-means clustering and neural networks, though the latter appear to be more accurate than the former.

Once edge detection has been performed, feature extraction is undertaken through segmentation which separates the image into different regions of brightness homogeneity. A feature may be defined as a tuple $f = \langle F, x, y \rangle$ where $F = $ image region (e.g., 7×7 typically) and $x, y = $ feature centroid coordinates. Thresholding generates a binary edge map of light/dark binary values which separates pixels into two groups—background and object. Global thresholding is used for scenes where the objects vary significantly from the background by using a single constant intensity threshold value. Local thresholding is used in situations involving intensity images where contrasts may not be high and the threshold value depends on local average intensity. To cope with variable illumination, edge detection thresholds must be variable. Dynamic thresholding depends on pixel position so that proximate point intensities determine the threshold value. A histogram provides for dynamic thresholding (Figure 6.15).

The threshold can be estimated using a histogram that encodes the frequency of intensity in an image. In simple cases, the histogram has two peaks, one for background intensity and one for object intensity. From the histogram, the mean gray level is computed from which the threshold can be computed. The intensity level at which to set the threshold is the rare intermediate intensity level forming the valley between the two peaks. Small objects are not easy to extract from their backgrounds by thresholding because the histogram may be comparatively flat and many objects may have different intensities. The means of the object(s) and background can be computed to generate further thresholds. A good choice of threshold reduces variance in pixel values within a group. For multiple small objects that defy thresholding, entropy estimates offer a more robust solution [467]:

$$H_{\text{obj}} = -\sum_{i=1}^{s}\sum_{j=1}^{t} \frac{P_{ij}}{\sum_{i=1}^{s}\sum_{j=1}^{t} P_{ij}} \ln\left(\frac{P_{ij}}{\sum_{i=1}^{s}\sum_{j=1}^{t} P_{ij}}\right) = \ln\sum_{i=1}^{s}\sum_{j=1}^{t} P_{ij} + \frac{-\sum_{i=1}^{s}\sum_{j=1}^{t} P_{ij}\ln P_{ij}}{\sum_{i=1}^{s}\sum_{j=1}^{t} P_{ij}}$$

$$H_{\text{bgd}} = -\sum_{i=s+1}^{m}\sum_{j=t+1}^{m} \frac{P_{ij}}{\left(1 - \sum_{i=1}^{s}\sum_{j=1}^{t} P_{ij}\right)} \ln\left(\frac{P_{ij}}{\left(1 - \sum_{i=1}^{s}\sum_{j=1}^{t} P_{ij}\right)}\right)$$

$$= \ln\left(1 - \sum_{i=1}^{s}\sum_{j=1}^{t} P_{ij}\right) + \frac{-\sum_{i=1}^{m}\sum_{j=1}^{m} P_{ij}\ln P_{ij} + \sum_{i=1}^{s}\sum_{j=1}^{t} P_{ij}\ln P_{ij}}{\left(1 - \sum_{i=1}^{s}\sum_{j=1}^{t} P_{ij}\right)}$$

where $P_{ij} = $ probability of pixel intensity (i,j). The entropy criterion is given by:

$$\phi(s,t) = H_{\text{obj}} + H_{\text{bgd}}$$

where the optimum threshold satisfies $\hat{\phi}(s,t) = \max \phi(s,t)$.

Lines are grouped into contours, contours into surfaces, and surfaces into objects. Image contours usually correspond to definite physical events—shadows, depth discontinuities, etc. If S_l, S_r are areas enclosed by contours C_l and C_r and (A_l, B_l), and (A_r, B_r) are the center of the mass coordinates of contours:

$$\frac{S_l}{S_r} = \frac{1 - A_l p - B_l q}{1 - A_l p - B_r q}$$

There are a number of different approaches to the representation and recognition of visual objects in terms of classification capability. Categorization involves not

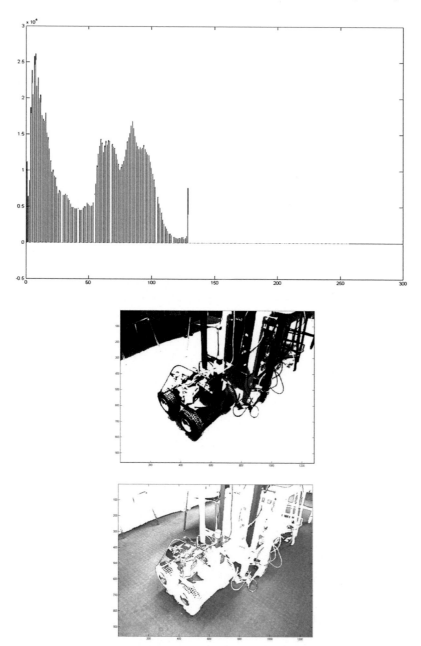

Figure 6.15. (top) Histogram separating (center) foreground from (bottom) background [credit Adam Mack, Carleton University].

only a specification of an object such as its geometric shape, but also its resemblance to the other known objects. Shapes may be identified using generalized angles which facilitates rapid pattern matching [468]. The polygonal pixel chain approximating the curve of an object boundary is characterized by a sequence of turning angles. Hence, any object shape may be represented by a series of turning angles α_i and edge lengths l_i. A line model has a polar form:

$$r = x \cos \alpha + y \sin \alpha$$

where α = angle between x-axis and line normal and r = perpendicular distance of line to origin. The covariance matrix of line parameters is given by:

$$\cos(r, \alpha) = \begin{pmatrix} \sigma_r^2 & \sigma_{r\alpha} \\ \sigma_{r\alpha} & \sigma_\alpha^2 \end{pmatrix}$$

Common clustering algorithms to reject outliers include RANSAC (random sample consensus) but superior performance is offered by the Hough transform. The Hough transform is a popular line extraction algorithm which is robust to outliers. It identifies clusters of features with a consistent interpretation such as contours of edges. The Hough transform can be used to detect the alignment of points of high gradient corresponding to straight line elements which may be grouped into lines (i.e., the simple Hough transform finds regular patterns such as lines in 2D datasets). Artificial objects often have straight or circular edges which project as straight and elliptical boundaries. The Hough transform is a global method for linking such edges. It maps data into parameter space and a search is made in this parameter space for data clusters. It finds infinite straight or curved lines on which the image edges lie, and the portion of the line between two ends is the line segment. If we take a point (x, y) in an image, all straight lines that pass through that point have form: $y = mx + c$, or equivalently, $c = -mx + y$. Each line with a different gradient and intercept corresponds to one of the points on the line in (m, c) parameter space. If two pixels p and q in (x, y) space lie on the same single line, there is an intersection of the two lines p and q in (m, c) parameter space (i.e., all the pixels of a single line lie at a single point in (m, c) parameter space. Every straight line (x, y) in Cartesian space is transformed by the Hough transform into a peak (m, q) in parameter space. The Hough transform transforms a pixel into a parametrized curve in parameter space. A point x, y is transformed into a straight line by: $y = mx + c$ where $c = y - mx$ of (m, c) parameter space. Similarly, two points may be mapped into (m_1, m_2) parameter space through $y_1 = m_1 x_1 + c_1$ and $y_2 = m_2 x_2 + c_2$. A point in (x, y) can also be represented as a curve in (r, θ) parameter space. For curved lines, the line may be represented as:

$$r = x \cos \theta + y \sin \theta$$

Circles may be represented as:

$$r^2 = (x - a)^2 + (y - b)^2$$

Hence, (m, c) space may be represented as a 2D accumulator array and (a, b, r)

space as a 3D array. A general ellipse requires a 5D parameter space. The search for the elements of the accumulator array produces large values where they correspond to lines in the original edge map. A randomized Hough transform (RHT) may be used for curve detection where n pixels are randomly selected to be mapped into the parameter space [469]. RHT offers the advantage of efficient computational overhead. The radon transform performs a similar operation to the Hough transform by transforming into polar space (ρ, θ). A straight line may be represented in polar coordinates as:

$$\rho = x \cos \theta + y \sin \theta$$

The radon transform is given by:

$$R(\rho, \theta) = \int_{-\infty}^{\infty} \int_{-\infty}^{\infty} g(x,y) \delta(\rho - x \cos \theta - y \sin \theta) \, dx \, dy$$

where $g(x,y) = $ function to be transformed, $\delta(x,y) = $ Dirac delta function, $\rho = $ distance of straight line from origin, and $\theta = $ normal angle. The generalized Hough transform extends its applicability to arbitrary shapes based on sinusoids but does not perform well with natural objects. Furthermore, it is computationally expensive but robust to noise. The generalized Hough transform provides vectors of the form:

$$f(x) = A \sin(wx - \phi) - c$$

where $A = $ amplitude of sinusoid, $w = 2\pi/\tau = $ angular velocity, $\tau = $ period, $\varphi = $ phase, and $c = $ axis offset of sinusoid

$$H(A, w, \phi, c) = \int_{-\infty}^{\infty} g(x, [A \sin(wx - \phi) - c]) \, dx$$

$$= \int_{-\infty}^{\infty} \int_{-\infty}^{\infty} g(x,y) \delta[y - A \sin(wx - \phi) - c] \cdot dx \, dy$$

If non-zero, (x, y) fits the sinusoid.

To cope with more general, non-elliptical closed contour shapes, active contour models may be used (they can be used for potential fields) [470]. The snake is an energy-minimizing spline guided by external forces and influenced by image forces that pull it towards features such as lines and edges. It is a 2D elastic model attached to edge features in the image. The snake contour is shrunk onto the object boundary by snake contraction forces along the energy gradient. Such methods may be used for tracking moving objects in successive images [471].

This bears similarities to the elastic net approach to the NP-hard traveling salesman problem by minimizes the length of paths [471, 471]. However, using an energy measure, a closed circular path is elongated non-uniformly until it passes through each city. This expansion is controlled by a scale parameter K corresponding to a maximum *a posteriori* Bayesian estimate.

A series of control points are connected by straight lines—snake position is parametrized by the coordinates of control points defining a deformable curve $v(s)$. The active contour is specified by the number and coordinates (x, y) of each of those control points. A deformable curve $v(s,t)$ with spatial index s and time t is a function of its coordinates (x,y):

$$v(s,t) = (x(s,t), y(s,t))$$

An energy function is used to reduce the energy of the active contour by moving control points:

$$E_{contour} = E_{int} + E_{ext}$$

Equivalently,

$$F_{ext} = -\nabla E_{ext} \quad \text{and} \quad F_{ext} + F_{int} = 0$$

Internal forces are potential forces defined as the negative gradient of a potential function while external forces are pressure forces. Internal (elastic) energy is dependent on intrinsic properties such as curvature and length. External energy depends on externally defined constraints such as edges. The motion of the active contour is dependent on simulated forces. Elastic energy is proportional to total length via the sum of squares of distances between adjacent control points:

$$E_{el} = K_1 \sum_{i=1}^{n} (x_i - x_{i-1})^2 + (y_i - y_{i-1})^2$$

where $K_1 =$ constant dictating degree of elasticity and $x, y_{i,i-1} =$ control point coordinates i and $i-1$. More generally,

$$E = \frac{1}{2} \int_s K_1 \left| \frac{dv}{ds} \right|^2 + K_2 \left| \frac{d^2v}{ds^2} \right|^2 - \lambda |\nabla^2 I| \cdot ds$$

which is minimized by steepest descent. The first term is a tensioning smoothing force which is strongest at high curvature. The second term is a rigidity term. The third external term is derived from image intensities (following preprocessing with a Gaussian mask operator) and leads the active contour to step edges. A variant is the balloon which has an additional normal term to maximize its area while smoothing the contour. This generates an elastic force on the ith control point:

$$F_{el} = -2K_1 [(x_i - x_{i-1}) + (x_i - x_{i+1}) + (y_i - y_{i-1}) + (y_i - y_{i+1})]$$

Control points are pulled towards their nearest neighbors so that the force is towards the average position (i.e., moved inwards while smoothing). Outlying points are pulled in fastest and highest curvatures are smoothed fastest. The external energy function is defined as the negative of the sum of intensity levels of pixels. Minimizing the function moves the contour towards the brightest parts of the image:

$$E_{ext} = -K_2 \sum_{i=1}^{n} I(x_i, y_i) \rightarrow F_{ext} = \frac{K_2}{2} (I(x_{i+1}, y_i) - I(x_{i-1}, y_i) + I(x_i, y_{i+1}) - I(x_i, y_{i-1}))$$

Control points are pulled in the direction of the intensity gradient. After the internal force is applied, the external force alone is applied to close the active contour. Active contours can be implemented which average over more pixels or even more sophisticated variants. Snakes are a special case of generalized Bayes criterion minimization [472]. Total energy is composed of image energy, internal energy to ensure a smooth snake contour, and external energy:

$$E_{\text{total}} = \frac{1}{2} \int (E_{\text{int}} + E_{\text{image}}) \cdot ds$$

where

$$E_{\text{image}}(x, y) = -\gamma |\nabla I(x, y)|^2$$

$$E_{\text{int}}(v(s)) = \alpha(s) \left| \frac{\partial v}{\partial s} \right|^2 + \beta(s) \left| \frac{\partial^2 v}{\partial s^2} \right|^2 = \alpha_i |v_i - v_{i-1}|^2 + \beta_i |v_{i-1} - 2v_i + v_{i+1}|^2$$

where α, β, γ = weighting constants. Hence,

$$E_{\text{total}} = \sum_{i=1}^{n} \alpha |v_i - v_{i-1}|^2 + \beta |v_{i-1} - 2v_i + v_{i+1}|^2 + \gamma(-|\nabla I(x_i, y_i)|^2)$$

The minimum points of total energy are solutions. The gradient operation in image energy may be performed by convolution of the image with a Sobel operator:

$$S(x, y) = I(x, y) \otimes H(-x, -y) = \sum_i \sum_j h(i, j) I(i + x, j + y)$$

where

$$h_x = \begin{pmatrix} -1 & 0 & 1 \\ -2 & 0 & 2 \\ -1 & 0 & 1 \end{pmatrix}, \quad h_y = \begin{pmatrix} 1 & -2 & -1 \\ 0 & 0 & 0 \\ -1 & 2 & 1 \end{pmatrix}$$

Snakes have been applied to path planning for serpentine manipulators whereby the energy-minimizing curve approximates to the manipulator shape [473]. Active contours have problems with concave boundaries that require strong pressure forces but this can make the contour unstable. To overcome this, gradient vector flow field (GVF) snakes have been proposed in which external force is not a negative gradient of a potential function [474]. GVF fields are related directly to optic flow fields of the form:

$$f(x, y) = \begin{pmatrix} u(x, y) \\ v(x, y) \end{pmatrix}$$

which minimizes energy

$$E = \int \int k(u_x^2 + u_y^2 + v_x^2 + v_y^2) + |\nabla I|^2 |f(x, y) - \nabla I|^2 \, dx \, dy$$

An object is completely determined by its boundary curve. Encoding the shape of image regions has been implemented using shape-encoding filters that receive a 2D object boundary matrix and output identified features based on a mathematical model of shape [475]. Segmentation involves grouping pixels together into regions of similar intensity values. The scene is partitioned into regions of approximately uniform brightness corresponding to surfaces or texture regions that are locally consistent by iteratively subdividing it. For real world scenes, region growing is preferred over segmentation since it does not require straight boundaries such as curves and is well suited to texture segmentation and analysis—a merging process continues with segments growing until a uniformity condition holds for the region by adding in similar neighboring pixels. Adjacent regions are then successively merged when their boundary contrasts or texture constraints are low until only boundaries of strong contrast remain. A region is delineated by its closed boundary which may be specified by parametric curves, its center and radii, or a vector pair. Object recognition is based on geometric shapes encoded as features invariant to rotation, translation, and scaling such as perimeter, area, and moments. Global features such as area, perimeter, centroid, perimeter radii, first and second moments, and number and area of holes may be extracted. The area of an object in the image is defined by the number of pixels in the object image. The perimeter of the object is defined by the number of steps required to follow its border. The extent of the object is defined by the length of the line in that direc-tion—the greatest extent is its diameter. The chief difficulty is that natural terrains are not easily characterized by standard sets of features readily extractable from images. The extraction of geometric moments invariant to mathematical opera-tions such as reflection, rotation, etc. are defined by [476, 477]:

$$m_{pq} = \int_{-\infty}^{\infty} \int_{-\infty}^{\infty} x^p y^q I(x,y) \cdot dx \cdot dy$$

$$= \sum_i \sum_j x^p y^q I(x,y) \quad \text{in discrete form}$$

where $p + q =$ order of moments. A more robust version is given by:

$$m_c^{p,q} = \sum_x \sum_y x^p y^q \delta(c - I(x,y)) w(x,y)$$

where $\delta(x) =$ Kronecker delta and $w(x,y) =$ weighting function that emphasizes the center over edges. The object center of mass (centroid) is given by:

$$\bar{x} = \frac{m_{10}}{m_{00}}, \quad \bar{y} = \frac{m_{01}}{m_{00}}$$

Central moments about the center of gravity are given by:

$$m_{pq} = \sum_i \sum_j (x - \bar{x})^p (y - \bar{y})^q I(x,y)$$

Hence, the moment of inertia defined with respect to the center of mass (Figure 6.16) is:

$$J = \int\int r^2 I(x,y)\, dx\, dy = \int\int (x \cos\theta + y \sin\theta - \rho)^2 I(x,y)\, dx\, dy$$

The orientation angle is defined by the orientation of the axis passing through an object such that the second moment of the object about that axis is minimal

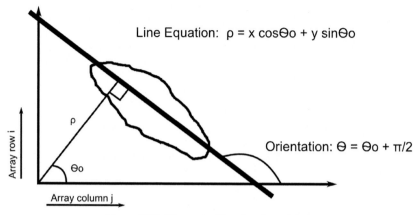

Figure 6.16. Moment of inertia of an object.

(axis of least inertia) ($\rho = x\cos(\theta - \pi/2) + y\cos(\theta - \pi/2)$):

$$\theta = \tfrac{1}{2}\tan^{-1}\left(\frac{2m_{11}}{\sqrt{m_{20}^2 - m_{02}^2}}\right)$$

where

$$m_{20} = \iint (x - \bar{x})^2 I(x,y)\, dx\, dy$$

$$m_{11} = \iint (x - \bar{x})(y - \bar{y}) I(x,y)\, dx\, dy$$

$$m_{02} = \iint (y - \bar{y})^2 I(x,y)\, dx\, dy$$

The principal axis is the line through the centroid about which the moment of inertia is least. Normalized central moments are given by:

$$\eta_{pq} = \frac{m_{pq}}{m_{00}^{\gamma}}$$

where $\gamma = (p + q)/2$, m_{00} = area. Object eccentricity is given by:

$$\varepsilon = \left(\frac{\mu_{02}\cos^2\theta + \mu_{20}\sin^2\theta - \mu_{11}\sin 2\theta}{\mu_{02}\sin^2\theta + \mu_{20}\cos^2\theta - \mu_{11}\cos 2\theta}\right)^2$$

where $\mu_{20} = m_{20} - \mu\bar{x}^2$, $\mu_{02} = m_{02} - \mu\bar{y}^2$, and $\mu_{11} = m_{11} - \mu\bar{x}\bar{y}$. Moments are invariant to rotation, scaling, and translation. Zero to second-order moments enable calculation of perimeter, outline area, centroid (center of mass if mass distribution is even), angle to major axis, elliptical cross section, area/(perimeter)^2 (this measures the compactness of an object or its inverse complexity), and area/

(thickness)^2 (elongatedness). The second and third moments of inertia are typically sufficient for distinguishing between natural landmarks. Moments represent a class of Fourier power spectrum techniques.

6.6   TERRAIN IMAGING

Color variation cannot be used as the primary distinguishing characteristic for the geological identification of rocks on Mars as rock color variation is small. However, the degree of variation of color and texture (quantified by Shannon entropy) may be used to indicate the distance to an obstacle on the basis that such variation is reduced at closer distance [478]. This method may be used to complement optic flow (discussed later). There are a number of 3D extraction "shape from x" methods [479] where x = shading (image intensity variation)/stereo (binocular depth perception)/motion (image flow)/texture (surface marking distribution)/etc. which require generic assumptions about the physical world which are usually justified (e.g., equipolarity, rigidity, etc.) [480]. Surface orientation is usually represented as the surface normal vector. Shape parameterization requires two parameters, $p = \partial z/\partial x$ and $q = \partial z/\partial y$ describing a surface patch S whose image is point (x, y) of intensity $I(x, y)$ where $I(x, y) = R(p, q)$ and R = surface orientation reflectance—this is the image irradiance equation. The surface normal vector is given by:

$$\frac{(p, q, -1)}{\sqrt{p^2 + q^2 + 1}}$$

Shape from motion is the problem of determining object shape based on egomotion—reconstruction through least-squares estimation and extraction of a Jacobian through singular value decomposition (SVD) is very sensitive to errors such as image noise and image parallax degradation with object–camera distance [481]. Shape from shading recovers shape from gradual variations in image shading and assumes that surfaces are Lambertian with only diffuse reflectance and locally smooth variation [482]. For a Lambertian surface, image intensity is given by:

$$I(x, y) = \frac{\rho(1 + pp_s + qq_s)}{\sqrt{1 + p^2 + q^2}\sqrt{1 + p_s^2 + q_s^2}}$$

where ρ = albedo (assumed constant), $(p, q, -1)$ = object surface normal, and $(p_s, q_s, -1)$ = light source direction. Shape from shading has been used for terrain analysis [483]. It is being explored to complement shape from stereo for the ExoMars rover mission. An object is distinguishable from the background by its characteristic pattern of gray levels representing its visual texture.

   Terrain classification would add greater functionality to autonomous rover navigation and, in particular, for automated traction control. Texture is an innate property of a surface defining its microstructure and its relationships. It is important in defining the 3D surface structure of an object. Projection introduces

Table 6.2. Fractal dimension of different terrains.

Terrain	Fractal dimension D
Flat plane	2.0
Rolling countryside	2.1
Old eroded mountain range	2.3
Young rugged mountain range	2.5
Stalagmite-covered plane	2.8

distortion clues—distance effect (objects closer to the image plane appear larger), position effect (angle between line of sight and image plane distorts the image), and foreshortening effect (angle between line of sight and surface normal distorts the image). Texture comprises small elements (texels) whereby surface orientation is determined from the uniformity and differences in neighboring texels. Texture gradients are variations in projected surface texture encoding information about the surface orientation and distance [484]. Surface slant angle σ is related to the gradient of texture:

$$\tan \sigma = \frac{\nabla \rho}{3\rho}$$

where $\rho = \rho_0(d^2/\cos \sigma) =$ measured texture density, $d =$ egocentric radial distance, and $\rho_0 =$ surface texture density with constraints on texture uniformity and regularity. No one approach is superior in all situations suggesting that multiple, integrated approaches should be adopted, each providing visual cues at multiple resolutions [485, 486]. Most natural objects require fractal description [487]. Fractals are ubiquitous in nature—they have dimension D greater than their topological dimension (Table 6.2).

Roughness can be quantified through the Hausdorff–Besicovitch dimension 0 (rough) $< h < 1$ (smooth). Fractal functions model 3D natural surfaces effectively because many physical processes generate fractal surfaces. A fractal has a Hausdorff–Besicovich dimension larger than its topological dimension (e.g., mountains have variable dimensions from 2.3 for an old worn mountain range to 2.5 for a young rugged mountain range). It thus corresponds to the concept of roughness of texture but different textures can have the same fractal dimension [488]. This makes it less suitable for characterizing texture. However, natural textures can be modeled as Brownian fractal noise whose local differences have a Gaussian probability distribution function with self-similar randomness at all scales. The power spectrum of fractal noise is given by:

$$P(w) = kw^{-2H-1}$$

where $H = T + 1 - D$, $T =$ topological dimension of space $= 2$ for images, and $D =$ fractal dimension. The image intensity of a fractal surface will also be fractal.

Automated terrain analysis using multispectral imagery incorporates texture data
derived from specifically selected channels including near-infrared wavelengths (a
function of chemical and physical composition) to segment the terrain into
distinctive regions [489]. Texture is defined from the mean and standard deviation
values of each $n \times n$ window of the image where n depends on the feature. Texture
gradient refers to the increasing fineness of visual texture with depth (especially
useful for outdoor scenes). Texture gradient is described by a 2D discrete Fourier
transform of an image window such that:

$$F(n,m) = \frac{1}{p^2}\sum_{x=0}^{p-1}\sum_{y=0}^{p-1} I(x,y)e^{-2\pi i(xn+ym)/p}$$

where $p =$ square window size. The power spectrum is defined as:

$$P(n,m) = \sqrt{F_{\text{Re}}^2(n,m) + F_{\text{Im}}^2(n,m)}$$

The phase spectrum is defined as:

$$\psi(n,m) = \tan^{-1}\left(\frac{F_{\text{Im}}(n,m)}{F_{\text{Re}}(n,m)}\right)$$

The power spectrum is invariant to translation but not rotation. Transforming the
power spectrum from cartesian to polar coordinates $P(r,\phi)$ provides directionality
information which at any given r is a 1D function $P(\phi)$. Peaks in $P(\phi)$ indicate
texture directionality while a uniform function indicates non-directionality of
texture. Depths derived from the texture gradient are only relative so distance is
computed from a projection function:

$$R = k\int P \cdot dy$$

where $k =$ proportionality constant (which requires calibration) and $y =$ vertical of
image. To compute a texture gradient image, a $(2k+1) \times (2k+1)$ window is
centered on each pixel where $k =$ region size of interest. Both k and n are set for
each image according to the nominal range of terrain class boundaries. The
texture gradient at pixel i for an $n \times n$ window is defined as:

$$\max\sqrt{[(\mu_i - \mu_{i+1})^2 + (\sigma_i - \sigma_{i+4})^2]}$$

where $\mu =$ mean of image intensity, and $\sigma =$ standard deviation. The confidence
associated with each classification may be computed using a Bayesian measure of
spectral classes.

The difference between two textures globally may be detected in their first-
order statistics (probability distribution) of local features comprising elongated
blobs of given orientation and size (textons) [490]. Wavelet transformation may be
regarded as texton decomposition where each texton equates to a particular
wavelet function. An extension to the texton concept is the 3D texton which has
an associated appearance vector that characterizes radiance distribution under

different lighting and viewing conditions [491]. Texture is defined by its response
to a bank of orientation and spatial frequency selective linear filters forming a
small set of prototype response vectors (textons). In particular, surface normal
variations offer clues to texture—specularities, ridges, grooves, shadows, occlusions, etc. which may occur at different heights and orientations. Robust
classification of 3D textons requires many images under different lighting and
viewing conditions in order to disambiguate features. Clustering involves unsupervised classification of patterns (features) based on their statistical similarity [492].
Similarity measures include the Minkowski metric (a special case of which is the
simple Euclidean distance) and Mahalanobis distance, the latter based on the
covariance matrix. From this, squared error may be used to perform clustering—
the k-means algorithm uses the squared error criterion with low computational
complexity. The K-means clustering algorithm finds the local minimum of a sum-
of-squares distance error:

$$e = \sum_{i=1}^{N} \sum_{j=1}^{K} q_{ij} |x_i - c_j|^2$$

where

$$q_{ij} = \begin{cases} 1 & \text{if } |x_i - c_j|^2 < |x_i - c_k|^2 \text{ for } k = 1, \dots, K \\ 0 & \text{otherwise} \end{cases}$$

The k-means algorithm may be implemented neurally with the Kohonen net which
offers superior performance in solution quality. Textons may be learned through
Bayesian-type methods [493]. Textons may be expressed as joint probability
distributions represented by the frequency histograms of rotationally invariant
filter responses [494]. The filters comprise a bank of Gaussians, derivatives of
Gaussians, and Laplacian of Gaussians at different orientations and scales (e.g.,
the MR8 filter bank comprises an edge and bar filter both at six orientations and
three scales, a Gaussian, and a Laplacian of Gaussian filter). Outputs from the
orientation-sensitive filters (edge and bar) are collapsed by selecting only the
maximum responses. Filter responses are clustered using the k-means algorithm
into textons which represent exemplar filter responses. Textons are used to label
each filter response in the training set during learning. Histograms of texton
frequencies are built corresponding to image models. Textons from different
texture classes are combined to form the texton dictionary. New histograms are
compared with the prelearned library using a nearest neighbor classifier to select
the closest model based on a χ^2 distance metric. Classification through the k-
nearest neighbor classifier uses a database of labeled features to classify new
features. It looks at the k nearest neighbors in the database and assigns the most
representative category of the k nearest neighbors with the smallest distance to the
unknown feature. However, it has been suggested that Markov random fields
(MRFs) supplant filter banks for texton representation [495]. Rather than using
individual pixels, an $n \times n$ pixel square neighborhood is adopted excluding the
central pixel. A 3×3 neighborhood gives good classification accuracy with
maximum performance achieved with a 7×7 neighborhood.

6.7   STEREOVISION

Stereopsis is the process whereby distance (depth) is visually reconstructed from binocular (or more) images [496]. Distance to the object can be computed from corresponding pixels in the two images and camera parameters. Views from binocular vision are slightly disparate horizontally (baseline) enabling depth to be recovered. The use of stereoscopic binocular vision from two differently positioned cameras allows 3D range estimation. Features are extracted in the two images, matched with each other, and the disparity computed from the positional differences between matched features. Stereovision is based on the disparity between the two images for which the two images equate thus:

$$I_{\text{left}}(x + \tfrac{1}{2}d(x)) = I_{\text{right}}(x - \tfrac{1}{2}d(x))$$

where $d(x) =$ disparity at point x. Three-dimensional information is derived from the disparity between corresponding regions in the two images from the two cameras. This angular disparity between the two images together with a known separation distance between the cameras enables computation of range. There are three primary problems to be solved in stereovision corresponding to three steps for the measurement of stereo disparity—calibration, correspondence, and reconstruction [497]. Calibration determines the camera system parameters and the selected location on the scene in one image. The correspondence problem defines the identification of corresponding points in the images of the two cameras so that the images may be matched (i.e., the same location must be identified in the other image). Correspondence comprises registering points in the two images that are projections from the same object. Thus, disparity between two image points can be measured (reconstruction). The two cameras should be mounted such that their sensing axes are parallel to camera optical axes. However, the possibility of using converging cameras has been investigated; they employ a least-squares estimation procedure [498]. Several constraints may be applicable [499]. Corresponding points must share the same elevation y (equipolar constraint). Stereoscopic vision imposes constraints on vergence (pan/yaw) and requires the cameras to have the same pitch (tilt) orientation and no roll.

The use of two images separated by a binocular parallax (baseline) allows intersection of the two projections to determine distance (Figure 6.17). Each point on the surface generates corresponding points in each of the two images. The intersection of light rays determines the location of the surface according to geometric constraints. Stereo configuration is usually through a parallel camera system of focal length f separated by intercamera distance b (baseline) with their optical axes parallel to each other. A point in a world coordinate system (x, y, z) projects onto point (x_L, y_i) in the left image and (x_R, y_i) in the right image:

$$x_L = f\frac{\left(x + \dfrac{b}{2}\right)}{z}, \quad y_i = f\frac{y}{z}, \quad x_R = f\frac{\left(x - \dfrac{b}{2}\right)}{z}$$

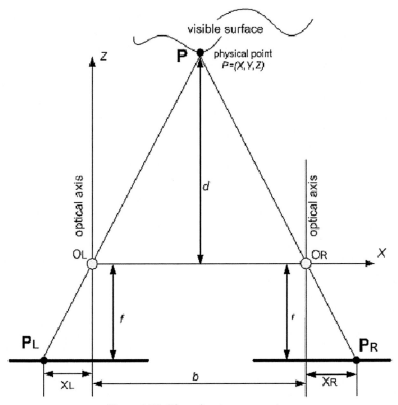

Figure 6.17. Binocular stereo geometry.

where $z =$ depth of object relative to the camera. From the midpoint (X, Y, Z) in cyclopean coordinates:

$$X = \frac{b\,(x_L + x_R)}{2\,(x_L - x_R)}, \quad Y = \frac{b}{2}\frac{by}{x_L - x_R}, \quad Z = \frac{bf}{x_L - x_R}$$

where $d = x_L - x_R =$ stereo disparity. Hence, range Z is dependent on disparity d which is the difference in the x-coordinates for corresponding points in the two images. Evidently, baseline distance b determines the accuracy of depth estimates but for a rover vehicle this will typically be limited. The coordinates (x_R, y_i, x_R) refer to coordinates in disparity space defined as $x_L - x_R$. Disparity is computed to extract the Z position of the feature in 3D space. Distance is inversely proportional to disparity; disparity is proportional to the baseline. This assumes parallel optic axes—vergence generates a more complex formulation. A point $P(X, Y, Z)$ is projected onto the stereo plane defined by the two convergent optical axes ($Y = 0$)

defining depth as:

$$Z = b \cdot \frac{\tan(\alpha - \gamma)\tan(\beta + \delta)}{\tan(\alpha - \gamma) + \tan(\beta + \delta)}$$

where $\alpha, \beta =$ vergence angles, $\gamma = \tan^{-1}(x_L/f_L) =$ angular disparity, $\delta = \tan^{-1}(x_R/f_R) =$ angular disparity, $x_R = x_L + d$, $f_L, f_R =$ focal lengths (in pixels) of cameras, and $b =$ baseline. Distance to the fixation point with zero angular disparity is given by:

$$Z_{fp} = B\frac{\tan\alpha\,\tan\beta}{\tan\alpha + \tan\beta}$$

Active camera control allows self-calibration of camera focal lengths and vergence angles. Camera focal length may be estimated by moving the camera pair between two fixation points:

$$\theta = \tan^{-1}\left(\frac{x}{F}\right)$$

where $\theta =$ camera rotation and $x =$ displacement of fixation point on image plane. Hence, 3D information is derived from the disparity between corresponding regions in the two images from the two cameras. This angular disparity between the two images together with a known separation distance between the cameras (baseline) enables the computation of range.

The reconstruction problem involves determining the 3D structure from the disparity map. Disparity vectors are computed for a grid of points. The binocular assumption implies that the optic centers of both cameras and the baseline between them define the equipolar plane of projection. Corresponding pixels lie on equipolar lines which limits computational requirements. A sliding window is placed over left and right images for each value of the disparity vector. There are several algorithmic approaches for determining similarity between images for correspondence. Mismatches due to poor spatial discrimination grow when uncertainties increase as the search window size increases (typically 10–20 pixels in size) but too small a window will yield false positives. A cross correlation algorithm computes a matching score for each pixel between the two cameras. The best match provides correspondence between the two images. Cross-correlation methods of matching are sensitive to noise and illumination differences. Errors can be reduced by assuming that the terrain is locally continuous. Multiple cross-correlation algorithms based on the vector inner product detect patterns in multiple images. Stereovision should be based on area correlation rather than edge correlation as natural scenes are rich with texture but deficient in straight edges. The correlation coefficient describes the similarity between features and is given by:

$$c_{ij} = I \otimes S_{ij} = \sum_{k=1}^{m}\sum_{l=1}^{n} I_{kl}S_{kl}$$

There are several basic stereovision algorithms:

$$C_{\text{NCC}} = \frac{\sum\limits_{u,v}(I_i(u,v) - \bar{I}_i)(I_2(u+d,v) - \bar{I}_2)}{\sqrt{\sum\limits_{u,v}(I_i(u,v) - \bar{I}_1)^2(I_2(u+d,v) - \bar{I}_2)^2}} \qquad \text{normalized cross-correlation}$$

$$C_{\text{SSD}} = \sum\limits_{u,v}(I_1(u,v) - I_2(u+d,v))^2$$

$$= \sum\limits_{i=-n}^{n} I_1^2(u,u) + \sum\limits_{i=-n}^{n} I_2^2(u+d,v)$$

$$- 2\sum\limits_{i=-n}^{n} I_1(u,v)I_2(u+d,v) \qquad \text{sum of squared differences}$$

$$C_{\text{NSSD}} = \sum\limits_{u,v}\left(\frac{I_1(u,v) - \bar{I}_1}{\sqrt{\sum\limits_{u,v}(I_1(u,v) - \bar{I}_1)^2}} - \frac{I_2(u+d,v) - \bar{I}_2}{\sqrt{\sum\limits_{u,v}(I_2(u+d,v) - \bar{I}_2)^2}}\right)^2$$

normalized sum of squared differences

$$C_{\text{SAD}} = \sum\limits_{u,v}|I_1(u,v) - I_2(u+d,v)| \qquad \text{sum of absolute differences}$$

SAD is the most computationally efficient approach but it is not robust. With SSD, as the correlation (last term) between the two images increases so the Euclidean distance decreases. A normalized correlation quantifies similarity which is invariant to scaling. If the two images are regarded as vectors (either line or array images), the correlation may be regarded as an inner product of the form:

$$\langle I_1, I_2 \rangle = |I_1||I_2| \cos\theta$$

where θ = angle between I_1 and I_2. Correlation is maximized when the angle between the vectors is zero. A similar argument applies to 2D images so that filter convolutions are a form of correlation:

$$I_1(x,y) \cdot I_2(u,v) = \sum\limits_{i=-n}^{n}\sum\limits_{j=-n}^{n} I_1(x,y)I_2(u,v)$$

Normalized cross-correlation is the most commonly used approach for finding the highest correlation between matching points:

$$C_{ij} = \frac{\sum\limits_{i}^{n}\sum\limits_{j}^{m}(I_1(i,j) - \bar{I}_1)(I_2(i+d,j) - \bar{I}_2)}{\sqrt{\sum\limits_{j}^{n}\sum\limits_{j}^{m}(I_1(i,j) - \bar{I}_1)^2(I_2(i+d,j) - \bar{I}_2)^2}}$$

where

$$d = \text{baseline}$$

$$\mu_1 = \frac{1}{mn} \sum_i^n \sum_j^m I_1(i,j)$$

$$\mu_2 = \frac{1}{mn} \sum_i^n \sum_j^m I_2(i,j)$$

$$\sigma_1^2 = \frac{1}{mn} \sum_i^n \sum_j^m (I_1(i,j)^2 - \mu_1(i,j)^2)$$

$$\sigma_2^2 = \frac{1}{mn} \sum_i^n \sum_j^m (I_2(i,j)^2 - \mu_2(i,j)^2)$$

This utilizes mean pixel intensity and variance of pixel intensity in each image. The maximum score indicates the best match. An efficient sum-updating technique may be adopted to reduce computational burdens [500]. Each pixel is assigned a 3D position based on the computed range. To further reduce computational requirements, a subset of pixels is chosen randomly for analysis. Correlation also provides a best-fit matching criterion for comparison with prototype templates [501] (e.g., Moravec correlation):

$$c_{ij}(f) = \frac{\sum_{k=0}^n \sum_{l=0}^m f_{k,l} I_{i+k,j+l}}{\frac{1}{2} \sum_{k=0}^n \sum_{l=0}^m f_{k,l}^2 + \frac{1}{2} \sum_{k=0}^n \sum_{l=0}^m I_{i+k,j+l}^2}$$

In template matching, template image I of p pixels contains m features to be matched with a target image T of q pixels containing n features. This requires correlations above a given threshold to be computed across the entire image. Total match of template T and image I (for a minimum of N feature matches):

$$\mu(T,I) = \begin{cases} \frac{1}{|Q|} \sum_f c(f) & \text{if } |Q| > N \\ 0 & \text{otherwise} \end{cases}$$

In fact, there will be a distribution of candidate disparity values. One approach is to estimate disparity independently (statistically uncorrelated) for each image pixel [502]. This models a joint a priori distribution of the disparity field with noise modeled as stationary Gaussian white noise with zero mean.

$$I_l(x) = I(x) + n_l(x)$$
$$I_r(x) = I(x + d(x)) + n_r(x)$$

where $d = $ disparity. Measurements are intensity differences (errors) between two images in small windows around pixels being matched x_0:

$$e(x_0 + \Delta x_i; d) = I_r(x_0 + \Delta x_x - d) - I_l(x_0 + \Delta x_i)$$

$$e(x_0, d) = (e(x_0 + \Delta x_1; d), \ldots, e(x_0 + \Delta x_n; d))^T$$

where Δx_i = pixel indexes of window and n = window size. The joint probability density function of noise to be maximized (thereby minimizing $e^T e$):

$$f(e|d) = \frac{1}{(2\pi)^{n/2}\sigma} e^{-(1/(2\sigma^2))e^T e}$$

where $\sigma = \sqrt{\sigma_r^2 + \sigma_l^2}$. Maximum likelihood estimation (MLE) then treats disparity as an unknown deterministic parameter. Bayesian estimation is suitable when probabilistic a priori information is available about the disparity field, so posterior probability of disparity d given measurements e is:

$$f(d|e) = \frac{f(e|d)f(d)}{\sum_i f(e_i|d_i)f(d_i)}$$

where $f(d)$ = prior probability function of d. Optimal estimate of d is given by maximum a posteriori probability $f(d|e)$. The prior probability function of d derives from external sensors independent of images. In the simplest case, this may be assumed to be uniform over minimum and maximum disparity limits:

$$f(d|e) = \frac{f(e|d)}{\sum_i f(e_i|d_i)} = K \cdot f(e|d)$$

Maximizing $f(e|d)$ provides the best estimate of disparity d. More complex models use Gaussian models of the prior probability function of d. Stereovision algorithms output a 3D point cloud that may be input directly into an autonomous navigation algorithm (Figure 6.18).

An alternative solution proposed for the correspondence problem of stereovision uses bandpass complex Gabor filters that are sensitive to local shifts between images indicated by local phase differences to generate highly accurate disparities, but it suffers from difficulties at high-depth boundaries and occlusions [504]. This involves convolving left and right images with a complex Gabor filter and the difference in phase between points in each image indicates the shift. Alternatively, a Fourier transform approach translates the translation in the spatial domain into a proportional translation in the frequency domain exhibited as a phase difference. A Cepstral filter may be used to measure the disparity. The two images are phase-shifted versions of each other:

$$R(x, y) = L(x, y) * \delta(x - x_d)(y - y_d)$$

where $x_d = h/(2\pi)\Delta\theta^i$, $y_d = w/(2\pi)\Delta\theta^j$, $\Delta\theta^i = \theta_R^i - \theta_L^i$ = phase difference between two images (horizontal), and $\Delta\theta^j = \theta_R^j - \theta_L^j$ = phase difference between two

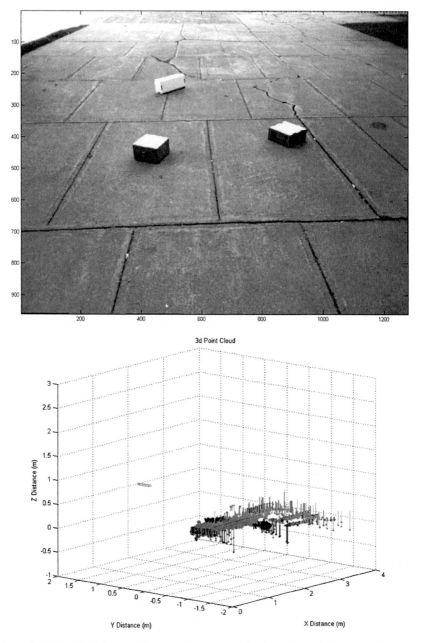

Figure 6.18. Rectified image of an environment with obstacles and the 3D point cloud extracted through stereovision [credit Adam Mack, Carleton University].

images (vertical). Disparity along the x direction is given by:

$$d_x = \frac{M}{2\pi}(\theta_R^i - \theta_L^i) + \frac{m}{2\pi}\Delta\theta$$

where $\Delta\theta = $ phase angle error and $M = $ window x dimension. Hence, the error term can be reduced by reducing the window size. Despite these alternatives, cross correlation is almost universally adopted for stereovision.

Binocular vision also includes vergence control. Vergence motions are based on disparity cues to compensate for motion in depth. Vergence control involves servoing the tilt of a binocular vision system to fixate at a selected point—this is achieved by decreasing binocular disparity between the two images [505]. The fixation point is selected from one camera, its pose controlled to keep its optical axis pointing at it. The pose of the other camera is determined by the disparity between the two images. The control of vertex provides fixation on a target by ensuring that the optical axes of the two cameras intersect at this point. By gazing at the target, optical flow is reduced. Vergence controls the angle between the optic axes of each stereo camera [506]. Horizontal disparity $d = x_l - x_r$ is used to measure the vergence angle such that:

$$d = G_r(\theta_r - \theta_r^0)$$

where $G_r = $ sensitivity, $\theta_r = $ vergence angle, and $\theta_r^0 = $ vergence angle for centered target such that $d = 0$. Target disparity between stereo image pairing is given by:

$$S(I_{x,y}^r, I_{x,y}^l) = \sum (I_{x,y}^r - I_{x,y}^l)^2$$

The minimum correlation function is given by:

$$C(I^r, I^l, d) = \int (I_{x+d}^r, I_{x,y}^l)$$

A parallel neural network may be tuned to be sensitive to discrete values of disparity. Image coordinates are related to pan θ_p and tilt θ_t angles by:

$$x = G_p(\theta_t^0, \theta_v^0)(\theta_p - \theta_p^0)$$
$$y = G_t(\theta_t^0, \theta_v^0)(\theta_t - \theta_t^0)$$

For static targets, image velocities are related to pan w_p and tilt w_c velocities:

$$v_x = G_p(\theta_t^0, \theta_v^0)w_t$$
$$v_y = G_t(\theta_t^0, \theta_v^0)w_t$$

The extraction of multiple cues from images provides greater robustness in recovering 3D structure than from a single cue [507]. Two successive pairs of stereo images invoke a requirement for matching four images, one set for stereo and the other for image flow.

The overall vision-processing chain involves a number of modules that may be combined in different ways (Figure 6.19). The MER stereovision sequence was as

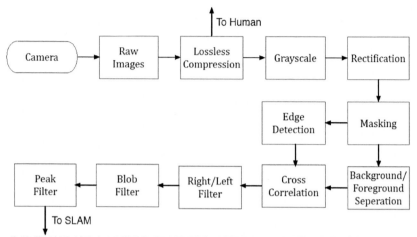

Figure 6.19. Typical rover vision-processing chain [credit Adam Mack, Carleton University].

follows:

(i) 1,024 × 1,024 pixel images were averaged down to 256 × 256 pixels;
(ii) images were rectified so that horizontal alignments matched;
(iii) difference of Gaussian filter was applied to remove noise;
(iv) 7 × 7 windows were correlated between images to establish and eliminate outliers;
(v) 3D coordinates were computed for matches.

The results from these image-processing algorithms have been impressive. This algorithm takes 5 s to compute on a R3000 12 MHz processor. Real time stereoscopic vision processing requires high processing powers which can only be supplied by multiprocessing architectures [508]. Much of the processing is concerned with repetitive low-level algorithms—matrix computations, convolutions, and FIR filtering—to which dedicated DSP (digital signal processors) can be applied. DSP processors such as the 32-bit Motorola DSP96002 include an on-chip single-cycle multiplier/accumulator, integer ALU and data address generator for computation and multiple memories, and data/address buses for support functions for 100 MFLOPS performance. A passive range pattern sensor system has been proposed which comprises an optical stereo system with two lenses and one CCD imager that measures the range of a scene projected onto the CCD [509]. Although this reduces the hardware required for stereovision, the short baseline will introduce difficulties for image correlation algorithms. The high computational cost of real-time stereovision lends itself to implementation on a field-programmable gate array (FPGA) [510]. Successful parallel/pipelined implementation on a single FPGA has included stereomatching between the two rectified images using correlation, disparity mapping, subpixel estimation using parabola fitting, and spike removal using thresholds.

7

Rover vision—advanced capabilities

In this chapter, we progress beyond the basics of rover vision to introduce some more advanced approaches to vision processing as it underlies the development of greater enhancements to rover missions than any other facet. Often, these require considerable computational effort which may be traded with dedicated hardware. Neuromorphic vision sensors are biomimetic chips that process images rapidly at the focal plane in parallel [511]. For instance, such an approach has been proposed for planetary landers [512]. Alternatively, a log-polar vision sensor with decreasing resolution towards the periphery may be employed to implement active vision though this is unlikely in the context of planetary exploration (this is relatively easily done in software) [513]. The most likely eventuality is that computational resources onboard planetary rovers will evolve greater capacities to enable these greater performances.

7.1 VISUAL ODOMETRY

Although an advanced technique, visual odometry has been employed and validated on the MER missions. Stereo visual odometry is a vision-based motion estimation technique that estimates the egomotion of a stereo camera pair through its environment in all six degrees of freedom [514]. Visual odometry tracks visual features—it uses the stereo navigation cameras from which image features were extracted in sequential stereoscopic image pairs. Visual odometry compares stereo image pairs from the stereo cameras to self-localize and correct for the in-accuracies of wheel odometry. The 3D positions of these features were computed through stereo-matching. These features were tracked through successive frames to determine the distance moved. The MER rover Opportunity was driven uphill over 19 m distance and wheel odometry underestimated the travel distance by 1.6 m due to slippage (i.e., ~10%). Visual odometry reduced localization error to

© Springer-Verlag Berlin Heidelberg 2016   263
A. Ellery, *Planetary Rovers*, Springer Praxis Books,
DOI 10.1007/978-3-642-03259-2_7

~1%. Visual odometry is essential under conditions of high slip on inclined and sandy surfaces where slippage is much higher. It involves taking stereoscopic images and processing with 3D stereovision at time t and consecutive stereo pair images at time $t + 1$ within a traverse segment between two rover locations. Pixels are matched between the two images of both stereoscopic pairs at the different times. Pixel-based stereovision provides a set of 3D points in Cartesian space. 3D-point-to-3D-point matches are established and used to recover velocity and position values between time t and $t + 1$. For precise motion estimates, the number of pixel matches should be large. Points can be tracked over more than two stereo frames and used in an estimation filter to limit error growth from local estimates. This is a simplified version of simultaneous localization and mapping (SLAM). This process thus involves (Figure 7.1):

(i) acquiring an image pair from the left and right cameras;
(ii) finding features in images (using Harris or related operators);
(iii) computing the 3D coordinates of features;
(iv) matching features between different consecutive or viewpoint images;
(v) computing motion between viewpoints.

Successive image pairs allow tracking the motion of interesting terrain features in 2D pixel and 3D world coordinates. A final motion estimate is determined through a maximum likelihood estimator [515]. This involved tracking the relative motion of terrain features between successive stereo images. Features were selected on the basis of their matchability between images using an interest operator which quantified distinctiveness such as the Harris corner detector which is rapidly computed. To maintain an even distribution of features, a minimum distance between features is defined. To reduce computational costs, pixels may be merged into larger cells to which the corner detector is applied. The feature with the strongest corner response in each grid cell was selected as a candidate feature until a fixed number of features had been defined. Each feature's 3D position is computed by stereo-matching. A normalized cross-correlation algorithm along an equipolar line around a 3×3 neighborhood provided the best stereo match. The 3D positions of features were determined by intersecting rays projected from camera models. Errors and inaccuracies ensured that they do not usually intersect at a point. The shortest distance between two rays indicated the goodness of the stereo match.

Position of a feature with respect to each camera is given by:

$$P_1 = C_1 + m_1 \hat{r}_1$$

where $C_1 = x_1, y_1, z_1 = $ position of camera 1 and $m_1 = |P_1 C_1| = $ magnitude of vector r_1 from camera to feature

$$P_2 = C_1 + m_2 \hat{r}_2$$

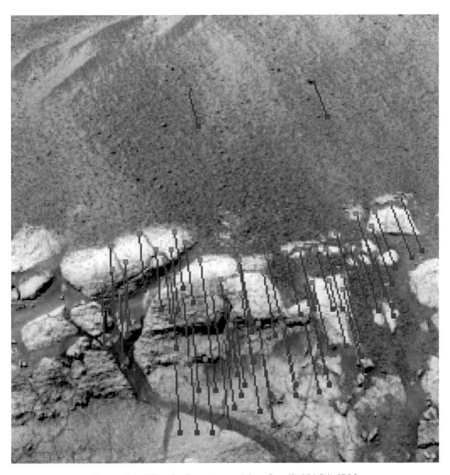

Figure 7.1. Visual odometry on Mars [credit NASA JPL].

where $C_2 = x_2, y_2, z_2 = $ position of camera 2 and $m_2 = |P_2C_2| = $ magnitude of vector r_2 from camera to feature. This gives:

$$m_1 = \frac{b\hat{r}_1 - (b \cdot \hat{r}_2)(\hat{r}_1 \cdot \hat{r}_2)}{1 - (\hat{r}_1 \cdot \hat{r}_2)^2}, \quad m_2 = (\hat{r}_1 \cdot \hat{r}_2)m_1 - b\hat{r}_2, \quad P = \frac{(P_1 + P_2)}{2}$$

where $b = C_1 - C_2 = $ stereo baseline. The covariance of each feature is computed. After the rover moves, a second pair of images is acquired and projected onto the new images based on wheel odometry. A cross-correlation–based search locates the features in the second image pair and stereo-matching extracts the depth coordinate based on affine transforms [516]:

$$x_2 = ax_1 + by_1 + c \quad \text{and} \quad y_2 = dx_1 + ey_1 + f$$

where (a, b, c, d) = unknown affine transform coefficients determined by minimizing a merit function of the form:

$$m = \sum (I_1(x_1, y_1) - I_2(x_2, y_2))^2$$

Motion estimation proceeds in two steps—first, a least-squares estimate determines motion rapidly to eliminate outliers. Stereo-matching tracked features on the second image pair yields their new 3D positions. If the motion is accurately measured, the difference between the initial and final 3D positions will lie within an error ellipse. A maximum likelihood estimator completes the process more accurately:

$$P_i = RP_{i-1} + T + \varepsilon$$

where P_i = current feature position, P_{i-1} = previous feature position, R = three rotation angles of the rover, T = three translation coordinates of the rover, and ε = position error modeled as a Gaussian distribution. This is achieved through minimization of

$$\sum_{i-1} A_j^T W A_j$$

where $A = P_i - RP_{i-1} - T$ and W = inverse covariance matrix of ε. The maximum likelihood estimate of parameter θ is the value $\hat{\theta}$ that maximizes the likelihood function $f(y|\theta)$. If Gaussian statistics apply, the least-squares estimate provides a maximum likelihood estimate. From Bayes theorem:

$$f(\theta|y) = \frac{f(y|\theta)f(\theta)}{f(y)}$$

where $f(y)$ = true probability density function and $f(y|\theta)$ = model probability density function. The Kullback–Leiber information distance represents relative entropy between two probability distributions:

$$I_{\mathrm{KL}} = \sum_{i=1}^{n} f(y_i) \log \frac{f(y_i)}{f(y_i|\theta)}$$

The Fischer information matrix is defined by:

$$M = \left\langle \left(\frac{d}{d\theta} \log f(y|\theta) \right)^2 \right\rangle$$

Gaussian error models are propagated into pose estimation to improve precision. The forward kinematics of the rover provides rover motion given wheel, steering, and kinematic lever positions and rates. Inverse kinematics computes wheel velocities and steering angles to generate the required rover motion. The algorithms are specific to the rocker-bogie configuration with 16 DOF enabling derivation of wheel Jacobians related to rover velocity [517]. An indirect extended Kalman filter is used to fuse inertial measurements with relative pose from visual odometry to determine the state vectors of vehicle attitude, vehicle position, and rates. Optical tracking can be achieved through a mean shift tracking algorithm [518]. It tracks within a rectangular search window representing the most probable target in a consecutive image based on visual feature distribution. The target may be defined by color distribution or texture. It uses an iterative gradient ascent (mean shift) algorithm to find a maximum similarity peak between the target and candidate image—it thus resembles a traditional cross-correlation algorithm. The Bhattacharyya coefficient is a Bayesian-like metric that measures similarity. For a normal distribution, the Bhattacharyya distance is given by:

$$d = \tfrac{1}{8}(\mu_2 - \mu_1)^T \left(\frac{\Sigma_1 + \Sigma_2}{2}\right)^{-1} (\mu_2 - \mu_1) + \tfrac{1}{2} \ln\left(\frac{|\tfrac{1}{2}(\Sigma_1 + \Sigma_2)|}{\sqrt{|\Sigma_1||\Sigma_2|}}\right)$$

The first term gives the class separability between the two means of each distribution, while the second term gives the class separability of covariances of the mean distribution. The chief problem was that visual odometry required considerable computation time using the onboard processor so it was used only on short drives with high slippage (on slopes and during wheel dragging). This favors the use of optic flow which may be used to perform motion detection, object segmentation, time to collision, and focus of expansion computations [519].

The range of rover-based stereovision is limited by the short baseline distance between stereoscopic cameras. Wide-baseline stereovision was accomplished by using two images captured by the rover at different positions [520]. It involves acquiring consecutive NavCam stereo pairs within a traverse segment between two rover locations. This allowed longer range depth estimation. This is achieved through cross-correlation matching between images with maximum likelihood estimation to achieve highly accurate localization. This is a form of active vision fusing both short and long-baseline viewpoints to reduce visual ambiguities [521]. A network of continuously linked images over several frames is built to track the motion of terrain features. Some forms of visual odometry rely on optic flow in tracking ground point features (Figure 7.2).

Visual odometry has been successfully applied to the Mars Exploration Rovers to correct for wheel odometry errors under high-slip conditions [522, 523]. Wheel encoder and inertial measurement unit data became unreliable for navigation, necessitating the use of visual odometry which involved the use of stereo images from the NavCam to compute an update to the 6 DOF rover pose. The MERs experienced slippery slopes with high inclinations up to $30°$ during which slippage reached 100%. This involved tracking the relative motion of terrain features between successive stereo images. These features were selected on the basis of distinctiveness based on the use of a Harris or Fortner corner detection operator. It required at least 60% overlap between adjacent images (i.e., 75 cm and/or $18°$ steer drive steps). Each step required 2–3 min of computation time on the MERs' 20 MHz RAD6000 onboard computers so visual odometry was used only sparsely. Thus, visual odometry requires features in the environment which can be sparse on flat terrain (though slip is low in such environments).

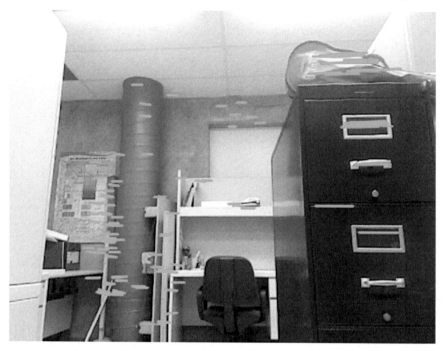

Figure 7.2. Visual odometry–based feature flow using a Kinect sensor [credit Pablo Molina, Carleton University].

The Meridiani Planum was challenging for visual odometry due to its lack of features and texture. However, such terrain is generally flat with reasonable wheel odometry and attitude measurement accuracy. Slopes, where visual odometry become more useful, are generally characterized by distinctive features. Similar arguments apply to rocky environments. Visual odometry has been used several times (e.g., on Opportunity from sol 133 to 312 within the 130 m diameter Endurance crater). It used visual odometry for 11% of its total traverse over its first 5.95 km (compared with 28% using stereovision). The Gusev crater site where the MER rover Spirit landed was more suited to visual odometry by virtue of its rock distribution and it was employed more extensively on Spirit particularly to compensate for wheel dragging. It used visual odometry for 16% of its first 4.80 km of traverse (compared with 21% using stereovision). In both cases, the remainder of each traverse was conducted as blind drives. Similarly, in both cases, visual odometry was primarily employed on sloping terrain where slippage was high. The chief problem was that visual odometry required considerable computation time using the onboard processor so it was used only on short drives with high slippage (on slopes and during wheel dragging).

7.2 OPTIC FLOW–BASED NAVIGATION

Planetary rovers utilizing traditional autonomous navigation systems based on stereovision and mapping must stop in order to perceive and plan its trajectories. They currently possess limited onboard computational resources which limits their effective traverse speed. A typical navigation profile is based on a stop–scan–drive approach. It will adopt a stop–move strategy in which it images the scene while halted, performs map building, and plots a trajectory to follow. Environmental data about the scene are acquired by stereovision and/or LIDAR scanning while stationary. This provides the basis of digital environment mapping (DEM) of the terrain. Once the trajectory has been computed, the rover will follow it for a fixed distance according to the nature of the terrain (nominally 5 m). These motions are executed blindly until the next perception cycle. On completion of the drive, the rover stops and repeats the process. This current stop–move method every 5–10 m proposed for ExoMars and other rovers is inefficient and potentially hazardous. The rover is "blind" during the traverse—any errors in the plotted trajectory, uncertainty in sensory data, or other unobserved hazard could render the rover damaged resulting in mission loss. Visual measurement of rover motion would compensate for this deficiency—frame transfer CCDs that allow charge image motion may be used to implement time delay and integration (TDI) mode imaging [524]. Rover motion imposes image blurring which degrades image quality. To compensate for blurring, an estimate of the image point spread function is required to allow image deconvolution (e.g., the Richardson–Lucy algorithm maximizes the image model likelihood function). One way to achieve this is by using a high-resolution camera and a low-resolution camera, which imposes a mass penalty [525]. Optic flow provides the capability of continuous sensing during rover motion [526]. As a robotic rover is a moving vehicle, motion detection allows the determination of vehicle motion. Optical flow field methods provide the means for motion detection and have applications in active vision particularly for smooth pursuit, optokinetic response, and vestibular optical response. Optic flow may be implemented to perform obstacle detection through inverse perspective mapping [527, 528]. This transforms image motion in world coordinates into frontoparallel motion of observer-referenced coordinates. An obstacle is then defined as an entity that rises above the horizontal plane of the observer's 2D path. Flying insects such as bees use optic flow for navigation purposes including obstacle avoidance, dead reckoning, and landing. They rely on measurement of visual motion generated by egomotion which generates an optomotor response to keep the insect on a straight line course. Optic flow may be implemented through a behavior control methodology involving two basic behaviors—collision avoidance and goal pursuit. Optic flow is considered to be one of the most important aspects of vision given its fundamental use in biological organisms [529]. Optic flow is directly applicable to rover navigation but has yet to be deployed in that capacity [530]. Optic flow can also be used for slip sensing—when a downward-pointing camera moves in the direction of motion, an image of the ground and surroundings sweep backward forming an optic flow field. Optic flow provides the capability of continuous sensing during rover motion [531].

Optic flow is defined by the pattern of velocities of brightness patterns in an image due to relative motion between the observer and objects in the environment. Optic flow is a projection of 3D motion enabled by movement of a sensor moving relative to a static scene. 2D optic flow is the 2D motion of points in an image which represents the perspective projection of 3D object motion onto an image plane. Projection of point $P = (X, Y, Z)$ onto an image is:

$$p = (x, y) = \left(\frac{fX}{Z}, \frac{fY}{Z} \right)$$

Optic flow results from the projection of object movement in 3D (with 6 DOF) onto a 2D image plane. Optical flow as a 2D velocity field can be used to extract 3D motion of the objects in the visual field. An equivalent 3D scene flow was described by Vedula et al. (2005) [532]. Optical flow methods generate a vector field representing the motion of an image's pixels. Associated with each pixel is a velocity vector, and the vector field of all pixels of the image forms the optical flow field. The assumption is that the motion field and optic flow field coincide as spatial variations in intensity correspond to physical features on visible 3D surfaces though, in general, this is not necessarily true [533]. For the condition to be true, self-motion must be at sufficiently high speeds to make environmental object motion negligible—this will be so in static planetary environments. There are certain general local assumptions—uniform illumination, Lambertian surface reflectance, and pure translation parallel to the image plane. Discontinuities in the optic flow pattern enable segmentation of the image into regions corresponding to different objects. Continuity of the motion field is ensured by applying the constant brightness constraint which states that a moving pixel's brightness is constant over time between images. Optic flow is a gradient-based approach that relates motion to image brightness through a differential equation (x defines two image spatial coordinates x, y):

$$I(x, t) = I(x + \delta x, t + \delta t)$$

Taylor series expansion yields:

$$I(x, t) = I(x, t) + \nabla_x I \cdot \delta x + \delta t I_t + O^2$$

where $\nabla_x I = \partial I / \partial x = $ spatial gradient of intensity and $I_t = \partial I / \partial t = I_i - I_{i-1} = $ time gradient of intensity. Total temporal derivative along the trajectory of an image point is given by differentiating:

$$\frac{dI}{dt} = \frac{\partial I}{\partial t} + \nabla I \cdot v = -K\rho(x) \frac{d(n, r)}{dt}$$

where $\nabla I = $ spatial gradient between images, $\partial I / \partial t = $ local temporal intensity difference between images, $v = \binom{u}{v} = $ translational motion field in motion direction (optic flow), $\rho(x) = $ constant albedo of image point, $n = $ surface normal,

r = surface radiance, and K = scaling factor. It is assumed that $n \cdot r$ is constant:

$$\frac{d(n \cdot r)}{dt} = 0$$

Hence, horizontal translation of a point from one image to another is given by:

$$\frac{dI}{dt} = \frac{\partial I}{\partial x} u + \frac{\partial I}{\partial y} v + \frac{\partial I}{\partial t} = 0$$

where $u = dx/dt$ and $v = dy/dt$. Hence,

$$\frac{dI}{dt} = 0$$

This is the smoothness constraint such that the optic flow field varies only slowly over single objects [534]. This gives the optic flow constraint equation assuming constant brightness in the direction of the image gradient (normal component of optic flow):

$$\nabla I v + \frac{\partial I}{\partial t} = 0, \quad \text{that is,} \quad v = -\frac{\partial I/\partial t}{\nabla I}$$

Typically, images are smoothed prior to computation of the derivatives but such smoothing may introduce blurring. Optic flow is a gradient-based approach that determines small local disparities between consecutive images by determining the differential equation relating motion and brightness. Optic flow (u, v) at image point $P(x, y)$ may be extracted assuming a standard camera model. Transformation from world to image coordinates is given by:

$$x = \frac{fX}{Z}, \quad y = \frac{fY}{Z}$$

where (X, Y, Z) = Cartesian world coordinates, f = focal length of camera, and (x, y) = image coordinates. This assumes that there are no scaling factors to be considered. Now,

$$\frac{dP}{dt} = v + w \times P$$

where $v = (v_x, v_y, v_z)$ and $w = (w_x, w_y, w_z)$. Hence [535, 536]:

$$u(x, y) = \frac{1}{Z}(-f v_x + x v_z) + w_x \left(\frac{xy}{f}\right) - w_y \left(f + \frac{x^2}{f}\right) + w_z y$$

$$v(x, y) = \frac{1}{Z}(-f v_y + y v_z) + w_x \left(f + \frac{y^2}{f}\right) - w_y \left(\frac{xy}{f}\right) - w_z x$$

Or, more compactly, we can approximate the flow field by an affine mapping (ignoring focal length for convenience) [537]:

$$\begin{pmatrix} u \\ v \end{pmatrix} = \begin{pmatrix} w_x xy - w_y x^2 + y w_z \\ w_x y^2 - w_y xy - x w_z \end{pmatrix} + \begin{pmatrix} \dfrac{v_x - x v_z}{z} \\ \dfrac{v_y - y v_z}{z} \end{pmatrix}$$

The optic flow field has both translational and angular velocity components. The translational component has a radial structure expanding/contracting from a focus of expansion at location

$$(x_0, y_0) = \begin{pmatrix} \dfrac{f v_x}{v_z}, \dfrac{f v_y}{v_z} \end{pmatrix}$$

The focus of expansion (FOE) defines points on the motion field from which flow vectors appear to originate. The focus of contraction (FOC) defines points where flow vectors disappear. They are both in the field of view for pure translation but lie outside the field of view for pure rotation. Mixed rotation and translation cause the relative positions of FOE and FOC to vary. The focus of expansion which lies at the center of the radial flow pattern determines the direction of movement. Determination of the focus of expansion of the optic flow field based on translational velocity is important for approach to targets. The rotational component is a linear combination of three weighted angular velocity components. Flow circulation computes rotational parameters from which the focus of expansion can be computed [538]. Circulation about a contour is related to the curl of the flow field at the centroid of the contour which allows angular velocities to be computed. Optic flow may be exploited above and below the horizon in different ways—below the horizon provides x and z measurements while above the horizon provides rotation [539]. The horizon offers a high-contrast transition. If the camera faces the tangent of its path (i.e., faces forward without attitude motion), then:

$$v_x = v_y = w_x = w_y = 0$$

Then:

$$\begin{pmatrix} u \\ v \end{pmatrix} = \begin{pmatrix} w_y x^2 / f \\ w_y xy / f \end{pmatrix} - \begin{pmatrix} \dfrac{x v_z}{z} \\ \dfrac{y v_z}{z} \end{pmatrix}$$

If $v_z = 0$ (no vertical motion):

$$\begin{pmatrix} u \\ v \end{pmatrix} = -\frac{f}{z} \begin{pmatrix} v_x \\ v_y \end{pmatrix}$$

this is parallel in the direction of translational motion.

Estimating the visual flow field is underconstrained due to the aperture problem such that flow is determined only along the image gradient. With a centering reflex, the vertical component of flow is negligible so horizontal flow is given by [540]:

$$u = -\left(\frac{dI/dt}{dI/dx}\right)$$

This is particularly suitable for rovers which will have vertical displacements induced by the terrain. Only normal velocity $v_\perp$ perpendicular to the viewing direction—

$$\frac{I_t}{\sqrt{I_x^2 + I_y^2}}$$

can be extracted—this is the aperture problem as parallel motion to the viewing direction cannot be computed to extract 3D object velocity components:

$$v_\perp = \frac{-I_t \nabla_z I}{|\nabla_z I|^2}$$

Optic flow is created by the translational and rotational motion of the observer $(v, w)^T$ with respect to an object at a distance d and offset by θ (Figure 7.3):

$$v_\perp = -w + \left(\frac{v}{d}\right)\sin\theta$$

where $w =$ angular velocity of robot rotation (yaw), $v =$ rover true translational velocity, $\theta =$ angle between obstacle and direction of motion, and $d =$ line-of-sight distance between rover and object. Closer features produce larger flows. Optic flow is maximum when $\theta = 90$ (tangential) and essentially zero when $\theta = 0$ (normal) so optic flow cannot be used to detect obstacles in the direction of motion (except through looming). Hence,

$$d = \frac{v \cdot \sin\theta}{v_\perp + w}$$

In order to compute distance, rotational motion should be minimized (blowflies enforce this by undergoing rapid discrete rotations) so $w = 0$. Spatial computa-

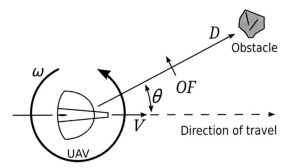

Figure 7.3. Ground or aerial rover traveling at velocity past an obstacle.

tions are complex to compute so an alternative temporal formulation may offer greater efficiency [541, 542]:

$$\Delta\theta = \frac{v\Delta t \sin\theta}{d}$$

This is the principle of motion parallax. The background is differentiated from objects in the foreground such that the apparent angular velocity of the background at distance nd (nominally assumed to be infinite) is given by:

$$\Delta\theta = \frac{v\Delta t \sin\theta}{nd}$$

Similarly, for flying vehicles, apparent angular velocity $\dot\theta$ of a point on the ground is given by the equation of optic flow [543]:

$$\dot\theta = \frac{v}{d}\sin\theta$$

where v = vehicle speed, d = altitude, and θ = trajectory angle to horizontal. Optic flow can be used as a rate gyroscope measurement of yaw in that pure rotation about the vertical axis results in opposing optic flows of the same magnitude in either eye. In insects, navigation occurs through a centering response such that a collision-free path between obstacles is followed by equalizing optic flow on left and right obstacles when optic flow error is given by [544]:

$$e = v_L + v_R = v\left(\frac{1}{z_R} - \frac{1}{z_L}\right)$$

where v_L, v_R = left and right optic flow velocities, v = vehicle forward velocity, and z_L, z_R = horizontal projections of moving objects onto images. The vehicle may be centered between obstacles to balance retinotopic visual flows using a proportional controller of the form

$$\theta^d = k\left(\frac{\partial I^{\max}}{\partial t_{\text{left}}} - \frac{\partial I^{\max}}{\partial t_{\text{rught}}}\right) + \theta_{\text{current}}$$

This is valid as long as rotational speed is much lower than translational speed. Optic flow has low computational complexity $\sim O(n)$ where n = horizontal resolution.

When a camera moves forward in the direction of motion, an image of the ground and surroundings sweep backward forming an optic flow field (Figure 7.4). This flow field radiates outward from a FOE and converges at a rear FOC. This forms a particular distribution of motion stimuli on the camera—from this retinal velocity field information about collisions can be inferred. For pure translational observer motion, the heading direction coincides with the focus of expansion defined as the spatial location from which all velocity vectors diverge. Flow field divergence within the field of view may be used as a cue for obstacle avoidance while the robot is traversing the terrain [545]. An obstacle in motion towards the robot will generate an expanding image of positive divergence whereby the distance to the obstacle is inversely proportional to the divergence

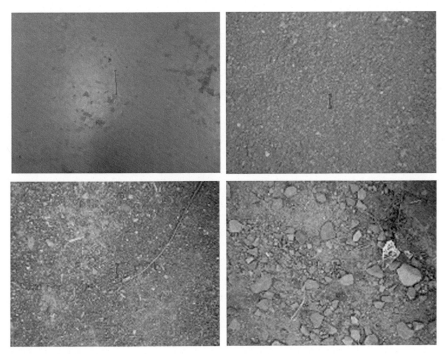

Figure 7.4. Optic flow measurements for rover velocimeter on (top left) cement surface, (top right) asphalt surface, (bottom left) soil surface, and (bottom right) rocky surface [credit Adam Mack, Carleton University].

magnitude. Considering only horizontal optic flow, the focus of expansion corresponds to the point from which vectors alter direction. By definition, the focus of expansion may be defined as the zero-crossing either side of which velocity vectors diverge. The image may be projected onto an image sphere of radius r. Directional divergence of image $I(x, y)$ at distance r_ϕ in the ϕ-direction may be defined as:

$$\frac{\partial I_\phi}{\partial r_\phi} = \cos^2 \phi \frac{\partial I_x}{\partial x} + \sin^2 \phi \frac{\partial I_y}{\partial y} + \sin \phi \cos \phi \left(\frac{\partial I_x}{\partial y} + \frac{\partial I_y}{\partial x} \right)$$

where $I = I_{\text{perpendicular}} + I_{\text{parallel}}$, $\partial I_\perp / \partial r_\phi = v_\perp / z$ for perpendicular motion (expansion as a consequence of approach), $\partial I_{\text{para}} / \partial r_\phi = (\tan \beta v_{\text{para}} / z)(\cos^2 \phi \cos \alpha + \cos \phi \sin \phi \sin \alpha) \rightarrow 0$ for parallel only motion, $\alpha = $ tilt angle of projection of surface normal, and $\beta = $ slant angle between surface normal and the z axis. Any object on a collision course will generate $\partial I_\phi / \partial r_\phi > 0$ over a significant area of the image. Relative speed is determined from uniform expansion of the image away from the focus of expansion. In fact, time to collision can be estimated robustly for arbitrary motion by measuring normal velocity around a closed contour from

Gauss' divergence theorem. For a circular contour of radius r:

$$\int_c v \cdot n \cdot ds = \frac{2\pi r_0^2}{r}$$

where $n =$ unit normal vector Hence, time to collision is based on measurement of radial motion assuming constant velocity as:

$$\tau = \frac{Z}{v} = \frac{2\pi r^2}{\displaystyle\int_c v \cdot n \cdot ds}$$

where $r =$ radius of circular image contour C, $v =$ velocity field, and $n =$ normal vector to contour. Distance to the nearest point on an approaching object may be determined by:

$$Z_{max} \leq \max\left[\left|\frac{2v}{\nabla}\right|, \left|\frac{vr}{r\nabla}\right|\right]$$

where $r =$ radius of the sphere, $\nabla = \cos^2 \phi(\partial I_x/\partial x) + \sin^2 \phi (\partial I_y/\partial y) + \sin \phi \cos \phi((\partial I_x/\partial y) + (\partial I_y/\partial x))$, and $\varphi =$ intensity divergence direction. Time-to-contact estimates of the form $t = d/v$ may be derived to detect head-on collisions by analyzing the optic flow velocity difference of objects between left and right peripheral cameras (similar to insect visual navigation) [546]. Time to contact represents the duration before a rover collides with an obstacle assuming a relative constant velocity. This can be computed directly from optic flow. However, several such samples are required for a robust estimate. We can also perform motor tasks on the basis of estimating the time to collision [547]:

$$\tau = \frac{\theta}{d\theta/dt}$$

where $\theta =$ object's angle of orientation and $d\theta/dt =$ rate of expansion of image. For an object moving directly towards the camera at a constant speed:

$$\tau = \frac{b}{D(d\delta/dt)}$$

where $b =$ baseline distance, $d\delta/dt =$ rate of change of relative disparity, $D =$ current distance between moving object and object's initial location, and $\delta =$ current angle of viewpoint of camera. Hence:

$$\frac{d\delta}{dt} = \frac{b}{D\tau}$$

The rate of change of relative disparity increases with distance for any given τ.

For flying vehicles, optic flow can be used for altitude measurements from downward-pointing imagers at low altitude. Optic flow may be used when hovering with image velocity kept at zero everywhere. It is also used as an insect landing strategy in that optic flow on the landing surface is kept constant while forward speed is maintained proportional to vertical speed. By keeping w

constant, an increase in speed invokes ascent. Similarly, decreased forward speed yields descent. On landing, a constant descent angle can be defined:

$$\alpha = \tan^{-1}\left(\frac{1}{w\tau}\right)$$

On approaching a landing site, front-to-back expansion of the visual world promotes landing behavior; back-to-front motion inhibits landing behavior [548]. The time to collision is given by the rate of expansion:

$$\tau = \frac{\theta}{\dot{\theta}}$$

where θ = visual angle between surface and focus of expansion, $\dot{\theta}$ = radial component of optic flow = $(v \cdot \sin\theta)/d$ during pure translation, and d = distance to the point P. Forward collision response to head-on collision occurs by measuring the relative rate of expansion. A collision avoidance strategy used by insects is the body saccade response in which the insect's body is turned away from high image velocity regions. A balancing strategy control law defines the centering response from optic flow across the whole FOV:

$$\Delta(F_L - F_R) = K\left(\frac{\sum w_L - \sum w_R}{\sum w_L + \sum w_R}\right)$$

When the agent is translating, motion parallax indicates that closer objects generate faster motion across the retina than objects farther away. In addition, closer objects take up more of the FOV. The agent turns away from the side of greater flow. In order to achieve soft docking with the target, a form of PD control may be used:

$$\Delta(F_L + F_R) = K(\dot{\tau} + c)$$

where $c = 0.5$. In order to extract optic flow vectors, features must be tracked over several image frames in order to sample the image flow field. There are several ways to estimate image velocity—intensity-based differential (gradient) methods, frequency-based filtering, and correlation-based methods [549, 550]. Differential methods compute image velocity from spatiotemporal derivatives of image intensities such that $dx/dt = (dI/dt)/(dI/dx)$, but these show poor sensitivities despite being quite robust to noise. Correlation methods spatiotemporally cross-correlate filtered signals that have been pre-smoothed with a Gaussian filter. This is to reduce their sensitivity to temporal aliasing but they are sensitive to illumination and stimulus contrast although they are robust to noise.

Edge-based filtering may be used. Discontinuities in motion occur at the boundaries of objects to enable object differentiation. This minimizes the square of the gradient of optic flow velocity:

$$\left(\frac{\partial u}{\partial x}\right)^2 + \left(\frac{\partial u}{\partial y}\right)^2 \quad \text{and} \quad \left(\frac{\partial v}{\partial x}\right)^2 + \left(\frac{\partial v}{\partial y}\right)^2$$

An equivalent form is the sum of squares of the Laplacians of optic flow:

$$\nabla^2 u = \frac{\partial^2 u}{\partial x^2} + \frac{\partial^2 u}{\partial y^2} \quad \text{and} \quad \nabla^2 v = \frac{\partial^2 v}{\partial x^2} + \frac{\partial^2 v}{\partial y^2}$$

The simplest assumption is that both Laplacians are zero which defines the smoothness constraint. We may approximately differentiate thus:

$$I_x \approx \tfrac{1}{4}\left(I_{i,j+1,k} - I_{i,j,k} + I_{i+1,j+1,k} - I_{i+1,j,k} + I_{i,j+1,k+1} - I_{i,j,k+1} + I_{i+1,j+1,k+1} - I_{i+1,j,k+1}\right)$$

$$I_y \approx \tfrac{1}{4}\left(I_{i+1,j,k} - I_{i,j,k} + I_{i+1,j+1,k} - I_{i,j+1,k} + I_{i+1,j,k+1} - I_{i,j,k+1} + I_{i+1,j+1,k+1} - I_{i,j+1,k+1}\right)$$

$$I_t \approx \tfrac{1}{4}\left(I_{i,j,k+1} - I_{i,j,k} + I_{i+1,j,k+1} - I_{i+1,j,k} + I_{i,j+1,k+1} - I_{i,j+1,k} + I_{i+1,j+1,k+1} - I_{i+1,j+1,k}\right)$$

The Laplacians of (u, v) may be approximated by:

$$\nabla^2 u \approx \kappa(\bar{u}_{i,j,k} - u_{i,j,k}) \quad \text{and} \quad \nabla^2 v \approx \kappa(\bar{v}_{i,j,k} - v_{i,j,k})$$

where $\kappa = 3$ and

$$\bar{u}_{i,j,k} = \tfrac{1}{6}\left(u_{i-1,j,k} + u_{i,j+1,k} + u_{i+1,j,k} + u_{i,j-1,k}\right)$$
$$+ \tfrac{1}{12}\left(u_{i-1,j-1,k} + u_{i-1,j+1,k} + u_{i+1,j+1,k} + u_{i+1,j-1,k}\right)$$
$$\bar{v}_{i,j,k} = \tfrac{1}{6}\left(v_{i-1,j,k} + v_{i,j+1,k} + v_{i+1,j,k} + v_{i,j-1,k}\right)$$
$$+ \tfrac{1}{12}\left(v_{i-1,j-1,k} + v_{i-1,j+1,k} + v_{i+1,j+1,k} + v_{i+1,j-1,k}\right)$$

This defines a weights mask:

$\frac{1}{12}$	$\frac{1}{6}$	$\frac{1}{12}$
$\frac{1}{6}$	-1	$\frac{1}{6}$
$\frac{1}{12}$	$\frac{1}{6}$	$\frac{1}{12}$

To minimize the sum of errors for the rate of change of brightness and departure from smoothness:

$$e^2 = \int\int (\alpha^2 e_s^2 + e_b^2)\, dx\, dy$$

where $\alpha^2 =$ weighting factor $\to 0$ and $e_b = I_x u + I_y v + I_t \to 0$

$$e_s = \sqrt{\left(\frac{\partial u}{\partial x}\right)^2 + \left(\frac{\partial u}{\partial y}\right)^2 + \left(\frac{\partial v}{\partial x}\right)^2 + \left(\frac{\partial v}{\partial y}\right)^2} \to 0$$

From calculus of variations,

$$I_x^2 u + I_x I_y v = \alpha^2 \nabla^2 u - I_x I_t$$
$$I_x I_y u + I_y^2 v = \alpha^2 \nabla^2 u - I_y I_t$$

Using the Laplacian approximation:

$$(\alpha^2 + I_x^2)u + I_x I_y v = (\alpha^2 \bar{u} - I_x I_t)$$

$$I_x I_y u + (\alpha^2 + I_y^2)v = (\alpha^2 \bar{v} - I_y I_t)$$

To solve for (u, v):

$$(\alpha^2 + I_x^2 + I_y^2)(u - \bar{u}) = -I_x(I_x \bar{u} + I_y \bar{v} + I_t)$$

$$(\alpha^2 + I_x^2 + I_y^2)(v - \bar{v}) = -I_y(I_x \bar{u} + I_y \bar{v} + I_t)$$

Flow velocity that minimizes error lies in the direction towards the constraint line that intersects the line perpendicularly. If all errors and weightings are zeroed, we obtain:

$$(I_x^2 + I_y^2)(u - \bar{u}) = -I_x(I_x \bar{u} + I_y \bar{v} + I_t)$$

$$(I_x^2 + I_y^2)(v - \bar{v}) = -I_y(I_x \bar{u} + I_y \bar{v} + I_t)$$

Solving this for a complete image by Gauss–Jordan elimination will be computationally costly. The iterative Gauss–Seidel method was suggested to compute the next set of velocity estimates (u_{i+1}, v_{i+1}) from previous estimates (u_i, v_i) based on derivatives yielding:

$$u_{i+1} = \bar{u}_i - \frac{I_x(I_x \bar{u}_i + I_y \bar{v}_i + I_t)}{(\alpha^2 + I_x^2 + I_y^2)} \quad \text{and} \quad v_{i+1} = \bar{v}_i - \frac{I_y(I_x \bar{u}_i + I_y \bar{v}_i + I_t)}{(\alpha^2 + I_x^2 + I_y^2)}$$

Alternatively, filtering using a spatiotemporal Laplacian of a Gaussian convolution kernel $\partial \Delta G / \partial x$ applies a center–surround field horizontally to find zero-crossings. Spatial frequency filtering involves convolving the image with a receptive field operator representing the Laplacian of the Gaussian to generate a filtered image:

$$v_\perp = \frac{\partial \nabla^2 G \cdot I / \partial x}{\partial \nabla^2 G \cdot I / \partial t}$$

where $G(x, t) = u(\alpha/\pi)^{3/2} e^{-\alpha |x|^2 + u^2 t^2} = $ Gaussian function and $\alpha, u = $ envelope shape control parameters. The receptive field function determines response selectivity (such as image rotation sensitivity during motion) to model elementary motion detectors found in insects [551]. To reduce computational load, measurements are provided only at zero-crossings of illuminance.

Correlation-based methods are similar to those used in stereovision in that correlation windows are used to search matching features between image frames from different viewpoints based on maximum similarity. They are best used after coarse estimates of motion have been extracted to reduce the computational burden of searches in cross-correlation methods. Typically, edges are first extracted using nonlinear operators (the Hough transform is the commonest) which are tracked and then velocity is estimated at those edges. Optic flow can be computed as a disparity field between two images using a correspondence algorithm (hence, optic flow can be used for stereovision). It is assumed that disparity varies smoothly over small windows of n pixels, $p_1, \ldots, p_n$, so disparity may be estimated using a least-squares algorithm such that:

$$v = (A^T A)^{-1} A^T b$$

where

$$A = \begin{pmatrix} \nabla_x I(p_1) \\ \nabla_x I(p_2) \\ \vdots \\ \nabla_x I(p_n) \end{pmatrix} \quad \text{and} \quad b = \begin{pmatrix} \dfrac{\partial I}{\partial t}(p_1) \\ \dfrac{\partial I}{\partial t}(p_2) \\ \vdots \\ \dfrac{\partial I}{\partial t}(p_n) \end{pmatrix}$$

A sum-of-squared-differences measure can be used to ascertain the maximum correlation with a weighting function found by least-squares minimization. The match strength may be defined as:

$$M(x, y, u, v) = \sum_{i,j} \phi(I_1(i,j) - I_2(i + u, j + v))$$

This approach is robust to noise. A Kalman filter approach allows new measurements to be integrated with existing estimates. Unfortunately, correlation-based methods are sensitive to occlusions which are characteristic of unstructured environments.

Frequency-based approaches generally model the orientation selectivity of biological models similar to the way in which a Fourier transform extracts motion. The Fourier transform of optic flow is defined as:

$$\hat{I}(k, w) = \hat{I}_0(k)\delta(v^T k + w)$$

where $\hat{I}_0(k) = $ Fourier transform of $I(x, 0)$ and $\delta = $ Dirac delta function. This yields the optical flow constraint equation: $v^T k + w = 0$. A spatiotemporal Gabor filter (Gaussian function multiplied by a sine and/or cosine wave) may be used to extract spatiotemporal orientation:

$$G(x, t) = \frac{1}{(2\pi)^{3/2}\sigma_x\sigma_y\sigma_t} e^{-(x^2/2\sigma_x^2 + y^2/2\sigma_y^2 + t^2/2\sigma_t^2)} e^{i(2\pi f_0 x \cos\theta + 2\pi f_0 y \sin\theta)}$$

where $\sigma = $ standard deviation of the Gabor filter (defines scale) and $f_0 = $ central frequency with peak amplitude response

$$\iint \left(u\frac{\partial I}{\partial x} + v\frac{\partial I}{\partial y} + \frac{\partial I}{\partial t} \right) \times G(x - x_0, y - y_0) \, dx \, dy = 0$$

Assuming (u, v) is constant, this gives:

$$u\frac{\partial I^* G}{\partial x} + v\frac{\partial I^* G}{\partial y} + \frac{\partial I^* G}{\partial t} = 0$$

A Gabor filter bank with least-squares estimation at multiple resolutions has been demonstrated for optic flow processing [552]. In this case, the Gabor filter was broken down into three successive stages: modulation, low-pass Gaussian filtering, and demodulation of the form:

$$y(k) = x(k)(Ke^{-(k^2/2\sigma^2)}e^{i2\pi f_0 k}) = ((x(k)e^{-i2\pi f_0 k}) * Ke^{-k^2/2\sigma^2})e^{i2\pi f_0 k}$$

Maximum detectable velocity is defined by:

$$|v_\perp| < \frac{\sigma}{2\pi\sigma f_0 + 1}$$

Gabor filters are likely to be computationally intensive unless greater computational resources become available on rovers such as FPGAs.

These approaches have been evaluated [553]:

(i) phase-based methods using orientation-sensitive filters are the most accurate but are the most computationally intense (Fleet and Jepson);
(ii) differential methods based on gradient constraints are the least accurate (Horn and Schunk);
(iii) differential methods based on weighted least-squares fitting are computationally simple as well as being accurate (Lucas and Kanade).

Optic flow algorithmic computational complexity has been assessed by its processing time [554] (Table 7.1).

Assuming an $M \times N$ image, the standard Lucas–Kanade method requires 10^5 $M \times N$ operations per frame and 16 $M \times N$ bytes of frame memory. Its complexity will drop significantly by using feature-based approaches. The combined accuracy and efficiency of the Lucas and Kanade algorithm lent itself to real time implementation on FPGAs [555]. This system achieved a throughput of 2,857,000 pixels per second equating to almost three frames per second for $1,024 \times 1,024$ images (Table 7.2). Thus differential methods based on weighted least-squares fitting appear to be optimal given onboard computational resource restrictions.

To implement optic flow, the simplest type of camera to be considered is the omnidirectional vision camera which simulates the functionality of hemispherical optic flow vision employed by insects [556]. Catadioptric imagers use a conical mirror and single camera without imposing the distortion inherent in wide-angle fish-eye lenses. However, such imaging cameras are unlikely to be proposed for planetary missions due to their restricted utility. Optic flow from a pair of side-looking cameras to avoid side obstacles and stereovision from a pair of forward-

Table 7.1. Sample execution times for tree sequence translation (in min:s) [from Bober and Kittler, 1994].

Horn	Uras	Anandan	Lucas	Fleet
8:00	0:38	8:12	0:23	30:02

Table 7.2. Lucas and Kanade algorithm processing times on a LEON processor.

Image size	Instructions per frame (million instructions)	Memory (MB)	Time on LEON (100 MIPS) (s)	Frame rate on LEON (FPS)	Time on COTS (600 MIPS) (s)	Frame rate on COTS (FPS)
256×256	6.9	1	0.069	14.5	0.012	83.3
512×512	27.5	4	0.275	3.6	0.046	21.7
$1{,}024 \times 1{,}024$	110	16	1.1	0.9	0.18	5.5

MIPS = million instructions per second; FPS = frames per second.

looking cameras to avoid frontal obstacles may be combined to provide the advantages of both [557]. In fact, a single pair of forward-facing fish-eye cameras can implement stereovision at the central portion of the image with optic flow at the periphery of the image. For pure translational motion, the heading direction coincides with the focus of expansion defined as the spatial location from which all velocity vectors diverge. Considering only horizontal optic flow, the focus of expansion corresponds to the point at which vectors alter direction. By definition, the focus of expansion may be defined as the zero-crossing either side of which velocity vectors diverge. Time to impact before collision can be computed by optic flow without estimating the focus of expansion. Optic flow is computed as:

$$\frac{dI}{dt} = \frac{d\nabla I}{dt} = 0$$

where I = image intensity of point (x, y). Global optic flow is measured and comprises two components. A rotational component of flow field V_r may be computed from proprioceptive measurement of camera rotation and focal length. This may be subtracted from global flow velocity to yield translational flow to give the time to impact:

$$t = \frac{d}{v_t}$$

where d = distance of point (x, y) from focus of expansion and v_t = translational flow velocity. However, the focus of expansion can be eliminated using stereo disparity. Depth is measured with respect to the baseline bisecting the axis perpendicular to the baseline. Optic flow is measured parallel to the optic axis of a camera. The two reference frames are related by:

$$Z(x, y) = Z^{\text{of}}(x, y)h(x)$$

where $h(x) = \sin \alpha + (x/F_l) \cos \alpha$. This allows a robust time to impact to be computed as:

$$t = \left(1 + \frac{x}{F_l \tan \alpha}\right) t^{\text{of}}$$

The chief problem is that optic flow data are sparse and reliable only in regions of high contrast. However, a single coarse measure of contrast may be defined as:

$$m = \frac{I_{\text{obj}} - I_{\text{back}}}{I_{\text{obj}} + I_{\text{back}}}$$

to provide a first pass to find regions of high contrast. Furthermore, image prefiltering increases the accuracy of velocity estimates. This favors the use of filtering masks. The smoothness constraint is not generally applicable for complex outdoor environments. Again, this favors approaches that extract edges. Generally, optic flow is best computed hierarchically at multiple resolutions using a coarse-to-fine strategy, particularly for sparse environments. However, given the ubiquity of cross-correlation algorithms used for stereovision, this represents the most readily implemented approach for planetary rovers. Optic flow may be used for tracking other robots to aid in the maintenance of distributed formations of multiple robots in the absence of communication [558]. It has been suggested that appearance variation cues may be used to augment optic flow to estimate the proximity and relative motion of objects in the visual field [559]. This is based on the premise that images have smaller variation in texture and color (due to fewer objects) at closer distances than farther away (where more objects are in view). However, this is not particularly reliable at close quarters where this entropy of texture increases due to increased resolution. In order to reduce the computational requirements of optic flow, image bulk flow has been proposed based on a weighted centroid of the image [560].

The most common use of structure from motion (SfM) is in 3D reconstruction and is probably most suited to the scientific phase of rover operations. Recovery of 3D structure from image sequences (structure from motion) may be implemented through optic flow although motion parameters may be computed without using optic flow. The determination of SfM is effectively equivalent to stereo vision with a single camera. SfM requires 3D motion of the camera to be deduced using optic flow (or other methods). Recovery of 3D shape from motion due to the changing image is uniquely determined from the 2D transformation. The 3D positions of the matched points using standard stereo equations can then be implemented. SfM assumes object rigidity to derive the 3D structure that projects onto the changing 2D image. To infer the 3D structure of object surfaces requires more detailed measurement of relative motion [561, 562]. Three distinct views of four points on the object in motion are sufficient to compute a unique 3D structure. If, however, a continuous velocity field is measured over the complete image, then the acceleration field can be computed and a unique 3D structure can be recovered but it requires imaging over a reasonable time. To reduce sensitivity to error, dense measurements of spatial and temporal derivatives of the image are required. There are two types of SfM approaches: direct or feature based. The most common is the feature-based Kalman filter approach while the other approaches are too computationally expensive to be implemented in real time. Real time performance places a limit on the number n of landmarks due to the computational time complexity of the EKF being $O(n^3)$ where n is the number of features. The computational complexity for SfM with, say, 30 features would involve 10^5–10^7 FLOPS [563]. This method may require lower computational power than dense optical flow calculation in a sparse environment such as the Martian flats where there are only a few features. However, in a complex environment with significant numbers of rocks, a large number of visible rocks will cause a significant increase in computational complexity. SfM has some other disadvantages. A change of camera internal geometry (e.g., focal length) will produce errors unless it is recalibrated. It requires a good feature-tracking algorithm, perhaps based on an optical flow algorithm. There is little doubt that optic flow–based methods offer a vision-based behavior control approach for reactive obstacle avoidance which will be essential for autonomous navigation. The question becomes exactly how will it be implemented in the future. Optic flow–based vision inspired by insect vision has been proposed as part of a bio-inspired engineering of exploration systems (BEES) approach [564]. A variation on this is to use a single catadioptic camera with a conical mirror to provide a 360° horizontal FOV and 70° vertical FOV without distortion—such an approach would be ideally suited to implementing optic flow–based navigation. Optic flow Reichardt detectors may be adopted to implement parallel processing [565]. A VLSI visual motion chip has been developed which implements the Adelson–Bergen algorithm in hardware to generate local motion sensing [566]. Similarly, both FPGAs and microcontrollers have been configured for optic flow functions [567, 568]. Alternatively, an orientable centrally foveated vision system may be implemented to reduce image-processing requirements [569]. Potential fields are also compatible with optic flow methods in which motion generates an optic motion field offering potential synergies between these two navigation techniques [570].

There are several commonly used metrics in visual autonomous navigation [571]. Visual threat cue (VTC):

$$VTC = -R_0 \frac{dR/dt}{R(R - R_0)} \quad \text{for } R > R_0$$

where R = range between observer and target and R_0 = minimum distance. The visual field around a moving agent defines safe travel without collision. The visual field associated with VTC is similar to the optical flow field but is independent of rotational velocity. If $VTC > 0$, the object lies in front of the agent. A variation of VTC is temporal VTC (TVTC):

$$TVTC = \frac{d(VTC)}{dt} = \frac{2RR_0\dot{R}^2 + RR_0^2\ddot{R} - R_0R^2\ddot{R} - R_0^2\dot{R}^2}{R^2(R - R_0)^2}$$

Local spatial tone variation gives a visual pattern of texture; blurring of textual detail is characterized by dissimilarity in image intensity values which is measured by image quality measurement (IQM):

$$IQM = \frac{1}{|D|} \sum_{x_i}^{x_f} \sum_{y_i}^{y_f} \left(\sum_{-L_c}^{L_c} \sum_{-L_r}^{L_r} |I(x,y) - I(x+p, y+q)| \right)$$

where $I(x,y) =$ image pixel intensities, $x_i, y_i =$ initial x, y coordinates, $x_f, y_f =$ final x, y coordinates, $L_c, L_r =$ positive integer constants, and $D = (2L_c + 1)(2L_r + 1)(x_f - x_i)(y_f - y_i)$. Relative temporal changes in IQM are given by:

$$\frac{1}{IQM} \frac{d(IQm)}{dt} = -\frac{R_0}{R(R - R_0)} \frac{dR}{dt}$$

Hence, VTC can be measured using IQM.

7.3 ACTIVE VISION

Active vision has yet to be used on board a planetary rover but autonomous rover navigation will require such a capability. In order to extract scientific data during the traverse phases of rover missions, the mast-mounted panoramic camera may be panned to search for opportunistic scientific targets. This requires automated camera control during the rover traverse. The use of active vision offers the potential for reducing computational effort by devoting computational resources to only relevant segments of an image. Active vision indicates that internal representation does not form a complete world model of the environment but is focused on task-based knowledge [572]. Traditional vision-processing problems are ill posed and rely on unrealistic assumptions such as Lambertian reflectance, smooth surface, uniform illumination, and noise-free conditions [573]. Active vision requires image positioning through pan/tilt/focus actuation of the camera system. It involves interleaving rapid camera movements over the visual field in search of target objects (saccades) and slow camera movements to track such objects (smooth pursuit)—this is an exploratory activity. It involves an active observer which controls its sensory apparatus to improve the quality of perceptual sensing such as maintaining visual fixation on a moving target. Active vision has two aspects: goal-irected image acquisition ("what" component of object identification) and control of image acquisition ("where" component of object location)—it is the goal direction that allows sensory selection; in this way, active vision provides the basis for goal-directed expectations to direct low-level processing [574]. Active vision may be used to direct the robot's navigational tasks as a form of behavior control driven by camera control. Active vision requires the use of inertial measurement (for attitude measurement) and pan/tilt cameras (for visual tracking). Active vision is a paradigm that suggests the integration of controlled actuation (such as through zoom/pan/tilt lenses) may convert an ill-posed, nonlinear, and unstable problem from a passive observer to a well-posed, linear stable solution for an active observer by utilizing multiple information sources

[576]. Active vision allows solutions to a number of specific vision problems such as shape-from-*x*. Shape from shading assumes surfaces to be smooth with Lambertian (isotropic) reflectance, point source illumination, and noise-free images. Shape from contour and shape from texture suffer from similar problems of assumption. Moving the camera in a known translation and rotation reduces the reliance on such assumptions by providing multiple viewpoints. Orienting the eyes (gaze stabilization) to track static or moving objects in the environment provides the core of active vision. Unlike cameras that adopt uniform resolutions over the visual field, the human retina exhibits variable visual acuity that decreases monotonically with increasing eccentricity from the fovea [577]:

$$r = \frac{2\pi}{3n}e$$

where r = visual field radius, e = retinal eccentricity, and n = number of pixels at eccentricity e. Active vision exploits the 60:1 difference in spatial resolutions between the central fovea of 2–4° and the periphery of the 160° field of view of the human eye. The use of directed foveated vision considerably reduces the computational requirements of the visual system. It filters information extracted from the environment. Active vision provides greater flexibility for automated camera control on board a planetary rover. Active vision provides the basis for animate vision in which gaze control is employed to rapidly aim the narrow field of view of the fovea of the eyes to track targets in the visual field [578]. Active vision involves many biologically inspired mammalian capacities: vestibular ocular reflex (VOR), optokinetic response (OKR), smooth pursuit, and saccades. Optic flow for motion detection has applications in active vision particularly for smooth pursuit, optokinetic response, and vestibular–ocular response. Active vision includes a number of mammalian vision features controlling foveated eye movement [579]:

(i) *smooth pursuit*—this is slow eye movement at 1–30°/s which attempts to stabilize the image of a moving target on the retina;

(ii) *saccades*—these are rapid ballistic step-like open-loop eye repositioning movements lasting 30–100 ms in the range 1–40° with accelerations of ~40,000°/s², peak velocities up to 500–1,000°/s, and frequencies of 3–4 Hz;

(iii) *fixations*—saccades are separated by approximately constant fixations that average 330 ms in duration (with high variability) but random fluctuations and drift corrupt fixations if there is no visual feedback;

(iv) *VOR*—this maintains eye fixation during head rotations by counterrotating the eyes with typically unity gain and very low latency lasting ~10 ms so is the fastest visual reflex with open-loop control driven by the otoliths (linear) and semicircular canals (angular);

(v) *OKR*—this stabilizes gaze during sustained head movement driven only by optic flow with a response time of ~100 ms; it is believed that the smooth pursuit mechanism and the optokinetic response system are aspects of the same mechanism;

(vi) *vergence*—eyes move in opposite directions with a maximum range of 15° and velocity of $<10°/s$ when they alter their fixation targets (focusing) at different distances;

(vii) *microsaccades*—rapid random eye movements of $<1°$;

(viii) *drifts*—slow random eye movements $<0.1°/s$;

(ix) *tremors*—high-frequency ~ 30–$150\,$Hz eye oscillations $<30\,$arcsec.

Active vision involves purposeful gazing and the fixation of attention. Stabilization of gaze with respect to the environment is the key to active vision—this requires extraction of image velocity (optic flow). By gazing at the target, optic flow is reduced. The use of inertial information in stabilization is also essential to measure the rotational velocity of the head as an independent source of information. This also reduces the amount of computation required as inertial data are orders of magnitude less than visual data. Two cameras with foveated vision which are able to pan and tilt synchronously, each with independent saccading ability at 3 Hz and convergence control in the yaw axis, could be used. Human saccades reach up to $700°/s$ at a frequency of $\sim 3\,$Hz. Intervening fixations between saccades last $\sim 300\,$ms. However, it is not clear that the camera's inertia and articulation motors could match those of the human eyeball (or even if such rates are desirable). Fast saccadic eye movements move the fovea between different targets in the visual field while active gaze control keeps the fovea on a specific target. The saccade overrides the other eye control systems while saccades occur. Our eyes are constantly in motion characterized by saccades interspersed by relatively steady fixations with vergence control [580]. During fixation, retinal slip is used to stabilize the retinal image. Active vision is required if visual fixation is to be kept on a moving target. Target selection concerns the where-to-look problem so that gaze can be directed appropriately. Selection of a point in the visual field for visual fixation is goal dependent but subject to certain constraints:

(i) visual sequences are centrifugally ordered (i.e., sequential fixation points are proximate to each other and closer potential targets are selected for fixation);

(ii) upward eye motion is preferred over downward movement;

(iii) specific left or right-eye rotations are favored over mixed directions;

(iv) eye fixations of $\sim 250\,$ms tend to concentrate near the vertices and corners of polygonal shapes;

(v) saccades are directed to areas of high information density such as facial features (or content words during reading [581, 582]);

(vi) if peripheral movement occurs a saccade is directed towards the moving agent.

Active vision continually interacts with the environment in that frequent saccades rapidly move the eye over different regions of the visual field. Gaze is rapidly altered sequentially across the visual field rather than employing a uniform image of the entire field. Saccadic and smooth pursuit motions use retinal position and velocity errors to compensate for the lateral motion of visual targets. Gaze searches for information in the visual scene. In general, visual searches are NP-complete if unbounded but, if bounded by expectations, these searches become polynomial or linear in complexity. The visual search pattern is fractal to match the characteristics of natural scenes (e.g., trees, clouds, etc.). A fractal has a Hausdorff–Besicovitch dimension that is greater than its topological dimension. It describes the fragmentation or roughness of the figure. The repeating fractal search pattern of the eye occupies greater space than the 1D line but less than the 2D plane—the greater the fractal fine structure, the greater the fractal dimension of the visual coverage. The eye searches one area with small movements (with a fractal dimension of 1.9) before jumping (with a fractal dimension of 1.1) to another area which is searched with similar small movements. The average fractal dimension of total eye movement is robust at 1.5 [583]. Animal-hunting patterns follow a similar pattern which is more efficient than purely random trajectories. Some 10–15% of saccades are regressions to previously fixated regions. The more informative a region of the visual field the greater the number of fixations. Visual information is acquired only during fixations as information acquisition is suppressed during the saccades themselves masking any blurring effects. Fixation points are generally selected on the basis of high information regions [584]: high contrast, high spatial frequency, vertices, symmetry axes, proximity to previous fixation point, etc. Fixation duration tends to be longer for information-rich targets such as high spatial–frequency changes and contrasts [585]. Furthermore, the sudden appearance or disappearance or movement of peripheral stimuli often invokes saccades towards that location. The presence of a target in the visual field may generate an estimate of the log-likelihood of a target being present which increases until it reaches a threshold when a saccade is initiated. This may be cast into a Bayesian framework. Posterior likelihood is given by the prior likelihood plus the effect of observational evidence [586]:

$$\log L_{\text{post}} = \log L_{\text{prior}} + \log \frac{p(E|H_{\text{tar}})}{p(E|H_{\text{ref}})}$$

where E = observational evidence, H_{ta} = hypothesis of target, and H_{ref} = reference hypothesis. Bayes theorem computes the a posteriori probability from the a priori probability which is not dependent on camera parameters:

$$p(x_i|z_i, a_i) = \frac{p(z_i|x_{i+1}, a_i)p(x_i)}{p(z_i|a_i)}$$

This may be computed iteratively with the former a posteriori probability feeding into the next iteration as the new a priori estimate:

$$p(x_{i+1}) = p(x_i|z_i, a_i) = \frac{p(z_i|x_i, a_i)p(x_i|z_{i-1}, \ldots, z_0)}{p(z_i|a_i)}$$

Probability density functions are estimated during training. This comprises the prior contextual information. Randomness is an important component of visual saccades suggesting a probabilistic aspect [587]. This can provide a mechanism for active vision in selecting camera parameters (e.g., focal length, pan/tilt angles, and field of view). Maximization of mutual information based on Bayesian computations may be used to select actions for visual viewpoint selection based on the prior and the likelihood function which reduces uncertainties in the state of the environment [588]. Mutual information places the receptive field of the camera over the area of interest. It defines how much uncertainty is reduced in x if observation o (image quality) is made which depends on action a (selected camera parameters)—it is given by:

$$I(x_i; z_i | a_i) = H(x_i) - H(x_i | z_i, a_i) = \int_{x_i} \int_{z_i} p(x_i) p(z_i, a_i) \log \left(\frac{p(z_i | x_i, a_i)}{p(z_i | a_i)} \right) dz_i \, dx_i$$

where

$$H(x_i) = - \int_{x_i} p(x_i) \log(p(x_i)) \, dx_i$$

$$H(x|z) = \sum p(x|z) \log p(x|z)$$

$$= \text{entropy of posterior}$$

Contextual information is critical to enabling object search in natural scenes which may be captured within a Bayesian framework [589]. Attention (and so foveation) is guided by the global scene context. There are two parallel computations— bottom-up local features such as parsed objects (saliency) and top-down global features of the image with scene context which influence the early stages of visual processing (within 150–200 ms of image onset). Contextual information (prior) such as the co-occurrence of objects is particularly important in object detection. These factors, but especially contextual information, allow prediction of the next region of the image to be fixated. Fixation occurs on the image location that has the highest probability of containing the target object given the contextual information (i.e., by estimating $p(O, X|L, G)$, the probability of the presence O of the target object at location X given local measurements $L(X)$ and a set of features G). From Bayes theorem:

$$p(O, X|L, G) = \frac{p(L|O, X, G) p(X|O, G) p(O|G)}{p(L|G)}$$

Local features are computed through saliency while global features are computed through information-theoretic measures.

Vestibular–ocular reflex (VOR) capability with smooth pursuit of targets would provide robust static and moving target tracking. The use of vision within a servo control loop—visual servoing [590]—requires extraction of sufficient relevant information from the scene (contours or regions) for execution of a given task compatible with the closed-loop rate of the control algorithm such as motion rate control [591]. Despite saccadic eye movements, there is egocentric perceptual stability of objects in the visual world from the observer's perspective [592, 593]. This is afforded by tracking retinal projections during successive fixations. Visual servoing is an extension of the active vision paradigm [594]. A hybrid affine visual-servoing approach may be adopted [595]. An object's surface may be defined by:

$$\xi(x, y) = \frac{\partial \xi}{\partial x} x + \frac{\partial \xi}{\partial y} y + c$$

A projection point P in camera coordinates and the object frame is given by:

$$\begin{pmatrix} p_x \\ p_y \end{pmatrix} = \lambda \begin{pmatrix} x_c \\ y_c \end{pmatrix} + T \begin{pmatrix} x_0 \\ y_0 \end{pmatrix}$$

where $\lambda = f/z_c$ and

$$T = \lambda \begin{pmatrix} c\alpha c\beta c\gamma + s\alpha s\gamma & c\alpha c\beta s\gamma - s\alpha c\gamma \\ s\alpha c\beta c\gamma - c\alpha s\gamma & s\alpha c\beta s\gamma + c\alpha c\gamma \end{pmatrix}$$

where $\alpha = $ tilt, $\beta = $ slant, and $\gamma = $ orientation. Visual servoing includes both self-localization and closed-loop navigation control [596, 597]. Splines may be used to compute the desired trajectory to the next saccade target from the current position from which a Moore–Penrose pseudoinverse derives the eye motion rate. The visual servoing control law has the form:

$$\dot{\theta} = -J^+ e$$

where $e = \Delta q = $ Cartesian image coordinate variation, $J^+ = (J^T J)^{-1} J^T = $ Moore–Penrose pseudoinverse of the Jacobian, and $J = \partial q / \partial \theta = $ Jacobian matrix of image coordinates with respect to joint angles. The Moore–Penrose pseudoinverse can be computed by the single-value decomposition algorithm. Pursuit tracking requires gaze holding to provide feedback signals. The process has three stages—rest, vergence stabilization, and pursuit.

Camera model is given by the perspective equation:

$$x_u = f \frac{x}{z} \quad \text{and} \quad y_v = f \frac{y}{z}$$

where $(x_u, y_v)^T = $ image coordinates. In fact, image points are transformed thus:

$$u = s_x x_u + u_0 \quad \text{and} \quad v = s_y y_v + v_0$$

where $(u_0 v_0)^T = $ image center and $s_i = $ scale factor. We need to alter camera angles θ_r and θ_l to keep the target in the center of the image.

Camera horizontal turning angle for vergence control is given by:

$$\Delta\theta = \tan^{-1}\left(\frac{\Delta u}{s_x f}\right)$$

where $\Delta u = u - u_0 = $ horizontal disparity. Angle and distance to the fixation point is given by:

$$\theta_n = 90 - \tan^{-1}\left(\frac{h}{p}\right)$$

where $\quad h = d_v \tan \theta_r = d_l \tan \theta_l, \quad d = d_l + d_r = \text{baseline}, \quad p = d_l - b/2, \quad$ and $\quad d = \sqrt{h^2 + p^2}$. When $\theta_l = \theta_r$, $d = b/2 \tan \theta_l$ to move the camera:

$$\theta_l = \tan^{-1}\left(\frac{h}{d_l}\right) \quad \text{and} \quad \theta_r = \tan^{-1}\left(\frac{h}{d_r}\right)$$

The similarity operator is given by:

$$S(x,y) = \sum_{i=-n}^{n} \sum_{j=-m}^{m} [P(i,j) - I(x-i, y-j)]^2$$

where p = search pattern and I = image. This effectively implements gaze holding through smooth pursuit and vergence control [598].

Human oculomotor control operates with a time delay. All oculomotor behaviors follow velocity commands except saccades which are position based— this is compatible with optic flow computations. Smooth pursuit, VOR, and OKR are the core capabilities. VOR is based on feedback measurement of head rotation by the vestibular system. VOR counteracts the effects of head movement by generating opposing eye movements up to $\sim 500°/\text{s}$. The optokinetic response (OKR) is similar but is controlled by the vision system. VOR has shorter latency (~ 15–$30\,\text{ms}$) than OKR (~ 80–$100\,\text{ms}$) which uses retinal slip (velocity) feedback to augment VOR. OKR counteracts high image slipping up to $80°/\text{s}$ due to body movements. VOR and OKR cooperate to minimize retinal slip error by control- ling head movements to ensure perceptual stability [599]. VOR maintains the eye's orientation in the opposite direction to head movement so that the target appears static within the visual field. It is possible to merge data from inertial sensors with data from stereoscopic vision for automatic camera pan–tilt control emulating the human VOR used for head stabilization [600]. VOR evokes compensatory eye movements when the head moves to stabilize visual fixation which is operated with an open-loop (feedforward) component for rapid response up to $5\,\text{Hz}$. Angular VOR responds to angular head movements while translational VOR responds to linear head movements. VOR gain is the ratio of eye to head velocity, typically close to unity. OKR provides a delayed closed-loop system (with no feed- forward component) to correct errors in VOR but operates more slowly ($<1\,\text{Hz}$). The pursuit system to track moving objects in the foveal region also acts to stabil- ize eye movement and provide feedback to VOR [601]. Smooth pursuit enables tracking of small moving objects up to $40°/\text{s}$ to keep it on the fovea. The pursuit system operates at higher frequency than OKR. Thus there are interactions between different eye movement control systems, in particular between VOR and OKR [602]. Stabilization of gaze with respect to static or moving targets is a fundamental requirement of an active visual system. Retinal image flow is counter- acted by head/eye reflexes from optokinetic and vestibular–ocular reflexes. Even during fixed gaze, however, there are small involuntary drifts/(microsaccades) a few arcminutes in amplitude with $\sim 0.6\,\text{s}$ periods.

Active vision requires the use of inertial measurement (for attitude measurement) and steerable pan–tilt cameras (for visual tracking). Oculomotor smooth pursuit may be driven by optic flow field velocity whereby the eyes respond to a target by moving opposite to the flow field. Eye movements respond to image velocity information to generate smooth pursuit tracking while responding to eye position information in generating oppositely directed saccades when the edge of the eye orbit is reached. Optic flow thus allows implementation of active vision for camera control. Reliance on attitude measurements to supplement image velocity measurement significantly reduces the computational complexity of reliance on optic flow alone. The use of inertial information to measure head movements is the basis of the vestibular–ocular reflex (VOR) which generates a feedforward signal based on angular head velocity [603]. An onboard inertial navigation system would best be implemented on the camera masthead but this would be unlikely for accommodation reasons. However, inertial measurements of the masthead may be estimated from inertial measurements of the rover and pan–tilt angle measurements (assuming that the mast is rigid). VOR may be implemented on a multi-DOF manipulator/pan–tilt mast mounting a camera system. The feedforward component comprises desired pan–tilt angles, angular velocities, and angular accelerations. These may be converted into the equivalent joint torque values [Lynch, B., 2010 private communication]. This may be combined with joint motion feedback (retinal slip cannot be used for visual feedback) to allow accurate tracking of visual objects while the rover is in motion.

The world (inertial) coordinate system W is the reference frame which which the rover body B has a location at position p_0 and an orientation R, P, and Y (Figure 7.5). Camera frame C is mounted on the end effector of an n-degree manipulator arm including pan–tilt angles α and β. The camera's x-axis is directed towards the target. The camera coordinate system relative to the end effector is defined by the pan tilt rotation matrices:

$$R_C^{(N)} = \begin{bmatrix} \cos\alpha & -\sin\alpha & 0 \\ \sin\alpha & \cos\alpha & 0 \\ 0 & 0 & 1 \end{bmatrix} \begin{bmatrix} \cos\beta & 0 & \sin\beta \\ 0 & 1 & 0 \\ -\sin\beta & 0 & \cos\beta \end{bmatrix}$$

$$R_C^{(N)} = \begin{bmatrix} \cos\alpha\cos\beta & \sin\alpha & \cos\alpha\sin\beta \\ \sin\alpha\cos\beta & \cos\alpha & \sin\alpha\sin\beta \\ -\sin\beta & 0 & \cos\beta \end{bmatrix}$$

The camera coordinate system in world coordinates is defined by the pan–tilt rotation matrix postmultiplied by the end effector orientation matrix in world coordinates:

$$R_C^{(0)} = R_N^{(0)} \begin{bmatrix} \cos\alpha\cos\beta & -\sin\alpha & \cos\alpha\sin\beta \\ \sin\alpha\cos\beta & \cos\alpha & \sin\alpha\sin\beta \\ -\sin\beta & 0 & \cos\beta \end{bmatrix}$$

The camera is directed along the x-axis of the camera coordinate system to the

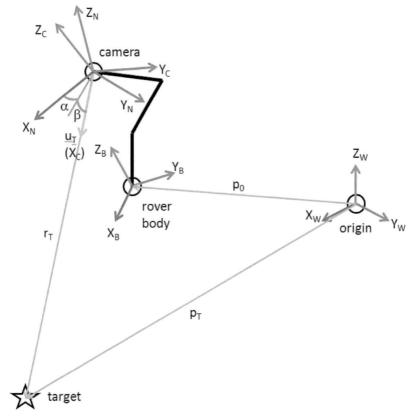

Figure 7.5. Coordinate frames for a rover, its camera mounting, and the visual target with respect to inertial coordinates [credit Brian Lynch, Carleton University].

target object with unit vector u_T:

$$u_T = R_N^{(0)} \begin{bmatrix} \cos\alpha\cos\beta & -\sin\alpha & \cos\alpha\sin\beta \\ \sin\alpha\cos\beta & \cos\alpha & \sin\alpha\sin\beta \\ -\sin\beta & 0 & \cos\beta \end{bmatrix} \begin{bmatrix} 1 \\ 0 \\ 0 \end{bmatrix} \quad \text{and} \quad u_T = R_N^{(0)} \begin{bmatrix} \cos\alpha\cos\beta \\ \sin\alpha\cos\beta \\ -\sin\beta \end{bmatrix}$$

The required camera pan–tilt angles can then be determined:

$$\begin{bmatrix} \cos\alpha\cos\beta \\ \sin\alpha\cos\beta \\ -\sin\beta \end{bmatrix} = (R_N^{(0)})^T u_T$$

The end effector orientation matrix is dependent on the rover body orientation matrix:

$$R_B = \begin{bmatrix} \cos\psi & -\sin\psi & 0 \\ \sin\psi & \cos\psi & 0 \\ 0 & 0 & 1 \end{bmatrix} \begin{bmatrix} \cos\theta & 0 & \sin\theta \\ 0 & 1 & 0 \\ -\sin\theta & 0 & \cos\theta \end{bmatrix} \begin{bmatrix} 1 & 0 & 0 \\ 0 & \cos\phi & -\sin\phi \\ 0 & \sin\phi & \cos\phi \end{bmatrix}$$

where φ, θ, ψ = roll–pitch–yaw angles. The position of the target with respect to the camera's position is given by:

$$r_T = p_T - p_0 - p_c^{(0)}$$

The position of the camera with respect to the rover's body in world coordinates is determined by the manipulator kinematic configuration:

$$p_C^{(0)} = \sum_{i=1}^{N} R_i^{(0)} L_i \quad \text{where} \quad R_i^{(0)} = R_B \sum_{j=1}^{i} R_j$$

The target unit vector is given by:

$$u_T = \frac{r_T}{d_T}$$

where $d_T = (r_T^T r_T)^{1/2}$. Hence, camera pan–tilt angles can be determined readily [604]. The time rates of change variables are determined through differentiation.

The agile eye is a spherical 3 DOF camera-orienting system based on three mutually perpendicular circular arc links to provide camera-swiveling capability within a compact unit [605]. It is a kinematic variant on the Stewart platform class of parallel manipulators with applications to manipulator wrists [606]. A VLSI retina chip that directs currents through resistors to generate different inhibitory strengths at different locations in the retina through shunting inhibition has been demonstrated [607]. Exploiting small-amplitude shifts of the optical axis similar to biological microsaccades in electronic imagers would enable increased resolution beyond the limits of photoreceptor spacing. Such a camera should be capable of generating saccades every 50–100 ms with an accuracy of $1°$ over a complete range of $\sim 60°$. To rotate the visual axis, the imager and a focusing lens may be fixed while a rotating microprism grating in front of the focusing lens provides orientability. Prism orientation would be controlled through a smooth pursuit algorithm to maintain the target in the center of the FOV. A saccadic exploration algorithm operates periodically to alter the focus quasi-randomly.

Many of the complexities associated with stereovision may be overcome using active stereovision through the control of camera focus, zoom, and orientation which provides redundant information with depth from focus as well as depth from vergence [608]. Binocular systems are more efficient at gaze stabilization than monocular systems as vergence control allows consistent depth information to be extracted [609]. Eye vergence movements are also controlled by disparity measurements. Defocusing the image ensures that the lens law is satisfied:

$$\frac{1}{Z_{\text{object}}} + \frac{1}{Z_{\text{image}}} = \frac{1}{f}$$

thereby obtaining monocular estimates of Z_{object}. In fact, this will have a range of values given by [610]:

$$z_{\text{upper}} - z_{\text{lower}} = \frac{2ACz_{\text{est}}f(z_{\text{est}} - f)}{A^2f^2 - C^2(z_{\text{est}} - f)^2}$$

where $A =$ aperture area, $C =$ diameter of circle of confusion (blur) $= d(Z_{\text{upper}} - Z_{\text{lower}})/z_{\text{image}}$, $d =$ lens diameter $= f/F$, $z_{\text{upper}} - z_{\text{lower}} =$ image plane displacement from focus, $f =$ focal length, and $F = F-$ number. To implement a basic VOR–OKR system for a binocular system, each eye must maintain its gaze at a particular point P in the world when the head rotates in a horizontal plane such that the instantaneous eye angle is given by [611–613]:

$$\theta_e = \tan^{-1}\left(\frac{v_b \times v_g}{v_b \cdot v_g}\right)$$

where $v_b =$ vector from eye position to mid-baseline and $v_g =$ vector from eye position to gaze point P. From geometry:

$$v_g = (\tfrac{1}{2}b\cos\theta_h - a\sin\theta_h, d - (a\cos\theta_h + \tfrac{1}{2}b\sin\theta_h))$$
$$v_b = (\tfrac{1}{2}b\cos\theta_h, -\tfrac{1}{2}b\sin\theta_h)$$

The angle of each camera is given by:

$$\theta_e = \tan^{-1}\left(\frac{a - d\cos\theta_h}{(b/2) - d\sin\theta_h}\right)$$

where $\theta_h =$ head rotation angle, $d =$ (fixation) distance of fixation point from head center of rotation, $b =$ interocular distance (baseline), $a =$ distance from head center of rotation to mid-baseline. Differentiating gives the VOR relation:

$$\dot{\theta}_e = \left(\frac{d^2 - dZ}{(b/2)^2 + a^2 + d^2 - 2dZ}\right)\dot{\theta}_e = G_{\text{VOR}}\dot{\theta}_h$$

where

$$Z_l = (a\cos\theta_h + \tfrac{1}{2}b\sin\theta_h) = z \text{ coordinate of left eye}$$
$$Z_r = (a\sin\theta_h - \tfrac{1}{2}b\sin\theta_h) = z \text{ coordinate of right eye}$$

This maintains binocular fixation. Ideally, G_{VOR} should equal 1.0 but motor saturation limits may require reducing this by extending the fixation distance. Similarly, eye velocity for vertical motion may be derived for pan/tilt motions. For translational rather than rotational VOR:

$$\dot{\theta}_e = \left(\frac{(d - a)}{(\tfrac{1}{2}b - x)^2 + (d - a)^2}\right)v_x \quad \text{from the perspective of the left eye}$$

Assuming symmetrical convergence, a similar expression will be valid for the right eye but in the opposite direction. By measuring head angular velocity and slewing the eyes open loop, the amount of image velocity due to retinal slip is reduced substantially (and so computation). This residual ocular velocity from 3D object motion $(v_x, v_y, v_z, w_x, w_y, w_z)$ is given by:

$$u(x,y) = f_x \left(\frac{\frac{x}{f_x} v_z - v_x}{Z} + w_x \frac{xy}{f_x f_y} - w_y \left(1 + \frac{x^2}{f_x^2} \right) + w_z \frac{y}{f_y} \right)$$

$$v(x,y) = f_y \left(\frac{\frac{y}{f_y} v_z - v_y}{Z} + w_x \left(1 + \frac{y^2}{f_y^2} \right) - w_x \frac{xy}{f_x f_y} - w_z \frac{x}{f_x} \right)$$

This provides closed-loop control based on the OKR that results from retinal slip:

$$\dot{\theta}_e = G_{\text{AVOR}} \dot{\theta}_h + G_{\text{OKR}} \binom{u}{v} \quad \text{in the angular case}$$

$$\dot{\theta}_e = G_{\text{LVOR}} v_h + G_{\text{OKR}} \binom{u}{v} \quad \text{in the linear case}$$

As the horizontal component counters rotations about the vertical y axis and a zx plane crossing the center of the image $(x, y) = (0, 0)$:

$$u(0,0) = f_x \left(-\frac{v_x}{z(0,0)} - w_y \right)$$

where $w_y = (1 - G_{\text{VOR}}) \dot{\theta}_h$ = rotational velocity of the eye about its rotation axis, $v_x = (b/2 \cos \theta_e - a \sin \theta_e) \dot{\theta}_h$ = vector translation along frontal image plane, and $z(0,0)$ = vector from camera $(0,0)$ to gaze point P:

$$\frac{u(0,0)}{f_x} = -\frac{v_x}{z(0,0)} - w_x$$

where $w_x = w_y + \Omega_y$ an Ω_y = head rotational velocity. The output angles from the cameras should be bandpass filtered to compensate for thermal drift, mechanical vibrations, and electronic noise.

Fixation distance can be determined from [614]:

$$d = b \frac{\cos \theta_r \cos \theta_l}{\sin \frac{1}{2} (\theta_l + \theta_r) \cos(\theta_l - \theta_r)}$$

This may be used to adaptively adjust camera pointing. Optic flow has a limited range of image velocities so visual stabilization only is limited and requires supplementation by inertial stabilization. Furthermore, if the rotation axis of the camera and head are coincident and the fixation point lies at infinity, inertial information suffices to stabilize gaze. However, this is not the case in a binocular camera–head system as the camera and head rotation axes are different—rotation of the head generates rotation and translation of the cameras. In this case, visual information is also required. Image comparison between two consecutive images may be used to stabilize a pan–tilt camera by computing the Euclidean distance between images to return the camera to its original position:

$$d(I_i, I_j) = \sqrt{\sum_{k=1}^{h \times w} \sum_{l=1}^{c} [I_i(k,l) - I_j(k,l)]^2}$$

where $h, w =$ search window height and width. The global minimum of the Euclidean error surface defines the desired actuator signals. Rather than identifying specific features (the search window) to track across the target image, it is possible to use representative pixels from the search image that applies to generalized images [615]. A number of options may be considered which involve selecting a representative fraction of the image including a predefined selection of distributed pixels from the entire image and a random selection of pixels representative of the complete image both of which require selection of around 50% of the image pixels but is robust to many different types of images. This eliminates the need for edge finding and feature identification. The Papanikolopoulos and Khosla (1993) visual tracking algorithm incorporated consideration of a delay factor [616]. It has been suggested that a predictive component based on the Smith predictor rather than using pure open-loop methods would add significant robustness to gaze control and visual tracking and compensate for delays [617, 618]. Such a Smith predictor controller would have the form:

$$\bar{C} = \frac{C}{1 + CA(1 - z^{-\tau})}$$

where $C(z) =$ delay-free controller, $z^{-T} =$ delay, and $A(z) =$ plant. In smooth pursuit, retinal slip is used as feedback to control eye movement. Smooth pursuit is highly accurate in tracking and requires a three-component system: a feedback controller (subject to significant delays from retinal slip), the target velocity predictive controller, and an inverse oculomotor model [619]. Accurate tracking by the predictor requires an accurate inverse oculomotor model. A Kalman filter may predict target locations when supplemented by a Smith predictor to cope with time delays. Dias et al. (1998) used a Kalman filter to predict the future position, velocity, and acceleration of the target to facilitate pursuit and target tracking [620]. This is motivated by the fact that binocular vision-based primate smooth pursuit is based on predictive encoded signals from visual feedback to drive pursuit to compensate for the $\sim$100 ms delays in OKR [621]. To ensure that mechanical stop limits are not transgressed, a nonlinear comfort function may be implemented:

$$\theta_{av} = \frac{\theta}{(\theta - \theta_{max})^2}$$

where $\theta =$ average pan–tilt angle and $\theta_{max} =$ maximum pan–tilt limits. Visual feedback of retinal slip may be modeled as [622]:

$$f = \frac{d}{dt} \sqrt{(\theta_h^2 + \theta_v^2)}$$

where $\theta_h, \theta_v =$ horizontal and vertical eye displacement error. The Robinson model on the integration of smooth pursuit with eye velocity feedback and vestibular feedforward inputs suggests that vestibular inputs are critical to providing predictive capability to maintain retinal image stabilization [623]. Smooth pursuit may be summed with VOR–OKR. One of the most sophisticated models of oculomotor gaze stabilization is that of Shibata and Schaal (2001) based on vestibular–cerebellar modeling through feedback error learning [624]. An estimate of motor error is given by the output of a feedback controller based on a reference model (i.e., a model-referenced adaptive controller). Image stabilization prevents self-motion from interfering with visual functions. Indeed, closed-loop cancellation of retinal slip to provide retinal stabilization suggests that image acquisition for autonomous navigation no longer requires the rover to be static. An extended Kalman filter–trained neural network has been employed to implement a VOR-like smooth pursuit system for a 3 DOF rover camera mast [625]. This neural model constitutes a forward model of the arm-mounted camera dynamics to track targets in the camera visual field while the rover is in motion. If there is deviation error from the desired tracking performance, a feedback control system is used; it is also used to train the forward model. The combination of forward and feedback control yields superior camera tracking than feedback control only (Figure 7.6).

Neural fields have been proposed to determine the next saccade target through a competitive dynamical neural network [626]. Visual flow input was fed into a saliency map in which a winner-takes-all strategy selected the next saccade target.

Optic flow may be adapted for use in active binocular vision to track targets [627]. Motion detected through optic flow and structure from stereovision with active vision camera control may be employed to support visual navigation and obstacle avoidance [628]. Binocular disparity may be combined with optic flow to obtain a relative depth map of the scene. Integration of multiple methods of vision enable solutions to some key problems such as explicit calibration and robustness to noise. Binocular disparity for each image pair over time may be combined with monocular optical flow to generate 2.5D scene representation in terms of time to impact and relative depth. Active gaze control to stabilize camera rotations will provide significant robustness to these methods by ensuring that image flow corresponds to motion flow. Active gaze control and visual servoing using potential field maps rather than image data provide robustness during rover maneuvers. Redirecting gaze through rapid saccades does not permit vision processing but a cyclopean eye is assumed to look forward to the target and both eyes have symmetric vergence. Optic flow provides the basis for smooth pursuit assuming that the target is not far from the fixation point (within the foveated region) generated by the saccade. The velocities of the target generated are used as inputs to a Kalman filter to generate updated velocities to control the camera position. A point (X_c, Y_c, Z_c) in cyclopean coordinates moves with a velocity given by:

$$\begin{pmatrix} v_x^C \\ v_y^C \\ v_z^C \end{pmatrix} = - \begin{pmatrix} w_x \\ w_y \\ w_z \end{pmatrix} \times \begin{pmatrix} X_c \\ Y_c \\ Z_c \end{pmatrix} + \begin{pmatrix} v_x \\ v_y \\ v_z \end{pmatrix}$$

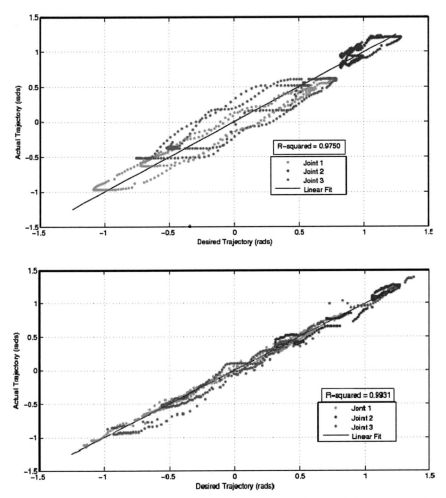

Figure 7.6. Deviation of camera from a desired pointing trajectory during pursuit using (a) feedback control only and (b) both feedforward and feedback control [credit Jordan Ross, Carleton University].

In terms of retinal coordinates:

$$\begin{pmatrix} v_x \\ v_y \end{pmatrix} = -R_{\text{ret}} \begin{pmatrix} v_x^C \\ v_y^C \\ v_z^C \end{pmatrix}$$

where R_{ret} is the 3×2 rotational matrix that maps from cyclopean coordinates to retinal coordinates of left/right cameras. Retinal image motion flow disparity is given by:

$$\Delta = v_l - v_r$$

Assuming that the target is verged with equal vergence angles (i.e., $X_c = Y_c = 0$), only horizontal motion disparity is relevant:

$$\Delta = \begin{pmatrix} \Delta_x \\ \Delta_y \end{pmatrix} = \begin{pmatrix} \dfrac{v_z f \sin 2\theta}{Z_C} \\ 0 \end{pmatrix}$$

where $f =$ same focal length for both cameras. This gives the time to contact,

$$t_c = \frac{Z_C}{v_z} = \frac{(f \sin 2\theta)}{\Delta_x}$$

where $Z_C = \frac{1}{2} b \tan \theta$ and $b =$ baseline So,

$$\Delta_x = \frac{4(v_z f \cos^2 \theta)}{B}$$

Differentiating:

$$\frac{\partial Z_C}{\partial t} = \frac{b}{2 \cos^2 \theta} \frac{\partial \theta}{\partial t} = v_z = \frac{b \Delta_x}{(4 f \cos^2 \theta)}$$

This gives:

$$\frac{\partial \theta}{\partial t} = \frac{\Delta_x}{2f}$$

This is the angular velocity (smooth pursuit) of vergence which maintains the target on the moving object steady.

   Self-localization and mapping (SLAM) [629] using active closed-loop control has been investigated in conjunction with active vision. The key is to use cameras with foveated vision to predict the expected position of a feature. This allows redetection of features over time to reduce drift but the approach was too artificial for planetary rover applications [630]. An autonomous exploration strategy has been developed which seeks out regions of the environment that match its internal representations using a gaze-planning strategy [631].

8

Autonomous navigation—behaviors and architectures

All space missions involve the control of mission operations from a ground station mediated through a communications channel. Rover missions are no exception. For planetary missions, important constraints are imposed by the nature of the terrain, limited amount of energy available, limited computational resources, limited communication with human operators, and the nature of scientific targets and their location. Fully autonomous control through autonomous task planning is not yet conceivable nor, indeed, is it desirable—there will always be a need for human supervision. Telerobotics for space or planetary missions is characterized by the remoteness of the operational environment. Indeed, for lunar missions, direct teleoperation is feasible due to the short separation distance and the gravitiational locking of the Moon's orbit around the Earth. For more distant missions, such as Mars, teleoperation becomes impractical. However, fully autonomous control through autonomous task planning is not yet conceivable, nor indeed, is it desirable—there will always be a need for human supervision. Teleoperation must proceed at a much faster speed than autonomous rovers in order to reduce the human operator workload (the Lunokhods traversed at 1–2 km/h). An important aspect of human–machine interfaces in teleoperation is the behavior of the human operator. Human reaction response speed is determined by the Fitts'–Hick's law [632]:

$$T = k \log_b N + c \log_2 N$$

where N = total amount of information received, b = number of alternative actions available, and k and c are empirical constants. Although human reaction speed is typically 150–200 ms, it is variable, with auditory responses time being $\sim$50 ms faster than visual response times. Fitts' law also gives the total duration T of a movement by [633]:

$$T = a + bI = a + b \log_2 \left(\frac{2d}{w} \right)$$

© Springer-Verlag Berlin Heidelberg 2016
A. Ellery, *Planetary Rovers*, Springer Praxis Books,
DOI 10.1007/978-3-642-03259-2_8

where $I = \log_2(2d/w) = $ index of difficulty of the motion, $d = $ distance moved, $w = $ target width, and a and b are empirical constants. The index of difficulty has several different definitions beyond the Fitts' definition above:

$$I = \log_2\left(\frac{d}{w} + \frac{1}{2}\right) - \text{Welford}$$

$$I = \log_2\left(\frac{d}{w} + 1\right) - \text{MacKenzie}$$

Schaub's version of Fitts' law is often used to describe how movement time T varies in terms of the difficulty of the motion and has the form [634]:

$$T = a + (b + cG)I$$

where $G = $ gain to account for learning and a, b, and c are system information capacity parameters.

Human–machine interfaces must deal with fault management under dynamic conditions and this requires certain trends [634]: (i) the interface should act as an external memory aid to relieve demands on the operator at the cost of flexibility if necessary; (ii) the interface should not distract the operator from the task with too many windows or displays; (iii) critical events should be displayed and highlighted in an integrated manner.

8.1   TELEROBOTICS

In general, space-based teleoperation will be an extension of remote control of robots through computer networks [635–637]. Teleoperation in hostile remote environments is characterized by low situational awareness, constrained bandwidth with time delays, and the need for highly skilled operators. Low situational awareness results from restricted camera viewpoints, uncontrolled illumination often giving poor visual definition, and a lack of real-time tactile feedback. For the Moon, it is plausible to utilize teleoperation to control a rover: video channels for teleoperation from the ground generally required high data rates of at least 120 Mbps at high video frame rates without compression but this can be reduced to 250 kbps without compression for near-real time imaging. Telerobotics is the distributed control of a remote robot (such as a planetary rover) and a separately located human operator (such as Earth) through a communication channel characterized by a time delay. Computers at the Earth-based ground station are not subject to computational resource limitations inherent on the remote rover so they can accommodate significant processing functions. Teleoperation is a cybernetic man–machine interface system designed to augment and project human senses and dexterity across physical distances. It is feasible only for short communication distances such as lunar surface operations. Teleoperation involves reflecting at a distance the physical motions of the human operator (HO). Telerobotics involves

varying degrees of sophistication according to the degree of supervision at higher
levels of control given by the HO [638]. There are three levels of telerobotic
control—manual teleoperation, supervised telerobotic control, and autonomous
control. Control is effected by sending command signals to the remote rover and
receiving telemetry feedback on its state. The master environment is that of the
HO while the slave environment contains the remote worksite joined by a communications link between them. Humans are primarily visual information–
processing animals, so interfacing is primarily through visual feedback which
guides the HO in the conduct of the task. Direct teleoperation commonly requires
a number of different viewpoints including global stereoscopic views (i.e., multiple
images). Similarly, graphical/pictorial representation of data is the best way of
displaying it. To ease operator burden, graphical virtual models are required with
increased autonomy from the ground operator. Telemetry data rate dominates the
communications bandwidth particularly for images. Bandwidth constraints limit
the bit rate for telemetry, particularly for video-imaging. High-compression
factors are required for images. There are a number of ways to reduce the image
transmission rate—low-resolution color images combined with high-resolution
black-and-white images can yield significant data rate reductions. Predictive displays graphically simulate the robot and the target motions in real time. The 3D
graphics models (e.g., Wireframe or 3D models) of the robot and objects in the
environment are overlayed over the 2D scene TV images. The teleautonomous
approach separates the remote and control station systems in a manner that is
applicable only to pure kinematic formulations in free space environments such as
rovers [639]. The predictive display is effective for free motion while shared
compliance control is effective during contact operations [640].

Robonaut is one of the most sophisticated robots in existence—it is currently
in service on ISS. It is an anthropomorphic robotic upper torso designed for EVA
(extravehicular activity)-equivalent tasks similar to those of a human astronaut.
It comprises two 7 DOF arms with a reach of 0.7 m (roll–pitch–roll–pitch–roll–
pitch–yaw configuration), two five-fingered hands, a head mounted onto a 3 DOF
orientable neck (pan–tilt–verge), a torso and a 6 DOF grappling leg. Each arm
has a shoulder, elbow, and wrist assembly in a roll–pitch–roll–pitch–roll–pitch–
yaw configuration. Each hand has 14 DOFs and includes a forearm, a 2 DOF
wrist mounting a 12 DOF hand (two 3 DOF fingers, one 3 DOF thumb, two
1 DOF fingers, and a 1 DOF palm). The hand is mounted onto a 5 DOF arm—
the forearm mounts Robonaut's 14 motors/harmonic drives, 12 circuit boards, and
the wiring harness. In total, it has 42 DOFs. The hands are designed to handle
EVA tools including power tools. Each finger possesses a six-axis force sensor and
has force sensors at each joint to detect applied forces. Each finger has a grasping
force of 2 kg while each arm as a whole can lift 9.5 kg. It is limited to exerting
90 N for safety reasons and can reach, orient to, and manipulate objects
including flexible objects. The 150 kg Robonaut 2 implements a total of 350
sensors. The head comprises two binocular stereo cameras with a fixed verge at
arm's length. The head sits on a pan–tilt neck with a pitch axis below the
camera frame to permit forward translation of the neck. It has been proposed to

mount it onto a planetary rover chassis in a "centaur" configuration to impart sophisticated handling capabilities [643]. Robonaut's torso mounts 38 PowerPC processors running VxWorks connected through a virtual machine environment (VME) backplane. The Robonaut telerobotic system architecture adopts full immersion telepresence using a stereovision helmet and a pair of haptic datagloves appropriate only for on-site astronauts controlling teleoperated rovers (Figure 8.1) [644].

The trend in human–computer interfacing is to maximize transparency between the human operator and the remote site. Transparency minimizes the cognitive load on the human teleoperator. This is the principle behind virtual reality—an extension of predictive graphics—generates simulated environments in real time through multiple sensory channels to provide the sense of human operator "immersion" within that environment [645]. Immersion attempts to enhance the transparency between the HO and the robot by creating the illusion of telepresence—situation awareness that the HO is physically present in the remote work environment through a computer-generated graphical environment. VR engines are typically multiprocessor graphics stations through which the user interacts via input/output devices—3D trackers, force-sensing datagloves, stereo head-mounted displays, and 3D sound. Output from the simulated world is through graphical visual feedback and tactile feedback. In order to retain the sense of immersion, high frame rates are required ($\sim$30 Hz) which may be accomplished by using variable detail at lower resolution for distant and/or non-foveated regions of the environment. Remote camera orientation may be controlled by the operator's eye movements monitored by infrared reflections from the pupils [646].

The use of brain–computer interfaces using 2D pointer control for spacecraft navigation has been explored [646]. It was based on EEG measurement of event-related potentials (P300) in the operator's brain to generate control vectors for a 2D screen pointer. A trained support vector machine learns the training neural–behavioral vectors from several operators concurrently to provide robust responses. This approach could potentially be applied to rover control for the Moon.

The maximum time delay for efficient teleoperation is $\sim$0.2–0.5 s. When longer transmission delays occur, move-and-wait strategies are usually adopted by the operator. Time delays introduce phase lag which reduces system stability margins. However, any time delay and limited bandwidth limitations degrade the possibility of telepresence [647]. There are a number of approaches for accommodating time delay loops during teleoperation [648]. A graphical representation of the remote robot—a phantom robot—such as a wireframe surface or 3D model may be overlaid on a video image of the real robot. The graphical overlay is registered with the position of its real counterpart and is used to predict remote motion. Predictor displays provide the basis for estimating the future state of the rover in real time and are particularly valuable for long time delays >0.5 s. Most involve adding the current state and time derivative states into a model with estimated future control signals based on current control signals. The Smith predictor attempts to remove the effects of time delay from a closed-loop control system.

Figure 8.1. Robonaut in (top) ISS configuration and (bottom) centaur configuration [credit NASA].

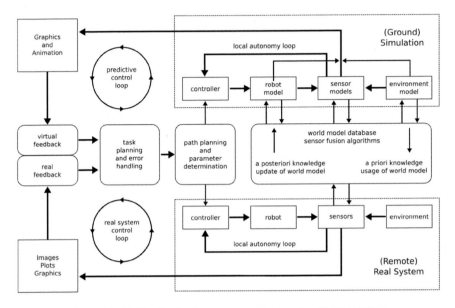

Figure 8.2. DLR telerobotic systems architecture [credit DLR ESA].

Shared control for advanced telerobotic operation with a mixture of automation and teleoperation is desirable (i.e., supervised autonomy). Generally, supervised control is adopted in which high-level commands are sent to the remote rover and closed-loop servo-control is performed by the rover at the remote site. This is different from the traditional approach of ground control of a spacecraft where low-level time-tagged commands are transmitted which affords little flexibility. Repeatedly executed tasks introduce the opportunity for high degrees of preprogrammed autonomy (e.g., automatic control of camera position and orientations based on camera position would significantly reduce the operator's workload since continuous work on the task is possible without interruption for camera adjustment). By limiting data transfer to higher levels of control reduces the required human information-processing rate by a factor of $\sim10^7$ to ~50–100 bps (comparable with reading at 40 bps and speech at 10 phonemes/s) [651]. DLR telerobotic system architecture is one that can accommodate shared control (Figure 8.2) [652, 653]:

The human operator acts as the mission planner and provides overall supervision. Supervised autonomy allows simulation and modification of task commands prior to being sent for actual execution. A task is essentially defined as a sequence of commands. Command sequences are typically composed of a fixed repertoire of command types that can be parameterized and sequenced to generate a wide range of behaviors. Most commands have a basic "MOVE (object) TO (destination)" syntax with concatenated sequences defining the traverse or kine-

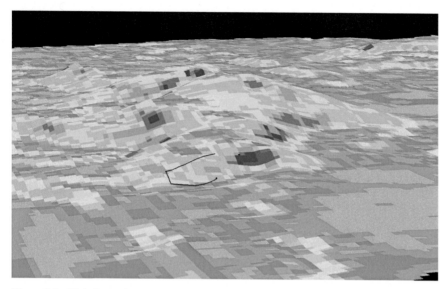

Figure 8.3. Digital terrain map (DTM) construction of Columbia Hills for MER traverse planning [credit NASA JPL].

matic requirements. Command sequences are generated as the ground control station uploads new goals and intermediate waypoints to the rover during communication windows. Desired waypoints are selected daily by the ground control staff for guiding the rover's autonomous navigation. The command line gives the user direct access to computer operations including scripts of commands which offer great power and flexibility, but this power and flexibility can be dangerous. The primary purpose of human computer interfacing is to hide the low-level operations of the computer. The most common generic approach is WIMPS (Windows Icons Menus Pointers) in which menus in particular restrict user operations on the computer.

The human operator requires a view of the remote environment on the planetary surface. Divergent performance from nominally commanded motion should invoke automated reflexes either to halt the motion or correct it—such event-invoked safety reflexes are critical for robustness [654]. A 3D digital elevation map (DEM) of the environment is constructed automatically on the ground from the panoramic stereoscopic images from the lander [655, 656] (Figure 8.3).

Beagle 2 was to be controlled through commands generated by the operations planning team uploaded from a ground control center. The commands were to be uploaded to Mars Express (once every 1–4 days) or Mars Odyssey (periodically) which would then be relayed to Beagle 2 on the surface during the overhead pass. In particular, Mars Odyssey resided in a Sun-synchronous polar orbit providing twice daily contact windows. The commands were to cover surface activities for up to four days. The Mars Express/Beagle 2 mission control center had overall responsibility for the mission and resided at ESOC which controlled the communications link to Mars Express. The early phases of the landed mission were to be conducted through Mars Odyssey while Mars Express maneuvered into its operational orbit. Operations planning were to proceed through the construction of a DEM of the landing site using a virtual reality representation from Envision Software. The first activity was to be a wide-area panoramic image of the landing site using the wide-angle mirror after the lander lid was opened but prior to deployment of the ARM. Primary surface targets on deployment of the ARM were rocks, particularly below the weathered oxidized rind and the subsurface to depths of up to 1.5 m. Following analysis by the ARM instruments, samples were to be analyzed by the GAP instrument.

Sojourner's ground control station was supported by a Silicon Graphics Onyx 2 graphics workstation for the graphical user interface based on the WIMPS philosophy. The operator wore goggles to provide 3D stereo-imaging and controlled the rover using a joystick which allowed a firm grip and easy movement but had less cursor guidance than a mouse. American teleoperative controllers have generally been dual joystick–based while European teleoperative controllers have been of the space mouse–type. Other options include a spaceball which can be twisted and pushed, the Polhemus tracker wand whose position/orientation is transmitted to the cursor, and the DataGlove whose position/orientation and gesture is registered. Symbolic darts were used to indicate the direction of the pathways for the rover to follow. Human operators on Earth specified waypoints in 3D maps of the landing site generated from stereo images. Waypoints were planned and generated by the ground station teleoperator while navigation between designated waypoints was conducted autonomously by the rover. The ground controller explored and evaluated potential paths between chosen sites while minimizing a weighted sum of power consumption, risk of tipping over, traverse time, risk of slippage, etc. Waypoint coordinates were uploaded to command the rover to move to those coordinates using a simple software navigation algorithm:

```
If no hazard exists
  Move forward and steer towards the goal
Else if there exists a hazard on the left
  Turn in place to the right until no hazard is detected
Else if there exists a hazard to the right
  Turn in place to the left until no hazard is detected
```

All in all, Sojourner traversed 104 m enabled by 114 commanded movements uploaded from the ground. The MERs were controlled through a graphical user interface, the "Virtual Dashboard", to send telecommands to the remote rover. Command sequences were planned through a 3D terrain model of the Martian environment based on stereo images from the MERs' stereo cameras.

It is envisaged that the ExoMars ground station will implement the DREAMS (Distributed Robotic & automation Environment for Advanced Missions Specification & Supervision) system which will interact with the onboard autonomous navigation system of the ExoMars rover [657, 658]. DREAMS is an integrated distributed multiuser systems architecture for specifying, validating, monitoring, and controlling robotic systems in space or planetary surfaces from the ground. It was designed for general robotic operations in space and has heritage from its predecessors JERICO, FAMOUS, SPARCO, MARCO, and CESAR. The FAMOUS (Flexible Automation Monitoring and Operation Users Station) robotic control system allows six hierarchical levels of abstraction—compound task, simple task, action, actuation, device, and physical level. CESAR (Common controller for European Space Automation and Robotics) is a flexible, modular, and open robotic control architecture for servicing and planetary mission scenarios built on the earlier SPARCO (Space Robot Controller) [659]. It is composed of a master robot control unit (RCU) which performs most of the high-level tasks and a set of slave servo-control unit (SCU) modules. DREAMS implements space robot control by defining robotic operations hierarchically at the mission level, task level, and action level [660]. A mission is a concatenation of tasks which in turn are comprised of actions. It is expected to be extended to include a task-planning tool (TAPAS), a formal mathematical specification and verification tool (MUROCO) [661], and an advanced teleoperation feedback capability (TELEMAN which is not relevant for ExoMars). A library of actions are con-catenated into tasks which in turn are assembled into plans. FORMID (formal robotic mission inspection and debugging tool) is a production system which implements logical functions of actions for the DREAMS architecture. Each pro-duction rule has a set of preconditions which define invocation events and postconditions which define the effects of actions. Exception handling is essential for critical systems like planetary rovers. Actions can be combined sequentially and in parallel to create complex tasks. The A-DREAMS (advanced DREAMS) ground control station has been further developed for ESA to support its space robotics applications [662]. It adopts an offline programming environment for planning robotic tasks based on 3D dynamic modeling while providing tele-operation and interactive autonomy with automated calibration through hardware-in-the-loop validation and haptic/visual feedback. It has a three-layered hierarchy that decomposes mission objectives into tasks, actions, and plans outputs. It covers three core operational phase functions [663]: (i) preparation and validation station (including ground calibration); (ii) robot control, monitoring, and command station; (iii) payload monitoring and command station (based on SCOS 2000, the standard spacecraft control system implemented at ESOC). Of most importance to ExoMars (where teleoperation is impossible) is the CNES Autonomous Rover Navigation System. DREAMs will implement the CNES Autonomous Rover Navigation Software Simulator as a ground emulation facility, a visualization tool, and a terrain testbed for physical validation. The rover receives the goal location (heading and distance, waypoint coordinates, etc.) from Earth identified by the ground personnel from the previous communication session images. This goal may be visible but its distance is not computable from the cameras or beyond visible range if taken from orbital imagery. The rover then embarks on its traverse segment until it has reached its goal identified by the ground station. During the traverse, it may interrupt its progress due to a priority override such as a communication window or the path to the goal is blocked by obstacles.

The DEM defines a cartesian grid of cells each of which may be classified using a Bayes classification scheme into different terrain types of flat, rough, and unknown according to measured attributes such as texture. The ground station displays the local planetary terrain in 3D and the operator selects a sequence of locations (waypoints). These images are taken from the onboard stereo-imaging system of the rover. A joystick is used to position a 3D graphical model of the rover at the waypoints. A straight line trajectory defines the nominal path. A sequence of commands is then uploaded to the rover. This emplaces the high processing demands on the ground. Although the DEM is commonly constructed on the ground from downloaded images, the adoption of onboard DEM construction would reduce the downloaded data requirement to Earth by 20–100 times [664]. It has been proposed that stereoscopic vision may evolve into 3D video where 3D telepresent scene representation offers the possibility of look-around capability [665]. This has heritage from 3D computer graphics. It requires remote cameras to be horizontally configured with a small distance between their optical axes (similar to human binocular baseline), optical axes to have small convergence, and each camera to have human-level color sensitivity. Algorithmically, this will require stereo depth analysis, motion estimation, shape analysis, and texture analysis.

The Mars Outpost comprises a human-inhabited research base on Mars supported by in situ resource utilization factories and extensive robot support of rovers, balloons, hoppers, etc. to explore larger regions of Mars. Humans within the relative safety of the Habitat can teleoperate the vehicles to perform large-area reconnaissance. Another alternative is to deliver a human crew to Mars orbit and deploy robotic explorers to the Martian surface which are teleoperated by the human crew in orbit [666]. This approach offers reduced cost and human risk while minimizing the possibility of forward and backward biological contamination.

The age-old argument of robot versus human spaceflight is periodically dusted off for recapitulation. The two approaches are completely complementary. There is a synergistic and efficient division of labor between man and machine and this division will evolve in symbiotic fashion, the machine acquiring greater skills to carry out more onerous tasks, leaving the human to deal with more sophisticated and demanding tasks [667, 668]. Consistent, repetitive, and routine tasks requiring high precision, speed, and repeatability in hostile environments are ideally suited to automation as machines do not tire or become inattentive over long periods. They can also be deployed for monitoring multiple complex systems and fault detection which can overwhelm human operators. Humans are most suited to non-repetitive tasks with high variability of presentation requiring high manual dexterity, robust pattern recognition in noisy uncertain environments, and flexible judgments based on lifetime learning and dealing with unexpected or difficult events. Humans are adaptable and capable of making decisions based on experience while robots cannot yet achieve such capabilities. Robots are limited in their autonomous capabilities by technological development—however, as technology advances, so robots will be able to subsume more complex functions from humans, freeing them for more cognitively challenging tasks. Robots are required for precursor survey missions, crew support on planetary surfaces, and for projecting human capabilities to remote, hostile environments. Robotic field assistants can be used to carry tools and equipment, act as remote teleoperated scientific platforms for astronaut scientists, and act as astronaut transport vehicles [669, 670]. The HERRO (human exploration using real-time robotic operations) concept is premised on the division of labor between human and robotic exploration in which humans in orbit around Mars teleoperate surface rovers deployed on the Martian surface [671]. Not landing humans onto the surface of Mars substantially reduces the cost and risk to manned Mars exploration. The manned Mars orbit vehicle may remain in Mars orbit for one or more years or humans may be deployed to a Phobos base of operations but restricted to less than an 8 h period (i.e., a maximum of a 4 h near-equatorial overpass). Teleoperating in situ rovers by the crew is enabled by the short distance between the surface and Mars orbit. An elliptical orbit with a half-sol period of 12 h 20 m at an altitude of 30,000 km allows a response delay of less than 100 ms. Multiple rover teams may be controlled during the orbital overpass at different locations on the surface. However, although the requirements for rover autonomy are much relieved, Mars surface operations are limited by the actuation capabilities of the deployed rovers (nominally a fleet of rovers with varying capabilities). We are concerned with two aspects of autonomous behavior—autonomous navigation which operates during rover traverse phases and autonomous science which operates during the scientific instrument deployment, sample acquisition, processing and measurement phase. In all aspects of autonomy, failure mitigation is a significant consideration. Execution of the actions may be regarded as logical effects—changes to the state of the environment as a result of those actions—initiated by preconditions defined by the state of the environment as detected by the sensors. An important subset of actions is reaction to failure and rescheduling (Table 8.1).

Consideration of human psychology is essential in the context of human–machine interfaces such as frame analysis [671]. Cognitive processing is not a mechanical process but a reconstructive one involving situated knowledge, cultural background, and personal skills. Humans indulge in cognitive frame jumping which continually reconstructs personal perspective and perception—the interface mediates that process in a structured manner so its design is crucial.

Robots in the future will begin to develop ever more sophisticated social relationships with humans based on social etiquette including the communications of affect (emotions) [672]. The development of such robotic social skills will vastly expand the range of human–robotic interaction beyond spoken dialog to include subtle social cues. Such capabilities however have yet to make their debut but do offer the promise of vastly increasing the amount of information transferable between human and robot. Fundamental to this capability will be automated face

Table 8.1. Failure mitigation actions.

1	Effects of failure of executed actions	Retry action as immediate reschedule
2	Precondition check for executable action failure (resource availability changes)	Reschedule existing executable actions
3	Precondition for executable action failure	Plan repaired via insertion of extra actions
4	New task presented so scheduled actions now infeasible	Plan/replan executable actions at lowest level
5	Existing task sequence infeasible— resource changes or other threats or opportunities	Reschedule at task level
6	Mission goals faced by threat or opportunity	Generate new tasks and plan for them

recognition. Face recognition is a specialized form of image processing that often exploits Karhunen–Loève transforms, singular value decomposition, or neural networks to extract features of high curvature [673]. They rely on conventional preprocessing such as Canny's edge detector or Hough's feature detector both of which are based on intensity gradient measurements. Gabor wavelet decomposition has shown promise for feature extraction. The 2D Gabor transform and its Fourier transform are given by:

$$g(x, y; u, v) = e^{\left[-\left(\frac{x^2}{2\sigma_x^2} + \frac{y^2}{2\sigma_y^2}\right) + 2\pi i(ux + vy)\right]}$$

where $\sigma_{x,y} = $ Gaussian spatial width and $(u, v) = $ complex sinusoid frequency

$$G(u, v) = e^{[-2\pi^2[\sigma_x^2(u-u')^2 + \sigma_y^2(v-v')^2]]}$$

8.2 ROVER SURFACE OPERATIONS

There is an up to 40 min signal delay between Earth and Mars precluding the possibility of telerobotic operation for Mars operations and beyond so a high degree of autonomous navigation and control for a planetary rover is required. Furthermore, line of sight communication windows are limited (e.g., Mars Pathfinder had only two 5 min communication windows to Earth per day). There exists a tradeoff between communications that require power to transmit data and computations that require power to process data. It is desirable to minimize communications requirements due to communications bandwidth constraints which favors onboard computational processing of information. Communication limitations dictate the degree of autonomy required between communication windows—this essentially defines the surface operational mission for rovers. Furthermore, such communication windows are primarily concerned with downlinking of telemetry and science data (uplinking of commands based on the previous down-linked telemetry generally has a much smaller footprint than telemetry). There is however a tradeoff between onboard computation which does not impose band-width requirements for communication nor time delays in reactivity but does impose requirements for onboard computational processing and data storage. Onboard data handling is a limited resource imposing major limitations on performance. For instance, the 180 kg Zoe rover platform used six high-end onboard computers with one devoted to self-localization, two devoted to onboard autonomous navigation, and two devoted to motor control [674]. This level of computing resources is impractical for a planetary rover at least currently. For early and current rover missions, it was therefore desirable to minimize onboard computational requirements for rover navigation.

The Martian day (sol) is close to 24 hours in length similar to (but not exactly the same as) that on Earth. MER surface operations are particularly instructive. They included: (i) communications sessions (determined by the orbiter overhead pass typically up to 1 h maximum) for uploading images; (ii) battery charging sessions (typically conducted close to midday); (iii) image acquisition prior to each surface traverse segment; (iv) surface traverse segments; (v) scientific measurement cycles. Initial localization of both MERs was performed through two-way X-band Doppler signals between Earth and Mars and two-way UHF Doppler signals between the rover and the Odyssey orbiter [675]. The MERs were commanded once per Martian day using prescheduled sequences of precise movement com-mands for that day. There were primarily three types of command: rover imaging and scientific measurements, robotic arm positioning commands, and rover loco-motion commands [676]. These could be low-level commands that specified actuated motions precisely, higher level directed primitives that specified path arcs, and finally autonomous path selection. Directed driving primitive commands were executed most quickly (up to 120 m/h) but had greater risk as there was no look-ahead facility except the AutoNav VisOdom facilities that could detect forward obstacles at speeds up to 10 m/h. AutoNav provided terrain imaging and closed-loop hazard avoidance for distances over which the terrain was visible (nominally 5–10 m). MERs were restricted to specific times of day for surface operations [677]. The rover was activated in mid-morning (around 9.00 AM), shut down at midday for 1–2 h of battery charging (11.30–12.30), and finally shut down mid-afternoon (around 3.00 PM). Image acquisitions, traverses, target approach, and scientific experiments are thus limited to 5 h per day. For a 100 m/sol traverse, this required a minimum traverse speed of 20 m/h but the MERs devoted specific days to each function. A day of drive traverse was usually limited to ~50 m. A target approach day was limited to 10 m. Similarly, scientific instruments were deployed on science days. Furthermore, the energy available decreased over time due to dust deposition on upper-deck solar arrays (0.3% per day) over the nominal 90-day mission life (25% reduction) though MERs significantly exceeded this during the extended mission due to clearing dust from the solar arrays as a result of intermittent dust devils. During sleep, almost all devices were powered down with battery charging, thermal control, and computer timer only being powered. Sleep occurred through the night from 3 PM until 9 AM except during orbiter overflight periods at 4 PM and 4 AM. The rover operations cycle was 4 days in length, one day for panoramic image capture, one for target definition by the ground crew which was uploaded, one for traverse to the ground-selected target accompanied by imaging at higher resolution and precise positioning, and finally one for science measurements. Hence, considerable planning had to be performed for communications scheduling, science planning, power scheduling, etc. which could not be predefined before flight. The maintenance of power and its scheduling was particularly important and represented a basic facet of self-sufficiency which has to be incorporated in any planetary rover's behavior to ensure its survival 678].

Clearly, the ground-planning horizon severely limits the scientific productivity of the rover. On MER, daily plans were dependent on the successful completion of the previous sol's activities. Complications highlighted included dead-reckoning errors and unanticipated obstacles during rover traverse, contact forces generated by the robotic arm, abrasion tool grinding process, etc. which could not be effectively modeled. Note that these are all functions associated with robotic interaction with an unmodeled environment. Currently, the ground control station plays a critical role for 3D environmental modeling (based on maps constructed from downloaded images), rover chassis dynamics modeling, wheel/soil modeling, and trajectory planning. Autonomy implies implementing such functions on board the rover with the ground station acting in a supervisory role (supervised autonomy).

8.3 BEHAVIOR-BASED AUTONOMY

The U.S. has pioneered the use of behavior-based control on planetary rovers for reflexive obstacle avoidance [679]. European research in planetary rover navigation and control has focused on top-down trajectory planners using a centralized world model. However, in recent years, these approaches have been converging. The first U.S. planetary rover—the Pathfinder Sojourner microrover—had extremely limited computational resources [680]. The rationale for the behavior control approach was to minimize the energy cost of computation by employing simple programs that requires <40 kB EEPROM memory for storage and <1 MIPS processing loads [681]. Behavior control is an autonomous robot control methodology inspired by animal ethology rather than human capabilities [682, 683]. An example is the Braitenberg approach in which complex behaviors such as fear and aggression can be generated by little more than simple hardwired connections between simple sensors and actuators [684]. Robot behavior is seen not so much as purposive but as traversing between points in an environment while avoiding obstacles—adaptable behavior involves no logical reasoning. The simplest form of animal behavior is the reflex—a rapid, stereotypical response triggered by specific stimuli. Reflexes are generally mediated by direct connections between sensory and motor neurons with no intermediate neural processing. All nervous systems are organized into movement patterns. The simplest pattern is the open-loop simple reflex. More complex reflex patterns are closed loop (e.g., grasp reflex). These latter modifiable reflexes underlie preadaptations for learning motor skills. Reactive robots act and in the process determine the sensory patterns they receive from the environment. There are several tenets to the physical grounding hypothesis that underlies behavior control [685]:

(i) situatedness—robots are agents in the real world so the world is its own best model (intelligent behavior does not require explicit world models as simulated world models are incomplete representations);

(ii) embodiment—robots have bodies with sensors and actuators with which they interact with the real world directly (i.e., meaning is grounded in the physical environment) (intelligence does not require explicit symbol manipulation);

(iii) intelligence—intelligence is a property of the robot's behavior in surviving in the real world through dynamic interaction with its environment;

(iv) emergence—intelligent behavior emerges from the robot's interactions with the real world.

Situatedness implies that the agent has a perspective in being embedded in the real world—its behavior is based on its interaction with its environment. Given that plan-based robots suffer from the combinatorial explosion problem, plans are not appropriate. Intelligent behavior is emergent from such processes without central control where global communications occur through interaction with the environment—this is the principle of parsimonious design [686]. In addition, the principle of ecological balance requires that there must be a match between the complexity of the sensors, actuators, and control system. This is closely related to the animat approach which emphasizes the importance of operation in complex natural environments without deliberative planning [687]. Embeddedness in the physical world may be crucial to the quest for artificial intelligence as it must cope with the dynamic uncertainty of the real environment [688].

Behavior control was proposed to overcome the slowness of robotic planning algorithms—the Stanford cart moved 1 m every 10–15 min due to the high processing overhead [689]. Planner-based control strategies employ a centralized world model updated from sensor fusion data, while reactive control strategies employ reactive condition–action rules linking sensors and actuators directly without sensor fusion into a central world model representation. The generation and maintenance of a single monolithic world model representation is computationally intensive. Behavioral control involves decomposing autonomous intelligent behavior into multiple modular task-based competences (behaviors)—each modular behavior represents a goal-oriented control law which encapsulates a specific simple task characterized by a situation–action rule with preconditions (sensory patterns) and postconditions (action) modeling a stimulus–response reflex. Thus, each reflexive behavior is connected directly to relevant sensors and actuators without any intervening symbol processing (such as centralized world models). The sensors and actuators serve to couple the robot directly to the environment which stresses the fundamental importance of situatedness and embodiment. Each behavior requires only task-relevant data to function (egocentric invariance). Each is invoked by a highly specific subset of sensory patterns and generates a specific motor response (i.e., only a small segment of the environment is filtered for perceptual analysis which is determined by the sensory transduction process). Basic behaviors include: Avoid_Obstacles (level 0), Move_Forward (level 1), and Turn_Towards _Target (level 2). As the robot moves around in the environment, sensory patterns will vary according to distance from objects, viewing conditions, etc. Each behavior functions asynchronously and runs in parallel. This may be regarded as a form of command fusion without the need for world models or complex sensor fusion. Higher level complex global behavioral patterns emerge which are not reducible to the individual simpler primitive behaviors comprising them [690, 691]. It is through interaction with the world that multiple behaviors emerge without centralized control—global communications occur through the environment without any intervening world model on the basis that "the world is its own best model" [692–695]. The total number of unique behaviors emergent from different primitive behaviors may be assumed to be a binomial function in that i possible combinations of n items taken at time k is given by [696]:

$$i = \sum_{k=0}^{n} \frac{n!}{k!(n-k)!}$$

Coupling between behaviors may be temporal or spatial. Alternative orderings include hierarchical coupling of primitive behaviors by adding temporal ordering. As each behavior requires only limited sensory data, the control system can react rapidly without delays incurred by sensor fusion and world modeling. The problem of autonomous behavior is decomposed vertically into number of levels of task competence based on behavioral units rather than functional modules. The control system is built bottom-up incrementally into layers of competence and each level of competence includes the lower level as a subset. These task-level behaviors are built up from basic locomotion tasks to increasing levels of sophistication in an incremental fashion but each behavior encapsulates its own perception and actuation functions. The hierarchical arrangement of behaviors is shown in Figure 8.4 and Table 8.2. Only the first three behaviors are typically implemented in behavior control with scaling to more complex behaviors being less successful due to the progression from reactive to purposive functions which imply the need for world modeling. A typical obstacle avoidance procedure is to back up one vehicle length, turn 90° away from the obstacle, move forward by one vehicle length, and turn forward again by 90°. Essentially behavior control implements a large CASE statement in which tests for each are specific sensor patterns generated by the environment. The action component of the CASE statement specifies which behaviors are activated.

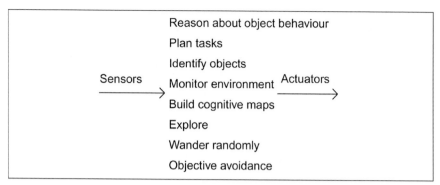

Figure 8.4. Behavior control layering of competences.

Action selection is a significant problem in behavior control as different invoked behaviors can result in conflicting behaviors (such as cyclic or freezing behaviors) which requires arbitration and selection (conflict resolution). Simple behaviors may be combined in a number of different ways including (weighted) parallel/distributed vector summation, spreading activation, hierarchical suppression, and sequential invocation [697]. The subsumption architecture is a structured

Table 8.2. Subsumption architecture.

Level	Module	Effect
0	Collide, runaway	Remains stationary until a moving obstacle approaches it
1	Wander randomly, object avoidance	Generate new headings, periodically accepts a force vector input from level 0 and suppresses output from runaway module
2	Goal-directed exploration	Finds a corridor to a specified goal at a distance and moves towards it (likened to exploratory behavior for the provision of information)
3	Build cognitive maps	Plan routes between landmarks
4	Monitor environment	Monitors dynamic changes in the environment
5	Identify objects	Identifies objects in the environment and reasons about tasks to be performed on them
6	Plan tasks	Formulates plans to change the state of the world as required
7	Reason about object behavior	Reasons about object behavior in the world and modifies plans accordingly

and layered approach to fusing simple task-achieving behaviors in a fixed prioritorization scheme of higher behaviors taking priority over lower behaviors thereby resolving conflicts [698, 699]. It is a hierarchical approach that adopts fixed prioritorization for a winner-takes-all selection of behaviors—higher level behaviors suppress the behaviors of lower layers thereby eliminating conflict through priority (i.e., higher level behaviors can subsume low-level behaviors). One instance of the subsumption architecture is the logic-based subsumption architecture in which each behavior is implemented as non-monotonic logic statements based on circumscription [700]. Subsumption has been implemented as networks of augmented FSMs (finite state machine of registers and timers with output wires) with asynchronous message passing of 8-bit numbers over local combinatorial circuits to generate actuator outputs [701]. Behavior control based on the subsumption architecture has been applied to robot manipulators through 15 independent behaviors grouped into 6 competences: cradle, grip, path, park, skim, and local [702]. Fixed prioritorization such as subsumption limits the behavioral flexibility generated. Behaviors may be weighted according to a learning algorithm and thresholding behaviors eliminate the possibility of chaotic behavioral responses. Fuzzy logic may be used to implement behavioral if–then rules augmented with fuzzy truth values to generate fuzzy behaviors. Selection of behaviors in the form of metarules may itself be fuzzy which implements a set of graded preferences. Context-dependent blending of behaviors allows different behavior patterns to be expressed: IF obstacle_close THEN avoid_obstacle; IF ~obstacle_close THEN go_to_target. A partly closed obstacle will yield a blend of both behaviors. In this way, complex goal-oriented behaviors can be constructed from simple local fuzzy rules. The number of fuzzy rules is $O(K^n)$ where $K =$ number of fuzzy predicates and $n =$ number of input variables. The complexity of the fuzzy rule set can become large. This may be alleviated through hierarchical structuring. The distributed architecture for mobile navigation (DAMN) is a behavior-based approach that employs a central command arbiter which generates actions by combining (fusing) outputs from behavior modules according to a voting scheme [703]. Rosenblatt and Payton (1989) introduced a variation on the subsumption architecture which adopted a finer-grained task-oriented behavioral suite organized in a connectionist-type, highly structured architecture [704]. Behaviors are assigned in two layers—an obstacle avoidance layer and a follow path layer. Behaviors are taken to represent specific concepts—each unit receives multiple-weighted input from other units and computes a single activation level output. Each link is weighted according to its priority. Each unit computes a weighted sum of votes received from each active behavior $0 < w < 1$. The sum is normalized by dividing by the sum of weights. Units with the highest activation are selected for command execution. The system implements a gradient field to avoid obstacles and pursue goals with commands being based on compromise arbitration.

Bayesian approaches to robot path planning were initially applied to obstacle avoidance in the context of decision theory [705]. Action selection is a fundamental aspect of intelligent behavior and is dependent on both the current state of the agent and that of its environment. The sensorimotor system—sensors and actuators that interface with the environment—is subject to noise. This introduces the necessity for estimation of these states. These state estimates must be combined with assessments of the potential rewards or costs of different actions. Barnes et al. (1997) extended the notion of a behavior to include a measure of its utility to generate behavior output in the form [706]:

$$y = \begin{pmatrix} f_r(s) \\ f_u(s) \end{pmatrix}$$

where f_r = response function based on sensory data s and f_u = utility function based on sensory data s. Utility may be regarded as equivalent of a gain within a schema. This is similar to Rosenblatt (1997) who introduced a set of behaviors that compute the utility of world states. Bayesian decision analysis may be used to determine the next sensing action to be performed [707]. A choice is offered by a set of actions $a \in A$ with probability distribution $p(x|a)$ over a set of possible states of the world X. The outcome of each action in each possible state $a(x)$ is characterized by a utility function for each action. Utilities may be combined to produce the maximum expected utility of performing selected action a estimated through Bayesian probabilities [708]:

$$U(a) = \sum_c U_c(a(c)) \cdot p(c|a,e)$$

where $U(c)$ = expected utility for each consequence c of action a and $p(c|a,e)$ = probability consequence c will occur given the observed evidence e and the adoption of action a. This allows optimal action selection of the maximum utility through the use of Bayes theorem to reason about uncertainty.

Behavior control substantially reduces the mass required to support computational resources on board a rover requiring only $\sim$0.1–1 MIPS and 10 kB memory (within the capabilities of a MC6811 microcontroller) compared with traditional planners that require $\sim$100–1,000 MIPS for real time performance [709–712]. The Rocky series of microrover prototypes (precursors to Sojourner) possessed only limited computational resources. They adopted a three-layered software architecture with each layer passing information down the control hierarchy: the lowest layer implemented motor control of speed and direction, the second layer implemented trajectory monitoring and obstacle avoidance, and the third layer implemented high-level sequencing control to move the robot through a series of x, y waypoints specified by the human operator. The GoTo_XY command was the principle autonomous command for traversing between waypoints. Rocky III/IV/VII platforms implemented ALFA, a behavior programming language which implemented a network of computational behavior modules that intercommunicated in a dataflow architecture [713, 714]. Each module was a computational state space machine that computed a transfer function between inputs and outputs. Each layer interfaced the layer above through computational channels that accept a number of inputs and combines them into a single output. Interrupt handling provided the means for reflex events within the software control loop.

ALFA allows higher layers to pass information to lower layers (in a subsumption architecture) through dataflow-based communications channels. The Rocky navigation algorithm was as follows [715]:

```
1. LOCALISATION
   Measure global rover position from lander
2. WAYPOINT
   Set new reference position from task queue
3. TURN-TO-GOAL
   If position error is small
     Goto 1
   Else
     Turn in place towards goal
4. OBSTACLE-DETECT
   Measure terrain in front of rover
5. TURN-IN-PLACE
   If obstacles center or left and right
     Turn nominal rotation right
   Goto 4
     If obstacles left/right
       If previous obstacles left/right
         Goto 6
       Else
         Turn nominal rotation left
         Goto 4
6. THREAD-THE-NEEDLE
   If obstacle center
     Move total alley length straight backwards
     Goto 4
   Else if obstacles left or right
     Move nominal translation straight forward
     Increment total alley length
     Obstacle_detect
     Goto 6
   Else if obstacle clear
     Move nominal translation straight forward
     Goto 4
7. LOOP-TO-GOAL
   If orientation error is small
     Move nominal translation straight forward
   Else if orientation error is medium
     Set turn radius to large
     Move nominal translation forward
   Else if orientation error is large
     If position error is medium
```

```
   Goto 3
 Else
   Set turn radius to small
   Move nominal translation forward
 Goto 4
```

Such behavior control methods were implemented as part of the autonomous control systems on Sojourner and the MERs. Sojourner's behavior control software used 90 kB of EEPROM for autonomous navigation between waypoints selected by the ground station operator. Onboard software implemented a single control loop which performed scheduled functions such as housekeeping and command handling. Rover motion commands were sequenced into a batch command file and uploaded daily to Sojourner during daily communications windows. Command sequences were based on combinations of high-level language of the form Go_To_Waypoint_X_Y (move in a straight line for a specific distance as measured by odometry), Turn (turn a specific angle measured by a yaw gyroscope), and Update_Position (estimate position within a Cartesian reference with the lander as its origin). The Pathfinder lander camera generated images of the rover from a known reference position augmented by stereoscopic images from the Sojourner rover itself. These images provided the basis for the selection of waypoint goal locations which were uploaded from the ground station during communication window sessions. Sojourner was commanded to move between these goal locations while implementing a form of reflexive behavior control to avoid hazards in its commanded path (i.e., navigation between the waypoints to the goal location was conducted autonomously by the rover). The real time onboard behavior control algorithm autonomously navigated between waypoints while avoiding obstacles. The control loop incorporated interrupt handling for reflex events such as bumper contact. GoTo_XY was the basic command for traversing between waypoints while avoiding obstacles. It computed an estimate for the XY position and heading of the rover to determine the steering angle. Along the way, onboard sensors detected the presence of obstacles to provide corrective setpoints. If an obstacle was detected by proximity sensors to the right/left, the rover stopped, turned left/right, and drove forward a short distance. For Sojourner, at the end of each day's operations, the lander camera imaged the rover and its surrounding terrain for the ground controller to select the waypoints for the following day's traverse. During the autonomous traverse, the detection of hazards invoked avoidance behavior until the vehicle passed the obstacle whereupon its original traverse was resumed. The main rover control loop was executed every 2 s until shutdown occurred (due to diminishing power or timeout). If a problem occurred which the rover could not solve autonomously (such as excessive tilt), it stopped and entered safe mode to await further instructions from the ground—it was this mode of operation that caused so many problems with Sojourner's transit across the rock garden. The ground station initiated the error recovery action to be implemented when error alarms were invoked.

It has been suggested that the behavior-based approach can be adapted to make predictions, plans, and goals without a central deliberative world model [716]. However, in reflexive behavior control, the complexity of behavior is a reflection of the environment rather than of the agent so the current state of the world completely determines the actions of the rover. Simple control rules may generate complex dynamics through interaction of the agent within its environment. They make limited use of internal representations of the external environment (world models). Therein lies the limitation of behavior control. It has been proposed that behaviour control may be expanded to enhance its task complexity capability to include goal-directed behaviors by structuring the environment through the addition of new features (markers) but this would only be suitable for infrastructure construction tasks on planetary surfaces [717, 718]. Gershenson (2004) performed some interesting experiments to assess approaches for controlling robots—rule based, knowledge based, behavior based, neural network based, and Braitenberg based—and each indicated limitations and strengths but none was demonstrably superior [720]. This suggests the use of multiple approaches. In particular, the behavior control mantra that the world is its own best model which eschews world models altogether does not provide sufficient capability for planetary rovers. Reactive approaches have been likened to Watsonian behaviorism in which human behavior is characterized as a reaction to stimuli and the internal processes of cognition are illusory [721]. Internal representation is however necessary to generate different behaviors for the same sensory patterns (i.e., memory)—this provides the basis of cognition. For instance, on sol 22, the Sojourner microrover on Mars was commanded to back up to a specific rock (Souffle) and emplace its arm on the rock for spectrometer measurement integration followed by a traverse to another rock. Unfortunately, it stopped short of the target rock and performed its measurements in mid air before subsequently heading off for its next target. The loss of memory of its previous traverse lost a scientific opportunity.

8.4 INTELLIGENT ROVER ARCHITECTURES

Behavior control has significant limitations in that it is not readily scalable to more complex behaviors. It is limited to comparatively simple tasks and can suffer from cyclic behavior due to limited sensory horizons to stimuli. Behavior control essentially treats autonomous navigation as a Markov chain without reliance on previous experience. The lack of memory in reactive behavior limits the flexibility of response introduced by dependence on previous inputs that require the implementation of internal states (i.e., the incidence of memory states [722]). This allows implementation of goal-directed behavior—such goals may include scientific instrument placement and sample acquisition, one of the raisons d'être of the planetary rover. Furthermore, intelligent autonomy requires high-level decision making to provide predictive capability to anticipate the outcomes of behavior [723]—this allows goal-oriented actions to be implemented.

Blackboard systems use multiple knowledge sources to analyze different aspects of a complex problem. They provide a uniform system which integrates a number of diverse, specialized, and independent knowledge sources that communicate through a common global database (blackboard). The use of a common memory between multiple knowledge sources provides the means for sharing results when subproblems cannot be solved by independent non-communicating agents. The blackboard is a high-level operating system that integrates the control of distributed component knowledge sources according to its "master" plan. An executive program controls the activation of each knowledge source in turn when an event occurs. System goals determine the plan that specifies the tasks to perform and in which order according to priorities and time constraints. The blackboard architecture has become the standard robot controller architecture [724]. It has commonly been used to integrate multisensory data at a symbolic level. This is the level of representation of the world model. The blackboard is a multiagent system with a shared central control unit—it is suited to multiple hierarchical levels of control including high-level reasoning that supervises multiple lower-level behavior modules. It comprises a distributed set of independent but cooperating knowledge sources that monitor the shared blackboard and activate themselves based on the state of the blackboard to contribute to a problem-solving process. Different specialists contribute to different aspects of the plan which are posted to the blackboard and incorporated into the plan. All domain-specific data (including partial solutions) are posted onto the blackboard by each knowledge source for accessibility allowing the broadcast of events to all knowledge sources via the blackboard. A master control program examines the blackboard and schedules the component sub-solutions into composite solutions. The blackboard is a common centralized global data structure that may be hierarchically partitioned and represents the problem domain. It supports inter-knowledge source communication and acts as a shared memory to knowledge sources. More formally, the blackboard is a tuple $\langle X, P, B, I, T \rangle$ where X = set of blackboard data objects, P = set of blackboard object states, B = set of specialized knowledge sources, I = initial values of blackboard data objects, and T = problem-solving communication channel between knowledge sources [725]. The blackboard is a centralized database that allows hypothesize-and-test processes. The system maintains a set of hypotheses based on its world model and uses them to focus processing effort on expected events. Each knowledge source either applies hypotheses or tests them on the blackboard. Knowledge sources can modify posted plans based on subsequent contributions from other sources. Executive knowledge sources apply inference techniques to generate solution elements on the blackboard and knowledge sources respond to, generate, and modify solution elements on the blackboard. The blackboard system is then both data and goal driven. Each behavioral module comprises a sensorimotor function [726]. An activity is defined by a specific set of features which invokes a specific set of actions (a sensorimotor function). A set of solution elements on the blackboard constitutes a partial plan. Complementary or alternative plans can coexist and one or more partial plan can be merged to form more complete plans. Each agent develops partial interpretations and hypotheses based on their incomplete data. Solutions can be constructed by the aggregation of mutually constraining partial solutions on a blackboard using hypothesize-and-test strategies. Such results sharing based on different perspectives is data directed. Partial hypotheses are proposed and tested for plausibility at each stage of the processing. Results sharing facilitates solutions to problems that cannot be subdivided into subtasks. The blackboard supervisor provides great flexibility in the instantiation of behavior modules and arbitrates between them. The blackboard may be represented as a topological map of the scene using potential fields and modules as separate expert systems dedicated to mobile robotic behavior.

The integration of high-level planning and low-level reactive capabilities has been regarded as essential [727, 728]. The Maes action selection dynamics concept combines goal-driven (top-down) and event-driven (bottom-up) behaviors. Behaviors were linked into a semantic network in which goals, perceptions, and behvior units were connected by excitatory and inhibitory links between current and goal situations. Information flow through the network creates a spreading activation across behavioral units through cause–effect links. The highest activated behavior network is selected for execution. This reflects biological cognitive architectures [729]—an evolutionarily old reactive layer with dedicated sensorimotor loops and an evolutionarily newer deliberative layer that provides for planning future actions. Both levels are augmented by learning capabilities. While the reactive layer is based on fast parallel operations for immediate action reflexive behaviors, the deliberative layer is based on slower serial processing for decision making and planning. This is the rationale behind the goal-oriented CogAff architecture which passes reactive processes through a variable threshold attention filter on their way to the deliberative layer at the second level [730]. The attention filter directs attention and processing resources to prioritorized goals. There is a need to represent the world in autonomous navigation in the form of spatial memory for goal-directed behavior [731, 732]. This requires integration of event-driven and goal-directed activities [733, 734]. Payton et al. (1990) [735] stressed the need of integrating high-level planning with lower-level reactive behaviors by suggesting a hierarchical approach as a development from Rosenblatt and Payton (1989). The lowest level was essentially behavior based with the higher level system being a map-based route planner. The internalized plan was represented as a gradient field imposed on a cellular map (similar to a potential field). The gradient field was subjected to the A* search algorithm to calculate the route cost though the gradient field which was used as "advice" rather than as a unified representation. Most autonomous navigation architectures comprise four major components overseen by a system executive that coordinates the major components—a sensing/perception system, a path planner, an execution-monitoring system, and a vehicle control system to plot a rover trajectory through a coarse-resolution map of the terrain [736]. It is possible to add a third higher layer beyond the reactive and goal oriented—meta-management—which is concerned with internal actions for self-knowledge, self-modification, self-monitoring, and self-evaluation in order to improve performance.

Robot control architectures are commonly hierarchically structured into a pyramid of decision levels in order to cope with their complexity [737, 738]. The hierarchy is a method of decomposing complex systems into smaller interrelated subsystems nested into levels until some lowest level is reached. Decomposition into hierarchical modules is essential to analyzing large and complex systems for efficient information flow [739–741]. The OASIS (operations and science instrument system) architecture is a hierarchically layered architecture for remote robot control. NASREM is a commonly cited reference model intelligent robot architecture comprising a set of modules with a number of specific components—sensory perception with attention focusing, deliberative and reactive behavior generation (planning and control), world modeling (environment mapping), and value judgment (cost–benefit analysis) supported by a knowledge base (reasoning) [742]. NASREM is a model-based architecture that is also hierarchical [743–745]. The three-layer hierarchy of execution–coordination–organization reflects increasing levels of cognitive abstraction characterized by decreasing precision and increasing temporal arch. Thus, in any intelligent control architecture, a minimum of three control levels is required—a supervisor/decision level for planning of tasks to achieve goals (e.g., temporal logic planner), a coordination level for coordinating each task of the plan (e.g., partial plan scheduler), and an executive level for controlling motor actions (e.g., behavior control reflexes) [746, 747].

There are thus three levels to the navigation architectures commonly adopted—this is based on the principle of "decreasing precision with increasing intelligence" [748]. The servo-level implements the servo-control law and primitive behavioral routines. The middle executive level coordinates servo-level primitives according to the tasks required with conflict resolution through a procedural reasoning system (e.g., obstacle avoidance). The highest level is the planner which outputs the temporal sequences of tasks required from a description of the state of the world and the desired goals. Each hierarchical level is defined by spatial and temporal decomposition into finer resolution down the hierarchy—hence, the control bandwidth decreases at each higher level. Chatila (1995) suggested a three-layer control hierarchy [749]—a reflexive mode, a 2D model mode (a regional map), and a 3D planner mode (a map model). The lowest functional level comprises modules incorporating basic skills such as image processing, obstacle avoidance, motion servo-control, etc. These modules are essentially numerical algorithms that have a high cycle rate ~ 10–$100\,Hz$ and are associated with elementary motion commands. The executive level controls and coordinates the execution of the functional modules according to the task at hand and interfaces between the decisional and functional levels. This trajectory generation level also implements reactive behaviors. The highest decision level with a much slower cycle rate $>10\,s$ typically implements logical reasoning based on symbol processing in order to generate plans and supervise their execution. The path planner generates task subgoals (generating go-to-goal commands) subject to time and energy constraints based on a global map. The common structure comprises a perception system that monitors the environment, an action system that implements physical actions on the world (which is monitored by the perception system), and a decision/control system that defines the goal-oriented actions to be performed under given environmental conditions (plan). This is the iterative perception–control–action cycle. This requires certain basic functions to be performed—environmental data acquisition, environment modeling (map building), self-localization (with respect to the map), motion generation (to the goal), motion execution and monitoring, and control based on feedback error. The LAAS architecture follows this general scheme but includes a requests control level as its executive level. The requests control level checks requests sent to the functional modules and resource usage, acting as a filter according to the current state of the system.

JPL's CAMPOUT (Control Architecture for Multirobot Planetary Outpost) is a behavior control–based architecture for organizing modular behaviors for reactive behaviors in multiple planetary robots while incorporating higher-level task planning through subgoaling (Figure 8.5) [750]. It is based on a generalized, three-layer hierarchical and distributed behavior control approach for coordinating teams of mobile robots without using a centralized planner [751, 752]. It is a hierarchical hybrid reactive/deliberative architecture with high-level task planning and decomposition within finite resources and constraints with lower-level behaviors for reactive control. It implements no centralized planning or control but is highly distributed. In particular, physical constraints imposed by cooperative tasks such as the transport of large extended objects require tight coordination over rugged terrain. CAMPOUT (Figure 8.5) implements its behaviors as finite state machines (encoded as simple if–then rules) with a hierarchy of

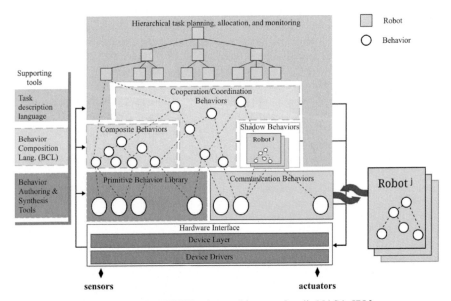

Figure 8.5. CAMPOUT robot architecture [credit NASA JPL].

primitive behaviors with composite behaviors built from more basic ones. Group behaviors are composed by coordinating behaviors across multiple robots. Priority-based arbitration is achieved by assigning one behavior as dominant for a single control cycle to suppress lower-priority behaviors (subsumption) [753]. Alternatively, command fusion combines the outputs from multiple behaviors representing the consensus through maximum weighting of votes or simple Boolean logical operators. In addition, a number of fuzzy metarules for dealing with context are implemented:

```
IF (obstacle is close) THEN avoid collisions
IF NOT (obstacle is close) THEN follow target
```

Behaviors communicate through environment interaction or through direct communication. There are two main single-robot behaviors—Avoid_Obstacle and Goto_Target—which together generate Safe_Navigation composite behavior. There are two main group behaviors—Assume_formation and Approach_target. Assume_formation turns the formation toward their deployment target area while Approach_target uses a visual target to traverse to the target. Active compliance between the robots is through implicit communication through the shared payload (as determined by external and internal disturbances) and explicit communication.

The middle layer of CAMPOUT is BISMARC (Biologically Inspired System for Map-based Autonomous Rover Control) based on a neural behavior control architecture that resembles DAMN [755, 756] (Figure 8.6). BISMARC has two

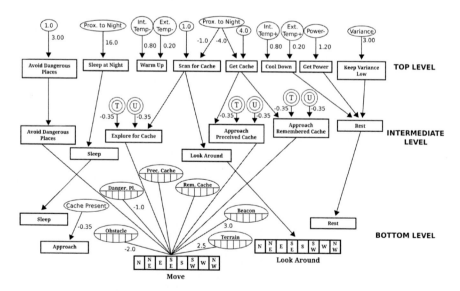

Figure 8.6. BISMARC action selection network [credit NASA JPL].

levels—the first level generates motor action based on stereo-imaging while the second level uses weighted action plans to drive motors. Behaviors are organized with weighted links between them which can be combined through weighted summation. The behaviors are either generic (avoid obstacles) or task specific.

BISMARC uses a map-based long-term memory referenced to external landmarks corresponding to obstacles and goals while short-term memory is represented by the current perceptual occupancy map. The long-term map is updated probabilistically from the short-term map as the rover traverses. BISMARC has been adapted for multirobot coordination [757]. JPL's CASPER (Continuous Activity Scheduling Planning and Replanning) is an extension of ASPEN (Automated Scheduling and Planning Environment) which plans and schedules activities according to resource constraints and current goals [758]. It uses a search algorithm with a temporal logic representation to create plans from a set of current goals, a current state, a current plan, and state projections of the plan into the future. The CASPER basic algorithm is as follows:

```
Initialise P to the null plan
Initialise G to the null goal state
Initialise S to the current state
Given current plan P and current goal G
    (i) update G to reflect new goals and delete goals no longer
        required
   (ii) update S to the new current state
  (iii) compute conflicts on (P, G, S)
   (iv) apply conflict resolution planning to P within resource
        bounds
    (v) release relevant near-term activities in P to Executor
        for execution
   (vi) goto 1
```

Plans are generated continuously through iterative repair which classifies conflicts and resolves them through plan modifications. It also includes a real-time control system to monitor task execution. In a similar vein, the Remote Agent planning system deployed as the Deep Space 1 probe autonomous flight control system has been adapted to planetary rover traverse planning [759]. Its primary roles were for autonomous navigation, power planning, and failure recovery. The Remote Agent architecture interfaces to a real-time control system and itself comprises three major reasoning components—a ground-based temporal logic-based planner/scheduler that generates a schedule from high-level goals, an onboard smart executive that decomposes the schedule into lower-level rover commands subject to mission constraints, and an onboard model-based diagnosis and reconfiguration system that monitors the rover state and generates recovery plans if failure occurs. The problem with planning approaches is that they are not robust to changing or uncertain environments and require extensive processing which, despite being suitable for in-space flight control, is less suited to planetary rovers.

Interaction with the real world involves real time interaction, uncertain knowledge, and environmental complexity. Plans cannot be fixed sequences of abstract actions but must be strategies that define successive actions based on the current situation. In fact, even reactive systems involve internal symbolic representations of objects in the environment and their functional relevance [760]. Internal symbolic systems referring to objects or the state of the environment can be interfaced to sensors and actuators. Often such internal representation is essential in providing context to situated actions. Hybrid techniques of reactive and symbolic planning systems will be necessary.

9

Autonomous navigation—self-localization and mapping (SLAM)

Durrant-Whyte (2001) classified five critical technologies for the realization of autonomous mobile field robots which are highly applicable to planetary rovers [761]: (i) terrain mobility; (ii) self-location; (iii) local navigation; (iv) global navigation; (v) communications. All of these issues are associated in one form or another with autonomous navigation. Navigation is the process of traversing terrain in order to reach a given goal (location goals or scientific goals). Goals are typically locations of scientific interest where the rover will acquire and analyze samples in situ. Goals may be a coordinate location, a visual target, or event based. Autonomous navigation on remote planetary surfaces is one of the most challenging requirements for planetary missions. Autonomous navigation of robotic rovers through hostile planetary terrain requires the ability to sense the environment, plan and traverse a safe course through that environment towards its goal while reacting to unexpected situations robustly. Navigation involves four main processes: (i) perception of the environment; (ii) self-localization with respect to landmarks; (iii) path planning; (iv) path traversal. The simplest form of navigation is dead reckoning in which the displacement of the rover is measured relatively through odometry but odometry is subject to cumulative drift errors. Absolute position measurements are necessary to recalibrate relative sensors from drifting. External sensor references will be required every 15 min or so. Proprioception (odometry) alone is insufficient for accurate navigation so exteroceptive (vision) sensors are required to provide feedback from environmental features [762]. In fact, internal proprioception is inherently noisy while external exteroceptive sensing suffers from perceptual aliasing in that similar sensory data may result from different locations. Proprioceptive data must be calibrated with exteroceptive data while exteroceptive data must be disambiguated by proprioceptive data. The navigation architecture for the Kapvik microrover is shown in Figure 9.1.

© Springer-Verlag Berlin Heidelberg 2016
A. Ellery, *Planetary Rovers*, Springer Praxis Books,
DOI 10.1007/978-3-642-03259-2_9

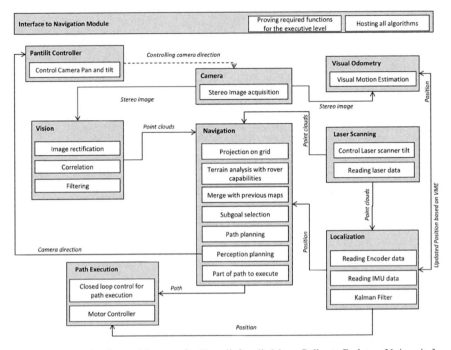

Figure 9.1. Navigation architecture for Kapvik [credit Marc Gallant, Carleton University].

9.1   DIGITAL ELEVATION/TERRAIN MODEL (DEM/DTM)

Unlike behavior-based approaches to robot behavior, a map-based approach requires higher-level cognitive capacities—a map is an internal representation of the real environment. The requirement for planning implies the need for some kind of representational structure which in the case of autonomous navigation is an internal map. Robust robotic navigation requires the use of neighboring and distant reference landmarks including a compass (such as a Sun reference), and the use of canonical rather than arbitrary paths to disambiguate sensor signals. There are two approaches to robot navigation—metric and topological. Topological maps are graph-based representations of the environment in which each cell represents a specific place while connecting edges between the nodes provide routing information. Topological mapping encodes a place relationally. In topological maps, object positions are stored in a local reference frame with relative numerical data with respect to distinct places. A topological map may thus be enhanced by metric information such as distances. The environment is represented graphically with nodes corresponding to distinct places or landmarks. A topological map is a specialized example of a spatial semantic hierarchy and may be regarded as Brooks' behavior-based competence level 3. In metric maps, the positions of objects are stored in a common global reference frame with absolute numerical data. A metric map that requires accurate environmental representations may also be included but generally involves large data storage requirements and slow processing. Biological strategies suggest the use of topological rather than geometric maps (as metric maps are error prone) [763]. Lemon and Nehmzow (1998) also suggest that for navigation, topological maps rather than geometric maps should be adopted [764]. Topological representation comprises a graph with nodes that are defined by connectivity to other nodes. The nodes represent geographical regions while arcs represent the adjacency of these regions to each other (i.e., traversability). It represents non-ranging information that may be important for navigation such as landmarks that impose structure to the environment. Most topological maps are coarse grained wherein each node represents a landmark while arcs represent physical adjacency. This provides a dynamic world model of the environment. Global planning may be implemented topologically with local metric mapbuilding connected by the topological map [765]. Relative distances may be computed from the Mahalanobis distance which incorporates variances and covariances of position uncertainty:

$$d = (\hat{x}_k - x_k)(H_k P_k H_k^T)^{-1}(\hat{x}_k - x_k)^T$$

The Mahalanobis distance works well for cluttered environments. Places are defined through localization by local landmark configurations (distances and directions) and/or spectral signatures [766]. A topological map is a sparser representation and easier to build than a metric map [767]. However, it suffers from difficulties in the recognition of previously visited places. Topological maps may be implemented using simple associative memories. Metric maps have higher resolution than topological maps but suffer from high spatial and temporal complexity. It is possible to integrate both grid-based metric and topological maps to obtain the accuracy of metric maps and the efficiency of topological maps, the topological map being generated from partitioning of the grid-based map into a small number of regions [768]. A geometric map may simply represent only the physical locations of objects rather than their properties as only positions are useful for navigation. However, the process of simplification, selection, and abstraction of certain environmental features reduces the fidelity of the representation of the environment. The combination of directional landmarks and topological navigation offers greater efficiency than metric navigation [769]. Metric maps are computationally expensive to construct while topological maps store only neighborhood relationships between landmarks but not distances. A hybrid approach to SLAM based on combining metric and topological maps provides an efficient method [770]. The environment may be based on a topological map while rover pose submaps within the global topological map are metric. One example is the biologically inspired RatSLAM which uses a global world–referenced topological map in conjunction with a series of local robot-centered obstacle maps [771]. Insects also use two primary strategies for navigation—path integration for long-range navigation and visual odometry for homing in close to the target [772]. We do not cover these biologically inspired approaches to navigation further.

Robust mobile robot navigation requires the use of visual landmarks for self-localization by triangulation [773]. This implies that stereoscopic vision is required to provide depth information (perhaps supported by LIDAR). Planetary terrains generally have sufficient texture to allow the use of pixel-based stereovision alone without the use of LIDAR, however. Stereoscopic vision involves triangulating between two identical cameras with the same focal length, a horizontal baseline, and parallel optic axes to extract depth. It is important that the cameras are calibrated to compensate for alignment errors to ensure the accuracy of depth extraction. Triangulation requires a minimum of three distinguishable and widely distributed landmarks to uniquely identify the robot pose in six dimensions of translation and orientation (i.e., it requires an omnidirectional or steerable camera). By taking a bearing on a landmark, the back bearing gives the self-location. A unique location through line intersection is provided by the three landmarks while a fourth adds redundancy. It is preferable that the spectral signature of a landmark be definitive (from multiple viewpoints) and unique to a particular place. Stereo-matching is enabled by the construction of a disparity image from the image pairs using cross-correlation algorithms to detect matches between the images. The stereovision algorithm computes a correlation match score for all the pixels and selects the best match. When several pixels have similar matching scores, mismatches may be eliminated under the assumption of a locally continuous surface. The rover is thus required to construct a 3D geometric map of the local environment (its terrain) around the rover in order to localize itself relative to visible landmarks. "Landmark" stands for a unique sensory perception, and mechanisms exist to identify such landmarks autonomously and without prior definition [774]. The geometric world model is computed primarily from visual data so landmarks must be readily identifiable by vision sensors. They should have sufficient contrast with the background. Naturally occurring boulders can be used such as local peaks supplemented by using the lander as an artificial landmark. The range of reliable obstacle identification is given by

$$R = \frac{hn}{\phi n_{\min}}$$

where h = minimum obstacle size, n = number of scan lines/frame, and $n_{\min}$ = minimum number of frame lines for reliable image. Accurate range data limit the look-ahead distance to a 10–15 m range which defines the limits of world model fidelity. The image frame rate imposes a maximum speed to the rover

$$v \leq \frac{\Delta s}{2t}$$

where Δs = difference between upper and lower boundary of the FOV and t = time between frame changes. Thus, image generation rates must be consistent with the rover traverse speed and low frame rates are sufficient for navigation. Stereo cameras are assumed to be mounted on a mast-mounted pan–tilt assembly with a minimum height of 1.5 m. The mast should be mounted forward on the rover deck to minimize obscuration of the terrain in front of the rover. Point clouds are the

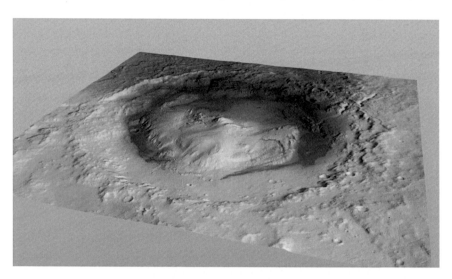

Figure 9.2. DEM construction of Mt. Sharp in Gale Crater [credit NASA JPL].

simplest form of map representation in 3D global coordinates. They are readily converted into cellular grids. After cross-correlation, data are constructed as a rectangular grid representing a set of 3D points corresponding to each image pixel with an altitude. Thus, stereo-imaging is used to build a local topographic map of the robot's environment. Any planetary rover will be required to self-localize and navigate in an a priori unknown environment based on a map of its environment [775]. The map is generally a rover-centered cellular map derived from stereo-scopic images (e.g., the Mars Pathfinder landing sites) [776] (Figure 9.2). The digital terrain model (DTM) comprises a digital elevation map (DEM), triangular mesh model, texture map, and thematic map.

A DEM is a discretized rectangular occupancy grid of (x, y) cells of the form $z = f(x, y)$ defining the geometry of the terrain [777]. The world is thus represented as a regular Cartesian grid of rectangular horizontal–vertical cells populated with data derived from its vision (and other) sensors. Rough terrain imposes constraints on traverse such as stability, obstacle negotiation height, chassis articulation, etc. Each cell thus represents a measure of navigability which may be classified as traversable (1), non-traversable (0) or unknown (*) depending on the height of obstacles within the cell (defining terrain difficulty scores). Thus a traversability map can be derived from the DEM by merely adopting such cell labeling [778]. The resolution of each cell depends on the distance from the rover. In fact, resolution may be represented in an iterative, hierarchical approach (e.g., 1 m resolution for 64×64 m map, 0.25 m resolution for 16×16 m map, and 0.125 m resolution for 18×18 m map. Each cell includes the average height and slope of the terrain which will constrain the pose of the rover (limited by its kinematics). A DEM is constructed from depth computed stereoscopically at each pixel using maximum cross-correlation matching with an $n \times n$ kernel window. It is estimated that construction of a DEM of the terrain generated on board the rover will reduce the data transmitted to Earth by 20–100 times [779]. A DEM is triangulated between pixels to create a geometric mesh model. A meshed 3D surface is formed using functions of the form:

$$z(x, y) = \sin(y + a) + b \sin(x) + c \cos(d\sqrt{y^2 + x^2})$$
$$+ e \cos(y) + f \sin(f\sqrt{y^2 + x^2}) + g \cos(y)$$

Texture-based terrain meshes are then added—this is characterized as the variance in elevation z of all points in the grid of radius r. Regions with similar slope are grown and merged. Terrain slope is defined as:

$$k = \frac{z_2 - z_1}{\sqrt{(x_2 - x_1)^2 + (y_2 - y_1)^2}}$$

A thematic map encodes the regularity of the terrain according to soil hue and color consistency. Each cell in the occupancy grid is labeled as traversable, non-traversable, or uncertain depending on its occupation state (i.e., obstacles). Finally, false color is added to delineate forbidden regions such as obstacles (red), loose soil regions (orange), and safe regions (green). Each cell includes an uncertainty measure of the data which may be classified using a Bayes classification scheme into different terrain types of flat, rough, obstacle, and unknown based on different images according to measured attributes such as texture [780, 781]. Tarokh et al. (1999) proposed a cell impedance measure of traversability of the form [782]:

$$\rho = \tfrac{1}{2}(\delta + \kappa)$$

where δ = normalized surface roughness due to obstacle (height $\times$ areal footprint) and $\kappa = 1/r$ = normalized surface vertical curvature (jaggedness). A fuzzy set representation of uncertainty may be adopted to represent the possibility of occupancy [783]. For instance, $\pi = 0$ for certain emptiness, $\pi = 1$ for certain occupation, $\pi = 0.5$ for total uncertainty. Seraji (1999) suggested a fuzzy measure of traversability based on terrain slope and roughness but it appears to have no obvious advantages over other such measures [784]. Classification of the traversability of each cell is made more difficult as soil parameters are difficult to measure online. Gat et al. (1994) proposed generating a continuous gradient measure that propagates information through spreading activation from the goal [785, 786]—this bears some similarities to the potential field' approach to navigation. Typically, the actual final position will differ from the assigned goal limited by the rover's self-localization capability (determined by its sensors).

Local maps may be merged incrementally with a previously constructed global terrain map that is built incrementally image by image over time to ensure robust self-localization with long-range navigation. Such map matching proceeds by attempting to establish a statistical correspondence between the local image and the global model through a search algorithm but emphasizing recent local data so that errors on localization do not accumulate. Merging may be achieved using Bayes analysis: the probability that a cell is occupied is conditional on sensor measurements. Dempster–Shafer probabilistic fusion which differentiates between imprecision and confidence produces superior results at the expense of greater computational complexity [787]. In this case, each cell is characterized by three values—elevation z, precision p, and confidence c on the elevation value. For each of n cells of the map:

$$z_{i+1} = \frac{c_i z_i + c_0 z_0}{c_i + c_0}; \quad p_{i+1} = \frac{c_i p_i + c_0 p_0}{c_i + c_0}; \quad c_{i+1} = c_i + c_0$$

where $c_0 = 1/n = $ current confidence measure and $p_0 = $ current precision measure. Alternatively, updated local maps may be compared with a previous global map of the environment based on sensor values through the principle of maximum likelihood estimation [788, 789]. Local map L is comprised of n features $\{l_1, \ldots, l_n\}$ while the global map is comprised of m features $\{g_1, \ldots, g_m\}$. A distance (which may be a Euclidean, Hausdorff, or Mahalanobis metric) between feature l_i in the local map and feature g_i in the global map is defined as

$$d_{ij}^X = \text{dist}(X(l_i), g_j)$$

The distance from the feature in the local map to the closest feature in the global map is given by

$$D_i^X = 1 \leq \min_j \leq m | d_{ij}^X$$

The likelihood function for robot position X is given by:

$$L(X) = p(X) \sum_{i=1}^{n} p(D_i^X)$$

where $p(D_i^X) = k_1 + (k_2/\sigma\sqrt{2\pi})e^{-(D_i^X)^2/2\sigma^2} = $ probability density function of each feature and $p(X) = $ a priori probability of position X. $L(X)$ is selected to yield the maximum likelihood of position X. This technique was robust to outliers, noise, and occlusions. The chief difficulty with a global map is its high memory storage capacity requirements which may render them infeasible.

9.2 MER ROVER NAVIGATION

During the early phases of the MER missions, the control strategies adopted were highly conservative. However, during the extended phases of their missions, greater levels of autonomy were implemented in part due to the more benign environments compared with Viking and Pathfinder (Figure 9.3).

Stereovision proceeded by first reducing the $1{,}024 \times 1{,}024$ pixels of the CCD raw image down to 256×256 pixels by averaging thereby reducing the computational requirement by eight times. A difference of Gaussian filter was

Figure 9.3. Three Martian terrains: (top left) rocky and hostile, (top right) duney and hostile, and (bottom) relatively flat and benign [credit NASA JPL].

applied to each averaged pixel to reduce noise through high-pass filtering. The filtered images were fed to a cross-correlation algorithm which used a 7×7 search window to find corresponding pixels in each image through disparity estimates. The cross-correlation algorithm computed a similarity score to find the maximum matching pixels between the two images. Cross-correlation could be run in the reverse direction: right window on left image and left window on right image as a check. The disparity map took around 30 s to compute on the MER flight processor. From this, a camera model was applied to extract the range. To minimize computation, a sliding sum was used. A peak filter was applied which required a similarity score to be higher than for surrounding pixels. The final check was to eliminate outliers using a threshold (blob) filter. A 3D geometric model of the environment was constructed as a pie-shaped wedge from the rover HazCams with a width of $115°$ and range of 0.3–3 m. The MER Cartesian grid represented an area 10×10 m subdivided into cells. Each cell, typically the size of a rover wheel (20×20 cm), represented whether a rover-sized object centered on the cell would encounter obstacles in adjacent cells. Each cell was initially populated with a median value of 0.5. Step and roughness hazards were computed from the elevation difference between adjacent pairs of cells—if the step or roughness hazard was below $1/3h$ (h = rover clearance height), the cell was classed as traversable. Pitch hazard was computed from the slope between adjacent cells— for traversability, this should not exceed the maximum pitch angle. Border hazard was defined by an untraversable or unknown cell classification in an adjacent cell. All these hazards were weighted according to uncertainty and summed to define a final value. From the resultant tesselation grid model where each cell represented an (x, y, z) coordinate, navigable (x, y, z) waypoints were selected. The waypoints were continuously connected with high-traversability cells. Sensor readings were placed in the grid using probability profiles which attached a certainty factor to the existence of objects in each grid cells. This map-building process took 10.0 s on a R3000 12 MHz processor.

A set of potential trajectory arcs were generated and evaluated on their successful goal completion. MER navigation was based on extracting geometric properties of the local 3D environment through grid-based estimation of surface traversability applied to local terrain (GESTALT) [790]. GESTALT uses a configuration space representation of the environment comprising arcs to assess the risks of different trajectories. The circular arcs were based on a global probabilistic model (Figure 9.4). Every iteration updated the terrain model with an interest

Figure 9.4. Set of circular arcs generated using GESTALT [credit NASA JPL].

value (inverse cost to reach goal) and risk (traversability measure) of a family of circle arcs. The circle arcs that maximized the interest/risk ratio were selected.

GESTALT determined the direction of the rover by generating trajectory arcs until it reached its goal. The trajectory comprised a set of line segments that joined points where course changes occurred. A desired trajectory was selected and motion commands were issued to drive and steer the rover through a specified 0.3 m trajectory segment—the MERs stopped every 0.3 m to reacquire new images. The trajectory was decomposed into elementary wheel torques and speeds for execution. The traverse segment was executed as measured by the wheel odometers. Between image acquisitions while static, the MERs did not utilize their imagers while on the move relying rather on tilt sensors, motor current monitoring, and wheel and lever odometers. The odometry estimate of the travel distance was fused with the yaw rate gyro measurement of heading. The chief limitation of the MER approach was its lack of long-range navigation sometimes requiring it to backtrack. This was due to excessive reliance on HazCams.

A variation on the circular arc approach is the nested arc approach for rough terrain [791]. A nesting depth of 20×2 m long arcs for the first depth level and 10×2 m long arcs for the second level gave good results. For each arc, a set of discrete regularly spaced positions are evaluated according to the chassis attitude and mechanism internal angles generated by geometric placement of the rover on the terrain. This allows risk to be quantified as a quadratic function of these angles so that large angles are increasingly penalized. The maximum interest/risk arc is selected then using the A* algorithm. The MER navigation process required three or more days and a large number of ground operators to accomplish a single task. A premium is therefore placed on developing greater autonomy. There is a need for higher performance rovers with more complex capabilities to operate autonomously.

9.3   CNES/LAAS AUTONOMOUS NAVIGATION SYSTEM

While the MER navigation system is primarily reactive, the CNES/LAAS approach is premised on the assessment that deliberation is required for highly autonomous missions. The CNES/LAAS autonomous navigation system follows a traditional approach to path planning and navigation based on stereovision from stereoscopic imaging cameras (and optionally a laser rangefinder) to build a 3D geometric DEM [792, 793]. Like the MER algorithms, the CNES autonomous navigation algorithms are run while the rover is stationary to acquire and process its stereo images. Like the MER algorithms, it is assumed that stereo cameras have a minimal resolution of 256×256 pixels each encoded with 8 bits (256 color levels). Correlation-based pixel set matching has been used [794–796]. This generates a disparity map from which triangulation data create a range map that may be converted into global coordinates. Tracking between sequences of stereo frames is achieved by selecting those pixels to track with high accuracy determined by an error model based on correlation strength with each of its neighbors [797].

Typically, image maps are limited to <10–20 m look-ahead distance (depending on the camera height and obstacle distribution), significantly greater than the 0.3 m adopted for the MERs, but difficult rock conditions can reduce this to around 2 m. The DEM has a resolution of 5×5 cm characterized by an elevation parameter—thus is much finer resolution than for wheel size adopted by the MER algorithm but 3D resolution decreases with distance from the rover.

The traversability of each cell is determined by the maximum step height of the rover and the maximum rover transverse/longitudinal tilt angle by evaluating the configuration of the rover axles. Each cell is analyzed by evaluating the configuration of virtual rover axles and the internal chassis configuration at each cell. The maximum tilt angle is limited by the maximum slope of traverse (given by soil traction rather than static stability) (Figure 9.5). The middle of a rover axle is placed on each cell of the terrain grid to determine if it exceeds maximum step height and maximum slope. If any of the axle orientations yield an impossible configuration beyond the maximum tilt angle, that cell is classified as untraversable. These factors determine the difficulty score used to label the cell. Safety margins are applied but their size must be carefully regulated in order not to eliminate potential paths. Obstacle sizes are expanded to account for localization and trajectory errors which grow between image acquisitions. Account must also

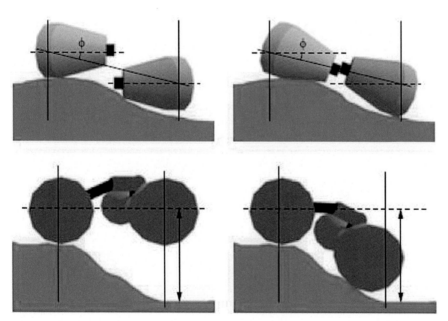

Figure 9.5. Traversability for the Marsokhod rover using the CNES/LAAS algorithm [credit CNES-ESA].

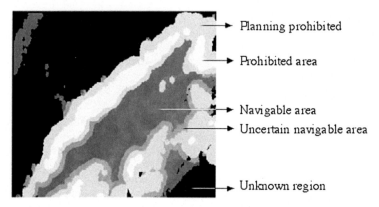

Figure 9.6. CNES autonomous navigation map [credit CNES-ESA].

be taken to label farside blind regions behind an obstacle which is dependent on camera height.

The goal is defined on the navigation map close to the border and subgoals are selected within known regions of the map (Figure 9.6). The CNES algorithms use a two-tiered graph search algorithm in which an A algorithm is used to find a coarse-grained path from the current position which is subdivided into subgoals (defining waypoints) to which the A* algorithm is used to find the optimal path. Each subgoal is generated near the border of known and unknown regions in a navigable region accessible by the current position. The A* path-planning algorithm computes the optimal path to the subgoal.

At each subgoal, the rover directs its stereo cameras to maximize its knowledge of the terrain towards the goal but away from non-traversable regions. As the attitude of the rover affects the line of sight, at each stop it is measured. This allows definition of the heading to the next stopping point to reacquire new images and maps towards the goal target. This generates a trajectory of straight/ curved line segments and heading changes to be executed defined by a sequence of waypoints. The CNES autonomous navigation system generates a continuous safe navigation path. The CNES approach is similar in spirit to that of the MERs but gyroscopes are not used to estimate rover heading. The CNES/LAAS approach to rover navigation recommends a global multilayered composite environmental representation of terrain to serve different navigation functions of reflexive and deliberative behavior [798, 799].

The CNES autonomous rover navigation system has certain characteristics that require consideration [Bertrand, R., 2004, private communication]:

(i) the calibration procedure takes no account of distortion models but a table-based approach has been tested though not implemented (this method interpolates tables stored on board using a bilinear interpolation algorithm);

(ii) the stereo camera pair must be identical and have the same focal length;
(iii) the stereo camera baseline is horizontal to the ground plane to within 0.5 pixels and the optical axes of both cameras must be parallel (no toe-in) compared with the MER pancam toe-in of 1°;
(iv) stereovision is based on two hierarchical levels of processing including subpixel interpolation to increase speed and reduce mismatching;
(v) a perception-planning algorithm determines the suitable direction of the line-of-sight direction relative to the ground (from the configuration and attitude/position of the rover, camera pan and tilt angles are computed);
(vi) the disparity-filtering algorithm is based on image segmentation;
(vii) a DEM algorithm retains maximum and minimum values rather than averaged (maximum expected errors are incorporated into obstacle size as additional radius);
(viii) there now exists a much faster algorithm, the D* algorithm (dynamic A*), than the A* algorithm which will substantially reduce the time for this computation.

The CNES algorithms have undergone trials in the JPL Mars Yard on the JPL FIDO rover platform which was equipped with a PC104-based Pentium CPU running at 133 MHz on the VxWorks 5.3 real time operating system [800, 801]. Although mechanically capable of 5 cm/s traverse, the limited computational resources on board FIDO limited it to 1 cm/s. This may be compared with the MER 20 MHz RAD6000 RISC processor with 8 kB cache and 128 kB RAM on the VxWorks operating system. The 512×486 pixel FIDO NavCam stereo pair generated a DEM grid size of 251×251 cells, each 50 mm in size, within 1.53 s.

The CNES AutoNav algorithms required a total memory size of 3 MB of RAM but 5 MB allows margin. A minimum of 100 kB non-volatile memory is required to store the executable code. Each raw image stere opair requires 200 kB storage suggesting that storage requirements will amount to 100 MB/km assuming a stereo pair every 2 m. Were just the DEM stored and images discarded, this may be reduced to 5 MB. Image sizes dominated the computer processing time according to the cube of their size—a 256×256 image takes 64 times less than a $1,024 \times 1,024$ image. The CNES algorithms involving stereovision and map building comprised 5% of the traverse time on FIDO (i.e., 2.2 s to process 256×243 images). The current European baseline space-rated computer processor is the 100 MHz LEON processor—this takes 3.3 s to run the CNES stereovision and DTM-building algorithm assuming a 256×256 image stereo pair excluding framegrabbing from the camera to the computer. The time to compute the D* algorithm to generate a trajectory is 2.0 s on the LEON processor. Additional functions including visual odometry add up to 6.5 s with a worst-case computation time of 8.25 s on the LEON processor. Nominally, additional locomotion traverse time adds up to 13.0 s for 2 m traverse segments with visual odometry performed every 0.5 m. Safeguarded mode during close approaches requires navigation to be performed every 0.5 m (6.5 s processing time) followed by traverse of 8.5 s with visual odometry performed every 0.5 m. An alternative may be to utilize black-

and-white images (3-bit encoding) for one camera while retaining color for the other (8-bit encoding); however, this may introduce matching difficulties. Recent interest in COTS hardware suggests the use of PC-104 stack architectures which may be implemented through a Linux-based embedded Pentium PC processor board interfaced to a flash memory card. There is also promise of massively increased performance offered by the use of FPGA (field programmable gate array) in-space systems.

The CNES software is somewhat out of date as there have been substantial developments in autonomous navigation techniques since it was developed:

(i) Taking account of cell resolution reduction with distance allows labeling using a Bayesian classifier on the basis of texture.
(ii) Integration of wheel odometry (estimate of distance traveled), gyroscopic data (estimate of R, P, Y angles), and Sun vector (estimate of rover heading) to provide an update of the 6 DOF of the rover state. This may be fused with stereo-imaging to generate an estimate of the rover location through Kalman filtering.
(iii) The inclusion of soil parameters to compensate for slip on rough terrain has been attempted by modeled soil–wheel terrain interaction through a Coulomb relation based on friction only (and soil contact viscoelastic properties modeled as a compressive-only spring damper) rather than a Mohr–Coulomb relation based on cohesion and friction [802]. Furthermore, periodic imaging of rover tracks to measure sinkage will provide data on soil properties to support rover traction control. Local geology of importance to rover traverse includes identification of rocks, ridges, troughs, etc.

The CNES system is one of the major contenders for implementation on the European ExoMars rover.

A key component in visual SLAM is bundle adjustment, which represents a considerable computational overhead by virtue of matrix multiplications and Jacobian computations. Bundle adjustment searches for the 3D positions and camera-pointing directions to minimize projection error with maximum likelihood. This is formulated as a nonlinear least squares problem with the error defined as the square of the difference between n observed 3D landmark locations and the projection of these positions onto the image plane. This has a computational complexity of $O(n^3)$ where $n =$ number of landmarks. The Levenberg–Marquardt or Gauss–Newton algorithm is usually adopted for this bundle adjustment nonlinear least squares problem through a series of linear approximations. The former is favored over the latter if the initial estimate is far from the real solution. There are several approximations to reduce computational load such as restricting the procedure to a subset of the global map. The objective function to be minimized finds camera poses and visual feature locations x that maximize the likelihood of measurements z:

$$F(z|x) = \frac{1}{2}\sum_{i=1}^{n}(z_i - h_i(x))^T R_i^{-1}(z_i - h_i(x))$$

where h = observation model and R_i = observtion covariance matrix. A first-order Taylor series expansion defines the Jacobian:

$$J(x)^T R_i^{-1} J(x) = -J(x)^T R_i^{-1}(z_i - h_i(x))$$

where $J = \partial h/\partial x$ = Jacobian and $H = J^T RJ$ = Hessian matrix. Bundle adjustment is considered a computationally demanding process so it is not suitable for planetary rover deployment unless restricted to local maps.

9.4 KALMAN FILTER–BASED SLAM

According to Brooks [803]: "The key problem is in trying to build a model of the world ... The problem is in correlating the current readings from the sensors with the existing (partial) world model. The original readings used to build the existing model were noisy and introduced uncertainties in the representation of the world. The new readings also include noise. Furthermore, if the robot has moved between sensor readings then there is uncertainty in how the coordinate systems of two (or more) sets of sensor readings are related ... If more than one type of sensor is used, there is also the problem of fusing the different classes of data into a single representation." These are the problems addressed in simultaneous self-localization and map building (SLAM) (Figure 9.7).

SLAM may be implemented using a Bayes filter with a motion estimation step (prediction) and a measurement update step (correction). The belief represents the probability of being at a current position while the action update determines a new belief after every movement. The sensor update then computes the new belief after measurement. The key to dealing with the uncertainty for planning purposes is the Bayesian framework such as Kalman filtering [804]. The Kalman filter is a generalized iterative version of the Wiener filter (and more generally Bayesian estimation procedure). Any robot navigation system that accounts for uncertainty in its localization will perform better than one that does not. SLAM attempts to provide estimates of the relative location of the rover between landmarks. The execution of a navigation trajectory must be correlated with position estimates from sensors to ensure tracking of that trajectory. A planetary rover navigation system may be based on state estimation through wheel odometry, inertial measurement sensors, and Sun sensor resembling the approach used for attitude control systems of spacecraft [805]. Control of spatial orientation during locomotion and navigation is essential and relies on the integration of multisensory (specifically visual, proprioceptive, and vestibular senses) data in egocentric body coordinates and the monitoring of deviations from planned trajectories from internal models in world coordinates [806]. The approximate positions and orientations of the rover are known from odometric data. This constrains the map-matching search by providing an estimate of position and an estimated expected image. This estimate can be matched against the actual visual image from the vision sensors. The process of construction and accuracy of representation comprises the core issue of the SLAM problem. SLAM is based on using conditional probability dis-

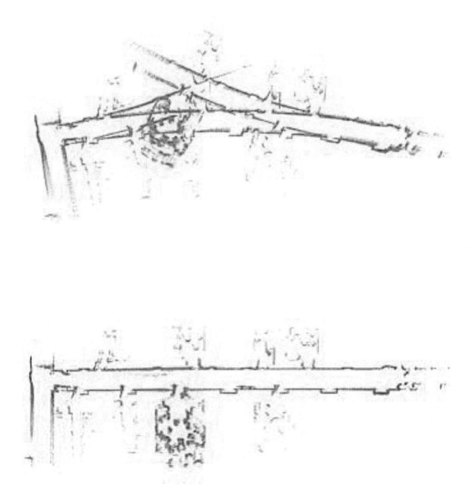

Figure 9.7. Effect of localization error with and without SLAM [reproduced with permission from Montemerlo, M., and Thrun, S. (2007) *FastSLAM: A Scalable Method for the Simultaneous Localization and Mapping Problem in Robotics*, Springer-Verlag, New York].

tributions to blend information from world models (exploitation) and sensor measurements (exploration). The world model comprises a conditional probability distribution so SLAM involves the propagation of a probability density function over time which represents estimates of the rover's position in its environment. Belief concerning the current state x_k of the robot may be defined probabilistically (see Appendix):

$$\text{Bel}(x_k) = p(x_k|u_1, z_1, \ldots, u_k, z_k)$$

$$= \eta p(z_k|x_k, u_1, z_1, \ldots, u_k)p(x_k|u_1, z_1, \ldots, u_k) - \text{Bayes theorem}$$

$$= \eta p(z_k|x_k)p(x_k|u_1, z_1, \ldots, u_k) - \text{Markov assumption}$$

$$= \eta p(z_k|x_k) \int p(x_k|u_1, z_1, \ldots, u_k, x_{k-1})p(x_{k-1}|u_1, z_1, \ldots, u_k) \, dx_{k-1}$$

$$= \eta p(z_k|x_k) \int p(x_k|u_k, x_{k-1})p(x_{k-1}|u_1, z_1, \ldots, z_{k-1}) \, dx_{k-1} - \text{Markov assumption}$$

$$= \eta p(z_k|x_k) \int p(x_k|u_k, x_{k-1}) \, \text{Bel}(x_{k-1}) \, dx_{k-1}$$

Markov localization is defined by $\text{Bel}(x_{k-1}), u_k, z_k$ with initial $\eta = 0$ which must be updated to generate a new belief. The predicted belief is the product of the old belief and the sensor model:

$$\overline{\text{Bel}}(x_k) \leftarrow \int p(z_k|x_k) \, \text{Bel}(x_{k-1})$$

This predicted belief is then convolved with the motion model:

$$\text{Bel}(x_k) \leftarrow \int p(x_k|u_k, x_{k-1}) \, \overline{\text{Bel}}(x_k) \, dx$$

Mapping becomes a maximum likelihood estimation problem. The Kalman filter may be represented as a dynamic Bayesian network where the hidden variables represent the states [807]. The Kalman filter computes the probability of the hidden state given all past observations, $p(X_i|y_i, \ldots, y_0)$. The Kalman filter allows tracking objects with simple states, but representing more complex internal states requires a network of states. A Bayesian network can accomplish this.

SLAM is typically presented as a Kalman filter [808] or other recursive estimation problem which captures both geometric modeling and measurement uncertainty. The most commonly used method for data fusion [809, 810] in SLAM is the extended Kalman filter (EKF) which gives an optimal estimate to a noisy linearized system in a least squares error sense [811]. Indeed, the Kalman filter extends the least squares estimate—the most probable value of x that minimizes the sum of squares of residuals—to time-variable quantities [812]. The Kalman filter integrates information on the system dynamics and its noise charac-teristics, and the sensory system and its noise characteristics. All previous measurements are accommodated but their contribution decreases exponentially with time. It is an optimal integrator that minimizes the mean square error of the estimate with a history of sensory feedback. It thereby maintains an optimal estimate of the state of the system given a system model and measurements. The recursive nature of the Kalman filter lends itself to computational traction with minimal training. There exists uncertainty in the system state defined as a probability distribution function over its state. The Gaussian distribution function is fully characterized by its mean and covariance, respectively:

$$\hat{x} = \langle x \rangle \quad \text{and} \quad P(x) = \langle \bar{x} \cdot \bar{x}^T \rangle$$

where $\bar{x} = x - \hat{x}$. It is a stochastic technique in that the mathematical models and sensors are corrupted by Gaussian noise. The Kalman filter uses the statistical properties of sensory measurement assuming Gaussian noise to provide optimal estimations [813]. It combines all available information about measurement data, prior knowledge of dynamics, and statistical analysis of uncertainty to estimate the posterior of the robot state and map of its environment. It assumes a known state space model of estimated variables and an assumption of Gaussian process and measurement (white) noise (with zero mean and constant variance). It uses system measurements, system dynamics, and noise models as inputs, and outputs filtered estimates of the system state and innovation (difference between predicted and observed measurements). It is an optimal recursive estimate of the robot's state based on a system model into which all the measurement data are fused. The Kalman filter is thus an observer-based approach that estimates the system state from the observed inputs and outputs of the system. It uses a weight determined by the covariance matrix which calculates error in the state estimate. The Kalman filter projects the current state and error covariance forward in time to estimate the a priori state (predictor) and new measurements are incorporated into the a priori estimate to improve the a posteriori estimate (corrector) [814]. Previous a posteriori estimates are used to project new a priori estimates. The Kalman filter performs the three stages—prediction, observation, and update—recursively. "Recursively" implies that it uses only the latest measurements without the need to store previous measurements or estimates—it cycles through prediction and correction. This recursive property has allowed the Kalman filter to supplant the Wiener–Kolmogorov filter which operates on all its data directly. The Kalman filter extends the least-squares estimate to time-variable quantities and is a generalization of the Wiener filter and the ARMA filter. If the noise is Gaussian, the Kalman filter produces an optimal minimum variance Bayesian state estimate. It computes the mean of the posterior conditional probability density function given the prior statistics of the state and the statistics of measurement.

The Kalman filter is used to estimate state variables recursively in two steps. The first step is a prediction of the conditioned state vector $\hat{x}(k + 1|k)$ given the previous state estimate $x(k)$ based on a dynamic model. This is the a priori estimate which is used to generate a prediction of the measurement. The second step is an updated measurement $\hat{z}(k + 1|k)$ to correct the a priori estimate into the a posteriori estimate based on past observations. The difference between the estimated and measured values defines the residual error $\bar{x}_k = \hat{x}_k - x_k$ which must be reduced to zero. Robot pose is defined by a state vector, and a covariance matrix provides the means for maintaining a pose estimate through extended Kalman filter updates. As the robot moves, its state is predicted using dead reckoning (odometry and inertial measurements) and the covariance model. State

estimation of rover position and attitude is based on fusing multiple sensory data into a global map [815]. The best estimate of robot position is computed by fusing the position prediction with all the sensory measurements. Map building must be periodically supported by external calibration from external sensor measurements. The chief problem with 3D terrain mapping is to minimize errors which tend to accumulate. Uncertainty is introduced by the sensing process which is noisy and thereby contributes to a lack of reliability in the estimates. Rover position is represented as a probability distribution over a finite set of discrete positions. Confidence estimates define the probability that the cell is occupied by an obstacle, initially an a priori probability which is subsequently updated through a sensor model. Alternatively, fuzzy logic may be used to define occupancy values [816]. Sensory data are integrated with the map to update it using Bayesian or Kalman filtering methods [817]. This map cannot be too large otherwise it will require extensive computational storage requirements. Kalman filter-based SLAM algorithms have a computational complexity $\sim O((2N)^2)$ where $N =$ number of landmarks due to matrix Riccati recursion in computing P. A wavelet-based compression algorithm can reduce storage requirements by 90–95%. Frame rates are typically low $<1\,\text{Hz}$ (normal video rates are at 25–30 Hz). Vehicle roll and pitch may be estimated by least squares fitting of stereo range data points to a plane and surface roughness may be estimated as the chi-squared residual of the fit to generate a traversability value for each cell as the minimum of the three values [818]. A 2D discrete wavelet transform was used to represent local terrain frequency by Pai and Reissell (1998) [819] to create multi resolution smoothed representations of rough terrain. As the robot traverses the environment, environmental landmarks are extracted from vision and/or laser scan sensors and corresponded to the map. Subsequent observations of these landmarks are fused with measurements from odometric and inertial sensors to update the robot's localization estimates. The Kalman filter may also be used to integrate images from two different frames in stereovision and/or image tracking [820]. Tracking techniques such as the Kalman filter requires frequent updates to track a small number of features between images. Dissanayake et al. (2001) showed that this recursive property of Kalman filtering provides convergence of errors to zero over time with successive observations of landmarks. This occurs as long as cross-
correlations of errors in landmark estimates are maintained within the map covariance matrix. One of the problems with the Kalman filter for target tracking is its sensitivity to the model. The use of multiple models suggests the use of Markov model filters (particle filters). The Kalman filter is more efficient than Markov localization but the latter is more robust to noise and error.

The rover begins its traverse within an unknown environment and incrementally builds a map of this environment using its sensors while simultaneously using the map to self-localize. The dynamic model uses internal measurements to generate its predictive pose of the robot. Error models of inertial navigation, three-axis gyroscope, and tri-axial accelerometer rate measurements are incorporated in the extended Kalman filter for estimating the position and orientation of the vehicle [821]. The rover state vector is defined by its planar position (x, y) and heading (yaw) angle ϕ relative to a global reference frame (though a 5 DOF state (x, y, R, P, Y) is more general) (i.e., $x(k) = [x(k), y(k), \theta(k)]$) [822–824]. The robot state vector defines the position and orientation of the robot; control inputs are steering angle and translational velocity. The measurement model relates sensory measurements to the vehicle state through the innovation covariance. The approach presented here assumes that rover state x is estimated from wheel odometry (θ) and then updated by attitude measurements (φ). Attitude is defined by roll (R), pitch (P), and yaw (ϕ) angles, the latter defining the steering heading. Application to a robotic vehicle uses kinematic equations to predict the vehicle state from velocity v and steering ϕ inputs assuming a Cartesian reference frame centered midway between the rear wheels:

$$\dot{x} = v \cos \phi, \quad \dot{y} = v \sin \phi \quad \text{and} \quad \dot{\theta} = \frac{v \tan \phi}{l}$$

where $l =$ wheelbase. Rover state is modeled as generalized robot position:

$$x(k) = \begin{pmatrix} x(k) \\ y(k) \\ \phi(k) \end{pmatrix}$$

The Kalman filter provides a prediction estimate of rover state with minimum covariance of the form assuming a nonlinear system. It maintains an estimate of the state x of a system and a covariance estimate P of its uncertainty. Robot position is determined by integrating odometric information $u(k) = [d(k), \varphi(k)]$ with other sensory data (x, y, φ). Systematic errors may be incorporated in the state model (e.g., slip errors):

$$x(k) = (x(k), y(k), \phi(k), s_{\text{left}}(k), s_{\text{right}}(k))^T$$

Control inputs are given by vehicle velocity and steering commands:

$$u(k) = \begin{pmatrix} v(k) \\ w(k) \end{pmatrix}$$

The Kalman filter is used to recursively estimate the state of the rover system which evolves as a process model that is observed by the measurement model. Kalman filters are highly dependent on the model used. The state vector of the system includes the position and orientation of the rover $x_v(k)$ and position of landmarks $p(k)$ for up to n landmarks. Suitable landmarks include points of maximum curvature that is invariant to viewpoint, resolution, occlusion, ambiguity, and noise [825]. The raw image is convolved with a Gaussian filter to smooth it and identify points of maximum curvature as landmarks where curvature is given by:

$$\kappa = \frac{1}{r} = \kappa(x, y) = \frac{\partial^2 y / \partial x^2}{[1 + (\partial y / \partial x)^2]^{3/2}} = \frac{\dot{x}\ddot{y} - \dot{y}\ddot{x}}{[\dot{x}^2 + \dot{y}^2]^{3/2}} = \dot{x}\ddot{y} - \dot{y}\ddot{x}$$

where $r =$ radius of curvature. This is convolved with a Gaussian kernel to yield

new curvature estimates. Curvature is defined as dominant if it is greater than
the two nearest points either side of it. As smoothing scaling increases, only the
largest dominant extrema survive. If the landmarks are stationary, they are
assumed noiseless in the map-building process:

$$p_i(k+1) = p_i(k) = p_i = \text{constant for } 0 \le i \le n \text{ landmarks}$$

Hence,

$$x(k) = [x_{\text{rob}}(k), p_1, \ldots, p_n]^T$$

In discrete form, the system model describes changes in robot state $x(k)$ due to
control inputs $u(k)$ as:

$$x(k+1) = F[x(k), u(k) + n(k)]$$

where $n(k) = $ process noise,

$$F = \begin{pmatrix} x(k) + d(k) \cos \phi(k) \cos(\theta(k) + \phi(k)) \\ y(k) + d(k) \cos \phi(k) \sin(\theta(k) + \phi(k)) \\ \theta(k) + d(k) \sin \phi(k)/l \end{pmatrix}$$

$\varphi = $ steering angle, $\theta = $ angle between robot heading and x coordinate, and
$l = $ distance between forward and rear axles. For a linear system, the system
dynamics is modeled as a first-order differential state space equation:

$$x(k+1) = A(k)x(k) + B(k)u(k) + v(k)$$

where $A(k) = $ state transition matrix, $B(k) = $ control matrix, $u(k) = $ control input
vector, and $v(k) = $ process noise error vector with covariance matrix Q. In an
extended Kalman filter, the dynamics are linearized about a nominal trajectory
and a Taylor series expansion approximation about the mean is used:

$$\hat{x}(k+1) = f(\hat{x}(k), v(k)) \equiv f(\hat{x}) + \frac{\partial f(x)}{\partial x}\hat{x} + \cdots \approx F(k)\hat{x}(k) + w(k)$$

$$\begin{pmatrix} x(k+1) \\ y(k+1) \\ \theta(k+1) \end{pmatrix} = \begin{pmatrix} x(k) + d(k) \cos(\theta(k) + \phi(k)) \\ y(k) + d(k) \sin(\theta(k) + \phi(k)) \\ \theta(k) + d(k) \tan \phi(k)/l \end{pmatrix} u(k) + w(k)$$

where $d(k) = v(k)\Delta t = $ distance. Error covariance P represents the uncertainty of
the state estimate:

$$P(k+1) = A(k)P(k)A^T(k) + Q(k)$$

where $Q(k) = \langle v(k)v(k) \rangle = $ the process noise covariance matrix. The covariance
matrix is defined as:

$$P = \begin{pmatrix} \sigma_x^2 & \sigma_{xy} & \sigma_{x\theta} \\ \sigma_{xy} & \sigma_y^2 & \sigma_{y\theta} \\ \sigma_{x\theta} & \sigma_{y\theta} & \sigma_\theta^2 \end{pmatrix}$$

Off-diagonal elements are covariances that are defined by correlation coefficients:

$$\rho_{ij} = \frac{\sigma_{ij}}{\sigma_i \sigma_j} = \frac{\langle \bar{x}_i \bar{x}_j \rangle}{\sqrt{\langle \bar{x}_i^2 \bar{x}_j^2 \rangle}}$$

where $-1 \leq \rho_{ij} \leq 1$ so that $\sigma_{ij} = \rho_{ij}\sigma_i\sigma_j$. The mean and covariance yield a normal distribution according to the maximum entropy principle:

$$p(x) = \frac{1}{\sqrt{2\pi P}} e^{-\frac{1}{2}(x-\bar{x})^T P^{-1}(x-\bar{x})}$$

The measurement (sensor) model describes how measured observables (output) z relates to the state vector x:

$$z(k) = H(k)x(k) + w(k)$$

where $H(k) = $ measurement observation Jacobian relating sensor output z to state $x(k) = \nabla h\hat{x}$ and $w(k) = $ measurement white noise with covariance matrix R. The measurement model gives range distance $r(k)$ and relative bearing $\gamma(k)$ to landmarks measured by each sensor (such as vision or LIDAR):

$$z(k) = h(k) + w(k) = \begin{pmatrix} r(k) \\ \gamma(k) \end{pmatrix}$$

where

$$w(k) = \text{measurement noise}$$

$$h(k) = \frac{y_i - y(k)}{x_i - x(k)} - \theta(k) = \text{measurement function}$$

$$r(k) = \sqrt{(x - x_r(k))^2 + (y - y_r(k))^2} + w_r(k)$$

$$\gamma(k) = \tan^{-1}\left(\frac{y - y_r(k)}{x - x_r(k)}\right) - \phi(k) + w_\gamma(k)$$

$$\phi = \text{orientation of robot}$$

$$\begin{pmatrix} x_r(k) \\ y_r(k) \end{pmatrix} = \text{location of sensors on the vehicle}$$

$$x_r(k) = x(k) + a\cos\phi(k) - b\sin\phi(k)$$

$$y_r(k) = y(k) + a\sin\phi(k) + b\cos\phi(k)$$

$$(a, b) = \text{sensor offsets from vehicle rear axle}$$

$$\frac{\partial h}{\partial x} = \frac{-(y - y(k))}{(x - x(k))^2 + (y - y(k))^2}$$

$$\frac{\partial h}{\partial y} = \frac{-(x - x(k))}{(x - x(k))^2 + (y - y(k))^2}$$

Assuming that all rover maneuvers are based on Ackermann steering, the state equation for rover motion is given by:

$$\frac{dx}{d\theta} = \begin{pmatrix} \dfrac{dX}{d\theta} \\ \dfrac{dY}{d\theta} \\ \dfrac{d\phi}{d\theta} \end{pmatrix} = \begin{pmatrix} r\cos\phi \\ r\sin\phi \\ \dfrac{r}{b}u \end{pmatrix} + w(\theta) = f(x,u) + w(\theta)$$

where

$$x = \begin{pmatrix} X \\ Y \\ \phi \end{pmatrix} = \text{rover position}$$

and where $\theta = (\theta_r + \theta_l)/2 = $ average rover wheel odometry measurements for straight and steered traverses, $r = $ wheel radius, $b = $ half-distance between wheel axles, $u = (d\theta_l - d\theta_r)/(d\theta_l + d\theta_r) = $ control input, $d\theta_l, d\theta_r = $ average differential wheel rotations of left/right wheels, and $w(\theta) = $ process noise-modeling slippage as random. For turning in place maneuvers,

$$\frac{dX}{d\theta} = \frac{dY}{d\theta} = 0$$

Estimation error covariance matrix $P(\theta)$ for the extended Kalman filter evolves over time and is defined by:

$$\frac{dP}{d\theta} = F(x)P(\theta) + P(\theta)F(x)^T + Q$$

where $F(x) = $ current transition equations and $Q = $ covariance matrix of process noise $ = \mathrm{diag}(Q_{XX}Q_{YY}Q_{\phi\phi})$. Update equations to compute the Jacobian of F are given by:

$$P(k+1|k) = \nabla F P(k|k) \nabla F^T + Q(k)$$

where

$$\nabla F = \frac{\partial F}{\partial x} = \begin{pmatrix} 1 & 0 & -d(k)\cos\phi(k)\sin(\theta(k)+\phi(k)) \\ 0 & 1 & d(k)\cos\phi(k)\cos(\theta(k)+\phi(k)) \\ 0 & 0 & 1 \end{pmatrix} = \text{Jacobian of } F$$

The probability of a priori estimate $\hat{x}(k+1|k)$ is conditioned on all prior measurements $z(k+1|k)$ according to Bayes rule. The a posteriori state estimate reflects the mean of the state distribution while the a posteriori estimate error covariance reflects the variance of the state distribution:

$$\hat{x}(\theta_{i+1}) = \hat{x}(\theta_i) + K(\theta_{i+1})[z(\theta_{i+1}) - h(\hat{x}(\theta_i))]$$

where $K(\theta_{i+1}) = P(\theta_i)H(\hat{x}(\theta_i))^T [H(\hat{x}(\theta_i))P(\theta_i)H(\hat{x}(\theta_i))^T + R]^{-1} = $ Kalman gain, $\theta_{i+1} = $ new measurement, $\hat{x}(\theta_{i+1}) = $ updated rover state, $\hat{x}(\theta_i) = $ rover state estimate from integrating state equations, $H(\hat{x}(\theta_i)) = $ measurement Jacobian at estimated state, and R = measurement noise covariance matrix of yaw sensor. The measurement stage computes the innovation (or residual) defined as the difference between predicted and actual observation $z(k+1)$. Prediction error (innovation) is given by:

$$V(k+1) = z(k+1) - H(k)\hat{x}(k+1|k) = z(k+1) - \hat{z}(k+1|k)$$

Innovation covariance matrix S_{k+1} is computed thus:

$$S(k+1) = \nabla h P(k+1|k)\nabla h^T + R(k+1)$$

where $\nabla h = \partial h/\partial x$ = Jacobian of h and $R(k) = \langle w(k) \cdot w(k)^T \rangle$ = measurement error covariance matrix = $tr(\sigma_{ii}^2)$. Generally, it is important to characterize Q and R accurately—this can usually be estimated. If noise is uncorrelated (i.e., different sensors), R will be diagonal. The RMS noise level of a sensor or discretization errors in ADC can be squared to give variance as elements of matrix R. Q is more difficult and is often initially set randomly. Generally, the values in Q and R determine the emphasis placed on the plant model and measurements. The state estimate and state estimate covariance are updated to provide a filtered estimate of the state vector at time $(k+1)$ given by the update equation:

$$\hat{x}(k+1|k+1) = A(k)\hat{x}(k|k) + u(k) + K(k+1)v(k+1)$$
$$= \hat{x}(k+1|k) + K(k+1)V(k+1)$$
$$= \hat{x}(k+1|k) + K(k+1)[z(k+1) - h(k+1|k)]$$

where $z(k+1) = y(k+1) - H(k+1)\hat{x}(k+1|k)$ = measurement residual. The state estimate of covariance determines how much the actual measurement deviates from the prediction:

$$P(k+1|k+1) = P(k+1|k) - K(k+1) \cdot S(k+1) \cdot K^T(k+1)$$

where

$$K(k+1) = P(k+1|k)\nabla h^T S(k+1)^{-1}$$
$$= P(k+1|k)H^T(k)S^{-1}(k+1) = \text{Kalman filter gain}$$
$$= P(k+1|k)H^T(k) \cdot (H(k+1)P(k+1|k)H^T(k+1) + R(k+1))^{-1}$$

Hence,

$$P(k+1|k+1) = (I - K(k+1)H(k+1)^T)P(k+1|k)$$

The Kalman filter updates an estimate based on the Kalman gain that defines the relative weighting of the correction accorded to noisy sensory measurements compared with the a priori estimate based on the internal model. The weighting is proportional to the prior covariance in the state estimate and inversely proportional to the conditional covariance of the measurement. Thus, gain is determined by the relative variance of the model and measurement errors giving it its optimality characteristic. A reduction in computational processing may be achieved by using a quaternion representation of orientation. Odometry includes motor angular position sensors, motor tachometers, and ground Doppler LIDAR/

radar. Inertial measurement sensors generally consist of an integrated platform of three gyroscopes, a triaxial accelerometer, and the addition of tilt sensors [826]. Gyroscopic rate sensor drift bias may be estimated through a linear Kalman filter by incorporating it in the state equation to be estimated. The approach outlined treats slippage as a Gaussian noise source. This assumes a linear model but the extended Kalman filter may be applied to nonlinear systems. Linearization is performed twice to generate state and measurement Jacobians. EKF assumes Gaussian noise and non-Gaussian noise can be approximated as the weighted sum of several Gaussian probability density functions. The Kapvik microrover incorporated an EKF (later replaced by a CKF) algorithm interfaced to the CNES/LAAS stereovision mapping algorithms. The EKF tends to underestimate state covariance and does not perform well with highly nonlinear systems. Arras et al. (2001) applied the Kalman filter to line segments extracted from images [828]. The Kalman filter may have a large state vector that may require high memory and processing demands. Maintenance of the covariance matrix has a computational complexity $O(n^2)$ where $n =$ number of features in the map. Matrix inversion has a complexity of $O(n^{2.4})$ while matrix multiplication has a complexity of $O(n^2)$. The main computational bottleneck is matrix inversion the size of which is dependent on the number of landmarks. The Kalman filter provides the basis for multiple target tracking and it is in this application that the data association problem is well known [829] but which similarly applies to SLAM involving multiple landmarks [830] (Figure 9.8).

It has been suggested that removing old landmarks as the rover progresses to new territory relieves this problem [831]. In particular, there will arise inconsistencies between observations and features in the map. As SLAM is a nonlinear problem, there is no guarantee that estimated covariances will match estimation errors. Although the Gaussian noise assumption in Kalman filters is erroneous, the extended Kalman filter performs well. Robustifying the Kalman filter has been proposed using either a heavy-tailed non-Gaussian observation error or state error (but not both) [832]. Non-Gaussian statistics derived from maximum a posteriori (MAP) Bayesian approaches may be adopted [833, 834]. It has been suggested that the construction of independent rover-centric local maps rather than incremental merging into an absolute reference frame global map reduces error growth [835, 836. It has also been suggested that a mixture of expert Kalman filters with different system parameters (including non-Gaussian statistics) regulated by a gating network forming a parallel Kalman filter bank offers online adaptability [837]. These Kalman filters yield different estimates that can be weight-summed to yield an optimal estimate, the weighting being determined from Bayes rule:

$$\hat{x}_{\text{opt}} = \sum_{i=1}^{l} \hat{x}_k(\alpha_i) w_i$$

where $l =$ number of Kalman filter modules, $\hat{x}_{k+1}(\alpha_i) =$ Kalman filter i output, $\alpha_i =$ parameter of interest, $w_i = p(\alpha_i|z_k) \equiv (p(z_k|\alpha_i)p(\alpha_i))/\sum_{j=1}^{l} p(z_i|\alpha_i)p(\alpha_i)$, $z_k =$ measurement inputs, and $p(z_k|\alpha_i) = (1/(2\pi|S_k|)e^{-\frac{1}{2}v_k^T S_k^{-1} v_k})p(z_{k-1}|\alpha_i)$. Alterna-

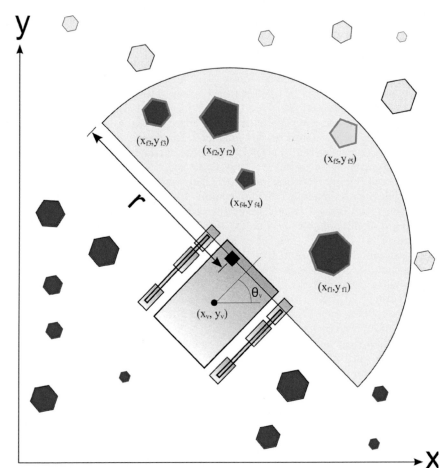

Figure 9.8. Simulated laser scanner–based SLAM with data association to distinguish between old and new objects [credit Rob Hewitt, Carleton University].

tively, fuzzy logic rules have been used to enhance the Kalman filter approach by adjusting the weighting factor to prevent divergent behavior [838]. However, these methods are computationally intensive. Kalman filters have been implemented in FPGA (field programmable gate array) hardware form for fast matrix multiplies providing a processing speed up of three times that of a transputer implementation [839]. The use of the Connection Machine of 64k processors connected in a hypercube network reduces the computational complexity of the Kalman filter to $O(n \log n)$ from $O(n^3)$ [840]. An alternative approach may be the artificial neural network controller that implements function approximation and may implement Kalman filter-like behavior. The neural network can augment the extended

Kalman filter to learn unmodeled dynamics which trains online [841, 842]. One
approach would be to use a neural network to approximate uncertainty in the
system model of an extended Kalman filter (EKF) for SLAM [843].

9.5   UNSCENTED FILTER–BASED SLAM

The unscented Kalman filter (UKF) offers many advantages over the traditional
Kalman filter including applicability to nonlinear systems and no requirement for
Gaussian distributions [844–846]. The unscented filter uses linear weighted regression of nonlinear models without Taylor series expansion linearization. It has
some of the advantages of particle filtering by incorporating aspects of the particle
filter within the Kalman framework. It does not require computation of the
Jacobian. To avoid computation of the Jacobian matrix, approximate solutions to
integral solutions of mean and covariance matrices may be used—numerical
integration methods include Gauss–Hermite quadrature, unscented filter variant to
the Gauss–Hermite quadrature approximation, and Monte Carlo methods [847].
The unscented transform is essentially a truncated form of a Gaussian–Hermite
quadrature approximation that may be input to a Kalman filter. Monte Carlo
integration is attractive for large-dimension problems such as multitarget tracking.
The UKF technique represents a good performance compromise with reduced
computational complexity for short running times. Nonlinear filters such as particle filters and unscented filters are superior to the EKF but usually at the cost of
computational complexity [848]. A better approximation to a nonlinear function is
provided through the unscented Kalman filter. For the UKF, a minimal set of
sample (sigma) points are selected which approximate the probability distribution
function of the state (i.e., an ensemble of sigma points with the same mean and
covariance as the true probability distribution function). It is easier to perform a
nonlinear transformation on a single point than through a general nonlinear function. Using the matrix square root of covariance of the state of dimension n, $2n$
sigma points are selected to undergo nonlinear transformation. These sigma points
capture different properties of a given distribution. The nonlinear function is
applied at each of these points to produce a transformed sample from which
predicted mean and covariance are computed. This resembles the Monte Carlo
approach but the samples are not drawn at random, rather they are selected on
the basis of the information they hold. The sigma points are spread out to ensure
accurate representation of the probability distribution function. The mean and
covariance of the probability distribution function are approximated as the mean
and covariance of these points. These form the a priori estimate. We initialize the
state and covariance matrix:

$$\hat{x}_0^+ = \langle x_0 \rangle \quad \text{and} \quad P_0^+ = \langle (x_0 - \hat{x}_0^+)(x_0 - \hat{x}_0^+)^T \rangle$$

in which $2n$ sigma points x_i are formed approximating a Gaussian distribution
describing the statistics of the state gradient (where $n =$ state space dimension):

$$\bar{x}_{k-1}^i = \bar{x}_{k-1}^+ + \tilde{x}^i$$

where $i = 1, 2, \ldots, 2n$, $\tilde{x}^i = (\sqrt{nP_{k-1}^+})_i = i$th column of $(\sqrt{nP})$ such that $(\sqrt{nP})^T(\sqrt{nP}) = nP$, and $\tilde{x}^{n+i} = -(\sqrt{nP_{k-1}^+})_i$ for $i = 1, 2, \ldots, n$. The sigma points are located at the mean and symmetrically along the covariance matrix. They are propagated through the state model to form the a priori estimate:

$$\bar{x}_k^i = f(\bar{x}_{k-1}^i, u_{k-1})$$

$$\bar{x}_k^- = \frac{1}{2n} \sum_{i=1}^{2n} \bar{x}_k^i$$

$$P_k^- = \frac{1}{2n} \sum_{i=1}^{2n} (\bar{x}_k^i - \bar{x}_k^-)(\bar{x}_k^i - \bar{x}_k^-)^T + Q_k$$

The sigma points are propagated through the nonlinear measurement function to form the a posteriori estimate:

$$\bar{x}_k^i = \bar{x}_k^- + (\sqrt{nP_k^-})^T$$

$$\bar{x}_k^{n+i} = \bar{x}_k^- - (\sqrt{nP_k^-})^T$$

$$\bar{z}^i = h(\bar{x}_k^i) \quad \text{for } i = 1, 2, \ldots, 2n \quad \text{and}$$

$$\bar{z}_k = \frac{1}{2n} \sum_{i=1}^{2n} z_k^i$$

The weight on each sigma point apart from the zeroth point is the same. These samples approximate the first three moments of the Taylor series. The predicted measurement, innovation covariance matrix, and cross-covariance matrix may be computed:

$$P_{zz} = \frac{1}{2n} \sum_{i=1}^{2n} (\bar{z}_k^i - \bar{z}_k^-)(\bar{z}_k^i - \bar{z}_k^-)^T + R_k$$

$$P_{xz} = \frac{1}{2n} \sum_{i=1}^{2n} (\bar{x}_k^i - \bar{x}_k^-)(\bar{z}_k^i - z_k^-)^T$$

Cross-covariance corresponds to the Jacobian term. Updating involves computing a posteriori estimates that are the same as for the Kalman filter:

$$K_k = P_{xz} P_{zz}^{-1}$$

$$\bar{x}_k^+ = \bar{x}_k^- + K_k(z_k - \bar{z}_k)$$

$$P_k^+ = P_k^- - K_k P_{zz} K_k^T$$

As the algorithm is recursive so state and covariance estimates are input to the filter recursively. The UKF has a similar computational complexity to the EKF (i.e., $\sim n^3$ where $n =$ problem dimension). Unscented transformation gives an equivalent performance to a third-order Taylor expansion compared with first order for Taylor linearization of the extended Kalman filter. The chief advantage is that there is no requirement for large Jacobian matrices to be computed. However, the UKF can suffer from the curse of dimensionality. The cubature Kalman filter uses efficient methods of integration (cubature rules) which avoids matrix inversion and provides greater numerical accuracy and stability. The cubature Kalman filter is a Bayesian filter under Gaussian conditions valid for nonlinear systems. It uses a spherical–radial cubature rule to numerically compute the mean and covariance of the state variables in a nonlinear Bayesian filter [849]. According to Bayes rule, posterior density of the current state is given by:

$$p(x_k|D_k) = \frac{p(x_k|D_{k-1})p(z_k|x_k, u_k)}{p(z_k|D_{k-1}, u_k)}$$

where $D_{k-1} = (u_i, z_i)_{k-1} = $ input–output pair history up to $k - 1$:

$$p(x_k|D_{k-1}) = \int p(x_{k-1}|D_{k-1})p(x_k|x_{k-1}, u_{k-1})\, dx_{k-1} = \text{predictive density}$$

$$p(z_k|D_{k-1}, u_k) = \int p(x_k|D_{k-1})p(z_k|x_k, u_k)\, dx_k$$

$p(x_k|x_{k-1}, u_{k-1}) = $ state transition function derived from

state equation $x_k = f(x_{k-1}, u_{k-1}) + v_{k-1}$

$p(z_k|x_k, u_k) = $ measurement likelihood function derived from

measurement equation $z_k = h(x_k, u_k) + w_k$

For a system of n state variables, the cubature KF uses $2n$ cubature points to numerically compute standard Gaussian integrals rewritten as spherical–radial third-degree summations:

$$I_n(x) = \int f(x)N(x, \mu, P)\, dx \approx \sum_{i=1}^{2n} w_i f(\xi_i)$$

where $\xi_i = \sqrt{n}[1]$, $w_i = 1/(2n)$, $i = 1, 2, \ldots, 2n$, and $n(\ldots) = $ Gaussian probability density function. The set of cubature points (ξ_i, w_i) are used to numerically compute the Gaussian integrals.

Square root CKF improves robustness against numerical instabilities by using a square root of the posterior density function of the form $P_0 = S_0 S_0^T$. It factorizes matrices into lower triangular and diagonal matrices of the form $X = LDL^T$ for the covariances. Its computational complexity is similar to that of the CKF or EKF but without using Jacobians. The time updates are thus:

(i) Compute cubature points for $i = 1, 2, \ldots, 2n$:

$$x_{k-1|k-1} = \hat{x}_{k-1|k-1} + S_{k-1|k-1}\xi$$

(ii) Compute propagated cubature points for $i = 1, 2, \ldots, 2n$:

$$x_{k|k-1}^* = f(x_{k-1|k-1}, u)$$

(iii) Compute predicted state:

$$\hat{x}_{k|k-1} = \frac{1}{2n} \sum_{i=1}^{2n} x^*_{k|k-1}$$

(iv) Compute square root of predicted error covariance:

$$S_{k|k-1} = Tri([x^*_{k|k-1} \quad S_{Q,k-1})$$

where $P_{Q,k-1} = S_{Q,k-1} S^T_{Q,k-1}$, $Tri = QR$ factorization, and $x^*_{k|k-1}$ = weighted centered matrix.

A cubature UKF algorithm was developed at Carleton University and implemented on the Kapvik microrover to augment the CNES/LAAS stereovision-based navigation algorithms by adding a sensor fusion capability (Figure 9.9).

9.6 SLAM BY BAYESIAN ESTIMATION

The self-localization aspect introduces the problem of belief regarding its position within the map over an extended distance and time. A single hypothesis belief representation posits a single unique point in the map. However, this position has an uncertainty modeled as a statistical distribution. A standard Bayesian update rule computes the posterior probability of occupancy of each cell of the environment map based on the prior probability of occupancy and current measurements. For a Gaussian distribution, this is represented by mean and standard deviation parameters. This single position must be updated over time as the robot moves with the appropriate uncertainty. The Kalman filter can be used to track the position over time using a Gaussian probability distribution. During every perception–action cycle, only the mean and standard deviation of the Gaussian probability representing the position belief are required to be updated. The Kalman filter is efficient in using Gaussian probability but can get lost. The Kalman filter is a single hypothesis method in representing only a single mode of probability density function, but multiple Kalman filter banks can implement multiple Gaussian hypotheses. In a multiple belief representation, a set of robot positions is maintained. Each position is marked with an uncertainty represented by a statistical distribution. The robot's multiple beliefs can accommodate significant changes in environmental information. Markov localization uses probability distributions over all possible robot positions—any probability density function may be used including multimodal distributions. The robot's belief is represented as separate probability distributions for each robot position in its map. The probability of each cell must be updated during every perception–action cycle. This update mechanism computes the belief state when new sensory information is incorporated in the prior belief state—Bayes theorem may be used. The prior describes the structure of the world in terms of a probability density function. Markov localization can recover from ambiguous situations (kidnapped robot situation) because the robot tracks multiple beliefs about its position.

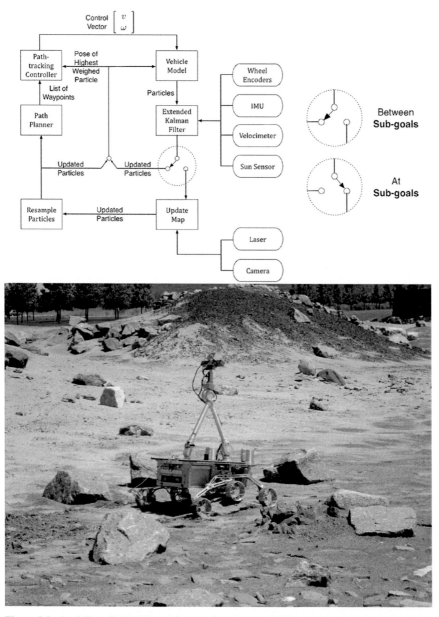

Figure 9.9. (top) Kapvik SLAM architecture based on an EKF algorithm [credit Rob Hewitt, Carleton University] and (bottom) EKF was subsequently updated to a CKF that was subjected to successful field trials on Kapvik at the CSA's Mars Yard [courtesy of Ian Sinclair, MPB Communications].

Multiple hypotheses impose high computational cost. If a continuous environment is tessellated into a grid map with discretized probability values at each cell, this comprises the Markov approach. This is computationally more tractable. A fixed resolution of the map comprises an occupancy grid within which each cell is regarded as occupied by an obstacle (non-traversable) or empty free space (traversable). As the map grows, map memory requirements grow. This favors a topological representation which does not include geometric information. When several alternative hypotheses exist, the consequences of each one can be determined and experimentally tested. The hypotheses effectively predict probabilistically what the observable consequences will be. A probability distribution is associated with the set of hypotheses reflecting the degree of belief in each. If the predicted events are observed, this increases the belief in that predictive hypothesis. Hypothesis H is perceived as causing evidence E. Hypotheses must be mutually exclusive and exhaustive. Bayes theorem is the simplest way of updating the probability distribution. It is used to fuse observational and motion measurements [850–853]. Bayesian inference fuses multisensory data through probability estimates that the state is correct [854]. Bayesian methods do not require the assumption of Gaussian noise, being dependent merely on the observations.

Particle filters represent beliefs of the location of the robot in the environment. They represent the unknown non-Gaussian probability distribution as a set of weighted particles. The initial probability that the robot is at a specific position at a specific time comprises the prior probability. Bayes rule is used to compute the robot's new belief regarding its position based on its prior belief and new evidence from sensor measurements. Bayes filters estimate the localization state x (pose) of the rover from sensor measurements of its environment (range measurements, camera images, odometry, etc.). Bayes filters assume that the environment is characterized as a Markov chain such that past and future data are independent and that the future state is determined only by the current state. The objective is to compute the probability of a given set of hypotheses on the basis of observational evidence. A set of prior probabilities describe the probabilities of each hypothesis prior to sensory measurement. The prior probability is effectively an internal model of previous trials but is primarily subjective. The joint probability of the observable and the hypothesis is the likelihood of making observations when the robot is at a given location defined by the map model. Sensory evidence adjusts this prior probability such that it becomes a posterior probability. The Bayesian method models the robot's belief about its location. Bayes theorem is used to estimate posterior probability density (belief) over state space x_i conditional on sensory data s_i [855–858]:

$$\text{Bel}(x_t) = p(x_t|s_{0,...,t}, m)$$

where x_t = state at time t, $s_{0,...,t}$ = sequence of sensory data from time 0 up to time t, and m = world model (environment map). Bayes theorem effectively weights the combination of the internal model and noisy sensory likelihood. There are two types of sensory data—exteroceptive data y_i that are referenced to environmental properties such as camera images and proprioceptive data u_i that are internal state measurements such as odometry:

$$\mathrm{Bel}(x_t) = p(x_t | y_t, u_{t-1}, y_{t-1}, u_{t-1} \ldots, u_0, y_0)$$

where $s_{0,\ldots,t} = (y_t, u_{t-1}, y_{t-1}, \ldots, u_0, y_0)$. From the sequence of outputs, it is possible to infer the dynamics of the system. Bayes filters (Markov localization) estimate belief recursively initialized with a uniform distribution over the state space:

$$\mathrm{Bel}(x_t) = \frac{p(y_t | x_t, u_{t-1}, \ldots, y_0) p(x_t | u_{t-1}, \ldots, y_0)}{p(y_t | u_{t-1}, s_{0,\ldots,t-1})}$$

Measurements y_t are independent of past measurements given state x_t (Markov):

$$p(y_t | x_t, u_{t-1}, \ldots, y_0) = p(y_t | x_t) \quad \text{or} \quad p(y_t | x_t, u_{t-1}, \ldots, y_0) = p(y_t | x_{t-1}, u_{t-1})$$

Hence,

$$\mathrm{Bel}(x_t) = \frac{p(y_t | x_t) p(x_t | u_{t-1}, \ldots, y_0)}{p(y_t | u_{t-1}, s_{0,\ldots,t-1})}$$

Integrating at time $t - 1$:

$$\mathrm{Bel}(x_t) = \eta p(y_t | x_t) \int p(x_t | x_{t-1}, u_{t-1}) \mathrm{Bel}(x_{t-1}) \, dx_{t-1}$$

where $\eta = 1/(p(y_t | u_{t-1}, s_{0,\ldots,t-1})) =$ normalization constant. If beliefs are represented by Gaussians, this defines the Kalman filter. Three probability distributions are required—initial belief (assumed uniform if the initial position is inknown), next state (motion) probability $p(x_t | x_{t-1}, u_{t-1})$ which represents a probabilistic kinematics model, and perceptual likelihood $p(y_t | x_t)$ which represents a probabilistic sensor model, all of which depend on the sensors. Assuming time independence, the probability density $p(y, x)$ is a random mixture of variables that model the response of the sensor (convolved with Gaussian noise) to given environmental properties. The probabilistic representation allows multiple hypotheses to be maintained weighted by sensor evidence. The optimal sample size is 100–500. Since the state space is continuous belief update is non-trivial so particle filters approximate belief with a set of weighted discrete state space samples (particles) where importance weights p_i sum to unity:

$$\mathrm{Bel}(x) = \{x_i, p_i\}_{i=1,\ldots,n}$$

Belief is approximated by:

$$\frac{p(y_t x_t) p(x_t | u_{t-1}) \mathrm{Bel}(x_{t-1})}{p(y_t | s_{0,\ldots,t-1}, u_{t-1})} \approx p(x_t | x_{t-1}, u_{t-1}) \mathrm{Bel}(x_{t-1})$$

where

$$p(y_t | x_t) \sim \frac{1}{p(x_t | x_{t-1}, u_{t-1}) \mathrm{Bel}(x_{t-1})} \frac{p(y_t | x_t) p(x_t | u_{t-1}, x_{t-1}) \mathrm{Bel}(x_{t-1})}{p(y_t | s_{0,\ldots,t-1}, u_{t-1})}$$

This particle filter approach offers much reduced memory requirements than grid-based maps with higher resolution. To alleviate the problem of slow convergence under conditions of low-noise sensors, a mixture of distributions may be used.

One approach to estimating the posterior probability function (belief) is to use maximum likelihood estimation over a sequence of occupancy grid-based maps (x, y) each with occupancy $m = (0, 1)$ with increasing likelihood:

$$\text{Bel}(m_{xy}) = \eta p(y_t | m_{xy}) \sum_{m^{xy}=o}^{1} p(m_{xy} | m_{xy,t-1}, u_{t-1}) \text{Bel}(m_{xy,t-1}) = \eta p(y_t | m_{xy}) \text{Bel}_{t-1}(m_{xy})$$

$$= \eta \frac{p(m_{xy} | y_t) p(y_t)}{p(m_{xy})} \text{Bel}_{t-1}(m_{xy})$$

Markov decision processes may be applied to rover navigation in modeling state transitions from statistical measures from observed variables. They provide control in stochastic environments that issue rewards from the environment. Markov localization is based on representing the robot position as a probability distribution over the positions of each grid in the environment according to sensor data. It is a multiple hypothesis method using histograms to track position uncertainty. It maintains a density over the set of all possible positions of the robot in its environment which is updated whenever the robot gains new sensory data. It is well suited to dynamic environments. Markov models are random process models where the probability of change depends only on the current state (Markov property)—they tend towards stationarity. The Markov chain is a memoryless random process whose future state depends only on the current state. Markov localization is a variant of Bayesian methods and uses a discrete approximation of the prior probability distribution. When the state is not completely observable, we must provide a model of observations. This is provided by Bayes rule. The hidden Markov model (HMM) is the simplest type of Bayesian network which may be used to recognize robotic behaviors where observations are viewed as signals [859, 860]. They represent a system as a set of discrete states and evolution of the system is dictated by state transitions with a probability distribution. Bayes theorem is used to estimate the probability of the hidden states given the observables.

Global localization from scratch requires multimodal representations of hypotheses concerning the state (position) of the rover (thereby accommodating the kidnapped robot problem). This may be represented as a mixture of Gaussians in which each hypothesis is tracked by an EKF. However, for nonlinear systems, a general non-Gaussian probability distribution function representing rover state (position) must be approximated. This is achieved using recursive versions of Monte Carlo algorithms based on point mass representations of probability densities (i.e., particle filters). Particle filter algorithms approximate desired a posteriori distributions as a set of particles that are random samples of states distributed according to the Bayesian estimation of probability distribution. Such particle filters represent beliefs by a weighted set of samples (particles) according to the probability distributions of possible locations [861]. The probability distributions are not Gaussian to represent multiple positions. For non-Gaussian systems, particle filters are appropriate. There are several particle filters: particle filters that require resampling (to prevent particle dispersion by replacing low-probability particles with high-probability particles) including the sequential importance sampling (SIS) particle filter and the bootstrap particle filter which is similar to the SIS method except in its update; particle filters that do not require resampling include the Gaussian particle filter which may also utilize Monte Carlo, Gauss–Hermite, and unscented approximations. Monte Carlo methods are useful for stochastic modeling and are not restricted to Gaussian probability density functions [862]. The Gauss–Hermite particle filter has the shortest runtime. A particle filter is a Markov chain Monte Carlo algorithm that approximates the belief state [863]. Particle filters are a variant on Bayes filters that maintain a set of hypotheses about robot pose and object locations. Randomized sampling using particle filters (Monte Carlo algorithms use sampling to approximate probability density functions) involves representing a random sample of possible positions rather than all possible positions in the environment. This reduces the computational overhead of Markov localization. Weighting this sample set with the probability values increases samples with higher probabilities. The set of particles is characterized by a state and weight:

$$s(k) = (x_i(k), w_i(k))$$

where $w_i(k) = p(z(k)|x_i(k))$. Once the probability distribution function of robot pose is known, the pose may be estimated from the weighted mean (typically). After a few iterations, position ambiguity is reduced—most particles have very small weights which can be ignored as they do not represent the pose. When the sample size drops below a percentage of particles, resampling is required. Particles with small weights are eliminated and replaced with duplicated particles with high weights. The particle filter samples and updates estimates of the maximum posterior probability density for localization.

A Markov chain Monte Carlo method is a stochastic simulation used to generate a sample from a complex distribution such as a Bayesian posterior distribution. In Monte Carlo–based particle filters, the posterior probability density of the state is represented by a set of particles randomly drawn from the probability density function (PDF) [864]. Monte Carlo methods evolve these particles over time—the posterior distribution is updated with repeated observations of the system. Samples are selected in proportion to the likelihood thereby focusing on highly likely states—this property enhances the efficiency of the particle filter. The use of samples reduces computational memory requirements over other probability representations and allows representation of multimodal distributions. The particle filter is recursive and so comprises two stages [865]—each particle is weighted according to an existing system model to predict the next state PDF and then updated by the weight reevaluated according to current observations. The sequential importance sampling (SIS) filter is a recursive Monte Carlo–based approach on which most particle filters are based. Posterior probability density is represented by a large set of random samples with associated weights. The particle filter is given by:

(i) Initialization by generating samples (particles) of state vector x_i

$$s_k = (x_k^i, w_k^i)$$

to approximate predicted density $p(x_{k-1}|z_{k-1})$.

(ii) Update importance weights w_k^i of each sample by the likelihood of x_k given measurements z_k:

$$w_k^i = w_{k-1}^i p(z_k|x_k^i) = w_{k-1}^i p(z_k - h(x_k^i))$$

The weight signifies the quality of that specific particle. This is normalized by:

$$w_k^i = \frac{w_k^i}{\sum_i w_k^i} \quad \text{to ensure that} \quad \sum_{i-=1}^n w_k^i = 1$$

Weights are chosen according to the importance sampling principle to approximate posterior density $p(x_k|z_{k-1})$. The state estimate comprises the weighted sum of all particles: Let

$$\hat{x}_k = \sum_{i=1}^n w_k^i x_k^i$$

Sample-based posterior density converges to the true posterior density at a rate $1/\sqrt{n}$ as $n \to \infty$.

(iii) The variance of particles increases over time causing particle divergence. Resampling selects with a high degree of probability those samples with high likelihood. Resampling implements the sequential importance resampling (SIR) filter derived from the SIS algorithm which consists of sampling $(s_1, \ldots, s_n)$ with replacement from (s_i, p_i) where p_i = probability weighting function. N samples are replaced with probability

$$w_k^i = \frac{1}{N} \quad \text{when } N = \frac{1}{\sum_i (w_k^i)^2} < N_{th}$$

where $N_{th} = \frac{2}{3}N$ (threshold) nominally. To reduce the effects of divergence, a large N_{th} is required to increase the size of N. Resampling occurs when N falls below the threshold—resampling involves eliminating small weights to concentrate on particles with high weights. Resampling involves the generation of a new set of particles from an approximate representation of posterior density $p(x_k|z_{k-1})$. SIR uses a discrete approximation while the regularized particle filter (RPF) uses a continuous approximation to prevent sample impoverishment whereby all particles are too tightly localized.

(iv) The prediction phase simulates the state equation and the iteration cycles once again to (ii).

Particle filters give superior results than EKF for SLAM applications [866]. Large numbers of particles are required as the state dimension increases but this comes at a severe computational cost. To reduce the computational burden, the number of particles may be reduced but this introduces sample impoverishment so that the particle distribution approximates the probability distribution function poorly [867]. It has been proposed that the sample number may be decreased during self-localization based on the Kullback–Leibler distance which defines the error between the true and sample-based posterior [868]:

$$K(p, q) = \sum p(x) \log \left(\frac{p(x)}{q(x)} \right)$$

These Monte Carlo–based methods are computationally expensive. Monte Carlo methods are often intractable for large sample sizes—in this case, a multivariate Gaussian approximation may be used. Particle filters are difficult to implement in real time due to the large number of particles. However, if the process model or observation model is linear, Rao–Blackwell simplifications may be used to reduce the computational load. The Rao–Blackwell particle filter maintains posterior probabilities over all vehicle and environment states and is typically used for SLAM. Comparison of the performance of particle filters and extended Kalman filters implies that the EKF is less accurate but faster than the particle filter [869]. Bayesian methods offer some advantages over Kalman filters in complex sensor modeling within complex environments—the Kalman filter assumes zero-mean Gaussian noise and cannot cope with systematic errors and highly nonlinear systems [870]. Gutman et al. (1998) [871] compares Kalman filtering with the Markov localization method [872] and found that the Kalman filter is more efficient and accurate but that Markov localization is more robust to noise. Particle filtering, however, is computationally intensive.

9.7 FASTSLAM

The SLAM problem may be represented as a tree-like dynamic Bayesian network in which rover state x_k evolves as a function of rover control inputs u_k. Each landmark measurement z_k is a function of landmark position θ_k and of rover state x_k. The covariance matrix grows quadratically with the number of features in the map. For a robot in a 2D map with n features, the covariance matrix will have $(2n + 3)^2$ elements. In the particle filter, a finite set of particles (sample states) is created representing probability distributions—this allows representation of belief about the state of the system. High-probability-state space contains large numbers of particles while low-probability regions contain few or none. Typically, 2,000–5,000 particles are required for SLAM. A probabilistic weighting weights each particle according to the likelihood of the belief. With sufficient particles, the particle filter can represent complex multimodal probability distributions. However, the number of particles can scale exponentially with the number of states to be estimated, limiting particle filters to low-dimensional problems. SLAM computes posterior distribution over the robot path by means of a map

$$p(x_i, \theta | z_i, u_i, n_i)$$

where x_i = measured robot position, θ = measured landmark positions in the map, z_i = measurement sequence, u_i = sequence of control inputs, and n_i = data association variables. FastSLAM is premised on the decomposition of robot localization from the landmark estimation problem. This reduces the computational burden by decoupling the robot path and environment map. FastSLAM uses particle filtering to estimate robot pose which has low dimensionality and then uses EKFs to estimate the state of the environment which is potentially of much larger dimensionality (i.e., FastSLAM is thus a hybrid between particle filtering and the EKF). This Rao–Blackwellianization involves the use of a hybrid particle and Kalman filter approach to compensate for small particle numbers N and reduce their divergence. The trajectory is represented by the weighted samples of a particle filter while the map of landmarks is constructed using an extended Kalman filter. This assumes that the rover's path and individual landmark measurements are conditionally independent. This FastSLAM scales only logarithmically $O(N \log K)$ with the number K of landmarks in the map [873, 874]. Each of the N particles estimating the rover path possesses K independent Kalman filters for K landmark locations. Robot state is given by a nonlinear model $p(x_i|x_{i-1}, u_i)$ while sensor measurements are determined by a nonlinear measurement model $p(z_i|x_i, \theta, n_i)$. The FastSLAM problem involves determining the locations of all landmarks θ and rover positions x_k from measurements z_k and control inputs u_k by calculating the posterior. Hence, FastSLAM posterior probability is partitioned into robot localization and independent landmark positions:

$$p(x_k, \theta|z_k u_k, n_k) = \eta p(z_k|x_k, \theta_n, n_k)p(x_k|x_{k-1}, u_k)p(x_{k-1}, \theta|z_{k-1}, u_{k-1}, n_{k-1})$$

Hence,

$$p(x_k, \theta|z_{k-1}, u_k, n_k) = p(x_k|x_{k-1}, u_k)p(x_{k-1}, \theta|z_{k-1}, u_{k-1}, n_{k-1})$$

The rover pose dynamic model is given by $p(x_k|x_{k-1}, u_k)$ while the measurement model is given by $p(z_k|x_k, \theta_n, n_k)$ where n_k = landmark correspondence mapping index (data association variables) and θ_n = observed landmark position. The data association problem is represented by mapping n_k between observations and landmarks. Whereas the Kalman filter maintains only a single data association based on maximum likelihood, the representation of multiple data associations is more robust. FastSLAM assumes that the posterior may be factored into one rover path posterior and the product of K individual landmark posteriors conditioned on those paths [875]:

$$p(x_k, \theta|z_k, u_k, n_k) = p(x_k|z_k, u_k, n_k)\prod_{i=1}^{K}p(\theta_i|x_k, z_k, u_k, n_k)$$

This allows use of a particle filter for the rover's path and K independent Kalman filters for each particle to represent landmark positions. Uncertainties in different map landmarks are correlated only through rover path uncertainty—if the rover's path were certain, landmark errors would be independent of each other. A set of particles $s_k = (s_1^i, s_2^i, \dots, s_k^i)$ is generated representing the posterior $p(x_k|z_k, u_k, n_k)$ in which samples are estimated recursively from the posterior $s_k^i \sim p(s_k|s_{k-1}^i, u_k)$

where s_{k-1}^i = posterior estimate of rover location at time $k-1$. The new particle distribution—the proposal distribution—is given by $s_k^i \sim p(s_k|z_{k-1}, u_k, n_{k-1})$. N particles are generated in this manner. FastSLAM resamples this set of particles to form a new particle set obtained by sampling this distribution with probability proportional to importance weight:

$$w_k^i = \frac{p(s_k^i|z_k, u_k, n_k)}{p(s_k^i|z_{k-1}, u_k, n_{k-1})} \approx \int p(z_k|\theta_n^i, x_k^i, n_k)p(\theta_n^i)\, d\theta_n$$

The resulting sample set is distributed according to the approximate desired rover position posterior $p(x_k|z_k, u_k, n_k)$. Conditional landmark position estimates $p(\theta_i|x_k, z_k, u_k, n_k)$ are made through EKF, one for each rover pose particle s_k. The EKF uses a linearized version of the measurement model

$$p(z_k|x_k, \theta_k, n_k) = h(x_k, \theta_n, n_k) + v_k$$

where v_k = Gaussian measurement noise with zero mean and covariance R_k. It also uses a linearized version of the rover motion model

$$p(x_k|x_{k-1}, u_k) = f(x_{k-1}, u_k) + w_k$$

where w_k = Gaussian process noise with zero mean and covariance P_k. Hence, there are nk EKFs to be computed. Importance weight may be reduced to:

$$w_k^i \approx \frac{1}{|2\pi Q_k^i|^2} e^{-\frac{1}{2}(z_k - \hat{z}_k^i)^T Q_k^{i-1}(z_k - \hat{z}_k^i)}$$

where $Q_k^i = H_k^{iT}\Sigma_{k-1}^i H_k^i + R_k$, H = first-degree Taylor coefficient of the measurement function, and Σ_k^i = measurement covariance. The particle filter approximates the Bayesian filter estimate of robot pose [876]. Each particle representing vehicle pose has a set of independent EKFs to estimate the positions of features in the map. The posterior estimate is thus factored into the product of simpler terms—based on separate estimation of robot path, the positions of environmental features in the map are independent of each other. Each environment feature is tracked by a set of EKFs based on the particle filter–generated robot path. There are two fastSLAM algorithms—version 1.0 is the foundation on which version 2.0 offers some improvements. Particles in the distribution are weighted relative to the true distribution defined by Bayes rule depending on innovation covariance:

$$w_i = \frac{1}{\sqrt{|2\pi Z_k|}} \exp(-\tfrac{1}{2}(z_k - \bar{z}_k)^T Z_k^{-1}(z_k - \bar{z}_k))$$

where $Z_k = H_k P_k H_k^T + R_k$, $P_k = (I - K_k H_k P_{k-1})$, $K_k = P_{k-1}H_k^T(H_k P_{k-1}H_k^T + R_k)^{-1}$, and $H_k = \partial h/\partial\lambda$. A new set of particles are sampled with probabilities in proportion to weights. This resampling maintains a constant number of particles while eliminating unlikely candidates yet not compromising particle diversity. The difference between FastSLAM 1.0 and FastSLAM 2.0 lies in their proposal distribution and so in their importance weights: unlike FastSLAM 1.0, FastSLAM 2.0 includes the most recent measurement z_k rather than only the rover model in its proposal for sampled positions [877]. This is less vulnerable to situations in which process noise is larger than measurement noise which introduces sample rover positions with low measurement likelihood. FastSLAM 2.0 offers greater robustness and improved performance over FastSLAM 1.0. FastSLAM 2.0 extracts its samples from posterior $s_k^i \sim p(s_k|s_{k-1}^i, u_k, z_k, n_k)$ rather than $s_k^i \sim p(s_k|s_{k-1}^i, u_k)$ as used in FastSLAM 1.0. The proposal distribution for FastSLAM 2.0 is now given by

$$p(s_{k-1}|z_{k-1}, u_{k-1}, n_{k-1})p(s_k|s_{k-1}, z_k, u_k, n_k)$$

and importance weight by

$$w_k^i = \eta p(z_k|s_{k-1}^i, z_{k-1}, u_k, n_k)$$

Particle weight is slightly modified where

$$Z_k = H_k Q H_k^T + H_k P_k H_k^T + R_k$$

Unscented FastSLAM uses the unscented filter to update state estimates using measurements of features in the map [878]. Linear approximations to nonlinear functions are not required as updates to state estimates are performed using the unscented filter. Unscented FastSLAM does not require computation of Jacobians that are used in versions 1.0 and 2.0 of FastSLAM. It is more accurate than FastSLAM 2.0 with a marginal increase in computational cost.

The Kapvik microrover used EKF estimation prior to LIDAR scans and the FastSLAM algorithm to make corrective updates to rover and map states. This accommodates limitations in the frequency of LIDAR scans available. Prior to each scan, a set of particles is propagated forward through a kinematic model (a priori). Kalman measurement update from odometry, IMU, and Sun sensor is applied to each particle. This process can occur a number of times prior to each LIDAR scan. Following the LIDAR scan, a cubature FastSLAM 2.0 updates the state using LIDAR measurements. Uncertainty in the pose estimate can be used to indicate when the LIDAR scan may be appropriate. A neural network was used to classify measurements and perform data association. In SLAM, one of the more difficult problems relates to data association (i.e., correspondence between measurement data and features in the geometric map of the environment). Data association involves association of observations with targets—this is challenging particularly for moving targets. The stability of SLAM is critically dependent on data association which must solve the perceptual aliasing problem. To solve the data association problem, n_k is usually chosen to maximize the likelihood of measurement z_k (maximum likelihood estimator) by minimizing the Mahalanobis distance. The dissimilarity between vectors x and y may be quantified using the Mahalanobis distance:

$$d_M = \sqrt{(z_k - \bar{z})^T Z_k^{-1}(z_k - \bar{z})}$$

where Z = state covariance and z = measurement. A variant on this is based on an energy metric for a grid map representation to be minimized in determining data association matches between measurements (equivalent to the Mahalanobis

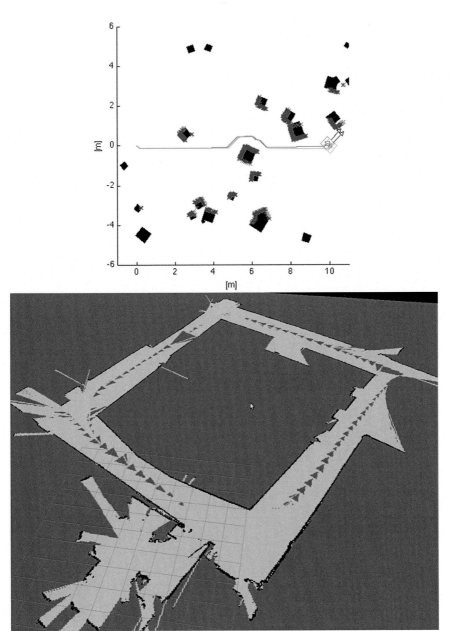

Figure 9.10. FastSLAM navigation (top) simulated through an obstacle field [credit Rob Hewitt, Carleton University] and (bottom) indoors around a physical circuit [credit Pablo Molina, Carleton University].

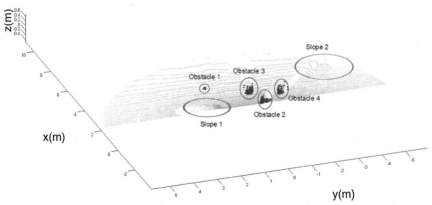

Figure 9.11. FastSLAM 2.0 obstacle map detecting rover during Petrie Island tests [credit Rob Hewitt, Carleton University].

distance in maximum likelihood estimation) [879]. Data associations are selected to maximize the probability of sensor measurement z (maximum likelihood estimator). Any time there is more than one possible data association, multiple hypotheses are generated and represented within a branching tree. The state that

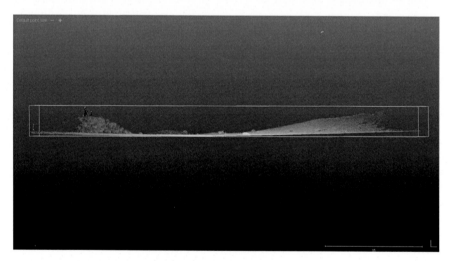

Figure 9.12. FastSLAM 2.0 terrain map generated during rover-based LIDAR trials at CSA's Mars Yard: (top) planar and (bottom) elevated views [credit Rob Hewitt, Carleton University].

maximizes the posterior probability is selected. The unscented Kalman filter (UKF) has been applied to the SLAM problem [880]. The UKF chooses $2n + 1$ sigma points with weights w_i for $i = 1, \ldots, n$:

$$\sigma_i = \hat{x}_{l|k} + \left(\sqrt{(n + \kappa)P_{l|k}}\right) \quad \text{with } w_i = \frac{1}{2(n + \kappa)}$$

where $l = k$ or $k + 1$ and $\kappa =$ degree of freedom such that $n + \kappa = 3$

$$\sigma_{i+n} = \hat{x}_{l|k} + \left(\sqrt{(n + \kappa)P_{l|k}}\right) \quad \text{with } w_{i+n} = \frac{1}{2(n + \kappa)}$$

The UKF involves computation of the square root of the state covariance matrix at each time step. Another variant on FastSLAM uses the Kalman filter for the linear part of the state model and the particle filter for the nonlinear part by partitioning the state equation (e.g., applying the particle filter to rover position estimates only while position derivatives are estimated by a Kalman filter [881]). The use of a visual vocabulary constructed from invariant features represents another possibility where scenes are represented as a collection of symbolic attributes chosen from a vocabulary of symbols [882]. Carleton University has implemented a FastSLAM autonomous navigation algorithm on its Husky test vehicle (Figure 9.10).

A neural network approach has been adopted that classifies LIDAR data in a terrain map for the Kapvik rover [883, 884] (Figure 9.11). The neural network automatically classifies laser scanner data in a simulated environment to enable creation of an obstacle grid map [886]. Neural network weights are trained by a Kalman filter that has been applied to the data association problem with an output given by [887]:

$$y_i = f(w_i, u_i) + v_i$$

where $w_i = w_{i-1} + \omega_i =$ state model, v_i and $w_i =$ measurement and state noise, and $f(\cdot) =$ neural network. FastSLAM 2.0 was deployed to estimate rover pose to input to the neural network. The mean and variance in obstacle height is also input to the neural network. The neural network was trained using a Kalman filter training algorithm on a simulation model of Martian terrain with random rock distributions but tested in real world environments. Because LIDAR scans are time consuming, the localization algorithm between LIDAR scans was a modified version of FastSLAM 2.0 [888]. Despite training in a simulated environment, the system performs well in realistic outdoor environments (Figure 9.12).

10

Rover path planning

Path planning for a rover will be online as orbiter maps will lack the resolution required for determining local rock distributions. Once the SLAM process on board the rover has created a map representing its environment, a critical capability in rover autonomous navigation is generation of the trajectory to the goal (Figure 10.1). This is the realm of path planning and execution. Several approaches are in use with differing degrees of sophistication but the most flexible is the potential field representation.

10.1 PATH PLANNING BY SEARCH ALGORITHM

The Bug algorithm is the simplest obstacle avoidance scheme in which the robot follows the contours of each obstacle in its path to circumnavigate it. Bug2 varies this by following the obstacle contour but departing it once it has direct line of sight to the goal. TangentBug adds range sensing and a local tangent graph that approaches optimal paths. The tangent graph comprises all line segments in free space connecting the initial position, goal, and obstacles. WedgeBug is modeled on the sensory field of view being wedge shaped [889]. The WedgeBug algorithm based on the TangentBug algorithm has been applied to the planetary rover navigation task. While TangentBug assumes omnidirectional viewing, WedgeBug takes account of the sensor-limited FOV and implements gaze control. A sensor detects object ranges within a wedge $W(x, v)$ of radius r and angle 2α centered in the direction of velocity v. WedgeBug is complete, correct, and uses a world model and onboard sensors for motion planning. TangentBug has a small world model of only sensed obstacle endpoints. The planner has two functions—motion-to-goal (MtG) mode and boundary following (BF) mode which interact incrementally. MtG is the typically dominant behavior and directs the robot towards the goal. If the robot encounters a local minimum in $d(x, T)$ corresponding to a blocking

© Springer-Verlag Berlin Heidelberg 2016
A. Ellery, *Planetary Rovers*, Springer Praxis Books,
DOI 10.1007/978-3-642-03259-2_10

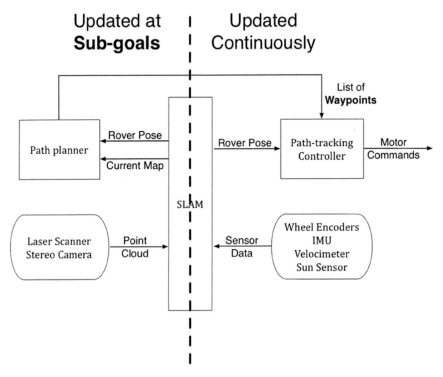

Figure 10.1. Path planner and its relation to SLAM [credit Marc Gallant, Carleton University].

obstacle, the planner switches to BF to skirt the boundary of the obstacle until the robot is back on a new subpath to the goal. The RoverBug algorithm is an extension to WedgeBug for map building which has reduced computational requirements [890–893]. It has been implemented on JPL's Rocky 7 rover testbed. Local information is used to build a local map from obstacle endpoints—obstacle boundaries are represented as contours and endpoints comprise obstacle vertices. By storing only boundary endpoints in its world model, computational resources are saved. It is based on two modes of operation—MtG (move-to-goal) and BF (boundary following)—which interact to ensure convergence. MtG is the dominant mode and a graph is constructed of all free space line segments connecting the start position, all object vertices, and goal positions. Obstacle contours are grown to account for the rover's physical size to form configuration space obstacles. The rover moves to reduce its distance from the goal directly through free space (direct mode) or along obstacle boundaries (sliding mode). This algorithm generates the shortest path to the goal through an A* search of the local graph. Rowe and Richburg (1990) described a new approach to path planning based on optical ray tracing and Snell's law with light ray paths obeying Fermat's principle in following a path between two locations in minimum time [898]. For dense obstacle fields, it requires an obstacle avoidance path-planning technique so is not suited to a planetary rover mission within a rock field.

Path planning for topological maps is straightforward as it requires simple graph searching while metric maps require discretization first [899]. There are a number of exhaustive search algoriths such as breadth first (Dijkstra's algorithm), A*, or D* (dynamic A*) to traverse graphs. The graph search uses a cost function in determining the energy consumption in following a given path. The most basic graph search method for finding the shortest route through the graph is Dijkstra's algorithm which maintains a working set of partial paths and updating the shortest distance at each node:

```
for all cell nodes v do dist(v): =∞ end for;
dist(S): =0;
Paths: ={[S]};
while Paths ≠ {} do
  select P from Paths;
  t: =head(P);
  for each neighbour n of t do
    newdist:=dist(t)+length(edge(n,t));
    if newdist < dist(n) then
      dist(n): =newdist;
      Paths: =Paths U {P+n};
    endif;
  endfor;
endwhile;
```

The algorithm allows specification of breadth-first, depth-first, or best-first searches. The best-first search expands the fewest paths, the depth-first search is the simplest, and the breadth-first search is a compromise. The Dijkstra algorithm is fine for theoretical purposes but is of limited value in mobile robotic navigation. The "find-path" problem requires determination of a continuous path for an object from an initial location to a goal location while avoiding obstacles. The Configuration or C–space approach recasts the problem by keeping the robot away from obstacles that are modeled as convex polygons. Each convex polygon representing an obstacle is expanded to accommodate the dimensions of the robot into a C-space obstacle while the robot itself is shrunk to a point. An alternative method is the generalized cylinders approach which involves fitting 2D generalized cylinders to the space between obstacles which may be stacked in 3D. The obstacle-free space path between obstacles can then be computed. The shortest path should include rotations to avoid expanded obstacles. The traditional approach to search the digital grid map for collision-free trajectories is to use an A* search algorithm on the grid-based terrain map. The A* algorithm is an extension of Dijkstra's algorithm augmented with heuristics. It expands each vertex to generate a search tree with computed costs from the start to the current position

and from the current to the goal position. This graph-traversing best-first A*
algorithm can be used to find the minimum cost solution (i.e., a minimax
principle). It is a branch and bound graph search algorithm that searches for the
minimum cost path through a tree based on a cost-to-go measure. It is effective
at dealing with average case situations though it fares less well under worst-case
scenarios. Given any intermediate node N between start node S and goal node G,
the shortest straight line path between N and G and from S to N can be
computed. By expanding the most promising paths first, the optimal path for the
minimum length between S and G:

```
for all nodes v do dist(v): =( end for;
dist(s): =0;
shortest: =∞;
Paths: ={[S]};
while Paths({} do
  select P from Paths with minimum minlength(P);
  t: =head(P);
  for each neighbour n of t do
    newdist: =dist(t)+length(edge(n,t));
    if newdist < dist(n) then
      dist(n): =newdist;
      Paths: =Paths U {P+n};
    endif;
  endfor;
endwhile;
```

The A* algorithm uses an evaluation function for ordering nodes. Path segments
can be identified using the graph-traversing A* algorithm in constructing free
space corridors representing straight line trajectories between obstacles based on a
distance-traveled cost function. The A* algorithm is an optimal path-planning
algorithm and requires a grid-based representation. It is imperative to use a metric
that minimizes energy consumption and risk. Each arc segment has the following
cost function [900]:

$$C(A_i) = K_1 C_d(A_i) + K_2 C_n(A_i) + K_3 C_m(A_i)$$

where $C_d(A_i) = LS_{\max}((1 + \Delta z/L)/(H(S_{\max} - \Delta z/L)) =$ distance and slope metric,
$C_n(A_i) = \min(n_k) =$ ground characteristic metric, $C_m(A_i) = ((1/N) \sum_{i=1}^{N} \rho_v(P_i)/$
$H(\rho_0 - \min(\rho_v(P_i))) =$ mobility metric, $\Delta z =$ height difference arc segment,
$L =$ length of segment, $S_{\max} =$ maximum slope, $H(x) = 0$ if $x < 0$ or 1 else,
$A_i = $ A* segment, $n_k =$ ground roughness, $\rho_0 =$ limit of curvature, and $N =$ pixel
volume of segment. Straight line path segments may be found by a two-step A*
algorithm at coarse resolution, then finer resolution to generate safe paths—this
two-step approach is computationally more tractable than a single A* search. If
an obstacle is detected, a local path is derived around the obstacle back to its
straight line trajectory. Schiller (2000) introduced a metric for path planning

based on the minimum time for traverse and vehicle stability [901]. A three-stage optimization process began with determining the maximum speed along a path represented as a smooth cubic B-spline mesh between control points for which the vehicle is dynamically stable. Parameter optimization and search for the best path proceeds normally. The A* algorithm is restricted to movement along the edges connecting vertices—for an eight-neighbor grid, paths are limited to $\pi/4$ steering increments. The A* algorithm may be guided by route cost based on distance to be navigated, vertical hazards, and turning maneuverability [902].

A more rapid approach is to use a D* (dynamic A*) algorithm which offers more efficient replanning than A* [903] (Figure 10.2). It is based on cell decomposition whereby each grid-based world model is built as local information is gathered by the sensors. The D* algorithm can plan paths in unknown or changing environments efficiently rather than replanning from scratch. This is particularly suited to Mars exploration where the environment is unknown initially until sensory feedback is available. The algorithm uses sensory data to build a spatial grid of traverse data and plans paths that avoid obstacles while minimizing the distance traveled. It provides planning with a reactive capability. The problem space is represented as a directed graph of robot state nodes connected by directional arcs representing costs. The D* search algorithm forms the core of the distributed architecture for mobile navigation (DAMN) [904]. The D* algorithm, like the A* algorithm, is sound, complete, and optimal but is far more efficient being some 200 times faster to compute than the A* algorithm. The D* algorithm can plan paths in unknown, partially known, and changing environments in an efficient, optimal, and complete manner [905]. It resembles the A* algorithm except it can dynamically change cost parameters during generation of a path if an unknown obstacle is encountered. Unlike the A* algorithm, it does not recompute a new path on failure but incrementally repairs paths as new information is gained. D* grid-based paths are labeled showing the cost of moving between adjacent cells, an estimate of the cost to attain the final goal, and a pointer in the goal direction. Every state x except the goal state has a backpointer to the next state to represent the path to the goal. Every state except G has a backpointer $b(X) = Y$ to the next state and the series of backpointers represents the path to the goal. The cost of traversing from state Y to X is $c(X, Y)$. D* initially computes the optimal path to all locations in the robot's environment. Similar to A*, it searches a map of traversability cost values to find the least-cost path to the goal. Both D* and A* maintain an OPEN list of states initially populated with the goal state of zero cost. The state with the maximum path cost on the OPEN list is repeatedly expanded propagating path cost computations to its neighbors. The OPEN list is populated with the minimum path cost until the optimum cost to all cells in the map is computed. The robot then follows the pointers marking out the optimal path. Whilst moving, if an unexpected obstacle occurs, then the optimal path needs to be replanned. In the case of the A* algorithm, this means beginning from scratch. For the D* algorithm, the obstacle state is updated with a very high cost, and adjacent cells are placed into the OPEN list to compute new paths to the goal. D* recomputes the optimal path by "repairing" the plans

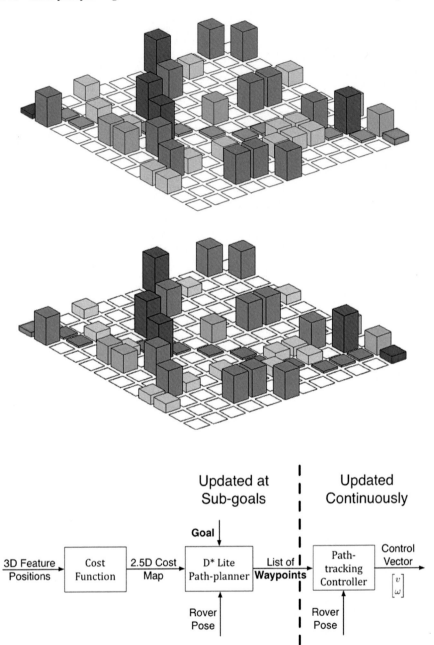

Figure 10.2. D-star lite graph search on discrete field in opposing directions; D-star lite architecture implementation [credit Marc Gallant, Carleton University].

incrementally. D* maintains an OPEN list of states and operates through two functions: PROCESS-STATE to compute optimal path costs to the goal, and MODIFY-COST to change the arc cost function and enter affected states on the OPEN list. PROCESS-STATE is repeatedly called until the robot's state is removed from the OPEN list. The robot proceeds to follow backpointers in the sequence until it either reaches the goal or finds an error in the cost function. MODIFY-COST then corrects the error. The major components of the D* algorithm are as follows:

Function: PROCESS-STATE()

```
L1   X=MIN-STATE( )
L2   if X=NULL the return -1
L3   k_old=GET-KMIN( )
L4   DELETE(X)
L5   #Reduce h(X) by lowest-cost neighbour if possible
L6   for each neighbour Y of X:
L7     if t(Y)=CLOSED and h(Y)≤k_old.
L8       and H(X)>h(Y)+c(Y,X) then
L9         b(X)=Y
L10        h(X)=h(Y)+c(Y,X)
L11    endif
L12  endforeach
L13  #Process each neighbour of X
L14  for each neighbour Y of X:
L15    #Propagate cost to NEW state
L16    if t(Y)=NEW then
L17       b(Y)=X
L18       h(Y)=h(X)+c(X,Y)
L19       p(Y)=h(Y)
L20       INSERT(Y)
L21    endif
L22    else
L23       #Propagate cost change along backpointer
L24       if b(Y)=X and h(Y)(h(X)+c(X,Y) then
L25         if t(Y)=OPEN then
L26           if h(Y)<p(Y) then p(Y)=h(Y)
L27           h(Y)=h(X)+c(X,Y)
L28         endif
L29         else
L30            p(Y)=h(Y)
L31            h(Y)=h(X)+c(X,Y)
L32         endelse
L33         INSERT(Y)
L34       endif
L35       else
```

```
L36        #Reduce cost of neighbour if possible
L37        if b(Y)≠X and h(Y)>h(X)+c(X,Y) then
L38         if p(X)(h(X) then
L39           b(Y)=X
L40           if t(Y)=CLOSED then p(Y)=h(Y)
L41           h(Y)=h(X)+c(X,Y)
L42           INSERT(Y)
L43         endif
L44         else
L45           p(X)=h(X)
L46           INSERT(X)
L47         endelse
L48        else
L49         #Set up cost reduction by neighbour if possible
L50         if b(Y)≠X and h(X)>h(Y)+c(Y,X) and
L51          t(Y)=CLOSED and h(Y)>k_old then
L52           p(Y)=h(Y)
L53           INSERT(Y)
L54         endif
L55        endelse
L56      endelse
L57    endelse
L58 endforeach
L59 #Return k_min.
L60 return GET-KMIN
L61 endfunction
```

Function: MODIFY-COST()
```
L1   #Change the arc cost value
L2   c(X,Y)=cval
L3   #Insert state X on the OPEN list if it is closed
L4   if t(X)=CLOSED then
L5     p(X)=h(X)
L6     INSERT(X)
L7   endif
L8   #Return k_min.
L9   return GET-KMIN( )
L10  endfunction
```

An extension to D*—focused D*—focuses on cost updates to reduce computational costs [906]. A heuristic biasing function effectively reduces runtime two to threefold. The Life in the Atacama (LITA) project is a robotic astrobiology approach to support robotic life detection through autonomous traverse based on the D* algorithm. The algorithm is based on estimation of the battery energy required to reach the goal as the cost function [907]. The Theta* algorithm is a variant on the A* algorithm with similarities to the D* algorithm but with close-to-minimal length paths with fewer steering changes [908]. It is not restricted to vertex edges and can generate paths in any direction.

Multiple goals may be accommodated by propagating a search wavefront from each goal. The wavefronts meet at a Voronoi edge forming a potential ridge defining a decision. A Voronoi diagram comprises edges formed by lines from all points equidistant from two or more obstacles. Voronoi diagrams require representation of a global grid. The nodes thus lie equidistant from two or more obstacles joined by lines/surfaces. These connected nodes represent a roadmap at a maximum distance from obstacles. The roadmap tends to maximize the distance between the robot and the obstacles in the map but this does not necessarily yield optimal paths. Geometric approaches that decompose objects into convex polygons are NP-hard suffering exponential combinatorial explosion and require significant simplifying assumptions in order to make them tractable [909]. Such combinatorially explosive computations cannot be resolved by increasing processing resources but require algorithms that focus on specificity rather than general algorithms [910]. Another alternative is to use a mesh of (cubic) *B*-spline curves which are parametric curves based on blending functions [911]—however, they are not well suited to rugged terrains.

10.2   PATH PLANNING BY POTENTIAL FIELD

The construction of repulsive force fields around obstacles and an attractive force field at the goal location imposes an artificial potential field on the individual rover. The potential field represents a particularly powerful single representation scheme for low-level reactive and high-level path planning by combining behaviors at each level in a sensor fusion-type process independent of sensory modality [912, 913]. Kweon et al. (1992) advocated the artificial potential field as a means of sensor fusion representation [914]. Each sensor generates its own potential field model of any objects and multiple sensor data may be combined as the superposition of repellant forces in the force field [915]. The potential field technique can incorporate behavior control while implementing a form of cognitive map. Level 0 collision avoidance involves sensing repellant forces around objects. Avoidance behaviors cause motion away from certain stimuli and attractive behaviors cause motion towards other stimuli. Each behavior may be mapped onto a potential field where repulsive forces protect objects and attractive forces guide the rover towards specific locations. Both level 0 obstacle avoidance and level 1 generalized wandering invoke random exploration behavior while avoiding obstacles. Generalized wandering behavior is an exploratory mode of behavior necessary for the construction of cognitive maps of the environment as the basis for navigation in complex environments. The addition of level 2 goal-directed exploration invokes searches for free spaces between obstacles towards target locations. Command fusion is implemented through the potential field mechanism. Plans require logical reasoning and representation capability to solve new tasks.

Path planning generates a trajectory between an initial point and a goal location. Reactive obstacle negotiation is required during execution of the plan. An implicit plan that integrates reactive planning may be represented as a gradient field towards a goal to guide actions [916]. The commonest form of weighted fusion is vector summation which is used in potential fields. Although treated as a behavior control method, the gradient field may be regarded as an internalized plan which defines the flow vector from the current location to the goal location [917]. Flow fields form ridges as wavefronts from multiple goal sites meet at a Voronoi edge— the goal selected depends on path history with respect to the ridge. A grid-based terrain map may be used to generate artificial potential fields but the potential field method does not require prior discretization. Occupancy maps may be improved by incorporating virtual force fields—each cell occupied by an obstacle exerts a repulsive force while the goal exerts an attractive force. The potential field effectively adds a path-planning component to a reflexive approach which can be performed while on the move.

The advantages of the potential field include the generation of smooth trajectories, low computational overhead, and the ability to cope with a dynamic environment. Real time obstacle avoidance and local path planning may be achieved by using the artificial potential field method in the low-level control laws [918]. It generates direct commands that are sent to the servocontroller which generates PWM signals that control the rover's servomotors. Collision avoidance is fundamentally a fast-response capability appropriate to real time control which implements dynamic steering using local feedback information to generate collision-free paths. The potential field may be incorporated as part of a computed torque control law which is used for manipulator control on a rover-mounted manipulator [919]. Indeed, a potential field similar to a postural force model of human arm motion has been proposed in which limb posture results from equilibrium in the elastic length–tension properties of the muscles [920]. Potential fields have also been applied to the problem of deploying a mobile sensor network on planetary surfaces where each mobile sensor (rover) and obstacle are treated as repulsive nodes so that the network spreads out [921]. They have been applied to hyper-redundant serial link snake-like manipulator path planning [922], to telerobotic interfaces to enhance human–robot performance [923], and to online robotic learning of appetitive and aversive stimuli whereby the approach–avoid gradient vector is used as a reward–punishment signal in reinforcement learning such as Q-learning [924]. There are a number of potential field variants including scalar electrostatic potential fields based on Kirchoff's current law developed by means of resistor networks that model the environment [925, 926]. Obstacles may be modeled as insulating solids and start/goal positions as current source–sink such that current is defined as $j = -\nabla U$ [927]. Such electrostatic approaches are not considered further.

The artificial potential field generates a gradient field across a map of the environment from the current point to the goal location. The potential field must have a global minimum at the goal point and high potential around the obstacles. The rover is considered to move through a field of forces whereby the goal position provides an attractive force for the robot and obstacles generate repulsive forces at their surfaces. The rover is attracted to the goal position and repelled by obstacles by an artificial force generated by the artificial field potential. This comprises a direction-based vector based on the negative gradient of the potential field, $F = -\nabla U$. Obstacles exert repulsive forces $F_o = \sum F_{oi}$ while the goal exerts an attractive force F_g. Sensors such as vision sensors provide data concerning the shape and location of the obstacles. Generally, a minimum distance is defined between the agent and the obstacle. The rover's dimensions may be added to those of the obstacles allowing the rover to be shrunk to a point or, alternatively, the rover's surface may be treated as an obstacle. The chief characteristic of the artificial potential field where $F(q) = \nabla U(q)$ is the linear additivity of independent vector fields—the principle of superposition of forces [928]. This vectorial addition is reliant on field irrotationality

$$\text{curl}\left(\frac{\partial U}{\partial q}\right) = \nabla \times \left(\frac{\partial U}{\partial q}\right) = 0$$

Most robot navigation systems assume that the robot traverses a 2D surface without terrain variation. This enables consideration of only translational position without regard to orientational degrees of freedom. The potential field is modeled as the vector sum of potential energy generated by an attractive field at the goal position and a repulsive field at all obstacles (Figure 10.3):

$$U(q) = U_r(q) + U_a(q)$$

where $U_a =$ attractive potential and $U_r =$ repulsive potential. The potential gradient is found by differentiating the potential function. The resultant force field

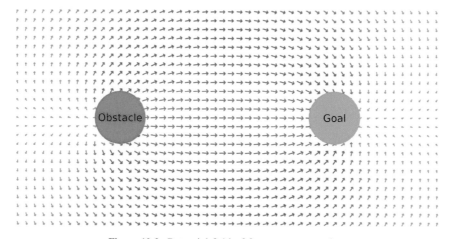

Figure 10.3. Potential field of forces representation.

determines the motion of the rover:

$$F(q) = F_r(q) + F_a(q)$$

where $F(q) = -\nabla U(q) = \partial U/\partial q$, F_a = attractive goal force, and F_r = repulsive obstacle forces.

The resultant force due to the potential gradient determines the motion of the rover which directs the rover towards the goal but away from the obstacles following a path of potential minimum (closely related to energy minimization). The vector field indicates the direction for the rover to follow (i.e., it is a gradient descent algorithm). These forces may be incorporated directly into a control law to generate actuator torques. Rimon and Koditschek (1992) suggest the addition of a dissipative function as a damping factor to damp out oscillations [929]:

$$F(q, \dot{q}) = -\nabla U(q) + D(q, \dot{q})$$

where D = dissipative field vector. This forces a loss of energy along the trajectory which allows close approach to obstacles for tasks such as docking, parts mating, etc. Ge and Cui (2000, 2002) suggest that the attractive potential field be defined in terms of both position and velocity of the form [930, 931]:

$$U(q, \dot{q}) = K_p |q_g - q|^m + K_v |\dot{q}_g - \dot{q}|^n$$

Addition of the rover velocity damping term results in smoother trajectories, places constraints on how fast an object can be approached, and reduces the undesirable effects of local minima. The chief disadvantage of the potential field method is the existence of local minima for certain obstacle configurations such as U-shaped convex configurations. This occurs where:

$$\nabla U = \frac{\partial U}{\partial x}\hat{x} + \frac{\partial U}{\partial y}\hat{y} = 0$$

Trapped states can be detected (by no progress towards the goal or limit cycling behaviors) and bring about reflexive recovery procedures such as simulated annealing and/or altering the parameters of the field (e.g., α—see later).

For the goal point, the simplest potential is an attractive quadratic well (where $K_v = 0$, $m = 2$). Goal force is given by:

$$F_g(q) = -\nabla U_g = -K_p(q_g - q)$$

where K_p = proportional gain, q_g = generalized goal position vector with respect to reference coordinates, and q = current position vector with respect to reference coordinates. The resultant force determines which direction the rover should move (i.e., a vector characterized by a magnitude and a direction). In polar coordinates, this would have the form:

$$F(r) = -K_p \frac{\Delta q}{r^\sigma} \quad \text{such that } F = \sqrt{F_y^2 + F_x^2} \text{ and } \theta = \tan^{-1}\left(\frac{F_y}{F_x}\right)$$

In Cartesian coordinates, the attractive potential is given by:

$$U_g(q) = \tfrac{1}{2}k_p(q_g - q)^m$$

This attains its minimum value of zero when $q = q_g$. This potential field varies with the exponent of distance from the goal. If $0 < m \leq 1$ the potential is conical and when $m = 2$ it is quadratic. Higher values of m yield steeper potentials (e.g., m can be varied to escape local minima). An attractive field that grows quadratically with distance represents a parabolic potential well:

$$U_g(q) = \tfrac{1}{2}k_p(q_g - q)^2$$

This is a proportional control law so damping is added for stability:

$$F_g(q) = k_p(q_g - q) + k_v\frac{dq}{dt} \quad \text{such that} \quad k_p^{\text{obj}}(q_g - q) > k_v^{\text{obj}}\dot{q}_{\text{max}}$$

This is similar to the virtual impedance approach [932]. A linear attractive force may be represented as a conic well potential of quadratic form whose gain scales as the potential well:

$$U(q) = \begin{cases} ke^2 & \text{for } |e| = |q_g - q| < s \\ 2ks|e| - ks^2 & \text{for } |e| = |q_g - q| \leq s \end{cases}$$

This generates a centrally attractive force field at large distances which was a minimum value of zero when $q = q_g$. A generalized potential field is a function of both the position and velocity of the rover. It is possible to combine these potential wells to shape them so that attractive linear behavior exists far from the goal position which transitions to a parabolic potential well close to the goal position [933]:

$$U_g(q) = K_g\sqrt{(q_g - q)^2 + w^2}$$

where $w = d_m\sqrt{1/([F_g^{\text{max}}(q)]^2)}$, $d_m = \text{limit distance}$, and $F_g^{\text{max}} = \text{maximum}$ attractive force. This gives:

$$F_g(q) = -K_p\frac{(q_g - q)}{\sqrt{(q_g - q)^2 + w^2}}\frac{\partial(q_g - q)}{\partial q}$$

The potential field method may be extended to include a rotational component which assumes that repulsion is a function of distance from the obstacle and orientation of the rover relative to the obstacle (i.e., moment on the rover). The steering angle is thus:

$$\theta = k \times M_\theta = k \times (p \times F_p) \cdot k$$

The Cartesian potential field has zero curl while the rotational potential field has zero divergence and both may be added vectorially:

$$F(q) = \sum_{i=1}^{n}\phi_i + \sum_{i=1}^{m}\psi_i$$

where $\text{curl }\phi_i = 0$ and $\text{div }\psi_i = 0$.

Obstacles may be described by the composition of geometric primitives such as a cylinder or ellipsoid:

$$F_o = \sum_{i=1}^{n} F_{oi}$$

due to the superposition property of the potential field (Figure 10.4). The repulsive potential field should follow obstacle surface contours with a limited range of influence and have spherical symmetry at large distances. It and its derivative should also be smooth. However, this makes it difficult to define repulsive force parameters to obtain constant behavior with respect to obstacles at far and close distances. There are a number of candidates including spherically symmetric repellors whose inverse cube decays with radial distance. As far as obstacles are concerned, the repulsive field generates a barrier at the convex polygonal obstacle surface which falls off rapidly with distance from the surface. GNRON (goal not reachable with obstacles nearby) is a general repulsive potential field given by:

$$U_0(r) = \begin{cases} \frac{1}{2} K_r \left(\frac{1}{q} - \frac{1}{q_0} \right)^2 q_g^m & \text{if } q \le q_0 \\ 0 & \text{if } q > q_0 \end{cases}$$

where q_t = distance from goal. The FIRAS (force involving artificial repulsion from the surface) repulsive potential field is a special case of GNRON of the form:

$$U_0(r) = \begin{cases} \frac{1}{2} K_r \left(\frac{1}{q} - \frac{1}{q_0} \right)^2 & \text{if } q \le q_0 \\ 0 & \text{if } q > q_0 \end{cases}$$

where q_0 = limiting range of potential field influence = $q/(1 - (\sqrt{m/s})v_{max} q_{max})$, q = shortest distance from rover to obstacle, m = shortest distance from object center to object surface, K_r = potential field scaling factor = constant, q_{max} = clearance width of rover, and v_{max} = maximum speed of rover. The FIRAS repulsive obstacle force is given by:

$$F_0(x) = -\nabla U_0 = \begin{cases} K_r \left[\frac{1}{2} \left(\frac{1}{q} - \frac{1}{q_0} \right) \left(\frac{1}{q} \right)^2 \dot{q} \right] & \text{if } q \le q_0 \\ 0 & \text{if } q > q_0 \end{cases}$$

where F_0 is the force inducing artificial repulsion from the surface of the obstacle. The FIRAS repulsive field can generate local minima either side of the object when combined with an attractive potential well. One way to avoid this is to impose spherical symmetry to avoid local minima generation. This implies the need for a repulsive field that follows the object surface but decays to spherical symmetry away from the object. However, polyhedral objects do not necessarily have a spherically symmetric form and the imposition of spherical field symmetry may reduce the amount of free space available for the rover to traverse. Haddad et al. (1998) proposed a more sophisticated variant from FIRAS for their repulsive

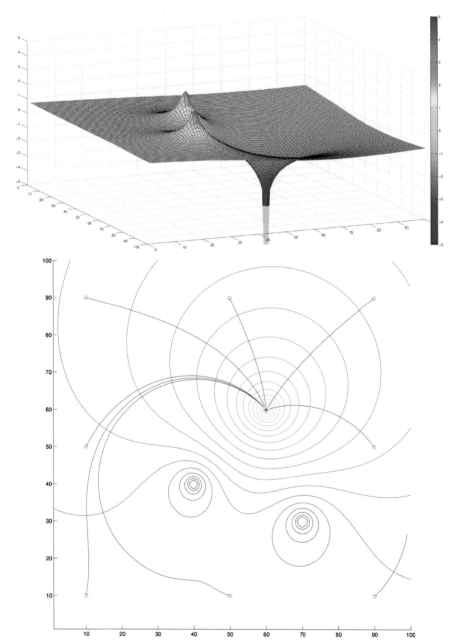

Figure 10.4. (a) and (b) Potential field representation of an obstacle field; (c, overleaf) typical terrain map generated for path planning [credit Cameron Frazier, Carleton University].

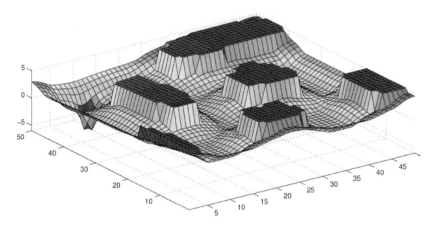

potential of the form:

$$U_0(q) = \begin{cases} \alpha^2 \left[\frac{1}{2}k_1 \left(\frac{1}{p} - \frac{1}{p_0} \right)^2 \right] + \frac{1}{2}k_2(p - p_0)^2 & \text{for } p < p_0 \\ 0 & \text{otherwise} \end{cases}$$

where $\alpha = |\sin(\Delta\theta)|$ and $\Delta\theta$ = relative angle between rover direction and nearest cell segment. This comprises summed quadratic and linear behaviors such that the quadratic dominates near the obstacle to prevent collision and linearity dominates near the limit distance to allow preavoidance of the obstacle. The limiting distance may be defined dynamically:

$$p_0(\phi) = p_m \left(\frac{1 + e^{\cos\phi}}{1 + e} \right)$$

where p_m = security distance corresponding to p_m in the static case and ϕ = angle between rover direction and heading of the closest point of the obstacle. The Yukawa repulsive potential has the form:

$$U(q) = A \frac{e^{-\alpha q}}{q}$$

which has high repulsion at short distances and exponential decay at longer distances. Parameter α determines the rapidity of decay and A is the overall scale factor. An approach function that provides zero artificial force when contact is made is given by:

$$U(q) = \begin{cases} \frac{A}{q} e^{-\alpha q} & \text{for } q \geq 1 \\ A \exp(-\alpha q^{1+(1/\alpha)}) & \text{for } 1 > q \geq 0 \end{cases}$$

Local minima (singularities) may be overcome using a background noise component to produce a low-magnitude random force vector to remove the rover from undesirable equilibrium points. A Gaussian form of repulsive field adds noise to eject the rover from potential minima. The most common obstacle potential is based on the Gaussian probability distribution:

$$U_{\text{Gauss}} = C_{\text{Gauss}} e^{-\sum_{i=1}^{n} \sum_{j=1}^{n} \frac{1}{\sigma_{ij}}((x_i - a_i)(x_j - a_j))}$$

where $C_{\text{Gauss}} =$ obstacle size constant $= K_1 R_0$, $\sigma_{ij} =$ obstacle shape constant $= K_1 R_0$, $x_{ij} =$ current coordinates, $a_{ij} =$ obstacle coordinates, and $R_0 =$ obstacle radius. Alternatively,

$$U_{\text{power}} = \frac{C_{\text{pow}}}{\sum\limits_{i=1}^{n} |x_i - a_i|^N}$$

where $C_{\text{pow}} =$ obstacle size constant and $N =$ exponent controlling decrease rate. The power of the repulsive field determines its dropoff rate. A low exponent suggests that obstacles may be detected at farther distances giving them a wider berth. This may also act to avoid potential local minima flanked by obstacles. Objects may be represented by superquadratics in conceptual space which are 3D geometric primitives derived from parametric equations of the form:

$$f(\eta, w) = \begin{pmatrix} a_x \cos^{\varepsilon_1} \eta \cos^{\varepsilon_2} w \\ a_y \cos^{\varepsilon_1} \eta \sin^{\varepsilon_2} w \\ a_z \sin^{\varepsilon_1} \eta \end{pmatrix}$$

where $(a_x, a_y, a_z)^T =$ superquadratic lengths of the superquadratic axes of the object, $\varepsilon_1 =$ object's longitude form factor, and $\varepsilon_2 =$ object's latitude form factor. The superquadratic approach avoids the problem of potential minima [934]. Superquadratic potential contours are used as 3D primitives to generate repulsive potential fields of any arbitrary shape but with limited range of influence. Superquadratics are superellipsoid shapes derived from the quadratic equation with trigonometric functions of the form [935]:

$$f(x, y, z) = \left[\left(\frac{x}{a_x} \right)^{2/\varepsilon_1} + \left(\frac{y}{a_y} \right)^{2/\varepsilon_2} \right]^{\varepsilon_1/\varepsilon_2} + \left(\frac{z}{a_z} \right)^{2/\varepsilon_1}$$

The superquadratic becomes square when form factors are <1 and rounded when close to 1 (e.g., $(0.1, 0.1)$ results in a box, $(0.1, 1.0)$ results in a cylinder, and $(1.0, 1.0)$ results in an ellipsoid). Superquadratics provide a mechanism of approaching obstacles without collision. The potential function is finite at the obstacle surface rather than infinite. This ability to approach obstacles is essential for robotic deployment devices to touch and interact with rocks of scientific interest. A superquadratic representation of objects with motion in 6 DOF space (p_x, p_y, p_y, R, P, Y) has been proposed as an intermediate-level metric representation format between 3D geometric (reactive) and symbolic (linguistic) levels of situation–action pairs for robot control to unify vision and artificial intelligence [937, 938]. Such a superquadratic may be represented compactly as a vectorial symbol encoding its object parameters and its location in the form:

$$m = (a_x, a_y, a_z, \varepsilon_1, \varepsilon_2, p_x, p_y, p_z, R, P, Y)$$

This may be generalized to a time-variable function $m(t)$ which is defined as a "knoxel". A superquadratic is a deformable parametric surface to model the isopotential surface:

$$U(q) = \left[\left(\frac{x}{a} \right)^{2n} + \left(\frac{b}{a} \right)^2 \left(\frac{y}{b} \right)^{2n} \right]^{1/2n} - 1$$

A superquadratic may be used to model a repulsive field with an object represented by at least one superquadratic:

$$U(x, y, z) = \left[\left(\frac{x}{a_x} \right)^{2n} + \left(\frac{y}{a_y} \right)^{2n} \right] + \left(\frac{z}{a_z} \right)^{2m} = 1$$

This is an ellipsoid function whose axes values a are determined by the dimensions of the object plus a margin to account for rover dimensions, typically defined as half the rover's largest dimension. For example, a surface n-ellipsoid parallelipiped of dimensions a, b, c has the form:

$$\left(\frac{x - x_0}{a} \right)^{2n} + \left(\frac{y - y_0}{b} \right)^{2n} + \left(\frac{z - z_0}{c} \right)^{2n} = 1$$

Similarly, a surface n-cylinder of cross section (a, b) and length $2c$ has the form:

$$\left(\frac{x - x_0}{a} \right)^2 + \left(\frac{y - y_0}{b} \right)^2 + \left(\frac{z - z_0}{c} \right)^{2n} = 1$$

where $n = 1/(1 - e^{-\alpha K})$ and $\alpha =$ adjustable parameter. Typically, $n = 4$ gives a good approximation. The ellipsoid function may be modified to alter its shape with distance to become elliptical at a large distance but close to the object contour close to the object:

$$K = \left[\left(\frac{x}{a_x} \right)^{2n} + \left(\frac{a_y}{a_x} \right)^2 \left(\frac{y}{a_y} \right)^{2n} \right]^{2n} - 1$$

where $n = 1/(1 - e^{-\alpha K})$ (m is identical in the 3D n-ellipsoid case) and $\alpha =$ adjustable parameter. By altering the a_x, a_y, a_z values, the n-ellipsoid can be deformed into almost any arbitrary shape (e.g., altering a_z will adjust the repulsive field height). This may be used to form the repulsive potential field by means of a Yukawa potential with $1/K$ dependence:

$$U(K) = \frac{S e^{-\alpha K}}{K}$$

where $S =$ scaling factor and $\alpha =$ gradient steepness factor. This may be used for the attractive field which must incorporate damping to absorb kinetic oscillations and have a Gaussian form at the surface so no artificial forces are experienced on contact with the object. This suggests an attractive field of the form:

$$U(K) = \begin{cases} \dfrac{Ae^{-\alpha K}}{K} & \text{for } K \geq 1 \\ Ae^{-\alpha K^{k+\frac{1}{a}}} & \text{for } k > K \geq 0 \end{cases}$$

Warren (1989) suggested an expression for determining a potential within an obstacle—this may be of use for crossing obstacles [939]:

$$U_{in} = U_{max}\left(1 - \frac{r_{in}}{r_{max}}\right) + U_s$$

where U_{max} = maximum potential, r_{in} = interior distance from center of obstacle, r_{max} = radius of obstacle to surface, and U_s = additional penalty potential for crossing the obstacle. To limit potential influence outside the boundary,

$$U_{out} = \frac{1}{2}U_s\left(\frac{1}{1 + r_{out}}\right)$$

where r_{out} = distance from obstacle boundary. Borenstein and Koren (1989) suggest employing a histogram grid representation of cells that are defined by certainty values to indicate confidence that an obstacle exists within the cell [940]. Grid resolution Δs must be related to sample period T by:

$$\Delta s > Tv_{max}$$

where v_{max} = maximum velocity. They recommend velocities to be limited to provide a damping effect if the rover is heading towards an obstacle:

$$v = \begin{cases} v_{max} & \text{for } F_0 = 0 \\ v_{max}(1 - |\cos\theta|) & \text{for } F_0 > 0 \end{cases}$$

where $\cos\theta = (v_x F_{0x} + v_y F_{0y})/|v||F_0|$. One proposal for action selection is to use m-out-of-n weighted consensus voting such that an action is selected if it has received m-out-of-n votes—majority voting is a special case in which $m = n/2$. If each behavior can vote for more than one action, such plurality will select the correct action. Koren and Borenstein (1991) highlight that potential field navigation can yield oscillations when the rover is traveling at high speed while narrowly flanked on both sides by obstacles necessitating slow speeds [941]. The vector field histogram method bears some resemblance to potential field methods [942]. It uses a 2D Cartesian grid (i.e., a histogram grid) of cells as a world model representing the environment. Each cell represents the probability of an obstacle. Detection of an object or free space adjusts the cell value incrementally according to the uncertainty level. Each active cell within the certainty grid exerts a virtual repulsive force in proportion to its certainty value (based on a probability estimate of confidence) and inversely proportional to its distance from the vehicle of the form:

$$F_{ij} = \frac{lF_{max}p_{ij}}{d_{ij}^n}\left(\frac{\Delta q}{d_{ij}}\right)$$

where $l =$ rover frontal cross section, $d_{ij} =$ scalar distance of cell from rover, $p_{ij} =$ cell occupancy certainty value, $n = 2$ for inverse square force, and $\delta q =$ vector distance of cell from rover. A virtual attractive force of constant magnitude is applied to the rover:

$$F_g = F_c \left(\frac{\delta q}{d_g} \right)$$

where $F_c =$ attractive force at goal point, $d_g =$ scalar distance of goal to rover, and $\delta q =$ vector distance of goal to rover. Superposition of the repulsive and attractive forces applies as with other potential field approaches. A vector field histogram creates a small local potential field map around the rover. The histogram grid may be defined on directional sectors to find open spaces between obstacles and generate a steering heading (and speed) closest to the target away from obstacles. Each cell in the current local region exerts a repulsive force of the form:

$$F_r = K_p c_{ij} \left(\frac{w}{d_{ij}} \right)^2 \left(\frac{r}{d_{ij}} \right)$$

where $d_{ij} =$ distance between vehicle center and cell ij, $c_{ij} =$ certainty value of cell, $w =$ vehicle width, and $r = \sqrt{\Delta x^2 + \Delta y^2} =$ distance between current location and cell. A constant virtual attractive force acts towards the target of the form:

$$F_a = K_{pr} \left(\frac{r}{d} \right)$$

where $d =$ distance between rover and target. This gives a steering command to the motors, the stability of which is determined by the ratio between the repulsive force constant and the attractive force constant. Furthermore, a smoothing interpolation polynomial is required between cells to reduce discrete changes. Labrosse (2006) [943] has modeled insect-like navigation comprising a series of short-range homing maneuvers specified by environmental snapshots along the route without feature extraction but based on color distribution distance to compute steering gradients. Although color discrimination is not considered reliable enough for planetary rovers, image-based homing would be a useful capability for certain types of rover traverses such as planetary base operations. A series of snapshot images at target locations, each associated with a motion vector, would allow precise navigation between those locations repeatably [944]. Desert ants use visual stimuli to identify goal locations by attempting to match the current retinal image of local landmarks with a stored one (snapshots) for proximity navigation close to the navigation target.

An alternative representation of the potential field is the directional vector field which defines a field including a fixed point attractor (goal) and repulsive obstacles [945, 946]. This field may be represented by a direction vector field defining the turning rate as a form of taxis:

$$\dot{\phi} = \frac{1}{\tau} \cdot \sin(\phi - \psi) + n(t) + \lambda_i(\phi - \psi_{\text{obs}(i)}) \exp \left(\left| \frac{(\phi - \psi_{\text{obs}(i)})^2}{2\sigma_i^2} \right| \right)$$

where $\psi(t) =$ goal point attractor direction, $\psi_{\text{obs}(i)}(t) =$ obstacle direction,

$\phi(t)$ = vehicle heading direction, τ = relaxation timescale of motion (related to velocity), $n(t)$ = gaussian noise, R_{obs} = distance from obstacle, d_{veh} = vehicle width, r_{obs} = obstacle radius, w_i = normalizing weight, $\sigma_i = \tan^{-1}(\tan \Delta\theta/2 + r_{veh}/(r_{veh} + R_{obs}))$ = repulsion range, $\lambda_i = \beta_1 \exp(-R_{obs}/\beta_2)$ = repulsion strength, $\psi_{obs} = \sum_i(\sin^{-1}((R_{obs} + d_{veh})/r_{obs}))$, and $\beta_{1,2}$ = global strength/global decay rate. A polar histogram represents the probability of an obstacle through the polar angle—from this a steering direction can be determined. A cost function is applied to all free openings and the lowest cost direction is selected. Huang et al. (2006) [947] have defined a potential field over steering angles based on the angular acceleration of rover heading $\ddot{\phi}$ which is a development of the Fajen and Warren (2003) model. It is defined by [948]:

$$\ddot{\phi} = k_g(\phi - \psi_g)(e^{-c_1 d_g} + c_2) - b\dot{\phi} + \sum_i k_0(\phi - \psi_{0_i})(e^{-c_3|\phi-\phi_{0_i}|})e^{-c_4 d_{0_i}}$$

where $I_G = k_g(\phi - \phi_g)(e^{-c_1 d_g} + c_2)$ = goal acceleration, $I_{O_i} = k_0(\phi - \phi_{O_i})(e^{-c_3|\phi-\phi_{0_i}|})$ $(\tan(\theta_i + c_5) - \tan(c_5))$ = obstacle acceleration, $c_5 = (\pi/2) - 2\tan^{-1}(r_0/(r_0 + r_r))$ = obstacle clearance, $\theta_i = 2\tan^{-1}(r_{O_i}/d_{O_i})$ = obstacle angular width, b = artificial damping coefficient, and $r_{0,r}$ = obstacle/rover radius. This only provides control of steering. Turn acceleration is dampened in proportion to turn rate through the damping coefficient. The goal term pulls the vehicle toward the goal with strength according to the stiffness parameter k_g, decay rate c_1, and minimum goal acceleration c_2. The obstacle term pushes the vehicle away from obstacles under the influence of repulsion from strength parameter k_0 and angular decay rate c_3. An earlier version that had a distance decay rate c_4 and assumed obstacles of zero size was replaced with obstacles of finite size determined by parameter c_5. The exponentials ensure that field influences drop rapidly away from obstacles. The potential field changes as the rover moves. Range to obstacle is not required as obstacle repulsion is based on size. Velocity control may be based on obstacle density:

$$v = v_{max}e^{-k_v \sum I_{O_i}} - \varepsilon$$

Using the default parameters, a statistical run comprising 100 runs was performed over a distance of 12.7 m at different speeds up to 10 cm/s over different types of Martian terrain.

The steering potential field controller encountered severe difficulties with the MPF rock distribution due to the large rock sizes encountered—there is a balance to be struck between halting for safety and the probability of collision which determines the navigability of rocky terrain [949, 950] (Figure 10.5). The implementation of both Cartesian and polar potential fields emulates the biological implementation of place and head direction cells in the hippocampus [951]. Artificial potential fields may also be used to represent search strategies similar to animal foraging [952]. For instance, the error signal may act as a linear or polynomial restoring force to drive an agent to its goal which comprises the equilibrium point of the force field [953]. Similarly, the potential approach may be augmented with additional capacities to enhance its flexibility [954]:

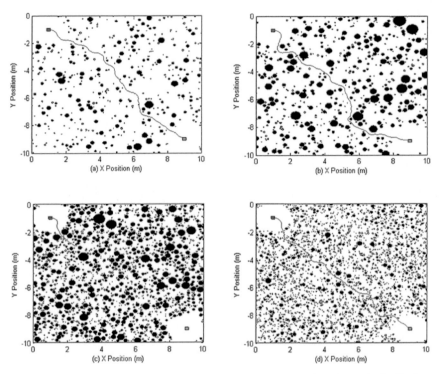

Figure 10.5. (a) VL1 landing site rock distribution; (b) VL2 landing site rock distribution; (c) MPF landing site rock distribution: $L = 20.0$, $s = 11.0$; (d) MER-A landing site rock distribution: $L = 40.0$, $s = 24.0$ [credit Adam Mack, Carleton University].

(i) exponential speed throttling according to encountered obstacle forces—this is the risk force that opposes obstacles in proportion to obstacle repulsion but exponentially to the inverse approach velocity:

$$F_{risk} = -\varepsilon K_{risk} F_{obs}$$

where $\varepsilon = e^{-c_d \cdot v}$, $K_{risk} =$ proportionality factor, $c_d =$ decay constant, and $v =$ vehicle velocity;

(ii) exponentially decaying tangential forces perpendicular to obstacle forces—this is the tangential force that avoids local minima conditions when attractor and obstacle lie along the travel path—a bias force is introduced to allow passage through tight passageways:

$$F_{tan} = \varepsilon \gamma K_{tan} R F_{obs}$$

where $K_{tan} =$ tangential force gain, $R = \begin{pmatrix} 0 & -1 \\ 1 & 0 \end{pmatrix}$, and

$$\gamma = \begin{cases} RAND(1,-1) & \text{for } v < v_{tan} \\ SGN(\lambda) & \text{otherwise} \end{cases}, \quad \lambda = SGN(v)$$

The first acts as a buffer and the second as a conduit force to propel the rover through narrow passages.

The potential field method is prone to entrapment in local minima that occur when obstacle repulsive forces balance goal attractive forces (particularly in cluttered environments). This generates oscillatory behavior under these conditions. This is particularly true in the case of concave objects and narrow corridors between obstacles. These Lagrange points occur when the rover is within the range of short-range influence of the obstacle potential field and the long-range attractive goal potential field such that

$$|F_{goal}| = |F_{obstacle}|$$

These points will trap the rover, particularly in cluttered environments. It is then necessary to impose a global approach to overcome a local minimum or alternatively add a probabilistic Monte Carlo element, or techniques such as simulated annealing, or using a grid solver that determines a potential field that satisfies the Laplace equation. Another mechanism for avoiding local minima is to ensure that

$$|\theta_t - \theta_0| < 90$$

where θ_t = rover–target direction and θ_0 = travel direction. A higher level global planner (e.g., C-space/free space approaches) derives a series of biasing interim points along a path that avoids local minima. Global planners work by searching for acceptable paths using a cost function based on the shortest distance (e.g., the A* or D* algorithm). Within the schema representation, local minima may be overcome using a background noise component to produce a low-magnitude random force vector to remove the rover from undesirable equilibrium points. The noise schema in conjunction with the Avoid_Obstacles schema generates exploratory behavior. Kim and Khosla (1992) introduced harmonic functions to eliminate local minima based on the theory of incompressible flow fields such that flow velocity is defined as [955]:

$$v = -\nabla U$$

Since the fluid is incompressible,

$$\nabla v = 0$$

Harmonic functions that are solutions to Laplace equations can implement potential field-based path planning without suffering from local minima. Substitution gives:

$$\nabla^2 U = \sum_{i=1}^{n} \frac{\partial^2 U}{\partial q_i^2} = 0$$

This is the Laplacian function whose solutions are harmonic functions that define fluid flow without vorticity. As far as harmonic functions are concerned, obstacles have potential $U = 1$ while goal regions have potential $U = 0$. These harmonic functions are calculated over a discrete Cartesian grid. Now, the diffusion equation models a disturbance (perturbation) due to the introduction of another fluid into a different fluid medium:

$$\frac{\partial U}{\partial t} = -\lambda \frac{\partial^2 U}{\partial q_i^2} = -\lambda \nabla^2 U \quad \text{such that} \quad \lim_{t \to \infty} \frac{\partial U}{\partial t} = \nabla^2 U$$

This limit is a generalized form of the Laplace equation which states that, for a steady-state condition, the second derivative of potential must be zero:

$$\nabla^2 U = \frac{\partial^2 U}{\partial x^2} + \frac{\partial^2 U}{\partial y^2} + \frac{\partial^2 U}{\partial z^2} = 0$$

Equivalently, in polar coordinates (assuming no orientation):

$$\nabla^2 U = \frac{\partial^2 U}{\partial r} + \frac{n-1}{r} \frac{\partial U}{\partial r}$$

This may be integrated to yield:

$$\frac{\partial U}{\partial r} = \frac{c}{r^{n-1}}$$

If $n = 2$:

$$U = k + c \log(r)$$

If $n > 2$:

$$U = \frac{c/(2-n)}{r^{n-2}} + k$$

where $r = \sqrt{q_1^2 + q_2^2 + \cdots + q_n^2}$. The scalar potential U of the Laplace function is harmonic through the field so does not suffer from local minima problems. By definition, any potential minimum must have a positive gradient on one side and a negative gradient on the other (i.e., a non-zero second derivative that contradicts the Laplace equation). The goal point is defined as the low potential and the Laplace function to it applied so it is the singular minimum. For example, the goal and obstacles may take the form (for $n =$ obstacles and 0 signifies the goal) with exponent $n = 2$:

$$U_i(q) = \sum_{i=0}^{n} \frac{\lambda_i}{2\pi} \log r$$

where $r = \sqrt{(x_i - x)^2 + (y_i - y)^2 + (z_i - z)^2}$ and $\lambda_i =$ strength parameter < 0 (source) or > 0 (sink). For complex environments, it is not possible to solve the Laplace function globally, so finite elements forming a mesh must be applied with a discrete Laplace difference equation applied to each node. The path of steepest descent to the goal is defined by:

$$p_{i+1} = p_i + \Delta p \frac{\nabla U}{|\nabla U|}$$

where $p_i =$ current path position, $\Delta p =$ path increment, and $\nabla U =$ potential gradient vector. Under certain circumstances, there may be situations in which control actuations should be restricted only when the rover strays a certain angle from the steepest descent path:

$$\frac{p \cdot (-\nabla U)}{|p||\nabla U||} < \cos \theta \rightarrow \text{no control action}$$

$$\text{otherwise} \rightarrow \text{control action}$$

If $\theta = \pi/2$, control action is invoked only if the potential is increasing—this will allow wandering until the path deviates such that the potential gradient enforces a control action down it. The digitized Laplace method suffers from considerable computational processing and storage requirements making it less suited to real time control. This method has been applied to freeflyer spacecraft navigation and proximity operations with other spacecraft whereby the potential function is treated as a Lyapunov function for real-time control to determine the required Δv thrusting maneuvers in the form [956, 957]:

$$v + \Delta v = -k \frac{\nabla U}{|\nabla U|}$$

where $v = $ current orbital velocity, $\Delta v = $ required velocity change, and $k = $ velocity shaping function. This is equivalent to extradimensional bypass (i.e., many local minima become saddle points in higher dimensional space [958]). The most significant problem with this approach to potential field navigation is its computational requirements which, given the severe limitations on onboard computing resources as global knowledge of the topology of the environment is required, make it infeasible for planetary rovers.

10.3 PATH PLANNING USING SIMULATED ANNEALING

Potential fields provide a generalized environmental medium by virtue of their conceptual proximity to energy measures. Environmental complexity may be quantified using potential energy metrics across both the space and time dimensions. We would expect the potential field configuration to be smooth and rolling rather than flat and pocked (this does not necessarily have to resemble the scale of the actual terrain), so simulated annealing would be expected to perform well in overcoming local minima [959]. Simulated annealing is a multiparameter optimization method for finding the minima or maxima for functions with many independent variables such as the NP-complete traveling salesman problem modeled on the physical process of annealing [960]. It is not a deterministic mechanical procedure in the definition of an algorithm but is more a strategy resembling other soft-computing methods such as genetic algorithms and offers similar performance [961]. Simulated annealing is a local search algorithm using statistical mechanics principles to find minimal cost solutions by minimizing energy. If the current position is defined by E, a neighboring position with energy E' is randomly selected until escape is established [962]. Simulated annealing provides a dynamic gradient due to generation of a disturbing perturbation at the goal which propagates through the field—in the case of simulated annealing, the perturbation is a heat source. A low-energy state implies a highly ordered state such as a crystal lattice. Simulated annealing may be used to eject from local minima analogous to phase states during cooling. For optimization problems, simulated annealing can produce solutions using effective temperature as the control variable. This technique is an optimization method for finding minima or maxima for functions with many independent variables such as the NP-hard traveling salesman problem. It corresponds to maximum a posteriori (MAP) estimation by maximizing entropy across all probabilistic equilibrium states (the MAP problem is solved using the Viterbi algorithm). Defect-free crystals are grown through annealing, heated to a high temperature, and cooled slowly until the material freezes. In the gas state, each atom is weighted by its Boltzmann probability

$$p = e^{-E_i/k_B T}$$

where E_i = energy of atom, k_B = Boltzmann constant, and T = absolute temperature. This generates a normal Boltzmann distribution of energies. At low temperatures, however, when the ensemble begins to solidify, ground states become dominant with their configurations being weighted by $e^{-N/2}$. The probability density function may be modeled:

$$p(q, t) = \frac{1}{\sqrt{2\pi}\sigma} e^{-\left(\frac{q^2}{2\sigma^2 t}\right)}$$

where $\sigma^2 t$ = variance which determines the range of the random walk. At high temperatures, the molecules of a liquid can occupy a continuum of energy states based on its temperature. As cooling ensues, they gradually lose energy and solidify to form a crystal which represents the lowest global minimum energy state. Annealing involves melting the material at a high temperature and then lowering the temperature very slowly and keeping it close to the freezing point for sufficient time to allow the system to achieve thermal equilibrium. In crystal growth, rapid cooling (quenching) at exponential rates such as at

$$T_n = \left(\frac{T_i}{T_0}\right)^n T_0$$

causes the system to get out of equilibrium and the solid will form a glass with no crystalline order but with metastable, locally optimal structures far from the globally optimal highly ordered crystalline state. If the system is cooled rapidly, an amorphous polycrystalline state is generated with local minima of higher energy states. Iterative algorithms and techniques for optimization problems are equivalent to this process of rapid quenching from high to low temperatures where the cost function plays the role of energy. However, if a slow-cooling annealing schedule is adopted, the system will retain thermal equilibrium during slow cooling and will tend towards a Boltzmann distribution without entering non-equilibrium metastable states. The annealing schedule raises the temperature and cools slowly to a low temperature to ensure global minimum is reached. Slow cooling ensures that sufficient time is spent at each temperature to ensure thermal equilibrium. For optimization problems, simulated annealing can produce solutions using effective temperature as the control variable. The simplest slow-cooling schedule is

$$T_n = \alpha T_0$$

where $\alpha < 1$ but this is unlikely to be optimal as the sampled mean and variance of cost values modeled as Markov systems at the current temperature allow the cooling schedule to adapt. This generates a global ground state with a defect-free highly ordered crystalline state. By only allowing solutions that decrease the cost function, only metastable solutions will be found. So, simulated annealing allows uphill steps in the energy cost function to jump out of local minima. Solutions that yield $\delta E < 0$ are accepted, but so are those with $\delta E > 0$. At higher temperatures, the probability of energy increase is high due to the randomness of the energy distribution, but at lower temperatures this probability is small. The probability of a solution with $\delta E > 0$ is

$$P(\delta E) = e^{-\frac{\delta E}{k_B T}}$$

A random number generator $0 < \text{Rnd} < 1$ value is generated: if $\text{Rnd} < P(\delta E)$, the solution is retained, otherwise it is rejected. This procedure avoids getting stuck in local minima since the stochastic component allows uphill transitions. Simulated annealing finds the global minimum by allowing occasional uphill jumps in energy to eject from local minima using the Boltzmann probability distribution:

$$p(E) = \frac{1}{kT} e^{-E/kT}$$

where $T = $ temperature, $k = $ Boltzmann constant, and $E = $ energy cost function. This generates a normal Boltzmann distribution of energies. High temperatures correspond to gradient descent algorithms while low temperatures correspond to stochastic searches. Simulated annealing is essentially an adaptive divide-and-conquer strategy and provides an interesting model of the way that evolution by natural selection operates. This generates a global ground state with a defect-free highly ordered crystalline state. Hence, it is inherently parallel offering the potential for fast computation corresponding to maximum a posteriori estimation by maximizing entropy across all probabilistic equilibrium states.

10.4   PATH PLANNING USING MOTOR SCHEMAS

The potential field technique may be implemented through behavior-based schemas. Schemas are psychological constructs encapsulating a generalized motor control program [963]. Schemas are generic specifications of appropriate patterns of behavioral actions codified in an organized data structure. They provide the basis for the integration of action, perception, and cognition. Motor schemas are knowledge structures (similar to but distinct from frames [964]) which include motivations for actions. Behavior schemas may be regarded as the integration of perceptual and motor schemas. Sensorimotor schemas are coarse-grained recursive modules of structured knowledge that encapsulate procedural knowledge about physical actions based on sensory data—they thus encapsulate sensorimotor behavior in the form of condition–action rules. Behaviour schemas are defined by the context C (preconditions) for invocation and behavioral action outputs A (postconditions) as goal-directed control trajectories (i.e., if–then rules). They can be anticipatory in modeling condition–action–expectation rules. They select and attend to specific stimuli for invoking their instantiation. Their instantiation is controlled by their activity level which may be determined by motivational and contextual factors. The generalized schema can be used in novel situations and combined into a multitude of different sequences of motor actions. The schema-based architecture is distributed and may cooperate or compete or both (competitive cooperation). They may excite or inhibit each other. Furthermore, instantiation of several schemas can generate emergent behaviors. Combining the potential field representation with reflexive behavior-based control provides a rich form of implementation of autonomous navigation. With the potential field approach, schemas can be fused (action fusion) simultaneously and directly. Each behavior may be mapped onto a potential field where repulsive forces project from objects and attractive forces guide the rover towards specific locations [965]. Schemas are fully compatible with the behavior-based and potential field–based approach to navigation. Integration of reactive and planned systems may also be achieved through behavior schemas that link physical motions to abstract action commands [966]. The potential field is not computed globally rather it is computed locally based on current sensory data, so the rover undergoes a continuous process of discovery through its camera observations of its immediate terrain. However, by virtue of the limitations of viewpoint, there will be shadow zones on the leeward side of obstacles which cannot be observed. However, the assumption of symmetry may be made until observation determines otherwise.

   Arkin et al. (1987, 1990a, 1990b) used a form of behavior-based control whereby behaviors are packaged into independent modules (schemas) which encapsulate all knowledge on how these behaviors are to be implemented [967–969]. Primitive behaviors were implemented as multiple concurrent and independent motor schemas capable of communicating and coping with conflicting data—this AuRA (Autonomous Robot Architecture) adopts sensorimotor schema-based planning that allows encapsulation of the perception–action process within a framework that is compatible with top-down expectation/attention and bottom-up reactive behavior mechanisms. The AURA architecture combines hierarchical plans with reactive control through schema theory. Each schema is an individual basic behavior pattern and in combination they are operated concurrently. A planner generates a plan and decomposes it into a set of schemas, the outputs of which are combined. The schema is effectively a generic computing agent. Expectations are provided by a priori knowledge about the environment and constraints on its capabilities. Hence, the schema is more encompassing and flexible than simple behavioral rules. They have gains that reflect activation levels which may be tuned through learning. AuRa possesses five major subsystems [970]:

(i) perception subsystem—this module implements external sensing and short-term memory (dynamic short-term world model);
(ii) cartographic subsystem—this module implements long term memory (essentially static a priori knowledge);
(iii) planning subsystem—this module implements a planner and executor;
(iv) homeostatic control subsystem—this module implements survival-related goals based on internal sensors modeled on the endocrine system (e.g., energy levels);
(v) motor subsystem—this module implements motor commands generated by the planner (this is the vehicle-dependent subsystem).

Short-term memory is a blackboard-type mechanism that is transitory and based on current sensory data; long-term memory is a static knowledge base structure. Potentially, learning mechanisms may be implemented to transfer persistent abstract data from short-term to long-term memory. In AuRa, the motor schema is the basic unit of behavior from which complex behaviors can be constructed. They specify appropriate patterns of behavioral actions codified in an organized data structure some of which resemble the operational modes—motion-to-goal and boundary following—of the RoverBug algorithm that generates a wedge-shaped terrain map in C-space [971]. Each motor schema contains parameters affecting the attractive–repulsive gain value, sphere of influence, and minimum range. Example motor schemas include:

- ```
  Move_to_goal: |v| is given a fixed gain value
     v direction lies towards the goal
     Gain=1.0
  ```
- ```
  Avoid_static_obstacles: Gain=4.0
     Sphere-of-influence=3.0
  ```
- `Move_forward`
- `Stay_on_path`
- `Dock_with_station`
- ```
  Move_randomly: |v| is given a fixed gain value
     v direction is random
     Gain=0.1
     Noise persistence=2 steps
  ```
- `Move_uphill`
- `Move_downhill`
- `Maintain_attitude`

Schemas are invoked by specific patterns of sensor data processed through Hough transforms for recognition—commonly used to extract edges. They provide expectations and anticipations for the appropriate motor actions—this is action-centered perception which models the potential for action inherent in specific objects or situations (Gibsonian affordances). Expectations provide a focus of attention and tracking to direct sensors to specific perceptual events (similar to active vision). Each schema outputs a single motor vector based on the current
  ```

position. Primitive schema behaviors may be combined to yield more complex behaviors such as obstacle avoidance, etc. Output vectors from each activated motor schema are summed and normalized to give speed and steering actions. Motor schemas act as a dynamic cooperative network of active behavioral schemas that change as perceptual sensing of the environment changes—this is Arbib and Liaw's (1995) Principle 1 [972]. In animal ethology, these networks of schemas implemented neurally coevolve to yield successful behaviors [973]. Reactive navigation of a rover emerges through instantiation of these motor schemas. Each schema outputs a potential force field vector. The `Avoid_ Obstacle` schema is given by:

$$
|v| = \begin{cases} 0 & \text{for } d > S \\ \dfrac{S-d}{S-r} \times G & \text{for } r < d \leq S \\ \infty & \text{for } d \leq r \end{cases}
$$

where S = sphere of influence of obstacle from obstacle center, r = radius of obstacle, G = gain, and d = distance of rover from obstacle center. The v direction is defined along the line from the rover to the centre of the obstacle moving away from the obstacle. An alternative version is an exponential function:

$$
|v| = \begin{cases} 0 & \text{for } d > S \\ (e^{\frac{S-d}{S-r}} - 1) \times G & \text{for } r < d \leq S \\ \infty & \text{for } d \leq r \end{cases}
$$

The `Move_to_Goal` schema is simple: $v = G$ in direction of goal; noise may be added thus: $N = G$ in random directions that persist for N steps. The `Path_ Following` schema is given by:

$$
v = \begin{cases} P & \text{for } d > w/2 \\ \dfrac{d}{w/2} \times G & \text{for } d \leq w/2 \end{cases}
$$

where w = path width, P = off-path gain, G = on-path gain, and d = distance of rover from path center. Each schema has an activation level that acts as a threshold for instantiation and this activation level may be controlled by altering the gains which provides a form of attention focusing. These schemas (behaviors) may be combined through disjunctive chaining and parallel conjunction (non-conflicting) to form complex controllers. These control trajectories resemble artificial potential fields that represent goals and pathways but the process of combination differs from weighted superposition. Fuzzy logic allows the use of logical operators for the combination of elementary behaviors. Fuzzy logic can represent both symbolic and numerical aspects of reasoning to integrate between high-level and low-level planning. If–then rule associations may be learned through reinforcement [974]:

$$
w_{i+1} = w_i + \eta r_i \bar{\mu}_i
$$

where η = learning rate and $r_i = +1.0$ to -1.0 = reinforcement signal. Each schema is independent and activated in parallel without layering and their potential field outputs are summed according to their weights. Summing requires communication to resolve conflicting forces and this may be accomplished through a blackboard. This essentially provides a world model/cognitive map which is a blackboard that uses behavior control modules as its parallel-invoked knowledge sources [975].

Motor schemas drive the rover to interact with its environment to satisfy the goals generated from a planning system. A neural-inspired schema has been developed based on behavior being resultant from the concurrent activity of interacting behavioral units [976]. Each schema is implemented as a single neural net forming a network of neural nets that interact with each other. Perceptual schemas are specialized to sensory inputs, motor schemas are specialized to action outputs, and sensorimotor schemas integrate sensory inputs and action outputs. Lyons and Hendricks (1995) also use a schema approach to integrating planning and reactive robotic systems in which the schemas represent a network of concurrent interacting schemas [977]. The implementation of local spatial memory to avoid the past offers a further development in the avoidance of local minima [978]. Spatial memory of the environment may be defined as a 2D array of integers representing the number of times each cell has been visited. This resembles the laying down of pheromone trails by ants to aid foraging. Spatial memory is implemented within Avoid_Past and Past_Mapper schemas. Past_Mapper stores and updates the array but does not contribute a motor output. Avoid_Past compares the map with the current location and computes a motor vector (similar to a gradient computation in image processing) to move away from previously visited areas, the repulsive force being dependent on the number of visitations. Avoid_Past gain (typically $G = 3.0$) must be higher than that of the Move_to_Goal schema. The longer the rover resides in a local minimum, the more times it logs the location as visited increasing the repulsive field generated until it escapes. A decay mechanism may be added to ensure that the rover is not repelled by the global minimum.

Moorman and Ram (1993) extended the schema-based potential field approach by including a simple form of case-based reasoning whereby successful previous behaviors are favored under similar conditions which led to that success [979]. In particular, case-based reasoning defines gain values to determine the weighting and variations in variables such as sphere of influence, thresholds, distance definitions, etc. based on most successful strategies for similar situations in the past. This requires storage of instantiations of schemas in a case library to generate more specific cases:

- Clear_Field: increase goal gain and reduce noise;
- Ballooning: increase obstacle gain to increase wander;
- Squeeze: reduce obstacle gain and increase goal gain for narrow passages;
- Hugging: increase boundary-following gain;
- Randomize: increase noise levels to increase random wander;
- Repulsion: decrease goal gain to negative values to move away from it temporarily.

This type of addition allows the method to overcome box canyon potential minima situations. Arkin and Balch (1997) added a deliberative hierarchical planner that used a free space graphical map searched by the A* algorithm to generate route plans [980]. Reif and Wang (1999) introduced spring-based force control laws rather than inverse square force control laws [981]. Potential fields are consistent with behavior control methods similar to the "feelforce" module of the subsumption architecture that summed repulsive fields from detected objects [982]. Potential field representation is fully consistent with the dynamical approach to robotics advocated by Beer [983, 984], in which emphasis is placed on robot–environment interaction rather than the processing of information. The potential field approach is consistent with dynamical systems theory and the analysis of attractor states [985]. The schema-based potential field approach to autonomous navigation lends itself to modular implementation that maximizes its flexibility and upgradability. Specific modules may be created that serve specific action-oriented perceptive and motor functions such as traction control while negotiating obstacles. Motor schemas generate appropriate actions according to the situation and are implemented as potential field vector generators. Each schema links perceptual situations to physical motor actions. Recognition of objects and planning emerges from the interaction of multiple schemas that convert appropriate motor actions into more complex motor behaviors. The integration of high-level planning and low-level reactive architectures is considered essential for true autonomy [986]. The key is action selection between different behaviors through the fusion of behavior commands rather than sensor fusion. They used a connectionist network framework with multiple weighted fine-grained behaviors which computed an activation level that determines whether an output is generated. The basic obstacle avoidance and directed navigation behaviors were divided into finer scale behaviors—trajectory selection and trajectory speed (for obstacle avoidance), and goal pursuit and gradient speed (for directed navigation) behaviors. Action selection and composition are defined by weights on the links between behaviors. The (potential) gradient field was used as an internalized plan to propagate a search wavefront simultaneously from multiple goals to generate the planned trajectory. Nikolos et al. (2003) used the inverse potential field force as part of a fitness function for a genetic algorithm for path planning with obstacle avoidance [987]. Behavior schemas provide the basis for integrating both reactive and planning capabilities [988]. Goals and behaviors can thus be combined into compositions of schemas to yield complex control schemes. They suggest that multivalued logics (such as fuzzy if–then rules) may be used to implement more abstract control schemas that interact through conjunction and chaining to form composite behaviors.

10.5   EXPLORATION BY MULTIROVER TEAMS

Planetary exploration may be enhanced through the deployment of cooperating teams of rovers. Multiple rovers operating in cooperative teams offer improved performance capabilities in achieving tasks beyond that of single rovers, more rapid task completion, greater mission redundancy and graceful degradation to faults inherent in the redundancy offered by multiple platforms, and enhanced scientific measurement mapping. As far as scientific exploration is concerned, the capacity for distributed measurement is a particular attraction. Multirobot coordination is required when there are interdependencies between different activities [989]. Key capabilities include distributed cooperative control [990, 991]. Hierarchic social entropy is an information-theoretic measure of a robot group's behavioral diversity (number and size of types of agent) [992]. Larger groups are more efficient at tracking environmental changes due to their greater capacity for social interactions which facilitates rapid allocation of individuals to different tasks [993]. The control of a cooperative team of N robots to achieve spatially distributed tasks may be modeled dynamically as a series of differential rate equations dN/dt but this has limited utility for planetary application [994]. Although not directly applicable to planetary exploration, RoboCup has several similar goals in attempting to implement robot teams in a dynamic game with standardized parameters [995]. The planetary exploration task, however, does bear many similarities to the problem of foraging for food [996] (in this case, searching for targets of scientific interest). Cooperative map construction of an unknown environment during multirover exploration is a highly useful and fundamental capability [997, 998]. Maximum likelihood methods may be used to determine the relative positions of all rovers from their previous positions and their distances from each other. One example of a collective localization approach involves a single centralized Kalman filter that exploits sensor data from distributed robots shared through a communications network [999].

There are several types of organizational control approaches relevant to multirobot teams. In centralized control, a single leader agent is responsible for the control of all other follower agents. The chief advantage is that global information, planning, and decision making may be implemented. Distributed control implements locally autonomous control independently in each agent which is commonly favored over centralized control [1000, 1001]. Explicit communication is not required for cooperative behavior if decentralized control is adopted [1002, 1003]. Hierarchical control is decentralized among agents but involves a centralized chain of command. A variation on this is to implement a degree of local autonomy but with higher level coordination. The problem of communication falls into three classes [1004]: (i) limited interrobot interaction through the environment itself without explicit communication (e.g., behavior-based approaches); (ii) local interrobot interation through passive sensing of each other without explicit communication (e.g., CEBOT—cellular robotic system); (iii) explicit communication between robots either directly or through broadcast (e.g., task allocation through the contract net protocol). All sociality within groups of interacting agents emerges from individual behaviors [1005]. Behavior-based approaches to group coordination have often been proposed as superior to model-based approaches but there is little demonstrative evidence for this assertion [1006]. However, distributed artificial intelligence methods focus on negotiation and cooperative coordination of multiple agents through direct communication

but these techniques are often slow. There are three design principles required for
simple behavior-based systems to realize complex cooperative goals using multiple
rovers: (i) minimization of resources required for the task at hand; (ii) statelessness
defining the degree of reactivity necessary without implementation of internal
states; (iii) tolerance to uncertainty.

Basic individual behaviors provide the basis for higher level group behaviors
that minimize interference between individuals generated by competition for
shared resources. The most common form of social relation that minimizes interference is the dominance hierarchy. Interference may also be minimized using
other social rules. Behavior-based methods do not require a centralized conroller
but rely on the interactions between individual agents. Social behavior-based
control is a decentralized control approach for multiple rovers: goal behavior is
location attractive, avoid behavior is obstacle repulsive, follow behavior is attractive to other rovers, slow behavior reduces speed to prevent rover collisions, and
find behavior which is exploratory and is the lowest priority behavior. Follow and
slow behaviors are specific to interrobot interactions. Flocking behavior in simulated birds (boids) may be maintained using a combination of simple behaviors
including [1007]:

(i) repulsive collision avoidance by maintaining a minimum distance from the
nearest boids;
(ii) match velocity to adjacent boids;
(iii) stay close to the flock center through attractive forces.

A more sophisticated minimalist set of basic behaviors for multirobot group
behaviors include (Obstacle) Avoidance, (Herd) Following, Aggregration,
Dispersion, Homing, and Wandering [1008–1011]. Each behavior is invoked by
environmental conditions as measured by sensors. Avoidance (of obstacles) and
Following (targets) imply the ability to discriminate between the two. These
behaviors may be combined by summation or sequencing to generate more
complex emergent behaviors. One of the simplest composite behaviors is `Safe_
Wandering` which comprises Avoidance and Wandering. Flocking and Foraging
are two more complex higher-level behaviors: (i) Flocking results from the parallel
summation of Avoidance, Aggregation, and Wandering (the addition of Homing
implements goal-directed migratory flocking); (ii) Foraging results from temporal
switching between Avoidance, Dispersion, Following, Homing, and Wandering.
Unlike Flocking, movement in formation requires the group to maintain a specific
shape (such as a column, spearhead, or diamond) [1012, 1013]. Navigation in formation may be implemented through a system of nearest-neighbor tracking which
by definition requires interrobot communication [1014]. Information foraging
theory is premised on the search for information rather than food with a performance metric based on time minimization or resource maximization [1015]. Holling's
disk equation quantifies the rate of information gain:

$$R = \frac{\lambda G - s}{1 + \lambda H}$$

which is maximized at

$$R = \frac{G}{H}$$

where λ = rate of information encountered during search (depending on environmental richness), G = information intake rate, H = information-processing cost rate, and s = search cost rate (power consumption). From this, Charnov's marginal value theorem states that a rover should remain in an information patch as long as the marginal rate of gain within the patch is greater than the overall rate of gain when navigation time of $1/\lambda$ is factored in (i.e., the prevalence of information patches). Optimal foraging theory has been applied to human food-gathering practices that maximize the net rate of return of energy per search time [1016].

Potential fields of force with attraction/repulsion can be used to coordinate fleets of rovers and exhibit swarm intelligence and formation behaviors through self-organizing principles [1017]. In this case, the potential force field has zero intensity for distances r (distance between surfaces of rover i and j) larger than a predefined threshold (replacing $1/r^i$ or exponentially decaying functions)—this eliminates the incidence of local minima. For collision avoidance, the short-range repulsive field is given by:

$$F_{\text{rep}}(r) = \begin{cases} \left(\tan\left[\frac{\pi}{2}\left(\frac{r}{\sigma} - 1 \right) \right] - \frac{\pi}{2}\left(\frac{r}{\sigma} - 1 \right) \right) & \text{for } 0 < r \leq \sigma \\ 0 & \text{for } r > 0 \end{cases}$$

where σ = maximum range parameter. This has been proposed for direct application to coordinate space vehicle fleets [1018]. Such social potential fields can be implemented to reflect social relations between rovers such that any rover's motion is the result of artificial forces imposed by other rovers and obstacles. A variation on this is schema-based behavior control which offers dynamic and flexible fusion of basic perception–motor modules through weighted summation and temporal sequencing as part of potential field representation [1019]. Schemas such as Obstacle_Avoidance, Move_To_Goal, and Maintain_Formation as well as noise can be combined to yield multirobot formation behavior towards a target [1020]. The Maintain_Behaviour schema determines interrobot separation distance and formation speed. Social potential fields can exhibit group behaviors such as clustering, guarding, escorting, patrolling, etc. Potential field modeling interaction forces between individuals of a flock have been used to model flock behavior in response to predator attack [1021]. They can also charac-terize a game-theoretic utility function representing an appropriate mechanism for multiagent cooperation [1022]. Neural fields have been proposed as a means to coordinate the movement of multiple rovers in formation [1023]. Neural fields bear some similarities to a discretized polar potential field and provide for neural implementation of behaviors as the attractors of dynamical systems [1024, 1025]. Bird flocking may be characterized by Steer_and_Avoidance behavior com-posed of more basic collision avoidance, velocity matching, and flock centering.

Although single-ant behavior can exhibit chaotic behavior [1026], social insects
have been used as models for collective intelligence using simple agents with
simple behavioral rules [1027–1029]. Several types of coordinated group behavior
are readily replicated using simple algorithms [1030]: path planning, brood sorting,
nest building, task differentiation, flocking, and hunting. Five simple mechanisms
suffice to generate group behavior without centralized control or the use of explicit
communication: common goal-finding behavior with obstacle avoidance, follow
herd behavior, environmental cues (matched filters) to control behavior transitions, task-specific behavior instantiation when within group, and stimulation by
group broadcast [1031]. The use of adaptive logic networks (binary Boolean
neural networks) allows sophisticated construction tasks to be performed [1032].
The swarm behavior of multiple rovers may be implemented through virtual pheromone messaging implementing a distributed computing approach. Bacterial cells
similarly generate and detect signal molecules (such as N-acyl-homoserine lactone
in gram-negative bacteria) which provide the medium for coordinated action
amongst bacterial cells once a critical population density is reached (quorum
sensing) [1033]. Quorum sensing is involved in a wide range of microbial activities
such as microbial biofilm differentiation, swarming attacks on prey, virulence
factor production, and fruiting body differentiation [1034] Pheromones are used
by ants and termites to communicate and coordinate their joint activity simulating
a shared blackboard. Pheromones emit chemical messages in all directions through
diffusion. Artificial pheromone messaging (or beacons) between neighboring
robotic agents may be adopted where distances are measured from received signal
power to allow gradient following similar to a potential field gradient [1035]. Pheromone gradients provide information about the location of neighboring agents.
Virtual pheromones may be characterized by a field of view defined by a beamwidth and vector direction to implement directional messaging. They propagate
locally generating a local gradient and invoke simple attraction–repulsion behaviors. Gradient following is the most basic of behaviors that guides rovers to a
particular location or region. This model of self-organization emerging from local
interactions of individual organisms with simple behaviors is inspired by social
insects (e.g., termites) which build extremely complex structures without global
control but which emerge from individual insects' limited behavioral response to
local pheromone concentrations [1037, 1038]. Termite pathways are determined by
the superposition of random walk and chemotactic behavior. Ants can solve a
number of mathematical problems such as shortest path, traveling salesman, and
ant colony optimization. For instance, scout worker ants can measure irregular
spaces much larger than their sensing range in order to assess the suitability of
cavities as nests [1039]. They use Buffon's needle algorithm (geometric method for
determining π) to estimate the spatial area of potential nests by secreting pheromones while exploring the periphery of the nest with random excursions into the
nest interior [1040]. Buffon's needle algorithm exploits random dropping of a
needle onto lined paper with spacing the same as the needle length—the number
of times the needle touches a line provides an estimate of π ($= 2N/n$ where
N = total number of drops and n = number of times the needle touches a line).

For ants, over multiple visits, they lay an initial pheromone trail and measure the intersection frequency of each subsequent random exploration path.

The emergence of complex social behaviors in insects requires three properties: (i) complex temperospatial patterns from homogeneous initial conditions; (ii) existence of multi-stable states due to amplification of random initial heterogeneity; (iii) existence of bifurcations in dynamic response to small parameter changes. There are many variations in termite mound design according to local conditions but all are designed to maintain constant environmental conditions. Termites build structures through self-organization based on chemical cues from pheromones [1041]. Self-organization involves the emergence of collective spatiotemporal patterns from the collective interactions of simple individual insects. It involves both positive and negative feedback, multiple interactions between agents, and the amplification of random fluctuations (such as natural gradients in temperature or humidity). There are several properties of self-organization: (i) positive feedback (amplification); (ii) negative feedback to counterbalance positive feedback); (iii) amplification of fluctuations; (iv) multiple interactions. This creates the spatio-temporal structures in which several stable states can coexist and in which bifurcations can occur. The ant or termite randomly transports and drops soil pellets impregnated with pheromones to build pillars. The probability of an ant releasing its pellet at a distance r from an arbitrary origin forms a Gaussian ring of width σ and $r = \sigma + \rho$ from the origin:

$$dp = \frac{1}{\sqrt{2\pi}} e^{-(r-\rho)^2/2\sigma^2} \, dt$$

One fluctuating pheromone source may initially be randomly favored exceeding a critical size. This becomes amplified through recruitment of other ants due to the greater attraction, thereby reinforcing the pheromone signal. Ants or termites are attracted over larger distances by stronger hormone concentrations that become further reinforced. Social insect constructions built by decentralized simple creatures exhibiting only simple stimulus–response behavior can be highly complex— termite mounds exhibit a complex internal alveolar structure. Complex ant nests are constructed through the simple rule-following behavior of individual ants by cumulatively depositing small pellets of soil. Their stimulus–response nervous systems generate attractive/repulsive behaviors that interact to form morpho-genetic patterns [1042]. The soil pellet is impregnated with pheromones that diffuse into the air forming a pheromone gradient. This attracts other ants and stimulates them to deposit more soil pellets close to the original pellet increasing the concentration of pheromones. These locations will grow into pillars as more ants deposit pellets. The nest begins with the construction of a peripheral wall broken by entrances [1043]. The nests of the African termite subfamily Macro-termitinae include brood chambers, pillars, bulwarks, spiral cooling vents with ribs and vanes, fungus gardens, and royal chambers. Termites acquire and deposit soil pellets masticated with pheromone that is deposited at locations where phero-mone concentrations are maximum. The termite queen continuously emits pheromones. Convective airstreams driven by the thermal output of termite bodies drives pheromone molecules in specific directions. The pheromone concentration gradient is described by:

$$\delta C = k_1 P - k_2 C + D_C \nabla^2 C$$

where C = pheromone concentration, k_1 = amount of pheromone emitted per unit of deposit per unit time, P = amount of active deposit, k_2 = pheromone decay constant, and D_C = diffusion coefficient. The construction process requires a minimum critical density of termites below which no construction occurs. This may be described by a Turing-like dynamic between short-range excitation (positive feedback) and long-range inhibition (negative feedback) of activity:

$$\frac{\partial c(x,t)}{\partial t} = \nu \left(k_d a + \frac{\alpha_1 a \phi_c}{\alpha_2 + \phi_c} - \frac{\alpha_3 \rho c}{\alpha_4 + \phi_c} \right)$$

$$\frac{\partial a(x,t)}{\partial t} = -\nu \left(k_d a + \frac{\alpha_1 a \phi_c}{\alpha_2 + \phi_c} - \frac{\alpha_3 \rho c}{\alpha_4 + \phi_c} \right) + D \nabla^2 a$$

where a = concentration of ants, c = concentration of pheromone-laden soil, ν = ant velocity, k_d = pheromone-laden soil deposition rate per ant, ρ = density of non-soil-carrying ants, and D = diffusion coefficient. The second and third terms represent density-dependent soil-dropping and removal rates. This has been applied to automatic construction of structures with robotic agent behavior determined by pheromone rules based on concentration which decays with distance according to an inverse square law [1044]. The ant colony system is a distributed algorithm that is capable of solving the traveling salesman problem which is known to be NP-hard [1045, 1046]. It involves a set of cooperating agents using indirect communication mediated by a pheromone signal that involves a form of reinforcement learning [1047]. Agent cooperation without central arbitration has also been modeled through emotions that implement conflict resolution to modify individual behaviors to monitor goals represented as motivations [1048, 1049]. It has been proposed that aggressive competition will reduce incidence interference and enhance team efficiency [1050]. "Fight" (based on the fear threshold) implements conflict resolution between multiple robots attempting to acquire the same resource. This breaks deadlocks through invoking "brave" behavior (greater than threshold) which continues the standard directed navigation behavior while "panic" (less than threshold) behavior invokes a random dash to open space. The fear threshold may be determined through a dominance hierarchy, random assignation, proximity to the resource (goal), or a function of past history of successful aggressions (either positive or negative). The latter exhibits either dominance hierarchy (increasing aggression with success) or turn taking (decreasing aggression with success).

Ant behavior—in particular, the prey–transport task by ant swarms—inspires the box-pushing task for groups of cooperative rovers [1051–1053]. Worker ants are recruited by pheromones to cooperate in the transport of large prey [1054]. For the box-pushing task, the behavior modules are similar to the standard behaviors in a subsumption architecture: Find to move the rover forward, Follow to move the rover to its closest-neighboring rover (keeping formation), Slow to reduce the rover speed on approach, Goal to locate the rover towards the box, and Avoid (highest priority) to move the rover away from obstacles (including other rovers) [1055]. Stagnation and cyclic behaviors are potential problems associated with the lack of memory. Environmental cues are the key to controlling a chain of sequential single-step behaviors by a group of rovers to define a collective task. Environmental cues control the transition between each behavior for each rover. The environmental cues may be differentiated into Obstacle_ Detection_Sensor, Box_Detection_Sensor, and Goal_Location_ Sensor. These are the transitions between three transport states: Find_Box, Move_To_Box, and Push_To_Goal. An adaptive logic network is a binary neural network that may implement simple combinational Boolean logic to classify such environmental cues. The box-pushing task may be used to construct an arch through three steps [1056]: (i) construct a freestanding pillar of blocks; (ii) construct an adjacent identical pillar: (iii) place a beam over the top of the two pillars. This leads to the construction of termite nests through a series of simple building steps (essentially blind bulldozing), each transition invoked by specific stimulus cues. This underlies the medium of Grasse's stigmergy without direct communication. Stigmergy is the indirect communication between individual social insects in which a specific stimulus triggers a specific action which has a cascade effect on both the environment and other social insects. The construction of termite nests involves a combination of self-organization and stigmergy. Group behavior (allelomimesis) is mediated by (chemical or visual) communication in social animals—positive feedback amplifies fluctuations through behavioral reinforcement enabled by the recruitment of others (individuals in a group copy each other's behavior) [1057]. This self-organization process may be augmented by stigmergy in which nest construction depends on the nest structure and the reaction of ants to that structure. A specific local environmental configuration of a nest construction stimulates a specific response in the ant. This response is a specific building action that transforms the current stimulating configuration into another configuration—this target configuration in turn can trigger another building action. Since the ant's actions change those environmental stimuli, subsequent actions will be altered. Different configurations trigger different actions. Stigmergy refers to the mechanism whereby one individual's actions influence the successive actions of other individuals indirectly through changes in the local environment. The dynamically evolving shape of a construction stimulates ant responses that may transform the the construction into a new configuration. This type of mechanism of ant trail laying has been applied to routing and load balancing in telecommunications networks [1058]. More complex cooperative tasks are also possible [1059–1061].

Automated construction based on social insect building behavior has been demonstrated using small groups of walking climbing robots by depositing polymer foam to construct arches and walls [1062]. Swarms of robotic bulldozers can implemet blind bulldozing to emulate the creation of ant colonies [1063]. This is particularly applicable to site preparation through the clearing of debris from a region by plowing outwards. Stigmergy controls the bulldozing process through a simple finite state machine (collision avoidance—plow in straight line—finish and random turn) representation. Assming that each robot acts independently, the final nest size is given by:

$$r = \frac{F_{ext} + \sqrt{F_{ext}^2 + (F_{fr}\rho w r_0)^2}}{F_{fr}\rho w}$$

where ρ = density of material moved, F_{ext} = robot applied contact force, F_{fr} = surface friction force, and r_0 = initial nest radius. Another demonstration used simple prefabricated bricks to build structures using robots capable of moving in a plane, climbing up/down a step the height of one brick, and picking up and attaching one brick in front of the robot (using a simple arm/gripper) [1064]. The structure constructed from the bricks provided the basis for a 3D coordinate grid reference from the first brick laid. The approach essentially discretized the construction process—the square bricks possessed self-alignment and magnets for attachment. The robots could build brick staircases to climb to higher levels (using whegs) but were restricted to only local sensing of patterned bricks (through active infrared sensors for brick detection). Ultrasound sensors detect the structure and other robots. All communication was implicit through the environment exploiting stigmergy. Thus, robots added bricks to the structure according to the current configuration. Each robot is programmed with the same representation of the target configuration allowing a set of structural pathways to be derived. All paths begin with the brick at the origin and follow a spiral in the same direction around and on the structure, deploying the brick appropriately and exiting the structure to acquire another brick.

Within rover teams, the question of division of labor arises between generalists who can perform all tasks and specialists who can perform only a subset of tasks [1065]. This is similar to the Prisoner's Dilemma problem. The optimal distribution divides tasks equitably amongst large specialized groups with increased population size [1066]. Hence, increased specialization of tasks amongst a population of rovers is favored. Furthermore, increasing behavioral heterogeneity—division of labor—between members of a group reduces the required control effort to achieve a given task [1067]. Role assignment may be dynamic and may be as simple as assigning one rover as the leader while the others are assigned as followers [1068]. Dynamic allocation of tasks within a rover fleet attempting to satisfy multiple goals under uncertainty implies that the best allocation strategy will change with prevailing circumstances and associated noise levels. Such dynamic task allocation may be implemented by a market-based auction system through a shared blackboard [1069, 1070]. Other market trading-based approaches are also feasible that use supply-and-demand to coordinate the rovers—a market is globally efficient in representing a Pareto equilibrium while each agent acts to maximize its own profit. Three popular examples of such economic approaches are the auction [1071, the contract net protocol [1072], and the Clarke tax mechanism [1073]. Negotiation mechanisms such as auctioning may be implemented through the blackboard architecture. A conventional approach to cooperative transportation of a large object using multiple rovers may be based on a global path planner and a local manipulation planner [1074]. The global planner was based on a reduced configuration space (C-space) representation (reduced by constraining the object-relative posture during manipulation) searched using a local potential field while the constraints of object manipulation were incorporated as the cost function of an A* search.

To date, multiple coordinated rovers have yet to be deployed but future missions, particularly those concerned with preparing for and aiding human exploration missions will require multiple rovers for a diverse array of functions. Although Mars is the primary deployment site of interest here, similar arguments apply to lunar exploration [1075]. For instance, a lunar facility should include a power plant, mining facility, product-manufacturing facility, consumable extraction facility, orbital shipping industry, surface transport system, solar panel production facility, repair facilities, and warehousing. It is envisaged that initial deployment of infrastructure assets to the surface will be in the form of outposts that will evolve into fully functional bases. For civil construction, recommended minimum compressive strength is $1.75\,\text{N/mm}^2$. Lunar regolith is an efficient thermal insulator so that temperature at a 0.5 m depth is 240 K despite large ambient temperature variations—it can be plowed as a thermal insulator for a static planetary base. Inflatable structures shielded by regolith are commonly proposed though lunar concrete is also sometimes suggested. This requires extensive ground preparation, deployment of the inflatable textile, and pressurizing and rigidizing the structure. Many approaches to lunar base construction are founded on the use of prefabricated modules (such as derived from space station–type designs) emplaced onto the lunar surface, linked into a complex, and partially or fully buried in the regolith. This will require substantial site preparation. Finally, local regolith may be bulldozed over the top of the modules as a radiation shield. This requires the doming of the modules to be shallow or the implementation of a standoff envelope supported by beams and columns over the modules. This will require significant construction capabilities. The most fundamental tasks involve rover vehicles as multipurpose devices for manipulating and handling regolith. An outpost—be it static or mobile—is a human–robotic facility that can support medium-duration habitation (nominally 6 months). For example, an initial Mars base might be based on the 8 m diameter Mars Arctic Research Station at Haughton Crater which comprises two decks [1076]: an upper deck is used as the living area with sleeping and kitchen areas; a lower deck used for rover stowage and maintenance, life support systems, and science laboratories. This may grow into a 10-person base station operating for a nominal 1,200-day mission at the Martian pole which will consume 92 tonnes of material and 61 kW of power [1077]. This requires a surface area of $40\,\text{m}^2$ per person and a modular form is favored. Each module is a disk-shaped membrane structure of Kevlar/Mylar/MLI of mass $3\,\text{kg/m}^2$ inflated to a pressure of 70 kPa. Within the structure, rigid panels unfold to impart further structural rigidity. Power is supplied by a 50 kW gas-cooled particle bed nuclear reactor using Brayton power conversion distributed as

440 V AC at source then transmitted 0.5–1.0 km down transmission lines and stepped down to 220 V AC at the station. Each module generates up to 10 kW of thermal power which must be rejected by 270 W/m^2 radiator panels. Each module is an independent structure mounted onto three legs ending in a tracked drive unit for mobility. The station comprises a core module surrounded by seven peripheral modules arranged radially around it separating the work, recreation, and sleeping environments [1078]. Dedicated special purpose modules include sleeping quarters, crew galley/recreation, life support/gardens, science/medical laboratories, EVA/ rover dock, and consumables storage.

Beyond exploration rovers, rovers will be the key element in developing infrastructure on the Moon, Mars, asteroids, or other celestial bodies. Any lunar construction will require compatibility with low gravity, outgassing due to vacuum conditions, and severe and extreme temperature cycles. This should be accomplished without or with minimal use of working fluids while dealing with the ubiquitous problem of dust. Many of the core technologies are essentially robotic in nature illustrating the importance of robotics in supporting human colonization [1079]. The emphasis will be on fleets of rovers with differing roles to support the outpost and its activities. The base itself may require adjustable thermal shielding to cope with day–night temperature environments—for a static base, regolith may be used as thermal insulation. Simple bulldozing of the regolith is unlikely to be sufficiently stable so some sandbagging may be required. The outpost will be supported by a landing/launch pad separated by at least 0.5 km to prevent contamination of the outpost. To mitigate against plume generation, the pad should be installed on a prepared surface either by heating into glass or by regolith removal. This will necessitate transport facilities between the launch/landing site and the outpost presumably by rover prior to the establishment of any permanent infrastructure. All activities must be supported by a power plant with a capacity of 100–500 kW—adjacent if solar but with dust mitigation capabilities (also requiring storage facilities) or buried/separated from the outpost if nuclear. Solar power facilities may be deployed in a modular fashion but there will still be a requirement for power storage through batteries, fuel cells, or capacitor banks; nuclear power requires no such storage but requires large radiators and radiation emission mitigation to under 5 rem/yr. Lunar regolith may be used as radiation shielding with thicknesses of ~2–3 m. In either case, these facilities will require monitoring and maintenance. In situ resource utilization (ISRU) facilities for the manufacture of essential consumables will be essential to support any outpost or base facilities. Extraterrestrial mining and excavation must be performed primarily autonomously to minimize reliance of human intervention. This will involve a diverse range of tasks: site surveying, resource prospecting, planning pit construction, in situ resource utilization, material processing, process monitoring, and outpost construction. Typical rover activities will include continual site surveying during rover operations, terrain leveling, haulage, berm building, and trenching. Site surveying alone involves mapping the region to be mined in terms of elevation and pit layout and pit construction planning. Pit planning involves determining the optimal excavation pattern and schedule. The primary effects are likely to be in the geotechnics of the regolith and implications for the robotic handling of regolith [1080]. Mining involves rock breakage (drilling, blasting, and beneficiation), overburden clearance (excavation), and material handling (loading and hauling). Rock breakage has primarily relied on explosives to reduce digging forces for scoops and excavators—explosives were used for the Apollo 16 ALSEP system and are used extensively in squibs. Specific explosive charges of $0.12 \, \text{kg/m}^3$ are required for hard rock. Alternatively, electrical blasting through the rapid discharge of electrical energy from a bank of capacitors in an electrolyte will form an expanding plasma [1081]. Focused solar energy, lasers, and microwaves may be deployed for both rock fragmentation and commanition. Surface mining involves excavating loaders and haulers (which may be the same vehicle). Strip mining is the most applicable mining method. Mine locations, processing facilities, and outposts/bases are unlikely to be colocated invoking the need for significant rover haulage capabilities. Hauling involves transporting the excavated material to where it will be processed using conveyors, cable excavators, and scrapers. In any case, the simplest mining approach would be to excavate the loose surface regolith for processing. Production monitoring involves measuring the throughput in terms of weight or volume of material. After processing, the end product consumables must be stored and/or distributed to the outpost. A critical issue will be minimization of transportation costs to support the outpost (product of distance, speed of traverse, and mass of cargo) [1082]. All this refers to the operational needs of an outpost or base without consideration of the robotics requirements for its emplacement which will involve an even greater robotics challenge [1083]. This requires autonomous earthmoving rovers such as bulldozers, diggers, excavators, wheel loaders, etc. The simplest excavator comprises a rover-mounted manipulator with a bucket end-effector and/or bucket wheels. Excavation using a backhoe is commonly used in construction and involves loosening the soil and removing it by dragging the bucket and then lifting. This will require sophisticated impedance control algorithms capable of estimating end effector/soil interaction forces during soil digging and dragging [1084].

Clearly, coordination of multiple rovers by a single human operator will be challenging [1085]. This will favor the use of multiple autonomous rovers operating cooperatively with minimum human assistance. Teams of rovers will play a critical role in the construction of infrastructure on planetary surfaces as precursors to human mission outposts such as habitats (or as part of a more sophisticated robotic infrastructure) [1086, 1087]. Such tasks will involve site preparation, the deployment of hardware elements, and the servicing of such elements—this may include terrain conditioning and roadway construction, cooperative handling and transport of heavy loads, base station construction, deployment of habitat modules including inflatables, deployment of in situ resource utilization plants for propellant production, deployment of solar photovoltaic tent arrays, long-range exploration surveying, establishment of robotic science stations, and a host of infrastructure servicing and repair operations. The power system deployment scenario is considered to be one of the most challenging tasks. Power requirements for human infrastructure support amounts to around

100 kWe which requires a solar area of 5,000 m^2 with an overall mass of 3.5 tonnes (this compares favorably with a nuclear power plant of 5–7 tonnes). Human missions to Mars will require extensive precursor site preparation for infrastructure emplacement [1088, 1089]—site preparation tasks will involve leveling of rough soil and the clearing of rocks from a selected 50 × 100 m area. This will involve transportation activities including cooperative manipulation and transport of objects. This bears some resemblance to the standard multirobot box-pushing task (emulating rock clearance) but site preparation is far more complex and involves only a handful of rovers (nominally four or five). Multiple fleets of rovers must be coordinated dynamically and continuously within a constantly changing environment. JPL's CASPER (continuous activity scheduling planning execution and replanning) planning system is an extension to ASPEN (automated scheduling and planning environment) which supports a distributed control approach through a scheme of iterative repair [1090]. Two rover control architectures have been proposed to control teams of earthmoving bulldozer rovers— BISMARC [1091] and ALLIANCE [1092]—though others such as AuRA are also suitable. The ALLIANCE architecture is a behavior-based, distributed multirobot control scheme that implements a motivational mechanism (impatience and acquiescence) to enable adaptive action selection [1093–1095]. Motivations (impatience and acquiescence) are mathematically modeled to augment its behavior-based architecture to respond to a dynamic environment. Motivation controls the invocation of basic behaviors by imposing an increasing level of activation until it surpasses a threshold or the situation changes. This allows ALLIANCE to select between conflicting behaviors. There are six behavior sets: `Prepare_for_-Clearing_Pass`, `Clear_Path_to_Subgoal`, `Request_Help`, `Cooperative_Clear`, `Maintenance_Operations`, and `Return_to_Home_Base`. Motivations are associated primarily with drives and emotions. Emotions provide the basis for distinguishing between situations that may be harmful or beneficial without reasoning about causes. They are instrumental in providing goal-based progress in behavior-based control systems beyond the traditional limitations of finite state machines particularly associated with multirover cooperation which may suffer from interdependent cycling [1096]. An example would be a robotic assistant failing to resupply another rover (e.g., due to hardware/software failure, time delays, etc.) which will generate a task failure in the waiting rover. Behavior-based approaches to distributed control of rover teams for diverse tasks have been popular; for example, AVATAR (autonomous vehicle aerial tracking and reconnaissance) which implements a hierarchical behavior-based control system in which more complex behaviors emerge from more fundamental simple behaviors using an aerial rover and a fleet of ground rovers [1097]. Multiple small, cheap rovers ~100 g have been proposed to be deployed in large numbers to provide blanket areal coverage of the planetary surface [1098] but such proposals take no account of the mode of delivery. A ground rover swarm may be deployed over large areas of planetary surfaces from an aerial vehicle such as a balloon through low-altitude drop landings. They may coordinate with collective behavior to form a communications network infrastructure.

11

Robotic sample acquisition

The primary purpose of the robotic planetary rover is to deploy scientific instruments to targets of interest and acquire physical samples for in situ analysis by scientific instruments or for subsequent return to Earth. The first step in scientific analysis of soil and rock from planetary surfaces and subsurfaces in particular is sample acquisition. Sample acquisition is typically performed by robotic manipulator, abrasion, drilling, and burrowing. Each is considered in turn. The first step toward such acquisition is to understand the nature of the environment from which the sample is to be extracted. There are two types of regolith environments that are likely to be encountered (excluding ice): (i) regoliths from planets with atmospheres such as Mars which have been subjected to weathering influences but which have been spared microtektite bombardment; (ii) regoliths from planets or other small bodies which have not been exposed to weathering but have been subjected to eons of microtektite bombardment (e.g., the Moon). For the Moon, hollow drive-coring tubes were hammered vertically into the regolith by the Apollo astronauts to depths of 1 m—it took 50 hammer blows to penetrate to 70 cm. Rotary drill core tubes were able to penetrate to depths of up to 3 m at which point the soil became highly compacted due to eons of micrometeoroid bombardment [1099]. The regolith layer is estimated to be 10–30 m thick. This characteristic of a thin layer of loose regolith overlying a highly compacted layer is expected to occur for all atmosphereless bodies such as asteroids.

The Martian surface and near-surface regolith is believed to be saturated in oxidants such as hydrogen peroxide, metal oxides, peroxides, and superoxides. This region will thus be devoid of organic material—the Viking lander gas chromatograph/mass spectrometer detected no organics at the parts per billion level for heavy organics and the parts per million level for light organics. This upper limit is far lower than that expected from meteoritic influx of carbonaceous chondrites over the eons. It is believed that solar UV flux acting on the small

© Springer-Verlag Berlin Heidelberg 2016
A. Ellery, *Planetary Rovers*, Springer Praxis Books,
DOI 10.1007/978-3-642-03259-2_11

amount of water vapor in the lower atmosphere generates hydrogen peroxide and
other peroxides in the soil. In addition, the UV-induced Fenton's reaction may
occur forming hydroxyl radicals in the soil which rapidly oxidize organic compounds. UV flux at the Martian surface is 2.6 mW/cm^2 (around four times that on
Earth). Furthermore, the loss of atmospheric water constituents such as hydrogen
from Mars was due to the higher UV flux during the Sun's early main sequence
phase which led directly to the incorporation of oxidants into the soil [1100]. The
distribution of this oxidant layer is controlled through both molecular diffusion
and meteoritic impact gardening which suggests a depth of several meters, nominally an average oxidant extinction depth is ~3 m based on crater population, onset
of oxidizing conditions, and absorption of water [1101–1104]. Oxidants comprise
H_2O_2, O_3, and H_2O in the Martian atmosphere at ~10^{10}/cm^3 which diffuse into
the Martian soil. The diffusion may be modeled by Fick's law of linear diffusion:

$$\phi = \frac{dC}{dt} = -D\frac{\partial^2 C}{\partial z^2} = -D\frac{f}{q}\frac{dy}{dz}$$

where C = peroxide density (concentration) in regolith, Φ = UV flux, D = diffusion
coefficient, f = porosity, q = tortuosity, γ = mass density of pore vapor, and
z = vertical depth. The extinction coefficient is defined to be the point at which
peroxide density drops to 10^{-6} of its surface value. The regolith is fully oxidized
to a depth of 30 cm by diffusion. The onset of oxidizing conditions occurred after
the end of the heavy bombardment phase ~3.8 Gyr ago when the cratering rate
was 10^4 that subsequently. Oxidants cannot survive excess water conditions.
Hence, the mixing conditions for oxidant stirring by impact gardening would have
been reduced suggesting a $1/e$ oxidant depth of 0.5–0.85 m (i.e., absence of
oxidant below a depth of 2–3 m). Hence, astrobiological prospecting will require
accessing samples from subsurface depths below this [1105, 1106].

   Although Kapvik did not incorporate any subsurface drilling capability, the
Vanguard Mars rover concept did. The mode of subsurface penetration selected
for Vanguard was the ground-penetrating mole. In situ sensor heads may be
accommodated within the mole while the instruments themselves reside on the
rover. Such a device can penetrate to a nominal depth of ~3 m. By employing
three such devices, there is no requirement to return the moles back to the surface
once they have been deployed—they can emplace sensor heads of rover-mounted
instruments into the borehole. This significantly reduces the robotic complexity
involved in extracting the moles. The microrover can deploy each mole in turn to
provide a triplicate depth profile. This requires the use of remote-sensing
instrumentation that can exploit separation of the sensor head from the main
instrument to eliminate the need for taking soil samples—the primary instruments
would be a Raman spectrometer, an infrared spectrometer, and a laser plasma
spectrometer. Augmentation with thermal probes and/or magnetometers within
the moles would provide valuable in situ geophysical data. The use of a ground-
penetrating radar for drill site selection would ensure that submerged boulder
obstacles within the regolith could be avoided. Given that the Martian regolith

will be comparatively loose in comparison with soils on atmosphereless bodies, a percussive mole offers a capable mode of reasonable depth penetration.

11.1 SAMPLE ACQUISITION BY ROBOTIC MANIPULATOR

An essential tool in autonomous science is the robotic manipulator for surface sample acquisition. Robot manipulators must be designed with respect to their required performance—(i) positional accuracy and repeatability; (ii) velocity/acceleration limits (operating cycle time and speed of response); (iii) life expectancy; (iv) reliability/maintainability. The acquisition of samples may require consideration of contact forces on an unstructured environment. Autonomous control systems require sufficient adaptability to cope with these exigencies to work with any type of soil and deal with random obstacles [1107]. Modern rovers carry robotic manipulators that have revolute joints with electric motors at each joint to provide maximum workspace and controllable configuration. The minimal requirement for robotic manipulation is 3 degrees of freedom (DOF) with an elbow movement—base sweep, shoulder pitch, and elbow pitch similar to the PUMA 560/600 manipulator—but the addition of a wrist pitch gives much greater flexibility in manipulation. Rover-mounted manipulator systems are generally of low mass (~1–3 kg) and characterized by compact stowage to eliminate interference with rover operations. They are required to have high payload capacity of ~1 kg, large workspace, and high versatility [1108]. Rover manipulator arms may be of the following configurations with increasing order of complexity [1109]:

   (i) a 2 DOF scoop;
  (ii) a 3 DOF scoop with two proximal actuators;
 (iii) a 3 DOF scoop with two distal actuators;
 (iv) a 4 DOF scoop, flat gripper, and arched access to ground and utility tray;
  (v) a 4 DOF scoop enabling front stowage with scoop and curved gripper for instrument handling;
 (vi) a 4 DOF scoop, flat gripper, and linear access to ground and utility tray;
(vii) a 4 DOF scoop and curved gripper for instrument handling with front or side stowage;
(viii) a 5 DOF gripper with planar access to ground, vehicle, and horizontal utility tray;
 (ix) a 6 DOF gripper (minus end effector roll) with access to tilted utility tray.

A 6 DOF manipulator typically violates the mass and volume constraints imposed by rover design, so 5 DOF manipulators are commonly used without the roll degree of freedom. A deployable 5 DOF Beagle 2–like articulated robotic arm is mounted externally to the rover and usually includes a rock corer and a small close-up camera for ground and rover self-inspection. It resides in front of the rover to give a good range of access to samples. The rover must accommodate cold, external storage of samples that may be accessed by the robotic arm. A rock

corer drill is deployed with the manipulator. With the exception of Russian
missions (Luna and Venera), the first robotic manipulator used on a planetary
mission was on the U.S. Viking Mars missions. The Viking 1 and 2 landers
landed on the Martian surface in 1976 in the Chryse Planitia and Utopia Planitia
regions, respectively, on opposite sides of the planet.

The two Viking landers on Mars (1975) were the first space missions to utilize
articulated remote manipulators to dig Martian soil up to 2.5 m from the landers
to a depth of 10 cm (Figure 11.1). All onboard experiments required the acquisition of Martian soil which was carried out by the lander telescopic arm mounted
with a scoop. The recovered soil was deposited through a sieve in the lander for
use in onboard experiments. The deployable boom comprised parallel strips of
stainless steel which were kinked to impart rigidity along the length and which
could be rolled up by retracting the boom's metal ribbons. A trench was excavated
by the Viking arm to a depth of 23 cm. The Viking collector head lower jaw was
4.45 cm wide with a serrated tip. A backhoe 6.1 cm wide by 6.45 cm high lay
10.2 cm from the serrated tip. The upper jaw was activated by a solenoid capable
of vibrating the upper jaw at 4.4 or 8.8 Hz. The upper surface of the upper jaw
had 0.2 cm holes to separate coarse and powdery fractions. The collector head was
24.3 cm long and attached to a furlable boom. The collector head could be
inverted about its longitudinal axis and pivoted vertically from its base through an
angle of 10°. The Viking boom sample acquisition sequence was thus:

   (i) position sampling boom to the desired azimuth;
  (ii) extend the collector head to the desired length;
 (iii) lower the collector head to the surface;
 (iv) extend the collector head into the soil at 0.025 m/s by 0.1 cm with the jaws open;
  (v) retract the collector head with the jaws closed;
 (vi) elevate the arm and deliver the sample through the 0.2 cm openings in the upper
       jaw of the collector head;
(vii) dump the remaining coarse fraction.

Excavation force in the lunar surface by a digging blade of weight W with
acceleration $a = (a_x a_y)^T$ is given by [1110]:

$$F = \sqrt{F_x^2 + F_y^2}$$

where x and y are referenced with respect to the blade:

$$F_x = P \cos(\alpha - \delta) + F_s \cos \beta + (W/g)a_x$$
$$F_y = W + P \sin(\alpha - \delta) + F_s \sin \beta + (W/g)a_y$$

where P = passive earth pressure on blade, F_s = side friction force, α = blade
inclination angle, β = side friction inclination angle, and δ = friction angle
between soil and blase. Passive earth pressure P and side friction force F_s are both
functions of excavation depth, soil specific gravity, soil cohesion, soil friction

Figure 11.1. Viking lander (1976) and its soil sampler [credit NASA JPL/Smithsonian Air & Space Museum].

angle, and surface surcharge (i.e., terramechanics properties, especially soil cohesion). Hence, the ability to measure soil physical parameters enables control of
earth-moving operations. Reece's fundamental equation of earthmoving describes
the mechanics of a cutting blade moving in soil perpendicular to the tool face
(e.g., bulldozers, excavator buckets, etc.) [1111]:

$$F = b(\gamma z^2 N_\gamma + cz N_c + qz N_q)$$

where N = passive earth pressure coefficients [1112, 1113] such that

$$N_\gamma = \frac{(\cot \alpha + \cos \beta) \sin(\beta + \phi)}{2 \sin(\alpha + \beta + \delta + \phi)} = \text{passive pressure coefficient}$$

$$N_c = \frac{\cos \phi}{\sin \beta \sin(\alpha + \beta + \delta + \phi)} = \text{cohesion coefficient}$$

$$N_q = -\frac{\cos(\alpha + \beta + \phi)}{\sin \alpha \sin(\alpha + \beta + \delta + \phi)} = \text{friction coefficient}$$

$$\cot \beta = \frac{\sqrt{\sin(\alpha + \delta) \sin(\delta + \phi)/\sin \alpha \sin \phi} - \cos(\alpha + \delta + \phi)}{\sin(\alpha + \delta + \phi)}$$

$$q = \frac{\gamma gl \tan \beta \tan \alpha}{2(\tan \beta + \tan \alpha)} = \text{surface pressure}$$

where α = blade approach angle, β = soil failure angle, and δ = soil–blade friction
angle. The Lockheed-Martin/Viking model based on the Luth–Wisner model for
the excavation force required by a robotic arm based on a bucket wheel excavator
or, equivalently, an earth-moving blade is given by [1114, 1115]:

$$F_{\text{fric}}^{\text{horiz}} = \gamma gwl^{1.5} \alpha^{1.73} \sqrt{d} \left(\frac{d}{l \sin \alpha}\right)^{0.77} \left(1.05\left(\frac{d}{w}\right)^{1.1} + 1.26\frac{v^2}{gl} + 3.91\right)$$

$$F_{\text{fric}}^{\text{vert}} = \gamma gwl^{1.5} \sqrt{d}(0.193 - (\beta - 0.714)^2)\left(\frac{d}{l \sin \alpha}\right)^{0.777} \left(1.31\left(\frac{d}{w}\right)^{0.966} + 1.43\frac{v^2}{gl} + 5.60\right)$$

$$F_{\text{coh}}^{\text{horiz}} = \gamma gwl^{1.5} \alpha^{1.15} \sqrt{d} \left(\frac{d}{l \sin \alpha}\right)^{1.21} \left(\left(\frac{11.5C}{\gamma gd}\right)^{1.21} \left(\frac{2v}{3w}\right)^{0.121} \left(0.055\left(\frac{d}{w}\right)^{0.78} + 0.065\right)\right)$$

$$F_{\text{coh}}^{\text{vert}} = \gamma gwl^{1.5} \sqrt{d}(0.48 - (\alpha - 0.70)^3)\left(\frac{d}{l \sin \alpha}\right)\left(\left(\frac{11.5C}{\gamma gd}\right)^{0.41} \left(\frac{2v}{3w}\right)^{0.041}\right.$$

$$\left. \times \left(9.2\left(\frac{d}{w}\right)^{0.225} - 5.0\right) + 0.24\frac{v^2}{gl}\right)$$

where α = rake angle, d = digging depth, C = soil cohesion, w = tool width,
l = tool length, γ = soil specific mass, g = acceleration due to gravity, and v = tool
speed. Friction contributions are far greater than other contributions when
cohesion is low (e.g., lunar soils).

While the Viking manipulator was used for sample acquisition, some scientific instruments require a pointing capability to detect their target. The alpha proton X-ray spectrometer (APXS) of the Sojourner rover was deployed by a deployment arm which moved the sensor head along a fixed path until the head contacted a surface. Its flexible wrist and motor-driven parallel link arm aligned the head to within 20° of the rock surface (it can accommodate surfaces from horizontal to vertical). Signals from three pairs of LEDs and a phototransistor provided feedback on contact forces of ~0.5–3 N on the sample. The 185 kg Mars Exploration Rovers (Spirit and Opportunity) each carried a small manipulator arm—the Instrument Deployment Device (IDD)—to deploy the scientific instruments that comprised a rock abrasion tool (RAT), an APXS, a Mössbauer spectrometer, and a microscopic imager which were used for analysis of selected rock and soil targets [1116].

The IDD comprised a 5 DOF robotic arm mounted to the front of the rover stowed below the main rover body. It had a deployed reach of 0.75 m. Its configuration comprised two shoulder joints (azimuth/elevation), a single elbow joint, and a wrist pitch and yaw—yaw was used to point different instruments. It was secured at the elbow and wrist during launch, cruise, and entry descent and landing. It was released on landing by pyrotechnically released spring-loaded pins. The link housing and shafts were constructed from Ti alloy tubing. It was driven by five DC brush motors with three-stage planetary gears, magnetoresistive encoders, and rotary potentiometers mounted on the output shaft. They were wet-lubricated with grease and sealed using Teflon O-rings, labyrinth seals, and felt. Each joint was driven by Maxon RE020 motors and had electric heaters supported by thermistor measurements to maintain their temperatures above −55°C. Motors/gears were placed at shoulder azimuth and elevation, elbow, and wrist azimuth and elevation (Figure 11.2). The shoulder and elbow actuators also included harmonic drive gearing to reduce backlash while the wrist motors also included two additional planetary gear stages.

The IDD had an accuracy of 4 mm and 5° and a contact preload capability of >10 N. It had a mass of 4 kg including 1 kg of electrical cabling offering a payload capacity of 2 kg. The 3 m long cable harness comprised over 200 flat flexprint conductors routed through the arm terminating at each joint but mostly terminating at the end effector instruments through micro-D connectors. The IDD used the frontal wide-angle stereo pair hazard cameras for visual feedback of its deployment and proximity feedback switch sensors at each instrument for determining contact to terminate movement. The IDD was PID-controlled at a rate of 8 Hz and used a trapezoidal velocity profile for its commanded movements. Intermediate points were generated between the initial and final poses through a recursive bisection method. There were four basic behaviors—free space, guarded, retractive, and preloaded motions. The motor controller board used feedback from the encoders while the payload analog board used feedback from the potentiometers and proximity sensors. The flight control and VxWorks real time operating system software resided on the main RAD6000 computer which was interfaced through a VME backplane.

Figure 11.2. MER Instrument Deployment Device (IDD) [credit NASA JPL].

IDD kinematics were represented by means of a Denavit–Hartenberg matrix representation of its pose $(p_x\,p_y\,p_z\theta_{az}\theta_{el})$ with the end effector z-axis aligned along its boresight (approach vector) (Table 11.1). The shoulder had right $(+)$/left $(-)$ configurations, the elbow had up $(+)$/down $(-)$ configurations, and the wrist had up $(+)$/down $(-)$ configurations.

The kinematics of the IDD were as follows:

$$^{0}T_{ee} = {}^{0}T_{5}\,{}^{5}T_{ee}$$

where

$$^{0}T_{5} = \begin{pmatrix} c_1c_{234}c_5 - s_1s_5 & -c_1s_{234} & c_1c_{234}s_5 + a_1c_5 & a_1c_1 + a_2c_1c_2 + a_3c_1c_{23} + d_4s_1 - d_5c_1s_{234} \\ s_1c_{234}c_5 + c_1s_5 & -s_1s_{234} & s_1c_{234}s_5 + c_1c_5 & a_1s_1 + a_2s_1c_2 + a_3s_1c_{23} - d_4c_1 - d_5s_1s_{234} \\ s_{234}c_5 & c_{234} & s_{234}s_5 & d_1 + a_2s_2 + a_3s_{23} + d_5c_{234} \\ 0 & 0 & 0 & 1 \end{pmatrix}$$

$$^{5}T_{ee} = \begin{pmatrix} c_{ee} & 0 & s_{ee} & {}^{5}p_{eex} \\ 0 & 1 & 0 & 0 \\ -s_{ee} & 0 & c_{ee} & {}^{5}p_{eez} \\ 0 & 0 & 0 & 1 \end{pmatrix}$$

Table 11.1. IDD kinematic parameters.

Link	Link offset a_i	Link twist angle α_i	Link length d_i	Joint angle θ_i
1	a_1	$\pi/2$	d_1	θ_1
2	a_2	0	0	θ_2
3	a_3	0	0	θ_3
4	0	$-\pi/2$	d_4	θ_5
5	0	$\pi/2$	d_5	θ_5

where $c_i = \cos\theta_i$ and $s_i = \sin\theta_i$. The inverse kinematics may be derived thus:

$$^0p_5 = {}^0p_{ee} - {}^0a_{ee}|{}^5p_{ee}|$$

$$\theta_1 = \tan^{-1}\left(\frac{{}^0p_{5y}}{{}^0p_{5x}}\right) \pm \tan^{-1}\left(\frac{d_4}{\sqrt{{}^0p_{5x}^2 + {}^0p_{5y}^2 - d_4^2}}\right)$$

$$\theta_2 = \tan^{-1}\left(\frac{{}^1p_{4y}}{{}^1p_{4x}}\right) \pm \left[(c_1{}^0p_{5x} + s_1{}^0p_{5y})(\pm^2p_{3y})\cos^{-1}\left(\frac{({}^1p_{4x}^2 + {}^1p_{4y}^2 - a_2^2 - a_3^2)}{2a_2\sqrt{{}^1p_{4x}^2 + {}^1p_{4y}^2}}\right)\right]$$

$$\theta_3 = \tan^{-1}\left(\frac{\pm(c_1{}^0p_{5x} + s_1{}^0p_{5y})(\pm^2p_{3y})\sqrt{1 - \left(\frac{{}^1p_{4x}^2 + {}^1p_{4y}^2 - a_2^2 - a_3^2}{2a_2a_3}\right)^2}}{\left(\frac{{}^1p_{4x}^2 + {}^1p_{4y}^2 - a_2^2 - a_3^2}{2a_2a_3}\right)}\right)$$

$$\theta_4 = \tan^{-1}\left(\frac{{}^0a_{eez}}{c_1{}^0a_{eex} + s_1{}^0a_{eey}}\right) - \theta_2 - \theta_3$$

$$\theta_5 = \tan^{-1}\left(\frac{c_1c_{234}{}^0a_{eex} + s_1c_{234}{}^0a_{eey} + s_{234}{}^0a_{eez}}{s_1{}^0a_{eex} - c_1{}^0a_{eey}}\right) - \theta_{ee}$$

Although not rover mounted, the Beagle 2 manipulator was similar to a rover-mounted manipulator [1117, 1118] (Figure 11.3). The Beagle 2 lander mission to Mars was designed and constructed to include a 2.4 kg 5 DOF ARM (anthropomorphic robotic manipulator) with revolute joints. The generalized 4×4 Denavit–Hartenberg matrix was determined from the Beagle 2 manipulator kinematic parameters given in Table 11.2.

ARM was placed on the opposite side to the solar panels to maximize the sampling area available to it. The robotic arm mounted the PAW (position-

Figure 11.3. Beagle 2 lander with its 5 DOF ARM [credit ESA].

Table 11.2. Beagle 2 kinematic configuration.

Link	α_i (deg)	a_i (mm)	d_i (mm)	θ_i (deg)
1	0	0	d_1	θ_1
2	90	a_2	0	θ_2
3	0	a_3	d_3	θ_3
4	0	0	d_4	θ_4
5	90	a_5	0	θ_5

actuated workbench) at its wrist [1119]. All links were constructed from Ti alloy and joint sections from CFRP bonded to Ti end fittings. Carbon fiber tubes made up the upper and lower ARM segments. The ARM possessed 162 components in total massing a total of 2.108 kg. A thermal spacer under the base of ARM minimized heat leaks. The Beagle 2 robot arm pointed its PAW (which mounted six scientific instruments) towards rock samples. A panoramic camera was mounted farther up the arm on the elbow. Each joint was moved separately to minimize power consumption to ~1 W. The first vibration mode of the arm was at 180 Hz. It had a reach of 1.09 m giving it a workspace of $0.7 \, m^2$ (such a reach would have given it access to two to four rocks >10 cm in diameter) and a payload capacity of 2.7 kg. This was the mass limit for PAW which sported an integrated sensor package: X-ray spectrometer (XRS), Mössbauer spectrometer, microscope, rock corer/grinder, and mole. Each revolute joint comprised a brushed Maxon DC motor with 100:1 harmonic drive gearing whose joint excursion was measured by a potentiometer in the output shaft. Maximum joint speed was 0.6°/s drawing power of 1.2 W at a maximum of 100 mA at 12 V. ARM positioning accuracy was ±5 mm and ±5° imposed by joint potentiometer feedback which increased to ±2 mm and ±1° after imaging by the stereo cameras—the former accuracy sufficed for delivery of the sample to the inlet on the lander but the latter accuracy was required by several of the instruments (e.g., microscope and X-ray/Mossbauer spectrometers). A number of safe configurations were defined within the workspace. All movements of ARM were preprogrammed with several standard moves:

 (i) deploy from stowed position to a safe position;
 (ii) move mole to GAP (gas analysis package) inlet from safe position;
 (iii) move corer to GAP inlet from safe position;
 (iv) change instrument by rotating PAW;
 (v) slow final approach move so that contact with rock exerts a predetermined load.

Once a sampling location was determined, ARM aligned PAW and advanced towards its target until contact was made with an applied force of 5 N (its maximum applied force was 30 N). Undesired contact forces measured by a piezo-electric force sensor at the ARM/PAW interface would invoke stoppage within 100 ms. ARM's most complex manipulation was to bring samples to the gas chromatograph/mass spectrometer (GCMS) on the lander.

The Phoenix Mars Lander (2008) carried a 4 DOF 2.4 m long robotic arm and backhoe scoop [1120]. Its configuration was based on shoulder yaw and elevation, elbow pitch, and wrist pitch. The scoop's front compartment collected material from the front blade while the rear chamber collected material from the rasp cutting tool on the rear of the scoop. Material could be transferred between chambers by rotating the entire scoop about the wrist axis. The arm dug trenches in the Martian regolith and acquired samples for scientific analysis by a thermal evolved gas analyzer (TEGA) and a microscopy electrochemistry and conductivity analyzer (MECA). It also inserted a thermal and electrical conductivity sensor into the regolith and atmosphere at different heights to measure humidity. It panned its camera to take extensive images of its close environment, particularly of samples prior to delivery to the scientific instruments. The arm acquired and delivered a total of 17 samples to onboard scientific instruments.

The Kapvik microrover employs a 4 DOF manipulator which doubled as the panoramic camera mast. Its end effector is a scoop for sample acquisition. Like the Viking scoop, it is designed to acquire soil and small rocks for subsequent analysis or in a sample-and-fetch capacity (Figure 11.4). The robotic arm is augmented by an additional pan–tilt unit on which the panoramic stereo camera is mounted. This approach integrates the robotic arm and camera mast in an eye-in-hand configuration. A mobile manipulator comprising a manipulator mounted onto a mobile platform has a much larger workspace than a fixed base manipulator. The most common approach is to decouple the problem of controlling locomotion (position control) and manipulation (force control) into separate sub-problems with interaction effects accommodated through compliant manipulators [1122]. Resolved acceleration control and PI force control have been proposed as control mechanisms for mobile manipulators [1123]. It is recommended, however, that full compensation of the dynamic interaction effects of manipulator motion on the mobile platform are made [1124]. This would enable visual servoing of the rover-mounted manipulator. Visual servoing involves controlling a robot arm to grasp an object through vision sensors [1125, 1126]. Image data are used directly to control the robot's pose and motion relative to a target. The error function is defined in terms of measurables in the image (e.g., feature points, lines of orientation, etc.) and a control law is constructed which maps the error directly to robot motion. End effector velocity is defined by:

$$\begin{pmatrix} \dot{x} \\ \dot{y} \\ \dot{z} \end{pmatrix} = - \begin{pmatrix} 0 & -z & y \\ z & 0 & -x \\ -y & x & 0 \end{pmatrix} \begin{pmatrix} w_x \\ w_y \\ w_z \end{pmatrix} + \begin{pmatrix} v_x \\ v_y \\ v_z \end{pmatrix}$$

The error function is the vector difference between the desired and measured locations of the image points selected. The vector of k measured image features is given by:

$$s_k(t) = \begin{pmatrix} u(t) \\ v(t) \end{pmatrix}$$

where $u(t), v(t) =$ image point coordinates. Image feature velocity is given by:

$$\dot{s}(t) = Jv$$

where $J = \partial s / \partial r =$ interaction matrix (image Jacobian) and $v = \binom{v}{w} = \dot{r} =$ end effector/camera velocity. The image Jacobian relates image velocity to camera coordinates [1127]:

$$\dot{x} = zw_y - \frac{vz}{\lambda} w_z + v_x$$

$$\dot{y} = \frac{uz}{\lambda} w_z - zw_x + v_y$$

$$\dot{z} = \frac{z}{\lambda} (vw_x - uw_y) + v_z$$

where $\lambda =$ camera focal length $f \times$ scaling factor α. Projected image velocity

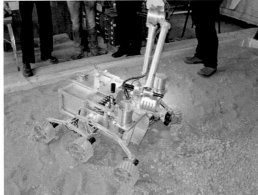

Figure 11.4. Kapvik acquiring rock samples using its manipulator-borne scoop [credit Ian Sinclair, MPB Communications].

relative to camera coordinates is given by:

$$
\begin{pmatrix} \dot{u} \\ \dot{v} \end{pmatrix} = \begin{pmatrix} \dfrac{\lambda}{x} & 0 & -\dfrac{u}{z} & -\dfrac{uv}{\lambda} & \dfrac{\lambda^2 + u^2}{\lambda} & -v \\[2ex] 0 & \dfrac{\lambda}{z} & -\dfrac{v}{z} & \dfrac{-\lambda^2 - v^2}{\lambda} & \dfrac{uv}{\lambda} & u \end{pmatrix} \begin{pmatrix} v_x \\ v_y \\ v_z \\ w_x \\ w_y \\ w_z \end{pmatrix}
$$

relative to the camera frame. Inversion requires use of the Moore–Penrose inverse least-squares solution. Resolved motion rate control can be applied using:

$$ e(t) = s(t) - s^d $$

The image Jacobian can be decomposed as:

$$ \dot{s} = J_v(u, v, z)v + J_w(u, v)w $$

Inversion of the image Jacobian is required to transform image motion into camera motion. A position-based control law may be determined:

$$ e_l = (x, y, z) - (x^d, y^d, z^d) $$

$$ e_\theta = \theta - \theta^d $$

The control law is given by:

$$ \tau = J^T k_p e - k_v \dot{e} $$

Image coordinates must be related to the robot's base coordinates. A variant on this visual servoing is servoing a camera-in-hand configuration [1128].

11.2 SAMPLE ACQUISITION BY CORER/GRINDER

A rock corer/grinder is required as rocks are often coated in a weathered rind—in the case of Martian rock this includes embedded dust and regolith which must be removed to expose fresh surfaces before analysis of rocks can be conducted. This may be similar to laminated desert varnish on Earth rocks that exhibit similar growth patterns to stromatolites [1129]. On Earth, the varnish comprises silica minerals and Fe/Mn oxides incorporating organics such as amino acids and even DNA, so represents a potential site for astrobiological investigation. Each MER carried a Honeybee low-power rotary coring drill—the 720 g rock abrasion tool (RAT) used a pair of high-speed percussion-based diamond-tipped grinding teeth to clean areas on rocks 40 mm in diameter and 5 mm deep between 30 min and 1 h (Figure 11.5). It was held against the sample rock by the IDD prior to deployment of the scientific instruments. RAT used three actuators to expose Martian rock: one was used to rotate the two grinding wheels at high speed—one wheel was equipped with two diamond teeth while the other had a brush to remove rock cuttings; another actuator moved the two grinding wheels around each other by

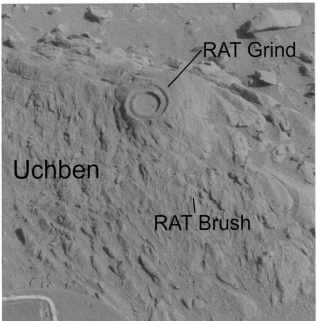

Figure 11.5. MER rock abrasion tool and its performance on Martian rock [credit NASA JPL].

rotating the head; the third actuator provided thrust into the rock. A ring surrounding RAT made contact with the rock. RAT used 50 Wh to grind a 40 mm diameter 5 mm deep hole in basalt rock to provide a pristine surface for examination by the MER APXS, Mössbauer spectrometer, and microscopic imager. The MER coring drill took rock samples up to 5 cm deep in rock 8–17 mm in diameter. The corer could drill vertically or up to 45° from the vertical. Once acquired, the sample could be examined by the arm instruments and/or placed in one of the compartments of the sample container. Once the sample had been acquired, the manipulator arm was retracted.

The Beagle 2 rock corer/grinder mounted onto the PAW of the robotic arm was to be used to abrade the dust-covered weathered rind of rocks to produce a flat 30 mm diameter fresh surface and provide access to internal pristine samples for scientific instruments [1130]. It was based on the holinser (holder + inserter) concept derived from dental forceps for handling inlays (tweezers are the most common laboratory tool used to grip objects of any shape up to 20 cm in diameter). The corer had contact dimensions of 15 to 25 mm while a larger grinder was to be used to pulverize rock for the GAP package. The rock corer/grinder was a motor-driven drill bit aided by a motor-driven cam mechanism hammer to provide a 1 Hz chiseling effect on the drill. The drill bit rotated at 60 rpm consuming 1.5 W. A sharp contact ring around the drill head had the purpose of maintaining high-friction contact with the rock surface with the Beagle 2 ARM providing 6–9 N thrust. ARM was required to move across the surface to provide the required 20 mm diameter area for the spectrometers and microscope. Drill time depended on rock hardness—porous rock would take around 15 min. The 348 g rock corer/grinder was to recover rock cores 2.5 mm in diameter and 7.0 mm deep (giving a mass of 60–80 mg assuming 3 g/cm^3 rock density) for each core sample (typically 75 mg assuming 3 g/cm^3 soil density).

Sonic/ultrasonic drilling is an aggressive technique that is particularly suited to methods that do not employ drilling fluid. A novel percussive approach to rock abrasion/shallow drilling is the ultrasonic/sonic driller/corer (USDC) comprising a piezoelectric ceramic stack in compression which drives an ultrasonic horn actuator [1131–1133] (Figure 11.6). An ultrasonic/sonic transformer comprises the drill/core column that converts 20 kHz ultrasonic frequency into drilling action. This was the resonant frequency of the horn. The ultrasonic horn actuator drives a 4 g proof mass which resonates between the horn and the 100 mm long drill stem (thickness 3 mm). The transformer comprises a free-floating drill stem which converts the ultrasonic 20 kHz drive frequency to a 60–1,000 Hz sonic wave. Vibrations of the horn excite the proof mass which resonates between the horn tip and the top of the drill stem at ~1 kHz. The proof mass jack-hammers the drill stem which fractures the rock. The coring bit creates a hole slightly larger than the drill bit diameter. The acoustic power delivered to the rock is high—the free mass impacts the top of the dill to deliver 0.1 GPa for 25–50 μs to the rock. The USDC has a mass of 800 g requiring a low preload force of 5–10 N while drawing only 5.3 W of power at a drill rate of 0.1 mm/s. It can drill through any hardness of rock including granite without blunting and can pull cores of various shapes. It

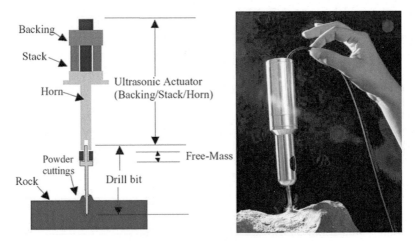

Figure 11.6. Ultrasonic/sonic driller/corer [credit NASA JPL].

has demonstrated its ability to procure 25 mm cores in basalt with a 3 N axial load and drill rate of 1 cm/min in sand or 1 mm/min in hardened regolith consuming only 3 W of power [1134]. The drill stem and free mass are the only two moving parts. The bit does not need to be sharp, so blunting of drill teeth is not a problem. The device does not suffer from drill walk during initiation and does not impose the large lateral forces and torques on any mounting platform typical of conventional rotary drills. Furthermore, conventional drills cause drill chatter of low-frequency perturbations ($\sim$2–10 Hz) on the drill platform necessitating massive structures. Ultrasonic drilling is fast and does not dull the drill bit but requires high power and undergoes significant energy dissipation in soft rock. It is not yet clear if this mode of drilling is scalable to greater depths within reasonable mass and power levels. Nevertheless, it offers similar rock-powdering capabilities to RAT.

An alternative is to replace the piezoelectric actuator with a solenoid that has a much longer stroke vibrated at high frequency using a sinusoidal input voltage and a restraining spring [1135].

11.3 SAMPLE ACQUISITION BY SUBSURFACE DRILL

The Viking lander investigation of Mars was restricted to 20 cm below the surface of the regolith due to the limits of its trench-digging experiments. Although the Martian subsurface is unknown, a popular model assumes a 1–2 m dust layer overlying a lava layer to a depth of 300 m, thereafter a permafrost layer to 2.5 km. Beyond this, a liquid water layer is expected. To penetrate even to the lower estimate of the permafrost layer at 300 m presents a considerable challenge.

However, there may be some water (probably ice) quite near the surface—samples acquired by Viking 2 beneath two rocks released 0.15–1.1% water when step-heated from 50–200°C compared with only 0.002% released from surface samples when heated in one step to 200°C [1136]. The soil beneath the surface and under rocks thus has much more adsorbed water at depths of 2.5–5.0 cm than surface soil. For subsurface samples, sample acquisition proceeds through robotic drilling. Ideally, to optimize selection of the drill site, ground-penetrating radar may be adopted augmented by an electromagnetic sensor to attempt to detect subsurface water prior to commitment to drilling. Surface mobility is essential in order to select optimal sites for subsurface sampling, preferably with a range of about a few kilometers. The depth of drill penetration dictates the types of samples that can be acquired. Drilling missions are generally classified according to their depth [1137]. Near-surface drilling penetrates down to a depth of 5–10 cm where geo-chemical sampling can be undertaken without difficulty; the near-subsurface can be considered the layer immediately beneath the potentially oxidized layer to obtain undegraded organic matter which could be as little as >2 m deep; shallow-depth drilling down to 200 m where organics could be cored from the biomolecu-lar fossil record is technically possible but highly challenging requiring a human presence and a significant infrastructure; deep drilling down to 5 km+ in search of liquid water and biospheric activity is currently beyond our extraterrestrial technology. Subsurface liquid water below the Martian surface has been modeled to exist at depths of 2.5–5.0 km at the equator and 8–13 km at the poles—unachievable currently for the foreseeable future but it is apparent that liquid water can exist locally at much shallower depths of ~100–200 m [1138]. Depths of 200 m have been considered but are unlikely to be achievable in a low-mass system [1139]. Evidently, deeper drilling implies greater stratigraphic access and data quality. Terrestrial drilling can achieve 12 km drilling depths but this requires a vast infrastructure and industrial-scale human teams. In general, the upper 1–2 m of a planetary surface is unconsolidated regolith/soil—this is typically exposed to the atmosphere (for atmosphere bodies) or the space environment (for atmo-sphereless bodies). However, it is expected that drilling to <1 m will yield no difference in scientific return than shallower trenching or sampling from a robotic manipulator. Hence, a drill is best deployed for deeper access beyond a depth of 1 m. For consolidated material that has not been subjected to turnover, a depth of up to 10 m is required. It is expected that drilling to 3–5 m will be required for robotic drilling in the near term with extension to 5–20 m being a longer term goal. Depths below 10 m are considered to be unlikely for robotic drilling yet the depth range 5–10 m is considered to have the highest scientific value [1140]. In the case of Mars, it is estimated that depths of 3+ m are required to access unoxidized material for access to pristine organic material. Bedrock from early magmatic resurfacing is not expected to occur until depths of 20+ m are reached (though more recent magmatic activity as evidenced on Mars gives lie to this assumption). In the near term, it is reckoned that depths below 5 m are unlikely to be attained (classified as shallow drilling), and depths of 1–2 m are currently projected for most near-term missions. In the case of the Moon, the maximum drilling depth is around 3 m due to high compaction of subsurface regolith due to eons of micro-tektite impact flux below this depth. Generally, it is considered for reasons of scientific replicability that a minimum of three to five boreholes should be drilled at a particular drilling site, though this may not be achievable within the constraints of a single robotic mission.

Robotic drilling on planetary surfaces such as Mars will be extremely challenging involving complex mechanical and control problems let alone the hostile mechanical environment characterized by low temperature and low press-ure. Unlike conventional drilling platforms, a planetary drill must be of low mass and operate at low power levels. Removal of cuttings usually requires drilling fluids that also cool the drill bit. Drilling muds should not be used as they can potentially destroy the integrity of the borehole environment. Furthermore, there are mass limitations and fluid-handling issues. The critical issue for Mars drilling is cuttings removal from the borehole where drilling fluids cannot be used. A helical auger on the external surface of the drill may be used to convey cuttings from a rotary drill bit. The chief problem is clogging of the augers with compacted cuttings. Low-pressure gas may be used as an air hammer to blow regolith out of the hole. This substantially reduces the required weight on bit to near zero. A combination of compressed atmospheric gas blasts and an auger may suffice aided by natural sublimation of exposed subsurface ice formations [1141]. Even small amounts of ice in the soil would generate sufficient water vapor to blow cuttings from the hole without sticking to the drill bit. Most planetary drilling scenarios assume an average drilling rate of 1 m/day into basalt, sediments, glasses, and ices with downhole thrust determined by weight on bit (this makes low-gravity envir-onments problematic). The subsurface terrain may vary from soft soil to hard rock (basalt) boulders to ice lenses. The Martian regolith comprises a range of soil types, the ground layer typically being unconsolidated eolian deposits from wind action. The soil varies from sandy soil and includes embedded gravel and rocks. Ice requires the lowest amount of drilling power while basalt requires the greatest drilling power with permafrost somewhere between. Ice within sediment acts as a cement and this may be an issue deep within the Martian regolith. Water ice increases in cutting strength from 25 MPa at 250 K to 60 MPa at 140 K—carbon dioxide ice has half the cutting strength of water ice at any given temperature [Garry, J., and Wright, I., 2007, private communication]. However, icy soils tend to maintain their borehole integrity in comparison with granular soils which are prone to collapse. For ice drilling, thermal melting is practical (e.g., electrothermal heating, hot water/steam jets, lasers, microwave heating, and thermochemical heating) though it is less efficient than mechanical cutting due to the high thermal capacity of water. Flame jets can drill in ice at a rate of 15 mm/s and frozen soil at a rate of 5 mm/s but power density is limited to 3 MW/m^2 due to the high heat capacity of meltwater. For pure solid ice at $-5°C$, a power density of 3 MW/m^2 gives a theoretical penetration rate of 10 mm/s—much lower than that achievable by mechanical drills. Mechanical drills, however, can yield to partial melting and refreezing of ice between the bit's teeth spaces—only the use of melting point depressants can counter this (salts, glycol, or methanol) but these represent contaminants. Hot-water jets pumped under pressure of ~1–10 MPa through a nozzle are far more efficient offering penetration rates of ~50–100 mm/s. Water from the hole is recirculated to a heater giving specific energies of 300 MJ/m^3 at 0°C. Drilling will have to cope with a wide range of deposits including hard rock such as basalt—indeed, if basalt can be drilled, the other materials should not be problematic. Soil porosity will decrease with depth according to:

$$\pi_z = \pi_0 e^{-z/K}$$

where $K = 2.82$ km on Mars and $\pi_0 = 40\%$ on the Mars surface (from Viking data). Martian surface soil is likely to contain adsorbed water and carbon dioxide in pore spaces. There may also be volatiles chemically bound in rock and regolith. This will require a robust design of the drill bit to suit multiple types of drilling medium including icy soils, sedimentary rock, sands, and volcanic rock such as basalt. Although steel is usually adopted for drills, Ti alloy offers reduced weight but tungsten carbide cutting edges could be adopted for frozen soils. Body-centric cubic structure metals cannot be used as they become brittle at low temperatures and composite materials suffer thermal fatigue due to thermal cycling. Diamond-impregnated bits comprise soft metal such as bronze impregnated with small diamonds dispersed in the metal. As the bronze is worn, diamonds are exposed to cut the rock. The bronze hardness must be matched to that of the rock to balance wear on the diamonds and the bronze. Discrete cutter bits have a few large teeth of tungsten carbide with a layer of diamond cutters on the forward edge. These give a much greater penetration rate than the former bit but suffer from greater wear. Tungsten carbide cutters and polycrystalline diamond cutters were compared at 60 rpm indicating that polycrystalline diamond cutters with a specific energy of 210 MJ/m^3 were superior. Both these drill types suffer from poor performance in ice and icy soils due to ice melting and refreezing which favors the use of negative rake angles. A hybrid approach uses two cutters—a positive rake diameter cutter and diamond-impregnated material. Diamond-impregnated drill bits in which polycrystalline diamond cutting elements are impregnated into soft metal such as bronze (ideally matched to slightly less than the rock hardness) will expose new cutting teeth as drilling proceeds. Such drills with limited weight-on-bit capability available from rover weight can drill into basalt but to a shallow depth of just a few centimeters [1142. To access the subsurface, some form of drilling is required whether it be by rotary screw drill, by linear impact percussion, or by ultrasonic drilling. Drilling comprises three processes—penetration into the ground, removal of displaced cuttings, and maintenance of borehole integrity. Percussive drilling chips and crushes rock while rotary drilling gouges and shears rock. Conventional drilling involves mechanical cutting and the use of rotary drag bit, rotary roller bit, percussive, and rotary percussive methods [1143]. Most drilling scenarios are based on continuous coring by means of rotary ultrasonic diamond core drilling, a rotary PDC (polycrystalline diamond compact) reamer, and downhole liner placement [1144].

There are a number of subsurface penetration techniques that may be deployed on planetary surfaces ranging from exotic methods such as laser/electron beam drilling to more traditional methods such as rotary drilling: (i) non-traditional drilling (e.g., laser, electron beam, microwave, etc.), which requires high power—a scarce resource on planetary surfaces; (ii) melting tip drilling, which requires high power and pollutes the environment by forming a glass casing; (iii) rotary drilling (e.g., Rosetta lander), which is the commonest method but suffers from the need to autonomously assemble the drill string beyond penetration depths of ~1 m; (vi) percussive drilling, which is the most viable in terms of power consumption. Modern drill technologies such as laser, electron beam, microwave jet, etc. require high power, about three to five times as much energy as conventional drills. Melting tip drilling is eminently suitable for penetrating ice such as found in the Martian polar caps or the Europan/Enceladus ice shell. A variation on this approach is the use of hot-water jets, currently the approach adopted for penetrating Lake Vostok in Antarctica. The most suitable power source for such approaches is an advanced radioisotope thermal generator (RTG). Due to water's high specific heat capacity, this represents an energetically costly approach. In general, it is not clear if a drilled hole will maintain its integrity against collapse which makes recovery of the drill tool at depth difficult at best or impossible at worst. If the soil has low cohesion, the borehole will collapse unless casing is placed down the hole. The two methods that are deemed suitable for planetary deployment are rotary drilling and percussion drilling [1146]. Percussive drilling is more efficient than rotary drilling. Percussion is less likely to be hindered by small buried rocks which will be displaced while larger rocks will be fractured by percussion.

Rotary drilling has the advantage of a substantial body of experience in the petroleum industry. Rotary drilling bores holes by applying thrust and torque to the drill bit. Rotary drilling rotates and pushes the drill rod and employs a cutting bit or a coring bit at the end of the drill rod. In terrestrial environments, it is generally used in conjunction with wash boring which cools and lubricates the drill tool—a technique that cannot be employed on planetary missions. Rotary drilling is an extremely versatile method capable of penetrating cohesive and cohesionless soils and rock. Rotary drilling requires two motors (or a linked gearing single motor) to provide rotary action and vertical thrust. The rotary drive is external to the drill stem with torque transmitted to the bit through the drill rod while thrust is usually provided by weight on bit. Short-flight augers comprise a helix of limited length with cutting teeth at the end of the flight. Continuous-flight augers comprise a helix along the entire length of the stem (which may or may not be hollow to allow sampling at any desired depth). Depth of penetration is limited to the length of the flight. Cutting involves two processes—parallel chiseling to the surface using drag bits and percussive indentation normal to the surface using roller bits. For parallel chiseling motion, the tool is characterized by three angles: rake angle (β_1), clearance angle (β_2), and included angle (β_3). For ice and frozen soils, the rake, clearance, and included angles should be 40–50°, 10–15°, 30–40° respectively. Each cutter follows a helical path. For normal indentation into brittle material, a crater is formed by spallation.

Rotary drilling is the most widely used mode of drilling but involves dulling the drill bit and requires significant force at the initiation of drilling. Rotating and moving parts used in drill strings are subject to single-point failure modes. For shallow depths to <20 m, a continuous drill pipe is often employed comprised of individual drill segments screwed and locked together as the drill progresses deeper. It is possible to attach a downhole drill unit to a wireline so that drill depth is not constrained. The wireline is deployed by a winch and reel. This requires that the drill unit anchors itself to the sides of the hole while a spring provides weight on bit thrust. This spring may be compressed by a motor when the anchor is disengaged for descent. Cuttings are conveyed upward by an auger to the surface. Rotary drills can recover core samples within its coring section which retain the morphology of the sediment without contamination from the surface. Once a core has been acquired, the drill system is winched to the surface and the core deposited. Terrestrial systems use extensive land support structures of high mass, though this can be dispensed with for modest depths. Local heating to 250–500°C can be generated by drilling which could cause mineralogical alteration. Rotary drilling suffers from bit dulling/breaking, bit jamming, high axial forces at drill initiation, the potential for drill walk, and high power requirements.

The use of rotating drill bits and drill strings is not power efficient. Percussive drilling operates through the hammer and chisel principle in which kinetic energy is converted into a stress pulse. It is suited to hard brittle rock that can be chipped into fragments. The bit is held in contact with the rock by a small bias force (<70 N/mm bit diameter). Between each blow, the drill bit is rotated so that a fresh region of rock is presented to the drill at each impact. Such indexing can be generated by a small rotation motor. Percussive drilling typically involves impact frequencies in the range 20–60 Hz but become inefficient at a depth of 10 m due to dissipation in the drill column. For deeper percussive drilling, downhole hammers are used which may be attached to a rotary drill. The mode of penetration is by alternately raising and dropping the tool through 1–2 m using the winch to provide free fall percussion. Percussive drilling is limited to maximum blow energies of 0.5 kJ per impact depending on drill construction. A typical blow energy to mass ratio lies in the range 1.5–5 J/kg. For cohesionless soils, a shell comprising a heavy steel tube with a flap valve at the lower end is used. For cohesive soils, a clay cutter comprising a steel tube with a cutting shoe at the lower end is used. To break up buried boulders, a chisel weighted with a metal sinker bar may be used. Percussive methods require no drilling fluid or lengthy drill string but it has a low rate of penetration in hard rock and makes the transport of cuttings difficult. Percussive drilling does not guarantee core sample integrity but is more efficient that rotary drilling. Vibration-assisted drilling at the soil's natural frequency to break cohesive lumps is commonly employed in terrestrial drilling. Ultrasonic drilling is fast and does not dull the drill bit but requires high power, undergoes significant energy dissipation in soft rock, and requires mud fluid for acoustic coupling to the rock. In general, percussion-based methods have difficulty with cuttings transport. Resonant acoustic drilling is an aggressive drilling technique. The drill rig has a sonic head that uses counterrotating weights to generate energy at 150 Hz which expand and contract an attached drill pipe generating a cutting action. It can drill two to three times faster than conventional rotary drills. However, this method is massive and power intensive.

Existing drilling methods require high axial forces, high power consumption, and heavy support infrastructures. The use of rotating drill bits and drill strings is not power efficient. Conventional drills suffer from bit jamming, bit breakage, and bit blunting. Rotating and moving parts used in drill strings are subject to single-point failure modes, particularly in the dusty Martian and lunar environments, which reduce the likelihood of mission success. Modern exotic drill technologies for hard rock such as high-pressure water jets, thermal laser/electron beam/microwave lances, explosive jets, electric arc, and plasma drilling, etc. require high power. High-pressure water jets evidently require extensive water resources which are likely to be absent (except at polar regions or on water-laden bodies like Europa). Explosive jets (shaped charges) can drill holes by generation of a detonation wave from a shaped charge with a cone angle of around 60°. This generates a deep and narrow hole with a broad crater at the surface. Penetration depth is inversely proportional to the square root of the target material. Stronger target material generates deeper and narrower holes. Flame jets using fuel/oxidant jets can penetrate almost any rock, particularly those with high quartz content, in which thermal spalling causes fragmentation and flush debris from the borehole. Electrothermal hot-point drills require power densities of 0.3–3 MW/m^2 for melting boreholes and form a fused borehole lining. These methods require high energy (e.g., the melting of basalt rock requires 100 GJ/m^3 of energy). Non-traditional drilling technologies (laser, electron beam, etc.) are power hungry and require three to five times as much energy as conventional drills.

All drilling systems must consider cuttings removal for which there are several techniques: continuous flight auger (screw conveyor), cyclic lift auger, gas circulation, or liquid mud circulation. Furthermore, cuttings removal will remove 70% of the heat generated during rock fracture to prevent overheating. A continuous flight auger for cuttings removal is the typical approach adopted though this increases the energy requirements substantially for deep drilling beyond ~10 m so this approach has limited depth capability. Furthermore, augers are limited to dry, powdery materials but choke when conveying wet, sticky material such as clay. Terrestrial drilling systems utilize liquid drilling mud to cool the drill bit, convey cuttings to the surface, and provide hydrostatic pressure against borehole collapse. Fluid is typically forced down the inside of the drill pipe and returns up the spacing between the drill string and the borehole wall. The transport of fluids is not an option for planetary drilling systems. Fluids such as muds are infeasible due to their propensity to contaminate and permeate into regolith. However, a mole can use circulating liquid xenon or carbon dioxide pumped from the surface through a thin tube to the mole which ejects it through an orifice. The fluid mixes with soil/ice samples to convey the samples to the surface like drilling fluid (mud logging). Carbon dioxide might originate from the Martian atmosphere which would require compressors on the surface. This is similar to wash boring which involves pumping water through hollow boring rods under pressure to be released as jets in narrow holes at the end of the rod to loosen the soil. Maintenance of borehole integrity against collapse (particularly in sandy soils) is typically achieved through insertion of metal or plastic piping (casing) or through high-density drilling fluid (mud). This will not apply to planetary operations. However, two epoxy fluids may be pumped down from the surface to the mole where it is mixed and extruded from the mole to create a hollow shaft that solidifies to form a hard epoxy material liner as the mole descends. These methods involve fluid handling and the use of pumps, valves, and hydraulic devices that are complex to control. For cuttings conveyance only, a compressed gas system may be viable—a gas mass flow rate of 0.032/0.22 g/s is required for 0.1/0.5 mm diameter drill cuttings using a 25 mm diameter drill. The gas supply may be generated from chemical decomposition or compressed atmospheric gases. The latter is impractical to collect and pressurize as drilling fluid. Furthermore, CO_2 heated to 250°C during drilling would cause mineralogical alteration. For Mars, drilling without fluid may be possible in that the permafrost ice may sublime on drilling the borehole, clearing the hole of cuttings. However, this is dependent on sufficient ice sublimation advocating the use of blasts of gas to clear the hole periodically [1147].

The primary role of the drill is to recover a core sample of subsurface material—a 1 cm diameter core to a depth of 1.5 m yields a 200 g core. The drill must be able to cope with the variable hardness of soil and rock and must be able to do so autonomously. Several drilling concepts have been applied to space missions. Most planetary drills are rotary drills employing a helical auger which despite their disadvantages are well understood but the most efficient approach is to use a mechanical percussion device that is highly power efficient. It has been suggested that a screw drill auger with a spiral edge may be mounted onto a rover body to provide a rigid platform to drill to a depth of 1–2 m at a penetration rate of 13 cm/s—to penetrate 1.5 m requires 1.15 kW of power [1148]. The auger would have to be housed in a metal tube to enable recovery of soil samples which are carried to the surface from the sub-surface. The auger cannot drill hard rock but does not require fluid for lubrication or cooling. It is driven by a rotation drive unit (with a torque of 7–70 Nm and 0.75 kW of power) and a translation drive unit (with a force of 1.75 kN and 0.4 kW of power) which have a combined mass of 260 kg. The drill head has a width of 1.4 m and length of 1.7 m. Drilling velocity is linearly dependent on rotation speed. A point worth noting is that integrating scientific instruments into a rotating drill is not trivial—it is only in recent years that wireline logging [1149, 1150] has been partially superseded by MWD (measurement while drilling) technology. Drilling must be feedback-controlled based on temperature measurement to prevent overheating of the soil which can cause ice to melt and organic material to be destroyed.

One of the simplest planetary drills flown was the single-shot DS2 microprobe sampling microdrill which was stowed in the forebody. This drill generated a thrust force of 45 N with a torque range of 0.226–0.282 Nm and rotation speed of 7.5 rpm. Rotary drilling does, however, require automated assembly of drill segments into a growing drill string to achieve any significant depth below 1 m. The 5 kg Deep Driller is a prototype mounted within the payload cab of a large Nanokhod—a mobile drilling platform comprising two parallel track bodies connected by a lifting bridge and sandwiching a rotatable payload cab housing the drill and sampling system—capable of drilling to depths of ~1–2 m [1151]. The mobile drilling platform is a tracked tethered vehicle that enables the drilling and sampling system to be delivered to multiple locations and to return to a lander sample port. The tiltable payload cab allows vertical drilling into the ground or horizontal drilling into a cliff. The rotary drill can autonomously penetrate hard or soft soils and can drill multiple holes. It incorporates three motors to drive the assembly process—a thrust motor to provide downward thrust to assemble the drill string without reliance on local gravity, a drill pipe carousel that rotates to select a pipe, and the tool bit carousel that rotates to select the tooling bit. The tip of the drill can drill, core, and retrieve samples for analysis at the lander. A drill pipe is selected by rotating the drill pipe carousel to align the selected drill pipe with the thrust unit. The thrust motor drives down the drill pipe to the drill string. The drill pipe mates with the drill string (initially the drill tool only) with a one-way clutch, building up the assembled drill string. The assembled drill string drills into the soil powered by the rotation and thrust motors. When the top end of the drill pipe is reached, a new drill pipe is attached to the growing drill string.

The 10–25 kg Deep Drilling Sampler was derived from the canceled DS4 Champollion comet lander sample acquisition and transfer mechanism (SATM) cryogenic drill to acquire samples at the surface with a depth of 20 cm, 1 m, and 10 m [1152]. It had a drill string attachment in the form of a linear feeder with multiple drill strings to allow drilling down to 10 m. The drill pipe was built up from sequential drill string segments as the drill penetrated the ground. Multiple drill string segments could be added to each other robotically during drilling to create a short drill tower. Motors inside the lead drill string moved a center pushrod which ejected core samples. A rotating shear tube sheared the core sample from the base rock. The thin cross-section of the drill string made it susceptible to breakage during drilling.

The Rosetta mission is currently on its way to place its 85 kg pentagonal lander with three folding legs onto the surface of a comet and recover multiple ice samples for in situ analysis. The Rosetta lander's SD2 (sampling drill) rotary drill system is designed to drill into the surface of a comet and collect ice samples at different depths up to 0.25 m [1153, 1154]. Two anchoring harpoons will provide sufficient reaction for drilling. The drill head includes a force/torque sensor to monitor the maximum vertical thrust of 20–50 N for closed-loop control. The Rosetta SD2 comprises three components in a package of mass 3.1 kg requiring only 5 W of power for rotation speed <100 rpm, torque <10 Nm, and thrust <100 N: a tool box containing the drill and sampler, a carousel for rotating microcontainer ovens to hold sample material, and a local control unit to drive the SD2 motors. The Rosetta 12 mm diameter drill/coring tool comprises a hollow outer tube with a helical auger that can rotate at <100 rpm. The drill head comprises polycrystalline diamond cutters to cut into the ice and soil. The bottom of the core sample once acquired is cut with two quarter-spheres while rotating the drill. The non-rotating inner tube thrusts down during drilling and is extracted while the outer tube remains in the borehole. The bottom of the core sample once acquired is cut with two quarter-spheres while rotating the drill. The drilling head is mounted with a force/torque sensor to monitor the maximum vertical thrust of 20–50 N. The toolbox containing the drill can rotate around an axis orthogonal to the ground. This rotation moves the drill over the slot in the ground plate to provide access to the surface over the drilling site. The drill has two degrees of freedom—one vertical translation to push the drill down and one rotational about its axis. After drilling, the bit is removed from the borehole to return the sample to the surface. The drill then penetrates to a predetermined depth using its thrust and torque drives after which it is retracted. The tool box rotates to place the sampler over the borehole where it is inserted into the borehole for sample collection. The sampler is retracted and the tool box rotated to place the sampler over a microcontainer on the carousel. The microcontainer and its sample are rotated for access to different scientific instruments. Each sample is transported to a thin-plate carousel for deployment onto a sample feed point which rotates to gain access to different experiment stations: microscope/infrared spectrometer, heating ovens, and evolved gas analyzers.

The DeeDri (Deep Driller) rotary drill is a compact integrated drill based on the Rosetta SD2 for simultaneous drilling and core sampling to 1–5 m depths in regolith. It employs shutters to enable sample coring [1155, 1156] (Figure 11.7). DeeDri is different in that the drill and the sampler are part of the same device. DeeDri comprises a 50 cm long cylindrical steel rod, cutting bits, and an internal sliding piston which creates a central sample chamber both to collect and eject soil samples. The carbure central bit is augmented by six radially mounted polycrystalline diamond blades. The central piston is withdrawn to create an internal volume for core collection. The DeeDri drilling tool prototype has a 35 mm diameter with sample collection capability and the corer has a diameter of 17 mm and a length of 25 mm. Its penetration speeds vary from 1 mm/min in travertine with a vertical thrust of 50 N (10 W power) and 0.4 mm/min in marble with a vertical thrust of 200 N (20 W power); it has a rotation speed of 125 rpm.

In order to accommodate limited storage volume on a planetary mission, DeeDri employs multiple drill rods that are automatically assembled into an extendible drill string to reach the required drilling depth. The drill segments are mounted in a carousel inside the drill box. A complex mechanism is used to automatically assemble the drill strings and to change tooling. To drive each carousel the assembly has three motors: the thrust actuator to provide downward thrust to assemble the drill string, a drill pipe carousel that rotates to select a pipe, and the tool bit carousel that rotates to select the tool bit. The tip of the drill can drill, core, and retrieve samples for analysis at the lander. A drill pipe is selected by rotating the drill pipe carousel to align the selected drill pipe into the thrust unit. The thrust motor drives down the drill pipe to the drill string. The drill pipe mates with the drill string (initially the drill tool only) with a one-way clutch, building up the assembled drill string. The assembled drill string drills into the soil powered by

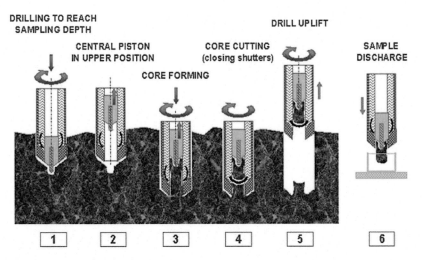

Figure 11.7. ExoMars drill prototype: (top) coring; (middle) core retracted; and (bottom) drilling and coring operation [credit ESA].

the rotation and thrust motors. When the top end of the drill pipe is reached, a new drill pipe is attached to the growing drill string.

The proposed ExoMars subsurface drill system is based on DeeDri and comprises a drill box and a 3 DOF drill-positioning unit of total mass 7.25 kg with volume of $175 \times 175 \times 540$ mm (Figure 11.8). Its basic requirements were to drill to a depth of 2 m and recover samples from surface rocks. It is to provide cores of 40 mm in length and 10 mm in diameter in soils with a hardness of 0–150 MPa implying a drill diameter of 25 mm. This depth required the use of three extension rods each of 575 mm length to be assembled onto the drill bit rod to automatically construct a drill stem of up to 2 m long. The first rod only has a sample collection facility while the additional rods are extension rods mounted onto a carousel. When the first rod is fully buried in the soil, it is disconnected from the mandrel and an extension rod inserted between it and the mandrel. The rods couple together with multiple electrical line connectors. This process is repeated until all the drill extension rods are connected to yield a depth of 2 m. For return of the sample to the surface, the drill rods are uplifted and disassembled in turn. This requires that the drill operations be observed using a camera at the bottom of the drill box. The drill tool and extension rods support optical fiber connection for in-tool instrumentation. It has to be capable of drilling

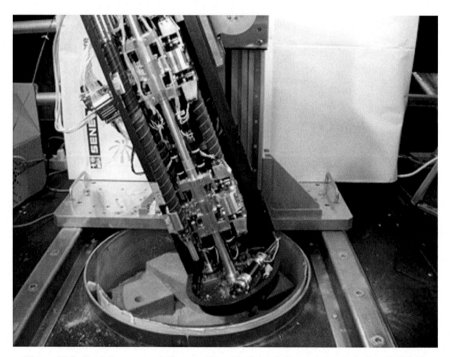

Figure 11.8. ExoMars rover drill assembly with three drill extension rods [credit ESA].

regolith and rock of unknown hardness up to 150 MPa hardness (equivalent to granite). It requires a thrust level of 100–200 N and a torque capability of 1.5–2.0 Nm utilizing 30–60 W of power (depending on material). The drill-positioning unit (plus associated electronics for power supply, microcontroller, motor drive boards, etc.) requires 3 DOF—two rotational (yaw/roll) and one translational (up/down) to place the drill box onto the soil. It can be jettisoned by pyrotechnic bolts if the drill becomes stuck. Particular attention must be paid to a dusty environment. Drilling is the most power-consuming operation during drill box deployment—this was estimated at 33 W for drilling in travertine at 1 mm/min at a rotation speed of 125 rpm (1.3×17 W for drilling $+ 4$ W for electronic interfacing $+ 3.5$ W for electronic computer boards).

A novel planetary drill mechanism has been proposed involving a tethered downhole motor drill which has low power consumption and eliminates the need for automated drill string assembly [1157]. A self-contained drilling mole descends while tethered to the surface—the unit comprises a pilot bit and a series of helical cutting auger sections. A series of helical augers of increasing size perform communition of the soil in excavating the borehole. It requires a surface tripod frame/winch to deploy and recover the drill into and from the borehole and an extendible sheath with a complex series of operations.

The requirement for autonomous assembly of drill strings is considered to be cause for concern with regard to reliability as a single-point failure mode in such systems. A bio inspired drilling approach has been proposed based on the mechanism of the ovipositor of the woodwasp *Sirex noctilio* (horntail sawfly) [1158–1160] (Figure 11.9). It deploys its ovipositor to drill into the bark of trees to lay its eggs. It comprised two half-sections: one side with teeth for wood cutting and the other with pockets to remove debris. The two sides operate reciprocally. The maximum pushing force of the drill is dependent on the buckling stress along the cylindrical drill length. It drills into wood across the grain to deposit its eggs while adhering to Euler end-load buckling constraints given by:

$$F_c = \frac{\beta \pi^2 EI}{L^2}$$

where $\beta =$ end condition constant $= 1$ for free joint or 2 for pinned joint, $E =$ Young's modulus, $I =$ moment of inertia, and $L =$ strut length. The ovipositor may be modeled as a circular cross section hollow tube with moment of inertia $I = (\pi(R_o^4 - R_i^4))/16$ and outer cross section of 0.1 mm. The first four teeth at the tip point proximally cut only on the upstroke (pull teeth). The following teeth are distally oriented and cut on the downstroke (push teeth). The spacing is such that push teeth on either side alternate with each other.

Euler end buckling provides the strongest constraint on the length/stiffness of the drill pipe. The ovipositor of the woodwasp *Megarhyssa nortoni* is longer and thinner than that of *Sirex noctilio* and is stabilized at the abdomen by a grooved guide that locks it to allow thrust to be exerted. The tip teeth face proximally so that initial fracture of the wood is through upward pulling in tension exerted by the abdomen muscles. The stiffness of wood is around 70 GPa with a 2% breaking

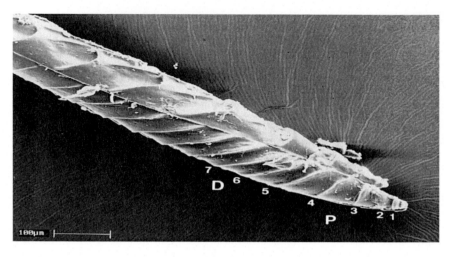

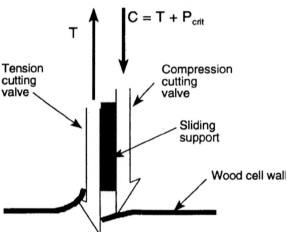

Figure 11.9. (top) Biological ovipositor of the woodwasp *Sirex noctilio*; and (bottom) ovipositor mechanism [reproduced with permission from Julian Vincent].

strain (i.e., fracture stress of 1.5 GPa). The two wall sections require one to be in tension to stabilize the ovipositor without buckling. The downward-pushing teeth are staggered so that cutting forces are distributed evenly in the power stroke. The pits are aligned to guide sawdust away from the cutting edges. *S. noctilio* rhythmically oscillates its body up and down in addition to the rapid mechanism of the ovipositor to aid in cuttings removal. The drilling rate is 1.5 mm/min.

A biomimetic drill inspired by the woodwasp ovipositor has been proposed for a planetary drill in that it overcomes the necessity for automatically assembling/disassembling drill strings under dusty conditions to provide greater mechanical reliability [1162, 1163]. The two sections combine teeth and pockets. The drilling force is generated between the two reciprocating sections so no external drilling force is required. Experiments conducted in condensed chalk, non-fired clay, and lime mortar at nine different power ratings demonstrated the viability of the bio-inspired drill—drilling speed increased with penetration depth as more teeth became engaged [1164]. Penetration speed was empirically determined by:

$$\nu = k\frac{P}{\sqrt{\sigma}}$$

where k = proportionality constant, P = input power, and σ = substrate compressive strength [1165]. The drill bit design was based on serrated half cones of increasing diameter. The edges of the cones provide gripping action. The drive mechanism is based on a simple but compact pin and crank [1166] (Figure 11.10). Piezoelectric motors provide closely matched oscillatory motion to that required over electric motors.

Coiled metal strips are pulled out of the reel due to reciprocation of the slider bars of the drive mechanism. A friction wheel roller with slots drives the metal strip to deploy and retract. The clips attached to the slider bar grip the teeth of the metal strips providing the force to pull the strips out of their roll canisters. Once fully extended, a solenoid is activated to switch the gear motor from the pin and crank to the reeling mechanism to allow the metal strips to be recoiled. Sample acquisition may be achieved by lining the inner core with angled bristles similar to a head of barley to convey particles upwards.

The biomimetic drill can also be used to recover water from ice pockets on asteroids, comets, or near-Earth objects (NEOs). Many NEOs are extinct comets comprising mostly water mixed with silicates and other rock-forming minerals (dirty snowball model). The drill tube descends as the working mechanism as well as acting as the borehole liner. Heat may be conveyed by an internal probe and melt the ice which may be pumped to the surface. Such water resources may also be used as thermal propellant (steam) to provide thrust for rockets or, alternatively, electrolyzed into hydrogen and oxygen for fuel—the hydrogen could be particularly valuable on the Moon where it is difficult to access.

The drill must be mounted in such a fashion not to interfere with rover locomotion. This involves mounting within a 3 DOF insulated drill box which may be stowed horizontally and sideways in front of the rover to maximize visibility but minimize interference with other rover systems. The drill is jettisonable in the event that the drill becomes jammed in the Martian soil to ensure rover mobility is not impaired. The shorter the drill, the easier its accommodation particularly if mounted vertically; conversely, the longer the drill, the fewer drill segments are required to drill to a given depth. Drill cameras enable monitoring of the drill emplacement and drilling process. Rotary drilling is generally used for depth sampling and coring—the integration of in situ scientific instruments into the drill collar is not a trivial problem. Measurement while drilling (MWD) tech-

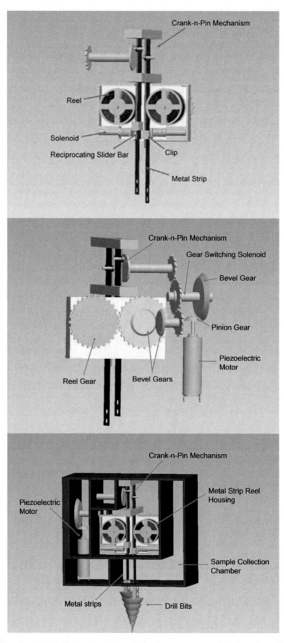

Figure 11.10. Biomimetic drill stowage and deployment mechanism [credit Tim Hopkins, Carleton University].

nology has only recently become common in replacing dedicated wireline-logging sondes in the terrestrial oil industry. The extraction of geotechnical data such as soil physical parameters is possible during drilling by exploiting measurement-while-drilling (MWD) techniques. The response of the drill differs according to the material through which it is drilling. This may be as simple as the difference between consolidated and unconsolidated material, or as complex as diferentiating between icy and dry regolith. The extraction of such data is complex involving learning/classification which may be achieved online if in situ chemical analysis can be conducted or offline if learning/classification are predetermined. The combination of the need for autonomous drill string assembly and the ubiqitous Martian dust offering a potentially hazardous situation for rotating parts suggests that rotary drilling to depth is not a reliable option for subsurface penetration of planetary surfaces.

Drill bit seismic emissions can help to predict hazards in advance of the drill bit [1167]. Acoustic energy is radiated both from a working drill as the drill penetrates the rock and from the axial force applied to the drill stem as the drill radiates compressional or primary waves into the surrounding rock. A roller core bit acts as the dipole source for P-waves radiating energy outwards. At the surface, an array of geophones in shallow holes to improve acoustic coupling can detect P-waves. Multicomponent accelerometers at the top of the drill string will detect axial vibrations up the drill string. Energy propagating through the rock formation travels more slowly than the axial vibrations of the drill stem, providing a time-shifted sensor signal. Time differences can be computed by cross-correlation methods. If the time for the axial vibration to travel up the drill is known, the time for seismic transmission from the bit to the surface can be computed. The energy propagated beyond the drill bit will often be reflected by impedance changes in the rock which allows look-ahead images to be generated.

MARTE (Mars Astrobiology Research & Technology Experiment) conducted in Río Tinto (Spain) was based on a partly automated auger coring drill assembly including a core sample-handling system and a number of scientific instruments [1168]. The sample-handling system comprised a core clamp to grip cores, a rock crusher, a double-bladed saw to expose core surfaces, and a linear rail to convey the clamped core sections to scientific instruments. The instruments included an imaging microscope and Raman specrometer for in-hole inspection and a visible/near-infrared spectrometer and life detection system (Signs of Life Detector—SOLID) for analysis of iron and sulfide-consuming chemoautotrophic bacteria.

Drilling into hard non-conductive material using concentrated microwave energy offers an alternative mode of drilling. This approach melts the localized material using a conventional magnetron source at high power density. Although microwave energy of 1 kW at 2.45 GHz is limited to a few centimeters in soil penetration depth, microwaves can be focused to mechanically assist drilling at much lower powers through rock fracturing. A microwave drilling head has been created comprising a coaxial waveguide that conveys microwave energy to a near-field concentrator that focuses microwave radiation into a small volume immediately at the end of the high melting point metal drill pin [1169–1171]. A molten

hot spot is generated by a thermal runaway process reducing the power requirements. This hot spot at temperatures up to ~1,500°C has a much smaller size than
the microwave wavelength (12 cm), typically ~1 cm. By using a sharpened tip at
the drill pin to increase the electric field concentration, this enables the use of
solid state microwave sources at lower power ratings ~10–100 W [1172–1174]. A
drill pin may be inserted into melt which cools and the pin retracted leaving a hole
with a glass casing. It has been tested in different hard non-conductive ceramics,
basalt, concrete, glass, and silicon showing its use in construction and machining.
Adapting for planetary exploration would not be difficult but the chief problem
would be corruption of the hole and any samples. More intriguingly, it was
proposed that microwave-induced breakdown spectroscopy of the slight plasma
generated might be feasible for low-cost mineral identification.

11.4   SAMPLE ACQUISITION BY MOLE

An alternative to drilling with a long drill string is to use burrowing robots—
moles—which offer a smaller size and greater flexibility of operation. The use of a
detached, self-propelled underground mole eliminates long drill strings and the
need for autonomous drill segment assembly. This approach allows collapse of the
hole behind the mole. A mole provides the capabilities of penetrators, but with
rover deployment they can be pinpointed to specific sites unlike atmospheric
penetrators due to inherent uncertainties in their landing ellipse. Moles do not
require external mounting platforms like drills. It is not clear, however, whether
they can be used for penetrating hard bedrock without significant power requirements, but they are ideally suited to penetrating compacted regolith. The ideal end
shape for penetration by percussive moles is a smooth cone rather than one that is
grooved, giving greater penetration for the same percussive force [1175, 1176].
The borehole is typically 7% larger than the diameter of the mole. Rotation of
the mole decreases penetration resistance which may be an option for increased
efficiency though power consumption tradeoffs would have to be studied.
Underground moles are an alternative drilling mechanism that is suited to loose,
unconsolidated, porous regolith with shallow-depth penetrations of <5 m but
cannot be used for hard soils. The Beagle 2 PLUTO (planetary underground tool)
mole was a variant on the percussive drilling approach developed by DLR in
Germany and was mounted at the back of PAW on a carrier tube (Figure 11.11)
[1177–1179]. It was a self-contained percussive device of length 0.365 m and
diameter 0.02 m and had a 60° conical tip. It was designed to acquire surface and
subsurface samples of Martian soil (though not rock) to a maximum depth of
1.5 m limited by soil density (compared with the Viking maximum depth of
20 cm). Its rear end was connected to the end of the Beagle 2 lander mechanical
arm by a 2.8 m long tether. The tether comprised Kapton-coated steel wire braid
through which there were two data signal paths (command/telemetry) and one
path for power from the lander. The mole travel distance was limited by the mass/
unit length which was determined by tension due to winch retrieval force (50 N).

Figure 11.11. Beagle 2 ground-penetrating mole [credit ESA].

An Al–Li launch tube against which the mole can react for initial soil insertion eliminated the need for external thrusting. The mole had a mass of 340 g (64 g outer casing, 220 g motor and 56 g mechanisms). The supporting equipment—95 g carrier tube, 375 g winch, and 80 g motor—added a further 550 g mass (860 g in total). The mole had an outer sleeve at the tip for generating core samples. PLUTO comprised an outer Ti alloy case that was driven into the soil by an internal self-penetration mechanism. The mole would be deployed downward into the soil which would involve ARM positioning PAW accurately to orient the mole vertically.

Within the tube, the shock mechanism comprised a sliding hammer restrained by an electric motor to compress a spring. A brushed electric motor pulled the internal proof mass against the compression of a spring. The spring was loaded by the motor through a gearbox and then released. The internal hammer then impacted the forward tip section of the mole causing soil displacement. Reaction to forward motion partially recompresses the spring. A sliding suppressor would absorb the reaction that was connected to a weaker brake spring to return the suppressor to its nominal forward position [1180, 1181]. A second shock from the internal hammer falling to its initial position causes additional forward motion. The spring was to be released repeatedly every 5 s to generate force at the tip of ~0.1 Nm to drive the mole ~2 mm per shock into the soil to a depth of 10 mm with no frictional resistance on the surface (i.e., over 1 h to reach a depth of 1 m depth ignoring soil compaction resistance changes), drawing power of only 2–5 W. Each shock was < 0.5 ms in duration giving high accelerations of ~8,000 g while friction acts over much longer periods (~300 ms) giving decelerations of <3 g. In hardened soil, this progression rate was much reduced (~1 mm/min). Friction increases with depth limiting the mole depth of penetration to 5–10 m depending on the soil properties. It was estimated that soil surface properties of density 2 g/cm^3, cohesion 8 kPa, and friction angle 34° would increase to 3 g/cm^3, 13 kPa, and 48°, respectively, at depth. Mole mass and length must be as small as possible to maximize shock efficiency. The dynamic coefficient of friction of the mole in soil was found empirically to be 0.35. The mole could also be deployed horizontally on the surface as the tether was unwound. In this way, if the mole encountered a rock, it would divert beneath it giving access to undisturbed pristine soil since the boulder emplacement shielded it from solar ultraviolet radiation.

It could reverse its hammering mechanism to backtrack to the surface and be reeled back to the arm at 1–5 mm/s to recover its 60 mg samples for analysis on the lander. The samples were to be acquired with a small sampling valve at the front tip of the mole. The mole launch tube rather than the arm itself suffered mole retrieval forces of 50 N in winching the mole in. The mole was tethered to a motor-driven winch mechanism at the rear of its carrier tube which could rewind the tether to retrieve the mole. Power and data were relayed through the tether to ensure that minimal electronics were required in the absence of thermal control. Return of the mole from the subsurface was aided by reverse hammering, allowing retraction back into the mole launch tube at the end of the robot arm from depths of up to 2 m. Peak power for the mole winch motor was 0.8 W while the mole drive motor used 1.2 W of power (peak power of 3 W) during soil penetration. The mole carried a temperature sensor to measure subsurface temperature at different depths. The tip incorporated a sampling device driven by the same motor capable of acquiring $0.24 \, cm^3$ of soil which was to be delivered to the GAP inlet on the lander by the robot arm once the mole was remounted in its launch tube. A piston drives the tip to open a cavity.

The DLR instrumented mole system is a development from the PLUTO mole and comprises a dual-mole configuration with a driven forward (tractor) mole housing the percussion mechanism trailing a payload mole housing the instruments of 300 g—both are connected by a short 3 mm diameter cable for electrical connection [1182]. Maximum penetration depth is dictated by the tether length to the lander (nominally 5 m). Total mole length is 415 mm comprising the 250×26 mm tractor mole and 158×26 mm trailed mole (with 150×24 mm instrument envelope). Total mole mass is 1 kg including a deployment mechanism with a total power requirement of 3 W.

NASA has also developed a percussive robotic mole, the Mars Subsurface Explorer (SSX) which acts similarly to the Beagle 2 mole but with some differences [1183]. It utilizes a spinning hammer on a screw thread. It is 1 m in length and 3 cm in diameter with a pointed tip in a mass of 5 kg. It requires 30–50 W of power to burrow at a rate of 5 m/day in hardened regolith and 100 W is sufficient to reach 5 km at 10 m/day over a mission lifetime of 500 days. It does not rely on a spring to store up internal energy like PLUTO. The pulsed electric motor shaft includes a long stretched screw thread and it is spun up at 10,000 rpm by the motor. A tungsten rotary hammer spins to the top of the shaft along the threads. When the motor is pulsed off, the shaft slows down and a ratchet immediately halts the shaft, engaging the screw thread. The hammer is still spinning at 10,000 rpm and spins down the screw thread to piledrive into the front of the mole with a force of 100 N or so enabling it to penetrate 10 m/day. The motor kicks in and the shaft begins spinning while the hammer, having been halted by its impact, spirals itself back up the screw thread for another pulse. It is reckoned that this ingenious percussion method is three times as efficient as other drilling methods. It can use liquid carbon dioxide compressed from the Martian atmosphere and pumped from the surface through a thin tube. The fluid mixes with soil/ice samples, acting as drilling mud to convey the samples to the surface. There is the possibility of using liquid Xe as drilling mud and/or epoxy fluids to extrude a solidifying hole casing. The use of fluids, however, introduces potential difficulties as they require complex fluid-handling capabilities such as pumps, valves, and hydraulics.

The chief problem with tethers is that there are high DC transmission losses along the tether imposing a high power requirement on initial transmit voltage but this is commonly limited to around 100 V DC limiting tether lengths to ~10–30 m. Power over Ethernet (PoE) transmits power and data at 48 V DC over Ethernet cables. If the tether is to bear load it must be reinforced. The Dante II tether comprised two 26 AWG coaxial conductors for two 384 kbps video channels, one 26 AWG twisted pair conductor for a 192 kbps Ethernet channel, four full duplex RS-232 serial channels, and one 18 AWG power conductor. An inner core of Cu AWG conductors was sheathed in a layer of load-bearing Kevlar interfaced by a 0.1 mm thick low-friction Gore-Tex fabric. The whole tether was encapsulated in an abrasion jacket. A number of tether winch mechanisms are possible. Tether respooling introduces a number of potential problems. The minimum bend radius determines the size of the winding drum. Tether stacking on the drum must avoid knifing whereby the lower layers of the tether are wound at low tension causing the outer layers which are under greater tension to force their way into lower layers. This can cause tether damage and interferes with payout. This can be avoided by winding the tether on the drum at 1/10 maximum expected tension. Non-uniformity of tether stacking will also lead to knifing. This can be avoided by using a sheath of high-friction coefficient. The winch may be mounted onto flexures that are compliant in the direction of pull but stiff in other directions to limit tether movement to a single direction. A set of pinch rollers that continuously apply tension to pull the tether from the drum by pinching it will eliminate slack in tether winding. Tether payout may be monitored by a potentiometer to measure tension during winching. The problems with tethered concepts are illustrated by the Dante II robot which used a 300 m tether to supply power and data communication wound on a winch drum which paid out under tension while rappelling. The maximum spooling speed was 3.5 rpm on a 32.3 cm diameter drum with a maximum static load of 7.7 kN. Dante II underwent an unfavorable tether angle while climbing out of Mt. Spurr in Alaska in 1994 which caused its tether to snag and break leading to the loss of the robot.

Tetherless moles are not depth-limited to the tether length but are required to be self-contained. The Inchworm Deep Drilling System (IDDS) mole is an advanced design for drilling within planetary surfaces [1184]. It is self-contained and powered by a radioisotope thermoelectric generator (RTG) eliminating the need for a tether. It is self-propelled, capable of burrowing to a depth of ~1–10 km, retrieving samples, and returning them to the surface. The 15 kg device comprises two symmetrical (fore and aft) segments of total length 1 m (extendable to 5 m) and diameter 15 cm. Each segment comprises a motor-driven rotary drill bit at its end and a set of three anchoring shoes. Feet on each segment grip the walls of the borehole to provide anchorage for drilling. The two sections are connected by a telescopic joint powered by a linear motor which contracts/expands the inchworm. Rotating auger flights along the body pass cuttings to the rear. The mode of operation involves anchoring the aft section in a deployment tube using the aft braking shoes while the fore section drills to the maximum length of the telescopic tube. Telescopic tubing serves to ensure the integrity of the hole between the fore and aft sections. The fore section anchors itself to the drilled hole by means of its braking shoes. The aft section releases its brake and slides down the hole to meet the fore section as the telescope contracts. The aft section engages its braking shoes and the fore section releases its brakes and resumes drilling. The propulsion cycle thus has the following form:

(i) shoes on the aft segment anchor to the borehole walls while the forward segment extends forward while rotating the forward drill bit;
(ii) shoes on the forward segment anchor to the borehole while the shoes on the aft segment disengage;
(iii) a linear actuator contracts pulling the aft segment forward;
(iv) shoes on the aft segment engage while shoes on the forward segment disengage.

This provides the facility for sample recovery. Using the same mode of propulsion it can return to the surface to deliver its samples by reversing its mechanical procedure with the aft section taking the lead. IDDS is initially deployed from a tube positioned on the surface of the soil. This is used to provide grip for the anchoring shoes so the forward segment can extend forward into the soil. The chief disadvantage of this device as with all mole devices is that it needs to return to the surface to deliver its samples. Although this represents a very advanced concept, it is massive and power hungry. A similar concept is the Space Worm drill which employs a two-segment borehole-anchoring mechanism [1185]. The major difference is that each segment comprises a forward section that can partially slide into a rear section. Furthermore, it employs a lead screw to convert motor rotation to linear motion to act as a linear actuator with a much higher mechanical advantage than an electromagnetic linear actuator.

It has been suggested that drilling to depths of 200–300 m might be achievable using electrically heated probes to melt through soil and rock creating a glass-lined borehole [1187]. The probe could be attached to a tether for power and data transmission and for probe recovery—this would require 70–100 W to penetrate 1 cm in 1 h. This eliminates the need for hole casing, drilling fluids, or debris ejection as the borehole wall is glassified during descent. A 1 m long thermal lance with a pointed ceramic tip heated by an electric heating resistance coil to 1500°C has been developed [1188]. The shaft of the lance is cooled by pumping gas from the cold Martian atmosphere to the rear of the lance. The tip melts the surrounding rock which cools into a glass casing to reinforce the hole. An unwinding spool of metal tubing carries power and coolant gas to the lance and gives it weight to push through the melting regolith. Such a method of drilling requires considerable electrical power. Furthermore, the high temperature of the thermal lance would chemically alter the mineral environment destroying the pristine borehole environment. In order to access permafrost, additional side-sampling drills or pyrotechnic bullets would be required to penetrate the glassy walls and end cap. The high power levels may require the use of nuclear sources, which has considerable political ramifications. This technique is massive, bulky, complex, untested, power-hungry, but yields the possibility of chemically reducing materials to be recovered.

11.5 DRILL PERFORMANCE MODELING

Theoretical prediction of drilling performance is notoriously difficult. Conventional rock properties such as hardness do not correlate well with drill penetration rate—this has motivated the use of new parameters such as rock quality designation or specific energy [1189] which are vague and not readily measurable. However, uniaxial compressive strength, tensile strength, and Young's modulus have the highest correlations with drill penetration rate [1190]. Drilling performance is dependent on both rock specific energy and rock compressive strength [1191]. The Mohr–Coulomb relation defines soil properties in terms of shear stress over the shear plane:

$$\tau = C + \sigma \tan \phi (1 - e^{-j/K})$$

where $C =$ soil cohesion giving shear stress at zero normal stress, $\sigma = P/A =$ normal stress, $\varphi =$ internal angle of friction giving the slope of shear stress to normal stress, $j = sx = (wp/2\pi) - v =$ ground deformation, $s = ((wp/2\pi) - v)/(wp/2\pi) =$ slip, $x = wp/2\pi$, $K =$ soil deformation coefficient, $w =$ rotation speed, and $p =$ drill rotor pitch. The unconfined compressive strength (UCS) of rock indicates the weight-on-bit-to-drill-bit cross section required $\sigma = W/A$ from the Mohr–Coulomb relation. The UCS of dry sandstone is 50 MPa with a range of 25–70 MPa while that for frozen sandstone is 50–110 MPa. For comparison, the UCS of basalt is 280–320 MPa. Specific energy (Nm/m^3) is the energy required to remove a unit volume of rock—it may be defined as:

$$SE = \frac{\sigma_c^2}{2G}$$

where $G =$ bulk modulus and $\sigma_c =$ rock compressive strength. Specific energy may be regarded as the product of drill torque and rock compressive strength. The specific energy for dry sandstone is 210 MJ/m^3 while that for frozen sandstone is 770 MJ/m^3. Uniaxial compressive strength including porosity is related to cohesion and friction angle by:

$$\sigma_c = \sigma_0 \left(\frac{A_c}{A_g}\right)$$

where $(A_c/A_g) = (1 - (b/a))^2 =$ grain-contact-area-to-unit-cell cross section, $a, b =$

grain/pore size, and $\sigma_0 = 2C/\tan(\pi/4 - \phi/2)$. Uniaxial compressive strength also increases with depth due to increased porosity:

$$\sigma_c = \sigma_0 \left(\frac{\rho_c}{\rho_0} \right)$$

Hence,

$$\tau = \left(C + \frac{A_c}{A_g} \frac{2C}{\tan\left(\frac{\pi}{4} - \frac{\phi}{2} \right)} \tan\phi \right) \left(1 - e^{-\left(\frac{wp}{2\pi} - v \right)/K} \right)$$

Alternatively, the Zelenin cutting force may be defined in terms of the number of impacts parameter. For rotary soil cutting, the Zelenin cutting force is given by:

$$F = Kr^{1.35}(1 - \cos\alpha)^{1.35}$$

where $K = 10N(1 + 0.026l)(1 + 0.0075\alpha)z$, $N =$ number of impacts $= 2.5$ for sandy soil, $l =$ cut width (radius of bit teeth), $\alpha =$ cutting angle relative to ground (20–90°, 30° being common), $z =$ teeth cutting coefficient $= 0.55$, and $r =$ radius of cut rotation (radius of bit teeth). The optimal cutting angle is 20° which reduces cutting forces by 25–35% compared with 45–60°. Teeth on the digging tool reduce the cutting forces required to break soil by 40%. The optimal tooth spacing is $a = 2b$ where $b =$ tooth width. A thorough review of soil excavation mechanics is given by Wilkinson and DeGennaro (2007) [1192]. In solid rock, there is no evidence of plastic failure in rock cutting, so it behaves like brittle material. Soft rocks, on the other hand, exhibit plastic behavior before failure. Failure occurs due to both tensile and compressive stresses in the rock which generates a crushed zone at the tool edge. Nishimatsu's theory of rock cutting gives axial thrust as [1193, 1194]:

$$F = F_0 + 4\sum_{i=1}^{m} \left(\frac{2Chd}{n+1} \frac{\cos\phi\sin(\phi' - \alpha)}{1 - \sin(\phi - \alpha + \phi')} \right)$$

where $F_0 =$ minimum thrust on bit to impact rock $= 1$ kN for basalt, $n =$ soil deformation coefficient $= 11.3-0.018\alpha$, $C =$ cohesion, $m =$ number of bits in tool, $h =$ cutting depth, $d =$ bit width, $\alpha =$ frontal rake angle of cutting edge, $\phi =$ soil friction angle, $\phi' =$ friction angle between rock and cutting tool $= 26°$ corresponding to a friction coefficient of 0.5 commonly, and $\tau_0 =$ initial torque to overcome friction. Tangential torque, according to Nishimatsu's theory of rock cutting, is:

$$\tau = \tau_0 + 2\sum_{i=1}^{m} \frac{2Chd}{n+1} \frac{\cos\phi\cos(\phi' - \alpha)}{1 - \sin(\phi - \alpha + \phi')} R_i$$

and power is:

$$P = \tau w = \left(\tau_0 + 2\sum_{i=1}^{4} \left(\frac{2Chd}{n+1} \frac{\cos\phi\cos(\phi' - \alpha)}{1 - \sin(\phi - \alpha + \phi')} R_i \right) \right) w$$

where $R_i =$ tool radius. These are subject to the conditions that $\alpha < \phi$ and $\alpha < \phi'$. Typical drill rates of 1–2 m/12 h are deemed acceptable performance. The drill penetration rate depends on the energy used to fracture the rock under the drill bit. Simple models for rotary and percussive drilling are, respectively, given by [1195, 1196]:

$$PR = \frac{kwW^2}{D^2\sigma_c^2}$$

where $k =$ drill parameter, $w =$ rotation speed, $W =$ weight on bit, $D =$ bit diameter, and $\sigma_c =$ rock compressive strength; and

$$PR = \frac{\eta Ef}{ASE}$$

where $E =$ energy per impact, $f =$ impact frequency, $\eta =$ energy transfer efficiency $= 0.7$ typically, $A =$ drill hole area, and $SE =$ specific energy. The prediction of drilling penetration rate (PRs) based on bit type, bit diameter, rotational speed, thrust, blow frequency, and flushing have been modeled by means of empirical relations. Weight on bit (thrust) is limited by the weight of the drilling platform (nominally the rover weight) unless an anchoring rig system is employed. Kahraman's (1999) equations are consistent with Rabia's (1985) analysis that drill penetration rate forms a power law of energy input to rock material strength [1197]. For rotary drills, the Kahraman equation is given by:

$$PR(m/min) = \frac{1.05W^{0.824}RPM^{1.690}}{D^{2.321}\sigma_c^{0.610}}$$

where $W =$ weight on bit (kg), $RPM =$ rotation speed (rpm) $= 100\text{--}300$ rpm nominally, $D =$ bit diameter (mm), and $\sigma_c =$ uniaxial compressive strength (MPa) $= W/A$. For percussive drills:

$$PR(m/min) = \frac{0.47f_{bpm}^{0.375}}{\sigma_c^{0.534}q^{0.093}}$$

where $f =$ blow frequency (bpm) and $q =$ quartz content (%). The quartz content of sandstone varies between 40 and 60% (average 50%) but quartz is absent on the Moon. Drill torque may be determined from the penetration rate:

$$\tau = \frac{1}{3}WD\sqrt{\frac{PR}{15f}}$$

where $W =$ weight on bit, $D =$ bit diameter, $PR =$ penetration rate, $f =$ impact frequency, and $RPM = 100\text{--}300$ rpm typically of Mars drills. Table 11.3 gives a sandstone example for rotary and percussive drilling and Figure 11.12 shows the results for several different materials [1198, 1199]:

At low speeds/frequencies, ceteris paribus, there is little difference in penetration rate but at high speeds/frequencies, rotary drills tend to outperform percussive drills. However, this comparison takes no account of energy consumption. Specific energy input (MJ/m^3), defined as the energy required to remove a unit volume of rock, can be computed from the penetration rate by assuming that

Table 11.3. Sandstone drilling performance for rotary and percussive drilling.

Rock type	Penetration rate (m/min)	Bit diameter (mm)	Weight-on-bit (kg)	Rotation speed (rpm)	Ultimate compressive strength (MPa)	Tensile strength (MPa)	Elastic modulus (MPa)	Impact strength	Density (g/cm³)	Quartz content (%)
Sandstone	0.41	165	1,493	72	70.5	5.5	13,855	75.8	2.56	40

Rock type	Penetration rate (m/min)	Bit diameter (mm)	Drill power (kW)	Blow frequency (bpm)	Ultimate compressive strength (MPa)	Tensile strength (MPa)	Elastic modulus (MPa)	Impact strength	Density (g/cm³)	Quartz content (%)
Sandstone	1.42	89	15.5	3,200	25.7	5.8	10,562	85	2.70	57

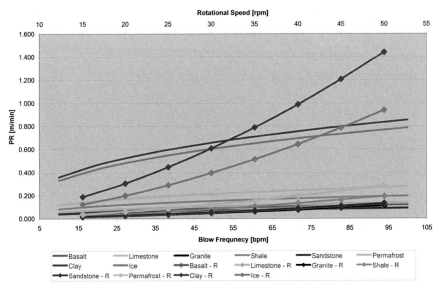

Figure 11.12. Comparison between rotary and percussive drilling models in different materials [credit Tim Hopkins, Carleton University].

the thrust energy is negligible) [1200]. Specific energy for basalt is $0.25–0.5\,\mathrm{GJ/m^3}$. For percussive drills, specific energy is given by:

$$SE = 4\eta\frac{P_{\mathrm{out}}}{d^2 PR}$$

where $PR = k((E_{\mathrm{in}})^a)/([(RIHN)\cdot(SH)]^b) = V/A = $ penetration rate, $\eta = $ energy efficiency $= 0.7$ typically, $d = $ bit diameter, $RIHN = $ rock impact hardness number, $SH = $ Shore hardness, $A = $ bit area, and $V = k((E_{\mathrm{in}})^a)/([(RIHN)\cdot(SH)]^b) = $ volume excavated. Rotary auger drills are limited to drilling to a depth of 100 m. For rotary drills, specific energy is given by:

$$E = 2.35\frac{W(RPM)}{d(PR)}\quad (\mathrm{MJ/m^3})$$

where $W = $ weight on bit, $RPM = $ rotation speed, and $d = $ borehole diameter. Drill time is given by:

$$PR = k(W/A)^a(RPM)^b(EFS)^c$$

where $EFS = $ effective formation strength; and

$$PR = \left(\frac{aS^2 d^3}{w^b W^2} + \frac{c}{wd}\right)^{-1}$$

where w = rotary speed and S = drilling strength. This yields a general equation:

$$\text{PR} = k(\text{RPM})^{a_1}(W)^{a_2}/S^b$$

which often reduces to

$$\text{PR} = k(\text{RPM} \cdot W)^a/S^b$$

The torque applied to move drill bits to new surfaces for cutting has an empirical form:

$$\tau = \frac{Wd}{3}\sqrt{\frac{\text{PR}}{15f\theta}}$$

where W = weight on bit, d = bit diameter, f = piston impact frequency, and θ = button diameter. A plot of auger torque against drill rotation speed (rpm) will yield the asymptotic rotation speed with minimum torque. From this, drill power may be determined using the Zacny–Cooper power model which comprises the power required to overcome sliding friction and that required to cut the rock:

$$P = \tau(\text{RPM})\left(\frac{2\pi}{60}\right) = P_{\text{cut}} + P_{\text{slide}} = \left(\sigma_c A\delta + \mu_d W \frac{d}{2}\right)\left(\frac{2\pi}{60}\text{RPM}\right)$$

where $A = (\pi/4)(d_{\text{out}}^2 - d_{\text{in}}^2)$ = cross section of drill, $\delta = 2\pi v/w$ = depth of cut per revolution, μ = coefficient of friction, $r = \frac{1}{2}(r_{\text{in}} + r_{\text{out}})$ = average drill radius, and $v = (2\pi/60)(\text{RPM})r$ = drill speed. In reality, drilling efficiency is determined by the ratio of drill power to the rate of material removal. The required characteristics of an auger system for cuttings removal are minimum rotation speed and minimum pitch height [1201]. The minimum rotational velocity (rpm) of an auger to initiate excavation is given by:

$$\text{RPM} = \frac{30}{\pi}\sqrt{\frac{2g}{d}\frac{\tan\alpha + \mu_{\text{ss}}}{\mu_{\text{rs}}}}$$

where d = auger diameter, μ_{rs} = rough surface friction between soil and borehole wall = 0.7 typically for soil, μ_{ss} = smooth surface friction between drill surface and soil = 0.3 typically for metal, and α = pitch angle of auger helix (20° typically). The minimum auger pitch height to prevent the auger from choking is given by:

$$h \geq \frac{\mu}{\cos i}\left(\frac{r_{\text{out}}^2 - r_{\text{in}}^2}{\mu_{\text{rs}}r_{\text{out}} - \mu_{\text{ss}}r_{\text{in}}}\right)$$

where $r_{\text{out}}, r_{\text{in}}$ = outer/inner radius of drill. A solid steel drill of 10 cm diameter can penetrate frozen silt or gravel at 10–30 mm/s. Thrust force is the product of rock compressive strength and tool area—its primary purpose is to maintain bit–rock contact. Drill torque is determined by the sum of tangential forces over the cutting face of the drill bit and tangential forces are proportional to contact pressure. Drill rotation torque requirements are proportional to the square of the bit diameter—typical torques of 35 kN/m are required for 75 mm diameter holes. Maximum thrust limits also correlate with the square of the bit diameter—maximum thrusts of 20 kPa are imposed for 75 mm diameter drill bits. Maximum drill rotation speed is inversely related to maximum bit diameter—typically, a maximum of 300 rpm for 250 mm diameter bits. Auger rotation is controlled so that the maximum linear speed of the outermost cutting bits are within 1–5 m/s. Penetration rates for augers in ice reach up to 20–80 mm/s and in frozen soil 10–30 mm/s at increasing rotation rates but for high coring quality the penetration rate should be bounded. For rotary drills, the power and the penetration rate are proportional to rotation speed. Drilling torque is not dependent on rotation rate and rotation rates are usually determined by the cuttings removal rate of the auger. For percussive drills, impact blow energy is inversely related to frequency of ~30 J for small drills. The product of impact frequency and blow energy defines the power delivered to the bit.

Typical electrical power requirements are around three times the mechanical power output for rotary drills (e.g., 60 W input for 20 W mechanical output). Percussive drills are more efficient with 40% of input power being delivered as mechanical output. Specific energy for basalt is 0.25–0.5 GJ/m^3 for a percussion drill or 0.6–1.0 GJ/m^3 for a rotary drill. As far as ice is concerned, specific energy for rotary drilling is 0.5–4 MJ/m^3 which reaches 2–6 MJ/m^3 for frozen soil and 2–20 MJ/m^3 for percussive drilling. For impacts at very high frequency, percussive drills become inefficient and act as melters at specific energies of 300 MJ/m^3. Specific energy for drilling correlates linearly with the uniaxial compressive strength of the rock. Typically, specific cutting energy/uniaxial compressive strength $E/\sigma_c = 0.05$–0.5 for rotary drilling and 0.25–1.5 for percussive drilling. However, the compressive strength of rock is a function of temperature and water content—frozen sandstone requires three times the power to drill than warmer dry sandstone [1202]. For Briar Hill sandstone under Martian pressure, a drill penetration rate of 80 cm/h requires 30 W for dry rock (specific energy of 210 MJ/m^3 and compressive strength of 43 MPa) and 100 W for water-saturated frozen rock (specific energy of 770 MJ/m^3 and compressive strength of 110 MPa). Similarly, dry rock requires much lower weight-on-bit values than frozen icy rock for the same penetration rates (e.g., 25 kg versus 60 kg for 80 cm/h). Minimum power of 5 W is required to overcome frictional resistance (with a zero penetration rate). Similarly, minimum threshold weights on bit required for drill penetration in dry sandstone and saturated frozen sandstone were 15 kg and 40 kg, respectively. Specific power required to drill is given by:

$$\frac{P}{Wv} = \left(\frac{1}{\sin \alpha} + \frac{\mu(((1 + h/d)^2 - 1)\cos^2 \alpha + 1)}{(1 + h/d)\cos \alpha - \mu \sin \alpha} \right)$$

where d = rotor diameter, h = blade height, α = helix angle, μ = coefficient of friction, v = speed, and W = weight on bit. These models allow automatic control of drilling performance as rock conditions change. Assuming a simple dynamic model of drilling, the drill tool interaction force can be related to drill penetration rate $\dot{x}$ and contact distance x:

$$F = c_1\dot{x} + c_2 x^a + c_3 = k_1 \dot{x}^b + k_2 x^a$$

where a, b = spring and damping exponents. Drilling models relate contact force nonlinearly with drill tool velocity, rock hardness, and tool bit temperature. The use of cutting tools induces both tool wear and significant energy consumption, much of it lost as heat. The heat generated due to tool action is given by:

$$q = F\nu$$

where F = tool shear force and ν = cutting speed. Cutting speed is the chief determinant on tool life τ through Taylor's relation:

$$C = \nu\tau^n$$

where C = empirical constant and n = exponent depending on cutting conditions. It is essential to avoid overheating of the core which implies that cooling must be employed. The low pressure of Mars straddles the triple point of water (0.63 kPa and 0°C). Below this pressure, subsurface ice will sublime removing cuttings from the borehole without the use of gas pumping. However, above this pressure, ice will melt which can potentially refreeze—this implies that intermittent stops must be applied to reduce overheating. Bit temperature measurement is essential to monitor temperature to limit drill bit heating. Alternatively, electrical resistivity measurement might be more appropriate to measure the incidence of liquid water directly given that temperature dependence may be affected by saltation. The penetration depth can be measured using lead screw switches. Force/torque feedback allows measurement of reaction forces to control the drill. Compliance control will be essential in such stiff environments of $\sim 10^5$ N/m. Neural network identification based on fully recurrent networks has been proposed for system identification with rapid adaptive tuning of the learning parameter such that it varies as $\eta \propto |\nabla E|^{-t}$ where E = energy function and t = time [1203].

The NASA Mars Analog Research & Technology Experiment (MARTE) has attempted a partially automated field test of robotic drilling at the Minas de Río Tinto in SW Spain (a chemoautotrophic biosphere based on iron and sulfide minerals as energy sources) involving a 10 m dray auger drill for core sampling, robotic sample handling, borehole inspection, and scientific instrument suite [1204]. This is a large and complex 10-axis rotary drill system 3 m high and 2.4 m in diameter designed to reach 10 m drilling depths without fluid. It represents to date the most complete simulation of robotic drilling and scientific analysis yet conducted. The drill comprises a 1.5 m first drill segment plus 10×1 m additional drill segments assembled autonomously. The borehole inspection system includes a Raman spectrometer and microscopic imager—this preselection has proven invaluable. The drill recovers its core sample which is then handed over to a core sample handling system. The core sample handling system receives the core sample by means of a multijaw 24 DOF clamp which is positioned beneath the drill and prepares the core by sawing it into sections—this newly exposed surface is conveyed by linear rail to a number of scientific instruments (microscope camera, visible/NIR spectrometer, and ATP luciferin–luciferase reagent-based detector). A rock crusher then powders the sample for analysis by astrobiological instruments (SOLID protein microarray using fluorescent labeling). Lessons learned from MARTE include the value of using compressed air to blow out cuttings which considerably reduces performance but ensures that coring is performed in the absence of compressed air circulationas well as the considerable cross-contamination during core sample handling due to dust generation during sawing and crushing.

11.6 SAMPLE HANDLING, PROCESSING, AND DISTRIBUTION (SHPD) DEVICES

Scientific instruments often require extensive processing of samples prior to analysis—this is particularly true for the latter classification of instrument. Furthermore, samples may be acquired faster than they can be processed and analyzed necessitating buffering. Samples usually require a sample acquisition mechanism such as a scoop, manipulator, corer/drill, etc. to take a sample of soil/ rock. It is these latter instruments that provide the most detailed interrogative analysis, particularly for astrobiological investigation. Sample handling and preparation generally requires a complex set of robotic devices. Most instruments require sample processing in order to enable extraction of good signals (e.g., grinding, crushing, sieving, sectioning, polishing, etc.). Different instruments require different sample forms which must be generated by:

 (i) comminution, grinding, and sieving;
(ii) sawing, sectioning, and polishing.

The commonest sample-processing requirement is for powdered samples (e.g., X-ray diffraction spectroscopy). Sample processing must be achieved remotely and robotically without contamination. Rock crushing may allow the release of inclusion fluids [1205]. Fluid inclusions may remain from former hydrothermal vent systems. Furthermore, recovered samples require protection from mechanical insult, excess temperature, and chemical alteration. Sample preparation and distribution provides a common functional integration system across scientific analysis instruments. This is far more efficient than individual instruments implementing their own sample processing. Beaty et al. (2004) defined the rationale for sample processing and distribution [1206]:

1. Improve accuracy, precision, and detection limits through adequate sample preparation (i.e., crushing, sieving, filtration, reagent addition, etc.).
2. Achieve synergy between scientific instruments with a diversity of relevant analyses.
3. Increase sample throughput with minimal cross contamination.
4. Analyze the appropriate samples with appropriate instruments using a triage system of low-level screening to determine which samples should be analyzed further with more sensitive equipment to optimize scientific return within resource constraints.

5. Sample analysis sequence should be logical with each analysis allowing decisions to be made for further analysis.
6. Enable new scientific instruments to be incorporated into the instrument suite through adequate sample processing.

Samples such as ice-bearing soils may require special handling methods. Given the need for general and more specific analyses, general analyses should precede more specific analysis. Beaty et al. (2004) also made certain recommendations to keep the engineering of the SPDS simple:

(i) Shared sample processing should be conducted only for instruments with common processing requirements—this is trivially obvious.
(ii) Dispose of analysed samples to reduce storage requirements.
(iii) Only one sample advanced analysis system should be activated at any one time.
(iv) Each instrument should be assigned a unique point in the sampling process flow.

Sample preparation must cut polished thin sections for optical microscopes, providing a smooth surface for an APXS (many instruments require a smooth flat surface to allow analysis), and grinding rock to fine powder of $<10\,\mu$ for pyrolysis in the GCMS. This will typically impose additional mass to the instrument payload. Clancy et al. (2000) proposed a six-instrument automated exobiology research facility for a Mars lander [1207]. It comprised a sample acquisition and delivery system, a sample preparation system, and a set of analytical instruments. The sample acquisition system comprised a robot arm, nanorover, and coring drill. The drill string could be mounted on the robot arm or in the payload cab of the Nanokhod rover. The sample preparation system comprised a cutter saw and polisher to cut and polish surfaces for optical examination. It also comprised a miller grinder to create $<10\,\mu$ sized powder for spectrometry. Analytical instruments included a microscope of high and low magnification, a Raman spectrometer imager, an infrared spectrometer, atomic force microscope, Mössbauer spectrometer, APXS, and GCMS.

For the ExoMars rover, the SHPD must be close to the drill to ease sample transfer to the entrance of the SHPD. It shall process Martian rock and soil samples for the scientific instrumentation. This involves crushing samples into a fine-grained powder which is then distributed to the instrument suite via a carousel of sample containers. The SHPD is externally mounted which allows sample cooling during the passage of samples into the SHPD. The SHPD must hold samples and prepare them for analysis through cutting/polishing, grinding/milling, and transferring them to different instruments. Multiple sample input ports to the SHPD ensure redundancy of access to scientific instruments. The sample holder is a tray-shaped container designed to hold a 1 cm diameter by 4 cm long sample. The mechanism for transport is based on a distribution carousel and a 3 DOF robotic arm. Samples are presented initially to optical instruments for non-destructive analysis. Samples may be ground into a fine powder for pyrolysis and chemical analysis; smooth polished sections are required for optical analysis; there is a mica window for Mössbauer spectroscopy. For ExoMars, the samples are first milled into a powder $\sim 100\,\mu m$ for subsequent analysis in a carousel of sample containers. The milling station may be similar to the NASA jaw crusher design capable of crushing basalt. Organics analysis and X-ray diffraction require finely powdered material. The final instruments are those that destroy the samples (such as the GCMS). The SHPD is divided into four sample-processing areas. The first sample area collects samples and transports them to the second processing area using a conveyor of container ovens which links all sample-processing stations. The second sample-processing area prepares samples by polishing them ready for optical examination by the color microscope, Raman/LIBS instruments, and the X-ray diffraction spectrometer. The third sample-processing area comminutes samples using a grinding mill in readiness for the optical instruments, life marker chip, oxidant sensor, Mars organics detector, and GCMS instruments.

The chief disadvantage of using the SHPD system is that it requires significant resources in terms of mass, energy volume, and complexity. The robotic complexity inherent in the SHPD system implies that there is increased risk of failure due to mechanical fault. The first issue that needs to be considered is that different instruments, particularly if different resolutions are involved, require differing volumes/masses of samples—this implies the need to divide samples such as core samples into smaller subsamples. In general, it is expected that acquired samples will be small ($\sim 1\,g$). The second issue concerns the order of analysis by different scientific instruments—clearly, those analyses that destroy samples should be performed after non-destructive sample analyses. There may be times when samples require storage as a result of the transport system delivering them faster than they can be processed and analyzed. Most scientific instruments require fresh exposed surfaces for analysis such that the sample analyzed is pristine. Some scientific instruments require flat surfaces (e.g., APXS). This may be achieved by sawing but polishing may also be necessary. Comminution of samples into differing degrees of fine powder may be required for some instruments (e.g., X-ray diffraction requires random finely divided powder samples). Some instruments require the evolution of volatiles which implies powdering to increase the surface area (e.g., mass spectrometers). Specific size fractions of the sample can be sieved.

SPADE (Sample Processing and Distribution Experiment) is a miniaturized NASA rock jaw crusher (milling station) and sample distribution system which generates rock fragments of varying size, sorts samples into fragments, and fines into sample bins mounted on a wheel (Figure 11.13) [1208]. This conveys rock samples to different scientific instruments for analysis of rock interiors and trap doors for removing rock products after analysis. The Rockhound jaw crusher comprises a fixed plate and moving plate forming a $20°$ wedge. Although the plates are constructed from hardened steel in the prototype, Ti alloy would be superior in terms of mass efficiency for a flight version. The moving plate rotates via a cam forcing the rock down as it compresses and crushes. A doubly geared high-speed motor with gear reduction of 60,000:1 drives the crusher delivering 28.25 Nm torque generating a force of 37,000 N at the top of the wedge. The rock crusher is 10 cm high and accepts rocks up to 5 cm in diameter. Each crusher plate

Figure 11.13. SPADE rockhound crusher prototype [credit NASA JPL].

cycle takes 3.5 min, an entire rock requiring several hours of operation. Crushed rock fragments smaller than the smallest gap (0.5 cm) between the plate and wedge fall are binned into size categories (in this case two—coarse and fine fragments with a separation threshold of 2 mm). A vibration-assisted mechanical sifting grate above one bin separates fine fragments from larger particles. Once bin capacity is achieved, the crusher is shut down automatically and the sample wheel moves to place another set of bins beneath the sorter. Trapdoors were used to remove samples from the sample bins and were found to be superior to sweepers. Further improvements include increasing the number of sample size bins using different individual sieves. Problems with crushing ice with the jaw crusher due to sliding extended ice-crushing time to 6 h.

Sample distribution requires the movement of samples to different scientific instruments. This generally involves three types of transport [1209]:

(i) robotic manipulator;
(ii) carousel;
(iii) microfluidics.

Carousels are useful for multi-instrument analysis while robot arms have the most flexible manipulation capabilities. Sample distribution is more efficiently achieved than rotating the instruments to get access to the sample—for this reason, a carousel (SD2) mounting sample containers, ovens that rotate under different instruments, was adopted on the European Rosetta mission. This approach does however impose major constraints on the configuration of the scientific instruments and works for a modest number of instruments. Gantry robots with 3–5 DOF such as ST Robotics' R16/R17 robotic arms of mass 20 kg are commonly used in the pharmaceuticals industry to handle chemicals in conjunction with sample carousels. Microfluidics is associated with lab-on-a-chip systems that require powdered samples. The carousel is rotated to emplace the sample so that each instrument can gain access to it sequentially. A cleaning station is incorporated to allow reuse of some of the containers. The carousel solution offers a smaller footprint area but greater height than the robotic manipulator. The carousel approach is more reliable. Both approaches are used in laboratory biochemical analysis. Such manipulators are designed to manipulate vials and tubes and are used extensively in robotic spotting of DNA microarrays. However, for SHPD systems, a 3 DOF manipulator should be sufficient to position samples for initial analysis prior to milling. A carousel may be used to mount a ring of sample containers in its rim (similar to the Rosetta SD2 system). These sample containers may be used to move the powdered sample for further analysis. A major issue is potential cross-contamination of samples.

The Mars Science Laboratory (MSL) sample acquisition and processing system is the most sophisticated system yet developed—this is the 8 kg CHIMRA (collection and handling for in situ Martian rock analysis) [1210] (Figure 11.14). For sample acquisition, it uses a 5 DOF robotic manipulator scoop for regolith samples and a rotary percussive drill for powdered samples. The first step is to bin the samples according to grain size to either 1 mm or sub-150 μm samples. There are two primary passages from two inlets (bulk regolith from the scoop and powered sample from the drill) that lead to a central reservoir. Thence, the samples pass through a 1 mm or 150 μ sieve, discarding the larger fraction in each case. One-way valves prevent samples from migrating to undesired locations within the passageways. To generate sample flow through the tubes, the CHIMRA assembly is vibrated as a whole while being reoriented with respect to gravity by the robotic arm through a specific sequence of rotations. Vibration is enabled by a spinning eccentric mass coupled to the natural frequency of the structure to generate shaking. Cleaning is accomplished through dilution and chemical flushing between samples. The sieves are attached to thwackers—mousetrap mechanisms based on latch-and-pawl devices that impulsively slam the sieve against a hardstop to prevent clogging. All internal areas of CHIMRA are visible by camera to aid troubleshooting.

In the future, as miniaturization develops, it is expected that micromanipulation will become more commonly employed in scientific instruments—indeed it is conceivable that micromanipulation of MEMS-based scientific instruments may provide reconfiguration of the scientific instrument suite

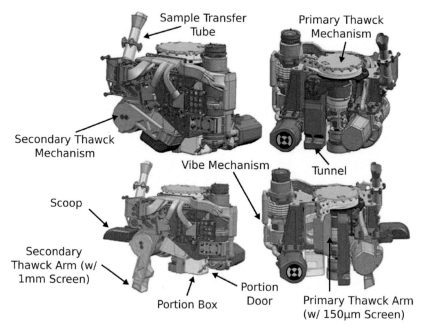

Figure 11.14. CHIMRA configuration on Curiosity [credit NASA].

according to need. Automated microdevice assembly has been proposed in which a micromanufacturing cell comprises a robotically configurable set of micro components [1211]. The key components are stereomicroscopic imagers, micro-grippers based on piezoactuators, and micropositioners exploiting electrostatic and capillary forces for adhesion and vibration-based release. The use of microfluidics is associated with the use of reagents with lab-on-a-chip instruments. Laboratory automation has been under development for DNA sequencing and data analysis in the form of a cubicle-sized laboratory—Adam [1212]. Adam performs experiments on microbial growth through experimentation to investigate the relationship between genes and their phenotypic effects (e.g., yeast enzymes). A liquid-handling robot emplaces microbial samples from a freezer into a growth medium on a test plate of wells. An incubator warms the test plate while a robotic arm moves each plate into an optical reader periodically to measure microbial growth. A centrifuge spins each plate to separate the microbial samples from the medium which is flushed away. Another robotic arm adds different types of growth media. This is the experiment phase—the incubator warms each plate and a robotic arm places plates into an optical reader to measure microbial growth. The data are analyzed using a knowledge-based expert system with heritage stemming from DENDRAL (mass spectrometer data), PROSPECTOR (mineral prospecting data), and BACON (general scientific laws). It generates hypotheses to test, designs experiments to test the hypotheses, and selects the simplest hypothesis.

12

Onboard autonomous science

Science-driven autonomy involves automating the detection of scientific events and their classification on board the rover. Such events may be static or transient, the latter allowing for scientific data acquisition while mobile. This will drive reductions in ground station operational costs, increased quality of science data through limited communications bandwidths, and enhanced intelligent scientific data processing. This also affords high compression of data. Planetary rovers have few opportunities for downlinking scientific data—typically, once or twice per day. Furthermore, uploading scientific objectives based on the previous day's images adds delays to scientific experiments. Implementation of a robotic geologist with onboard scientific decision-making capabilities would greatly enhance the scientific productivity of rover missions while reducing the need for the transmission of low-value raw data. The MERs did not possess "science" triggers (due to the likelihood of false positives) as all scientific objectives are selected by ground operators. Science targets can be identified visually up to 20 m in advance and the autonomous navigation system must bring the rover to within 1–2 m of the target. Scientific goals include:

(i) searching and characterizing rocks and soils indicative of water-based processes including water-deposited minerals, especially carbonates;
(ii) determining the spatial distribution and composition of minerals, rocks, and soils;
(iii) determining the nature of surface geological processes;
(iv) characterizing mineral assemblages and textures of different rock types and soils in their geological context;
(v) determining the geological context for possible biotic processes.

Within 1–2 m of the target, the rover is oriented to the target for fine-scale positioning (to within ±10 cm and 1° of the scientific target) (Figure 12.1).

© Springer-Verlag Berlin Heidelberg 2016
A. Ellery, *Planetary Rovers*, Springer Praxis Books,
DOI 10.1007/978-3-642-03259-2_12

Figure 12.1. Rover conducting rover-based science [credit NASA JPL].

This is to ensure that the robotic arm can perform final deployment of scientific instruments. JPL has developed an automated robotic arm instrument placement system for use on board planetary rovers [1213]. The initial position is 3 m from the scientific target and uses automated rover placement, collision-free manipulator planning, and vision-guided manipulation. The key process is camera handoff from NavCams to HazCams. The HazCams monitor the final positioning of the rover within the manipulator workspace. They track the target to generate collision-free manipulator trajectories to the target and monitor manipulator motions during trajectory execution using vision guidance. Automated target acquisition through vehicle–manipulator coordination involves the following steps [1214]:

(i) acquiring a stereo image pair;
(ii) the operator selecting a rock of interest;
(iii) identifying the 3D location of the rock based on target intensity differential, camera calibration, and stereo-imaging;
(iv) deriving the trajectory to the target;
(v) navigating to the target;
(vi) updating the terrain map with sensor data;
(viii) stopping when the target lies within the workspace of the manipulator;
(ix) deploying the manipulator to acquire a rock;
(x) vertical guarded move until soft contact is made;
(xi) force/torque feedback controlling target acquisition;
(xii) inserting samples into a sample box (predefined location).

Planetary rover autonomous science—the robotic scientist—is an important facility that requires autonomous navigation as part of its capability and the deployment of scientific instruments. Autonomous navigation and autonomous science phases essentially overlap right up to the final approach and positioning adjacent to the target. Scientific productivity is the primary performance metric for planetary missions including rover missions. It is essential to maximize the scientific productivity of rover missions while minimizing the necessity for intervention by the ground control team on Earth. It would thus be highly desirable that a rover mission perform at least some of its scientific analysis autonomously. This may involve acquisition of potential targets through visual analysis, selection of those targets according to priorities, scheduling the transmission of the most important scientific data to eliminate bandwidth bottlenecks, determination of the most appropriate actions such as instrument deployment, and integration of science planning with operational planning. Autonomous science capability is essential to prioritize options for action such as selection of scientific targets to be investigated, make scientific decisions such as selection of scientific instruments to be deployed, and relieve communication bandwidth bottlenecks in future rover missions. JPL's Onboard Autonomous Science Investigation System (OASIS) has been developed to use grayscale images to autonomously evaluate rocks and, on the basis of that evaluation, to acquire additional images [1215]. It integrates autonomous science acquisition and decision making with rover activity planning (within the CASPER rover planning system) including identification of new science targets within the constraints of rover resources [1216, 1217]. On triggering a science alert, it halts its current traverse and locates the rock that generated the alert. It approaches and takes additional scientific data on the rock and generates new science objectives. When measurements are complete, the rover then returns to its original traverse. Carleton University has been investigating opportunistic science in which a rover may autonomously select a previously unknown target during a traverse to a goal. On the basis of new priorities, the rover diverts to the new target and resumes its original traverse once its opportunistic investigation is completed or suspended. A variation on this theme is to use a scout rover, such as Kapvik, as a reconnaissance vehicle to a larger more expensive rover, such as ExoMars. The scout trailblazes and potentially explores riskier sites thereby reducing the risk to the main rover. Furthermore, it relays information about its traverse to the main rover to enhance the SLAM process.

12.1 VISUAL TARGET ANALYSIS

We briefly consider the use of vision in support of autonomous science—the techniques were explored earlier in considerable detail. The most fundamental task to be performed by a planetary rover is geological profiling in which the local region is examined to determine its geological features such as rocks [1218]. Selection of specific rock types of interest identified from specific features from imagery will be the primary modality of such science-driven autonomy. For instance, the ExoMars pancam is designed for geological mapping of scientific sites, identification of surface features and the characterization of different rocks and soil assemblages.

It currently comprises a stereopair of wide-angle cameras and a high-resolution camera (4° FOV) for close observations mounted onto a single optical bench. The detection of geological features is one of the ExoMars pancam roles using visual imagery and multispectral reflectance data from the wide-angle cameras and high-resolution camera. The two wide-angle cameras each have a 38.3° FOV with 50 cm baseline with 11 filter pairs—22 filters in all mounted onto two filter wheels, one per camera. Three filter pairs are broadband red–green–blue while 12 filters are narrowband geology filters over the range 440–1000 nm (determined by silicon detector sensitivity). Although most reflectance spectroscopy of minerals occurs at >1 μm, there do exist spectral features at <1 μm, but this necessitates the use of general multispectral filter sets in this spectral region in order to discriminate between minerals of interest. Under hydrothermal conditions, acidic aqueous interaction with basaltic material generates soils rich in sulfates, phyllosilicates, and ferric oxides with gypsum and zeolite veins. The six geology filter pairs were selected to optimize detection of such geological signatures of astrobiological significance—(i) sulfates, (ii) phyllosilicates, (iii) mafic silicates, (iv) ferric oxides, (v) all iron minerals, and (vi) all hydrated minerals [1217]. The set of 12 filters optimized for ferric oxide (440, 500, 530, 570, 610, 670, 740, 780, 840, 900, 950, 1000) offered excellent performance for iron oxides as well as the detection of hydrated minerals, sulfates, phyllosilicates, mafic silicates and carbonates. Principal components analysis was used on spectral data during field tests with the first three components containing most of the spectral parameters of astrobiological significance [1217]—red–blue ratio, 610 nm band, green–red slope, and green–red slope minus blue–red slope were especially useful, showing the importance of visible band data.

The Marsokhod rover has undergone field tests in California to demonstrate onboard rover intelligence in selecting science targets using visual data so that only scientifically interesting images are downloaded to Earth. Three algorithms have been proposed all based on finding edges in image intensity levels (thereby lending themselves to traditional edge-finding algorithms such as the Canny edge detector used in conjunction with snakes to determine boundary shapes): horizon detector, stratigraphic layer detector, and object/rock detector [1219]. The horizon detector differentiates the sky from the ground. Once the sky is segmented from the ground, it is then ignored. This is used to limit the search space for the other detector algorithms. The rock detector uses the solar vector as measured by the camera to predict the shadow size effects of rocks thereby locating them. This favors the detection of high-priority rock targets. The layer detector identifies stratigraphic layering as boundaries between different reflective layers. This is the key to understanding geological variations over time as part of geological history. Layers may be differentiated by texture due to grain size differences. In almost all cases, these involve edge detection algorithms such as the Canny edge detector at multiple scales. Although there were significant ambiguities, this demonstrated the utility in using automated pattern recognition to increase the scientific return from rover missions. Closed shapes may be classified as potential rocks. A visual discovery algorithm to extract different image intensities from the background based on the human visual cortical receptive field model using multiscale Gabor filters has been implemented [1220]. This has the advantage of detecting visually different objects from the background without a priori models. It is particularly suited to microscopic imaging. Gabor filters in conjunction with center–surround and corner-sensitive filters were used to identify objects. The center–surround filter identified bright or dark localized regions that differ from their surroundings.

Rocks are of specific geological interest. Their mineral constituents are chemical classes with distinctive physical properties such as color, hardness, density, luster, transparency, cleavage which may be used to classify minerals. The geological rock cycle begins with igneous rock of volcanic origin which may be intrusive (within the Earth's crust) or extruded (on the Earth's surface). Igneous rock may be weathered and eroded into fragments. The fragments are deposited as sediment and cemented through lithification into sedimentary rock. As well as clastic sedimentary rock formed from parent rock fragments, chemical precipitation of minerals in water form evaporates. Carbonates are primarily derived from biological marine fossils. If igneous or sedimentary rock are subjected to heat or pressure they are transformed into metamorphic rock with altered physical properties. Bodies of rock may be folded into arches (anticlines or synclines) or fractured into faults (normal, reverse, and strike–slip) by geological forces. Faults are classified according to the orientation of the major stresses—normal faults and grabens form when the maximum stress S is vertical giving an indication of previous geological context. Sedimentary facies define the set of characteristics of bodies of sedimentary rock that result from their depositional history indicative of the rock's original environment. These characteristics are defined by geometry, lithology, sedimentary structure, and fossils. The geometry of river channels, beaches, and glaciation render specific sedimentary geometries. The lithology is determined by color, texture, and composition. For example, oxidized Fe of red beds is indicative of arid or fluvial environments. Grain size, shape, and distribution, the determinants of texture, provide information on sediment formation energy—high-energy flows deposit larger pebbles than low-energy flows which deposit fine-grained sand and muds. Composition indicates chemical environments. Evaporites such as salt and gypsum indicate arid environments. Sandstones rich in feldspar indicate little weathering as feldspars are readily decomposed into clays by chemical weathering. The combination of red beds with feldspar would indicate arid environments. Oxygen isotope $^{18}O/^{16}O$ composition of sedimentary minerals indicates climatic temperature (higher temperatures preferentially evaporate the lighter isotope). Sedimentary structures include rippling formed by water, scouring indicating turbidity currents, etc. Sea level transgressions and regressions in marine environments have a strong influence on sedimentary facies. Tillites are lithified boulder clays of striated pebbles set in clay/silt deposited by glaciers indicating cold climates. Finally, fossils are of course the prize in Mars exploration but their recognition will be difficult at best and most likely ambiguous.

Visual feature discovery enables raw images to be processed on board and then discarded. Small rocks and rocks in the far field are difficult to detect. Rock extraction is based on albedo (reflectance), color, texture, size, and shape. Albedo gives information about the mineralogical composition of rock but this is affected by shadows and Sun angle. Quartz may be identified by its high albedo (e.g., two standard deviations above mean albedo). The size and shape parameters of rocks are estimated by fitting an ellipsoid to extract size, shape, and orientation which can indicate angularity and ruggedness. The core of visual feature extraction is texture analysis—it is expected that the images will have multiple filter channels across UV, visible, and NIR bands (hue, saturation, etc.). Texture may initially be analyzed through simple filters—Gaussian filter, Laplacian filters, and edge detection filters. Texture incorporates important geological infrmation. Visual images of rocks in planetary environments may be used to identify rocks on the basis of physical characteristics—grain size distribution, bedding, rippling, vesicles, etc. can aid in such identification. Vesicles are formed during the cooling of igneous rocks as gas bubbles escape. Sedimentary rocks are quite varied indicating differences in grain size texture from fine-grained clastics (such as shale) to medium-grained sandstones to conglomerates. Image resolution varies at different ranges for the same camera which is critical for object identification [1221]. Layering, texture, and identifying features are visible only at high resolution which implies either very close distance and/or high-quality imaging. Increasing distance loses high-frequency variations due to reduced resolution in detecting granularity (Figure 12.2). Increased resolution yields an exponential increase in extractable information.

A modified back projection algorithm may be used to create virtual geometric rock models from multiple viewpoints to allow planning prior to rover and/or manipulator movement (Figure 12.3).

We have selected a nominal standoff of 1 m with high-resolution imaging—it is unlikely that texture-based methods will be useful for identifying rocks beyond 5 m for which contour analysis would be more appropriate to indicate geological

Figure 12.2. Comparison of spatial resolution for rock identification; serpentine rock with texture analysis at 1, 2, 3, and 4 m distances [credit Adam Mack, Carleton University].

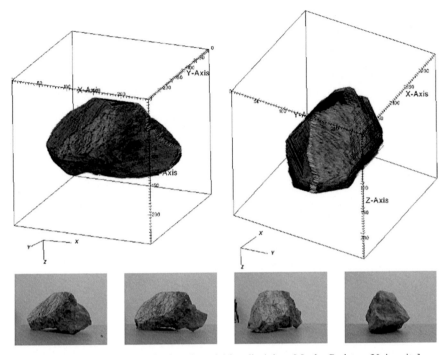

Figure 12.3. Serpentine rock virtual model [credit Adam Mack, Carleton University].

origin. We used multiple images of 30 different rocks classified into 5 fuzzy bins
[1222]. Prior to measurement, images are preprocessed into gray-tone spatial
dependence matrices (gray-level co-occurrence matrix—GLCOM) to provide
second-order feature extraction. GLCOM describes how often different neighboring pixels in four directions occur in an image. The 13 Haralick image parameters
are f_1 (angular second moment), f_2 (contrast), f_3 (correlation), f_4 (sum of squares
variance), f_5 (inverse difference moment), f_6 (sum average), f_7 (sum variance), f_8
(sum entropy), f_9 (entropy), f_{10} (difference variance), f_{11} (diference entropy), f_{12}
(information correlation 1), f_{13} (information correlation 2). The 14th Haralick
parameter f_{14} is rarely used [1223]. They are defined thus:

Angular second moment,
$$f_1 = \sum_{i=1}^{n} \sum_{j=1}^{n} p(i,j)^2$$

Contrast,
$$f_2 = \sum_{k=0}^{n-1} \sum_{i=1}^{n} \sum_{j=1}^{n} p(i,j) \quad \text{where } k = |i,j|$$

Correlation,
$$f_3 = \frac{\sum_{i=1}^{n}\sum_{j=1}^{n} p(i,j)p(i,j) - \mu_x\mu_y}{\sigma_x\sigma_y}$$

Sum of squares variance,
$$f_4 = \sum_{i=1}^{n}\sum_{j=1}^{n}(1-\mu)p(i,j)$$

Inverse difference moment,
$$f_5 = \sum_{i=1}^{n}\sum_{j=1}^{n}\frac{p(i,j)}{1+(i-j)^2}$$

Sum average,
$$f_6 = \sum_{i=2}^{2n} ip_{x+y}(i)$$

Sum variance,
$$f_7 = \sum_{i=2}^{2n}(i-f_8)p_{x+y}(i)$$

Sum entropy,
$$f_8 = -\sum_{i=2}^{2n}p_{x+y}(i)\log[p_{x+y}(i)+\varepsilon]$$

Entropy,
$$f_9 = -\sum_{i=1}^{n}\sum_{j=1}^{n}p(i,j)\log[p(i,j)+\varepsilon]$$

Difference variance,
$$f_{10} = \sigma(p_{x-y})$$

Difference entropy,
$$f_{11} = \sum_{i=0}^{n-1}p_{x-y}(i)\log[p_{x-y}(i)+\varepsilon]$$

Information correlation 1,
$$f_{12} = \frac{HXY - HXY1}{\max[HX, HY]}$$

Information correlation 2, $f_{13} = \sqrt{1 - \exp[-2(HXY2 - HXY)]}$

For example, angular second moment measures the homogeneity of an image. The 14 parameters are based on visual texture (f_1, f_2, and f_3), simple statistics of GLCOM transforms (f_4, f_5, f_6, f_7, and f_{10}), information theory (f_8, f_9, and f_{11}), and information correlation (f_{12}, f_{13}, and f_{14}). The f_3 (texture), f_5 (statistics), f_{10} (statistics), and f_{12} (information correlation), and/or f_{13} (information correlation) parameters were the most discriminatory but the entropy function f_9 (which appears to quantify grain size) and other inormation functions were found to be of little use due to the high degree of noise in rock images. However, the information correlations were entropy ratios that appeared to perform well. Haralick parameter computation is an efficient means of texture analysis to determine relative composition. The entropy measure provides good separation for only a few rocks but not most (Figure 12.4).

In combination with several other Haralick parameters, different rock types may be differentiated. This technique is computationally efficient but is limited—it offers no direction sensitivity. For such an enhanced capability, banks of Gabor

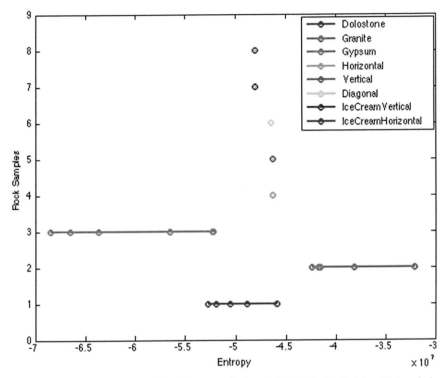

Figure 12.4. Entropy measure for different rocks [credit Helia Sharif, Carleton University].

filters which emulate mammalian visual processing offer a much richer form of information extraction but are computationally far more intense. We have used a bank of 80 Gabor filters to extract some basic textural properties of complex images, particularly regarding rock striations (unpublished work). Imagery may also be used for astrobiological investigation.

The cyborg astrobiologist is a wearable video/computer system to train an autonomous computer vision system making it capable of some of the decision making of a field geologist/astrobiologist [1224]. It performs low-level geological image analysis autonomously. It exploits the concept of metonymy in which the juxtaposition of different geological units is the key to scientific interest. This implies segmenting the image into different regions based on similarity within the unit and determination of the boundaries between them. A 2D histogram algorithm and peak search algorithm were used based on hue (color), saturation (hue purity), and intensity (brightness) bands (microtexture is to be included at a later date). Outcropping, faulting, and layering due to different textures, wetness, and color differences could be detected and marked as regions of interest but was subject to classifying shadow as interesting. This approach would not have

Figure 12.5. Martian blueberries [credit NASA].

detected the "blueberries" (sulfate salt minerals) on Mars as they would have been detected as background material (Figure 12.5). The detection of novelty may be based on color from a digital microscope in the field [1225]. The mean values of hue saturation and intensity are input to a Hopfield neural net which either relaxes to a familiar (previously known) state or remains at high energy (indicating novelty). Hue and saturation representing color are readily separable from intensity.

An autonomous science system must include other spectroscopic instruments as well as a vision system [1226]. Spectral imaging of rock, regolith, and dust provides the basis for rock identification. Spectral analysis is a necessary prerequisite to autonomous scene reconstruction particularly for identifying potential astrobiological targets evidenced by water-based activity (e.g., rounding of pebbles, evaporates, conglomerates, etc.). For example, the ExoMars panoramic camera [1227 is designed with 18 geological filters and 4 RGB filters based on the Beagle 2 design [1228] (compared with 12 for the Pathfinder and MER cameras [1229]).

The spectral imaging of rock, regolith, and dust provides the basis for rock identification (Figure 12.6). An autonomous feature detection algorithm in conjunction with an if–then rule-based system has been developed to detect carbonate absorption bands (at 2.33 and 2.5 µm) in near-IR reflectance spectra in the region 2,000–2,400 nm [1230]. Calcite (absorption band at 2,400 nm) and dolomite (absorption bands at 2,150, 2,000, and 2,350 nm) are typical carbonates with characteristic vibration modes of CO_2^{3+}.

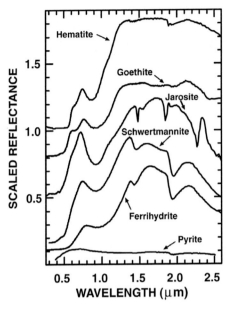

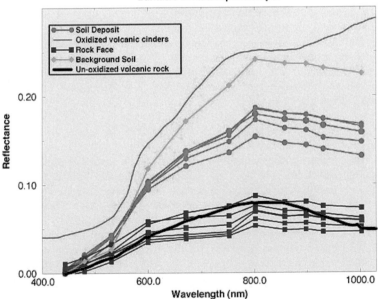

Figure 12.6. Electromagnetic reflectance spectra of different rock types typical on Mars [credit U.S. Geological Survey]; for comparison, APXS reflectance spectra of Martian rock Barnacle Bill [credit NASA].

Kalman filters have been applied to gamma ray spectroscopy which detects the concentration of radioactive minerals with subsurface depth [1230]. The Kalman filter state represents radioactive material concentration and measurement represents gamma ray flux. The same approach should be applicable to neutron measurements of hydrogen concentration below the surface of the Moon or Mars. On the MERs, however, attempts to correlate PanCam spectral imagery of rocks to APXS-measured elemental abundances and Mossbauer-measured abundances of Fe-bearing minerals have been only partially successful [1230]. This was due to the differences in sampling depth of the instruments and the effects of the surface varnish on imagery. This problem is not expected to occur if mini-TES were used in place of pancam imagery to provide deeper measurements or the use of LIBS in place of APXS/Mossbauer measurements to provide surface measurements. This illustrates the difficulty in correlating different scientific instrument data autonomously.

Classification is the next stage following visual feature extraction. Feature correlation computes a difference measurement between two image patterns which is typically thresholded. Large variations in images are common to render such methods difficult. Clustering involves organizing patterns into clusters so that patterns within a cluster are more similar to each other than those belonging to different clusters. Feature patterns may be reduced to low-dimensional vector spaces of all possible patterns allowing K-means clustering for image compression [1231]. Unsupervised classification allows the extraction of clusters of data points (e.g., the K-means algorithm allows identification of similar data). A maximum likelihood classifier computes the probability that data belong to a class. K-means clustering may use an adaptively changing normalized Mahalanobis distance metric to partition data samples into clusters. It is an example of an expectation maximization algorithm. The normalized Mahalanobis distance between a new feature x and the Gaussian centroid of a cluster μ (codebook of codevector prototypes) is given by:

$$M(x,\mu) = \tfrac{1}{2}(d \ln 2\pi + \ln|\Sigma| + (x-\mu)^T \Sigma^{-1}(x-\mu))$$

where d = vector space dimensionality and Σ = covariance matrix of cluster shape and elongation directions. The K-means clustering technique has been used in automated face recognition [1232]. Other kernel-based clustering methods include self-organizing maps and support vector machines [1233, 1234]. Ideally, rock classification would follow traditional geological patterns (Figure 12.7).

However, classification is not straightforward; rock color is corrupted by weathering and dust cover but rocks vary in their degree of dust covering. Fresh exposures allow rapid identification based on spectral reflectance. Texture in particular is deemed to offer enhanced geological classification—both different soil types and different rock types can be differentiated through texture. Rocks in particular are the key to geological investigation and give indications to geological history. Visual texture such as granularity has geological significance as in fluvial or evaporite weathering. Columnar jointing in volcanic rock, layering in sedimen-

Figure 12.7. Example sedimentary, metamorphic, and igneous rocks.

tary rock, conglomerations from breccias, etc. will yield textural signatures. There are, however, conditions in which geologically different textures may appear as similar visual textures. Ambiguities can often be dispelled by textural analysis at different scales. Certain minerals may be promising targets for astrobiology—evaporate deposits, hydrous minerals (carbonates, sulfides, etc.) indicative of hydrothermal vents, etc. Treated causally, each of n minerals (parents) causes the states of m attributes (child nodes).

Consider a choice between dolomite (a light-colored carbonate material) or kamacite (a dark-colored meteoritic material) [Gallant, M., unpublished material]. Conditional probability tables are given in Table 12.2.

If the color of a rock is dark then the probability that it is dolomite is given by (Figure 12.8):

$$p(D|C = \text{dark}) = \frac{p(C = \text{dark}|D)p(D)}{p(C = \text{dark})}$$

Color:
$$p_{\text{col}} = \left(\frac{0.85}{0.20}\right)0.05 = 0.21$$

Additional features such as texture may be incorporated. The incorporation of mineralogical estimates in the priors greatly improves classification accuracy.

Table 12.1. Some rock classification parameters.

Attribute	Parameter	Examples
Structure	Form	Planar, lenticular, nodular
	Orientation	Parallel, multiple, non-parallel
	Scale	Thick bedding, medium bedding, thin bedding
Texture	Luster	Earthy, pearl-like, vitreous, metallic
	Relief	Rough, striated, conchoidal, vesticulated, pitted, bumpy
	Scale	Thick bedding, medium bedding, thin bedding
	Shape	Spherical, tabular, bladed, geometric
	Roundness	Angular, rounded
	Size	Gravel, sand
Composition	Albedo	Low, medium, high
	Specularity	Low, medium, high
	Color	Reddish, greenish, black, white

Table 12.2. Conditional probabilities of dolomite and kamacite.

Color		
$p(C = \text{light})$	$p(C = \text{dark})$	
0.35	0.65	

Dolomite		
Color	$p(D = \text{false})$	$p(D = \text{true})$
Light	0.15	0.85
Dark	0.70	0.30
Kamacite		

Color	$p(K = \text{false})$	$p(K = \text{true})$
Light	0.90	0.10
Dark	0.35	0.65

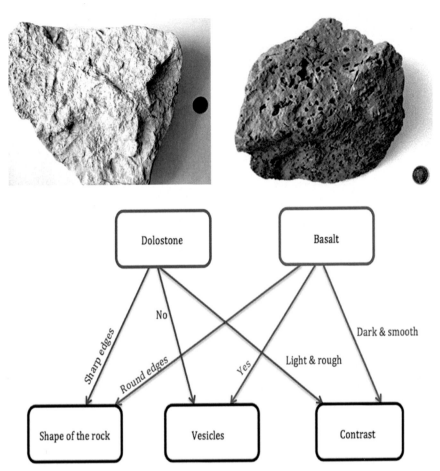

Figure 12.8. Overly simplistic Bayesian network correlating simple visual properties with discrete rock classes (dark basalt versus light dolostone) [credit Helia Sharif, Carleton University].

12.2 CLASSIFICATION BY BAYESIAN NETWORKS

The classification of scientific objects both a priori and novel through feature detection will be a prerequisite [1235]. Novelty detection requires identification of new data through automatic classification. Statistical approaches are based on modeling data according to its statistical properties [1236]. The construction of a probability density function for data of known classification allows assessment of the probability that new data belong to that class. Classification will require recognition of features based on comparison with predefined feature lists. It will require assessment of the scientific value of its observations through a scoring system. The most critical feature is to ensure low incidence of false positives requiring elimination of statistical outliers using weighted filters. The simplest means of achieving identification is by carrying out database searches using statistical methods in which observed spectra are compared with the library of spectral signatures from known samples. Correlation is indicated by maximum correlation coefficient r and minimum chi-squared criterion χ^2:

$$\chi^2 = \frac{1}{n-1} \sum_i \frac{(x_i - y_i)^2}{y_i}$$

where x_i = observed datapoint value, y_i = stored datapoint value, and n = number of datapoints

$$r = \frac{\sum_i (x_i - \bar{x})(y_i - \bar{y})}{(n-1)S_x S_y}$$

where

$$\bar{y} = \frac{\sum_i y_i}{n}, \quad \bar{x} = \frac{\sum_i x_i}{n}, \quad S_x = \sqrt{\frac{\sum_i (x_i - \bar{x})^2}{n-1}}, \quad S_y = \sqrt{\frac{\sum_i (y_i - \bar{y})^2}{n-1}}$$

The box plot constructed from lower-extreme, lower-quartile, median, upper-quartile, and upper-extreme values provides for outlier rejection. Outliers are rejected on the basis of their distance (more than 3 standard deviations) from the nearest neighbor in the class. It is assumed that data are from a known distribution such as a normal (Gaussian) distribution modeled on the basis of mean and covariance. However, extreme value theory may be used to consider abnormal values in the tails of distributions. The mixture of Gaussians approach models general distributions with parameters chosen by maximising the log likelihood of the data set.

The Bayesian classification approach was first developed for the Nomad rover for deployment in Elephant Moraine (Antarctica) to detect dark meteorites against the white surface of snow [1237, 1238]. Nomad was designed for autonomous meteorite searches and its science autonomy system was based on a three-layer hierarchy control architecture comprising a control layer, sequencing layer, and planning layer. The lowest level was the control layer for the robot's sensors and actuators. The second layer was the sequencing layer for coordinating primitive behaviors. The top level was the planning layer which generated plans to achieve tasks and optimize mission parameters such as energy consumption. Operating on this architecture was the scientific knowledge base. It incorporated a behavior control methodology with primitive behaviors including sensor calibration, sensor deployment, sensor data acquisition, sensor diagnostics, and sensor stowage. It operated in two modes: acquisition for acquiring new science targets, and identification for target classification. During target acquisition, the imaging color ratio between blue and green was used to identify rock as opposed to ice yet discriminate rocks from shadow. If the ratio blue/(blue + green) exceeded a threshold, it was classified as ice, otherwise rock. This then triggered the Bayesian classifier network for target identification. The first level of classification required for a Bayesian network was to differentiate between rocks, soil, sky, shadows, and everything else. This assumed that some form of prior scene segmentation (independently in hue saturation and intensity channels) had occurred—a 5×5 pixel area was commonly used initially and then merged into uniform regions. This multichannel segmentation was crucial to reliably detecting geologically relevant targets such as rocks. The Bayes network is used for classifying sensory data according to prior expert knowledge held in the scientific knowledge base. The Bayes network encoded statistical distributions of spectral features for 19 terrestrial rock types and 6 meteorite types assuming prior probabilities and measurements from a metal detector, color camera, and a spectrometer. From this, it computed the posterior probabilities of correct identification given the current sensor readings. It was designed to cope with different lighting conditions but was brittle to shadows. It was clearly limited in scope because a full geological classification system would require a database of 700 rocks/minerals using a more extensive measurement suite such as a visible/IR spectrometer. The science autonomy system successfully classified 42 samples from IR spectrometry with a 79% success rate. However, the identification of meteorites against a background of ice offer higher contrast than that expected on rocks and regolith on planetary surfaces, particularly if covered with the same type of dust (though IR detection based on differing thermal inertia may offer an alternative mode of classification).

Bayes theorem computes the posterior probability of hypotheses based on evidence and prior knowledge optimally from the previous time step (the Markov property) [1239]. Bayesian approaches estimate the posterior probability that the sample belongs to a class on the basis of multiple sensory evidence. Bayesian (maximum a priori likelihood) classification is a robust form of object classification. An event w defines an object belonging to a class $(w_1, \ldots, w_c)$. The probability that an object belongs to class w_i before measurement is defined—this is the a priori probability $p(w_i)$. The conditional probability that an object is in class w_i given measurement x is required $p(w_i|x)$. The conditional probability density that a member of class w_i will have feature x is also required $p(x|w_i)$. The conditional probability of class w_i relates to the conditional probability density of the measurement and the a priori probabilities of each class. Bayes rule defines the posterior probability in terms of prior probability given evidence x and minimizes the risk of misclassification:

$$P(w_i|x) = \frac{p(x|w_i)p(w_i)}{\sum_j p(x|w_j)p(w_j)}$$

If all classes are equally likely, $p(w_i) = 1$, then we have the decision rule: class w_i if $\forall j \neq i, p(x|w_i) > p(x|w_j)$. To represent $p(x|w_j)$, we use the Gaussian function:

$$p(x|w_i = \frac{1}{(2\pi)^{d/2}|C_i|^{1/2}} e^{-\frac{1}{2}(x-\mu_i)^T C_i^{-T}(x-\mu_i)}$$

where $d =$ dimension of vector x, $\mu_i =$ mean vector of training set X_i of samples in class $i = \mu_i = 1/N_i \sum_{x \in X_i} x$, $N_i =$ number of samples in X_i, and $C_i =$ covariance matrix $= 1/N_i \sum_{x \in X_i} (x - \mu_i)(x - \mu_i)^T$. We may use this formulation for recognizing objects by shape descriptors.

Example 12.1. An unknown rock feature is measured by an onboard microscope-based vision system to have a characteristic length of 2.5 μm. Empirically, a large number of dry inclusions have an average length of 2.1 μm with a length variance of 0.8 μm. Similarly, a large number of fluid inclusions have an average length of 2.8 μm and a length variance of 1.3 μm. Is the target more likely to be a dry inclusion or a fluid inclusion?

Apply the Gaussian assumption:

$$p(2.5|\text{dry inclusion}) = \frac{1}{\sqrt{2\pi(0.8)}} e^{-(2.5-2.1)^2/(2\times0.8)} = 0..404$$

$$p(2.5|\text{fluid inclusion}) = \frac{1}{\sqrt{2\pi(1.3)}} e^{-(2.5-2.8)^2/(2\times1.3)} = 0.338$$

Hence, the unknown object is more likely to be a dry inclusion rather than a fluid inclusion.

Example 12.2. Consider the problem of determining which of three phenomena a datapoint belongs to based on observations of features and knowledge of the feature frequencies associated with each phenomena. Consider that the three phenomena occur at the following rates: 81.9% A, 13.7% B, and 4.4% C. Assuming that $p(X_1|\theta_1) = 3.96 \times 10^{-9}$, $p(X_1|\theta_2) = 1.18 \times 10^{-8}$, and $p(X_1|\theta_3) = 1.91 \times 10^{-7}$, the prior probabilities for each phenomenon are 0.819, 0.137, 0.004. The posterior probability of a measurement belonging to each phenomenon:

$$p(X_1) = 0.819 \times 3.96 \times 10^{-9} + 0.137 \times 1.18 \times 10^{-8} + 0.044 \times 1.91 \times 10^{-9}$$

$$= 13.264 \times 10^{-9}$$

$$p(\theta_1|X_1) = \frac{0.819 \times 3.96 \times 10^{-9}}{p(X_1)} = 0.25$$

$$p(\theta_2|X_1) = \frac{0.137 \times 1.18 \times 10^{-8}}{p(X_1)} = 0.12$$

$$p(\theta_3|X_1) = \frac{0.044 \times 1.91 \times 10^{-7}}{p(X_1)} = 0.63$$

Hence, the third phenomenon is favored. Beyond simple texture analysis, multiple visual properties may be implemented in more complex Bayesian networks in order to classify rocks according to their depositional history (Figure 12.9).

The Bayesian network developed from the use of probability distributions to represent uncertainty in expert systems. Bayesian networks based on directed

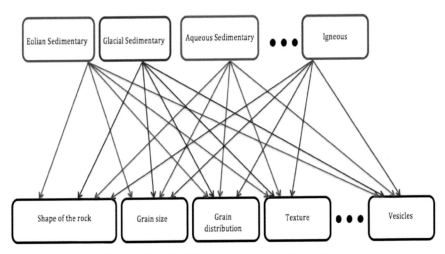

Figure 12.9. Simplified Bayesian network classifier.

acyclic graphs are to be differentiated from Markov random models which are
undirected models. A Markov random field defines a family of joint probability
distributions through undirected graphs. Energy minimization is equivalent to
maximizing probability. Bayesian networks are directed acyclic graphs that comprise nodes connected by directed links. The nodes represent random variables
and the arcs the dependent relationships between them. They can represent statistical dependencies between variables and thereby build complex probabilistic
models [1240]. Each root node comprises a prior probability table while each non-
root node comprises a conditional probability table. They thus represent complex
probability distributions in a propositional framework [1241]. These associative
networks essentially represent uncertain relationships and systems of beliefs. They
model statistical (causal) relationships between feature sets and data sets—each
node has a conditional probability distribution dependent on the states of its
immediate predecessor nodes. There are two types of Bayesian network—the
belief network that represents conditionally independent assertions and causal networks which also represent causal assertions [1242]. A belief network is a directed
acyclic graph representing the probability distributions of conditionally independent assertions. A directed acyclic graph constitutes a causal network, and a
causal network may be implemented through a Bayesian network [1243]. Bayesian
networks can model causal relations that are more complex than simple conditioning causal stimulus–response links including causal chains, multiple causes,
interactive causes, etc. [1244]. Causal connections are represented by directed
(cause–effect) links so a causal network is a belief network with non-root nodes
caused by parent nodes. In fact, the arcs represent direct causal links. The strength
of the causal link is encoded as a set of conditional probability distributions of
cause–effect relations. If C causes A and B—$p(A|C)$ and $p(B|C)$—then

$p(A \cap B|C) = p(A|C)p(B|C)$ so $p(A|C)p(B|C)p(C)$. The links are represented as causal probabilities under the Markov assumption as Bayes theorem which defines the probability of any parameter θ in a hypothesis model given data D:

$$p(\theta|D, H) == \frac{p(D|\theta, H)p(\theta|H)}{p(D|H)}$$

where $p(\theta|H) =$ prior probability of parameter θ assuming hypothesis H, $p(D|\theta, H) =$ likelihood function defining the probability of observation D conditional on the truth of parameter θ and hypothesis H, and $p(D|H) =$ evidence of model hypothesis H which normalizes Bayes theorem

$$= \int_{\theta} p(D|\theta, H)p(\theta|H)\, d\theta$$

Bayesian networks model uncertainties through probabilities which can be used for supervised learning of causal chains [1245, 1246]. Bayesian belief nets are probabilistic directed acyclic graphs in which nodes are randomized variables and arcs indicate conditional dependence among the variables. They represent causal relationships and allow the integration of logical inferencing with Bayesian probability. They are directed acyclic graphs with nodes in the network representing probabilistic sensory data selected by joint probability distributions. Evidence comprises the state of the nodes in the network which is propagated through the network through bidirectional links giving an overall joint probability distribution. The joint probability distribution can be encoded in the structure of the Bayesian network. The belief network computes the Bayesian belief on the basis of evidence: $\text{Bel}(H) = p(H|E)$. The root node represents the set C of mutually exclusive and exhaustive classes that are characterized by prior probabilities. Each node is a descendent of its causal parents. Bayesian nets can be used to model conditional dependences between belief and sensory evidence. They model abductive reasoning through computing conditional probabilities. If there are n nodes defined by binary random variables, there are 2^{n-1} joint probabilities. The Bayesian belief network classifies segmented regions into different categories. Bayesian networks may be used to classify minerals inferred from observations of features such as perimeter, dimensions, texture (surface roughness), and color (hue saturation and intensity) [1247]. These comprise the inputs to the Bayesian network that output a geological classification. The Bayesian network may serve as an analogue to knowledge structures of the human brain and are suitable models for expert systems. Prior knowledge may be incorporated readily in the topological structure of the network with probabilistic certainty factors. Bayesian nets offer a mechanism for combined low-level and high-level vision processing. Nodes represent random variables (propositions) while arcs represent direct probabilistic correlations between the linked variables—the strength of these dependences is quantified by conditional probabilities. Lack of arcs between variables indicates the probabilistic independence of nodes. The representation of conditional dependences and independences is an essential feature of the Bayesian network. Independencies are used to reduce the number of parameters required for probability distributions. Bayes theorem is used to determine the conditional probability tables for each node based on its predecessors. Conditional probabilities relate each node x_i (child) to its immediate predecessors (set of parents) π_i. A system of conditional probabilities associated with arcs corresponds to causal relations (i.e., the arc from parent A to child B indicates A causes B). The Bayesian network effectively provides a causal model of the world under consideration. Initial conditional probabilities are typically arranged in a conditional probability table that defines the node probabilities based on the states of the parent nodes. The Bayesian network may be represented as a triple $BN = (V, E, P)$ where $V =$ set of vertices, $E =$ set of edges, and $P =$ set of prior and conditional probabilities. The Bayesian net is used to compute posterior probabilities from prior probabilities as the primary mechanism for inferencing (which is NP hard for richly connected networks). The probability distribution function over n random variables $x_i, \ldots, x_n$ in a Bayesian network is given by:

$$P(x_1, \ldots, x_n) = p(x_n | x_{n-1}, \ldots, x_1), p(x_{n-1} | x_{n-2}, \ldots, x_1), \ldots, p(x_i) = \prod_{i=1}^{n} p(x_i | \pi_i)$$

Each node has its conditional probability distribution that represents the likelihood of its possible values based on those of its parent nodes. Conditional independence of x_k from all other variables except its parent variables x_{k-1} and x_{k-2} is given by:

$$P(x_k | x_{k-1}, \ldots, x_1) = P(x_k | x_{k-1}, x_{k-2})$$

The dynamic Bayesian network is a Bayesian network for modeling temporal stochastic processes. A network is defined for two or more consecutive time slices that are unrolled into a static network. Evolution of the hidden state of a system is given by:

$$X_i = A_i X_{i-1} + B_i W_i$$

where $W_i =$ stationary process white noise vector. The observation vector is given by:

$$Y_i = H_i X_i + J_i V_i$$

where $V_i =$ stationary measurement white noise vector.

The Bayesian network is trained using a set of F features each of which has M measurements associated with it. The measurements supply the evidence to update the probabilities of each node. The Bayesian network selects optimal actions to maintain an optimal balance between exploration and exploitation. It allows the use of prior knowledge while simultaneously integrating data from multiple sources. The Bayesian network's topological structure is unknown prior to learning from data. A Bayesian network reflecting probabilistic dependences may be constructed automatically from data sets based on assumptions about prior probabilistic knowledge [1248]. The idea is to learn the maximum a posteriori (MAP) Bayesian network structure that is equivalent to belief propagation. MAP extends the maximum likelihood (ML) by introducing prior probabilities with Bayes theorem to find the Bayesian structure that maximizes the log posterior

probability. The simplest prior distribution is a Gaussian distribution described by
a mean and covariance. Each node of the Bayesian network represents a random
variable $X = (x_1, \ldots, x_n)$ while the directed links between nodes represent joint
probability dependencies $p(x)$ such that:

$$p(x) = \prod_i p(x_i | x_j)$$

Bayesian networks define joint distributions over their nodes as conditional
Gaussian functions. In discrete nodes, the conditional distribution is defined by a
conditional probability table (CBT). For any node X_i with parents $X_{k_1}, \ldots, X_{k_n}$,
the conditional distribution is given by:

$$f(x_i | x_{k_1}, \ldots, x_{k_n}) = \frac{1}{\sqrt{2\pi\sigma^2}} e^{-1/(2\sigma^2)(x_i - u_i)^T (x_i - u_i)}$$

where $u_i = \mu_i + \sum_k w_{k_i}(x_k - \mu_k)$, w_{k_i} = arc weights (regression coefficients),
$\mu_i = (1/i)\sum_{j=1}^i x_j$, and $\sigma = (1/i)x_i - \mu_i\mu_i^T$. The joint Gaussian distribution over
all nodes may be represented as a potential function of the form:

$$\phi(x : p, \mu, \sigma) = p\, e^{1/(2\sigma)(x-\mu)^T (x-\mu)}$$

where σ = covariance matrix and $p = (2\pi)^{-|x|/2}|\sigma|^{-1/2}$ = normalizing constant so
that:

$$\int \phi(x; p, \mu, \sigma) = 1$$

Bayesian network learning from a training data set involves search for the
Bayesian network that yields the highest probabilities. Thus, the goal is to find
Bayesian network **B** that maximizes the (log) likelihood (expected gain) of data
$D = (X_1, \ldots, X_n)$:

$$\mathrm{LL}(B|D) = \sum_{i=1}^n \log p(x_i | \pi_i)$$

These scoring functions measure the goodness of fit of the network topology to
the data set. Such scores must be based on a conditional likelihood but this is
difficult to compute at large scales. Nominally, this begins with a topology
without links and then exhaustively applying link addition, link removal, and
reversal operations (cycles being illegal). As adding links almost invariably
increases log likelihood, overfitting can result. Several methods can alleviate this—
limit the number of parents for any node and penalize complexity by minimizing:

$$\mathrm{MDL}(B|D) = \tfrac{1}{2}m \log n - \mathrm{LL}(B|D)$$

where m = number of network parameters. Alternatively, the Bayesian Dirichlet
(BD) score is maximized:

$$\mathrm{BD} = p(B)p(D|B) = p(B) \prod_{i=1}^q \prod_{j=1}^r \frac{\Gamma(n'_{ij})}{\Gamma(n'_{ij} + n_{ij})} \prod_{k=1}^s \frac{\Gamma(n'_{ijk} + n_{ijk})}{\Gamma(n'_{ijk})}$$

where $\Gamma(\ ) = $ gamma function such that $\Gamma(x+1) = x\Gamma(x)$ and $\Gamma(1) = 1$, $q = $ number of parental states, $r = $ number of x_i states, $p(B) = $ prior probability of Bayesian structure, and $n_{ij} = \sum_{k=1}^{s} n_{ijk} = $ number of network states. Alternatively, the Bayesian information criterion (BIC) computes the likelihood of network B with a structural complexity penalty modeling data set D of size N:

$$BIC = \log p(D|B) - (\tfrac{1}{2} \log N) \dim(B)$$

where $p(B|D) = \sum N_{ijk} \log(N_{ijk}/N_{ij})$. Bayesian networks can be used as classifiers in which data are placed into predefined categories (class) on the basis of features or attributes. They use Bayes theorem to compute the relative likelihood of events and a priori data to improve the network. As the number of variables increases, the search for the set of all possible Bayesian networks is NP hard. The naive Bayes classifier—a special case of the Bayesian network where each attribute has only one parent (its class) such that all attributes of a given class are conditionally independent—is widely used for classification. The limitations of the naive Bayes classifier can be overcome using the tree-augmented naive (TAN) Bayes classifier which allows additional edges between attributes of any class to permit correlations between them [1249]. The conditional log likelihood of the class given the relevant attributes that measure the number of bits required to describe D based on probability distribution P may be maximized [1250]:

$$\mathrm{CLL}(B|D = \sum_{i=1}^{n} \log P(y_i|x_i)$$

where $B = $ number of parameters of network. Alternatively, conditional minimal description length is given by:

$$\mathrm{CMDL}(B|D) = \tfrac{1}{2} m \log n - \mathrm{CLL}(B|D)$$

There are no closed-form optimal estimates. The approximate conditional likelihood offers a solution to the problem of learning Bayesian network topologies [1251]. The Bayesian network may represent knowledge within an integrated inferencing system. Rather than computing exact posterior distributions in a large Bayesian network, a Markov chain model or mean field approximation may be used [1252]. The low values of Kullbach–Leibler cross entropy suggest that the Bayesian network can act as an accurate predictor extrapolated from prior knowledge [1253]. Bayesian networks provide the basis for probabilistic computing in representing the logic of unreliable nano-electronic circuits [1254]. Inadequacies of the Bayesian net in representing complex relationships may be overcome using multiple Bayesian nets [1255].

Bayesian networks comprise a useful mechanism for classifying rocks based on images, particularly in attempting to distinguish aqueous deposited from igneous rocks using grayscale contrast, texture, and vesicular nature (the latter may be partially diagnostic of igneous origin) (Figure 12.10) [1256]. Canny edge detection algorithms may be used to extract features [1257]. However, Bayesian classification of images of rocks processed using Haralick parameters have yielded accuracies of

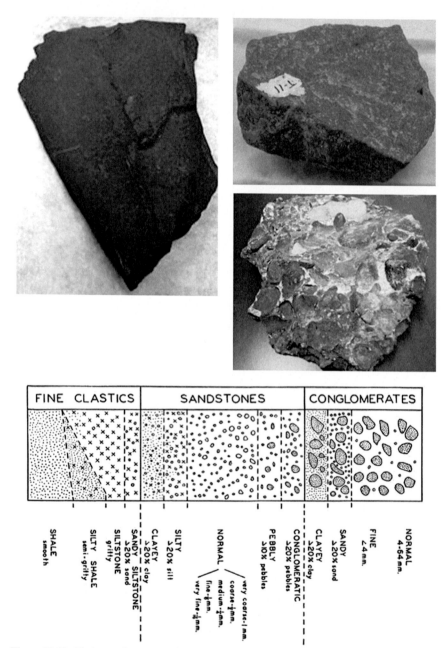

Figure 12.10. Shale, sandstones, and conglomerate based on grain size texture [credit Helia Sharif, Carleton University].

80% under laboratory conditions [1258]. A Bayesian network has been developed for classifying simple objects (tokens of rocks based on size and shape but not color) based on features extracted from image data [1259]. The Bayesian network classifies the visual features. An evaluation score based on the Bayesian feature classification is given by:

$$S = KD\sigma \sum_{i=1}^{n} w_i p_i$$

where K = scaling distance, $D = 2 \tan^{-1}(r/d)$ = feature angular diameter, r = feature radius, d = feature distance, σ = standard deviation of features, i = feature number, w_i = weighting factor, and p_i = Bayesian probabilities of feature i. The evaluation criterion is used to select the most appropriate image features and the path-planning algorithm plans the rover path to the new opportunistic target [1262]. The path-planning algorithm based on D*/θ* allows replanning to account for the opportunistic detection of salient targets [1263]. In particular, a small scout rover can be deployed to enhance the efficiency and reliability of a larger rover.

Principal components analysis (PCA) based on eigenvector decomposition of the covariance matrix identifies linear correlations between data. It reduces the dimensionality of high-dimensional data while retaining its statistical variability to preserve information. A Raman spectrometer example may be used to illustrate the principal components analysis approach [1264, 1265]. From the Raman spectrum $s(\lambda)$, concentration of a molecular component can be determined [1266]:

$$c = \sum_{\lambda} s(\lambda)b(\lambda)$$

where $b(\lambda)$ = characteristic intensity–concentration factor (from calibration). In general, $C = S \cdot B^T$ where B is determined from multivariate analysis. A number of approaches may be used to extract B including least-squares estimation and principal component regression. If a database of all chemicals in the mixture is known from experimentally determined pure spectra $S = C \cdot P$ then $C = S \cdot P^T (P \cdot P^T)^{-1}$ so $B^T = P^T (P \cdot P^T)^{-1}$. PCA chooses a subset of spectra (principal components) to characterize the range of all spectral variations. A linear combination of these principal components can be used to fit the measured spectra despite unknown molecules. Further useful developments would be in autonomous science whereby decision algorithms are applied to classify the data according to certain parameters. In logistic regression, the probability that sample i in the spectral data belongs to a particular category is given by the logistic/sigmoidal function:

$$p_i = \left(1 + \exp\left(\alpha + \sum_j w_j x_j\right)\right)^{-1}$$

where x_j = score associated with spectrum j, w_j = weighting coefficient, and α = constant offset. The likelihood function L defines the probability of observing a specific classification in a set of samples:

$$L = \prod_{i=1}^{n} p_i$$

In this way, scientific hypotheses can be generated from the analysis of spectral measurements. Commonly, singular value decomposition (SVD) is used:

$$X = U \cdot S \cdot V^T$$

where $S =$ non-negative diagonal matrix and $U, V^T =$ unitary matrices. More complex approaches to automated Raman spectroscopy using artificial intelligence methods have been described [1267]. A fuzzy logic controller based on fuzzy if–then rules and a fuzzy inference engine was used to filter noise from the Raman signal. A genetic algorithm optimization of spectroscopic polynomial functions is used to remove fluorescent background. The selection is from Lorentzian peak, Gaussian peak, and linear polynomial functions of the form:

$$y = y_0 + \frac{2A}{\pi} \frac{w}{4(x - x_0)^2 + w^2}$$

$$y = y_0 + A \cdot \exp\left(\frac{-(x - x_0)^2}{2w^2}\right)$$

$$y = A + Bx + Cx^2 + Dx^3 + Ex^4 + \cdots \quad \text{(nominally quartic)}$$

where $w =$ width (half-total height, $y_0 =$ amplitude offset, $x_0 =$ center of function, and $A =$ area. Normalization by standard normal variate analysis is applied to compensate for scattering:

$$\text{SNV}_i = \frac{(z_i - \bar{z})}{\sqrt{\sum_i^n (z_i - \bar{z})^2 / n - 1}}$$

where $z_i =$ intensity and $n =$ number of samples. Finally, principal components analysis is used to classify samples. Hidden Markov models are stochastic approaches for sequential data sets which contain a finite number of unobservable (hidden) states. The hidden Markov model approach has been proposed for the detection of novel rock features—hidden states estimate unobserved characteristics of the environment. Neural network–based approaches to novelty detection exploit their generalization capabilities [1268]. There are many types of neural networks—multilayer perceptrons, Kohonen self-organizing maps, radial basis functions, Hopfield nets, Elman networks, Boltzmann machines, adaptive resonance theory, grow-when-required networks, etc. Generalized radial basis functions can self-organize into a Bayesian classifier capable of novelty detection. Kohonen self-organizing maps, adaptive resonance theory, and Elman networks are effective classifiers.

12.3   SALIENCE AS A MEASURE OF INTEREST

Science on the fly refers to opportunistically identifying scientific targets during
traverses based on prioritized features [1270]. Such features may be predefined or
based on novelty or salience. Salience is a measure of the location with respect to
the background in terms of intensity contrast, color differences, texture differences,
orientation changes, and motion differences over multiple scales. There is also a
temporal component in that prior exposure and increasing exposure duration
reduces the visual attraction to the object or location in the visual field (habitua-
tion) [1271]. Salience thus measures uniqueness or rarity against the background
(i.e., an interesting location). It is a measure of unexpectedness:

$$S(x) = p(v_i)^{-1}$$

where $p(v_i) = [e^{-\frac{1}{2}(v_i-\mu)^T X^{-1}(v_i-\mu)}]/(2\pi)^{n/2}|X|^{1/2} = $ normal probability of feature
vector v_i. Thus, saccades may be directed towards locations of interest. The rover
must be able to detect and select regions of scientific interest in the surrounding
environment and plan trajectories to such regions. The salience values of scientific
targets are determined by appropriate range sensors such as vision and infrared
reflectance. The path planner selects the most salient objects in the map and computes the optimal route. Sensing strategies represent a development where task
requirements dictate information recovery from the environment [1272]. This
requires hypothesize-and-test approaches where hypotheses are generated about
object identities and poses which are assessed according to certain metrics and
new sensing configurations proposed to verify hypotheses based on matching
criteria. Most sensors have five degrees of freedom—three of position and pan–tilt
which define the viewing direction. The viewpoint is characterized by geometry,
radiometry, and field of view. In general, an actively slewed camera should follow
the rule of thirds in photography—an image must be divided into three segments
vertically and horizontally with the most interesting features of the scene placed
within the central rectangle. The photographic golden rule states that the horizon
should divide the frame into two regions with the larger ground region being 1.62
times the size of the smaller sky region. For task-driven sensing strategies, a priori
knowledge of the task is required. Sensor placement must account for sensor
uncertainty, visibility, occlusion, shadows, and contrast (illumination). The autonomous science facility has importance for the selection of camera fixation points if
active vision is employed through the "saliency" map. Target viewing points in the
visual field may be selected based on a number of criteria but attention shifting
may be determined by a saliency map such that each image pixel is assigned a
measure of saliency based on a weighted sum of low-level feature values [1273].

A saliency map is a 2D topographical map that provides the basis for visual
attention. It encodes the saliency (physical distinctiveness) of objects in the visual
environment [1274]. Salience is a measure of location difference with respect to the
background in terms of intensity contrast, color differences, texture differences,
orientation changes, and motion differences over multiple scales. Visual input is
decomposed into a set of topographic feature maps corresponding to a fixation

target. Different spatial regions compete to stand out (saliency) within each map
on the basis of their differentiability from the background. This may be accomplished using Gaussian filter pyramids at different scales using center–
surround operations [1275]. Multiple feature maps are created in parallel (e.g.,
intensity contrast, color contrast, motion direction and orientation-sensitive
filtering—marking out regions that are different from their backgrounds). Feature
intensity is computed through a center–surround arrangement based on on–off
contrasts with a fine-scale center c and a coarse-scale surround s. Colors aid in
highlighting the salient regions of different colors—color channels may be created
thus:

$$R = r - (g + b)/2 \qquad \text{(red)}$$

$$G = g - (r + b)/2 \qquad \text{(green)}$$

$$B = b - (r + g)/2 \qquad \text{(blue)}$$

$$Y = (r + g)/2 - (r - g)/2 - b \quad \text{(yellow)}$$

Different orientations may be computed by Gabor filters. A saliency map is a
pre-attentive process in which the most salient object is selected as the next visual
target through a winner-takes-all mechanism [1276]. The different local saliency
maps are integrated into a master saliency map that encodes the topography of
local maps to cover the visual scene. The chief advantage to selection of fixation
points is that attention can be dynamically allocated to high-value targets within
the restrictions of limited computational resources. Weights can be altered
dynamically to impose shifts of attention. To further reduce processing
requirements, a Laplacian pyramid with coarser-than-pixel resolution may be
implemented [1277]. Once foveated attention is directed to the region, finer resolution can be used. This kind of facility requires coordination with the autonomous
science facility in order to generate a saliency map. Visual fixations may be based
on informative regions of the scene—specifically, high spatial frequency regions
such as edges [1278]. This may be used to build a saliency map representing
regions where there is high contrast in color, intensity, edges, etc. over multiple
scales using a weighted combination. Priority may be defined as the product of
salience and relevance forming a priority map—the concept of relevance includes
top–down influences [1279]. This is directed by a scene schema comprising stored
knowledge about objects in the type of scene. A robotic model of reflexive attention based on points of intersection of lines of symmetry and edges has been
developed [1280]. Active recognition with viewpoint planning is required to search
for evidence to select between hypotheses that suggest the use of Bayesian classification [1281]. Bayesian networks can implement top–down contextual knowledge
while tracking moving objects in a visual scene for dealing with uncertainty [1282].
Prior probabilities initialize the network while subsequent evidence updates it.
A policy for gaze control may be implemented through a Bayesian framework on
the basis of local features in a visual scene integrated with goal-oriented top–down
processing [1283]. Goals impose contextual constraints on saliency which
maximizes task-specific information in observations over time due to establishing

dependencies between actions and observations. The probability of class c_j given observation o_{ik} (posterior probability) is to be maximized. Priors are supplied from top–down goals yet local features are essential to matching predictive top-level expectations. One such Bayesian model that integrates bottom–up salience with top–down attention is the contextual guidance model [1284]. It differentiates four items of information in Bayes rule:

$$p(O = 1, X|L, G) = \frac{1}{p(L|G)} p(L|O = 1, X, G)p(X|O = 1, G)p(O = 1|G)$$

where $1/(p(L|G)) = $ bottom–up salience, $p(L|O = 1, X, G) = $ top–down target knowledge, $p(X|O = 1, G) = $ task-dependent context-based priors, and $p(O = 1|G) = $ target presence probability. Gaze control may be implemented as a Bayes net inferencing system in which visual evidence is incorporated and propagated through the net [1285]. The figure–ground problem requires separation of objects from the background prior to recognition. The human visual system selects regions that automatically "pop" out from the scene based on uniqueness and high contrast. A saliency measure based on entropy is the key to selection of important regions of the scene. A uniqueness weight may be defined by

$$w(x) = \frac{x}{\sqrt{m}}$$

where $x = $ feature property and $m = $ number of local maxima that exceed a threshold. This weighting emphasizes only a few peaks to enable popout. The salient features may be computed by labeling a saliency map:

$$s = w(I) + w(\theta) + w(f)$$

Selection is through top–down target identification. This approach can be used to select easily redetectable landmarks for SLAM [1286]. Wavelet transforms may be used as salient point detectors with Daubechies 4 wavelets being slightly superior to the Haar transform which in turn is much superior to other approaches such as the Harris corner detector and its variants [1287]. The wavelet measures image variations at multiple scales and extracts smooth edges rather than just corners. Furthermore, extraction of both color and texture information at the salient points yields improved results [1288]. A saliency map has been demonstrated for rapid scene evaluation through the focusing of attention [1289, 1290].

Saliency-based maps for implementing attention have been proposed as a mechanism for autonomous spacecraft landing [1291]. The algorithm involved visual filtering with a Gaussian pyramid and center--surround operation at multiple scales to create several feature maps from which the saliency map is constructed. The most salient location is selected through a winner-takes-all competition (implemented through a competitive neural network). Performance is comparable with human landing site selection.

12.4 EXPERT SYSTEM KNOWLEDGE

The applications of expert systems are multitudinous: for medical diagnosis of bacterial infection (MYCIN), mineral prospecting (PROSPECTOR), electronic component troubleshooting (ACE), organic molecular analysis (DENDRAL), etc. The Multi-Rover Integrated Science Understanding System (MISUS) provides for autonomous hypothesis-directed science [1292]. MISUS is a planning/scheduling system with machine-learning capabilities. It combines planning with machine learning for autonomous scientific exploration onboard planetary rovers. It can classify different rock types using clustering methods to group rock types into classes by analyzing multispectral images. The planning/scheduling component determines rover activities to maximize the quality of scientific return. Similar to the Remote Agent planner, it uses an iterative repair approach to resolve conflicts and repair plans. It selects the most appropriate science opportunities for further observations depending on the expected utility of the observation and the expected cost of acquiring it. The expected utility depends on a figure of merit and the degree to which it will confirm/refute scientific hypotheses. MISUS employs machine learning to identify, evaluate, and select science goals. A distributed planner plans the sequence of rover activities required to realize these science goals and estimate the cost of doing so. Inductive inference may be implemented through rules that assign a recognized object to a class of objects according to its specific properties [1293]. The rules encode a priori descriptions that define the properties of the classes to afford generalization. Science data analysis is performed using clustering based on spectral mineralogical features used to classify rock types. Unsupervised clustering methods implement searches based on a mixture of Gaussian functions for similarity classes of spectral images of rock. Clustering is a distance-based classification process defined by similarity of visual texture and reflectance spectra amongst different rocks. These clusters are used to evaluate scientific hypotheses and to prioritize targets for further observations based on their scientific interest. This typically involves the use of self-organizing learning algorithms such as the Kohonen self-organizing map. Beyond this, scientific hypotheses may use rock-type classifications and spatial clustering to determine model geological processes that might have generated the data measured. As the system builds models of rock-type distribution, it assembles new observation goals which are passed to the rover planner (ASPEN).

Chien et al. (1999a, b) describe an autonomous drilling expert system originally designed for the canceled U.S. Deep Space 4/Champollion comet lander [1294, 1295]. It used high-level commands such as STEP, FFWD, MOVE-DRILL, START-DRILL, STOP-DRILL, TAKE-PICTURE, TURN-ON ⟨device⟩, etc. The planner modeled 11 resource operations including drill location, battery power, data buffer, and camera state, and 19 activities such as uplink data, move drill, compress data, take image, and perform oven experiment. There were 3 major classes of activity: drilling and material transport to the surface, instrument activity including imaging and testing in situ materials, and data uplink (drilling was the most complex). There were 3 separate drilling activities. Each drilling activity drilled a separate hole and acquired samples at three different levels during the process: at the surface, at a depth of 20 cm and at a depth of 1 m. Acquiring samples involved 5 separate mining operations after the holes had been drilled to remove 1 cm of material.

The proposed ASTIA autonomous science system for the ExoMars rover assigns attribute values rather than performing rock or mineral classification [1296, 1297]. It focuses on more global characteristics such as identification of geological boundaries and transitions between geological regions (e.g., identification of bedding planes). Segmentation is fundamental to differentiating between geological units. A scoring system—the science value score—was based on adding numerical values to structure, texture, and composition of geological structures and multiplying them by quality and bias factors. ASTIA is supported by a knowledge-based fuzzy expert system that determines the importance of identified rocks. A robotic scientist has visual data analysis integrated with a knowledge base of geological and biological attributes and a methodology for scientific assessment of potential targets as its basis. It requires using its a priori knowledge as context for its observations. Autonomous rovers must perform robotic scientific experiments and this requires the ability to interpret the results of scientific experiments to generate new hypotheses and the design of further experiments to test those hypotheses. The Mars Exploration Rovers downloaded images to the ground station for 3D terrain generation which allowed the use of a Science Activity Planner (SAP) to plan scientific activities [1298]. In particular, the Miniature Thermal Emission Spectrometer (mini-TES) that measures the infrared emission of soil and rocks produced data that were plotted as an array of spectral curves which could be overlaid on NavCam optical images. On the basis of the data, ground operators planned scientific activities such as image acquisition or rover trajectory to sites of scientific interest. This process currently takes four days to perform—it would greatly enhance scientific productivity if this could be reduced to one day. A critical issue to be considered is the sequence of scientific analyses performed on samples. It should follow a logical sequence of initial investigation to one that is more detailed at increasing degrees of resolution. For an ExoMars-type rover mission sporting a complex suite of scientific instruments, consideration must be taken of the degree of sample processing required and the effects on sample alteration or destruction:

Visual analysis to select sample → Acquire sample → Emplace sample to SPDHS inlet → Hand lens microscope imaging → High-resolution microscope imaging of selected regions → Raman spectroscopic analysis of selected regions

→ LIBS destructive analysis of selected regions

→ Grind sample into powder → X-ray diffraction measurement

→ Organics detection experiment → Life marker analysis

→ GCMS analysis

→ LIBS destructive analysis of selected samples

To make scientific decisions effectively, the rover must be programmed with background scientific information. The most basic form of knowledge base comprises if–then–else rules with uncertainty factors encapsulating domain knowledge. Geological features are often complex with geological classification based on structure such as bedding, texture based on grain size distribution and chemical composition. However, there exist expert systems that have been applied successfully to geological resource exploration (e.g., PROSPECTOR [1300]). MYCIN was an early expert system of 500 if–then rules for diagnosing infectious diseases with a 65% success rate compared with 80% for humans. This was considered to be a significant success. Expert systems are generally limited to around 5,000 production rules. The production rules contain logical AND and OR connectives but the elimination of OR connectives is desirable to enforce determinacy. Metarules can be used to modularize groups of rules that are relevant to certain situations to minimize search times. Rules should be ordered in sequences describing priority such that the first rule found that fulfills the conditions is fixed in preference to subsequent rules. Furthermore, the expert system must be intimately knowledgeable about its suite of scientific instruments, their function and operation. An inference engine must reason to generate hypotheses based on current observations in order to select scientific targets and the most appropriate scientific measurements to test the hypotheses.

12.5   ROBOTIC FIELD TRIALS OF AUTONOMOUS SCIENCE

The most comprehensive experiments in autonomous science (in astrobiology) have been conducted in Chile's Atacama Desert using the Zoe rover test platform [1302]. The 198 kg Zoe planetary rover testbed is a four-wheeled platform that tests robotic science in the field [1303, 1304]. Its primary mission is to demonstrate robotic astrobiology in the search for life during its autonomous traverses looking for water to maximize scientific data generation while operating within significant resource constraints. Based on the types of rocks that it images, it computes the possibility of discovering life. As on Mars, life in the Atacama Desert is likely to be sparsely distributed and possibly concentrated in small oases associated with highly localized microclimates. Zoe moved between defined goal positions at a speed of 0.5 m/s using both autonomous navigation and autonomous science planning. Zoe was equipped with four general purpose processors: two 2.2 GHz Pentium 4 processors for autonomous navigation, one 700 MHz Pentium 3 for sensor processing, and one 133 MHz AMD SC520 for power monitoring. Clearly, this is far in excess of what will be available to a planetary rover. The Atacama Desert is one of the most inhospitable regions of the Earth comprising primarily dry salt beds. It has been instrumental in the Life in the Atacama (LITA) project which uses standardized ecological transects in search of biological signatures employing its instruments for science on the fly (SOTF) [1305, 1306]. SOTF is a reactive behavior that uses a spectral signature (such as chlorophyll) obtained during traverse to trigger the rover to stop to collect further sample data within a planned standardized traverse of 180–210 m between standardized in situ sampling sites. In between endpoints of the transect, the rover stops for quick surveys every 30 m or so. SOTF was based on a Bayesian network classifier to interpret biosignatures and was successful 75% of the time, the failures being associated with the limitations the chlorophyll trigger has in detecting non-photosynthetic life. The Zoe rover deploys a plough to expose subsurface environments protected from the strong solar UV radiation. A mast-mounted pan–tilt panoramic stereo imager (400–700 nm) and a fixed forward NavCam were used to assess rock/soil morphology from a distance. A visible/near infrared spectrometer (350–2,500 nm) measures reflectance spectra from a variety of minerals (such as silicates, carbonates, sulfates, etc.) and organic materials such as chlorophyll. A thermal infrared spectrometer (similar to the mini-thermal emission spectrometer on the MERs) was used to measure emissivity spectra in the thermal infrared range (8–12 μm) for mineralogical information but its liquid nitrogen cooling precluded it from being mounted on the rover. A flash microscopic imager was supported by a fluorescent dye sprayer to detect bacterial colonies. The fluorescence imager could therefore detect chlorophyll. Other fluorescent dyes bind to DNA, proteins, lipids, and carbohydrates. Zoe computed the possibility of discovering life from an assessment of habitability derived from locally variable conditions using a simple Bayes rule computation of the probability of chlorophyll detection. Color images were generated by bandpass filtering at 630 nm (red), 535 nm (green), and 470 nm (blue). The imager provided the context for microscopic imaging. Zoe searched for chlorophyll, lipids, carbohydrates, proteins and DNA using an under-chassis fluorescence imager. The CCD-based fluorescent imager searched for fluorescence signals through 740 nm (infrared), 450 nm (blue), or 540 nm (green) bandpass filters characteristic of chlorophyll. While chlorophyll fluoresces naturally, DNA, proteins, lipids, and carbohydrates require stimulation by a light source tuned to specific absorption frequencies (Figure 12.9). Zoe used a lower power option—an Xe flashlamp provided full spectrum illumination that was bandpass filtered through a filter wheel to transmit only those absorbing frequencies.

In addition, Zoe had a facility for wet chemistry albeit limited in staining samples by spraying mild acetic acid and dyes to induce fluorescence in target molecules such as proteins and carbohydrates. The onboard science system involved making decisions on the selection of instruments to be deployed, but this facility is currently limited. It was found that increasing the diversity of sampling sites increases the probability of successful discovery even at the expense of reducing instrument deployment diversity. In fact, it was deemed preferable to begin with a reconnaissance phase and return to more promising sites later. However, in the context of a Mars mission, this is considered impractical in terms of mission time so a tradeoff would be required between sampling diversity and instrument suite diversity. Similarly, Nomad had similarly been deployed in the Atacama Desert to demonstrate the search for paleolife within evaporate deposits [1307]. In this case, aerial photography (emulating orbital imagery) was used to manually identify outcrops and lightly colored regions within a geology comprised of both volcanic and sedimentary deposits. In one rock sample thin section, an anomalous clast was identified through high-resolution multispectral imagery which was interpretable in three ways—fossil, chert nodule, or Fe-rich conglomerate clast. Selection of the correct hypothesis required the use of spectrometry which was not carried by Nomad. Carbonate identification may indicate a biotic source but is still ambiguous—in the case of the Earth, a thorough knowledge of geological history and dating provides the context for interpretation. This indicates that identification of fossils in situ on planetary surfaces may be fraught with difficulty unless sufficient contextual data can be generated.

As discussed by Cabrol et al. (2001) in the Atacama Desert, fossil discovery occurred during reconnaissance dominated by traverses rather than detailed site investigation indicating the utility of rover traverse during the discovery process (Figure 12.10). FIDO has been deployed in a more realistic scenario during scientific desert field trials in Nevada [1308]. Its Athena payload included a mast-mounted PanCam stereo camera pair for navigation and science imaging, front and rear-mounted HazCams for obstacle avoidance, an IR (1.3–2.5 μm range) spectrometer, color microscope, and rock-coring drill. Drilling was aided by a body-mounted belly camera.

12.6 GAS EMISSION SOURCE LOCALIZATION

Mars surface rover exploration in search of methane sources. though a rather specific mode of astrobiological search, may be integrated into the broader "robotic scientist" capacity, particularly that associated with visual analysis. Earth-based measurements of methane plumes are maximum during the summer over Elysium, Tharsis, and Arabia Terrae in the northern hemisphere indicating discrete sources [1309]. Methane has been confirmed by the Mars Express Planetary Fourier Spectrometer with a global aveage of 10 ± 5 ppb but varying between 0 and 30 ppb [1310]. However, the veracity of the methane measurements has been questioned [1311]. The Mars methane emission scenario has similarities to the problem of terrestrial odor detection [1312–1314]. The nature of localized methane sources is unknown as is the resolution and/or detectability of methane on the Martian surface (indeed, it is not clear whether they are real or artifactual). Methane emissions may result from non-biotic serpentine deposits or have a biotic origin (from either subsurface methanogens directly or the release of methane clathrates of biotic origin). There are several possible sources ranging from thermal evolution of near-surface methane clathrates or other organics (unlikely due to strong oxidants), to deeper sources such as Fischer–Tropsch-like reactions, carbonate reduction, serpentinization reactions, or even microbial methanogenesis. Chemolithoautotrophic methanogenesis involves the active reduction of CO and

CO_2 to methane [1315]:

$$4CO + 2H_2O \rightarrow CH_4 + 3CO_2 + 0.48 \, eV$$
$$4H_2 + CO_2 \rightarrow CH_4 + 2H_2O + 1.71 \, eV$$

Microbial sulfate reduction is a common source of methane on Earth. Sulfate reducers can often exploit other electron acceptors such as Fe(III). Both sulfur and iron are present in the Martian environment. In any case, if biotic in origin, the microbial population is very low. There are three major abiogenic hypotheses for the origin of methane. Volcanic emission would require recent Martian volcanism. The Mars Odyssey THEMIS (thermal emission imaging system) has mapped the Martian surface in the visible/infrared region at a resolution of ~100 m [1316]. No active thermal signatures have been detected, thereby discarding that hypothesis. Methane clathrate hydrates are stable at Martian temperatures and stability increases at shallower depths if methane is mixed with carbon dioxide [1317]. The gas hydrate model suggests that gas hydrates are formed when hydrocarbons and other gases are concentrated under high pressure and low temperature in the presence of water. They form van der Waals bonds within the cubic crystals of water ice. As Mars froze, the freezing point of the Martian cryosphere propagated downward incorporating subsurface methane as methane hydrates. Methane hydrates are stable at pressures above 140 kPa at 200 K—this corresponds to a depth of 15 m. Above this, methane hydrates are unstable and release methane. Carbon dioxide hydrates may also exist in the Martian subsurface but they are stable to shallower depths up to 5 m with confining pressure of 50 kPa at 200 K. Methane clathrate sources would require release due to geothermal heating which was discounted above. Serpentinites are rocks rich in serpentine formed from aqueous alteration of olivine at continental margins on Earth. Olivine is common on Mars so serpentinization is expected to be widespread [1318]. The weathering of serpentinite—a Mg and Fe-bearing phyllosilicate—by carbon dioxide yields carbonates:

$$\underbrace{2Mg_2SiO_4}_{\text{(olivine)}} + \underbrace{Mg_2Si_2O_6}_{\text{(pyroxene)}} + 4H_2O \rightarrow \underbrace{2Mg_3Si_2O_5(OH)_4}_{\text{(serpentine)}}$$

$$\underbrace{Mg_2SiO_4}_{\text{(olivine)}} + 2CO_2 \rightarrow \underbrace{2MgCO_3}_{\text{(magnesite)}} + \underbrace{SiO_2}_{\text{(quartz)}}$$

$$\underbrace{Mg_2SiO_4}_{\text{(olivine)}} + \underbrace{CaMgSi_2O_6}_{\text{(pyroxene)}} + 2CO_2 + 2H_2O \rightarrow \underbrace{Mg_3Si_2O_5(OH)_4}_{\text{(serpentine)}} + \underbrace{CaCO_3}_{\text{(calcite)}} + \underbrace{MgCO_3}_{\text{(magnesite)}}$$

Serpentine commonly contains ferrous iron which is oxidized during weathering to ferric iron releasing hydrogen:

$$3Fe_2SiO_4 + 2H_2O \rightarrow \underbrace{2Fe_3O_4}_{\text{(magnetite)}} + \underbrace{3SiO_2}_{\text{(quartz)}} + 2H_2(aq)$$
$$\underbrace{}_{\text{(fayalite)}}$$

$$6Fe_2SiO_4 + CO_2 + 2H_2O \rightarrow \underbrace{4e_3O_4}_{\text{(magnetite)}} + \underbrace{6SiO_2}_{\text{(quartz)}} + CH_4$$
$$\underbrace{}_{\text{(fayalite)}}$$

This hydrogen reacts with dissolved carbon dioxide to produce methane:

$$CO_{2(aq)} + 4H_{2(aq)} \rightarrow CH_{4(aq)} + 2H_2O$$

The coincidence of methane maxima over shield volcanoes favors the serpentiniza-tion hypothesis. The serpentinization process generates methane and hydrogen, which are potential biological fuels, so this model is not inconsistent with a biogenic model.

Unknown odor sources emit directional information that is corrupted by external biases and random noise. Scents are chemical particles that undergo spatially and temporally varying distribution with large voids and filaments of high concentration. The ubiquity and frequency of dust devils suggest that atmo-spheric mixing will be fairly thorough locally. A pump-driven snorkel may act similarly to a lobster's antennae to increase local intensity for measurement (depending on the wispiness of the vapor filament). Sampling measurements are necessary to provide point-by-point data on the intensity distribution. The first consideration will be the signal to be detected and the sensors used and their inte-gration times. Instantaneous odor intensity is different from the time-averaged distribution, the former being more readily and more rapidly detected and, indeed, this is what insects respond to [1319]. Chemical sensors are primarily chemically sensitive platforms in opposing pairs for directionality (odor compass). Tin oxide chemical sensors are commonly used in which sensor resistance is dependent on the concentration of reducing gases (such as methanol or ammonia) but they are not sufficiently specific for planetary exploration. In the case of Mars methane, a tuned laser absorption diode spectrometer (TDLAS) can measure methane (or even $\delta^{13}C$ excess in methane as a potential biomarker). Isotopic ratios will be much more challenging and more time consuming to measure than the presence and concentration of methane so the former will be unlikely. Measurement of the atmospheric forcing function—wind vector sensor in the case of scents—is essential. Wind vector measurements are de facto standard measurements on Mars missions (along with atmospheric temperature, pressure, and dust load measurements).

A physical model of the scent plume distribution over time is necessary to provide a predictive capability. This model must be stochastic in nature to account for uncertainties. Vapors emitted from the source are dispersed according to the Reynolds number of the flow. At low Reynolds number typical of micro-scopic environments, viscosity is dominant and vapors spread by diffusion. A fluid flow model may have the form:

$$\nabla(\rho V) = v_x \frac{\partial \rho}{\partial x} + \rho \frac{\partial v_x}{\partial x} + v_y \frac{\partial \rho}{\partial y} + \rho \frac{\partial v_y}{\partial y}$$

where $\rho = $ chemical density and $V = \sqrt{v_x^2 + v_y^2} = $ fluid flow velocity. The divergence of mass flux $\nabla(\rho V) > 0$ for a source and $\nabla(\rho V) < 0$ for a sink (i.e., this can be used to trace the plume mass flux gradient in a manner reminiscent of the potential field). In terms of chemical concentration C, the diffusion–advection–reaction equation models the dynamics of methane plumes in 2D in describing the time evolution of methane concentration C:

$$\frac{\partial C}{\partial t} = D_x \frac{\partial^2 C}{\partial x^2} + D_y \frac{\partial^2 C}{\partial y^2} - v_x \frac{\partial C}{\partial x} - v_y \frac{\partial C}{\partial y} + Q$$

where $D = $ molecular diffusion constants, $v = $ wind speed, and $Q = $ rate of methane production (source). The methane source is fixed and the wind vector V incorporates three processes—general plume movement (advection), puff mixing, and turbulent noise. The chemical concentration peaks at the source and decreases smoothly radially with a Gaussian distribution. Gaussian plume concentration may be modeled by:

$$C = C_{\text{max}} \exp\left(-\frac{y^2}{2\sigma_y^2}\right) \left(\exp\left(\frac{-(z-h)^2}{2\sigma_z^2}\right) + \alpha \exp\left(\frac{-(z+h)^2}{2\sigma_z^2}\right)\right)$$

where $C_{\text{max}} = Q/(2\pi V \sigma_x \sigma_y) = $ maximum plume concentration (centerline), $\sigma_i = \frac{1}{2} C_i x^{(2-n)/2} = $ mean plume width (dispersion coefficients) where $i = y$ or $z, n = 1$ gives a good match, $Q = $ odor release rate, $V = $ mean wind speed, $h = $ emission source height, and $\alpha = $ ground reflection coefficient. The Sutton model of the concentration of a Gaussian plume is simpler in form:

$$C(x, y, z) = C_{\text{max}} \exp\left(-\left(\frac{y^2}{2\sigma_y^2} + \frac{z^2}{2\sigma_z^2}\right)\right)$$

This may be simplified into a 2D form:

$$C(x, y) = \frac{Q}{\sqrt{2\pi} S_y u} \exp\left(-\frac{y^2}{2S_y^2}\right)$$

This model is valid for low Reynolds number fluids. Low-viscosity conditions allow measurement of chemical concentrations to apply a gradient descent to determine the direction to the source (i.e., chemotaxis). The artificial potential field approach lends itself to representing the odor localization problem as gradient minimization. Simple Braitenberg vehicle architectures with bilateral sensors coupled to bilateral wheels offer a reactive approach based on gradient-based source localization. However, although odor source tracking by animals bears some similarities to navigation through artificial potential fields, there is a dominant stochastic component that yields shifting goal positions with high uncertainty.

Odor localization involves searching for and finding the source of a volatile chemical emission in the environment guided by chemical sensing [1320]. The tracking of scents is fundamental across all phyla for the foraging of food or mates. Localization methods are inspired by nature such as mate localization by moths or food localization by lobsters and by bacteria such as *Escherichia coli*. All animal foraging strategies are variations on saltatory search in which periods of cruising for prey are interspersed with periods of waiting to ambush prey. Some animals favor one extreme over the other (e.g., hawks are cruisers while rattle-snakes are ambushers). However, active search is more efficient than waiting for odor patches to be encountered [1321]. The instantaneous odor distribution of a vapor plume will differ significantly from the time-averaged plume distribution. The key to measurement is the ability to integrate measurement by averaging instantaneous fluctuations. The most appropriate instrument for this type of gas measurement is laser-induced fluorescence which gives accurate concentration measurements. Four reactive chemotaxis algorithms have been evaluated—*E coli*, silkworm moth, dung beetle, and gradient-based ones [1322]. The bacterium *E. coli* is propelled by its flagella which can implement two types of motion—counter-clockwise flagella rotation resulting in smooth directed motion and random flagella rotation resulting in random tumbling [1323]. In detecting a nutrient con-centration gradient, it exhibits chemotactic motion towards the nutrient source along the gradient (i.e., smooth motion on detecting nutrients); if no nutrients are detected, random tumbling is implemented as a search strategy. *E. coli* uses a random walk in which randomly directed straight line runs are interspersed with random tumbles for reorientation. *E. coli* bacteria respond to low concentrations of attractants (aspartate) and can follow concentration gradients at a speed of 20–40 μm/s. Directional run behavior is instigated by flagella rotating counter-clockwise which form a coordinated bundle propeling the bacterium forward. Random tumbling behavior is instigated by flagella rotating clockwise causing uncoordinated motion. The ratio of directed to random propulsion increases with the attractant concentration. This approach is inefficient as a means of plume tracking except for plumes with no turbulence. The silkworm moth remains stationary and waits for the evolving odor plume to intercept it. On interception, it moves rapidly upwind and performs side-to-side searches of increasing ampli-tude. This approach is best suited to rapidly fluctuating plumes. The dung beetle employs a zigzag algorithm assuming that loss of signal indicates the plume edge. This is suited to intermediate plumes. Gradient-based approaches maintain straight line trajectories to follow the centerline of maximum concentration. This is an economical approach but requires directional bias to be effective. An artificial potential field is capable of modeling many foraging strategies effectively.

Gradient-based techniques cannot work with plumes characterized as intermit-tent patches driven by winds. Such trace gas flows occur in high Reynolds number fluids such that dispersion is dominaated by turbulence. Odors, unlike potential field pathways, are variable in intensity, sparse, shifting, uncertain, and ethereal.

At medium to high Reynolds numbers, turbulence is dominant over diffusion and vapors spread through background currents (advection) or temperature gradients (convection). A high-viscosity regime is representative of the Mars methane environment which is subjected to external forcings—winds. Chemical plumes form downflow from the source and meanders. Eddies due to turbulent advection disperse odors by stretching and folding over distances quantified by the Kolmogorov length $L = VT$ (a few centimeters or so) where $V =$ wind speed and $T =$ period of fluctuation. Chemical concentration decreases from the downflow centerline of the plume to crossflow plume edges. Turbulence generates chaotic trajectories for the fluid causing stretching and folding of concentration contours. The plume wanders almost randomly over a wide area forming filaments and patches. The plume becomes patchy with higher-than-average concentrations separated by voids. The combination of turbulence and diffusion increases the rate of mixing. Concentration gradients fluctuate over time rendering gradient descent approaches untenable. Odor is emitted at puff emission rate R with a finite lifetime of τ propagating due to diffusivity D and advected by wind current V. The odor plume may be modeled by an advection–diffusion equation thus [1324]:

$$D\nabla^2 U(r) + V \cdot \nabla U(r) - \frac{1}{\tau} U(r) - R\delta(r - r_0) = 0$$

where $U(r) =$ local scalar field (such as concentration) at r, $R(r) =$ odor emission rate at r, $D \approx \Delta V_{rms}^2 \tau =$ particle diffusivity, $\tau =$ odor trace lifetime, $V =$ advecting current velocity, and $\delta =$ Dirac delta function. The continuous solution is given by:

$$U(r|r_0) = \frac{1}{4\pi D |r - r_0|} e^{-\frac{V}{2D}(yy - y_0)} e^{-\frac{|r-r_0|}{\lambda}} \quad \text{where } \lambda = \sqrt{\frac{D\tau}{\left(1 + \frac{V^2 r}{4D}\right)}}$$

The discrete solution is given by:

$$U(r|r_0) = \left(\frac{e^{-\frac{|r-r_0|}{4Dt}}}{4\pi Dt} R(t) e^{\frac{1}{\tau}\left(1 + \frac{V^2 \tau}{4D}\right)t} \right) e^{-\frac{V}{2D}(y - y_0) - \frac{1}{\tau}\left(1 + \frac{V^2 \tau}{4D}\right)t}$$

Qualitative physical models have been proposed to model airflow in simplified (indoor) environments [1325]. However, qualitative models are of limited use and quantitative models offer greater effectiveness. A suitable approach is infotaxis where information plays the same role as concentration in chemotaxis [1326]. This strategy maximizes the expected rate of information gain which favors casting and zigzagging behaviors. Upwind zigzagging is a common search strategy in insects such as the silkworm moth in response to pheromones [1327]. This is enabled through neural circuits analogous to an electronic flipflop circuit which controls the zigzagging behavior.

The simplest approach is to track the plume along its length while garnering additional information from the environment such as wind vectors perhaps supported by a fluid flow model. An array of flow measurements enables extraction of the forcing flow environment to ensure tracking upwind. Most Mars missions carry wind vector sensors as part of their meteorological packages (which include pressure and temperature sensors). External flow measurements are essential to provide a history of measurements for Braitenberg approaches to gradient descent to work. Initial plume acquisition often involves a side-to-side raster scan of increasing magnitude perpendicular to the wind direction until complete circular motion is generated. Chemical plume tracing in turbulent fluids may follow a bowtie maneuver—the trajectory is aligned to 15° from the centerline of the plume in order to acquire the trace which is then reversed if the trace has not been acquired [1328]. This acquisition strategy may be readily adapted. In order to track an odor plume, it is necessary to bilaterally compare two separated vapor sensors either side of the vehicle [1329]. Similar arguments hold for a 3D environment. A more general search method would be an Archimedean spiral trajectory of increasing radius (as adopted by desert ants). Continuous time-recurrent neural networks have been evolved using a weighted fitness function to find odor sources by spiraling [1330]. The zigzag (dung beetle) algorithm involves movement at an intermediate offset angle α (typically between 30 and 60°) from the upwind bearing while monitoring odor concentration. Beyond 60°, the plume becomes wider than it is long. If the odor concentration falls below a certain threshold, the edge of the plume has been encountered. A new upwind measurement is undertaken and movement is offset at the opposing angle $(-\alpha)$ from the new upwind direction. This motion zigzags within the plume while making progress towards the source. Alternatively, the chemical cross gradient may be determined by sensor casting similar to lobster antennae deployment (e.g., RoboLobster). Casting (exploration) involves broader crosswind excursion and reduced upwind progression than zigzagging (exploitation). Alternatively, airflow may be forced to influence odor distribution in much the same way as the effect of wing flutter in the moth. A snorkel with an air pump may serve a similar function—this helps the TDLAS sensor in integrating concentration intensity. Both zigzagging and casting are strategies of infotaxis which maximizes the expected rate of information gain by following the odor trace (noisy message) emitted from the source (transmitter) to the searcher (receiver) [1331]. The rate of accumulation of information may be quantified as the rate of reduction of entropy. The posterior probability distribution of the unknown location of the source is updated iteratively through Bayesian inference from the history of odor encounters. Entropy diminishes more rapidly the closer to the source creating an information gradient. The best search strategy is to explore the environment and then exploit the acquired information to find the source [1332]. If the plume is lost, tracking is performed across-wind while using a predicted concentration gradient. This gradient is determined from the triangle formed by the last three edge points. This should track the upwind plume centerline. Fluctuations in the plume suggest that instantaneous measurements give transient information only. In this case, a potential field representation is applicable but with a variable goal attractor (chaotic attractor). The goal vector results from a linear combination of chemical concentration and airflow gradients. Looked at two-dimensionally, the time-averaged gas plume concentration at (x, y)

due to turbulent diffusion may be modeled by:

$$C(x,y) = \frac{Q}{2\pi K} e^{\left(-\frac{V}{2K}(r-\Delta r)\right)}$$

where $r = \sqrt{\Delta x^2 + \Delta y^2} = $ distance to the odor source, $\theta = \tan^{-1}(\Delta x/\Delta y) = $ upwind direction to the odor source, $\Delta r = \Delta x \cos\theta + \Delta y \sin\theta$, $Q = $ odor release rate, $K = \sqrt{D_x D_y} = $ turbulent diffusion coefficient, $D_i = $ diffusion coefficients, and $V = $ wind speed. If the diffusion coefficient is constant and known, puff dissipation time can be predicted from [1333]:

$$C(t) = \frac{Q}{4\pi t D}$$

A Kalman filter approach that weights measurements against a model may be appropriate here. The gas distribution model may be incorporated through the Kalman filter into the plume-tracking process. However, turbulent diffusion is only weakly realistic so the Kalman filter will be highly reliant on sensory measurements. The hidden Markov model (HMM) provides a stochastic approach to plume modeling in forming a source likelihood map (SLIM) [1334]. The environment is tessellated into a grid whose occupancy by an odor is determined probabilistically by the HMM. Plume detection events are separated by wide voids without any signal, and the probability of detection is dependent on the distance from the source. Hence, Bayes theorem may construct the posterior probability of the unknown location of the source given the measured detection events (particles) emitted by the source [1337]. The actual probability distribution must be estimated from the history of detection events—plume intermittency increases farther from the source. Hidden Markov models describe individual puffs in the plume from

Figure 12.11. Pioneer robot deployed at Jeffrey Mine (Asbestos, Quebec) [credit Ala Qadi, Carleton University].

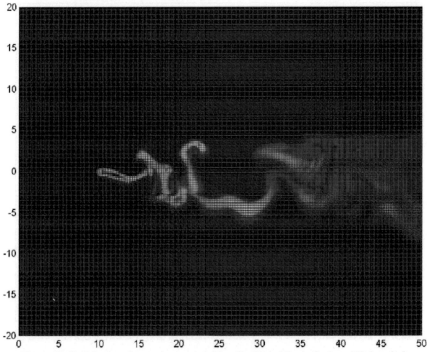

Figure 12.12. Mars methane plume model [credit Chris Nicol, Carleton University].

which backward integration through time allows the generation of a source like-lihood map (SLIM). Bayesian probabilities may be used to update the SLIM based on detection events from the sensors. Bayesian inference to predict likely odor source locations appears to provide a superior approach to SLIM than HMM [1338]. This is similar to the SLAM (self-localization and mapping) process which uses the Kalman filter in autonomous navigation. Plume tracking should be integrated with autonomous navigation—obstacles will affect odor distribution. However, the prior probability of cell occupancy is much lower for odors than for obstacles detected by vision or laser. Simulated Mars operations have been run at the Jeffrey Mine, in the city of Asbestos, Quebec using a Pioneer mobile robot carrying a small number of scientific instruments to search for serpentinite deposits (Figure 12.11) [1339, 1340]. This was based on the assumption that such deposits could be the source of methane emissions measured on Mars.

A Mars methane plume model including advection and diffusion has been developed to ascertain its fine structure (Figure 12.12). A rover surface exploration strategy based on a methane sensor and wind vector sensor has been investigated using a neural network with a Kalman filter learning rule [1342]. This type of exploration is suited to a scout rover role such as Kapvik supporting a larger rover such as ExoMars.

13

Case study: Robotic exploration of Europa

To illustrate the diversity of robotic vehicles that may be deployed in the exploration of the solar system, we consider a case study concerning a robotic lander to penetrate into the ice crust of Europa to access the subsurface ocean below (Figure 13.1). Exploration of Enceladus may be similar—one model of Enceladus incorporates a 10 km deep liquid water ocean some 30–40 km below the surface powered by hydrothermal activity in its porous silicate core. The gas plumes are generated from narrow fissures through the ice layer. This is likely to represent one of the most challenging robotic missions that could be undertaken in the forseeable future.

All aspects of a Europa mission will require high levels of autonomy due to the high communications delay and sporadic communication windows between Earth and Europa. In particular, surface, ice-penetrating, and submersible assets will be required to perform sophisticated robotic functions completely autonomously (Figure 13.2). Planetary protection will be a major design driver to minimize the probability of forward contamination of the Europan environment which imposes the necessity for exacting sterilization methods. Jupiter's magnetic field is 10 times more intense than that of Earth as Europa lies within Jupiter's main radiation belt. This represents the most hostile radiation environment in the solar system with a proton/heavy particle flux of 10^8–10^{10} particles/cm^2 s. The Europan environment is subjected to a high radiation exposure of 15 Mrad/yr. Total doses are expected to be 2.3 Mrad behind 5 mm of Al shielding and 953 krad behind 8 mm of Al shielding, particularly during transfer through the Jovian system yet radiation-hardened electronic devices are generally limited to 1 Mrad total dose which implies extensive use of spot shielding. A Europa cryobot/ hydrobot exploration mission may be preceded by a penetrator mission [1343]. Delivery of a network of seismometers onto the surface of Europa for P and S wave detection would enable determination of the ice crust and liquid mantle thicknesses. A more sophisticated surface lander on Europa would be highly

© Springer-Verlag Berlin Heidelberg 2016
A. Ellery, *Planetary Rovers*, Springer Praxis Books,
DOI 10.1007/978-3-642-03259-2_13

Figure 13.1. Europan ice crust is believed to overlie a liquid water ocean residing above a silicate mantle and Fe core [credit NASA].

Table 13.1. Europa landed segment.

	Dry mass (kg)	Power (W)
Lander	560–800	
Surface station	325–400	
Cryobot	110–150	1,000
Hydrobot	125–200	50–100

desirable after such an exploration mission. We assume that the Europa lander descends from a 100 km altitude orbit requiring a Δv of 2.2 km/s. It should possess a descent imager to survey the landing site before controlled landing and a panoramic imager for onsite environmental imaging. The Europa lander freefalls 10–20 m to prevent exhaust contamination of the Europan landing site. This requires telescopic legs with damping to enable vertical alignment of a drilling cryobot for deployment into the ice [1344].

The chief technological problems are perceived to be power, navigation, and communications.

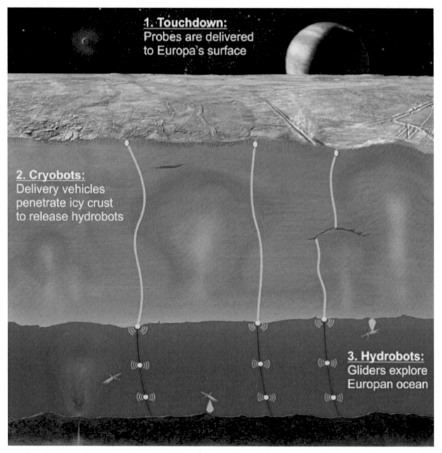

Figure 13.2. Mission profile of a cryobot–hydrobot to explore the subsurface of Europa [credit NASA JPL].

13.1 EUROPAN CRYOBOT

The subsurface exploration of Europa involves the use of a drill sonde or a mole device deployed through a borehole. Mechanical drilling into ice involves the use of a rotary or percussive drill head and requires reaction on the drill stem to push against the cutting force reaction. The required energy for mechanical cutting of ice is 1.2 kJ/kg but this will vary with depth as the ice becomes more consolidated. Deep boreholes as envisaged here introduce the possibility of jamming in long drill stems and borehole sidewall collapse due to the pressure differential between the borehole and the ice matrix. A Europan mission would likely be based on a cryobot—a 1–2 m long, 10–12 cm thick probe which when heated at the tip melts the ice. Allowing the ice to refreeze behind it minimizes the possibility of contamination. Such a cryobot would descend by gravity alone at a rate of 0.5–1.0 km/month. The nominal ice thickness of Europa is taken to be 10 km though it is expected that there will be regions as thin as 1–3 km. There is evidence that Europa may also be intermittently outgassing plumes of water vapor from its subsurface in the south pole indicating a thin icy crust. At 1 km depth, the pressure of the overlying ice is 1.2 MPa while that at 3 km depth is 40 bar (equivalent to 390 m water depth on Earth). Thermal ice penetration may occur through ice sublimation or ice melting—sublimation requires far more energy than melting. The latent heat of evaporation for water is 2,770 kJ/kg at low pressure across a wide range of temperatures compared with a latent heat of melting of 580 kJ/kg [1345]. Liquid water is displaced from the front of the probe to the rear providing pressure equalization. This also requires the walls of the probe as well as the front of the probe to be heated to prevent refreezing. Water would refreeze some 1.25 m behind it. There are two major approaches to thermal drilling of ice—the thermal lance (external water flow) and hot-water drilling (internal water flow). In both cases, gravity provides the basis for the cryobot's descent into the ice requiring a forward center of gravity. High-velocity fluid jets of ~100 MPa can penetrate almost any rock. The power requirement for a water jet is given by:

$$P \propto d^2 p^{3/2}$$

where $d^2 \propto q/p^{1/2}$ = nozzle diameter, $p^{1/2} \propto v$ = nozzle pressure, v = discharge velocity, and q = flow rate. High velocity and pressure require high power—70 MPa requires power/unit area of 3×10^4 MW/m^2. The energy required to melt 1 kg of ice at a temperature of 100 K is given by [1346]:

$$\Delta E = H_{\text{fusion}} + \int_{100}^{273} C(T) \cdot dT = 6 \times 10^5 \quad \text{J/kg}$$

where H_{fusion} = latent heat of fusion of ice = 3.3×10^5 J/kg and $C(T)$ = temperature-dependent specific heat of ice = $7.04T + 185$. If the water is to be evaporated, more energy is required ($H_{\text{vapor}} = 2.5 \times 10^6$ J/kg) making this a less viable approach. Stored chemical energy is not feasible and a long-lived nuclear-based power supply will be essential. Onboard power generation through radioisotope thermal generators (RTGs) suffers from a degradation in power output over time requiring complex thermal rejection through the early mission phases of orbit transfer. The design of the complete system is determined by the requirements of the cryobot probe. As the energy required for melting depends on the volume of water melted, it is advantageous to have as narrow a cross section as possible. In one month, 100 L of water may be melted and filtered using a power source of 20 W suggesting the use of a much higher thermal power source such as a radioisotope thermal generator which is capable of generating around 1 kW of thermal power. The descent rate (m/s) is given by:

$$\frac{dz}{dt} = \frac{P}{\Delta H \pi r^2}$$

where P = thermal power, r = borehole radius, and ΔH = enthalpy change from ice at temperature T to liquid water = $4.5 \times 10^8 \, \text{J/m}^3$ at $T = -100°\text{C}$. A combination of thermal heating to melt water and meltwater jetting offers a faster descent than that for pure heating only (1–2 m/h compared with 0.5 m/h). Hot water at $\sim 50°\text{C}$ is jetted downward which is recirculated aft to be pumped from intakes and reheated for jetting. The ice penetration rate is a function of temperature— assuming a drill rate of 1.5 m/h at $-15°\text{C}$, this drops to 0.9 m/h at $-55°\text{C}$, 0.5 m/h at $-120°\text{C}$, and 0.3 m/h at $-170°\text{C}$. In particular, melting will require higher power as there will be heat dissipation through thermal conduction of the ice but, once melted, water acts as a good thermal insulator. The energy required to melt a 13 cm diameter hole (assuming a 12 cm diameter probe) in ice for a 3 m long probe traveling at 2 m/h is 5 kW. This equates to 1 km every 21 days or 210 days for the complete traverse of 10 km. A 1.2 m long cylinder reduces the energy requirements to 1.0 kW with a lower speed of descent of ~ 0.5 km/month due to the lower weight. The chief advantage of melting is that the melt water can be sampled and filtered for deposits of biological material. Cryobots have also been proposed for melting into the Martian polar ice caps but the Europan deployment of cryobots is a much greater challenge. The integrated cryobot experimental (ICE) probe concept uses a combination of four thermal heaters and a central melt water pump. It is around 3 m in length constructed from a hollow Al cylinder with a copper melting head. There will almost certainly be a requirement to avoid obstacles during the ice descent but this will require active acoustic sensors to detect obstacles in advance in order to initiate differential heating of the cryobot surface. This will also require a spherical tip to allow a curved trajectory. On detecting and reaching the ice/water interface, the cryobot must anchor itself to the ice prior to deploying its hydrobot into the Europan ocean.

The JPL CHIRPS (Cryo-Hydro Integrated Robotic Penetrator System) mission architecture involves deploying a communications transmit/receive electronics package into a shallow depth of ice connected to the lander antenna through an electrical umbilical. The JPL CHIRPS cryobot has a mass of 22 kg, a length of 1 m, and a diameter of 12 cm. It uses four RTG-based thermoelectric heaters around the nose supported by a central hot-water jet using 1 kW of energy [1347]. Capillary fluid loops at the nose, midsection, and rear provide controlled water melting. Temperature, pressure, and inclinometer sensors monitor progress. A series of deployable transceivers are mounted at the rear for deployment. The cryobot uses two acoustic sonar sensors to detect obstacles and can steer up to 30° from the vertical using differential heating. Additional hot-water jets on its sides near the nose would greatly increase its descent rate. A lightweight compact nuclear power system would be ideal for Europan thermal probes similar to that proposed for a Martian polar ice cap cryobot [1348]. This concept—MICE (Mars Ice Cap Explorer)—uses warm-water jets powered by thermal energy from a miniaturized nuclear reactor to create a melt channel. The NEMO Europa melt probe uses the MITEE nuclear reactor's waste heat to melt Europa's surface ice [1349]. Enriched U-235 cermet fuel particles are embedded in a zirconium metal matrix. Cooled water flows into the thin-wall tubes arranged in a hexagonal grid forming the reactor core to remove fission-generated heat. Water acts as a moderator similar to a pressurized water reactor with an efficiency of 20%. The pressure is maintained at a minimum of 10 MPa to prevent boiling at the surface but this increases during the descent to ensure ejection of water. A small electrolyzer generates H_2 gas in the top of the melt probe to maintain bouyancy. The melt probe has a maximum thermal power of 500 kW giving an ascent/descent rate of 340 m/day. Reducing the thermal power to 200 kW reduces fuel consumption to 72 g U-235/year and allows the probe to make scientific measurements during the descent. Electric power output is nominally 10 kW but 20–30 kW can be generated. A similar reactor can be used to power the automatic underwater vehicle delivered to the Europa subsurface ocean to power its hot-water jets. Internal pressure is maintained above submarine pressure—at the ice/water interface at a depth of 10 km, the pressure is 120 atm increasing to over 250 atm at a depth of 10 km below the interface.

Typically, it is assumed that DC electric power and data communication to the cryobot can be supplied through a tether cable released aft of the subsurface explorer connected to the surface lander [1350]. The Dante II tether was capable of withstanding 7,700 N static and 13,600 N dynamic loads, respectively. The load-bearing tether comprised an inner core of Cu AWG conductors sheathed in a layer of load-bearing Kevlar interfaced by 0.1 mm thick low-friction Gore-Tex. The whole tether was encapsulated in an abrasion jacket. It provided two 26 AWG co-axial conductors for two 384 kbps video channels, one 26 AWG twisted pair conductor for a 192 kbps Ethernet channel, four full duplex RS-232 serial channels, and one 18 AWG power conductor. Its motor generated an output torque of 2.85 Nm at 2,000 rpm with a total gear reduction of 1246:1. Maximum spooling speed was 3.5 rpm on a 32.3 cm diameter drum with a maximum static load of 7.7 kN. Total winch weight was 169 kg (55 kg of which belonged to the 300 m tether of specific weight 0.186 kg/m). However, given the depth of the ice, tether-based communication and power transmission is not a reliable approach due to likely ice movement during tidal deformation which could shear the tether. Furthermore, there is a high DC transmission loss along tethers imposing a high power requirement on the surface lander. A 10 km long tensionable tether would also be potentially heavy. For communications, the periodic deployment of miniaturized microwave radio transceiver relays from the subsurface explorer may establish a communications link between the surface lander and the subsurface explorer [1351]. Radio waves travel through ice at a velocity of 168 m/s. These 0.1 m diameter quad dipole antennas will have to be able to supply sufficient RF output power of ~120 mW within a small, low-mass package. It has been suggested that three 1 W bismuth telluride radioisotope heating units could suffice supported by a 1.3 W ultracapacitor. The data rate will be low (~10 kbps) so image data will require long latency times for transmission. A shape memory alloy spring deploys an anchor into the ice to maintain the transceiver's position in the ice. A nominal range of 300 m between transceivers places a limit on the ice depth achievable—an ice temperature of $-15°C$ gives higher transmission distance than an ice temperature of $-5°C$.

13.2   EUROPAN HYDROBOT

Once the overlying ice has been penetrated, a hydrobot may be deployed. The most ambitious scientific goal for the Europan submersible would be to autonomously explore and locate hydrothermal vent sites at the seabed on a 100–200 km deep ocean. The simplest form of hydrobot would be a detachable section of the cryobot deployed while connected by a tether. The cryobot would anchor itself into the ice by freezing approximately 100 m from the ice/water interface. In this case, the detachable section is lowered into the liquid water layer from the cryobot by unspooling the tether. The hydrobot section would have to be negatively buoyant for descent and respooling would provide the ascent mechanism. By virtue of mass limitations, the tether length would be limited to 300 m in length giving it a range of 200 m into the water layer. However, as the deployed tether adds mass to the hydrobot, some form of buoyancy/ballast control in order to descend/ascend may be required. Streamlining with a fusiform (teardrop-shaped) body minimizes drag. Frictional drag is caused by friction between the skin and the boundary layer which generates flow shearing within the boundary layer. The fusiform shape keeps the boundary layer attached to the body. Submarines have been based on the fusiform shape, similar to dolphins with a rounded leading edge and slow tapering tail. Streamlining is defined by the fineness ratio of body length to maximum girth which is optimally in the range 3–8 for minimum drag for maximum volume. MASE (Miniature Autonomous Submersible Explorer) is a miniature (23 cm long by 5 cm diameter) torpedo-shaped autonomous submersible explorer concept designed for deployment into the subsurface of Europa. It extensively exploits miniaturization technology [1352]. It is designed for a deployment duration of 5 h limited by onboard battery power. It is tethered to the mothercraft (the cryobot) incorporating an optical fiber link. Its range is perceived to be 5–10 km—this requires an extremely long tether for optical fiber communication with the cryobot. The use of a tether severely restricts the depth of deployment—to reach the seabed at a depth of 100–200 km would necessitate utilizing an untethered autonomous submersible. There is little doubt, however, that maximum flexibility is afforded by adopting a free-swimming hydrobot with 6 DOF control similar to a submersible.

The first question that arises concerns whether freefall landers or powered submersibles should be used. Research on benthic communities in the deep ocean have been investigated using ballasted underwater landers which freefall-descend to the seabed at depths of 1–5 km and can remain on the seafloor for a considerable time under high pressure of ~600 bar [1353–1355]. Remote vessel sonar can be used for pulse activation of their onboard control programs. They carry out autonomous experiments using onboard instrument packages (such as microscopes, DNA sequencers, and mass spectrometers) and store data on board. They can be deployed to monitor the water column, track abyssal fish, analyze sediment, and undertake in situ chemical analysis of water, gas, and sediment. Once their task is completed, they drop their steel ballast and ascend to the surface for recovery. AUDOS (Aberdeen University Deep Ocean Submersible) is a representative example used to study the movements of deep-sea fish. It comprised an Al tubular frame with buoyancy weights mounted on which is an instrument payload including lights/cameras, acoustic Doppler current meter, and sonar.

Deep water offshore oil and gas production has been the major driver of suboceanic technology. Oceanographic submersibles may be adopted as an initial model for Europan submersibles despite the pressure at the base of Europa's 200 km deep ocean being around 3 kbar (three times that at the Mariana Trench). Tethered remotely operated vehicles (ROVs) can operate over significantly longer periods for seabed surveys and inspection, submarine cable/pipeline installation and maintenance, servicing of subsea oil platforms and military mine counter measures—they are often used as verification models for space robotic systems (e.g., the Neutral Buoyancy Vehicle, NBV, spaceflight simulation vehicle [1356]). The 1,000 kg Jason submersible is an ROV vehicle with high maneverability for detailed surveying and can trail a slave vehicle above the seafloor [1357]. Tethers—which are kept neutrally bouyant—are subject to significant loads transmitted from the motion of the surface ship so a tether management system connects through a heavy tether (typically, triple armored) ~10 km in length to the surface ship and also to the ROV through a lightweight tether ~1 km in length. Although some ROVs are little more than traditional CTD (conductivity, temperature, and depth) measurement platforms, many employ a tether with high-bandwidth optical fibers for telemetering video-imaging data and three copper cables for transmitting AC power over ~100 m. Nereus is a tethered robotic submersible capable of descending to 10 km depths—it successfully descended to the Mariana Trench in 2009. Tethers can present problems for station keeping due to variable ocean current-induced drag. Furthermore, the Japanese ROV *Kaiko* was lost by tether breakage in 2003 illustrating the single-point failure mode introduced by tethers.

Traditional submersible power systems have been based on lead–acid batteries, silver–zinc batteries, sodium–sulfur batteries, Li–ion batteries, solid polymer fuel cells, and closed-cycle diesel engines—the highest performance in terms of endurance results from solid polymer fuel cells [1358]. Such power systems will not provide the long-duration power required for long ranges >100 km. For a Europa submersible, the only viable options are radioisotope generators or a nuclear reactor—the latter has heritage in military submarines as pressurized water reactors but these are large and generate power output far in excess of that required. For smaller submersibles, the most appropriate power system would be primary batteries/fuel cells supported by RHU heat generation. The maximum range of a robotic submersible is given by [Bruhn et al., 2005]:

$$R = \frac{2\varepsilon}{3(P_h - P_s)}\left[\left(1 - \frac{\rho_h}{\rho_f}\right)V - M\right]\sqrt[3]{\frac{(P_h - P_s)\eta_m\eta_p}{\rho_f V^{2/3} C_D}}$$

where P_h = support system power, P_s = supply power, V = submersible volume, M = submersible mass, C_D = drag coefficient, ε = battery energy density, ρ_h = hull density, and ρ_f = fluid density. The propulsion power required at the propeller is determined by drag resistance [Schubak and Scott., 1995]:

$$P_{\text{prop}} = \tfrac{1}{2}\rho C_D V^{2/3} v^2$$

where ρ = seawater density, C_D = volumetric drag coefficient = 0.5 for a torpedo shape with a fineness (l/d) ratio of 5, V = vehicle volume, and v = vehicle speed.

Elimination of a tether decouples the dynamics of the submersible from that of the ocean surface. Autonomous underwater vehicles (AUVs) dispense with a tether and provide greater deployment flexibility without the need for continuous control or monitoring. Their reduced mission duration due to battery constraints has limited them to short-duration sea surveys but they are beginning to be deployed for more sophisticated manipulation tasks and long-duration observation tasks. Modern nuclear submarines are adopting greater automation—the Royal Navy's HMS *Astute* operates with a small crew of 98 and will not require refueling throughout its 25 yr life. Autosub is a 7 m long torpedo-shaped vehicle with a mass of 1.4 t (dry) or 3.7 t (wet). Its D-cell alkaline batteries give it a 500 km range over 6 days in low-power mode at depths up to 1 km.

Smaller AUVs have limited depth capabilities and ranges; a 50 kg AUV giving 100 m depth and 20–40 km being typical. Such AUVs are the workhorse for distributed sensor networks in providing both vertical and horizontal profiling capability. Gliders are buoyancy driven, similar to floating buoys, by pumping ballast but have wings that convert vertical sinking into forward motion. A tailfin rudder provides steering. They have extremely low power consumption. Propeller-driven AUVs however can move at much higher speed and against the current, unlike gliders, but have much shorter power duration. Seagliders are a new approach to AUV design with the capability of extreme long range, low hydrodynamic drag, and wide pitch control for high glide slopes of 0.2–3.0 [1359]. Seaglider comprises an Al alloy pressure hull rated to 1,000 bar and a fiberglass fairing to which wings, rudders, and trailing antenna are attached. The 52 kg Seaglider is 1.8 m long with a 1 m wingspan.

Seaglider carries ethyl ether oil that it can pump from inside its pressure hull into an exterior balloon to increase its displacement of seawater. The oil's density is determined by the ambient pressure:

$$\rho = \frac{\rho_0}{\left(1 - \dfrac{p - p_0}{E}\right)}$$

where E = bulk modulus.

Pumping oil into the exterior balloon causes it to ascend through the Archimedes principle:

$$F = mg - \rho g V$$

where V = displaced fluid volume.

Ocean gliders may use ocean temperature differences to draw energy—the Slocum glider uses temperature differences to alter its buoyancy thereby alternately descending/ascending. Deep Flight 1 is a winged torpedo that has two short wings to allow it to "fly" underwater but they are inverted to produce lift downwards for longer seabed operations up to 6,000 km with low power requirements. Biomimetic roboswimmers may be based on a number of different designs—fish, eels, seahorses, crabs, and worms—but fish were the most successful design resulting from genetic algorithm optimization [1360].

In 1960, Jacques Piccard and Don Walsh reached the Challenger Deep of the Mariana Trench in a steel bathyscaphe—*Trieste*—based on a bathyscaphe design for balloon ascent. Most military and research submersibles are restricted to depths of 400–500 m (though the NR-1 research submarine has a rated depth of 724 m) which provides access to most continental shelf regions. Most deep-sea submersibles are based on a pressure vessel hull with dry internal cabling. Pressure vessels must be designed to withstand internal/external pressures. They are typically single-walled structures designed to house an onboard power system, payload, and electronics. Material selection is complicated by deployment in the high-pressure corrosive environment of salt water. Al alloy has a higher strength-to-weight ratio than high-strength steel but is not generally used as it corrodes unless coated with Zn alloy. Ti alloys offer the highest strength-to-weight ratios of metals and accordingly most diving vessels are constructed from 7.5 cm thick Ti alloy. The *Triton* MRV adopted an Al alloy frame/pressure vessel with a depth rating of 3 km [1361]. The *Alvin* manned research submersible is rated to a maximum depth of 4.5 km, the average depth of terrestrial oceans. At a depth of 5 km, the pressure is 0.5 t/cm^2. The ROV *Jason* which explored the RMS *Titanic* wreck was rated to a depth of 7 km though the ROVs that capped the disastrous Deep Horizon drilling platform operated at a depth of 1.5 km. Most terrestrial AUVs are rated at a depth less than 3 km. The pressure on the seabed at a depth of 10 km is ~110 MPa requiring high-pressure resistance structures. The deepest terrestrial dive was made in 1960 by the manned bathyscaphe *Trieste*. Designed to withstand 1.2 kbar pressure, it reached a depth of 11.3 km in the Challenger Deep of the Mariana Trench near Guam in the western Pacific. *Trieste* employed a 12.7 cm thick steel passenger chamber. Today, Ti alloy has replaced steel. Alternative materials such as glass-reinforced plastics using S-glass can offer higher compressive strength for greater depth. Carbon fiber–reinforced composites offer the best tensile modulus. James Cameron's *Deepsea Challenger* was launched in January 2012. It was a 12 t vertically oriented torpedo-shaped submersible designed to achieve a maximum depth of 11 km at the Challenger Deep as rapidly as possible. It is constructed from carbon fiber composite and withstands 1,100 atmospheres. All the electronics were mounted into pressure-balanced oil-filled boxes. Autosub uses a central carbon fiber composite pressure hull rated to a depth of 1.6 km, which contains its power system, trim ballast, and buoyancy foam. The vehicle hull is constructed from glass-reinforced plastic panels mounted on an Al frame. For a single-walled shell, wall thickness is given by:

$$t = \sqrt{t_a^2 + t_r^2}$$

where $t_a = (n\pi r^2 p / 2\sigma \pi r) = (nrp/2\sigma) =$ axial   thickness, $t_r = (nrp/\sigma) =$ radial thickness, p = external pressure, r = segment radius, σ = material tensile strength, and n = factor of safety ~5. Hence,

$$t = \sqrt{\frac{5}{4}\frac{nrp}{\sigma}}$$

A more empirical form for shell thickness is given by [1362]:

$$t = kd\frac{p/\sigma}{1 \mp kp/\sigma}$$

where d = shell diameter, p = design pressure (MPa), σ = ultimate strength of vessel material (MPa), and k = dimensionless constant = 0.7 for metal or 2.15 for composite. A spherical hull provides the maximum strength-to-weight ratio but military submarines are constructed from a near-cylindrical pressure hull of steel cylindrical sections strengthened with steel girder ring sections blocked by end caps. A cylindrical form offers a better hydrodynamic form than a spherical form. Such torpedo-shaped submarines require rudders and bow/stern hydroplanes and three trim tanks (forward, aft, and midsection) for maneuvering. Hydrodynamic drag determines the drag coefficient on the basis of body shape:

$$C_D = \frac{2D}{\rho Av^2} \quad \text{and} \quad \text{Re} = \frac{\rho vl}{\mu}$$

where D = drag, ρ = fluid density, A = plan area of body, v = fluid velocity, l = linear body dimension, and μ = dynamic fluid viscosity. The Reynolds number is defined as the ratio of inertial to viscous forces. Bulkier designs use thrusters and propellers for maneuvering which offer higher maneuverability at the expense of speed. Compliant coatings may be used to delay hydrodynamic transition from laminar to turbulent flow, reduce skin friction drag, and suppress flow-induced noise for small submersibles [1363].

The first question that arises here concerns whether to adopt a dry or wet structure: dry systems are based on a pressure vessel capable of withstanding the external pressures of a deep water/ice environment; wet systems are based on flooding the internal cavity of the structure to maintain zero differential pressure from the external environment. In submarines, the pressure hull is surrounded by a flooded hydrodynamic hull which itself does not experience hydrostatic pressure differentials. Wet structures still require their electronics to be enclosed in dry conditions and sealed against high-pressure water infiltration. A wet structure will be essential for a Europa hydrobot where the ambient pressure will be high. Although wet cabling would be desirable, oil pressurant within the dry cavities, rubber sealing, end cap bulkhead connectors, and watertight compartmentation would protect against leakage. In the *Triton* MRV, all electrical junction boxes and robotic joints are filled with electrically insulating mineral oil, the pressure of which is equalized to the water pressure using flexible membranes.

The ROV must be neutrally buoyant for maximum maneuverability. Either side of the pressure hull there are large twin-bouyancy tanks that flood during dives—in a submarine the angle of descent is controlled by adjustable wings on the hull (trim). For ascent, air compressors pump air into the bouyancy tanks. Bouyancy spheres may be constructed from lightweight ceramic with high compressive strength to provide neutral buoyancy with low mass. This assumes that they are filled with air but synthetic foam may act as a buoyant material. The foam is comprised of tiny glass spheres encased in an epoxy resin matrix. Water may be pumped into and out of ballast tanks for sinking, rising, and providing trim.

Most AUV/ROV submersible propulsion operates through ducted propeller thrusters (though some are bio-inspired for low-power consumption). A propeller generates thrust by speeding up fluid flow through it from the vehicle speed v_{veh} to an output speed v_{op}. Thrust is a product of mass flow and the speed difference $(v_{op} - v_{veh})$. Power output is the product of thrust and vehicle speed. Power input is kinetic energy per unit time; that is:

$$P_{in} = \tfrac{1}{2}\dot{m}(v_{op}^2 - v_{veh}^2)$$

The Froude number defines propulsive efficiency, which may be given by:

$$\eta = \frac{2v_{veh}}{v_{op} + v_{veh}}$$

Now, $v_{op} > v_{veh}$ so $\eta < 100\%$. If $v_{op} \to v_{veh}$, a large mass flow is required for the smallest increase in speed. Hence, jet propulsion is inefficient as jets give reduced mass flow with higher output speed. Autosub uses a five-bladed propeller in its tail section in conjunction with control surfaces. Submersible attitude control will require three-axis thruster control in order to hover for extended periods over the worksite. This will require four horizontal and four vertical ducted propellers for vectored thrust. *Triton* has 4×380 mm diameter horizontal and 4×305 mm diameter vertical thrusters for vectored thrust at 3.5 knots and 2.0 knots, respectively. The NR-1 research submarine was able to reconnoiter the seabed on two retractable wheels in its base for deep-sea surveillance and recovery. The propulsion system must also implement maneuvering as well as cruising, which involves complex translational and rotational changes in direction and position [1364]. The combined cruise–maneuverability index may be defined through the Froude and Reynolds number [1365]:

$$\beta = \frac{Fr^4}{Re} = \frac{v^3}{L^3}\frac{\mu}{g^2}$$

where $Fr = v/\sqrt{gL}$ = Froude number, L = vehicle length, $Re = vL/\mu$ = Reynolds number, an μ = fluid kinematic viscosity. Biomimetic approaches to autonomous underwater propulsion may also be feasible [1366–1368]. Vortex propulsion is an approach to enable rapid acceleration and maneuverability similar to jellyfish and squid. A vortex ring forms when a quick burst of fluid is shot through an opening. The fluid spreads out as it is expelled and its front edges curl back. If fluid expulsion is rapid enough, the curling extends round until it travels forward again in the same direction as the original fluid burst. This vortex ring separates from the expelled fluid and carries off a large amount of momentum. This generates a reaction on the propulsor. Vortex rings can be created using a piston to push a column of fluid out into an ambient fluid. This enables tuning the speed of the expelled fluid to the speed of the vortex by controlling the accelerating piston and adjusting the diameter of the expulsion orifice. This requires that the piston chamber is four times longer than its width. A thruster constructed from a flexible plastic diaphragm dilates to ingest water which is compressed to push water back out again. This creates a vortex ring that accelerates the thruster. Although in an immature state of development, vortex thrusters with high oscillation can be used for attitude control and propulsion if used in arrays. Vehicle attitude control may be implemented through actuated movement of the vehicle center of mass relative to its center of buoyancy to control pitch and roll. Bouyancy control is based on a small high-pressure pump and internal piston powered by onboard batteries. Wings provide hydrodynamic lift to propel the vehicle forward as it sinks or rises. This is in contrast to aerodynamic gliders which use delta wings to produce downward lift with a small wing area for rapid descent. In both cases, trim may be adjusted by weight redistribution. Depth may be controlled by altering pitch and thrust.

Salt water is a strong absorber of radio frequency signals due to its high conductivity of 6 S/m. Most submarine radio communications have been based on extremely low frequency (ELF) radio limited to 10 characters/minute. Acoustic packet signaling under water offers shorter range but greater information rates. At depth, communications are restricted to low frequency (30 kHz) acoustic signals with a maximum range of 10 km. Deployment of a tether anchored to the base of the ice would provide one means for communication from the submersible. Furthermore, this would provide anchorage against high tidal currents expected in the Europan ocean. The elimination of tethers and their optical fibers for communication places reliance on underwater acoustic communication. Subsea communications suffer from multipath propagation effects which causes signal spreading. This may be alleviated by using spread spectrum methods. Ultrasonic signals can be transmitted up to 5 km but suffer from reflections from the ice, water surfaces, seabed, and thermoclines. This is characterized by low-bandwidth multipath propagation due to boundary reflections and high delay due to the speed of sound in water [1369, 1370]. This varies between 1,400 and 1,600 m/s (average 1,500 m/s) depending on the temperature, salinity, and pressure of the water. It decreases with decreasing temperature, pressure, and salinity, though the latter dependence is not strong (most of the terrestrial ocean lies within a salinity range of 33.8–36.8). The data rate is generally limited to around 100 bps. The layer where the terrestrial ocean temperature (in the range of 8–25°C) drops rapidly with depth (150–400 m in the tropics and 400–1,000 m in the subtropics) defines the thermocline below which ocean temperature is approximately constant at 3°C. Temperature change below a depth of 1,000 m is small remaining close to 4°C, beyond which the speed of sound increases with increased pressure. The minimum speed of sound occurs at around 1,000 m. Sound rays tend to bend towards the depth of the minimum sound speed channel and travel at that depth over large distances. The acoustic channel exhibits increased losses of amplitude and frequency with range—a long-range acoustic channel of several tens of meters is limited to a few kilohertz bandwidth. Bandwidth limitations are due to acoustic absorption in water above 30 kHz. This may be modeled as a Rayleigh fading channel. FSK transmission suffers from multipath propagation—as fading is correlated with closely separated frequencies associated with multipath spread, $\Delta f = 1/t_{mps}$, only frequency channels outside this range can be used to reduce the data rate further: FSK in the 20–30 kHz band is limited to a data rate of 5 kbps. To reduce or eliminate intersymbol interference, guard times must be introduced between successive pulses which reduces data throughput. The replacement of non-coherent modulation by phase-coherent modulation offers the possibility of increasing the data rate by an order of magnitude to ~30–40 kbps through the acoustic channel by increasing the efficiency of bandwidth usage [1371]. Phase-coherent PSK and QAM offer reduced multipath propagation—coherent DPSK can achieve 10–20 kbps data rates. The use of channel (error correction) coding can increase the reliability of information transfer substantially, whereas adaptive filtering using estimation further enhances coherent communication. Nevertheless, acoustic communication presents severe problems, particularly over long distances so the data rates supported will be low: ~1 kbps over a 1 km range and much less over a longer range (up to a depth of 100 km). Blue/green lasers may be used for communication—blue light penetrates water better than any other light frequency and, with sufficient power, signals can be detected by a receiver at almost any depth. Water absorption losses are given by:

$$L = e^{-kz} = e^{-\tau}$$

where $\tau = kz =$ optical diffusion thickness of water ~1–10, $k =$ diffusion attenuation constant, and $z =$ depth. Increased energy per pulse of transmission increases the depth of transmission. The transmission channel varies in quality over short time periods implying the need for a large margin. An alternative but risky strategy would be to deploy a free-swimming submersible and for it to return to its anchor point for transmission of its data.

Robotic submersibles require complex navigation and control systems. An inertial measurement unit (IMU) provides attitude measurement through three-axis accelerometers and three-axis laser/fiber optic gyroscopes. The yaw gyro determines the submersible heading while two tilt sensors can determine pitch and roll. Autosub uses a magnetic heading yaw sensor and pitch/roll gyroscopes for attitude determination. The Litton LN-200 system which includes both fiber optic gyros and micromachined accelerometers with a Kalman filter estimator can provide attitude and attitude rate sensing for vehicle stabilization essential for sidescan sonar [1372]. Gyros and accelerometers suffer from severe drift rates that require supplementation with periodic external reference calibration—this assumes that the seabed is within range. Typical navigation sensors are acoustic (namely, the forward-located sonar). Depth may be measured by a semiconductor pressure sensor rated to very high pressure. Altitude determination above the seabed requires an acoustic echo sounder with a range of up to 1 km (for use at depths close to the seabed). High-precision bathymetric surveys require narrow-beam scanning sonars that are used for external references to scan the environment—forward scanning and downward scanning from the front of the vehicle provides the main navigation system. Side-scanning sonar is widely used for 2D searches but is not suited to 3D bathymetry which requires phase sensitivity. The standard approach to acoustic navigation is to use a sonar beamwidth of 1–3° at 12 kHz which gives a range of around 10 km ± 10 m accuracy. Higher frequencies at 300+ kHz offer greater resolutions of ~10 cm but higher frequencies are attenuated more rapidly giving very limited range restricting it to shallow waters (i.e., depths <500 m). The *Jason* submersible uses 120 kHz broadband forward-looking and downward-looking sonar systems for navigation with a look-ahead range of 100 m and a resolution of ±1°/2 cm. It also uses 200 kHz side-looking sonar. High-frequency sidescan sonar provides seabed mapping through murky water. A 1,200 kHz acoustic Doppler scanning sonar provides a 30 m range for navigation with respect to the seabed. Multibeam Doppler sonar can improve underwater navigation—a minimum of three downward-looking high-frequency (~1,200 kHz) Doppler beams measure the apparent velocity of the seabed due to the vehicle's motion by active sonar pinging [1373]. The four pings generate a vector of velocities along the beam axes with an error rate that may be combined with lower frequency long-baseline scanning sonar through a Kalman filter to increase accuracy by generating a terrain map for self-localization. Bio-inspired echolocation is sophisticated, providing dolphins with the ability to identify objects [1374]. They use broadband impulsive clicks with a time–bandwidth product of unity. Peak frequencies and bandwidths range from 20 to 100 kHz. A miniaturized sidescan sonar system has been developed for use on miniaturized submersibles [1375]. The sonar element based on a PZT ceramic plate is 50 mm in length by 1.5 mm in width with an operating frequency of 666 kHz. This results in a beamwidth given by:

$$B(\theta) = 20 \log \left| \frac{\sin l}{l} \right|$$

where $L = (\pi/\lambda)l \sin \theta$ and l = transducer element dimensions.

CTD (conductivity, temperature and depth) measurements are commonly used to determine salinity, temperature, and pressure but may aid in the navigation process as additional sensor sources. In addition, a microfluidic sample system with a filter allows the acquisition of concentrated samples. Color and black-and-white cameras are often used with illumination sources to supplement sonar data but suffer from absorption and backscatter due to suspended particles limiting them to close-up imaging. Illumination should be provided that is sensitive to blue–green light. The advantage is high resolution to an accuracy of a few millimeters. *Jason* carried two 16-bit 1,024 × 1,024 CCD color cameras and one aft-pointed black-and-white camera supported by two strobe lights and five incandescent lights. Forward cameras may be supported by a laser striper to provide close-up imaging. Laser space–frequency difference scanning is an approach that can provide 3D data across a wide field of view without lighting and avoiding underwater backscattering [1376]. Space–frequency difference scanned signals at different times correspond to different observations at different positions. The main components are a laser scanner, a 2D imaging system, a 3D sensing system, and a signal-processing unit. A camera-controlled manipulator provides the opportunity to target samples. The *Triton* series of ROVs were equipped with 6 DOF robotic manipulators with 2+ fingered grippers, end effectors, or tooling. *Triton*, however, sported a 7 DOF manipulator with associated gripper and tools.

13.3 SUBMERSIBLE DYNAMICS

Maneuvering involves complex translational and rotational changes in direction and position [1377]. As a locomotion medium, water is an incompressible fluid with a high density. Any movement by an object will generate motion in the ambient fluid. The Navier–Stokes equations describe fluid motion for an incompressible fluid based on conservation of mass and momentum, respectively:

$$\nabla v = 0$$

$$\rho \left(\frac{\partial v_i}{\partial t} + v \nabla v_i \right) = \rho g_i + (\nabla \sigma)_i$$

The dynamics of underwater vehicles are highly nonlinear. It is generally assumed that fluid rotations due to vortices are negligible. Submersibles may be represented dynamically as 6 DOF systems. Dynamic control of position is a primary requisite for the submersible and the dynamics are nonlinear [1379]. The submersible must compensate for unpredictable disturbances such as currents and drag effects—if sensory processing and response times are more rapid than the unknown current dynamics then these can be compensated for. *Jason* employs a supervisory control system with emphasis on automatic control at the servo level. A minimum of one manipulator is used to acquire samples. The *Jason* manipulator has a payload limit of 15 kg and was the main instrument in the recovery of over 50 artifacts from the *Titanic* shipwreck. There will be significant vehicle–manipulator dynamic coupling. The 6 DOF dynamics of an AUV are given by [1380, 1381]:

$$\tau = BF = D(q)\ddot{q} + C(\dot{q})\dot{q} + H(\dot{q})\dot{q} + G(q)$$

where $D(q)$ = vehicle plus hydrodynamic (added masses) inertia matrix, $C(\dot{q})$ = Coriolis/centrifugal terms including hydrodynamic added mass, $H(\dot{q})$ = hydrostatic forces including drag and friction, $G(q)$ = gravity/buoyancy forces (nominally neutrally buoyant), F = thruster forces/torques to compensate for external forces, and B = thruster configuration matrix. External ocean currents must be added to this. This may be reduced to:

$$m\dot{v}_h = F_h + \eta F_{th} + W = 0$$

where v_h = vehicle heave, m = vehicle mass including added mass, η = thruster efficiency, $F_h = -k_d v_h - C_d v_h |v_h|$ = hydrodynamic damping, $F_{th} = k_1 w_p^2 - k_2 v_f w_p$ = thruster heave force, w_p = propeller revolution rate, v_f = water speed past propeller, and W = vehicle weight. The hydrodynamic parameters are unknown and significant external perturbations are normal. Flow is turbulent and even drag coefficients are dependent on the fluid Reynolds number. Thruster dynamics are typically nonlinear with properties of saturation, slew rate limits, friction, nonlinear propeller load, etc. The thruster acts as a sluggish nonlinear filter in which the response speed is dependent on the commanded thrust which may be compensated through a lead compensator followed by nonlinear cancellation [1382]. Added mass occurs as a result of the surrounding fluid accelerating when bodies move thereby generating additional inertial resistance. They are a function of vehicle shape and fluid density and may be modeled thus:

$$m_{\text{add}} = \gamma \rho A l$$

where $\gamma \geq 1$ = added mass coefficient, l = thrust duct length, and A = thruster cross section. If the body is symmetric and moves slowly, this added mass can be neglected. Hydrodynamic forces include both lift and drag due to ocean water velocities acting on the vehicle about the center of pressure. If there is a lever arm from the center of pressure to the center of mass, there will be a drag torque. Higher order skin friction due to viscous shear and vortex shedding are neglected in comparison with drag making hydrodynamic damping negligible. There will be unknown tidal effects but it is not clear as to their nature within Europa—however, they are expected to be significant due to tidal flexing and radioactive heat sources. We assume that ocean current velocity is negligible which allows application of a spacecraft–manipulator model. Separation of translation and rotation is achieved by the emplacement of vehicle thrusters relative to the vehicle center of mass and center of pressure. A minimum of three pairs of thruster jets require active control to stabilize the platform while a manipulator takes samples. Thruster propulsion is based on electric motor-driven ducted propellers. We also assume a freeflyer AUV controlled in all three rotational axes using a PID controller driving the thrusters for each axis. Integral control is essential to compensate for disturbances and unmodeled dynamics. A fixed gain PI control law may be applied to thruster control though gain scheduling may be used to compensate for unmodeled aspects of the vehicle and its environment [1383]. However, sliding mode control has been the most commonly applied control strategy for UUVs, which compensates for external perturbations [1384]. The system state is driven towards a switching surface within the state space of the system where it remains despite parameter changes and disturbances. The switching surface is given by:

$$s = \dot{\phi}_e + \lambda \dot{\phi}_e$$

where $\phi_e = \phi - \phi_d$ = system parameter error and λ = bandwidth parameter. Neural nets have also been applied to UUV control because of their ability to accommodate nonlinear dynamics [1385, 1386]. A variation on this relates to the control of manipulators mounted on underwater vehicles—indeed, most ROVs include one or more manipulators augmented by imaging cameras. Although a second arm may be used to anchor the ROV to the worksite, this is more complex than is immediately obvious as it requires a minimum of three degrees of freedom (yaw/pitch/telescope) to react to perturbing forces. Furthermore, strapdown laser/ fiber optic gyroscopes may be used to provide measurement of attitude at all three axes. The roll–pitch–yaw angles of attitude are defined by:

$$R = \begin{pmatrix} cYcP & cYsPsR - sYcR & cYsPcR + sYsR \\ sYcP & sYsPsR + cYcR & sYsPcR - cYsR \\ -sP & cPsR & cRcP \end{pmatrix}$$

Alternatively, a quaternion-based formulation may be used [1387]—however, AUV control during station keeping is primarily horizontal with a small variation in depth. *Jason* uses seven brushless DC motor-driven thrusters for maneuvering and translation at a maximum speed of 1 knot forward/vertically (from 260 N thrust forward/aft motors and 300 N thrust vertical motor) and 200 N thrust generating 0.5 knots laterally. Hydrodynamic forces on the vehicle are given by Morison's equation:

$$F_{\text{hydro}} = \tfrac{1}{2} C_D \rho A v^2 + C_M \rho u \dot{v} + \rho u \dot{v}$$

where A = cross section of vehicle, ρ = water density, v = velocity of vehicle with respect to water, $\dot{u}$ = water acceleration = 0 nominally, and C_D = drag coefficient. Coefficients $C_M \rho u$ comprise the added masses to the inertia matrix $D(q)$. In general, many of the hydrodynamic coefficients will be poorly known and unknown multidirectional currents will be present. Water currents will impose forces on each manipulator link which may be estimated as:

$$F_\parallel = \tfrac{1}{2} C_D A \rho v_\parallel^2 \quad \text{and} \quad F_\perp = \tfrac{1}{2} C_D A \rho v_\perp^2$$

where $v_\parallel, v_\perp$ = velocity components of water on each link, C_D = drag coefficient, and A = area perpendicular to movement. At low velocities, DC motor–driven thruster dynamics become important and suffer from poor control performance. Sliding mode controllers have commonly been applied to AUV vehicle control due to their robustness [1388] though many more recent controllers have been neural net–based for similar reasons [1389], and variations thereof [1390]. Even these approaches, however, often make simplifying assumptions suggesting that automatic control of AUVs is still an unsolved problem. We assume that hydro-dynamic forces and gravity/buoyancy effects can be feedforward-compensated through modeling akin to the computed torque control method. In addition, it is important to model thruster dynamics effectively for improved station-keeping capability. Such model-based control approaches are superior in performance to non-model–based approaches [1391]. Thin-foil propeller hydrodynamics are modeled based on axial fluid velocity and propeller rotational velocity. A brushless DC motor operating in current control mode has torque load given by:

$$I_{\text{mech}} \dot{\Omega} + B\Omega = k_t i - \tau_1 - \tau_{\text{fric}}$$

where $\tau_{\text{fric}} = k_v \Omega + k_s \, \text{sgn}(\Omega)$. A PID shaft speed control law is given by:

$$\tau = -k_p \Delta\Omega - k_i \int \Delta\Omega \, dt - k_d \Delta\dot{\Omega}$$

Generally, fluid momentum in the thrusters will generate a time lag in response to thrust generation by the propellers. The thrust propeller is mounted into a duct which increases thruster efficiency. Axial flow and cross flow effects reduce thruster force. Thruster dynamics are given by propellar angular acceleration and thrust, respectively:

$$\dot{w} = \frac{\tau}{\eta^2 p^2 \rho V} - \frac{\eta p A}{2V} w|w|$$

where ρ = fluid density, τ = thruster torque, w = propeller angular velocity, V = thruster volume, η = propeller efficiency, and p = propeller pitch (axial distance traveled by propeller blades/rad:

$$F_{\text{th}} = \gamma Q = A\rho\eta^2 p^2 w|w|$$

where $Q = \eta p A w$ = thruster volume flow rate, $\gamma = (A\Gamma)/V$ = fluid momentum/unit volume, $\Gamma = \rho V(Q/A^2)$ = fluid momentum, and A = thruster duct cross section area. It is assumed that the Europan submersible is small in size with rapid response to actuator commands for station keeping. It has been shown that the use of long-period triangular wave input commands provides a steady-state response by reducing water transient inertial response [1392]. The propeller will generate both lift and drag on the blade. Propeller shaft thrust is given by [1393]:

$$T = L \cos\theta - D \sin\theta$$

where $L = \frac{1}{2}\rho A v^2 C_L \sin(2\alpha)$ = lift perpendicular to incident flow, $D = \frac{1}{2}\rho A v^2 C_D \times (1 - \cos(2\alpha))$ = drag parallel to incident flow, $\theta = \tan^{-1}(v_p/v_t)$ = direction of incident flow, $C_L = 1.75$ commonly for thruster blades, and $C_D = 1.2$ commonly for thruster blades. Propeller shaft torque is given by:

$$\tau = 0.7r(L \sin\theta + D \cos\theta)$$

where r = propeller radius, A = thruster duct area, p = propeller blade average pitch angle, $\alpha = \pi/2 - p - \theta$ = blade angle of attack in fluid, and Ω = propeller angular velocity. Total fluid velocity with respect to the propeller:

$$v = \sqrt{v_a^2 + v_t^2}$$

where v_a = axial fluid flow velocity and $v_t = 0.7r\Omega$ = transverse axial flow velocity. Axial flow velocity varies over time as:

$$\dot{v}_a = \frac{\tau}{\rho A l} - \frac{K}{l} v_a^2$$

where K = axial flow form factor. This is a nonlinear system that requires inversion to determine the control law to control thrust based on the rotation speed of the propeller blades. As the propeller rotates, the fluid axial velocity changes through the blades. Axial thrust is related to the rate of change of fluid momentum through the thruster volume:

$$T = \rho Al\gamma \dot{v}_a + \rho A \Delta \beta v_a^2$$

where $\Delta\beta =$ differential momentum flux coefficient between inlet and outlet $=$ 0.2–2.0 empirically, $\gamma =$ effective added mass ratio, and $l =$ duct length. Controlling propeller velocity by means of axial flow velocity sensing will result in accurate thrust generation. The propeller suffers from several thrust loss mechanisms that are difficult to model accurately including axial water inflow (usually neglected), cross-coupling drag due to water inflow perpendicular to propeller axis, thruster–hull interaction due to friction/pressure losses, thruster–thruster interaction in neighboring thrusters and unsteady flow effects such as gusting [1394]. For deep submergence, we assume that these effects can be neglected. It is likely that some tethering will be required for a Europan explorer unless an acoustic communications link can be devised. Tethers can be damaged due to slacking and snap loads that can be modeled as a balance of forces [1395]:

$$\frac{\partial T}{\partial l} + W + F = ma$$

where $T = EA\varepsilon =$ tether tension, $EA = mc^2 =$ axial stiffness, $c =$ strain wave speed along tether, $F = -\frac{1}{2}(\varepsilon + 1)\rho_w \, dC_D \dot{v}^2 =$ hydrodynamic drag force, $W = \Delta\rho Ag =$ tether weight, $\Delta\rho = \rho_l - \rho_w =$ difference in density between tether and water, $m =$ tether mass/unit length, $\varepsilon =$ tether strain, $d =$ tether diameter, $C_C =$ tangential drag coefficient, and $E =$ Young's modulus of tether. Elastic displacement is given by:

$$u(s,t) = \sum_{j=1}^{n} \delta_i(t)\phi_{ij}(s)$$

where $\phi_{ij}(s) =$ shape function and $\delta_i(t) =$ time-dependent coefficient. Station keeping of the AUV must be maintained while the manipulator operates. There will be dynamic coupling between the vehicle and the manipulator which will disturb the vehicle's position and orientation. In ROVs this is typically achieved by the human operator who makes constant adjustments to counteract the reaction forces generated by the manipulator. Automated control of these effects would relieve the human operator and will be essential for an AUV such as the Europan submersible [1396]. The *Ranger* NBV (neutral buoyancy vehicle) is a zero-gravity simulator robotic system with four manipulators used to validate on-orbit servicing manipulator spacecraft (Figure 13.3).

Ranger possesses two 8 DOF arms with an R-P-R-P-R-P-Y-R configuration for EVA equivalence, a 7 DOF manipulator leg, and a 6 DOF manipulator to actuate a stereo camera pair which supports a telepresence control interface (virtual reality based). An important issue involves the use of force control for a manipulator that exerts further reaction forces on the vehicle disturbing station keeping, particularly during instances of loss of contact [1397]. This makes the use of an object-based or ground-based reference system untenable. End effector

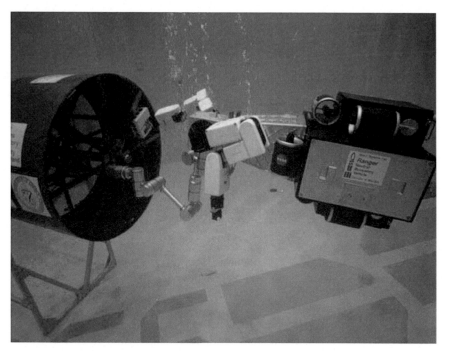

Figure 13.3. *Ranger* neutral buoyancy vehicle [credit NASA].

contact forces may be resolved at the joint level using the Jacobian transpose. The chief difficulty is that drag will be induced as there will be a lever arm along the manipulator from the center of mass of the vehicle to the center of pressure (assuming that the radius is much smaller than the length). We can assume that the center of pressure of each link resides at that link's center of mass and that the center of mass resides at the system center of mass. Hence, the virtual arm position represents the lever arm separation distance:

$$N_{\text{drag}_i} = 2 \int r_{\text{lever}_i} \times dF_i$$

where $F_i = $ drag force acting on each link. Hover control—similar to freeflyer control—avoids this disadvantage. The base of the manipulator is not fixed in the local inertial frame. The dynamics of such an underwater vehicle–mounted manipulator are similar to those of a spacecraft-mounted manipulator with several modifications [1398, 1399]:

(a) Added mass to account for the effective mass of fluid to be moved with the vehicle—added mass reflects the density of the fluid and results in delayed responses. Added mass is included in the inertial term only because it is dependent on acceleration. Fluid inertia is a function of body surface geometry and there are no principal axes: the fluid inertia matrix does not reflect that of the body itself. Added mass is defined as the derivative of fluid momentum:

$$f = -I_f \begin{pmatrix} \dot{w}_b \\ \dot{v} \end{pmatrix} - \begin{pmatrix} w_b & v \\ 0 & w_b \end{pmatrix} I_f \begin{pmatrix} w_b \\ v \end{pmatrix}$$

where $v_b =$ inertial body velocity, $\dot{v} = \dot{v}_b + w_b \times v_b =$ body acceleration with respect to fluid, $I_f = \begin{pmatrix} I_i & m_i r_i \\ m_i r_i & m_i \end{pmatrix} =$ body/fluid inertia matrix, and $w_b =$ body rotation.

(b) Fluid acceleration exerted on the vehicle by fluid motion

$$\begin{pmatrix} F \\ N \end{pmatrix}_f = \begin{pmatrix} m_f \dot{v}_f \\ m_f r \times \dot{v}_f \end{pmatrix}$$

where $v_f =$ fluid velocity and $m_f =$ mass of displaced fluid—only for a neutrally buoyant vehicle will the displaced fluid mass equate to the vehicle mass.

(c) Fluid viscous drag due to fluid viscosity which is primarily due to pressure rather than shear drag. Links are approximated as cylinders assuming that the radius is small compared with the length (this assumes that drag acts perpendicularly to the link axis and that parallel forces are negligible). In this case:

$$f_D = -\rho C_D r v^2$$

(d) Buoyancy force $f_B = -\rho g V$ where $V =$ displaced fluid volume—buoyancy force results in reduced apparent weight when immersed—for an Al body, body mass is two thirds its actual mass. Bouyancy equates to gravity forces $f = m\dot{v}$ under neutral buoyancy conditions. Furthermore, if the center of gravity and center of buoyancy coincide there is no buoyancy moment.

(e) Thruster dynamics are generated by a minimum of six thrusters—four vertical and four lateral.

(f) Fluid current fields that act as external hydrodynamic forces.

The Newton–Euler dynamics approach imposes $O(n)$ computational complexity while allowing explicit accommodation of additional hydrodynamic effects on manipulator control [1399–1401]. The control of robotic manipulators mounted on the Europan submersible during sampling of hydrothermal vent samples has been investigated [1402] based on an adaptation from space manipulator control algorithms [1403]. A modification of the computed torque control law can provide compensation for vehicle–manipulator coupling augmented by a robust nonlinear control law [1404]. A force–torque sensor at the base of the manipulator measuring the manipulator–vehicle coupling allows feedforward compensation of manipulator dynamics on the underwater vehicle [1405]. Minimization of a quadratic potential function of hydrodynamic drag exploits dynamic redundancy in these systems:

$$V(q, \dot{q}) = D^T W D$$

where $D = $ drag force [1406]. A computed torque model reference adaptive control law based on the Moore–Penrose pseudoinverse has been applied successfully [1407].

13.4 PLANETARY OCEANS GALORE

The resemblance between the Enceladus environment and that of Europa is obvious except that the former's subsurface ocean is much more accessible; however, scientifically it is of less astrobiological interest. Much of our earlier discussion applies either directly or with fewer more relaxed constraints. An alternative to the exploration of Europa would be investigation of one of Titan's hydrocarbon lakes (such as Kraken Mare) [1408]. There have been a number of large lakes identified on Titan including Ligeia Mare and Kraken Mare. The latter may be explored by submersible to determine its role in the global methane cycle. It could be deployed from a floating lander that would relay information to an orbiter via an X-band transponder. The submersible would deploy either as a free unit or tethered to the bottom of the lake. It would comprise two spheres connected by a cable: the lower sphere containing scientific instruments and the upper sphere the bus systems. During the descent at a rate of 1 m/s, the submersible would sample the lake at different depths. If a free submersible unit it would communicate with the surface lander via a VHF link through the hydrocarbon fluid. It would control its descent by means of sonar measurements of its environment. On reaching the bottom of the lake, the submersible would acquire samples of sediment for analysis for 30 days. On completion of this phase, the upper sphere of the submersible would return to the surface leaving the lower sphere of scientific instruments in situ. Buoyancy-driven gliders similar to the underwater glider Seaglider have been proposed for the exploration of Titan because of their capabilities for long-duration flight [1409]. These exploit the properties of aerial balloons with gliders. A variant on the Wave Glider unmanned surface vehicle may generate mechanical propulsion power from waves. In this scenario, a glider wing is attached by an umbilical to a surface float. As the float pulls up over a wave, the glider wing alters its angle of attack to generate thrust. During the downstroke, the glider wing reverses its angle of attack to generate further thrust. Wave oscillation is thus converted to horizontal movement. As gliders work by angle of attack, altering the position of the batteries allows the center of gravity to be shifted. A combination of gliding and propeller motion would give long-range and high-control precision, respectively. Even more speculative are flying submarines, a variant of seagliders with symmetrical wings whereby lift is provided by a nose-down pose when submerged (negative angle of attack) and by a nose-up pose when airborne (positive angle of attack).

A small probe has been proposed to propel itself to the Earth's core through a vast volume $\sim 10^8$–10^{10} kg of liquid Fe alloy migrating through a crack under its own gravity [1410]. The propagation velocity of the melt through a vertical crack of width d, horizontal dimension $w \gg d$, and vertical depth l in the Earth is given by:

$$v_p = \left[\left(\frac{\Delta\rho}{\rho}\right)\frac{gd^{5/4}}{\nu^{1/4}}\right]^{4/7} \approx 30d^{5/7}$$

where $\Delta\rho \approx \rho =$ difference in density between solid and melt rock, $g =$ acceleration due to gravity, and $\nu =$ kinematic viscosity of liquid $Fe = 10^{-6}\,m^2/s$. Crack pressure stress must equate to rock shear stress:

$$\frac{\mu d}{l} = \Delta\rho gl$$

where $\mu =$ rock shear modulus $= 10^{11}\,Pa$ and $l \approx \sqrt{\mu d/\Delta\rho g} \approx 1\,km \times d^{1/2}$ for a stress level of $3 \times 10^7 d^{1/2}\,Pa$ (i.e., $d = 0.1\,m$ with $l = 300\,m$). Assuming that $w = 300\,m$ also, this gives a volume of $10^4\,m^3$ of Fe (10^8 kg). The crack will propagate downwards at a rate of 5 m/s equating to it taking one week to reach the core. In order to initiate the crack, the energy required is given by:

$$E = \mu l w d \approx 10^{15}J \quad (Mt\ TNT)$$

(equivalent to a magnitude 7 earthquake). The probe would have a volume of $d^3 \sim 10^{-3}\,m^3$ and be constructed of a high melting point alloy in saturation equilibrium with the liquid Fe alloy melt. The probe would communicate through high-frequency acoustic (seismic) waves emitted to a ground-based detector. The power requirement to generate seismic oscillations is:

$$P = 4\pi\rho x^2 \left(\frac{w^4 d^4}{c}\right)$$

where $w =$ angular frequency, $A = (wd^2 x)/rc =$ seismic oscillation amplitude at the Earth's surface $= 10^{-13}(10^2\,Hz/f)$, $m, r =$ distance from seismic source, $c =$ wave propagation speed $= 10^4\,m/s$, and $x = 300(10^2\,Hz/f)^2\,\mu m$. Over the mission, this involves 10^8 cycles for frequency-modulated information transmission. The prospects of using such a technique for planetary exploration are a long way off.

13.5   SUBSURFACE OCEAN TERRESTRIAL ANALOGS

Europa has a partial terrestrial analog in Lake Vostok beneath the ice of East Antarctica similar in size to Lake Ontario [1411]. Antarctica is comprised of two major ice sheets: the East and West Antarctica ice sheets, comprising an ice volume of 30,000 km^3 of which 83% resides in the East Antarctic ice sheet. The ice thickness varies from 2.8 to 4.5 km, though it is currently thinning as a result of significant climate warming in the polar regions. The East Antarctic craton formed as part of the Paleozoic/Mesozoic supercontinent Gondwanaland until its breakup during the Jurassic/Cretaceous periods into Africa, East Antarctica, India, and Australia [1412]. However, Antarctic glaciation began some 33–14 Myr ago. Antarctica has some 150 buried freshwater lakes beneath its icy surface with an estimated total area of 35,000 km^3 and an average depth of 10–20 m [1413, 1414]. The largest is Lake Vostok with an area of 14,000 km^2 buried 3.7–4.2 km below the ice [1415]. It is 230 km long by 50 km wide with the characteristics of a rift valley. It has been isolated from the surface for $\sim$100,000–1,000,000 years, possibly as long as 15 Myr since the glaciation of Antarctica (i.e., dating back to the mid-Cenozoic period). The overlying ice represents refrozen lake water which suggests an age of $\sim$400,000 yr for the lake water itself. It may potentially harbor life with significant implications for Europa: photosynthesis cannot be supported at such depths but chemolithotrophic metabolism based on hydrogen, carbon dioxide, methane, iron, or sulfur is feasible.

Lake Vostok lies in an inactive tectonic rift valley (similar to Lake Tanganyika in East Africa) with steep side walls suggesting that there is significant geothermal energy heating of $\sim$55 mW/m^2. The surface temperature averages $-20°$C. The upper 3.5 km of ice originates from the glacial ice sheet, below which the ice originates from refrozen ice accretion from the lake water [1416]. The overlying glacial ice contains traces of potential nutrients such as sulfuric acid, nitric acid, methanosulfate acid, formic acid, salts, and minerals. VHF radar sounding is sensitive to ice/water interfaces and indicates that the lake is inclined at an angle of 0.0004°. There is evidence of a north–south water circulation within Lake Vostok due to the difference in lake thickness at each end. The depth of the lake water varies from -700 m in the south to -300 m in the north (average 500 m deep). Such convection could provide a transport mechanism for minerals and nutrients. Dissolved ions such as sulfate, sodium, chloride, and other ions would depress the local freezing point. Salinity has been estimated to be low $\sim$0.4–1.2% and the sedimentation rate is also estimated to be low. Accumulated sediment on the lake bed is estimated to be around 100–200 m thick. The water is expected to be oxygenated with dissolved oxygen from gas hydrates. Gas hydrates may exist in the lake and would be stable at high pressure $\sim$350 atm as a source of organic material—they have been recovered from cores above the lake [1417]. Dissolved carbon dioxide in the overlying ice indicates that biological oxidation of dissolved oxygen may occur in the lake.

Bacteria have been recovered from 200,000-year-old ice cores at 3 km depth in the Greenland ice sheet, their habitat being water nanofilms attached to microscopic clay grains from which they extract iron. Biological material in ice has a precedent—the permanent sandy ice covers (a few meters thick) of Antarctic lakes of the McMurdo Dry Valleys support complex microbial populations indulging in photosynthesis, nitrogen fixation, and decomposition [1418]. Furthermore, ice cores from 60 m into the Siberian permafrost suggest that bacteria can survive up to 3 Myr assuming they were deposited as aboriginal rather than later invaders. Lake Vostok was drilled to its maximum depth of 3.6 km in 2012 (equivalent to an ice core spanning 420,000 yr) though detailed analysis of the recovered samples has yet to be completed. The overlying ice cores from 3,541 to 3,611 m depth contained both prokaryotic and eukaryotic microbes at different depths. They exhibit concentrations of $\sim$10^3–10^4 bacterial cells/ml with low concentrations of growth nutrients at a depth of 3.6 km close to the subsurface lake water. Viable bacteria—particularly micrococci—have been recovered from overlying ice cores which are similar to modern terrestrial bacteria; indeed, the dominant population resides between 3.5 and 3.6 km depth. A proposed habitat comprises interconnected liquid veins along ice grain boundaries within which psychrophilic bacteria can survive using dissolved organic carbon [1419]. Subglacial lakes are thus expected to support habitats for microorganisms that may have invaded at the time of formation and subsequently adapted to the ambient conditions of high-pressure, low-temperature, permanent darkness, low-energy, and low-nutrient levels [1420].

Lake Ellsworth is a 10×10 km subglacial lake overlain by 3.4 km of ice near the Ellsworth Mountains in West Antarctica [1421]. Strong radio echo reflections with a smooth along-track variation indicate an ice–water interface with a gradient of 0.02 (Figure 13.4). It is located in a 1.5 km deep rift with a depth of 250–500 m indicated by airborne radar sounding. It is expected to be fairly representative of subglacial lakes generally and is estimated to be 150,000 years old. Lake Ellsworth is the target for exploration by the Subglacial Antarctic Lake Environments (SALE) Consortium [1422]. Hot-water drilling has been proposed to open a pres-

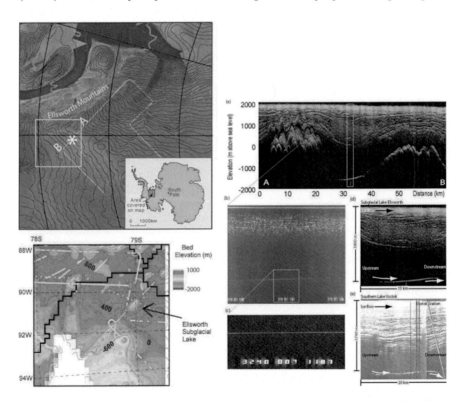

Figure 13.4. Lake Ellsworth survey [reproduced with permission from Martin Siegert].

surized borehole down which a robotic probe with scientific instruments would be winched by tether. This is similar in concept to a U.S. prototype Antarctic borehole ice probe [1423]. It is known that significant cavities of liquid water and rock debris can exist within the ice sheet. A hot-water drilling hose can create a 30 cm wide borehole 3.4–3.2 km deep to penetrate the underlying lake. The drilling system requires 2 MW of power from the burning of fuel to heat ice to 90°C. A 30 cm hole can be created to a depth of 3.4 km in 50 hours using 12,000 L of fuel (60 drums). This suggests an estimated pressure of 370 bar at 4 km depth. While drilling with the thermal lance, it is necessary to keep the fluid pressure in the hole close to the upward pressure of the underlying lake. A sterilized probe would be lowered through the borehole by winch through the water column to the lake floor. Water samples at different depths may be acquired by a set of tubes. Sonar imaging in conjunction with inertial sensors provide position measurement. The probe would acquire a sediment core of 2 m length for return to the surface. Equipment may be tested in the subglacial lake at Vatnajökull (Iceland). The chief issue is to sterilize the probe prior to its insertion into the lake to ensure that there is no contamination. Recovery of microorganisms would need to remain pressurized. However, recent evidence suggests that many of these lakes are connected through mobile ice streams. Unfortunately, initial attempts to penetrate Lake Ellsworth were unsuccessful—following the drilling of the initial hole of 300 m depth to create the initial water cavity, the main tube could not couple with the cavity and the mission was aborted. It is expected that a further attempt will be made in 2015. These terrestrial explorations indicate how challenging it is to penetrate through ice layers into subsurface water even with the benefit of human support and considerable technological infrastructure, which would not be available on Europa.

14

Future rover concepts

There are five classes of locomotion adopted in mobile robots that are applicable to planetary exploration rovers:

(i) wheels—rolling (e.g., automobile locomotion);
(ii) tracks—rolling/screwing (e.g., armored vehicle locomotion, rotary drill);
(iii) legs—walking (e.g., animal locomotion);
(iv) body articulation—crawling/sliding (e.g., snake undulation);
(v) non-contact locomotion—hopping/flying.

Biomimetic approaches to locomotion are many and varied but the end goal is to implement high-performance robotic explorers [1424–1427]. This includes the implementation of motive power through artificial muscles and feedback from quantum-limited sensors [1428]. Biomimetic robots are commonly used in animal neurethology as robots must interact and be tested with the real world environment subject to the same physical constraints, unlike computer models [1429]. Such robots can implement existence proofs of visual and other sensors, neural controllers, actuation mechanisms, taxis behaviors, etc. The lessons that need to be learned here include the implementation of sensor fusion, cooperativity, resonant mechanisms, open-loop control, online adaptivity, and limb sensors capable of detecting environmental changes. The BEES (Bio-inspired Engineering of Exploration Systems) approach seeks to exploit low-mass low-volume low-power robotic devices characteristic of the biological world with an emphasis on biological mobility and their control and navigation (such as neural control): legged locomotion, hopping, swimming, burrowing, and in particular flying [1430, 1431]. Biomimetics is also applicable to other robotic capacities relevant to planetary exploration [1432]: gecko-inspired climbing, spider-inspired jumping, strain sensors based on campaniform sensilla, peristaltic-based transport, and woodwasp ovipositor-based drilling. Bio-inspiration is not mere copying but an abstracting principle. Many biomimetic approaches are highly dependent on miniaturization brought about by MEMS-based microtechnology [1433]. A number of exotic missions to Titan have been proposed. TANDEM was a proposed ESA-led mission involving an orbiter for Titan and Enceladus with penetrators for Enceladus and a Montgolfier balloon and three entry probes for Titan. The Titan Mare Explorer was a proposed NASA mission to deliver a lake lander. The lake lander included a robotic submersible to sample lake sediments. An RTG-powered aerobot nominally cruising at an altitude of 10 km which can descend with a reelable tether to recover ground samples has also been suggested. However, no Titan mission concepts have been pursued, but this illustrates the versatility of planetary vehicles.

Biological animals locomote using legs, which provides better mobility on rugged terrain than wheeled vehicles—however, see McGeer (1990) for a mechanical relationship between wheeled and legged locomotion [1434]. Legged locomotion does not require continuous free pathways between discrete footfalls. Feet are required to overcome only compaction resistances at points of contact while wheels and tracks must overcome these forces continuously. Compaction resistance due to foot sinkage is given by [1435]:

$$R_c = \frac{l}{D} \frac{1}{(n+1)(k_c + bk_\phi)^{1/n}} \left(\frac{W}{l}\right)^{(n+1)/n}$$

where l = foot rectangular length and D = stride length. Resistance is reduced by increasing the stride length D. The slip relation for legs is similar to that for tracks with the foot length equating to track length. Legs use these resistive forces to aid movement rather than hinder movement. Legs decouple the ruggedness of the terrain from the movement of the vehicle body. Legged locomotion however requires considerable control, sophistication, and effort. An excellent review of biomimetic walking in insects as implemented robotically is given by Graham (1985) [1436]. A walking cycle has two phases—during the stance/power phase one or more legs support the weight of the body, while during the swing phase the body moved ballistically by swinging one or more legs. The power and swing phases are separated by transition points. The control of a legged vehicle over rugged terrain requires coordinated leg motion: gait is the coordinated response of leg motions. Gait generation is based on two rules [1437]: (i) never allow two neighboring legs to be raised from the ground simultaneously; (ii) a leg should step when its two neighboring legs have stepped more recently than it has—with reference to (i). Gait is defined as regular if all legs have the same duty factor β. Duty factor for a leg defines the fraction of cycle time spent on the ground propelling the body (i.e., the ratio of stance-to-stride duration, $\beta \geq 0.5$). The minimum β for a four-legged stable system is 0.75; the minimum β for a six-legged stable system is 0.5; the minimum β for an eight-legged stable system is 0.375 (i.e., the gain in stability is greatest for a six-legged system [1438]). The time for leg swing is given by:

$$\tau = (1 - \beta)T$$

so:

$$T = \frac{\tau}{1 - \beta}$$

Hence:

$$v = \frac{D}{\tau}\left(\frac{1 - \beta}{\beta}\right)$$

so:

$$v = 0.33\frac{D}{\tau} \quad \text{for four legs}$$

$$v = \frac{D}{\tau} \quad \text{for six legs}$$

$$v = 1.67\frac{D}{\tau} \quad \text{for eight legs}$$

So, speed increases with the number of legs but at the cost of greater complexity. As β decreases from 1.0 to 0.5, the robot speed increases while its stability decreases. When $\beta = 3/2$, we have a parallelogram gait where four legs are in contact with the ground at all times; when $\beta = 0.5$, we have a tripod gait where three legs are always in contact with the ground. Wave gait with a duty factor (ratio of support phase to total cycle time) β exceeding 0.5 provides good stability. The optimum duty factor β for a $2n$-legged wave gait is given by $3/n \leq \beta < 1$ which reduces to $1/2 \leq \beta < 1$ for a six-legged gait [1439]. This defines the gait adopted: wave gait ($\beta \sim 10/12$), quadruped gait ($\beta \sim 8/12$), tripod gait ($\beta \sim 6/12$). Tripod gait is fastest but the slow wave gait is stable to the loss of a leg. Gaits are selected by varying the phasing between leg oscillators. Central pattern generators which control gaits may be modeled as van der Pol oscillators coupled with each other in a network with each joint controlled by a single oscillator [1440, 1441]:

$$\ddot{x} + \mu(\alpha - x^{2n+2})\dot{x} + w^2 x^{2n+1} = 0$$

where $\alpha =$ oscillation amplitude and $w =$ oscillation frequency. The addition of a harmonic forcing function yields:

$$\ddot{x} + \mu(\alpha - x^{2n+2})\dot{x} + w^2 x^{2n+1} = \gamma \cos wt$$

Leg structure is optimized by having two inboard joint axes intersecting at right angles and a third joint axis parallel to the second with equal lengths for links 2 and 3. There are two ways to mount legs to the body: the first axis may be rotational or horizontal. The advantage of a vertical axis mounting is that the vertical joint undergoes the largest displacement about the axis parallel to the weight. However, this requires the kinetic energy of the leg to be absorbed at the end of each forward and back stroke. However, the hip joint and knee joint act in opposing directions acting alternately as driver and brake. The position of the feet may be defined through the Denavit–Hartenburg matrix repesentation with each foot following a trajectory in coordination with the other feet with respect to the body and conventional manipulator control algorithms applied [1442, 1443]. Legged vehicle velocity is given by:

$$v = \frac{D}{\beta T}$$

where D = leg stroke (step length) and T = gait period. Turning is achieved by using different stroke lengths on either side giving a radius of curvature and speed of turn:

$$r = b\frac{(D_l + D_r)}{(D_l - D_l)} \quad \text{and} \quad w = \frac{v}{\rho} = \frac{D_l - D_r}{2b\beta T}$$

where b = baseline between left and right feet. Power required to take a step is given by:

$$P = \alpha\frac{mv^3}{D}$$

where α = power coefficient, m = vehicle mass, and v = vehicle velocity. Biological legged locomotion may be characterized as a set of spring-loaded, inverted pendulums, which minimizes energy expenditure [1444]. In order to maintain stability in walking robots, sensor-based attitude control is required. The position of foot i in body coordinates is given by [1445]:

$$p_i = \begin{pmatrix} 1 & 0 & \gamma \\ \alpha\gamma & 1 & -\alpha \\ -\gamma & \alpha & 1 \end{pmatrix}$$

where γ = y-axis rotation and α = x-axis rotation. Weight distribution as measured by feet reaction forces alters during legged locomotion causing variations in body attitude. Force control is required to allow a legged robot to adapt to rugged terrain and different soils: (i) force sensing at each foot and body tilt is required to accommodate terrain height iregularity; (ii) force control in each leg is required to obtain smooth motion on soil to minimize slippage. Terrain response is given by:

$$F = kz^n$$

where F = vertical force and z = soil sinkage. For active compliance with position/force control at the end of the leg, the updated vertical leg coordinates are given by [1446, 1447]:

$$\dot{z} = \dot{z}^d + K_p(z^d - z) + K_f(f^d - f)$$

where $K_{p,f}$ = position and force gain constants <1. When insufficient force is sensed at the foot, the leg is lowered until a threshold is reached—this will maintain a constant body height. If the leg is lowered to the ground without detecting the ground, a foothold search is invoked with a widening horizontal motion. The simplest way to walk over soft soil is to use fixed locomotion cycles. When the leg changes from stance to swing, a foot contact sensor switches from active to inactive; when the leg changes from swing to stance, the foot contact sensor switches vice versa. The Newton–Euler dynamics may be defined by [1448]:

$$J_R^T \tau + J_T^T F = M\ddot{x} + H$$

where M = vehicle mass, J_R = rotational leg Jacobian, J_T = translational leg Jacobian, and $(F, \tau)^T$ = contact forces/torques on each leg. Constraints may be applied that include minimization of bending energy, load on the legs, and/or intermediate forces in the walking plane. An example of a sophisticated legged rover prototype is SCORPION [1449]. However, despite its high adaptability to rugged terrain, legged locomotion is complex in its control and highly power intensive due to the high number of drivable degrees of freedom [1450, 1451]. The pantographic design attempts to decouple vertical and horizontal DOF for simplicity of control. One of the early designs for a legged planetary rover was the 3 t six-legged Ambler robot which employed an ambling gait [1452]. Its legs decoupled horizontal and vertical motion. It required around 1.4 kW of power to generate a walking speed of 40 cm/min. Pantographic linkages are a popular choice for legged robot designs such as Dante II which was designed to explore active volcanic craters [1453]. Its walking frame had 8 pantographic legs arranged in two groups of 4 on the inner and outer frames. Pantographic legs amplify hip motion at the foot and allow large vertical foot strokes. There are examples of hybrid legged/wheeled vehicles such as Wheeleg which comprises two 3 DOF front legs for obstacle climbing and two rear wheels to bear most of the weight similar to a horse-drawn cart [1454, 1455]. Another example is Hybtor (hybrid tractor) which exhibits a "rolking" gait (combined rolling/walking) by coordinating its legs individually, each of which terminates in a single wheel [1456–1458]. A version of Hybtor called WorkPartner was designed for remotely controlled operations in forestry. WorkPartner's rolling gait is enabled by coordinating each of its four legs which terminate in a single wheel [1459–1461]. The wheels are locked initially to increase drive pull and then the wheels are actively rolled to generate a locomotion style similar to ski walking. The chassis includes 18 hinges, an actively articulated frame, and four legs. Hybrid locomotion as implemented in WorkPartner is to ensure high cross-country ability on terrain with complex relief and, at the same time, motion over a wide speed range (up to 12 km/h on hard surfaces) [1462]. Each leg has linear actuators with electric motors of 250 W power input. Similar motors are accommodated in the wheels. Its mass is 160 kg, its load-carrying capacity in walking mode is ∼60 kg, and it has overall dimensions of 1.4 m length, 1.2 m width, and 0.5–1.2 m height. The onboard power supply system consists of batteries (48 V) and a generator driven by a 3 kW combustion engine.

It has been suggested that walking machines are suitable for small planetary rovers as long as they are based on rigid frame hexapod designs such that vertical and horizontal motions can be decoupled thereby simplifying leg control [1463, 1464]. Certainly, legged locomotion has been a popular choice in microrobots (e.g., the eight-legged SCORPION, Figure 14.1). The 1.3 kg 6-legged Genghis walking robot capable of negotiating rough terrain was designed and built in 12 weeks by a small team [1465]. It comprised four onboard 8-bit processors (three

Figure 14.1. Eight-legged Scorpion robot [credit NASA].

for motor/sensor processing and one to coordinate the subsumption archiecture), 12 motors each with force feedback sensors, six heat-seeking pyroelectric sensors, two collision-detecting whiskers, and pitch/roll inclinometers. Each processor comprised an MIT Media Lab miniboard 2.0 with 8-bit Motorola 6811 neurochip CPU, 256 B of internal RAM, and 12 kB of programmable ROM. Robust walking behaviors were produced by a distributed system with little central coordination and it was capable of robust steering and target following. Each leg was attached at the shoulder joint with 2 DOF, each driven by a model airplane position-controllable servomotor. Each leg was controlled separately and rough terrain was compensated for by force monitoring in each leg. Whiskers anticipated obstacles and infrared sensors detected moving objects to which it was attracted. It adopted the behavior control strategy of simple sensorimotor behaviors organized in a fixed distributed layered subsumption architecture implemented as 57 finite state machines augmented with timers (ASM). As soon as a leg was raised, it automatically swung forward and then down. The act of swinging caused all the other legs to move backward so that the body moved forward. The process cycles through each leg in turn. All invoked output behaviors were summed vectorially (similar to a potential field representation) but higher level behaviors suppressed lower level behaviors. Each leg lift was controlled by 8 ASMs (i.e., 48 ASMs for leg control). The other ASMs were associated with whisker–front-leg interaction, forward–rear-leg balance, walking gait coordination, steering, and following. The control system was implemented in 8 modular task layers: Stand-up, Simple-walk, Force-balancing, Leg-lifting, Whiskers, Pitch-stabilization, Prowling, and Steered-prowling. Of the 57 ASMs, 48 were organized as 6 complete copies of an 8-machine control system for each leg, 2 for local behaviors connecting whiskers to the front legs, 2 associated with inhibitions of balance behaviors in the front and back leg pairs, 5 for "central" control (2 for walking, 1 for steering, and 2 for tracking of objects). Hence, complex behaviors emerged from a network of simple behaviors with little central control. The 2 kg Attila robot—a development from Genghis—employed 6 legs (24 actuators in total) armed with 150 sensors (14 types) and 9 microcontrollers plus 2 dedicated processors (for control and vision, respectively) connected through a serial I^2C bus and a solar array/battery power system. Many of the sensors were employed to serve legged locomotion though others were more generic such as the gyroscopically controlled camera and laser rangefinder. Genghis and its variants demonstrated the proof-of-principle of constructing lightweight robots with simple tasks which may be deployed in numbers to achieve more complex mission scenarios. Further developments in microrobots have been demonstrated such as the 2.5 kg microaerial autogyro VTOL vehicle, 6 kg microball-type freeflyer camera supported by full spacecraft subsystems, 100 g nanorover-based swarm, using micromachined components such as microcontrollers, micromotors, microcameras, and microaccelerometers/gyroscopes [1466].

Walking was eliminated early in the lunar exploration program due to the complexities of balancing, poor mechanical efficiency of walking, etc. [1467] For these reasons, we do not consider it further. Serpentine locomotion is highly versatile over rugged terrain or underwater [1468–1471] (Figure 14.2). There are four main snake gaits generated by its peristaltic movements: serpentine, side winding, concertina, and rectilinear. All the modular segments of a snakebot's body are identical, hinged together into a chain with each hinge powered by a single motor, gearbox, control electronics, and power source.

A robotic serpent can burrow into loose soil, stiffen itself like a mast, act as a robotic manipulator, or adopt a serpentine-traveling gait to transport itself across the surface such as sidewinding, coiling, or flipping. The chief disadvantage is that it suffers from poor energy efficiency. The critical property is that normal friction with respect to the body is much higher than tangential friction thereby reducing side slipping. This is difficult to achieve in practise, so many serpentine prototypes employ wheels to reduce ground friction (e.g., GMD-Snake [1472]). This rather defeats the object, so is not considered further.

Inchworm locomotion is an option for motion similar to that of segmented worms of the phylum Annelida [1473] such as the earthworm [1474]. An application was considered earlier (Inchworm Deep Drilling System). Inchworms move by peristalsis through the control of circular and longitudinal muscles around their fluid-filled body. The earthworm comprises a series of segments on which a pair of setae (bristles) are mounted. The worm crawls by elongating the body to push forward and by contracting to pull the rear part of the body. Circular muscles around the body provide radial shrinkage and expansion of the body. Longi-

Figure 14.2. Snakebot [credit NASA].

tudinal muscles along the body can contract and extend longitudinally. On expansion of the circular muscle, the setae are rigid preventing slippage by anchoring themselves to the soil. Contraction of the longitudinal muscles shortens the body and widens the girth of the segments. The circular muscles narrow the segment girth and elongate it at the same time. These muscles are operated in turn down the entire body length. Widening of several segments allows anchorage of the worm while the leading segments are elongated. The forward segments are then widened and the trailing end is contracted. Similarly, aquatic ribbon worms are capable of great elongation from a few centimeters up to ~30 cm. They have a constant volume body cavity—a hydrostatic skeleton—with a proboscis, a tube which extends by turning inside out. The membranous skin around the body of the ribbon worm incorporates embedded alternating left and right-hand helical fibers of collagen. They move by lengthening and shortening their circmferential and longitudinal muscles in peristaltic fashion. The peristaltic wave moves slower than the animal's forward velocity given by:

$$v_{\mathrm{per}} = qv(\Delta l/l)$$

where v = forward propulsion velocity, l = length of uncontracted section, Δl = length of elongated section, and q = fraction of worm's body that is extended. The squid can extend its tentacles by 50% in 15–30 ms through a similar process. A micro-inchworm actuator has been developed capable of moving at low (surface) speeds of 13 cm/min which may have utility in providing a modest sampling capability (e.g., for a penetrator) [1475, 1476]. Like an inchworm, it moves forward by alternately attaching and detaching part of its body and alternately actuating the moving part. The motor comprises a single push device (based on Terfenol-D magnetostrictive material) and two clamping devices (based on multi-stack piezoelectric actuators). The biomimetic Ver-Vite model was suited only to horizontal motion rather than drilling. A single shape-memory alloy actuator provides contraction while a coupled silicone bellows extends the robot body. This provides the basis for repetitive contraction and retraction. Two simple clampers are mounted onto the front and rear of the microrobot.

Climbing robots must be able to adhere to vertical surfaces while being able to locomote along those vertical surfaces [1477]. Mobility and grasping are implemented independently and the latter include suction, magnetic, gripping, rail-guided, and biomimetic adhesion. Vacuum suction is the commonest approach but it requires smooth non-porous surfaces. For scaling vertical shears such as cliff faces and deep canyons, climbing robots based on the gecko may be appropriate [1478]. Gecko foot hairs are based on micrometer-sized stalks ending in arrays of 100–200 nm width spatulae. Geckos use dry adhesion between their toe pads and the surface. Splitting the contact pad into many points increases the adhesion force. Arrays of such high-aspect-ratio structures adhere to any surface with a pressure-controlled contact area to maximize van der Waals forces. Dry adhesion results from arrays of tiny micro-spines of 200 μm in diameter attached to a number of toes on each leg. Gecko toe pads are lined with lamellae with multiple folded membrane-like structures made up of 3–5 μm setae which in turn are constructed from highly divided spatulae of a few nanometers. There are around 500,000 such bristles on each foot pad which split into around 100 M points of contact for each foot. Dry adhesion relies on van der Waals forces between the toe pads and the surface—assuming Hertzian contact (separation distance <2 nm), the adhesion force between a sphrical tip and a flat surface is given by the Johnson–Kendall–Roberts equation:

$$F = \tfrac{3}{2}\pi r \gamma$$

where r = hemispherical radius of micro-spine and γ = surface energy. The adhesion force for a gecko foot is 10 N (gecko body mass is 0.1 kg). The hairs—effectively cantilevers—should be stiff to prevent matting. The spines hook into asperites at the wall surface to provide frictional contact (e.g., Spinybot). Climbing based on synthetic gecko foot hairs requires nanomolding fabrication of complex microstructures [1479]. Thus far, synthetic bristles have been manufactured in polyurethane but not with the same density nor similar compliance to the biological analog. Spinybot is a three-limbed robot that maintains equilibrium while controlling contact slippage through force control [1480]. Planning of handholds for climbing may be based on traditional SLAM and path-planning methods in conjunction with handhold force feedback [1481].

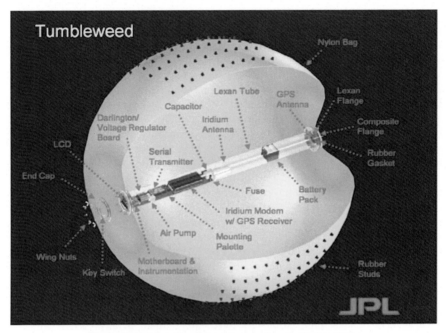

Figure 14.3. Mars Tumbleweed rover [credit NASA JPL].

The Tumbleweed rover is a large, wind-blown, inflated ball with an interior instrument payload. It was successfully deployed on a Greenland ice sheet over 130 km within 48 hours [1482] (Figure 14.3). A Mars Tumbleweed would exploit the Martian wind which has average speeds of 2–5 m/s (dynamic pressure of 0.03 N/m^2 for speeds of 2 m/s) with gusting up to 10–20 m/s (dynamic pressure of 0.8 N/m^2 for speeds of 10 m/s) [1483]. Tumbleweed has a spherical diameter of 6 m and a total mass of 5 kg with a drag coefficient of 0.5. The Greenland prototype comprises a 1.5 m diameter nylon bag within which resides a 1.2 m long tube containing instrumentation and electronics. Deployment on Mars would involve a 6 m diameter ball that can climb 1 m high rocks and 25° slopes. Furthermore, it may also be used as a parachute/airbag descent and landing system.

Hoppers are particularly useful for overcoming obstacles on low-gravity environments such as asteroids where tractive friction between the robot and the ground may be insufficient. Hop height is critical. Hopping is the only feasible means of exploring small bodies of the solar system as wheeled vehicles have a negative drawbar pull/weight ratio indicating that there is insufficient weight to provide traction. Hoppers have been proposed for asteroid exploration because of their low gravitational acceleration of ∼0.01 m/s^2 based on internal torquer wheels [1484]. A more conventional approach to asteroid/comet exploration is the use of a tripod of cold-gas thrusters in weak gravitational fields to generate ballistic trajectories by controlling the thrust direction [1485]. A ballistic hopper based on CO_2/metal rocket engines has been proposed for Mars [1486]. An alternative is to use a motorized spring (i.e., a single motor to drive a hopping leg [1487]). A linear spring may be compressed by a ball screw driven by the motor which is held in place by a lock–release mechanism. However, linear springs are subject to premature liftoff suggesting the use of a combined spring linkage system to generate nonlinear behavior [1488]. The pogo stick/foot is similar to legged locomotion in comprising four phases—two flight phases (ascent/descent) and two stance phases (spring compression/extension) [1489]. Jump height of the center of mass is determined by launch velocity. Peak launch height is given by:

$$y_{max} = -\frac{(\dot{y}_0 \sin\theta)^2}{2g} = \frac{E(\sin\theta)^2}{mg}$$

Maximum range is given by:

$$x = \dot{x}^2 \frac{\sin 2\theta}{g} = \frac{2E \sin 2\theta}{mg}$$

Maximum horizontal range occurs at a takeoff angle of 45° while maximum height occurs at a takeoff angle of 90°. The 50 kg PrOP-F on Phobos 2 and used hopping by means of a spring mechanism to propel the vehicle over 40 m. It comprised a bell-shaped body and a damper to reduce impact force. Hoppers are generally required to be small for higher efficiency [1490, 1491]. The MINERVA (Micro Nano-experimental Robot Vehicle for Asteroids) rover carried by the Hayabusa (MUSES-C) asteroid explorer spacecraft was 10 cm high by 12 cm in diameter. It employed a hopping mechanism powered by a dual DC motor-driven torquer. Its total mass was 588 g and it carried three CCD cameras, photodiodes, and thermometers. MINERVA was propelled on asteroid 25143 Itokawa by a rotating weight held within the robot, which generated a hopping motion at a speed of 9 cm/s (less than the escape velocity of 20 cm/s). It fired a 5 g pellet into the asteroid at 300 m/s to release a 1 g sample for capture by a sample catcher. Hopping is generally adopted by very small animals over scales where ground locomotion is inefficient because it minimizes energy costs for wide-area exploration in such cases. Bio-inspired hoppers are often based on grasshoppers, fiddler crabs, or fleas, which minimizes energy cost [1492]. Strain energy in biological systems such as insects is stored only when the spring mechanism is fully loaded waiting for energy to be released. The insect brings its legs or its body into the jumping position then loads the main spring using muscle power. This probably makes the system safer. Nature commonly uses bistable mechanisms. This is associated with separating the loaded configuration from the storage of strain energy. The mechanism is drawn over the center by the main spring and then the spring is loaded. When the system is fired, the trigger—characterized by low force but high mechanical advantage—allows the mechanism to move back over the center and the energy from the main spring is fed into the system. This has the advantage that there are no firing pins or hooks to jam or break and control is smoother and reliability improved. A further variation in locomotion was PrOP-M which skiied on two sledges each driven by two rotating levers.

Hovercraft have not been proposed for planetary surfaces primarily due to their high power requirements; however, they would be highly suitable for surface exploration of Titan or perhaps Venus. They can traverse obstacle fields, liquids, ice, snow, etc. They operate by creating a cushion of pressurized air to separate the vehicle from the surface. There are two types: the plenum chamber and the peripheral jet, primarily the former [1493]. In the plenum chamber, pressurized air is pumped into the chamber by a compressor to form the air cushion. The air pumped in should be equal to that leaking out of the airgap. The clearance height, however, is limited by the skirt diameter by the air-cushioning lift efficiency:

$$K = \frac{W}{F_l} = \frac{A}{2hlD}$$

where W = weight, h = clearance height, A = cushion area, l = cushion perimeter, D = discharge coefficient dependent on wall angle θ = 0.6 for an inward-pointing skirt, and L = lift force = $\rho V v$, and ρ = air density.

Wall angle θ	0	45	90	135	180
Discharge coefficient D	0.5	0.537	0.611	0.746	1.00

Lift is given by:

$$L = W = p_c A$$

where p_c = cushion pressure = 1.2–3.3 kPa for terrestrial hovercraft.

The velocity of escaping air is given by:

$$v_e = \sqrt{\frac{2p_c}{\rho}}$$

Airflow volume is given by:

$$V = hdDv_e = hdD\sqrt{\frac{2p_c}{\rho}}$$

where d = cushion perimeter.

The power required to sustain the air cushion is given by:

$$P = p_c V = hdDp_c^{2/3}\left(\frac{2}{\rho}\right)^{1/2} = hdD\left(\frac{W}{A}\right)^{3/2}\left(\frac{2}{\rho}\right)^{1/2}$$

There are also drag effects to be compensated including momentum drag, trim drag, skirt drag, and (on water only) wave drag. This is dominated by momentum drag:

$$R_d = \rho V v_{\text{rel}}$$

where v_{rel} = relative velocity of air. The additional power required to compensate for momentum drag is given by:

$$P = \tfrac{1}{2}\eta\rho V v_{rel}^2$$

where η = cushion system efficiency. Propulsion is typically achieved using propeller fans which may be differentially driven to steer. Fan size may be determined by the following:

$$\text{Fluid flow} \quad Q = kNd^3$$

where k = efficiency constant, N = fan speed (rev/s), and d = fan diameter

Pressure $\qquad p = k\rho N^2 d^2$

Power $\qquad\quad P = kN^3 d^5$

It is not clear how feasible hovercraft might be but they do introduce significant control challenges for small vehicles. Furthermore, ambient conditions (not on Mars where the winds are too feeble) may be suitable for the exploitation of planetary winds for the generation of electric power. Airflow may be converted into rotational shaft power through the aerofoil rotors of a wind-driven turbine. The kinetic power conveyed by the wind is dependent on air density ρ, cross section of flow A, and wind speed v:

$$P = \tfrac{1}{2}\rho A v^3$$

The power from the rotor is dependent on the lift and drag forces on the aerofoil which in turn are dependent on the windmill configuration that is quantified by the power coefficient c_p (typically ~0.5).

Lighter-than-air vehicles offer extended mission durations with long traverses as well as several advantages over aeroplanes or helicopters for planetary atmospheres. Powered airships are controllable offering precise flight paths for surveying large areas and a station-keeping capability to hover over sites of interest, even landing to deploy in situ scientific instruments [1494]. Most planetary aerobot designs are balloon based. Balloon-based systems are commonly used for terrestrial exploration and monitoring (Figure 14.4) [1495]. An extensive review of balloon-based aerobots is provided by Cutts et al. (1995) who discuss buoyancy control using reversible fluids to exploit phase changes [1496]. They provide an option for suborbital remote sensing and surface mapping at greater resolution than orbiters. It has been proposed that a solar-heated hot-air balloon (Solar Montgolfier) may be deployed on Mars like a parachute until it achieves buoyancy to reduce landing speed to <5 m/s [1497]. Balloons offer the capability of precise landing at scientifically interesting sites otherwise inaccessible to a surface rover (e.g., near a Martian gully). However, such sites may be subject to significant atmospheric turbulence. Balloons can be part of a multitiered architecture for comprehensive planetary exploration in conjunction with ground rovers by providing navigation support to the ground rovers [1498]. The DARE (Directed Aerial Robot Explorer) concept proposes using long-duration

Figure 14.4. Deployment of a Mars balloon [credit NASA].

autonomous balloons with trajectory control to deploy swarms of miniature probes over multiple target areas while acting as relays for microprobes [1499]. The trajectory control mechanism involves a hanging wing/sail at the end of a long tether which exploits the differences in wind vectors at the two different altitudes. Balloons may also be used to double as soft descent and landing systems providing a major mass saving [1500]. Two Venus superpressure balloons were deployed to 54 km altitude in the Venusian atmosphere in 1984 by the French–Russian–U.S. Vega project. They were inserted at the nightside near the terminator and one floated for 46.5 h returning meteorological data until battery power was exhausted. Balloons are commonly proposed for Titan, Mars, Venus, and even Jupiter, Saturn, Uranus, and Neptune. Generally, for Mars, Venus, and Titan superpressure balloons are envisaged while solar montgolfier balloons are envisaged for gas giants. The Mars 96 mission included a balloon concept which was to fly at an altitude of 4 km during daylight covering 2,000 km per day and land at night. The 5,500 m^3 balloon had a mass of 30.5 kg, enclosing 6.4 kg of gas for bouyancy, carrying a 15 kg gondola for scientific payload, and used a 13.5 kg guiderope for deploying instruments on the ground [1501]. For Titan and Venus, the mass of the balloon envelope would be small but for Mars the large volume of gas required demands a massive envelope. Titan in particular has been targeted as a potential planetary body for balloon exploration due to the high buoyancy-driven lift available from its thick atmosphere [1502, 1503]. This would require active heating due to the low solar flux of 1 W/m^2 at Titan. The chief problem is that balloons can control only vertical ascent/descent and are propelled by ambient winds.

There are several balloon options but superpressure balloons are the most favorable for planetary missions. Zero-pressure balloons are the simplest. They have an inner metallic coating that heats the interior with atmospheric gases. Using atmospheric gases as the float gas eliminates the need for compressed gas cylinders. Solar heating offers transport by day and landing of in situ science experiments to the surface at night at multiple locations. Pre-inserted volatile fluids such as methanol can act as the float gas when irradiated by solar radiation. Zero-pressure balloons require ballast unlike superpressure balloons. Superpressure balloons operate as near-constant volume systems [1504, 1505]. They comprise a sealed envelope filled with helium—hydrogen is excluded because it is more hazardous despite suffering reduced leakage—with a pressure greater than ambient pressure. This allows the balloon to maintain its shape and maximize lift (which is dependent on displaced volume). No ballast is required to maintain altitude as long as it is pressurized allowing ultra long-duration flight. The altitude range is determined by the amount of stretch that the balloon envelope material can withstand. Superpressure balloons have significant advantages over zero-pressure balloons by virtue of their long duration—they exploit venting of excess gas through a valve [1506]. NASA zero-pressure balloons have operated for 10–20 days in Antarctica but superpressure balloons offer the possibility of 100-day durations. Suitable construction materials include Mylar, polyethylene terephthalate (PET), and poly-phenyleneterephthalimide aramid (PPTA). Polyester laminates have low areal densities of $\sim$40–100 g/m^2 which can survive low temperatures of 77 K. Polyethylene films used for conventional zero-pressure balloons typically are a composite of polyester weave fabric with thin films of polyester and polyethylene. However, for superpressure balloons, Mylar film or nylon is more appropriate for the pressure vessel role as polyethylene films are tear sensitive. Lightweight fabrics such as Kevlar/Mylar/polyethene composite offer skin densities as low as 13 g/m^3. The major problem is in sealing the balloon against gas loss. The baseline balloon shape is spherical to maximize the volume-to-surface area ratio, though pumpkin (natural shape) balloons, commonly used for zero-pressure balloons, have been proposed using tendons to take meridional loads but these have not been successful [1507–1509]. Balloon technologies have been proposed for Mars aerial exploration but they require large diameters [1510]: a 11.3 m diameter by 6.8 m high superpressure pumpkin balloon made of polyethylene film and a 10 m diameter superpressure spherical balloon made of Mylar film is an example. They were deployed aerially at a speed of 40 m/s and inflated with He during parachute descent at 34 km altitude similar to a Mars deployment scenario. The pumpkin balloon inflated successfully while the spherical balloon only partially inflated and ruptured. The main limitation on balloon missions is electric power generation (particularly at night). Furthermore, they need to be anchored to the ground against wind. To act as a drilling platform, they need to exert sufficient force on the ground.

Gliding and powered flight offer greater control than achievable from balloon-based flight. The minimum wind gradient required for dynamic soaring is given by:

$$\frac{dv}{dh} = 2C_D v$$

where v = wind speed. For steady flight, the lift generated by a wing is given by:

$$L = \tfrac{1}{2}\rho v^2 C_L A$$

Drag is given by:

$$D = \tfrac{1}{2}\rho v^2 C_D A$$

For a wing with an elliptical lift distribution:

$$C_D = \frac{C_L^2}{\pi(AR)}$$

for small angles of attack. In steady powered forward flight, lift balances weight while thrust balances drag. Gliding involves zero thrust, so drag must be minimized. The gliding angle is minimized if drag is minimized—this occurs at a flight speed given by:

$$v_{\min} = \left(\frac{4K}{\rho^2 C_D} \frac{L^2}{Ab^2} \right)^{1/4}$$

where K = lift efficiency and b = wingspan. Time of flight for a glider is given by:

$$t = h\frac{C_L}{C_D}\sqrt{\frac{\rho C_L A}{2W}}$$

where h = initial altitude and $W = mg$ = weight. Powered flight is 10 times as energy intensive as ground locomotion (energy per unit time) but four times more efficient in ground coverage (energy per unit distance) on Earth. Powered planetary flyers involve similar technology to UAVs (unmanned air vehicles). Conventional aircraft use wings to reduce their thrust requirements:

$$T = W/(L/D)$$

where $L/D = 10\text{--}20$ typically. Aerodynamic forces are generated through fluid pressure:

$$F = -\int P\, dS$$

where P = surface pressure and dS = surface area of body. Shear stress due to fluid viscosity is given by:

$$\tau = \mu \frac{dv}{dx}$$

where μ = viscosity = 1.79×10^{-5} Ns/m^2 for air and dv/dx = velocity gradient. To minimize friction on a wing, the boundary layer of air at the surface should flow smoothly—laminar flow. Bernoulli's equation relates fluid pressure and fluid speed while neglecting shear stress:

$$P + \tfrac{1}{2}\rho v^2 = c$$

along streamlines tangential to flow direction. Air speed is commonly measured by a Pitot static tube that allows measurement of stagnation pressure P_0 in the inner tube and comparison with pressure P in the outer tube such that:

$$P_0 = P + \tfrac{1}{2}\rho v^2 \rightarrow v = \sqrt{\frac{2(P - P_0)}{\rho}}$$

There are a number of dimensionless constants used to characterize the performance of aeroplanes. The coefficients of wing lift and drag are given by:

$$C_L = \frac{L}{\tfrac{1}{2}\rho v^2 S} \quad \text{and} \quad C_D = \frac{D}{\tfrac{1}{2}\rho v^2 S}$$

where L = lift force, D = drag force, and S = planform area. Typically, the boundary layer exhibits disrupted airflow (turbulence), which typically accounts for a third of total air drag. Aircraft wings have a larger wingspan than a chord characterized by high aspect area $AR = l/c \gg 1$, where l = wingspan and c = chord length. The lift coefficient varies with the angle of attack: it increases linearly to a maximum at 10–20° and then drops at increasing angles of attack. Maximum lift occurs at the critical angle of attack (stall angle). Aeroplane wings are often cambered to generate more lift—upward bowing of 8% of the distance from front to back of the wing. The pitching moment coefficient for a wing is given by:

$$C_M = \frac{M}{\tfrac{1}{2}\rho v^2 S c}$$

where c = aerodynamic chord of the wing. The aerodynamic center is typically near $\tfrac{1}{4}$-chord aft of the leading edge. The pressure coefficient is given by:

$$c_p = \frac{P - P_\infty}{\tfrac{1}{2}\rho v^2}$$

The Reynolds number is given by:

$$\text{Re} = \frac{vc}{\nu}$$

where v = speed, ν = viscosity, c = chord, and ρ = density. Flow around the wing tips gives rise to trailing vortices that induce downwinds, which is detrimental. For Mars, long wingspans and high speeds of about Mach 0.5 are required due to the thin atmosphere. The DARPA micro air vehicle project goal is development of a 15 cm long vehicle under 150 g to fly a distance of 10 km in 2 h in winds of up to 50 kph. They generally have wingspans of ~350–400 mm and must be constructed of lightweight materials [1513]. Their flying environments are characterized by low Reynolds number ($\sim 10^4$–10^5) aerodynamics requiring thin-foil low-aspect-ratio wings with sweep angles as low as 40°—such wings can generate leading edge and tip vortices at low angles of attack [1514]. Powered aircraft for Mars deployment require extensive use of lightweight multifunctional structures including structural integration of solar cells and Li polymer batteries. Powered flyers for Mars have

been proposed despite the low atmospheric pressure of 7 mbar. The power requirement for flight is dependent on aerofoil performance and wing loading and is
given by:

$$P = Fv = (D + L)v = \tfrac{1}{2}\rho v^3 (A_s C_D + A_w C_L) = \sqrt{\frac{2W^3 C_D^2}{\rho A C_L^3}}$$

Long-duration flyers have also been proposed for Mars based on high-altitude
long-duration UAV technology (e.g., HARVE-2 for stratospheric flight [1515,
1516]). Such designs require very light structures with high-aspect-ratio wings
($AR = 19.3$ with span of 34 m and wing area of 60 m^2) with a lift coefficient of 1.2
for low Reynolds number operation at 26 km altitude for 48 h.

Most micro-air vehicles are based on rotary wings. MANTA (Mars Nano/
Micro-Technology Aircraft) was an autogyro concept that provided the basis for
aerial reconnaissance exploration with VTOL (vertical takeoff and landing) capability. A microcamera comprised the principal scientific payload. MANTA was
conceived to have a flight duration of 3 h with a maximum speed of 100 km/h. Its
characteristics are listed in Table 14.1 [1517].

Micro-UAVs are primarily hampered by the high mass of battery power they
need to carry which limits their flight durations (particularly for rotary wing craft).
The MASSIVA (Mars surface sampling and imaging VTOL aircraft) concept—
based on a ducted propeller in conjunction with Zagi 10-section fixed wings—has
a projected mass of 25 kg [1518, 1519]. It takes off vertically and has an operational altitude of 100 m. It utilizes two lift rotors of diameter 1.4 m and a fixed
wing area with high aspect ratio of 8 m^2 for level flight at a cruise velocity of
40 m/s. Lift requires 3.7 kW of power over 1 min. The approach involved using
two counterrotating rotors stacked vertically. The wing area was coated with
triple-junction thin-film solar cells supported by rechargeable Li polymer batteries.
Two small propellers would provide forward thrust. All structural and mechanical
components would be constructed from carbon fiber composite.

Table 14.1. Properties of MANTA.

Total mass (kg)	2.5
Rotor diameter (m)	1.5
Stowed dimensions (m)	0.8 × 0.5 × 0.38
Max mission duration (h)	3
Max speed (km/h)	100
Cruising speed (km/h)	60
Max range (km)	100
Glide path (power off)	1:6

Table 14.2. MASSIVA conceptual mass budget.

Subsystem	Mass (kg)
Propulsion	5.8
Power	4.7
Structure	9.5
Payload	4.0
Contingency	1.0
Total	25.0

Mesicopter is a helicopter measuring a few centimeters with a mass of 60 g designed to remain aloft on a planet such as Mars using an onboard power supply for its DC motors [1520]. Rotor thrust is given by:

$$T = 2\rho A v^2$$

For hovering, thrust is determined by weight:

$$v = \sqrt{\frac{W}{2\rho A}}$$

Each of the four blades of the rotor are 2.5 cm in diameter and 100 μm thick. Each rotor can generate a thrust of 20 g sufficient to support the mesicopter in hover. The rotors have a disk loading of <25 N/m². The motors use single-turn coils rather than multiple coils and a permanent magnet rotor. The speed controller is a PIC17 microcontroller implementing PWM motor drivers. While fixed wing flight has separate lift and propulsive devices, rotary wing and flapping flight involve a single integrated rotor system for both. Fixed wing aircraft cannot be scaled down as they would require high speeds for lift in a reduced Reynolds number environment. This requires around 150 W/kg specific power. Similarly, downscaled rotorcraft suffer from recirculation problems and high power requirements. The solid-state aircraft integrates aerofoil, energy sources, and control into the structure of the aircraft without moving parts for operation in the Venusian atmosphere to exploit its higher solar flux [1521]. It uses an ionic polymeric metal composite for both propulsion and control acting as an artificial muscle powered by thin-film photovoltaics and thin-film batteries to generate an electromagnetic field across the wing. It employs a high-frequency flapping motion of ~10 Hz and the flight profile of a hawk exploiting soaring for long periods.

Bio-inspired micro-air vehicles (MAVs) based on flapping wings offer the ability to glide and hover [1523]. MAVs may generate thrust through wing flapping or through conventional methods—a basic analysis of performance is given by Shyy et al. (1999) [1524]. Michelson and Naqvi (2003) recommend that air vehicle design should be bio-inspired rather than directly biomimetic [1525]—biological systems are often limited, difficult to replicate, and can be exceeded by technology. There are 10 design principles:

  (i) biomimetics is a good starting point;
 (ii) pure biomimetic approaches can yield non-optimal performance solutions;
(iii) sometimes optimal solutions result from traditional technology;
 (iv) biomimetic solutions may not be practical;
  (v) wing beating requires sufficient power—artificial muscles have yet to demonstrate sufficient efficiency to lift vehicles;
 (vi) biomimetic flapping is complex to control involving complex muscle structures and resilin energy storage;
(vii) flight stability control is complex involving flap angle, span, rate, and wing angle-of-attack modulation;
(viii) poor integration generates mass creep requiring extensive use of multi-functional structures;
 (ix) designs must capitalize on resonance in storing energy within their structures, which comprises ~35–38% of elastic energy storage compared with 10% efficiency for muscles;
  (x) average power density for current battery technology is marginal for small-scale flapping wing flight.

An efficient MAV has been developed by Jones and Platzer (2002) [1526]. This was a small biplane flapping wing configuration (for propulsion) mounted onto a large low-aspect-ratio fixed wing (for lift) [1527, 1528]. The vehicle has a large wing providing most of the lift and two small wings at the rear arranged in biplane fashion to provide propulsion. The trailing biplane arrangement of two opposed flapping wings in counterphase provides an enhanced ground effect. The two aft wings flap in counterphase at a fixed amplitude with a maximum frequency of 25 Hz. Increasing wing length l is the most efficient way to increase lift capacity due to its l^4 dependence. Aerodynamic power also favors inceasing wing length rather than flapping frequency at the expense of reduced flight speed. Their radio-controlled MAV model had a mass of 15 g with a wingspan of 30 cm and a speed of 2–5 m/s using Li polymer battery cells (with an energy density of 170 Wh/kg) for a 20 min flight time at a flapping frequency of 30 Hz [1529, 1530]. It had no control capability. The 50 g Entomopter is a candidate for Mars exploration at low Reynolds number with its 20 cm wingspan [1531]. It is capable of flapping flight based on the hawkmoth, ambulation, and swimming that is powered anaerobically by a reciprocating chemical muscle system. This power system operates by catalytic combustion and is regeneratively powered from chemical fuel sources offering 70 Hz flapping rates with 4 W of continuous power. Rather than using wing flapping with individual wing twisting, the X wing fore and aft wings pivot like seesaws about the central fuselage, 180° out of phase with each other while giving roll, pitch, yaw, and forward motion. The fuselage acts as a torsional

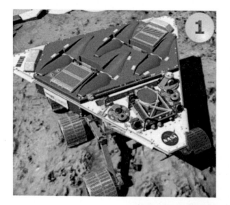

Figure 14.5. Entomopter deployed from a ground rover [credit NASA].

spring with a resonant frequency equal to the flapping rate. A pneumatically con-trolled airflow (generated as waste products from the reciprocating chemical muscle) over the wing through hollow ribs along the trailing edges provides circulation-controlled lift increasing lift by 500%. The Mars ExoFly offers a hovering capability and a range of 5–10 km at 70 km/h with a power consumption of 5–10 W. Entomopter may be deployed as a long-range scout from a ground rover (Figure 14.5) [1532].

The 3 g DelFly is a MAV with dual-flapping wings inspired by the dragonfly that is capable of 7 m/s forward flight from a 10 cm wingspan [1533, 1534]. The wing pairs are mounted vertically with respect to each other to exploit clap-and-fling aerodynamics in mid-flap. It exploits highly miniaturized rechargeable Li polymer batteries of mass ~1 g with 130 Wh/kg specific power, a miniaturized ~1 g DC Enterprises brushless motor with a 1.5 W power output, a miniaturized radio control system, and a miniaturized NTSC CMOS camera. Both of these concepts are currently advanced concepts only.

Appendix

Brief review of Bayesian and associated methods

Bayesian methods and their derivative such as Kalman filters appear as a consistent theme in rovers and robotics. This appendix provides a brief review. In a Markov process, each state transition occurs at discrete times according to a set of probabilities associated with each state. Any state can be reached from any other state (ergodic) determined by the state transitions. Transition probability is the posterior probability that the Markov chain will change from state i to state j. For a first-order Markov chain, the probabilistic state transition function is dependent only on the current and predecessor state. A Markov chain emits a sequence of observable outputs such that a Markov model is the five-tuple:

$$(\Omega_S, \Omega_O, O, S, S_0)$$

where Ω_S = finite set of states, Ω_O = finite set of observations of environmental states, O = observation probabilities, S = state transition probabilities defined by probability distributions, and S_0 = initial state distribution. In a Markov process, the current state and action must provide all the information available for predicting the next state. For a dynamic system, control input can be incorporated into the posterior:

$$p(x_k|u_{k-1}) = \int p(u_{k-1}|x_k, x_{k-1})p(x_{k-1}) \, dx_{k-1}$$

The Markov property assumed is that the current state (probability) is dependent only on the previous state (probability) (i.e., $p(x_k, x_{k-1}, x_{k-2}, \ldots) = p(x_k, x_{k-1})$):

$$p(z_k|x_k) = p(z_k|x_{1:k}, z_{1:k}, u_{1:k-1}) \quad \text{and} \quad p(x_k|x_{k-1}, u_{k-1}) = p(x_k|x_{1:k-1}, z_{1:k}, u_{1:k-1})$$

For a Markov decision process, policy π defines a sequence of functions that maps states to control actions. A partially observable Markov decision problem allows the state to be a function of all past observations. A Markov decision process is a

© Springer-Verlag Berlin Heidelberg 2016
A. Ellery, *Planetary Rovers*, Springer Praxis Books,
DOI 10.1007/978-3-642-03259-2

tuple:

$$(S, A, T)$$

where $S =$ finite set of states, $A =$ finite set of actions, and $T : S \times A \rightarrow P(S) =$ probabilistic state transition function subject to the Markov property that the state is dependent only on the previous state and the action taken [1535]. An optimal policy is one that maximizes expected accumulated reward R (i.e., a Markov decision process is given by a set of states S, a set of actions A, a probabilistic action model $P(S|S, A)$, and a reward function R. A discount factor γ controls the influence of rewards in the future. A Markov chain is a stochastic Markov process sequence which depends only on its previous state (i.e., $p(H_i|H_{i-1}, \ldots, H_0) = p(H_i|H_{i-1})$). It is defined by the initial probability distribution $p(H_0)$ and the transition probability $p(H_{i+1}|H_i)$. Belief networks are constructed from acyclic graphs of a large number of particles with probability distributions with the Markov property. Probabilistic state transitions model system noise. The relationship to Bayesian systems is evident. Since $p(x_k|z_k)$ represents the posterior probability of the state given the measurements, Bayes theorem gives:

$$p(x_k|z_k) = \kappa p(z_k, x_k) \int p(x_k|x_{k-1}) p(x_{k-1}|z_{k-1}) \, dx_{k-1}$$

where $p(z_k|x_k) =$ likelihood of state relative to measurements (current posterior), $p(x_k, x_{k-1}) =$ prior that defines the evolution of the state over time, $p(x_{k-1}, z_{k-1}) =$ previous posterior, and $\kappa =$ normalizing constant. Given the posterior probability of the state $p(x_k|z_k)$, it is possible to compute a maximum a priori (MAP) estimate of x_k.

In many cases, much information is hidden from observation (unobservable). The hidden Markov model (HMM) is a generalization of the Markov model in which there exist hidden states that are not directly observable. Outputs that are dependent on the hidden states are observable, which gives some information about the sequence of states. Independent components analysis, principal components analysis, mixtures of Gaussians, vector quantization, Kalman filters, and hidden Markov models are all variations of unsupervised learning [1536]. Estimating hidden states given observations and a model of fixed parameters comprises the filtering/smoothing problem. It is possible, nevertheless, to learn a few parameters that model observed data (identification). The basis of most learning algorithms is the iterative expectation maximization algorithm which maximizes the likelihood of observed data in the presence of hidden variables. Hidden Markov models are suited to linear statistical sequences—a finite n-state model describes a probabilistic distribution over an infinite number of possible sequences [1537]. The HMM is a double stochastic process that generates states which correspond to positions in a sequence. It is defined by a Markov chain comprising a finite number N of states, a finite number M of observables, and a finite set of state transition probability distributions. The HMM generates a sequence of observable outputs from hidden states that evolve according to transition probabilities. The observed outputs are generated from hidden states according to emission probabilities. The current state emits m symbols according to symbol emission probabilities B. The next state is determined by the state transition probability distribution A and the current state. From an initial state a sequence of states is generated according to state transition probabilities until an end state is reached. The sequence of states forms a Markov chain as the next state is dependent on the current state. These states, however, are hidden. Each state emits symbols according to the state's emission probability distribution. It is the symbol sequence that is observed from which the state sequence must be inferred. The state of the system is not directly observable, only state-dependent observables may be measured forming a discrete set. The HMM is defined by:

$S = (s_1, \ldots, s_q) = $ set of states

$\pi = p(s_i | t_0) = $ initial state probability distribution

$a_{ij} = p(s_j | s_i) = $ transition probability distribution of transitions between states

$Y \in S = $ set of all observable events corrupted by random noise

$B = p(x | s_i) = $ set of emission probability distributions of observations for each

state

The HMM generates a sequence of outputs from a system whose internal states evolve according to probabilistic rules (state transition function). The output sequence y_t (observations) depends on the current internal state x_t of the model through a probabilistic relation (output likelihood transition function). The HMM represents the simplest Bayesian network in which the state of the model is not directly observable but the output (which is dependent probabilistically on the state) is visible. Speech recognition may be characterized as a maximum likelihood decoding from strings of phonemes into strings of words using statistical models [1538]. Speech production may be characterized as a Markov source in which sequential state transitions are associated with a probability. An estimate of possible words minimizes the probability of word error through Bayes rule based on an a priori language (grammar) model. Speech models are based on hidden Markov models which generate symbol sequences [1539]. Given an observation sequence of symbols $O = (O_1, \ldots, O_r)$ and model $M = (A, B, q_0)$ where $q_0 = $ initial state, we compute maximum $p(O|M)$ based on the hidden state sequence $Q = (Q_1, \ldots, Q_r)$. The probability of the observed symbol sequence for a model is given by:

$$p(O|M) = p(O|Q, M) p(Q|M)$$

The solution to the HMM estimates hidden states given the observations and current model parameters. The HMM is trained to estimate π, A, and B through the Baum–Welch or Viterbi unsupervised learning algorithms, the latter being a simplified version of the former and so faster. The Baum–Welch algorithm is a maximum likelihood criterion estimation procedure used to calculate the probability of each state for the entire data sequence. The forward–backward algorithm forms the core of the Baum–Welch algorithm to maximize the conditional probability of the samples given in the model. The likelihood of the output is computed initially from the state transition probabilities through a forward pass. The forward probabilities are given by:

$$\alpha_i = b_i \sum_j a_{ij}\alpha_j$$

where $\alpha(i, i+k, q_i^d, q^{d-1}) = p(o_i, \ldots, o_{i+k}, q_i^d$ finished at $i+k|q^{d-1}$ started at i). The likelihood of the output is updated from the observation function through a backward pass. The backward probabilities are given by:

$$\beta_i = \sum_j a_{ji}\beta_j b_j$$

where $\beta(i, i+k, q_i^d, q^{d-1}) = p(o_i, \ldots, o_{i+k}, q_i^d$ started at i, q^{d-1} finished at $i+k$). The Viterbi algorithm is a dynamic programming algorithm to find the most probable state sequence of hidden states of the HMM, which results in a given observed sequence of outputs. It computes a maximum a posteriori (MAP) estimate of the sequence of states that maximize the probability of the state sequence given the data. It may be viewed as the shortest path through a trellis diagram. The Viterbi algorithm is used for decoding convolutional codes but is most commonly used in speech recognition—an acoustic signal represents an observed sequence of outputs while the word strings are the hidden cause generating the output sequence. Both the Baum–Welch algorithm and the Viterbi algorithm have a computational complexity of $O(nt^3)$ where $n =$ number of states and $t =$ length of observation sequence. Hierarchical hidden Markov models that use an extended Baum–Welch (forward–backward) algorithm are suited to multiscale structures such as language [1540]. Partially observable Markov decision processes may be employed to select optimal actions [1541]. In this case, a Markov decision process is a tuple $\langle S, A, T, R\rangle$ where $S =$ set of states of the world, $A =$ set of actions by the agent, $T : S \times A \rightarrow \pi(S) =$ state transition function mapping each world state to agent action through a probability distribution, and $R : S \times A \rightarrow R(s, a) =$ reward function for a given action in each state.

Bayes theorem defines a relationship between the a priori probability of a hypothesis, the conditional probability of an observation given the hypothesis, and the a posteriori probability of the hypothesis. Observational evidence is used to update the a priori probability of the hypothesis. A Bayesian inference about the unobserved states of a system is based on using a model of the system (prior knowledge) and measurements of observed states. Bayes theorem is used to update the (posterior) likelihood of a hypothesis given the prior likelihood estimate and additional observations (evidence):

$$p(H_i|E) = \frac{p(E(H_i)p(H_i)}{\sum_{j=1}^{n} p(E|H_j)p(H_j)}$$

where $\sum_{j=1}^{n} p(H_j) = 1$, $p(H_i|E) = $ a posteriori probability of hypothesis H_i being true given observations O, $p(H_i)$ = a priori probability of hypothesis H_i being true, and $p(E|H_i)$ = probability of observing O given that H_i is true (observation model or likelihood). Bayes theorem allows selection of the model that fits the prior knowledge and observables best (i.e., the highest posterior probability). The process of computing the posterior probability from the prior probability constitutes Bayesian learning as it classifies data according to the MAP probability. The MAP rule is used to classify instances of the class C:

$$\arg \max p(C = c) \prod_{i=1}^{n} p(E_i = p_i | C = c)$$

Now, a conditional probability defines the probability of A given event B and is defined by:

$$p(A|B) = \frac{p(A \cap B)}{p(B)}$$

Bayes theorem is a special case of Jeffrey's rule given by [1542]:

$$p(H) = p(H|E)p + p(H|\bar{E})(1 - p)$$

where $p = 1$ for Bayes rule (interpreted as event E occurring). Bayes theorem describes the classification of event H for observed features/states $E_1, \dots, E_n$ as a conditional probability:

$$p(H_i|E_j) = \frac{p(H \cap E)}{p(E)} = \frac{p(H_i)p(E_j|H_i)}{p(H_i, E_j)}$$

where $p(E|H)$ = likelihood and where

$$p(H_i, E_j) = p(H_i)p(E_j|H_i) = p(H_i) \prod_{j=1}^{n} p(E_j|H_i)$$

$$= \text{joint probability of observable } E \text{ and hypothesis } H \text{ being true}$$

If all prior probabilities are equal to each other (principle of indifference),

$$p(H_i) \prod_{j=1}^{n} p(E_j|H_i) = p(E)$$

Hence,

$$p(H_i|E_j) = \frac{p(H_i)p(E_j|H_i)}{p(E_j)}$$

Alternatively, Bayes theorem may be expressed in terms of its odds-likelihood form of the likelihood ratio [1543]:

$$\frac{p(H|E)}{p(\sim H|E)} = \left(\frac{p(E|H)}{p(E|\sim H)} \right) \left(\frac{p(H)}{p(\sim H)} \right)$$

where $O(H|E) = p(H|E)/p(\sim H|E)$ = posterior odds, $\lambda(H, E) = (p(E|H)/p(E|\sim H))$ = likelihood ratio, and $O(H) = (p(H)/p(\sim H))$ = prior odds (i.e., $O(H|E) = \lambda(H, E)O(H)$. Maximum entropy assumption distributes prior probabilities to maximize the entropy function [1544]:

$$H = - \sum_{ij} p(H \cap E) \log[p(H \cap E)]$$

In assigning probabilities as degrees of belief to events, we are using probability distributions across a set of possible worlds (models) and select those with maximum entropy. From information theory [1545, 1546], we can describe the gain in information from the prior information (model) updated with evidence (data). Information $I(M : E|C)$ about a model M (hypothesis) derived from the evidence E (data) given a background of contextual information C (prior) has certain properties. The entropy of posterior distributions of observations is given by:

$$I(t) = \sum_{o \in O} p(o, t) H(p(C|o, t))$$

where $H(p(C|o, t)) = \sum_{c \in C} p(c|o, t) \log_2(1/p(c|o, t))$, $p(o, t) =$ probability of observation o at time t, $H(p(C|o, t)) =$ Shannon entropy of posterior probability for observation o at time t, and $C =$ set of all mutually exclusive classes c.

The Bayesian problem is to determine the degree of belief in state $x_k = (x, y, \theta)^T$ at time k given observational data $z_{1,...,k}$ up to time k. Dynamic state estimation involves the construction of a (posterior) probability density function of the state of a system and its future evolution $p(x_k|z_{1,...,k})$ or, more compactly, $p(x_k|z_k)$. A posterior probability density function may be represented as a belief function:

$$\text{Bel}(x_k) = p(x_k|z_k, u_{k-1}, z_{k-1}, \ldots, u_0, z_0) = \eta p(z_k|x_k) \int p(x_k|x_{k-1}, u_{k-1}) \text{Bel}(x_{k-1}) \cdot dx_{k-1}$$

where $p(x_k|x_{k-1}, u_{k-1}) =$ system state dynamics, $p(z_k|x_k) =$ measurement model, and $\eta = p(z_k|z_{k-1}, u_k)$. Prediction comprises the integral term (predictive belief) while the update term is represented by normalizing constant η and the measurement term preceding the integral. This is the concept of the Bayes filter. The initial probability distribution function (the prior) $p(x_0|z_0) = p(x_0)$ is assumed known (without measurements). The Bayesian solution to compute the posterior distribution $p(x_k|z_k)$ of the state given past measurements is given by the Chapman–Kolmogorov equation:

$$p(x_k|z_{k-1}) = \int p(x_k|x_{k-1}) p(x_{k-1}|z_{k-1}) \cdot dx_{k-1}$$

This predictive phase is based on the motion model $p(x_k|z_{k-1})$ which assumes that the current state x_k is dependent only in the previous state x_{k-1} (Markov assumption). During the update step, measurement z_k (conditionally independent of earlier measurements z_{k-1}) is based on a measurement model. The measurement model is based on the likelihood $p(z_k|x_k)$ which quantifies the likelihood that the rover is at location x_k given observation of z_k. This may be used to update the prior to compute the posterior using Bayes theorem:

$$p(x_k|z_k) = \frac{p(z_k|x_k)p(x_k|z_{k-1})}{p(z_k|z_{k-1})}$$

where $p(z_k|z_{k-1}) = \int p(z_k|x_k)\,p(x_k|z_{k-1})\,dx_k =$ normalizing constant and $p(z_k|x_k) =$ likelihood function. This represents a Markov process of order 1 but in general this cannot be solved analytically. However, this may be repeated recursively. The prior $p(H)$ may be modeled as a Gaussian distribution with variance σ^2_{prior} and mean μ_{prior}. The likelihood $p(E|H)$ may similarly be modeled as a Gaussian distribution with variance $\sigma^2_{\text{likelihood}}$ and mean $\mu_{\text{likelihood}}$. The optimal estimate of a sensory measurement based on the means and variances of the prior and likelihood distributions is given by:

$$\hat{x} = \eta\mu_{\text{likelihood}} + (1 - \eta)\mu_{\text{prior}}$$

where $\eta = \sigma^2_{\text{prior}}/(\sigma^2_{\text{prior}} + \sigma^2_{\text{likelihood}}) < 1 =$ weighting between prior and sensory data. This is similar to the way in which the Kalman filter operates. The Kalman filter assumes that the prior probability distribution function $p(x_{k-1}|z_{k-1})$ is Gaussian and so the posterior $p(x_k|z_k)$ is Gaussian at all times. Gaussian statistics are fully describable through means and covariances (i.e., both the state and measurement models are Gaussian). In the case of a discrete state space, the posterior probability density function at $k - 1$ is given by:

$$p(x_{k-1}|z_{k-1}) = \sum_{i=1}^{n} p(x^i_{k-1}|z_{k-1})$$

If local linearization of a nonlinear function through a Taylor expansion is performed, the extended Kalman filter (EKF) approximation may be used. Typically, only the first term of the expansion is retained for computational complexity reasons. The unscented Kalman filter (UKF) implements a set of points selected from a Gaussian approximation which are propagated through the nonlinear function and then re-estimated.

The principle of maximum entropy is a generalization of the principle of indifference (equal probabilities between choices) which measures the degree of ignorance (lack of information) of a probability distribution [1547]. When faced with incomplete information, we assume that the probability distribution has maximum entropy based on the information available. The greater the distribution's concentration, the less the entropy (greater information content). Autoregressive models such as the Wiener filter represent a special case of maximum entropy. Maximum entropy methods are appropriate to Bayesian analysis. Bayesian approaches implement maximum entropy solutions to stochastic problems [1548]. Minimizing the error probability amounts to implementing the MAP that defines the most probable solution. If $p(x)$ is constant, this reduces to the maximum likelihood (ratio) estimator (MLE). Prior probability is given by

Bayes theorem:

$$p(x_i|z_{1:i-1}) = \int p(x_i|x_{i-1}, z_{1:i-1})p(x_{i-1}|z_{1:i-1}) \cdot dx_{i-1}$$

A correction step generates the posterior probability density function:

$$p(x_i|z_{1:i}) = p(z_i|x_i)p(x_i|z_{1:i-1})$$

Thus, the a posteriori probability density function may be estimated from Bayes rule defining the probability of x (state, hypothesis, or model) given z (measurement or evidence):

$$p(x|z) = \frac{p(z|x)p(x)}{p(z)}$$

where $p(z|x) =$ conditional probability of z given x is true (measurement likelihood), $p(x) =$ prior probability of x being true prior to observations, and $p(z) = \sum p(z|x)p(x) =$ normalizing prior probability of z. The posterior probability depends on the likelihood of the evidence given the hypothesis and its negation $p(z|x)$ and $p(z|\sim x)$ and the prior probability of the hypothesis $p(x)$. The posterior probability $p(x|z)$ defines the inverse model while $p(z|x)$ defines the forward model reasoning from state to sensor measurements. A recursive version is given by:

$$p(x|z_1, \ldots, z_k) = \frac{p(z_k|x, z_1, \ldots, z_{k-1})p(x|z_1, \ldots, z_{k-1})}{p(z_k|z_1, \ldots, z_{k-1})}$$

Given the influence of hypothesis x on evidence z, $p(z|x)$, Bayes theorem allows computation of the influence of evidence z on hypothesis x $p(x|z)$. Hypothesis x causes evidence z through $p(z|x)$ but Bayes theorem allows a reversal of this direction of inference in logic in a diagnostic fashion $p(x|z)$. Filtering recursively estimates the first two moments of x_i given $z_{i:i}$. The chief limitation to Bayes rule is that conditional probabilities are difficult to estimate and that assessment of prior probabilities before new evidence is considered is problematic. Furthermore, Bayes theorem cannot represent ignorance—if evidence is only partially in favor of a hypothesis, it must also partially support the negation of the hypothesis such that $p(H|E) + p(\sim H|E) = 1$. This is clearly counterintuitive and contradictory. Any probability not assigned to evidence in favor of the truth of proposition H is assigned to evidence in favor of $\sim H$. Alternatively, the Dempster–Shafer theory of evidence addresses some of the shortcomings of Bayes theorem [1549]. It is a generalization of Bayes theorem which overcomes the problem of assigning a $1 - P$ probability to a proposition's negation. It differentiates between a lack of probability from the evidence of plausibility. Bayes theorem assigns probability measures to events while the Dempster–Shafer theory assigns beliefs to propositions about events. The Dempster–Shafer theory quantifies the degree of support based on evidence. Like Bayes theorem, it is based on subjective probabilities (plausibility) to represent ignorance. Unlike Bayes thereom, a priori probabilities are not assigned under conditions of ignorance. In Dempster–Shafer theory, $p(\bar{A}) \neq 1 - p(A)$. This allows the Dempster–Shafer theory to assign belief to a proposition without assigning belief to its negation. Indeed, evidence is represented as a belief function rather than a probability function. It has been shown that modal and possibilistic logics are all subsets of Dempster–Shafer logic [1550]. Both modal and possibilistic logics are multivalued logics associated with a set of possible worlds made up of possibility and necessity functions:

$$\text{Poss}(X|A) = \sup[X(x) \wedge A(x)] \quad \text{and} \quad \text{Nec}(X|A) = \inf[1 - X(x) \vee A(x)]$$

such that

$$\text{Poss}(X|A) \geq (\text{Nec}(X|A)$$

Bayesian filtering involves integration over probability density functions that cannot be solved generally in closed form for non-Gaussian systems. This requires the use of approximations. If noise is assumed to be Gaussian, an extended Kalman filter (EKF) may be adopted—it requires only mean and covariance matrices for the Gaussian probability density functions. The Kalman filter is one of the simplest Bayes filters. It is modeled on a Markov chain of observed sensor measurements of a hidden Markov process corrupted by Gaussian noise. It resembles a hidden Markov model (HMM) but is restricted to Gaussian noise. The true state of the unobserved Markov process is conditionally independent of all earlier states given the immediately previous state:

$$p(x(t)|x(0), \ldots, x(t-1)) = p(x(t)|x(t-1))$$

Similarly, measurement at the current time is dependent only on the current state and is conditionally independent of all other states given the current state:

$$p(z(t)|x(0), \ldots, x(t)) = p(z(t)|x(t))$$

The probability distribution of the HMM over all states is given by:

$$p(x(0), \ldots, x(t), z(1), \ldots, z(k)) = p(x(0)) \prod_{i=1}^{t} p(z(i)|x(i))p(x(i)|x(i-1))$$

The Kalman filter attempts to estimate state x probabilistically:

$$p(x(t)|Z(t-1)) = \int p(x(t)|x(t-1))p(x(t-1)|Z(t-1)) \cdot dx(t-1)$$

The probability density function is a Kalman filter estimate. The Kalman–Bucy filter is a continuous time version of the Kalman filter. The Kalman filter is a maximum likelihood estimator (MLE) for Gaussian distributions. It can be used to construct a Gaussian approximation of the posterior density $p(x_i|x_i)$ with mean and covariance given by:

$$\hat{x}(t|t) = \hat{x}(t|t-1) + K(t)(z(t) - \hat{z}(t|t-1))$$

where

$$K(t) = P^{xz}(t|t-1)(P^{zz}(t|t-1))^{-1}$$

$$P^{xx}(t|t) = P^{xx}(t|t-1) - K(t)P^{zz}(t|t-1)K(t)^T$$

The Kalman filter is a recursive estimator and the state of the filter is represented by:

$\hat{x}(t|t)$ = a posteriori state estimate at time t based on observations up to time t

$P(t|t)$ = a posteriori error covariance (estimate of the accuracy of the state estimate)

The Kalman filter has two processes:

(i) The prediction process uses the previous state estimate to estimate the current state (predicted a priori estimate):

$$\hat{x}(t|t-1) = A\hat{x}(t-1|t-1) + Bu(t)$$

$$P(t|t-1) = AP(t-1|t-1)A^T(t) + Q(t)$$

(ii) The update process uses the predicted a priori estimate with the current observation (a posteriori estimate). Innovation/residual covariance:

$$R(t) = HP(t|t-1)H^T + I$$

Updated a posteriori estimates:

$$\hat{x}(t|t) = \hat{x}(t|t-1) + K(t)\tilde{y}(t)$$

where

$$\tilde{y}(t) = z(t) - H\hat{x}(t|t-1) = \text{measurement residual (innovation)}$$

$$K(t) = P(t|t-1)H^T R(t)^{-1} = \text{Kalman gain}$$

$$P(t|t) = \text{cov}(x(t) - \hat{x}(t|t)) = (I - K(t)H)P(t)$$

The optimal Kalman gain yields mean squared estimates. The extended Kalman filter has the form:

$$\dot{x} = f(x(t)) + w(t) = f(\hat{x}(t) + K(t)[y(t) - H\hat{x}(t)]$$

where $y(t) = Hx(t) + v(t)$. This may be linearized:

$$\delta\dot{x} = F(\hat{x}(t))\delta x(t) + w(t)$$

$$\delta y(t) = H\,\delta x(t) + v(t)$$

where $F = $ a Jacobian. Kalman filter covariance is given by:

$$\dot{P} = F(\hat{x}(t))P(t) + P(t)F^T(\hat{x}(t)) + Q(t) - K(t)HP(t)$$

$$K(t) = P(t)H^T R^{-1}(t)$$

The Kalman filter is an efficient recursive version of the Wiener filter. It is a technique for state estimation in the presence of noise and offers an optimal estimation solution. A priori information based on a dynamic model is predictive while measurement information is a posteriori. Observers (estimators) are used to predict the complete state of the system when the measurements are insufficient to determine the full state. It computes an a posteriori estimate of the system state using measurements as feedback to an a priori estimate based on a system model. Measurement feedback is based on external measurements which are compared with predicted measurements based on a measurement model. Predicted state and measurement difference are used to update the state. The Kalman filter is an observer of the form

$$\hat{x}_{k+1} = A\hat{x}_k + Bu_k + K(z_{k+1} - H(A\hat{x}_k + Bu_k))$$

For linear systems, the Kalman filter is an optimal estimation procedure. The extended Kalman filter approximates this process for nonlinear measurements. The extended Kalman filter involves four major computations: state propagation and update, covariance propagation and update. During state propagation, an optimal estimate of the state is propagated through a nonlinear dynamic system model. Similarly, state estimation error covariance is propagated. Sensor update utilizes the measurement equation that relates measurements to system states. Kalman filter gain is computed and multiplied by the innovation (i.e., the difference between the predicted and measured value of a variable) thereby generating a state estimate update. Kalman gain is used to update the error covariance estimate. The a posteriori state estimate of state x at time k (using a common shorthand notation) is given by

$$\hat{x}_k^+ = \langle x_k | z_k \rangle$$

and the a priori estimate of state x at time k given inputs up to step $k - 1$ is given by

$$\hat{x}_k^- = \langle x_k | y_k \rangle$$

The a priori estimate comprises the prediction phase while the a posteriori estimate comprises the correction. The first stage takes the a priori state estimate $\hat{x}_{k-1}$ from the previous times t_{k-1} and predicts forward to time t_k using time update equations. The state model for a nonlinear system is given by:

$$\dot{x}(t) = f(x(t), u(t), t) + w(t)$$
$$y(t) = g(x(t), t) + v(t)$$

This may be linearized and discretized into a discrete linear dynamic system with Gaussian noise:

$$x_{k+1} = A_k x_k + B_k u_k + w_k$$

where A = state transition matrix and w = noise with covariance Q

$$y_k = C_k x_k + D_k u_k + v_k \quad \text{or} \quad z_k = H_k x_k + v_k$$

where H = measurement matrix and v = noise with covariance R. The dynamics are linear and noise evolution is Gaussian so that the model represents a first-order Gauss–Markov random process. If the initial state is Gaussian then all future states are Gaussian. State estimate $\hat{x}_k$ is determined by the identity observer based on measurements:

$$\hat{x}_{k+1} = A_k \hat{x}_k + B_k u_k + K_k(\hat{y}_k - y_k)$$

$$\hat{y}_k = C_k \hat{x}_k + D_k u_k$$

We wish to drive estimation error to zero:

$$e_k = \hat{x}_k - x_k \to 0 \quad \text{as } k \to \infty$$

The estimator must have the form:

$$\hat{x}_{k+1} = A\hat{x}_k + Bu_k$$

$$e_{k+1} = x_{k+1} - \hat{x}_{k+1} = Ae_k$$

This cannot be guaranteed to be asymptotically stable and converge so feedback may be incorporated:

$$\hat{x}_{k+1} = A_k \hat{x}_k + B_k u_k + K_k(z_k - H_k \hat{x}_k)$$

$$e_{k+1} = x_{k+1} - \hat{x}_{k+1} = A_k x_k + B_k u_k - A_k \hat{x}_k - B_k u_k - K_k(z_k - H_k x_k)$$

$$= A_k e_k - K_k(H_k x_k - H_k \hat{x}_k) = (A_k - K_k H_k)e_k$$

where $K_k(z_k - H_k \hat{x}_k) =$ innovation and $K =$ gain This is an observer with eigenvalues $|\lambda_i| < 1$—a system is observable if the initial state can be uniquely determined by knowledge of inputs and outputs. The innovation is the difference between the predicted value and the measured value of a variable (i.e., this represents new information if the prediction is inaccurate). The first step predicts the robot position at time $k + 1$ based on its previous position and the control input $u(k)$:

$$\hat{x}(k + 1|k) = f(\hat{x}(k|k), u(k)) \quad \text{or} \quad x_k = f_{k-1}(x_{k-1}, u_{k-1}) + w_{k-1}$$

In the linear case, this a priori estimate becomes:

$$\hat{x}_k^- = A_{k-1} \hat{x}_{k-1}^+ + B_{k-1} u_{k-1}$$

The estimate is corrected when a measurement is made to form an a posteriori estimate:

$$\hat{x}_k = \hat{x}_k^- + K_k(C_k \hat{x}_k^- - y_k)$$

This observer is the current estimator with an estimated error of

$$e_{k+1} = (A_k - K_k C_k A_k)e_k$$

Observer gain may be chosen to minimize estimated error covariance:

$$P_k = \langle (\hat{x}_k - x_k)(\hat{x}_k - x_k)^T \rangle$$

Process and sensor noise is independent with Gaussian distributions and covariance matrices Q_k and R_k. Optimal observer gain is given by:

$$K_k = P_k C_k^T (C_k P_k C_k^T + R_k)^{-1}$$

where $P_{k+1} = A_k[P_k - P_k C_k^T (C_k P_k C_k^T + R_k)^{-1} C_k P_k]A_k^T + Q_k$. P_k is the solution to the discrete algebraic Riccati equation. The state covariance matrix is a measure of uncertainty in the state model and weighs the a priori estimate according to the model. The sensor covariance matrix is a measure of the quality of measurement and weighs the sensors according to their accuracies. First, a time update is performed to predict the a priori state estimate and state error covariance:

$$\hat{x}_k^- = \begin{cases} f(\hat{x}_{k-1}, u_{k-1}, k-1) & \text{(nonlinear a priori estimate)} \\ A_{k-1}\hat{x}_{k-1} + B_{k-1}u_{k-1} & \text{(linearized a priori state estimate)} \end{cases}$$

With knowledge of the process error, the a priori variance of the prediction can be computed:

$$P(k+1|k) = \nabla_p f \cdot P(k|k) \cdot \nabla_p f^T + \nabla_u f P(k) \cdot \nabla_u f^T$$

In the linear case, this becomes:

$$P_k^- = A_{k-1}P_{k-1}A_{k-1}^T + Q_{k-1} \quad \text{(a priori error covariance)}$$

The belief is represented by two parameters

$$\hat{x}(k+1|k) \quad \text{and} \quad P(k+1|k)$$

defining the Gaussian distribution. Then a measurement update is performed to generate a corrected a posteriori estimate and state error covariance:

$$K_k = P_k^- C_k^T (C_k P_k^- C_k^T + R_k)^{-1} \quad \text{(observer gain)}$$

$$\hat{x}_k = \hat{x}_k^- + K_k(y_k - g(\hat{x}_k^-, k)) \quad \text{(nonlinear a posteriori estimate)}$$

$$\hat{x}_k = \hat{x}_k^- + K_k(y_k - C_k\hat{x}_k^-) \quad \text{(linearized a posteriori estimate)}$$

$$P_k = (I - K_k C_k)P_k^- \quad \text{(a posteriori error covariance)}$$

The second step uses prediction $\hat{x}_k^-$ and measurement z_k to update the state estimate by means of posterior error covariance. Actual sensory measurements $z(k+1)$ are determined during the observation step. The robot position $\hat{x}(k+1|k)$ and a map are used to generate measurement predictions z_t:

$$\hat{z}(k+1) = h(z_k, \hat{x}(k+1|k)) \quad \text{or} \quad z_k = h_{k-1}(x_{k-1}) + v_{k-1}$$

where $h =$ coordinate transform between the world frame and the sensor frame. During matching, the best observations are matched with measurement predictions in the map. Matching is achieved through the innovation which is a measure of the difference between predicted and observed measurements:

$$v_{ij}(k+1) = z_j(k+1) - \hat{z}_i(k+1) = z_j(k+1) - h_i(z_t, \hat{p}(k+1|k))$$

Innovation covariance $P_{\text{in},ij}(k+1)$ is found by propagating the error:

$$P_{\text{in},ij}(k+1) = \nabla h_j \cdot P_p(k+1|k) \cdot \nabla h_i^T + P_{R,j}(k+1)$$

where $P_{R,ij}(k+1) =$ noise covariance of measurement. Validity of the corresponding match is given by the Mahalanobis distance:

$$v_{ij}^T(k+1) \cdot P_{\text{in},ij}^{-1}(k+1) \cdot v_{ij}(k+1) \leq d^2$$

where $d = \sqrt{v_k^T P_k^{-1} v_k}$. The best estimate is the best prediction before measurement plus a weighting (Kalman gain) times the difference between measurement and the best prediction. The Kalman gain minimizes the mean-squared error:

$$K(k+1) = P_p(k+1|k) \cdot \nabla h^T \cdot P_{in}^{-1}(k+1)$$

where ∇h = measurement Jacobian. In the linear case:

$$K_k = P_k^- H^T (H P_k^- H^T + Q)^{-1}$$

The state estimate may be updated:

$$\hat{x}(k+1|k+1) = \hat{x}(k+1|k) + K(k+1) \cdot v(k+1)$$

In the linear case:

$$\hat{x}_k = \hat{x}_k^- + K_k(z_k - H\hat{x}_k^-)$$

This estimate has error covariance given by:

$$P_p(k+1|k+1) = P_p(k+1|k) - K(k+1) \cdot P_{in}(k+1) \cdot K^T(k+1)$$

In the linear case:

$$P_k = (I - K_k H)P_k^-$$

The extended Kalman filter (EKF) linearizes the state equations about the current state estimate $x_{k-1} = \bar{x}_{k-1}^+$ using a Taylor expansion of the nonlinear state equation and then applies the standard Kalman filter. Superscripts "+" and "−" refer to estimates before and after measurement. The current state is estimated by projecting the previous state using the current input:

$$x_k = f_{k-1}(\bar{x}_{k-1}^+, u_{k-1}) + \frac{\partial f_{k-1}}{\partial x}(x_{k-1} - \bar{x}_{k-1}^+) + \frac{\partial f_{k-1}}{\partial w} w_{k-1}$$

$$= f_{k-1}(\bar{x}_{k-1}^+, u_{k-1}) + A_{k-1}(x_{k-1} - \bar{x}_{k-1}^+) + L_{k-1} w_{k-1}$$

$$z_k = h_k(\bar{x}_k^-) + \frac{\partial h_k}{\partial x}(x_k - \bar{x}_k^-) + \frac{\partial h_k}{\partial v} v_k$$

$$= h_k(\bar{x}_k^-) + H_k(x_k - \bar{x}_k^-) + M_k v_k$$

The covariance matrix P_k is bounded from below by the inverse of Fisher's information matrix J_k:

$$P_k = \langle (\bar{x}_k - x_k)(\bar{x}_k - x_k)^T \rangle \geq J_k^{-1}$$

where

$$J_k = \left\langle \left(\frac{\partial \ln p(z_k|x_k)}{\partial x_k} \right)^T \left(\frac{\partial \ln p(z_k|x_k)}{\partial x_k} \right) \Big| x_k \right\rangle = \text{Fisher information matrix}$$

and where $p(z_k|x_k)$ = joint conditional probability density function of z_k given x_k. For Gaussian assumption, this reduces to:

$$J_{k+1} = F_k^{-1T} J_k F_k^{-1} + H_k^T R_k^{-1} H_k \quad \text{where } F_k = \frac{\partial f(x)}{\partial x} \text{ and } H_k = \frac{\partial h(x)}{\partial x}$$

The covariance of the estimate is computed using the Jacobian to form the a priori estimate. The a priori estimate comprises the prediction phase from the system Jacobians:

$$P_k^- = F_{k-1}P_{k-1}^+ F_{k-1}^T + L_{k-1}Q_{k-1}L_{k-1}^T$$

where $F_{k-1} = \delta f/\delta x \cdot L_{k-1} = \delta f/\delta w$ and $Q_k = \langle w_k w_k^T \rangle$. The state and its covariance are propagated through the state model; after measurement a correction updates the state. Kalman filter gain is computed using Jacobians of the measurement model and covariances. Gain is multiplied by the difference between actual sensor measurements and predicted measurements based on the a priori estimate of the state (i.e., the Kalman gain weights measurements against the state model). A posteriori estimates are computed from a priori estimates and measurement data, the proportional weighting determined by the Kalman gain. This forms the posterior estimate. The a posteriori estimate comprises the measurement correction:

$$K_k = P_k^- H_k^T (H_k P_k^- H_k^T + M_k R_k M_k^T)^{-1} \quad \text{where } H_k = \frac{\partial h}{\partial k}, M_k = \frac{\partial h}{\partial v}, R_k = \langle v_k v_k^T \rangle$$

$$\hat{x}_k^+ = \hat{x}_k^- + K_k(z_k - H_k \hat{x}_k^-)$$

$$P_k^+ = (I - K_k H_k)P_k^-$$

The initial state covariance matrix is given by:

$$P_0^+ = \text{diag}(\sigma_i^2) = \text{diag}\langle (x_0 - \bar{x}_0^+)(x_0 - \bar{x}_0^+)^T \rangle$$

in the state space of dimension n. It may initially be set to large numbers to indicate high uncertainty. Q and R representing process and measurement noise are similar:

$$Q_k = \text{diag}(\sigma_i^2) \quad \text{and} \quad R_k = \text{diag}(\sigma_i^2)$$

where σ_i = noise variance. The UKF is applicable to non-Gaussian systems for which the sigma points representing the distribution are chosen algorithmically:

$$\chi_0(k|k) = \hat{x}(k|k) \quad \text{and} \quad W_0 = \frac{\kappa}{n+\kappa}$$

$$\chi_i(k|k) = \hat{x}(k|k) + (\sqrt{(n+\kappa)P(k|k)})_i \quad \text{and} \quad W_i = \frac{1}{2(n+\kappa)}$$

$$\chi_{i+n}(k|k) = \hat{x}(k|k) - (\sqrt{(n+\kappa)P(k|k)})_i \quad \text{and} \quad W_{i+n} = \frac{1}{2(n+\kappa)}$$

where $(n+\kappa) \neq 0$, $n+\kappa = k$, and $\sum_{i=0}^{P} W_i = 1$. Factor κ fine-tunes higher order moments of the approximation to reduce errors in those terms. Fourth and even higher order moments scale with $(n+\kappa)$ where the fourth-order moment defines kurtosis. The extended Kalman filter ignores these higher order terms. For Gaussian distributions, $n+\kappa = 3$. The process model is given by:

$$\chi_i(k+1|k) = f(\chi_i(k|k), u(k))$$

Predicted mean is given by:

$$\hat{x}(k+1|k) = \sum_{i=0}^{2n} W_i \chi_i(k+1|k)$$

Predicted covariance is given by:

$$P(k+1|k) = \sum_{i=0}^{2n} W_i [\chi_i(k+1|k) - \hat{x}(k+1|k)][\chi_i(k+1|k) - \hat{x}(k+1|k)]$$

The information filter is the algebraic equivalent form of the Kalman filter which explicitly expresses the effect of sensory measurements on the state estimate. It involves a canonical parameterization in which the inverse covariance defines the information matrix from which there are analog versions such as the extended information filter. Inverse covariance may be regarded as Fisher information— information at time $k+1$ is information at time k plus new information at time $k+1$ from z_{k+1}. The information filter replaces the states and covariance of the Kalman filter with information states and an information matrix (inverse of covariance) given by:

$$Y_{k|k-1} = P_{k|k-1}^{-1}$$

$$\hat{y}_{k|k-1} = Y_{k|k-1} \hat{x}_{k|k-1}$$

$$i_k = C_k^T R_k^{-1} y_k$$

$$I_k = C_k^T R_k^{-1} C_k$$

The information filter involves first an information measurement update:

$$\hat{Y}_k^+ = \hat{Y}_k^- + i_k$$

$$Y_k^+ = Y_k^- + I_k$$

This is followed by an information time update:

$$M_k = (A_k^{-1})^T Y_k^+ A_k^{-1}$$

$$Y_{k+1}^- = [I - M_k(M_k + Q_k^{-1})^{-1}]M_k$$

$$\hat{Y}_{k+1}^- = [I - M_k(M_k + Q_k^{-1})^{-1}](A_k^{-1})^T \hat{Y}_k^+$$

For instance, the information form of the square root cubature Kalman filter offers a simpler measurement update stage for multisensory fusion [1551]. We introduce random noise to the plant and measurement so the estimated state is computed as the average state

$$\hat{x}_k = \langle x_k \rangle$$

and the variance as

$$\sigma_k^2 = \langle (x_k - \hat{x}_k)^2 \rangle$$

So,

$$x_{k+1} = Ax_k + Bu_k + w_k$$

$$z_k = Hx_k + v_k$$

where w_k = process noise and v_k = measurement noise The Bayesian estimator attempts to find the conditional probability distribution of the state based on measurements $p(x_k|z_k)$. The a priori estimate is the PDF of x_k given all measurements prior to time k, $p(x_k|z_{k-1})$. Bayes filter allows estimation of a dynamic system's state from noisy observations. Uncertainty is defined by the probability distribution. Given new information from sensor measurements, the product rule gives the conditional probability of state x given measurements z:

$$p(x|z)p(z) = p(z|x)p(x)$$

The Bayes rule then follows to compute the posterior probability:

$$p(x|z) = p(z|x)\frac{p(x)}{p(z)} \quad \text{or} \quad p(z|x) = p(x|z)\frac{p(z)}{p(x)}$$

The assumption is that:

$$p(x_t|z_t) = \int p(x_t|x_{t-1}, z_t)p(x_{t-1}) \cdot dx_{t-1}$$

From this, the generalized Bayes filter is given by:

$$p(x_t|z_t) = \alpha_t p(z_t|x_t)p(x_t|z_{t-1})$$

where α_t = normalizing constant and

$$p(x_t|z_{t-1}) = \int p(x_t|x_{t-1})px_{t-1}|z_{t-1}) \cdot dx_{t-1} = \text{prior distribution}$$

$$p(x_{t-1}|z_{t-1}) = \frac{p(z_{t-1}|x_{t-1})p(x_{t-1})}{p(z_{t-1})} = \text{posterior distribution}$$

$p(z_t|x_t)$ = observation model

$p(x_t|x_{t-1})$ = state transition probability of first-order Markov process (system model)

This is the basis of the Markov assumption that the output is a function only of the previous state and the most recent measurements. The state model predicts the output of the system with the effects of noise (nominally Gaussian) accounted for. Bayesian estimation is a general approach to state estimation where the estimate is represented by a probability distribution function based on measurement data. The Bayes filter uses a prediction/update cycle to estimate the state of the system—sensory measurements update these hypotheses about the state of a dynamic system. The Bayes filter cannot in general be evaluated in closed form, so an approximation may be used such as the Kalman filter or particle filter—the current state of the robot is regarded as a probabilistic function of its previous state and the input.

References

[1] Space Exploration Engineering Group (SEEG), Department of Mechanical & Aerospace Engineering, Carleton University, Ottawa, Ontario, K1S 5B6, Canada.

[2] National Research Council (Commission on Physical Sciences, Mathematics, and Applications) (1999) *Scientific Rationale for Mobility in Planetary Environments*, National Academy Press, Washington, D.C.

[3] Vila, C., Savolainen, P., Maldonado, J., Amorim, I., Rice, J., Honeycutt, R., Crandall, K., Lundenberg, J., Wayne, R. (1997) "Multiple and ancient origins of the domestic dog," *Science*, **276**, 1687–1689.

[4] Brooks, R., and Flynn, A. (1989) "Fast, cheap and out of control: A robot invasion of the solar system," *J. British Interplanetary Society*, **42**, 478–485.

[5] Angle, C., and Brooks, R. (1990) "Small planetary rovers," *IEEE International Workshop on Intelligent Robots and Systems*.

[6] Ellery, A. (2004) "Space robotics part 3: Robotic rovers for planetary exploration," *Int. J. Advanced Robotic Systems*, **1**(4), 303–307.

[7] Mondier, J (1993) "Planetary rover locomotion mechanisms," *Proceedings of the International Symposium on Missions, Technologies and Design of Planetary Mobile Vehicles, Toulouse, September 1992*, CNES/Cipaduhs-Iditions, Toulouse, France (ISBN 2854283317).

[8] Chicarro, A. et al. (1998) "Scientific applications of robotic systems on planetary missions," *Robotics and Autonomous Systems*, **23**, 65–71.

[9] Zakrajsek, J., McKissock, D., Woytach, J., Zakrajsek, J., Oswald, F., McEntire, K., Hill, G., Abel, P., Eichenberg, D., and Goodnight, T. (2005) *Exploration Rover Concepts and Development Challenges* (AIAA 2005-2525, NASA TM-2005-213555), NASA, Washington, D.C.

[10] Wilcox, B., Litwin, T., Biesiadecki, J., Matthews, J., Heverly, M., and Morrison, J. (2007) "Athlete: A cargo-handling and manipulation robot for the Moon," *J. Field Robotics*, **24**(5), 421–434.

[11] Ellery, A. and Cockell, C. (2006) "Bio-inspired microrobots for support of exploration from a Mars polar station," in C. Cockell (Ed.), *Project Boreas: A Station for the Martian Geographic Pole*, British Interplanetary Society, London.

© Springer-Verlag Berlin Heidelberg 2016
A. Ellery, *Planetary Rovers*, Springer Praxis Books,
DOI 10.1007/978-3-642-03259-2

[12] Ellery, A., and the Rover Team (2006) "ExoMars rover and Pasteur payload Phase A study: An approach to experimental astrobiology," *International Journal of Astrobiology*, **5**(3), 221–241.

[13] Vago, J., Witasse, O., Baglioni, P., Haldemann, A., Gianfiglio, G., Blancquaert, T., McCoy, D., de Groot, R. and the ExoMars Team (2013) "ESA's next step in Mars exploration," *ESA Bulletin*, **155**, August, 13–21.

[14] Teshigahara, A., Watanabe, M., Kawahara, N., Ohtsuka, Y., and Hattori, T. (1995) "Performance of a 7 mm microfabricated car," *IEEE J. Microelectromechanical Systems*, **4**(2), 76–80.

[15] Fontaine, B. et al. (2000) "Autonomous operations of a micro-rover for geoscience on Mars," *Proceedings Sixth ESA Workshop on Advanced Space Technologies for Robotics & Automation Conference (ASTRA), December 2000*, ESTEC, Noordwijk, The Netherlands.

[16] Wilcox, B., and Jones, R. (2000) "MUSES-CN nanorover mission and related technology," *Proceedings of IEEE International Conference Robotics & Automation*, pp. 287–295.

[17] Newell, M., Stern, R., Hykes, D., Bolotin, G., Gregoire, T., McCarthy, T., Buchanan, C., and Cozy, S. (2001) "Extreme temperature (−170°C to +125°C) electronics for nanorover operation," *Proceedings IEEE Aerospace Conference*, Vol. 5, pp. 2443–2456.

[18] Mariorodriga, G. (1997) "Micro-rovers for scientific applications on Mars or Moon missions," *Preparing for the Future*, **7**(2), 16–17.

[19] Matijevic, J. (1997) "Sojourner: The Mars Pathfinder microrover flight experiment," *Space Technology*, **17**(3/4), 143–149.

[20] Shirley, D., and Matijevic, J. (1995) "Mars Pathfinder microrover," *Autonomous Robots*, **2**, 283–289.

[21] Matijevic, J. (1998) *Mars Pathfinder Microrover: Implementing a Low Cost Planetary Mission Experiment* (IAA-L-0510), International Academy of Astronautics, Johns Hopkins APL, Laurel, MD.

[22] Rover Team (1997) "The Pathfinder microrover," *Journal of Geophysical Research*, **102**(E2), 3989–4001.

[23] Qadi, A., Cross, M., Setterfield, T., Ellery, A., and Nicol, C. (2012) "Kapvik rover—systems and control: A micro-rover for planetary exploration," *Proceedings CASI ASTRO Conference, Quebec City, Canada* (Paper No. 84).

[24] Setterfield, T., Frazier, C., and Ellery, A. (2014) "Mechanical design and testing of an instrumented rocker–bogie system for the Kapvik micro-rover," *J. British Interplanetary Society*, **67**, 96–104.

[25] Marth, P. (2003) "TIMED integrated electronics module (IEM)," *Johns Hopkins APL Technical Digest*, **24**(2), 194–200.

[26] Ellery, A., Welch, C., Curley, A., Wynn-Williams, D., Dickensheets, D., and Edwards, H. (2002) "Design options for a new European astrobiology-focussed Mars mission—Vanguard," *Proceedings World Space Congress, Houston*, IAC-02-Q.3.2.04.

[27] Ellery, A., Richter, L., Parnell, J., and Baker, A. (2003) "A low-cost approach to the exploration of Mars through a robotic technology demonstrator mission," *Proceedings Fifth IAA International Conference on Low-Cost Planetary Missions* (ESA SP 542), ESTEC, Noordwijk, The Netherlands, pp. 127–134.

[28] Ellery, A., Cockell, C., Edwards, H., Dickensheets, D., and Welch, C. (2002) "Vanguard: A proposed European astrobiology experiment on Mars," *Int. J. Astrobiology*, **1**(3), 191–199.

[29] Ellery, A., Richter, L., Parnell, J., and Baker, A. (2006) "Low-cost approach to the exploration of Mars through a robotic technology demonstrator mission," *Acta Astronautica*, **59**(8/11), 742–749.

[30] Ellery, A., Ball, A., Coste, P., Dickensheets, D., Hu, H., Lorenz, R., Nehmzow, H., McKee, G., Richter, L., and Winfield, A. (2003) "Robotic triad for Mars surface and sub-surface exploration," *Proceedings Seventh International Symposium Artificial Intelligence Robotics & Automation in Space, Nara, Japan*.

[31] Apostopoulos, D., and Bares, J. (1995) "Locomotion configuration of a robust rapelling robot," *Proceedings IEEE International Conference Intelligent Robots & Systems, August 1995, Pittsburgh, PA*.

[32] Bares, J., and Wethergreen, D. (1999) "Dante II: Technical description, results and lessons learned," *Int. J. Robotics Research*, **18**(7), 621–649.

[33] Krishna, K. et al. (1997) "Tethering system design for Dante II," *Proceedings IEEE International Conference Robotics and Automation*, pp. 1100–1105.

[34] Northcote Parkinson, C. (1957) *Parkinson's Law and Other Studies in Administration*, Ballantine Books, New York.

[35] Causon, J (2007) "Six steps to efficient management," *Engineering Management J.*, **17**(5), 8–9.

[36] Collins, D. (1997) "Ethical superiority and inevitability of participatory management as an organisational system," *Organization Science*, **8**(5), 489–507.

[37] Beer, R. (1995) "A dynamical systems perspective on agent–environment interaction," *Artificial Intelligence*, **72**(1995), 173–215.

[38] Clark, A. (1999) "An embodied cognitive science?" *Trends in Cognitive Science*, **3**(9), 345–351.

[39] Muir, P., and Neumann, C. (1987) "Kinematic modeling of wheeled mobile robots," *J. Robotic Systems*, **4**(2), 281-340.

[40] Ellery, A. (2004) "Space robotics part 3: Robotic rovers for planetary exploration," *International J. Advanced Robotic Systems*, **1**(4), 303–307.

[41] Lessem, A. et al. (1996) "Stochastic vehicle mobility forecast using the NATO Reference Mobility Model," *J. Terramechanics*, **33**(6), 273–280.

[42] Ellery, A. (2015) *Space Technology for Astrobiology Missions*, Cambridge University Press, U.K. (in press).

[43] Novara, M., Putz, P., Marechal, L., and Losito, S. (1998) "Robotics for lunar surface exploration," *Robotics and Autonomous Systems*, **23**, 53–63.

[44] Greenwood, J., Itoh, S., Sakamoto, N., Warren, P., Taylor, L., and Yurimoto, H. (2011) "Hydrogen isotope ratios in lunar rocks indicate delivery of cometary water to the Moon," *Nature Geoscience*, **4**, 79–82.

[45] Starukhina, L. (2006) "Polar regions of the Moon as a potential repository of solar-wind-implanted gases," *Advances in Space Research*, **37**, 50–58.

[46] Boucher, D., Edwards, E., and Sanders, J. (2005) "Fully autonomous mining system controller for in-situ resource utilisation," *Proceedings Eighth International Symposium AI Robotics and Automation in Space, Munich, Germany* (ESA SP-603).

[47] Crawford, I. (2004) "Scientific case for renewed human activities on the Moon," *Space Policy*, **20**, 91–97.

[48] Bussey, D., McGovern, J., Spudis, P., Neish, C., Noda, H., Ishihara, Y., and Sorensen, S-A. (2010) "Illumination conditions of the south pole of the Moon derived using Kaguya topography," *Icarus*, **208**, 558–564.

[49] George, J., Mattes, G., Rogers, K., Magruder, D., Paz, A., Vaccaeo, H., Baird, R., Sanders, G., Smith, J., Quinn, J. et al. (2012) "RESOLVE mission architecture for

lunar resource prospecting and utilisation," *43rd Lunar and Planetary Science Conference Abstracts*, No. 2583.

[50] Horanyi, M., Walch, B., Robertson, S., and Alexander, D. (1998) "Electrostatic charging properties of Apollo 17 lunar dust," *Journal of Geophysical Research*, **103**(E4), 8575–8580.

[51] Liu, Y., and Taylor, L. (2008) "Lunar dust: Chemistry and physical properties and implications for toxicity," *NLSI Lunar Science Conference 2012* (abstract).

[52] Kruzeklecky, R., Wong, B., Aissa, B., Haddad, E., Jamroz, W., Cloutis, E., Rosca, I., Hoa, S., Therriault, D., and Ellery, A. (2010) *MoonDust Lunar Dust Simulation and Mitigation* (AIAA-2010-764033). American Institute of Aeronautics and Astronautics, Washington, D.C.

[53] Eimer, B., and Taylor, L. (2007) "Dust mitigation: Lunar air filtration with a permanent magnet system," *38th Lunar and Planetary Science Conference, League City, Texas*, No. 1338, 1654.

[54] Landis, G., and Jenkins, P. (2002) "Dust mitigation for Mars solar arrays," *Proceedings IEEE Conference Photovoltaic Specialists*, pp. 812–815.

[55] Calle, C., McFall, J., Buhler, C., Snyder, S., Ritz, M., Trigwell, S., Chen, A., and Hogue, M. (2008) "Development of an active dust mitigation technology for lunar exploration," *AIAA Space Conference and Exposition, San Diego* (AIAA 2008-7894), American Institute of Aeronautics and Astronautics, Washington, D.C..

[56] Clark, P., Curtis, S., Minetto, F., and Keller, J. (2007) "Finding a dust mitigation strategy that works on the lunar surface," *38th Lunar and Planetary Science Conference, League City, Texas*, No. 1338, 1175.

[57] Lauer, H., and Allton, J. (1992) "Mars containers: Dust on Teflon sealing surfaces," *Space International Conference Space III: Engineering, Construction and Operations in Space*, Vol. 1, pp. 508–517.

[58] Cong, M., and Shi, H. (2008) "Study of magnetic fluid rotary seals for wafer handling robots," *IEEE International Conference Mechatronics and Machine Vision in Practice*, pp. 269–273.

[59] Komatsu, G., and Baker, V. (1996) "Channels in the solar system," *Planetary and Space Science*, **44**(8), 801–815.

[60] Squyres, S. (1989) "Water on Mars," *Icarus*, **79**, 229–288.

[61] Squyres, S. (1984) "History of water on Mars," *Annual Review Earth and Planetary Science*, **12**, 83–106.

[62] Malin, M., and Carr, M. (1999) "Groundwater formation of Martian valleys," *Nature*, **397**, 589–591.

[63] Segura, T., Toon, O., Colaprete, A., and Zahnle, K. (2002) "Environmental effects of large impacts on Mars," *Science*, **298**, 1977–1980.

[64] Clifford, S. (1993) "Model for the hydrological and climatic behaviour of water on Mars," *J. Geophysical Research*, **98**(E6), 10973–11016.

[65] Goldspiel, J., and Squyres, S. (1991) "Ancient aqueous sedimentation on Mars," *Icarus*, **89**, 392–410.

[66] Forget, F., and Pierrehumbert, R. (1997) "Warming early Mars with carbon dioxide clouds that scatter infrared radiation," *Science*, **278**, 1273–1276.

[67] Carr, M. (1986) "Mars: A water-rich planet," *Icarus*, **68**, 187–216.

[68] Haberle, R. (1998) "Early Mars climate models," *J. Geophysical Research*, **103**(E12), 28467–28479.

[69] Pollack, J., Kasting, J., Richardson, S., and Poliakoff, K. (1987) "Case for a wet, warm climate on early Mars," *Icarus*, **71**, 203–224.

[70] Gulick, V., and Baker, V. (1989) "Fluvial valleys and martian palaeoclimates," *Nature*, **341**, 514–516.

[71] Head III, J., Hiesinger, H., Ivanov, M., Kreslavsky. M., Pratt, S., Thomson, B. (1999) "Possible ancient oceans on Mars: Evidence from Mars Orbiter laser altimeter data," *Science*, **286**, 2134–2137.

[72] Marquez, A., Fernandez, C., Anguita, F., Farelo, A., Anguita, J., and de la Casa, M.-A. (2004) "New evidence for a volcanically, tectonically, and climatically active Mars," *Icarus*, **172**, 573–581.

[73] Malin, M., and Edgett, K. (2000) "Evidence for recent groundwater seepage and surface runoff on Mars," *Science*, **288**, 2330–2335.

[74] Mangold, N., Allemand, P., Duval, P., Geraud, P., and Thomas, P. (2002) "Experimental and theoretical deformation of ice-rock mixtures: Implications on rheology and ice content of Martian permafrost," *Planetary and Space Science*, **50**, 385–401.

[75] Laskar, J., and Robutel, P. (1993) "Chaotic obliquity of the planets," *Nature*, **361**, 608–612.

[76] Touma, J., and Wisdom, J. (1993) "Chaotic obliquity of Mars," *Science*, **259**, 1294–1297.

[77] Ward, W., and Rudy, D. (1991) "Resonant obliquity of Mars," *Icarus*, **94**, 160–164.

[78] Jakosky, B., Henderson, B., and Mellon, M. (1995) "Chaotic obliquity and the nature of the Martian climate," *J. Geophysical Research*, **100**(E1), 1579–1584.

[79] Smith, H., and McKay, C. (2005) "Drilling in ancient permafrost on Mars for evidence of a second genesis of life," *Planetary and Space Science*, **53**, 1302–1308.

[80] Kargel, J., and Strom, R. (1992) "Ancient glaciation on Mars," *Geology*, **20**, 3–7.

[81] Melosh, H., and Vickery, A. (1989) "Impact erosion of the primordial atmosphere of Mars," *Nature*, **338**, 487–489.

[82] Jakosky, B., Pepin, R., Johnson, R., and Fox, J. (1994) "Mars atmospheric loss and isotopic fractionation by solar wind-induced sputtering and photochemical escape," *Icarus*, **111**, 271–288.

[83] Jakosky, B., and Jones, J. (1997) "History of Martian volatiles," *Reviews of Geophysics*, **35**(1), 1–16.

[84] Hollingsworth, J., Haberle, R., and Schaeffer, J. (1997) "Seasonal variations of storm zones on Mars," *Advances in Space Research*, **19**(8), 1237–1240.

[85] Chevrier, V., and Mathe, P. (2007) "Mineralogy and evolution of the surface of Mars: A review," *Planetary and Space Science*, **55**, 289–314.

[86] Bridges, J., and Grady, M. (2000) "Evaporite mineral assemblages in the nakhlite (martian) meteorites," *Earth and Planetary Science Letters*, **176**, 267–279.

[87] Halevy, I., Zuber, M., and Schrag, D. (2007) "Sulphur dioxide climate feedback on early Mars," *Science*, **318**, 1903–1907.

[88] Fairen, A., Fernandez-Remolar, D., Dohm, J., Baker, V., and Amils, R. (2004) "Inhibition of carbonate synthesis in acidic oceans on early Mars," *Nature*, **431**, 423–426.

[89] Hurowitz, J., McLennan, S., Tosca, N., Arvidson, R., Michalski, J., Ming, D., Schroder, C., and Squyres S. (2006) "In situ and experimental evidence for acidic weathering of rocks and soils on Mars," *J. Geophysical Research*, **111E**, E02S19.

[90] Knoll, A., and Grotzinger, J. (2006) "Water on Mars and the prospect of Martian life," *Elements*, **2**, June, 169–173.

[91] Parnell, J., Cockell, C., Edwards, H., and Ellery, A. (2003) "The range of life habitats in volcanic terrains on Mars," *Proceedings Third European Workshop on Exo/Astro-Biology* (SP-545), Madrid, Spain, pp. 81–84.

[92] Cabrol, N., and Grin, E. (2001) "Evolution of lacustrine environments on Mars: Is Mars only hydrologically dormant?" *Icarus*, **149**, 291–328.

[93] Ori, G., Marinangeli, L., and Komatsu, G. (2000) "Martian palaeolacustrine environments and their geological constraints on drilling operations for exobiological research," *Planetary and Space Science*, **48**, 1027–1034.

[94] Mancinelli, R., Fahlen, T., Landheim, R., and Klovstad, M, (2004) "Brines and evaporates: analogues for Martian life," *Advances in Space Research*, **33**, 1244–1246.

[95] Kovtunenko, V. et al. (1993) "Prospects for using mobile vehicles in missions to Mars and other planets," *Proceedings International Symposium on Missions, Technologies and Design of Planetary Mobile Vehicles, September 1992*, CNES/Cipaduhs Iditions, Toulouse, France.

[96] Burkhalter, B., and Sharpe, H. (1995) "Lunar roving vehicle: Historical origins, development and deployment," *J. British Interplanetary Society*, **48**, 199–212.

[97] Cowart, E. (1973) "Lunar roving vehicle: Spacecraft on wheels," *Proc. Institution Mechanical Engineers*, **187**(45/73), 463–491.

[98] Bekker, M. (1985) "The development of a moon rover," *J. British Interplanetary Society*, **38**, 537–543.

[99] Matijevic, J. (1997) "Sojourner: the Mars Pathfinder microrover flight experiment," *Space Technology*, **17**(3/4), 143–149.

[100] Shirley, D., and Matijevic, J. (1995) "Mars Pathfinder microrover," *Autonomous Robots*, **2**, 283–289.

[101] Matijevic, J. (1998) *Mars Pathfinder Microrover: Implementing a Low Cost Planetary Mission Experiment* (IAA-L-0510), International Academy of Astronautics, Stockholm, Sweden.

[102] Rover Team (1997) "The Pathfinder microrover," *Journal of Geophysical Research*, **102**(E2), 3989–4001.

[103] Mishkin, A., Morrison, J., Nguyen, T., Stone, H., Cooper, B., and Wilcox, B. (1998) "Experiences with operations and autonomy of the Mars Pathfinder microrover," *Proceedings International Conference Robotics and Automation, Leuven, Belgium*.

[104] Arvidson, R., Baumgartner, E., Schenker, P., and Squyres, S. (2000) "FIDO field trials in preparation for Mars rover exploration and discovery and sample return missions" (Abstract #6018), *Workshop on Concepts and Approaches for Mars Exploration, Johnson Space Center, Houston, TX*.

[105] Weisbin, C., Rodriguez, G., Schenker, P., Das, H., Hamayati, S., Baumgarter, E., Maimone, M., Nesnas, I., and Volpe, R. (1999) "Autonomous rover technology for Mars sample return," *Proceedings Fifth International Symposium: Artificial Intelligence, Robotics and Automation in Space* (ESA SP-440), pp. 1–8.

[106] Crisp, J., Adler, M., Matijevic, J., Squyres, S., Arvidson, R., and Kass, D. (2003) "Mars exploration rover mission," *Journal of Geophysical Research*, **108**(E12), 2.1–2.17.

[107] Erickson, J., Adler, M., Crisp, J., Mishkin, A., and Welch, R. (2002) "Mars exploration rover surface operations," *World Space Congress, Houston* (IAC-02-Q.3.1.03).

[108] Clarke, A. (1997) "Dynamical challenge," *Cognitive Science*, **21**(4), 461–481.

[109] Bogatchev, A. et al, (2000a) "Walking and wheel-walking robots," *CLAWAR 2000: Third International Conference, on Climbing and Walking Robots, Madrid, Spain October 2–4, 2000.*

[110] Bogatchev, A. et al. (2000b) "Mobile robots of high cross-country ability," *IFAC 2000 Conference.*

[111] Thianwoon, M. et al. (2001) "Rocker-bogie suspension performance," *Proceedings 12th Pacific Conference Automotive Engineering, Shanghai, November 6–9, 2001,* IPC2001D079.

[112] Mondier, J. (1993) "Planetary rover locomotion mechanisms," *Proceedings International Symposium Missions, Technologies and Design of Planetary Mobile Vehicles, September 1992,* CNES/Cipaduhs-Iditions, Toulouse, France.

[113] Wright, D., and Watson, R. (1987) "Comparison of mobility system concepts for a Mars rover," *Procedings SPIE Conference Mobile Robots II,* Vol. 852, pp. 180–186.

[114] Thakoor, S. (2000) "Bio-inspired engineering of exploration systems," *Proceedings NASA/DoD Second Biomorphic Explorers Workshop,* pp. 49–79.

[115] Colombano, S., Kirchner, F., Spenneberg, D., and Hanratty, J. (2003) *Exploration of Planetary Terrains with a Legged Robot as a Scout Adjunct to a Rover,* American Institute of Aeronautics and Astronautics, Washington, D.C.

[116] Genta, G., and Amati, N. (2004) "Mobility on planetary surfaces: May walking machines be a viable alternative?" *Planetary and Space Science,* **52,** 31–40.

[117] Santovincenzo, A. et al. (2002) *ExoMars 09* (CDF Study Report CDF-14A), ESA/ ESTEC, Noordwijk, The Netherlands.

[118] Clark, P., Curtis, S., and Rilee, M. (2011) "New paradigm for robotic rovers," *Physics Procedia,* **20,** 308–318.

[119] Hirose, S. et al, (1995) "Fundamental considerations for the design of a planetary rover," *IEEE International Conference Robotics and Automation,* pp. 1939–1944.

[120] Wright, D., and Watson, R. (1987) "Comparison of mobility system concepts for a Mars rover," *Proceedings SPIE Conference on Mobile Robots II,* Vol. 852, pp. 180–186.

[121] Ellery, A., Richter, L., and Bertrand, R. (2005) "Chassis design and performance analysis for the European ExoMars rover," *Trans. Can. Soc. Mech. Eng.,* **29**(4), 507–518.

[122] Schid, I. (1999) "Focussing terramechanics research towards tools for terrain development," *Proceedings 13th International Conference ISTVS, Munich, Germany, September 14–17,* pp. 799–808.

[123] Jain, A., Guineau, J., Lim, C., Lincoln, W., Pomerantz, M., Sohl, G., and Steele, R. (2003) "ROAMS: planetary surface rover simulation environment," *International Symposium Artificial Intelligence Robotics and Automation in Space, Nara, Japan.*

[124] Yen, J., Jain, A., and Balaram, J. (1999) "ROAMS: Rover analysis, modelling and simulation software," *Proceedings Fifth International Symposium on Artificial Intelligence and Automation in Space, Noordwijk, The Netherlands, June 1–3, 1999,* pp. 249–254.

[125] Bauer, R., Leung, W., and Barfoot, T. (2005) "Development of a dynamic simulation tool for the ExoMars rover," *Proceedings Eighth International Symposium on Artificial Intelligence, Robotics and Automation in Space (i-SAIRAS), Munich, Germany,* ESA SP-603.

[126] Patel, N., Ellery, A., Allouis, E,, Sweeting. M,, and Richter L (2004) "Rover mobility performance evaluation tool (RMPET): A systematic tool for rover chassis evaluation via application of Bekker theory," *Proceedings Eighth Advanced Space Technologies for*

Robotics and Automation (ASTRA), ESA-ESTEC, Noordwijk, The Netherlands, pp. 251–258.

[127] Apostolopoulos, D. (1996) *Systematic Configuration of Robotic Locomotion* (CMU Technical Report CMU-RI-TR-96-30), Carnegie Mellon University, Pittsburgh, PA; Apostolopoulos D (2001) "Analytical configuration of wheeled robotic locomotion," PhD thesis (CMU-RI-TR-01-08), Carnegie Mellon University, Pittsburgh, PA.

[128] Michaud, S., Richter, L., Patel, N., Thuer, T., Huelsing, T., Joudrier, L., Seigwart, S., and Ellery, A. (2006) "RCET: Rover chassis evaluation tools," *Proceedings ASTRA 2006*, ESA-ESTEC, Noordwijk, The Netherlands.

[129] Bickler, D. (1993) "New family of JPL planetary surface vehicles," *Proceedings International Symposium Missions, Technologies and Design of Planetary Mobile Vehicles, September 1992*, CNES/Cipaduhs-Iditions, Toulouse, France, pp. 301–306.

[130] Weisbin, C. (1995) "JPL space robotics: Present accomplishments and future thrusts," *ANS Sixth Topical Meeting on Robotics and Remote Systems, Monterey, CA*.

[131] Weisbin, C. (1993) "Evolving directions in NASA's planetary rover requirements and technology," *Proceedings International Symposium Missions, Technologies and Design of Planetary Mobile Vehicles, September 1992*, CNES/Cipaduhs-Iditions, Toulouse, France.

[132] McTamaney, L. et al. (1988) "Mars rover concept development," *Proceedings SPIE Conference Mobile Robots III*, Vol. 1007, pp. 85-94.

[133] Weisbin, C., and Rodriguez, G. (2000) "NASA robotics research for planetary exploration," *IEEE Robotics & Automation Magazine*, December, 25–34.

[134] Hacot, H. et al. (1998) "Analysis and simulation of a rocker-bogie exploration rover," *Proceedings 12th CISM-IFToMM Symposium RoMan Sy 98, Paris*.

[135] Volpe, R. et al. (1997) "Rocky 7: A next generation Mars rover prototype," *J. Advanced Robotics*, 11(4), 341–358.

[136] Miller, D. et al. (1993) "Experiments with a small behaviour-controlled planetary rover," *Proceedings International Symposium Missions, Technologies and Design of Planetary Mobile Vehicles, September 1992*, CNES/Cipaduhs-Iditions, Toulouse, France.

[137] Setterfield, T., and Ellery, A. (2010) "Potential chassis designs for Kapvik, a Canadian reconfigurable planetary microrover," *Proceedings ASTRO Conference, Toronto, Canada*, Paper 16.

[138]

[139] Setterfield, T., Frazier, C., and Ellery, A. (2014) "Mechanical design and testing of an instrumented rocker-bogie mobility system for the Kapvik micro-rover," *J. British Interplanetary Society*, **67**, 96–104.

[140] Bickler, D. (1993) "New family of JPL planetary surface vehicles," *Proceedings International Symposium Missions, Technologies and Design of Planetary Mobile Vehicles, September 1992*, CNES/Cipaduhs-Iditions, Toulouse, France, pp. 301–306.

[141] Hacot, H., Dubowsky, S., and Bidaud, P. (1998) "Analysis and simulation of a rocker-bogie exploration rover," reprint.

[142] Setterfield, T., and Ellery, A. (2012) "Terrain response estimation using an instrumented rocker-bogie mobility system," *IEEE Trans. Robotics*, **29**, 172–188.

[143] Iagnemma, K., Rzepniewski, A., Dubowsky, S., Pirjanian, P., Huntsberger, T., and Schenker, P. (2000) "Mobile robot kinematic reconfigurability for rough terrain," *Proceedings SPIE Conference Sensor Fusion and Decentralised Control in Robotic Systems III* (edited by G. McKee and P. Schenker), Vol. 4196, 413–420.

[144] Kubota, T. et al. (2003) "Small, lightweight rover 'Micro5' for lunar exploration," *Acta Astronautica*, **52**, 447–453.

[145] Siegwart, R. et al. (1998) "Design and implemention of an innovative micro-rover," *Proceedings Third ASME Conf and Exposition on Robotics in Challenging Environments (Robotics 98), Albuquerque, NM.*

[146] Estier, T. et al. (2000) "Shrimp: A rover architecture for long-range Martian missions," *Proceedings Sixth ESA Workshop on Advanced Space Technologies for Robotics and Automation (ASTRA), December 2000*, ESTEC, Noordwijk, The Netherlands.

[147] Estier, T. et al. (2000) "Shrimp: A rover architecture for long-range Martian missions," *Proceedings Sixth ESA Workshop on Advanced Space Technologies for Robotics and Automation (ASTRA 2000)*, ESTEC, Noordwijk, The Netherlands.

[148] Estier, T. et al. (2000) "An innovative space rover with extended climbing abilities," *Proceedings Fourth International Conference on Robotics for Challenging Situations and Environments, February 27–March 2, Albuquerque, NM*, pp. 333–339.

[149] Ellery, A., Richter, L., and Bertrand, R. (2005) "Chassis design and performance analysis for the European ExoMars rover," *Trans. Can. Soc. Mech. Eng.*, **29**(4), 507–518.

[150] Kucherenko, V., Gromov, V., Kazhukalo, I., Bogatchev, A., Vladykin, S., Manykjan, A. (2004) *Engineering Support on Rover Locomotion for ExoMars Rover Phase A* (ESROL-A Final Report ESTEC Contract No. 17211/03/NL/AG, ESA/ESTEC, Noordwijk, The Netherlands.

[151] Kucherenko, V., Bogatchev, A., and van Winnendael, M. (2004) "Chassis concepts for the ExoMars rover," *Proceedings Conference Advanced Space Technology for Robotics and Automation (ASTRA)*, ESTEC, Noordwijk, The Netherlands.

[152] Patel, N., Slade, R., and Clemmet, J. (2010) "ExoMars rover locomotion subsystem," *J. Terramechanics*, **47**, 227–242.

[153] Eisen, H., Buck, C., Gillis-Smith, G., and Umland, J. (1997) "Mechanical design of the Mars Pathfinder mission," *Proceedings Seventh European Space Mechanisms and Tribology Symposium* (ESA SP-410), ESTEC, Noordwijk, The Netherlands.

[154] Siegwart, R. et al. (1998) "Design and implementation of an innovative micro-rover," *Proceedings Robotics 98: Third Conference and Exposition on Robotics in Challenging Environments, Albuquerque, NM.*

[155] Lauria, M. et al. (2000) "Design and control of an innovative micro-rover," *Proceedings Fifth ESA Workshop on Advanced Space Technologies for Robotics and Automation (ASTRA)*, ESTEC, Noordwijk, The Netherlands.

[156] Richter, L., and Bernasconi, M. (2000) "Small wheeled rovers for unmanned lunar surface missions," *Proceedings Fourth International Conference on Exploration and Utilisation of the Moon (ICEUM 4)* (ESA SP-462), ESTEC, Noordwijk, The Netherlands, pp. 143–148.

[157] Richter, L., Bernasconi, M., and Coste, P. (2002) "Analysis, design, and test of wheels for a 4 kg-class mobile device for the surface of Mars," *Proceedings 14th International Conference*, International Society for Terrain–Vehicle Systems (ISTVS), Vicksburg, MS.

[158] Richter, L., Hamacher, H., Kochan, H., and Gromov, V. (1998) "The mobile subsurface sample acquisition and transport rover proposed for the Beagle 2 lander," *Proceedings Fifth ESA Workshop on Advanced Space Technologies for Robotics and Automation ASTRA 98*, ESA, Noordwijk, The Netherlands.

[159] Littman, E. et al, (1993) "Mechanical design of a planetary rover," *Proceedings International Symposium Missions, Technologies and Design of Planetary Mobile*

Vehicles, Toulouse, September 1992, CNES/Cipaduhs-Iditions, Toulouse, France (ISBN 2854283317).

[160] Chatila, R. et al. (1997) *A Case Study in Machine Intelligence: Adaptive Autonomous Space Rovers* (LAAS/CNRS Report 97463), Laboratory for Analysis and Architecture of Systems, Toulouse, France.

[161] Lacroix, S. et al. (2000) "Autonomous long range rover navigation in planetary-like environments," *Sixth ESA Workshop on Advanced Space Technologies for Robotics and Automation (ASTRA), December 2000*, ESTEC, Noordwijk, The Netherlands.

[162] Kemurdjian, A. et al, (1993) "Soviet developments of planet rovers in the period 1964–1990," *Proceedings International Symposium Missions, Technologies and Design of Planetary Mobile Vehicles, Toulouse, September 1992*, CNES/Cipaduhs-Iditions, Toulouse, France (ISBN 2854283317), pp. 25–43.

[163] Kemurdjian, A. et al. (1992) "Small Marsokhod configuration," *Proceedings IEEE International Conference Robotics and Automation, Nice, France, May*, pp. 165–168.

[164] Boissier, L. (1998) "IARES-L: A ground demonstrator of planetary rover technologies," *Robotics & Autonomous Systems*, **23**, 89–97; Boissier, L., and Maurette, M. (1997) "IARES: An onground demonstrator of planetary rover technology," *Preparing for the Future*, **7**(2), 12–13.

[165] Chatila, R. et al. (1997) *A Case Study in Machine Intelligence: Adaptive Autonomous Space Rovers* (LAAS/CNRS Report 97463), Laboratory for Analysis and Architecture of Systems, Toulouse, France.

[166] Chatila, R. et al. (1999) *Motion Control for a Planetary Rover* (LAAS/CNRS Report 99311), Laboratory for Analysis and Architecture of Systems, Toulouse, France.

[167] Eremenko, A. et al. (1993) "Rover in the Mars 96 mission," *Proceedings International Symposium Missions, Technologies and Design of Planetary Mobile Vehicles, Toulouse, September 1992*, CNES/Cipaduhs-Iditions, Toulouse, France (ISBN 2854283317).

[168] Schilling, K., and Jungius, C. (1996) "Mobile robots for planetary exploration," *Control Engineering Practice*, **4**(4), 513–524.

[169] McTamaney, L. et al. (1988) "Mars rover concept development," *Proceedings SPIE Conference Mobile Robots III*, Vol. 1007, pp. 85–94.

[170] Littman, E. et al. (1993) "Mechanical design of a planetary rover," *Proceedings International Symposium Missions, Technologies and Design of Planetary Mobile Vehicles, September 1992*, CNES/Cipaduhs-Iditions, Toulouse, France, pp. 345–359.

[171] Wettergreen, D., Bapna, D., Maimone, M., and Thomas, G. (1999) "Developing Nomad for robotic exploration of the Atacama Desert," *Robotics & Autonomous Systems*, **26**, 127–148.

[172] Shamah, B., Apostopoulos, D., Wagner, M., and Whittaker, W. (2000) "Effect of tyre design and steering mode on robotic mobility in barren terrain," *Proceedings International Conference on Field and Service Robots*, pp. 287–292.

[173] Apostopoulos, D., Wagner, M., Shamah, B., Pedersen, L., Shillcutt, K., and Whittaker, W. (2000) "Technology and field demonstration of robotic search for Antarctic meteorites," *Int. J. Robotics Research*, **19**(11), 1015–1032.

[174] Bodin, A. (1999) "Development of a tracked vehicle to study the influence of vehicle parameters or tractive performance in soft terrain," *J. Terramechanics*, **36**, 167–181.

[175] Wong, J., Garber, M., and Preston-Thomas, J. (1984) "Theoretical prediction and experimental substantiation of the ground pressure distribution and tractive performance of tracked vehicles," *Proc. Institution Mechanical Engineers D*, **198**, 265.

[176]

[177] Matthies, L. et al. (2000) "A portable, autonomous, urban reconnaissance robot," *Proceedings International Conference Intelligent Autonomous Systems.*

[178] Liu, Y., and Liu, G. (2009) "Track–stair interaction analysis and online tipover prediction for a self-reconfigurable tracked mobile robot climbing stairs," *IEEE/ASME Trans. Mechatronics*, **14**(5), 528–538.

[179] Bertrand, R. et al. (1998) "European tracked micro-rover for planetary surface exploration," *Proceedings Fifth ESA Workshop on Advanced Space Technologies for Robotics and Automation (ASTRA 98)*, ESTEC, Noordwijk, The Netherlands, p. 3.4-1.

[180] Bertrand, R., and van Winnendael, M. (2001) "Mechatronic aspects of the Nanokhod micro-rover for planetary surface exploration" reprint.

[181] Bertrand, R. et al. (2000) "Nanokhod micro-rover environmental capability requirements and design," *Proceedings Fifth ESA Workshop on Advanced Space Technologies for Robotics and Automation (ASTRA)*, ESTEC, Noordwijk, The Netherlands.

[182] Costes, N., and Trautwein, W. (1973) "Elastic loop mobility system: A new concept for planetary exploration," *J. Terramechanics*, **10**(1), 89–104.

[183] Costes, N. et al. (1973) "Terrain–vehicle dynamic interaction studies of a mobility concept (ELMS) for planetary surface exploration," *AIAA/ASME/SAE 14th Structures, Structural Dynamics and Material Conference, Williamsburg, VA, March 20–22*, Paper 73-407.

[184] Costes, N. (1998) "A mobility concept for Martian exploration," *Proceedings ASME Space Conference, Alburquerque, NM.*

[185] Ellery, A., and Patel, N. (2003) *Elastic Loop Mobility System Study for Mars Micro-rovers* (ESA-ESTEC Final Contract Report, Contract No 16221/02/NL/MV), ESA/ESTEC, Noordwijk, The Netherlands.

[186] Patel, N., Ellery, A., Welch, C., and Curley, A (2003) "Elastic loop mobility system (ELMS): Concept, innovation and performance evaluation for a robotic Mars rover," *International Astronautics Congress, Bremen, Germany* (IAC-03-IAA.1.1.05).

[187] Gee-Clough, D. (1979) "Effect of wheel width on the rolling resistance of rigid wheels in sand," *J. Terramechanics*, **15**(4), 161–184.

[188] Hetherington, J. (2005) "Tracked vehicle operations on sand: Investigations at model scale," *J. Terramechanics*, **42**, 65–70.

[189] Plackett, C. (1985) "Review of force prediction methods for off-road vehicles," *J. Agricultural Engineering Research*, **31**, 1–29.

[190] Maclaurin, B. (2007) "Comparing the NRMM (VCI), MMP and VLCI traction models," *J. Terramechanics*, **44**, 43–51.

[191] Godbole, R., Alcock, R., and Hettiaratchi, D. (1993) "Prediction of tractive performance on soil surfaces," *J. Terramechanics*, **30**(6), 443-459.

[192] Hetherington, J., and White, J. (2002) "Investigation of pressure under wheeled vehicles," *J. Terramechanics*, **39**, 85-93.

[193] Larminie, J. (1988) "Standards for the mobility requirements of military vehicles," *J. Terramechanics*, **25**(3), 171–189.

[194] Hetherington, J. (2001) "Applicability of the MMP concept in specifying off-road mobility for wheeled and tracked vehicles," *J. Terramechanics*, **38**, 63–70.

[195] Zuber, M., Smith, D., Phillips, R., Solomon, S., Banerdt, W., Neumann, G., and Aharonson, O. (1998) "Shape of the northern hemisphere of Mars from the Mars Orbiter Laser Altimeter (MOLA)," *Geophysical Research Letters*, **25**(24), 4393–4396.

[196] Parks, S., Popov, A., and Cole, D. (2004) "Influence of soil deformation on off-road heavy vehicle suspension vibration," *J. Terramechanics*, **41**, 41–68.

[197] Nikora. V., and Goring, D. (2004) "Mars topography: Bulk statistics and spectral scaling," *Chaos, Solitons & Fractals*, **19**, 427–439.

[198] Fischer, D., and Isermann, R. (2004) "Mechatronic semiactive and active vehicle suspensions," *Control Engineering Practice*, **12**, 1353–1367.

[199] Moore, H., and Jakosky, B. (1989) "Viking landing sites, remote-sensing observations, and physical properties of Martian surface materials," *Icarus*, **81**, 164–184.

[200] Moore, H. et al. (1999) "Soil-like deposits observed by Sojourner, the Pathfinder rover," *Journal of Geophysical Research*, **104**(E4), 8729–8746.

[201] Moore, H. et al, (1997) "Surface materials of the Viking landing sites," *Journal of Geophysical Research*, **82**(28), 4497–4523.

[202] Golombek, M. et al. (1997) "Overview of the Mars Pathfinder mission and assessment of landing site predictions," *Science*, **278**, 1743–1748.

[203] Golombek, M. et al. (1999) "Assessment of Mars Pathfinder landing site predictions," *Journal of Geophysical Research*, **104**(E4), 8585–8594.

[204] Golombek, M. et al. (1997) "Selection of the Mars Pathfinder landing site," *Journal of Geophysical Research*, **102**(E2), 3967–3988.

[205] Golombek, M., and Rapp, D. (1997) "Size–frequency distributions of rocks on Mars and Earth analogue sites: Implications for future landed missions," *Journal of Geophysical Research*, **102**(E2), 4117–4129.

[206] Jindra, F. (1966) "Obstacle performance of articulated wheeled vehicles," *J. Terramechanics*, **3**(2), 39–56.

[207] Wilcox, B., Nasif, A., and Welch, R. (1997) "Implications of statistical rock distributions on rover scaling," *International Conference Mobile Planetary Robots and Rovers Roundup, Santa Monica, CA, January 23–February 1*.

[208] Wilcox, B. et al. (1992) "Robotic vehicles for planetary exploration," *Proceedings IEEE International Conference Robotics and Automation*, pp. 175–180.

[209] Wilcox, B. et al. (1998) "Nanorover for Mars," *Space Technology*, **17**(3/4), 163–172.

[210] Ellery, A. (2005) "Robot–environment interaction: The basis for mobility in planetary micro-rovers," *Robotics & Autonomous Systems*, **51**, 29–39.

[211] Bekker, M. (1969) *Introduction to Terrain Vehicle Systems, Part 1: The Terrain* and *Part 2: The Vehicle*, University of Michigan Press, Ann Arbor, MI.

[212] Bekker, M. (1960) *Off the Road Locomotion*, University of Michigin Press, Ann Arbor, MI.

[213] Bekker, M. (1959) *Theory of Land Locomotion: The Mechanics of Vehicle Mobility*, University of Michigan Press, Ann Arbor, MI.

[214] Wisner, R., and Luth, H. (1973) "Off-road traction prediction for wheeled vehicles," *J. Terramechanics*, **10**(2), 45–61.

[215] Bekker, M. (1963) "Mechanics of locomotion and lunar surface vehicle concepts," *Automotive Engineering Congress 72*, Paper 632K, pp. 549–569.

[216] Plackett, C. (1985) "Review of force prediction methods for off-road wheels," *J. Agricultural Engineering Research*, **31**(1), 1–29.

[217] Moore, H., Bickler, D., Crisp, J., Eisen, H., Gensler, J., Haldemann, A., Matijevic, J., Pavlics, F., and Reid, L. (1999) "Soil-like deposits observed by Sojourner the Parthfinder rover," *Journal of Geophysical Research*, **104**(E4), 8729–8746.

[218] Rover Team (1997) "Characterisation of the Martian surface deposits by the Mars Pathfinder rover, Sojourner," *Science*, **278**, 1765–1767.

[219] Arvidson, R., Anderson, R., Haldemann, A., Landis, G., Li, R., Lindeman, R., Matijevic, J., Morris, R., Richter, L., Squyres, S. et al. (2003) "Physical properties and localization investigations associated with the 2003 Mars Exploration Rovers," *Journal of Geophysical Research*, **108**(E12), 11.1–11.20.

[220] Wallace, B., and Rao, N. (1993) "Engineering elements for transportation on the lunar surface," *Applied Mechanics Review*, **46**(6), 301–312.

[221] Gibbesch, A., and Schafer, B. (2005) "Multibody system modelling and simulation of planetary rover mobility on soft terrain," *Procedings Eighth International Symposium on Artificial Intelligence, Robotics and Automation in Space (i-SAIRAS), Munich, Germany* (ESA SP-603), ESA, Noordwijk, The Netherlands.

[222] Gibbesch, A., and Schafer, B. (2005) "Advanced modelling and simulation methods of planetary rover mobility on soft terrain," *Proceedings Conference Advanced Space Technologies for Robotics and Automation*, ESA/ESTEC, Noordwijk, The Netherlands.

[223] Komandi, G. (2006) "Soil vehicle relationship: The peripheral force," *J. Terramechanics*, **43**, 213–223.RED2

[224] Gee-Clough, D. (1976) "Bekker theory of rolling resistance amended to take account of skid and deep sinkage," *J. Terramechanics*, **13**(2), 87–105.

[225] Wong, J., and Preston-Thomas, J. (1983) "On the characterization of the shear stress–displacement relationship of terrain," *J. Terramechanics*, **19**(4), 225–234.

[226] Ishigami, G., Miwa, A., Nagatani, K., and Yoshida, K. (2007) "Terramechanics-based model for steering maneuver of planetary exploration rovers on loose soil," *J. Field Robotics*, **24**(3), 233–250.

[227] Wong, J.-Y., and Reece, A. (1967) "Prediction of rigid wheel performance based on the analysis of soil–wheel stresses, Part 1: Performance of driven rigid wheels," *J. Terramechanics*, **4**(1), 81–98.

[228] Shibley, H., Iagnemma, K., and Dubowsky, S. (2005) "Equivalent soil mechanics formulation for rigid wheels in deformable terrain, with application to planetary exploration rovers," *J. Terramechanics*, **42**, 1–13.

[229] Wong, J., and Huang, W. (2006) "Wheels vs tracks: A fundamental evaluation from the traction perspective," *J. Terramechanics*, **43**, 27–42.

[230] Wong, G. (2001) *Theory of Ground Vehicles* (Second Edition), John Wiley & Sons, New York.

[231] Wong, G. (1981) "Wheel–soil interaction," *J. Terramechanics*, **21**(2), 117–131.

[232] Wong, J. (1997) "Dynamics of tracked vehicles," *Vehicle System Dynamics*, **28**, 197–219.

[233] Jaeger, H. (2005) "Sand, jams and jets," *Physics World*, December, 34–39.

[234] Bak, P., Tang, C., and Wisenfeld, K. (1988) "Self-organised criticality," *Physical Review A*, **38**(1), 364–374.

[235] Boettcher, S., and Paczuski, M. (1997) "Broad universality in self-organised critical phenomena," *Physica D*, **107**, 171–173.

[236] Cerville, J., Formenti, E., and Masson, B. (2007) "From sandpiles to sand automata," *Theoretical Computer Science*, **381**, 1–28.

[237] Lindemann, R., and Voorhees, C. (2005) "Mars Exploration Rover mobility assembly, design, test and performance," *IEEE Int. Conf. Systems, Man, and Cybernetics*, **1**, 450–455.

[238] Wong, Z., and Reece, A. (1984) "Performance of free rolling rigid and flexible wheels on sand," *J. Terramechanics*, **21**(4), 347–360.

[239] Wong, J., and Asnani, V. (2008) "Study of the correlation between the performances of lunar vehicle wheels predicted by the Nepean wheeled vehicle performance model and test data," *Proc. Inst. Mechanical Engineers D (Automobile Eng.)*, **222**, 1939–1954.

[240] Richter, L., Bernasconi, M., and Coste, P. (2002) "Analysis, design, and test of wheels for a 4 kg-class mobile device for the surface of Mars," *Proceedings 14th International Conference International Society for Terrain–Vehicle Systems (ISTVS), Vicksburg, MS*.

[241] Richter, L. et al. (2000) "Wheeled mobile device for the deployment of surface and subsurface instruments and for subsurface sampling on planets," *Proceedings Sixth ESA Workshop on Advanced Space Technologies for Robotics and Automation (ASTRA 2000)*, ESTEC, Noordwijk, The Netherlands.

[242] Richter, L. et al. (1998) "Mobile micro-robots for scientific instrument deployment on planets," *Robotics & Autonomous Systems*, **23**, 107–115.

[243] Bauer, R., Leung, W., and Barfoot, T. (2005) "Development of a dynamic simulation tool for the ExoMars rover," *Proceedings Eighth International Symposium Artificial Intelligence, Robotics and Automation in Space (i-SAIRAS)* (ESA SP-603), ESA, Noordwijk, The Netherlands.

[244] Bauer, R., Leung, W., and Barfoot, T. (2008) "Experimental and simulation results of wheel–soil interaction for planetary rovers," *Proceedings International Conference on Intelligent Robots and Systems, Edmonton, Canada*.

[245] Nakashima, H., Fujii, H., Oida, A., Momozu, M., Kawase, Y., Kanamori, H., Aoki, S., and Yokoyama, T. (2007) "Parametric analysis of lugged wheel performance for a lunar microrover by means of DEM," *J. Terramechanics*, **44**, 153–162.

[246] Ding, L., Gao, H., Deng, Z., Nagatani, K., and Yoshida, K. (2011) "Experimental study and analysis on driving wheels' performance for planetary exploration rovers moving in deformable soils," *J. Terramechanics*, **48**, 27–45.

[247] Carrier III, W., Olhoeft, G., and Mendell, W. (1991) "Physical properties of the lunar surface," in G. Heiken, D. Vaniman, and B. French (Eds.), *Lunar Sourcebook: A User's Guide to the Moon*, Cambridge University Press, pp. 522–530.

[248] Bekker, M. (1962) "Land locomotion on the surface of planets," *ARS Journal*, **32**(11), 1651–1659.

[249] Richter, L., and Hamacher, H. (1999) "Investigating the locomotion performance of planetary microrovers with small wheel diameters and small wheel loads," *Proceedings 13th International Conference International Society for Terrain–Vehicle Systems, Munich, Germany, September 14–17*, pp. 719–726.

[250] Richter, L., Ellery, A., Gao, Y., Michaud, S., Schmitz, N., and Weiss, S. (2006) "A predictive wheel–soil interaction model for planetary rovers validated in testbeds and against MER Mars rover performance data," *Proceedings 10th European Conference International Society for Terrain–Vehicle Systems, Budapest, Hungary*.

[251] Yoshida, K., and Ishigami, G, (2004) "Steering characteristics of a rigid wheel for exploration on loose soil," *Proceedings IEEE/RST International Conference Intelligent Robots and Systems*, pp. 3995–4000.

[252] Yamakawa, J. et al. (1999) "Spatial motion analysis of tracked vehicles on dry sand," *Proceedings 13th International Conference International Society for Terrain–Vehicle Systems, Munich, Germany, September 14–17*, pp. 767–774.

[253] Wong, J., and Chiang, C. (2001) "General theory for skid steering of tracked vehicles on firm ground," *Proc. Institution Mechanical Engineers D*, **215**, 343–356.

[254] Wong, J., and Huang, W. (2006) "Wheels vs tracks: A fundamental evaluation from the traction perspective," *J. Terramechanics*, **43**, 27–42.

[255] Richter, L., and Hamacher, H. (1999) "Investigating the locomotion performance of planetary microrovers with small wheel diameters and small wheel loads," *Proceedings 13th International Conference International Society for Terrain–Vehicle Systems, Munich, Germany, September 14–17,* pp. 719–726.

[256] Braun, H., Malenkov, M., Fedosejev, S., Popova, I., and Vlasov, Y. (1994) *LUMOT: Locomotion Concepts Analysis for Moon Exploration* (Final Report, ESTEC Contract No. 141253), ESA/ESTEC, Noordwijk, The Netherlands.

[257] NASA (1995) *Selection of Electric Motors for Aerospace Applications* (Preferred Reliability Practice No. PD-ED-1229), NASA, Washington, D.C..

[258] Matijevic, J. (1997) "Sojourner: The Mars Pathfinder microrover flight experiment," *Space Technology,* 17(3/4), 143–149; Matijevic, J. (1996) *Mars Pathfinder Micro-rover: Implementing a Low Cost Planetary Mission Experiment* (IAA-L-0510), International Academy of Astronautics, Johns Hopkins APL, Laurel, MD.

[259] Oman, H. (2001) "Batteries for spacecraft, airplanes and military service: New developments," *IEEE Aerospace & Electronic Systems,* July, 35–44.

[260] Cross, M., Nicol, C., Qadi, A., and Ellery, A. (2013) "Application of COTS components for Martian surface exploration," *J. British Interplanetary Society,* 66, 161–166.

[261] Fuke, Y., Apostoulopoulos, D., Rollins, E., Silberman, J., and Whittaker, W. (1995) "A prototype locomotion concept for a lunar robotic explorer," *IEEE International Symposium Intelligent Vehicles,* pp. 382–387.

[262] Caldwell, D., and Rennels, D. (1997) "Minimalist hardware architecture for using commercial microcontrollers in space," *Proceedings 16th Digital Avionics Systems Conference,* pp. 1–8.

[263] Woodcock, A. (2002) "Microdot: A four-bit microcontroller designed for distributed low-end computing in satellites," US Air Force Institute of Technology Master's thesis, AFIT/GE/ENG/02M-28.

[264] Greenwald, L., and Kopena, J. (2003) "Mobile robot labs," *IEEE Robotics & Automation Magazine,* June, 25.

[265] Yilma, B., and Seif, M. (1999) "Behaviour-based artificial intelligence in miniature mobile robot," *Mechatronics,* 9, 185–206.

[266] Fraeman, M., Meitzler, R., Martin, M., Millard, W., Wong, Y., Mellert, J., Bowles-Martinez, J., Strohbehn, K., and Roth, D. (2005) "Radiation tolerant mixed signal microcontroller for Martian surface applications," *Proceedings 12th NASA Symposium on VLSI Design, Coeur d'Alene, ID,* pp. 1–6.

[267] Pollina, M., Sinander, P., and Habinc, S. (1998) "Microcontroller with built-in support for CCSDS telecommand and telemetry." *Proceedings First ESA Workshop on Tracking, Telemetry & Command Systems, Noordwijk, The Netherlands.*

[268] Wilcox, B., and Jones, R. (2000) "MUSES-CN nanorover mission and related technology," *Proceedings IEEE Conference,* pp. 287–295.

[269] Biesiadecki, J., Maimone, M., and Morrison, J. (2001) "Athena SDM rover: A testbed for Mars rover mobility," *Proceedings International Symposium Artificial Intelligence and Robotics in Space, Montreal, Canada.*

[270] Dote, Y. (1988) "Application of modern control techniques to motor control," *Proc. IEEE,* 76(4), 438–454.

[271] Kappos, E., and Kinniment, D. (1996) "Application-specific processor architectures for embedded control: Case studies," *Microprocessors & Microsystems,* 20, 225–232.

[272] Nilsson, K., and Johansson, R. (1999) "Integrated architecture for industrial robot programming and control," *Robotics & Autonomous Systems,* 29, 205–226.

[273] Illgner, K. (2000) "DSPs for image and video processing," *Signal Processing*, **80**, 2323–2336.

[274] Jameux, D. (2000) "Onboard DSP technologies applied to robotic applications," *Proceedings Sixth ESA Workshop on Advanced Space Technologies for Robotics and Automation (ASTRA), December 2000*, ESTEC, Noordwijk, The Netherlands.

[275] Grunfelder, S., and Kricki, R. (1999) "Buyer's guide to forward intersection for binocular robot vision," *Proceedings Fifth International Symposium Artificial Intelligence Robotics & Automation in Space* (ESA SP-440), pp. 649–654.

[276] Kukolj, D., Kulic, F., and Levi, E. (2000) "Design of the speed controller for sensorless electric drives based on AI techniques: A comparative study," *Artificial Intelligence in Engineering*, **14**, 165–174.

[277] Petersen, R., and Hutchings, B. (1995) "An assessment of the suitability of FPGA based systems for use in digital signal processing," *Proceedings Fifth International Workshop on FPGA Logic and Applications, August 1995, Oxford*.

[278] Biesiadecki, J., and Maimone, M. (2006) "Mars Exploration Rover surface mobility flight software: Driving ambition," *Proceedings IEEE Aerospace Conference, Big Sky, Montana*, pp. 1–15.

[279] Sotelo, M. (2003) "Lateral control strategy for autonomous steering of Ackerman-like vehicles," *Robotics & Autonomous Systems*, **45**, 223–233.

[280] Ishigami, G., Miwa, A., Nagatani, K., and Yoshida, K. (2007) "Terramechanics-based model for steering maneouvre of planetary exploration rovers on loose soil," *J. Field Robotics*, **24**(3), 233–250.

[281] Peynot, T., and Lacroix, S. (2003) "Enhanced locomotion control for a planetary rover," *Proceedings IEEE/RSJ International Conference Intelligent Robots and Systems*, pp. 311–316.

[282] Hacot, H., Dubowsky, S., and Bidaud, P. (1998) "Analysis and simulation of a rocker-bogie exploration rover," *Proceedings 12th CISM-IFToMM Sympowium RoManSy 98, Paris, France*.

[283] Tunstel, E. (1999) "Evolution of autonomous self-righting behaviours for articulated nanorovers," *Proceedings Fifth International Symposium Artificial Intelligence Robotics and Automation in Space* (ESA SP-440), pp. 341–346.

[284] Bernstein, D. (2001) "Sensor performance specifications," *IEEE Control Systems Mag.*, August, 9–18.

[285] Borenstein, J., Everett, H., Feng, L., and Wehe, D. (1997) "Mobile robot positioning: Sensors and techniques," *J. Robotic Systems*, **14**(4), 231–249.

[286] Borenstein, J. et al. (1996) *Where Am I? Sensors and Methods for Mobile Robot Positioning*, University of Michigan Technical Manual for Oakridge National Lab and U.S. Department of Energy.

[287] Boero, G., Demierre, M., Besse, P-A., and Popovic, R. (2003) "Micro-Hall devices: Performance, technologies and applications," *Sensors & Actuators*, **A106**, 314–320.

[288] Kapoor, A., Simaan, N., and Kazanzides, P. (2004) "System for speed and torque control of dc motors with application to small snake robots," *Proceedings IEEE Mechatronics and Robotics Conference, Germany*.

[289] Bell, J. et al, (2003) "Mars exploration rover Athena panoramic camera (pancam) investigation," *Journal of Geophysical Research*, **108**(E12), 4.1–4.30.

[290] Trebi-Ollennu, A., Huntsberger, T., Cheng, Y., Baumgartner, E., Kennedy, B., and Schenker, P. (2001) "Design and analysis of a sun sensor for planetary rover absolute heading detection," *IEEE Trans Robotics & Automation*, **17**(6), 939–947.

[291]

[292] Enright, J., Furgale, P., and Barfoot, T. (2009) "Sun sensing for planetary rover navigation," *IEEEEAC* (Paper No. 1340).

[293] Furgale, P., Enright, J., and Barfoot, T. (2011) "Sun sensor navigation for planetary rovers: theory and field testing," *IEEE Trans. Aerospace and Electronic Systems*, **47**(3), 1631–1647.

[294] Wehner, R., and Lanfranconi, B. (1981) "What do ants know about the rotation of the sky?" *Nature*, **293**, 731–733.

[295] Lambrinos, D., Moller, R., Lebhart, T., Pfeifer, R., and Wehner, R. (2000) "Mobile robot employing insect strategies for navigation," *Robotics & Autonomous Systems*, **30**, 39–64.

[296] Lambrinos, D., Kobayashi, H., Pfeifer, R., Maris, M., Labhart, T., and Wehner, R. (1997) "Autonomous agent navigating with a polarised light compass," *Adaptive Behaviours*, **6**(1), 131–161.

[297] Wehner, R. (2003) "Desert ant navigation: How miniature brains solve complex tasks," *J. Comparative Physiology*, **189A**, 579–588.

[298] Lambrinos, D., Muller, R., Lebhart, T., Pfeifer, R., and Wehner, R. (2000) "Mobile robot employing insect strategies for navigation," *Robotics and Autonomous Systems*, **30**, 39–64.

[299] Labhart, T., and Meyer, E. (2002) "Neural mechanisms in insect navigation: Polarisation compass and odometer," *Current Opinion in Neurobiology*, **12**, 707–714.

[300] Davis, J., Nehab, D., Ramamoorthi, R., and Rusinikiewcz, S. (2005) "Spacetime stereo: A unifying framework for depth from triangulation," *IEEE Trans. Pattern Analysis & Machine Intelligence*, **27**(2), 296–302.

[301] Cha, Y., and Gweon, D. (1996) "Calibration and range data extraction algorithm for an omnidirectional laser range finder of a free-ranging mobile robot," *Mechatronics*, **6**(6), 665–689.

[302] Jarvis, R. (1983) "Perspective on range-finding techniques for computer vision," *IEEE Trans. Pattern Analysis & Machine Intelligence*, **5**(2), 122–139.

[303] Hebert, M. (2000) "Active and passive range sensing for robotics," *Proceedings IEEE International Conference Robotics and Automation*, pp. 102–110.

[304] Kweon, I., and Kanade, T. (1992) "High resolution terrain map from multiple sensor data," *IEEE Trans. Pattern Analysis & Machine Intelligence*, **14**(2), 278–292.

[305] Krotkov, E., and Hoffman, R. (1994) "Terrain mapping for a walking planetary rover," *IEEE Trans. Robotics and Automation*, **10**(6), 728–739.

[306] Hebert, M., Krotkov, E., and Kanade, T. (1989) "Perception system for a planetary explorer," *Proceedings 28th Conference Decision and Control, Tampa, FL*, pp. 1151–1156.

[307] Langer, D., Rosenblatt, J., and Hebert, M. (1994) "Behaviour-based system for off-road navigation," *IEEE Trans. Robotics and Automation*, **10**(6), 776–783.

[308] Okubo, Y., Ye, C., and Borenstein, J. (2009) "Characterisation of the Hokuyo URG-04LX laser rangefinder for mobile robot obstacle negotiation," *Proceedings of SPIE*, **7332**.

[309] Miller, D., and Lee, T. (2002) "High speed traversal of rough terrain using a rocker-bogie mobility system," *Proceedings Fifth International Conference Exposition Robotics for Challenging Situations and Environments*.

[310] McLean, D., and Ellery, A. (2008) "Survey of traction control systems for planetary rovers," *CSME Forum (CCToMM), Ottawa University, Ontario, Canada*.

[311] Zaman, M. (2006) "High resolution relative localisation in a mobile robot using two cameras," *Proceedings Towards Autonomous Robotic Systems (TAROS) Conference, University of Surrey, Guildford, U.K.*

[312] Zaman, M. (2007) "High resolution localization using two cameras," *Robotics & Autonomous Systems*, **55**, 685–692.

[313] Bradshaw, J., Lollini, C., and Bishop, B. (2007) "On the development of an enhanced optical mouse sensor for odometry and mobile robotics education," *39th Southeastern Symposium on System Theory*, pp. 6–10.

[314] Jackson, J., Callahan, D., and Marstrander, J. (2007) "Rationale for the use of optical mice chips for economic and accurate vehicle tracking," *Proceedings Third Annual IEEE Conference Automation Science and Engineering, Scottsdale, AZ*, pp. 939–944.

[315] Barrows, G., Chahl, J., and Srinavasan, M. (2003) "Biomimetic visual sensing and flight control," *Proceedings Bristol UAV Conference.*

[316] Tunwattana, N., Roskilly, A., and Norman, R. (2009) "Investigations into the effects of illumination and acceleration on optical mouse sensors as contact-free 2D measurement devices," *Sensors & Actuators A: Physical*, **149**, 87–92.

[317] Bonarini, A., Matteucci, M., and Restelli, M. (2005) "Automatic error detection and reduction for an odometric sensor based on two optical mice," *Proceedings IEEE International Conference Robotics and Automation*, pp. 1675–1680.

[318] Hyun, D., Yang, H., Park, H., and Park, H-S. (2009) "Differential optical navigation sensor for mobile robots," *Sensors & Actuators A: Physical*, **156**, 296–301.

[319] Tan, C., Zweiri, Y., Althoefer, K., and Senevirante, L. (2005) "Online soil parameter estimation scheme based on Newton–Raphson method for autonomous excavation," *IEEE/ASME Trans. Mechatronics*, **10**(2), 221–229.

[320] Buckholtz, K. (2002) "Reference input wheel slip tracking using sliding mode control," *SAE World Congress, Detroit, MI* (Paper No. 2002-01-0301).

[321] Caponero, M., Moricini, C., and Aliverdiev, A. (2000) "Laser velocimetry: An application as smart driving agent for tracked vehicles," *Proceedings Fourth Russian Laser Symposium.*

[322] Ozdemir, S., Takamiya, S., Shinohara, S., and Yoshida, H. (2000) "Speckle velocimeter using a semiconductor laser with external optical feedback from a moving surface: Effects of system parameters on the reproducibility and accuracy of measurements," *Measurement Science & Technology*, **11**, 1447–1455.

[323] Charrett, T., Waugh, L., and Tatam, R. (2010) "Speckle interferometry for high accuracy odometry for a Mars exploration rover," *Measurement Science & Technology*, **21**(2), 025301.

[324] Wong, J-Y. (1971) "Optimisation of the tractive performance of four-wheel-drive off-road vehicles," *SAE Trans.*, **79**, 2238-2246.

[325] Wong, J-Y., McLaughlin, N., Knezevic, Z., and Burrt, S. (1998) "Optimisation of the tractive performance of four-wheel-drive tractors: Theoretical analysis and experimental substantiation," *Proc. Institution Mechanical Engineers*, **212D**, 285–297.

[326] Baumgartner, E., Aghazarian, H., and Trebi-Ollennu, A. (2001) "Rover localisation results for the FIDO rover," in: G. McKee and P. Schenker (Eds.), *Proceedings SPIE Sensor Fusion and Decentralised Control in Robotic Systems IV*, Vol. 4571..

[327] Ishigami, G., Nagatani, K., and Yoshida, K. (2009) "Slope traversal controls for planetary exploration rover on sandy terrain," *J. Field Robotics*, **26**(3), 264–286.

[328] Helmick, D., Cheng, Y., Clouse, D., Bajrachararya, M., Matthoes, L., and Roumeliotis, S. (2005) "Slip compensation for a Mars rover," *Proceedings IEEE/RSJ International Conference Intelligent Robots and Systems, Pittsburgh, PA.*

[329] Shibly, H., Iagnemma, K., and Dubowsky, S. (2005) "Equivalent soil mechanics formulation for rigid wheels in deformable terrain with application to planetary exploration rovers," *J. Terramechanics*, **42**, 1–13.

[330] Bekker, M. (1969) *Introduction to Terrain Vehicle Systems, Part 1: The Terrain* and *Part 2: The Vehicle*, University of Michigan Press, Ann Arbor; Bekker, M. (1959) *Theory of Land Locomotion: The Mechanics of Vehicle Mobility*, University of Michigan Press, Ann Arbor; Bekker, M. (1960) *Off the Road Locomotion*, University of Michigan Press, Ann Arbor.

[331] Yoshida, K., and Hamamo, H. (2002) "Motion dynamics and control of a planetary rover with slip-based traction model," *Proceedings IEEE International Conference Robotics and Automation*, pp. 3155–3160.

[332] Wong, G. (2001) *Theory of Ground Vehicles* (Second Edition), John Wiley & Sons, New York.

[333] Helmick, D., Roumeliotis, S., Cheng, Y., Clouse, D., Bajracharya, M., and Matthies, L. (2006) "Slip-compensated path following for planetary exploration rovers," *Advanced Robotics*, **20**(11), 1257–1280.

[334] Iagnemma, K., and Dubowsky, S. (2000) "Vehicle wheel–ground contact angle estimation: With application to mobile robot traction control," *Proceedings International Symposium Advances in Robot Kinematics*.

[335] Iagnemma, K., Shibley, H., and Dubowsky, S. (2002) "Online terrain parameter estimation for planetary rovers," *Proceedings IEEE International Conference Robotics and Automation*, pp. 3142–3147.

[336] Iagnemma, K., Kwang, S., Shibly, H., and Dubowsky, S. (2004) "Online terrain parameter estimation for wheeled mobile robots with application to planetary rovers," *IEEE Trans. Robotics*, **20**(5), 921–927.

[337] Shibly, H., Iagnemma, K., and Dubowsky, S. (2005) "Equivalent soil mechanics formulation for rigid wheels in deformable terrain with application to planetary exploration," *J. Terramechanics*, **42**, 1–13.

[338] Iagnemma, K., and Dubowsky, S. (2004) "Traction control of wheeled robotic vehicles in rough terrain with application to planetary rovers," *Int. J. Robotics Research*, **23**(10/11), 1029–1040.

[339] Iagnemma, K., and Dubowsky, S. (2004) *Mobile Robots in Rough Terrain: Estimation, Motion Planning, and Control with Application to Planetary Rovers* (Springer Tracts in Advanced Robotics (STAR) 12), Springer-Verlag, Berlin, Germany.

[340] Iagnemma, K., Rzepniewski, A., Dubowsky, S., and Schenker, P. (2003) "Control of robotic vehicles with actively articulated suspensions on rough terrain," *Autonomous Robots*, **14**(1), 5–16.

[341] Howard, A., and Seraji, H. (2001) "Vision-based terrain characterization and traversability assessment," *J. Robotic Systems*, **18**(1), 577–587.

[342] Shirkhodaie, A., Amrani, R., and Tunstel, E. (2004) "Soft computing for visual terrain perception and traversability assessment by planetary robotic systems," *Proc. IEEE Int. Conf Systems Man and Cybernetics*, **2**, 1848–1855.

[343] Sancho-Pradel, D., and Gao, Y. (2010) "Survey of terrain assessment techniques for autonomous operation of planetary robots," *J. British Interplanetary Society*, **63**, 206–217.

[344] Angelova, A., Matthies, L., Helmick, D., and Perona, P. (2007) "Learning and prediction of slip from visual information," *J. Field Robotics*, **24**(3), 205–231.

600   **References**

[345] Tunstel, E., Howard, A., and Seraji, H. (2002) "Rule-based reasoning and neural network perception for safe off-road robot mobility," *Expert Systems: Int. J. Knowledge Engineering & Neural Networks*, **19**(4), 191–200.

[346] Swartz, M., and Ellery, A. (2008) "Towards adaptive localisation for rover navigation using multilayer feedforward neural networks," *Canadian Aeronautics & Space Institute (CASI) ASTRO2008, Montreal, Canada* (Paper No. 74).

[347] Swartz, M., Ellery, A., and Marshall, J. (2015) "Simulation and analysis of a slip-adaptive rover navigation algorithm," submitted to *Int. J. Advanced Robotic Systems*.

[348] Helmick, D., Angelova, A., and Matthies, L. (2009) "Terrain adaptive navigation for planetary rovers," *J Field Robotics*, **26**(4), 391–410,

[349] Ojeda, L., Cruz, D., Reina, G., and Borenstein, J. (2006) "Current-based slippage detection and odometry correction for mobile robots and planetary rovers," *IEEE Trans. Robotics*, **22**(3), 366–378.

[350] Collins, E., and Coyle, E. (2008) "Vibration-based terrain classification using surface profile input frequency responses," *Proceedings IEEE International Conference Robotics and Automation*, pp. 3276–3283.

[351] Bajracharya, M., Tang, B., Howard, A., Turmon, M., and Matthies, L. (2008) "Learning long-range terrain classification for autonomous navigation," *Proceedings IEEE International Conference Robotics and Automation*, pp. 4018–4024.

[352] Kweon, I., and Kanade, T. (1990) "High resolution terrain map from multiple sensor data," *IEEE International Workshop Intelligent Robots and Systems*, pp. 127–134.

[353] Halatci, I., Brooks, C., and Iagnemma, K. (2008) "Study of visual and tactile terrain classification and classifier fusion for planetary exploration rovers," *Robotica*, **26**, 767–779.

[354] Komma, P., Weiss, C., and Zell, A. (2009) "Adaptive Bayesian filtering for vibration-based terrain classification," *Proceedings IEEE International Conference Robotics and Automation*, pp. 3307–3313.

[355] Brooks, C., Iagnemma, K., and Dubowsky, S. (2005) "Vibration-based terrain analysis for mobile robots," *Proceedings IEEE International Conference Robotics and Automation*, 3415–3420.

[356] Weiss, C., Tamimi, H., and Zell, A. (2008) "Combination of vision and vibration-based terrain classification," *Proceedings IEEE/RSJ International Conference Intelligent Robots and Systems*, pp. 2204–2209.

[357] Brooks, C., and Iagnemma, K. (2012) "Self-supervised terrain classification for planetary surface exploration rovers," *J. Field Robotics*, **29**(3), 445–468.

[358]

[359] Setterfield, T., and Ellery, A. (2013) 'Terrain response estimation using an instrumented rocker-bogie mobility system," *IEEE Trans. Robotics*, **29**(1), 172–188.

[360] Buehler, M., Anderson, R., and Seshadri, S. (2005) "Prospecting for in-situ resources on the Moon and Mars using wheel-based sensors," *Proceedings IEEE Aerospace Conference 2005*.

[361]

[362] Cross, M., Ellery, A., and Qadi, A. (2013) "Estimating terrain parameters for a rigid wheel rover using neural networks," *J. Terramechanics*, **50**(3), 165–174.

[363] Jain, A., and Dorai, C. (1997) "Practicing vision: Integration, evaluation and applications," *Pattern Recognition*, **30**(2), 183–196.

[364] Massaro, D., and Friedman, D. (1990) "Models of integration given multiple sources of information," *Psychological Review*, **97**(2), 225–252.

[365] Luo, R., and Kay, M. (1989) "Multisensor integration and fusion in intelligent systems," *IEEE Trans Systems Man & Cybernetics*, **19**(5), 61–70.

[366] Harmon, S. et al. (1986) "Sensor data fusion through a distributed blackboard," *Proceedings IEEE International Conference Robotics & Automation*, Vol. 3, pp. 1449–1454.

[367] Perlovsky, L. (2007) "Cognitive high level information fusion," *Information Sciences*, **177**, 2099–2118.

[368] Ernst, M., and Banks, M. (2002) "Humans integrate visual and haptic information in a statistically optimal fashion," *Nature*, **415**, 429–433.

[369] Hink, R., and Woods, D. (1987) "How humans process uncertain knowledge," *AI Magazine*, December, 41–51.

[370] Pang, D. et al. (1987) "Reasoning with uncertain information," *Proc. IEEE*, **134D**(4), 231–237.

[371] Hall, D., and Llinas, J. (1997) "Introduction to multisensor data fusion," *Proc. IEEE*, **85**(1), 6–23.

[372] Bloch, I. (1996) "Information combination operators for data fusion: A comparative review with classification," *IEEE Trans. Systems Man & Cybernetics*, **26**(1), 52–67.

[373] Henkind, S., and Harrison, M. (1988) "Analysis of four uncertainty calculi," *IEEE Trans. Systems Man & Cybernetics*, **18**(5), 700–714.

[374] Clark, D. (1990) "Numerical and symbolic approaches to uncertainty management in AI," *Artificial Intelligence Review*, **4**, 109–146.

[375] Hackett, J., and Shah, M. (1990) "Multisensor fusion: A perspective," *Proceedings IEEE International Conference Robotics and Automation*, 1324–1329.

[376] Knill, D., and Pouget, A. (2004) "Bayesian brain: The role of uncertainty in neural coding and computation," *Trends in Neurosciences*, **27**(12), 712–719.

[377] Bloch, I. (1996) "Information combination operators for data fusion: A comparative review with classification," *IEEE Trans. Systems Man & Cybernetics A: Systems & Humans*, **26**(1), 52–67.

[378] Duda, R., Hart, P., and Nilsson, N. (1976) "Subjective Bayesian network for rule-based inference systems," *Proc. AFIPS Computing Conf.*, **45**, 1072–1082.

[379] Wu, Y-G., Yang, J-Y., and Liu, K. (1996) "Obstacle detection and environment modelling based on multisensor fusion for robot navigation," *Artificial Intelligence in Engineering*, **10**, 232–333.

[380] Murphy, R. (1998) "Dempster–Shafer theory for sensor fusion in autonomous mobile robots," *IEEE Trans. Robotics & Automation*, **14**(2), 197–206.

[381] Yang, C., and Lin, C-F. (1993) "Multisensor data fusion for target recognition," *Proceedings First IEEE Aerospace Control Systems Conference*, pp. 118–121.

[382] Vasseur, P., Pegard, C., Mouaddib, E., and Delahoche, L. (1999) "Perceptual organization approach based on Dempster–Shafer theory," *Pattern Recognition*, **32**, 1449–1462.

[383] Zadeh, L. (1965) "Fuzzy sets," *Information & Control*, **8**(3), 338–353.

[384] Zadeh, L. (1978) "Fuzzy sets as a basis for a theory of possibility," *Fuzzy Sets & Systems*, **1**, 3–28.

[385] Zadeh, L. (1988) "Fuzzy logic," *IEEE Computer*, **21**(4), 83–93.

[386] Haack, S. (1979) "Do we need fuzzy logic?" *Int. J. Man–Machine Studies*, **11**, 437–445.

[387] Munakata, T., and Jani, Y. (1994) "Fuzzy systems: An overview," *Communications ACM*, **37**(3), 69–84.

[388] Shepard, R. (1987) "Towards a universal law of generalisation for psychological science," *Science*, **237**, 1317–1323.

[389] Shepard, R. (1984) "Ecological constraints on internal representation: Resonant kinematics of perceiving, imagining, thinking, and dreaming," *Psychological Review*, **91**, 417–447.

[390] Tong, R. (1977) "Control engineering review of fuzzy systems," *Automatica*, **13**, 559–569.

[391] Procyk, T., and Mamdani, E. (1979) "Linguistic self organising process controller," *Automatica*, **15**, 15–30.

[392] Daley, S., and Gill, K. (1986) "Design study for self-organising fuzzy logic controller," *Proc. Institution Mechanical Engineers*, **200C**(1), 59–69.

[393]

[394]

[395]

[396] Mamdani, E. (1977) "Application of fuzzy logic to approximate reasoning using linguistic synthesis," *IEEE Trans. Computers*, **26**, 1182–1191.

[397] Dodds, D. (1988) "Fuzziness in knowledge-based robotic systems," *Fuzzy Sets & Systems*, **20**, 179–193.

[398] Zadeh, L. (1989) "Knowledge representation in fuzzy logic," *IEEE Trans Knowledge & Data Eng.*, **1**(1), 89–100.

[399] Arzen, K.-E. (1996) "AI in the feedback loop: A survey of alternative approaches," *Annual Reviews in Control*, **20**, 71–82.

[400] Lee, C. (1990) "Fuzzy logic in control systems, Fuzzy logic controller: Part 1," *IEEE Trans Systems Man & Cybernetics*, **20**(2), 404–418.

[401] Lee, C. (1990) "Fuzzy logic in control systems, Fuzzy logic controller: Part 2," *IEEE Trans. Systems Man & Cybernetics*, **20**(2), 419–435.

[402] Ying, H. (1998) "General Takagi–Sugeno fuzzy systems with simplified linear rule consequent are universal controllers, models and filters," *J. Information Systems*, **108**, 91–107.

[403] Kosko, B. (1994) "Fuzzy systems as universal approximators," *IEEE Trans Computers*, **43**(11), 1329–1333.

[404] Pollatschek, M. (1977) "Hierarchical systems and fuzzy set theory," *Kybernetes*, **6**, 147–151.

[405] Wang, L-X. (1997) "Modelling and control of hierarchical systems with fuzzy systems," *Automatica*, **33**(6), 1041–1053.

[406] Cao, S., Rees, N., and Feng, G (1997) "Analysis and design for a class of complex control systems, Part I: Fuzzy modeling and identification," *Automatica*, **33**(6), 1017–1028.

[407] Cao, S., Rees, N., and Feng, G. (1997) "Analysis and design for a class of complex control systems, Part II: Fuzzy controller design," *Automatica*, **33**(6), 1029–1039.

[408] Floreano, D., Godjevac, J., Martinoli, A., Mondada, F., and Nicoud, J-D. (1999) "Design, control and applications of autonomous mobile robots," in S. Tzafestas (Ed.), *Advances in Intelligent Autonomous Systems*, Springer, Dordrecht, The Netherlands, pp. 175–189.

[409] Maiers, J., and Sherif, Y. (1985) "Applications of fuzzy set theory," *IEEE Trans. Systems Man & Cybernetics*. **15**(1), 175–189.

[410] Beliakov, G. (1996) "Fuzzy sets and membership functions based on probabilities," *Information Sciences*, **91**, 95–111.

[411] Hisdal, E. (1988) "Philosophical issues raised by fuzzy set theory," *Fuzzy Sets & Systems*, **25**, 349–356.

[412] Richardson, J., and Marsh, K. (1988) "Fusion of multisensor data," *Int. J. Robotics Research*, **7**(6), 78–96.

[413] Sasiadek, J. (2002) "Sensor fusion," *Annual Reviews in Control*, **26**, 203–228.

[414] DeSouza, G., and Kak, A. (2002) "Vision for mobile robot navigation: A survey," *IEEE Trans. Pattern Analysis & Machine Intelligence*, **24**(4), 237–267.

[415] Wong, H-S. (1996) "Technology and device scaling considerations for CMOS imagers," *IEEE Trans. Electron. Devices*, **43**(12), 2131–2142.

[416] Mansoorian, B., Yee, H-Y., and Fossum, E. (1999) "A 250 mW 60 frame/s 1,280 × 720 pixel 9B CMOS digital image sensor," *Proceedings IEEE International Solid State Circuits Conference*, pp. 312–313.

[417] Maurette, M. (2003) *CNES Autonomous Navigation, Basic Description and Preliminary Requirements: Rev. 1* (CNES Direction des Techniques Spatiale DTS/AE/SEA/ER/ 2003-009), Centre National d'Études Spatiales, Toulouse, France.

[418] Griffiths, A., Coates, A., Josset, J.-L., Paar, G., and Sims, M. (2003) "Scientific objectives of the Beagle 2 stereocamera system," *Lunar & Planetary Science*, **XXXIV**, 1609.

[419] Griffiths, A., Coates, A., Josset, J.-L., Paar, G., Hofmann, B., Pullan, D., Ruffer, P., Sims, M., and Pillinger, C. (2005) "Beagle 2 stereocamera system," *Planetary & Space Science*, **53**, 1466–1482.

[420] Eisenman, A., Liebe, C., Maimone, M., Schwochert, M., and Willson, R. (2002) "Mars exploration rover engineering cameras," available at *http://www-robotics.jpl.nasa.gov/ publications/Reg_Willson/MER_Cameras.pdf*

[421] Maki, J., Bell III, J., Herkenhoff, K., Squyres, S., Kiely, A., Klimesh, M., Schwochert, M., Litwin, T., Willson, R., Johnson, A. et al. (2003) "Mars Exploration Rover engineering cameras," *Journal of Geophysical Research*, **108**(E12), 12-1–12-24.

[422] Squyres, S., Arvidson, R., Baumgartner, E., Bell, J., Christensen, P., Gorevan, S., Herkenhoff, K., Klingelhofer, G., Madsen, M., Morris, R. et al. (2003) "Athena Mars rover science investigation," *Journal of Geophysical Research*, **108**(E12), 3.1–3.21.

[423] Bell, J., Squyres, S., Herkenhoff, K., Maki, J., Arneson, H., Brown, D., Collins, S., Dingizian, A., Elliot, S., Hagerott, E. et al. (2003) "Mars exploration rover Athena panoramic camera (pancam) investigation," *Journal of Geophysical Research*, **108**(E12), 4.1–4.30.

[424] Griffiths, A., Ellery, A., and the Camera Team (2007) "Context for the ExoMars rover: The panoramic camera (pancam) instrument," *Int. J. Astrobiology*, **5**(3), 269–275.

[425] Jameux, D. (2000) "Onboard DSP technologies applied to robotic applications," *Proceedings Advanced Space Technologies for Robotics and Automation (ASTRA) 2000*.

[426] Bajracharya, M., Maimone, M., and Helmick, D. (2008) "Autonomy for Mars rovers: Past, present and future," *IEEE Computer*, December, 44–50.

[427] Alexander, D., Deen, R., Andres, P., Zamani, P., Mortensen, H., Chen, A., Cayanan, M., Hall, J., Klochko, V., Pariser, O. et al. (2006) "Processing of Mars Exploration Rover imagery for science and operations planning," *Journal of Geophysical Research*, **111**, E02S02.

[428] Wagner, M., O'Hallaron, D., Apostoulopoulos, D., and Urmson, C. (2002) *Principles of Computer Design for Stereo Perception* (CMU-RI-TR-02-01), Carnegie Mellon University, Pittsburgh, PA.

[429] Kelly, A., and Stentz, A. (1998) "Rough terrain autonomous mobility, Part 1: A theoretical analysis of requirements," *Autonomous Robots*, **5**(May), 129–161.

[430] Kelly, A., and Stentz, A. (1998) "Rough terrain autonomous mobility, Part 2: An active vision, predictive control approach," *Autonomous Robots*, **5**, 163–198.

[431] Vergauwen, M. et al. (2000) "Autonomous operations of a micro-rover for geoscience on Mars," *Proceedings Sixth ESA Workshop on Advanced Space Technologies for Robotics and Automation (ASTRA), December 2000*, ESTEC, Noordwijk, The Netherlands.

[432] Stieber, M., McKay, M., Vukovich, G., and Petriu, E. (1999) "Vision-based sensing and control for space robotics applications," *IEEE Trans. Instrumentation & Measurement*, **48**(4), 807–812.

[433] Tsai, R. (1987) "Versatile camera calibration technique for high accuracy 3D machine vision metrology using off-the-shelf TV cameras and lenses," *IEEE J. Robotics and Automation*, **3**(3), 323–344.

[434] Barnes, D., Wilding, M., Gunn, M., Pugh, S., Tyler, L., Coates, A,., Griffiths, A., Cousins, C., Schmitz, N., Bauer, A. et al. (2006) "Multispectral vision processing for the ExoMars 2018 mission," *Symposium on Advanced Space Technologies in Robotics & Automation (ASTRA), ESTEC, Noordwijk, The Netherlands*.

[435] Ellery, A. (2000) *An Introduction to Space Robotics*, Springer/Praxis, Heidelberg, Germany/Chichester, U.K.

[436] Matthies, L., Maimone, M., Johnson, A., Cheng, Y., Willson, R., Villalpando, C., Goldberg, S., Huertas, A., Stein, A., and Angelova, A. (2007) "Computer vision on Mars," *Int. J. Computer Vision*, **75**(1), 67–92.

[437] Mallat, S. (1989) "Theory of multiresolution signal decomposition: The wavelet representation," *IEEE Trans. Pattern Analysis & Machine Intelligence*, **11**(7), 674–693.

[438] Mallat, S. (1996) "Wavelets for a vision," *Proc. IEEE*, **84**(4), 604–614.

[439] Nadernejad, E., Sharifzadeh, S., and Hassanpour, H. (2008) "Edge detection techniques: Evaluations and comparisons," *Applied Mathematical Sciences*, **2**(31), 1507–1520.

[440] Wells III, W. (1986) "Efficient synthesis of Gaussian filters by cascaded uniform filters," *IEEE Trans. Pattern Analysis & Machine Intelligence*, **8**(2), 234–239.

[441] Atick, J. (1992) "Could information theory provide an ecological theory of sensory processing?" *Network*, **3**, 213–251.

[442] Ullman, S. (1986) "Artificial intelligence and the brain: Computational studies of the visual system," *Annual Reviews Neuroscience*, **9**, 1–26.

[443] Sigman, M., Cecchi, G., Gilbert, C., and Magnasco, M. (2001) "On a common circle: Natural scenes and Gestalt rules," *Proc. National Academy Sciences*, **98**(4), 1935–1940.

[444] Kennedy, L., and Basu, M. (1997) "Image enhancement using a human visual system model," *Pattern Recognition*, **30**(12), 2001–2014.

[445] Canny, J. (1986) "Computational approach to edge detection," *IEEE Trans. Pattern Analysis & Machine Intelligence*, **8**(6), 679–698.

[446] Basu, M. (2002) "Gaussian-based edge-detection methods: A survey," *IEEE Trans. Systems Man & Cybernetics C: Applications & Reviews*, **32**(3), 252–259.

[447] Huntsberger, T., Aghazarian, H., Cheng, Y., Baumgartner, E., Tunstel, E., Leger, C., Trebi-Ollennu, A., and Schenker, P. (2002) "Rover autonomy for long range navigation and science data acquisition on planetary surfaces," *Proceedings IEEE International Conference Robotics and Automation*, pp. 3161–3168.

Molina Cabrera, P., and Ellery, A. (2015) "Towards a visual simultaneous localisation and mapping system for computationally constrained systems," submitted to *AJAA J. Aerospace Information Systems*.

[448] Alahi, A., Ortiz, R., and Vandergheynst, P. (2012) "FREAK: Fast retina keypoint," *Proceedings IEEE Conference Computer Vision & Pattern Recognition*, pp. 510–517.

[449] Schmid, C., Mohr, R., and Bauckhage, C. (2000) "Evaluation of interest point detectors," *Int. J. Computer Vision*, **37**(2), 151–172.

[450] Papardi, G., Campisi, P., Petkov, N., and Neri, A. (2007) "Biologically motivated multiresolution approach to contour detection," *EURASIP J. Advances in Signal Processing*, **2007**(71828).

[451] Geman, S., and Geman, D. (1984) "Stochastic relaxation, Gibbs distribution and the Bayesian restoration of images," *IEEE Trans. Pattern Analysis & Machine Intelligence*, **6**(4), 721–741.

[452] Konishi, S., Yuille, A., Coughlan, J., and Zhu, S. (2003) "Statistical edge detection: Learning and evaluating edge cues," *IEEE Trans. Pattern Analysis & Machine Intelligence*, **25**(1), 57–74.

[453] Carandini, M., Demb, J., Mante, V., Tolhurst, D., Dan, Y., Olshausen, B., Gallant, J., and Rust, N. (2005) "Do we know what the early visual system does?" *J. Neuroscience*, **16**, 10577–10597.

Geman, S., and Geman, D. (1984) "Stochastic relaxation, Gibbs distribution and the Bayesian restoration of images," *IEEE Trans. Pattern Analysis & Machine Intelligence*, **6**(4), 721–741.

Cross, G. (1983) "Markov random field texture models," *IEEE Trans. Pattern Analysis & Machine Intelligence*, **5**(1), 25–39.

Bello, M. (1994) "Combined Markov random field and wave-packet transform-based approach for image segmentation," *IEEE Trans. Image Processing*, **3**(6), 834–846.

[454] Haralick, R. (1979) "Statistical and structural approaches to texture," *Proc. IEEE*, **67**(5), 786–804.

[455] Haralick, R., Shanmugam, K., and Dinstein, I. (1973) "Textural features for image classification," *IEEE Trans. Systems Man & Cybernetics*, **3**(6), 610–621.

[456] Mack, A., and Ellery, A. (2015) "Autonomous science target identification using navigation cameras for planetary rovers," submitted to *Int. J. Advanced Robotic Systems*.

[457] Materka, A., and Strzelecki, M. (1998) "Texture analysis methods—a review," *COST B11 Report*, Institute of Electronics, Technical University of Lodz.

[458] Mallot, H. (1997) "Behaviour-oriented approaches to cognition: Theoretical perspectives," *Theory in Biosciences*, **116**, 196–220.

Weszka, J., Dyer, C., and Rosenfeld, A. (1976) "Comparative study of texture measures for terrain classification," *IEEE Trans. Systems Man & Cybernetics*, **6**(4), 269–285.

[459] Jain, A., Ratha, N., and Lakshmanan, S. (1997) "Object detection using Gabor filters," *Pattern Recognition*, **30**(2), 295–309.

[460] Xu, G., Ming, X., and Yang, N. (2004) "Gabor filter optimisation design for iris texture analysis," *J. Bionic Engineering*, **1**(1), 72–78.

[461] Wilson, R., Calway, A., and Pearson, E. (1992) "Generalised wavelet transform for Fourier analysis: The multiresolution Fourier transform and its application to image and audio analysis," *IEEE Trans. Information Theory*, **38**(2), 674–690.

[462] Jain, A., and Farrokhnia, F. (1990) "Unsupervised texture segmentation using Gabor filters," *IEEE International Conference on Systems, Man & Cybernetics*, pp. 14–19.

[463] Turner, M. (1986) "Texture discrimination by Gabor functions," *Biological Cybernetics*, **55**, 71–82.

[464] Lee, T. (1996) "Image representation using 2D Gabor wavelets," *IEEE Trans. Pattern Analysis & Machine Intelligence*, **18**(10), 959–971.

[465] Daugman, J. (1988) "Complete discrete 2D Gabor transforms by neural networks for image analysis and compression," *IEEE Trans. Acoustics, Speech & Signal Processing*, **36**(7), 1169–1179.

[466] Mikolajczyk, K., and Schmid, C. (2005) "Performance evaluation of local descriptors," *IEEE Trans. Pattern Analysis & Machine Intelligence*, **27**(10), 1615–1630.

[467] Tianxu, Z., Nong, S., Guoyou, W., and Xiaowen, L. (1996) "Effective method for identifying small objects on a complicated background," *Artificial Intelligence in Engineering*, **10**, 343–349.
Mallat, S. (1989) "Multifrequency channel decomposition of images and wavelet models," *IEEE Trans. Acoustics, Speech & Signal Processing*, **37**(17), 2091–2110.
Lee, T. (1996) "Image representation using 2D Gabor wavelets," *IEEE Trans. Pattern Analysis & Machine Intelligence*, **18**(10), 959–971.
Laine, A., and Fan, J. (1993) "Texture classification by wavelet packet signatures," *IEEE Trans. Pattern Analysis & Machine Intelligence*, **15**(11), 1186–1191.
Porat, M., and Zeevi, Y. (1988) "Generalised Gabor scheme of image representation in biological and machine vision," *IEEE Trans. Pattern Analysis & Machine Intelligence*, **10**(4), 452–468.
Porter, R., and Canagarajah, N. (1997) "Robust, rotation-invariant texture classification: Wavelet, Gabor filter and GMRF based schemes," *IEE Proc. Vision & Image Processing*, **144**(3), 180–188.

[468] Ozen, S., Bouganis, A., and Shanahan, M. (2007) "Fast evaluation criterion for the recognition of occluded shapes," *Robotics & Autonomous Science*, **55**, 741–749.

[469] Xu, L., Oja, E., and Kultanen, P. (1990) "New curve detection method: Randomized Hough transform (HRT)," *Pattern Recognition Letters*, **11**, 331–338.

[470] Kass, M., Witkin, A., and Terzopoulos, D. (1988) "Snakes: Active contour models," *Int. J. Computer Vision*, **1**(4), 321–331.

[471] Kim, W., Lee, C-Y., and Lee, J-J. (2001) "Tracking moving object using Snake's jump based on image flow," *Mechatronics*, **11**, 99–226.
Durbin, R., Szeliski, R., and Yuille, A. (1989) "Analysis of the elastic net approach to the travelling salesman problem," *Neural Computation*, **1**, 348–358.
Durbin, R., and Willshaw, D. (1987) "Analogue approach to the travelling salesman problem using an elastic net method," *Nature*, **326**, 689–891.

[472] Zhu, S., and Yuille, A. (1996) "Region competition: Unifying snakes, region growing, and Bayes/MDL for multiband image segmentation," *IEEE Trans. Pattern Analysis & Machine Intelligence*, **18**(9), 884–900.

[473] Herman, H., and Schempf, H. (1992) *Serpentine Manipulator Planning and Control for the NASA Space Shuttle Payload Servicing* (CMU Tech. Report CMU-RI-TR-92-10). Carnegie Mellon University, Pittsburgh, PA.

[474] Xu, C., and Prince, J. (1998) "Snakes, shapes and gradient vector flow," *IEEE Trans. Image Processing*, **7**(3), 359–369.

[475] Von Tonder, G., and Kruger, J. (1997) "Shape encoding: A biologically inspired method of transforming boundary images into ensembles of shape-related features," *IEEE Trans. Systems Man & Cybernetics, Part B: Cybernetics*, **27**(5), 749–759.

[476] Shen, D., and Ip, H. (1997) "Generalised affine invariant image normalisation," *IEEE Trans. Pattern Analysis & Machine Intelligence*, **19**(5), 431–433.

[477] Wood, J. (1996) "Invariant pattern recognition: A review," *Pattern Recognition*, **29**(1), 1–17.

[478] de Croon, G., de Weerdt, E., de Wagner, C., and Remes, B. (2011) "Appearance variation cue for obstacle avoidance," *Proceedings IEEE International Conference Robotics & Biomimetics*, pp. 1606–1611.

[479] Aloimonos, J. (1988) "Visual shape computation," *Proc. IEEE*, **76**(8), 899–916.

[480] Kanade, T., Binford, T., Poggio, T., and Rosenfeld, A. (1990) "Vision," *Annual Reviews Computer Science*, **4**, 517–529.

[481] Tomasi, C., and Zhang, J. (1995) "Is structure-from-motion worth pursuing?" *Proceedings Seventh International Symposium Robotics Research*, Springer-Verlag, New York.

[482] Zhang, R., Tsai, P-S., Cryer, J., and Shah, M. (1999) "Shape from shading: A survey," *IEEE Trans. Pattern Analysis & Machine Intelligence*, **21**(8), 690–706.

[483] Misu, T., Hashimoto, T., and Ninomiya, K. (1999) "Terrain shape recognition for celestial landing/rover missions from shade information," *Acta Astronautica*, **45**(4/9), 357–364.

[484] Stevens, K. (1981) "Information content of texture gradients," *Biological Cybernetics*, **42**, 95–105.

[485] Pankathi, S., and Jain, A. (1995) "Integrating vision modules: Stereo, shading, grouping and line labelling," *IEEE Trans. Pattern Analysis & Machine Intelligence*, **17**(8), 831–842.

[486] Cryer, J., Tsai, P-S., and Shah, M. (1993) "Integration of shape from X modules: Combining stereo and shading," *Proceedings IEEE International Conference Robotics and Automation*, pp. 720–721.

[487] Pentland, A. (1984) "Fractal-based description of natural scenes," *IEEE Trans. Pattern Analysis & Machine Intelligence*, **6**(6), 661–674.

[488] Chaudhuri, B., and Sarkar, N. (1995) "Texture segmentation using fractal dimension," *IEEE Trans. Pattern Analysis & Machine Intelligence*, **17**(1), 72–77.

[489] Bhanu, B., Symosek, P., and Das, S. (1997) "Analysis of terrain using multispectral images," *Pattern Recognition*, **30**(2), 197–215.

[490] Julesz, B. (1981) "Textons, the elements of texture perception and their interactions," *Nature*, **290**, 91–97.

[491] Leung, T., and Malik, J. (2001) "Representing and recognizing the visual appearance of materials using 3D textons," *Int. J. Computer Vision*, **43**(1), 29–44.

[492] Jain, A., Murty, M., and Flynn, P. (1999) "Data clustering: A review," *ACM Computing Surveys*, **31**(3), 264–323.

[493] Shotton, J., Winn, J., Rother, C., and Criminisi, A. (2006) "TextonBoost: Joint appearance, shape and context modeling for multi-class object recognition and segmentation," *Proceedings European Conference on Computer Vision*, pp. 1–15.

[494] Varma, M., and Zisserman, A. (2005) "Statistical approach to texture classification from single images," *Int. J. Computer Vision*, **62**(1/2), 61–81.

[495] Varma, M., and Zisserman, A. (2003) "Texture classification: Are filter banks necessary?" *Proceedings IEEE Computer Society Conference on Computer Vision and Pattern Recognition*.

[496] Poggio, G., and Poggio, T. (1984) "Analysis of stereopsis," *Annual Reviews in Neuroscience*, **7**, 379–412.

[497] Brown, M., Burschka, D., and Hager, G. (2003) "Advances in computational stereo," *IEEE Trans. Pattern Analysis & Machine Intelligence*, **25**(6), 993–1003.

[498] Krotkov, E., Henriksen, K., and Kories, R. (1990) "Stereo-ranging with verging cameras," *IEEE Trans. Pattern Analysis & Machine Intelligence*, **12**(12), 1200–1205.

[499] Brown, M., Burschka, D., and Hager, G. (2003) "Advances in computational stereo," *IEEE Trans. Pattern Analysis & Machine Intelligence*, **25**(8), 993–1008.

[500] van der Mark, W., Groen, F., and van den Heuvel, J. (2001) "Stereo based navigation in unstructured environments," *Proceedings IEEE Instrument and Measurement Technology Conference, Budapest, Hungary*.

[501] Binford, T. (1982) "Survey of model-based image analysis systems," *Int. J. Robotics Research*, **1**(1), 18–64.

[502] Matthias, L. (1992) "Stereovision for planetary rovers: Stochastic modelling to near real-time implementation," *Int. J. Computer Vision*, **8**(1), 71–91.

[503]

[504] Sanger, T. (1988) "Stereo disparity computation using Gabor filters," *Biological Cybernetics*, **59**, 405–418.

[505] Marefat, M., and Wu, L. (1996) "Purposeful gazing and vergence control for active vision," *Robotics & CIM*, **12**(2), 135–155.

[506] Bernardino, A., and Santos-Victor, J. (1998) "Visual behaviours for binocular track-ing," *Robotics & Autonomous Systems*, **25**, 137–146.

[507] Ho, A., and Pang, T-C. (1996) "Cooperative fusion of stereo and mission," *Pattern Recognition*, **29**(1), 121–130.

[508] Parkes, S. (1993) "Towards real-time stereovision systems for planetary missions," *Proceedings International Symposium Missions, Technologies and Design of Planetary Mobile Vehicles, Toulouse, September 1992*, CNES/Cipaduhs-Iditions, Toulouse, France (ISBN 2854283317).

[509] Takeno, J., and Rembold, U. (1996) "Stereovision systems for autonomous mobile robots," *Robotics & Autonomous Systems*, **18**, 355–363.

[510] Jin, S., Cho, J., Pham, D., Lee, K., Park, S-K., Kim, M., and Jeon, J. (2010) "FPGA design and implementation of a real-time stereo vision system," *IEEE Trans. Circuits & Systems for Video Technology*, **20**(1), 15–26.

[511] Mead, C. (1990) "Neuromorphic electronic systems," *Proc. IEEE*, **78**(10), 1629–1636.

[512] Orchard, G., Bartolozzi, C., and Indiveri, G. (2009) "Applying neuromorphic vision sensors to planetary landing tasks," *Proceedings IEEE International Conference Robotics and Automation*, pp. 201–204.

[513] Barnes, N., and Sandini, G. (2000) "Direction control for an active docking behaviour based on the rotational component of log-polar optic flow," *Lecture Notes in Computer Science*, **1843**, 167–181.

[514] Sunderhauf, N., and Protzel, P. (2006) "Stereo odometry: A survey," *Towards Autonomous Robotics*, University of Surrey, Guildford, UK.

[515] Cheng, Y., Maimone, M., and Matthies, L. (2006) "Visual odometry on the Mars Exploration Rovers," *IEEE Robotics & Automation Magazine*, **13**(2), 54–62.

[516] Li, R., Di, K., Matthies, L., Arvidson, R., Folkner, W., and Archinal, B. (2004) "Rover localization and landing site mapping technology for the 2003 Mars Exploration Rover mission," *Photogrammetric Engineering & Remote Sensing*, **70**(1), 77–90.

[517] Helmick, D., Cheng, Y., Clouse, D., Matthies, L., and Roumeliotis, S. (2004) "Path following using visual odometry for a Mars rover in high-slip environments," *IEEE Proc. Aerospace Conf.*, Vol. 2, pp. 772–789.

[518] Comaniciu, D., Ramesh, V., and Meer, P. (2000) "Real-time tracking of non-rigid objects using mean shift," *IEEE Conference Computer Vision Processing*.

[519] Campbell, J., Sukthankar, R., and Nourbakhsh, I. (2003) "Techniques for evaluating optical flow for visual odometry in extreme terrain," *Proceedings IEEE/RSJ International Conference Intelligent Robots & Systems*, Vol. 4, pp. 3704–3711.

[520] Olson, C., Matthies, L., Wright, J., Li, R., and Di, K. (2003) *Visual Terrain Mapping for Mars Exploration* (IEEE Paper 1176), Institute of Electrical and Electronic Engineers, Piscataway, NJ.

[521] Aloimonos, Y., Weiss, I., and Bandyopadyay, A. (1988) "Active vision," *Int. J. Computer Vision*, **1**(4), 333–356.

[522] Cheng, Y., Maimone, M., and Matthies, L. (2006) "Visual odometry on the Mars Exploration Rovers," *IEEE Robotics & Automation Magazine*, **13**(2), 54–62.

[523] Li, R. et al. (2005) "Initial results of rover localisation and topographic mapping for the 2003 Mars Exploration Rover mission," *Photogrammetric Engineering & Remote Sensing*, **71**(10), 1129–1142.

[524] Olson, G. (2002) "Image motion compensation with frame transfer CCDs," *Proc, SPIE*, **4567**, 153–160.

[525] Ben-Ezra, M., and Nayar, S. (2004) "Motion-based motion deblurring," *IEEE Trans. Pattern Analysis & Machine Intelligence*, **26**(6), 689–698.

[526] Ellery, A. (2007) *Optic-flow Based Autonomous Navigation for the ExoMars Rover* (PPARC Final Report, CREST programme), Particle Physics and Astronomy Research Council (now STFC), Swindon, U.K.

[527] Mallot, H., Bulthoff, H., Little, J., and Bohrer, S. (1991) "Inverse perspective mapping simplifies optical flow computation and obstacle detection," *Biological Cybernetics*, **64**, 177–185.

[528] Tan, S., Dale, J., Anderson, A., and Johnston, A. (2006) "Inverse perspective mapping and optic flow: A calibration method and a quantitative analysis," *Image & Vision Computing*, **24**, 153–165.

[529] Eklundh, J.-O., and Christensen, H. (2001) "Computer vision: Past and future," in R. Wilhelm (Ed.), *Informatics: 10 Years Back; 10 Years Ahead* (Lecture Notes in Computer Science), Springer-Verlag, Berlin, pp. 328–340.

[530] DeSouza, G., and Kak, A. (2002) "Vision for mobile robot navigation: A survey," *IEEE Trans. Pattern Analysis & Machine Intelligence*, **24**(2), 237–267.

[531] Ellery, A. (2007) *Optic-flow Based Autonomous Navigation for the ExoMars Rover* (PPARC Final Report, CREST programme), Particle Physics and Astronomy Research Council (now STFC), Swindon, U.K.

[532] Vedula, S., Rander, P., Collins, R., and Kanade, T. (2005) "Three dimensional scene flow," *IEEE Trans. Pattern Analysis & Machine Intelligence*, **27**(3), 475–480.

[533] Verri, A., and Poggio, T. (1989) "Motion field and optic flow: Qualitative properties," *IEEE Trans. Pattern Analysis & Machine Intelligence*, **11**(5), 490–498.

[534] Horn, B., and Schunck, B. (1981) "Determining optic flow," *Artificial Intelligence*, **17**, 185–203.

[535] Waxman, A., Kamgar-Parsi, B., and Subbarao, M. (1987) "Closed form solutions to image flow equations for 3D structure and motion," *Int. J. Computer Vision*, **1**, 239–258.

[536] Dias, J., Paredes, C., Fonseca, I., Araujo, H., Batista, J., and Almeida, A. (1998) "Simulating pursuit with machine experiments with robots and artificial vision," *IEEE Trans. Robotics & Automation*, **14**(1), 1–18.

[537] Barrows, G., Chahl, J., and Srinavasan, M. (2003) "Biomimetic visual sensing and flight control," *Aeronautical J.*, **107**(1069), 159–168.

[538] Sundareswaran, V. (1991) "Egomotion from global flow field data," *Proceedings IEEE Workshop Visual Motion*, pp. 140–145.

[539] Campbell, J., Sukthankar, R., and Nourbakhsh, I. (2004) "Techniques for evaluating optical flow for visual odometry in extreme terrain," *Proceedings IEEE/RSJ International Conference Intelligent Robots & Systems* (IRP-TR-04-06).

[540] Santos-Victor, J., and Sandini, G. (1997) "Embedded visual behaviours for navigation," *Robotics & Autonomous Systems*, **19**, 299–313.

[541] Camus, T. (1997) "Real-time quantized optical flow," *Real-Time Imaging*, **3**, 71–86.

[542] Dev, A., Krose, B., and Groen, F. (1997) "Navigation of a mobile robot on the temporal development of the optic flow," *Proc. IROS*, **97**, 558–563.

[543] Franceschini, N., Ruffler, F., and Serres, J. (2007) "Bio-inspired flying robot sheds light on insect piloting abilities," *Current Biology*, **17**, 329–335.

[544] Neumann, T., and Bulthoff, H. (2002) "Behaviour-oriented vision for biomimetic flight control," *Proceedings EPSRC/BBSRC International Workshop Biologically Inspired Robotics: The Legacy of W. Grey Walter, HP Labs, Bristol, U.K.*, pp. 196–203.

[545] Nelson, R., and Aloimonos, Y. (1989) "Obstacle avoidance using flow field divergence," *IEEE Trans. Pattern Analysis & Machine Intelligence*, **11**(10), 1102–1106.

[546] Coombs, D., Herman, M., Hong, T-H., and Nashman, M. (1998) "Real-time obstacle avoidance using central flow divergence and peripheral flow," *IEEE Trans. Robotics & Automation*, **14**(1), 49–59.

[547] Regan, D., and Gray, R. (2000) "Visually guided collision avoidance and collision achievement," *Trends in Cognitive Sciences*, **4**(3), 99–107.

[548] Borst, A. (1990) "How do flies land? From behaviour to neuronal circuits," *BioScience*, **40**(4), 292–299.

[549] Beauchemin, S., and Barron, J. (1995) "Computation of optical flow," *ACM Computing Surveys*, **27**(3), 433–467.

[550] Borst, A., and Egelhaaf, M. (1989) "Principles of visual motion detection," *Trends in Neural Sciences*, **12**(8), 297–306.

[551] Neumann, T., and Bulthoff, H. (2002) "Behaviour-oriented vision for biomimetic flight control," *Proceedings EPSRC/BBSRC International Workshop on Biologically-Inspired Robotics: The Legacy of W. Grey Walter*, pp. 196–203.

[552] Bruno, E., and Pellerin, D. (2002) "Robust motion estimation using spatial Gabor-like filters," *Signal Processing*, **82**, 297–309.

[553] Barron, J., Fleet, D., Beauchemin, S., and Burkitt, T. (1992) "Performance of optical flow techniques," *Proceedings IEEE Computer Society Conference on Computer Vision & Pattern Recognition*, pp. 236–242.

[554] Bober, M., and Kittler, J. (1994) "Robust motion analysis," *Proceedings IEEE Conference Computer Vision and Pattern Recognition, Seattle, WA*, pp. 947–952.

[555] Diaz, J., Ros, E., Pelayo, F., Ortigosa, E., and Mota, S. (2006) "FPGA-based real-time optical flow system," *IEEE Trans. Circuits & Systems for Video Technology*, **16**(2), 274–279.

[556] Spacek, L., and Burbridge, C. (2006) "Instantaneous robot motion estimation with omnidirectional vision," *Proceedings Towards Autonomous Robotic Systems Conference*.

[557] Hrabar, S., Sukhatme, G., Corke, P., Usher, K., and Roberts, J. (2005) "Combined optic-flow and stereo-based navigation of urban canyons for a UAV," *Proceedings IEEE/RSJ International Conference Robots & Systems*, pp. 3309–3316.

[558] Vidal, R., Shakernia, O., and Sastry, S. (2004) "Following the flock," *IEEE Robotics & Automation Magazine*, December, 14–20.

[559] De Croon, G., de Weerdt, E., de Wagter, C., and Remes, B. (2010) "Appearance variation cue for obstacle avoidance," *IEEE Trans. Robotics*, **28**(2), 529–534.

[560] Mack, A., and Ellery, A. (2009) "A method of real-time obstacle detection and avoidance using cameras for autonomous planetary rovers," *Towards Autonomous Robotic Systems (TAROS 09), Londonderry, U.K.*, pp. 112–118.

[561] Hildreth, E., and Koch, C. (1987) "Analysis of visual motion: From computational theory to neuronal mechanisms," *Annual Reviews of Neuroscience*, **10**, 477–533.

[562] Huang, T., and Netravali, A. (1994) "Motion and structure from feature correspondences," *Proc. IEEE*, **82**(2), 252–268.

[563] Soatto, S., and Perona, P. (1998) "Reducing structure-from-motion: A general framework for dynamic vision, Part 2: Implementation and experimental assessment," *IEEE Trans. Pattern Recognition & Machine Intelligence*, **20**(10), 11–17.

[564] Thakoor, S., Morookian, J., Chahl, J., and Zornetzer, S. (2004) "BEES: Exploring Mars with bioinspired technologies," *IEEE Computer*, September, 36–47.

[565] Sarpeshkar, R., Kramer, J., Indiveri, G., and Koch, G. (1996) "Analog VLSI architectures for motion processing: From fundamental limits to system applications," *Proc. IEEE*, **84**(7), 969–987.

[566] Higgins, C. (2002) *Airborne Visual Navigation Using Biomimetic VLSI Vision Chips* (Higgins Laboratory Technical Report), University of Arizona.

[567] Pudas, M., Viollet, S., Ruffier, F., Kruusing, A., Amic, S., Leppavuori, S., and Franceschini, N. (2007) "Miniature bio-inspired optic flow sensor based on low temperature co-fired ceramics (LTCC) technology," *Sensors & Actuators A*, **133**, 88–95.

[568] Aubepart, F., and Franceschini, N. (2007) "Bio-inspired optic flow sensors based on FPGA: Application to micro-air vehicles," *Microprocessors & Microsystems*, **31**, 408–419.

[569] Park, D-S., Kim, J-H., Kim, H-S., Parl, J-H., Shin, J-K., and Lee, M. (2003) "Foveated structure CMOS retina chip for edge detection with local light adaptation," *Sensors & Actuators A*, **108**, 75–80.

[570] Akishita, S., Kawamura, S., and Hisanobu, T. (1993) "Velocity potential approach to path planning for avoiding moving obstacles," *Advanced Robotics*, **7**(5), 463–478.

[571] Kunder, A., and Ravier, D. (1998) "Vision-based pragmatic strategy for autonomous navigation," *Pattern Recognition*, **31**(9), 1221–1239.

[572] Wood, S. (2004) "Representation and purposeful autonomous agents," *Robotics & Autonomous Systems*, **49**, 79–90.

[573] Aloimonos, J., Weiss, I., and Bandyopadhyay, A. (1988) "Active vision," *Int. J. Computer Vision*, **4**(1), 333–356.

[574] Bajcsy, R. (1988) "Active perception," *Proc. IEEE*, **76**(8), 996–1005.

[575]

[576]

[577] Sandini, G., and Tagliasco, V. (1980) "Anthropomorphic retina-like structure for scene analysis," *Computer Graphics & Image Processing*, **14**, 365–372.

[578] Ballard, D. (1991) "Animate vision," *Artificial Intelligence*, **48**, 57–86.

[579] Landholt, O. (2000) "Visual sensors using eye movements," in J. Ayers, J. Davis, and A. Rudolph (Eds.), *Neurotechnology for Biomimetic Robots*, Bradford Books, MIT Press, MA.

[580] Coombs, D., and Brown, C. (1991) "Cooperative gaze holding in binocular vision," *IEEE Control Systems*, June, 24–33.

[581] Liversedge, S., and Findlay, J. (2000) "Saccadic eye movements and cognition," *Trends in Cognitive Sciences*, **4**(1), 6–14.

[582] Rayner, K. (1998) "Eye movements in reading and information processing: 20 years of research," *Psychological Bulletin*, **124**(3), 372–422.

[583] Taylor, R. (2011) "Vision of beauty," *Physics World*, May, 22–27.

[584] Abbott, A. (1992) "Survey of selective fixation control for machine vision," *IEEE Control Systems*, August, 25–31.

[585] Henderson, J. (2003) "Human gaze control during real world scene perception," *Trends in Cognitive Sciences*, **7**(11), 498–504.

[586] Carpenter, R., and Williams, M. (1995) "Neural computation of log likelihood in control of saccadic eye movements," *Nature*, **377**, 59–61.

[587] Leopold, D., and Logothetis, N. (1999) "Multistable phenomena: Changing views in perception," *Trends in Cognitive Sciences*, **3**(7), 254–264.

[588] Denzler, J., and Brown, C. (2002) "Information theoretic sensor data selection for active object recognition and state estimation," *IEEE Trans. Pattern Analysis & Machine Intelligence*, **24**(2), 145–157.

[589] Torralba, A., Castelhano, M., Oliva, A., and Henderson, J. (2006) "Contextual guidance of eye movements and attention in real world scenes: The role of global features on object search," *Psychological Review*, **113**(4), 766–786.

[590] Richards, C., and Papanikopoulos, N. (1997) "Detection and tracking for robotic visual servoing systems," *Robotics & Computer Integrated Manufacturing*, **13**(2), 101–120.

[591] Epsiau, B., Chaumette, F., and Rives, P. (1992) "New approach to visual servoing in robotics," *IEEE Trans. Robotics & Automation*, **8**(3), 313–326.

[592] Schneider, W. (1998) "Introduction to 'Mechanisms of Visual Attention: A Cognitive Neuroscience Perspective'," *Visual Cognition*, **5**(1/2), 1–8.

[593] Bridgeman, B., van der Heijden, A., and Velichkovsky, B. (1994) "Theory of visual stability across saccadic eye movements," *Behavioural & Brain Sciences*, **17**(2), 247–292.

[594] Bernardino, A., and Santos-Victor, J. (1999) "Binocular tracking: Integrating perception and control," *IEEE Trans. Robotics & Automation*, **15**(6), 1080–1094.

[595] Conticelli, A.B., and Colombo, C. (1999) "Hybrid visual servoing: A combination of nonlinear control and linear vision," *Robotics & Autonomous Systems*, **29**, 243–256.

[596] Kosecka, J. (1997) "Visually guided navigation," *Robotics & Autonomous Systems*, **21**, 37–50.

[597] Cowan, N., Weingarten, J., and Koditschek, D. (2002) "Visual servoing via navigation functions," *IEEE Trans. Robotics & Automation*, **18**(4), 521–533.

[598] Días, J., Paredes, C., Fonseca, I., Araijo, H., Batista, J., and Almeida, A. (1998) "Simulating pursuit with machine experiments with robots and artificial vision," *IEEE Trans. Robotics & Automation*, **14**(1), 1–18.

[599] Buizza, A., and Schmid, R. (1982) "Visual–vestibular interaction in the control of eye movement: Mathematical modeling and computer simulation," *Biological Cybernetics*, **43**, 200–223.

[600] Lobo, T., and Dras, J. (2000) "Preserving 3D structure from images and inertial sensors," *Proceedings Sixth ESA Workshop on Advanced Space Technologies for Robotics and Automation (ASTRA), December 2000*, ESA/ESTEC, Noordwijk, The Netherlands.

[601] Miles, F., and Lisberger, S. (1981) "Plasticity in the vestibule-ocular reflex: A new hypothesis," *Annual Reviews of Neurosciences*, **4**, 273–299.

[602] Schweigart, G., Mergner, T., Evdokimidis, I., Morand, S., and Becker, W. (1997) "Gaze stabilization by optokinetic reflex (OKR) and vestibule-ocular reflex (VOR) during active head rotation," *Vision Research*, **37**(12), 1643–1652.

[603] Viollett, S., and Franceschini, N. (2005) "High speed gaze control system based on the vestibule-ocular reflex," *Robotics & Autonomous Systems*, **50**, 147–161.

[604] Murray, D., and Basu, A. (1994) "Motion tracking with an active camera," *IEEE Trans. Pattern Analysis & Machine Intelligence*, **16**(5), 449–459.

[605] Gosselin, C., and Hamel, J-F. (1994) "Agile eyes: A high performance three degree-of-freedom camera-orienting device," *Proceedings IEEE International Conference Robots & Automation*, pp. 781–786.

[606] Dasgupta, B., and Mruthyunjaya, T. (2000) "Stewart platform manipulator: A review," *Mechanism & Machine Theory*, **35**, 15–40.

[607] Maris, M. (2001) "Attention-based navigation in mobile robots using a reconfigurable sensor," *Robotics & Autonomous Systems*, **34**, 53–63.

[608] Ahuja, N., and Abbott, A. (1993) "Active stereo: Integrating disparity, vergence, focus, aperture and calibration for surface estimation," *IEEE Trans. Pattern Analysis & Machine Intelligence*, **15**(10), 1007–1029.

[609] Panerai, F., Metta, G., and Sandini, G. (2000) "Visuo-inertial stabilisation in space-variant binocular systems," *Robotics and Autonomous Systems*, **30**, 195–214.

[610] Ens, J., and Lawrence, P. (1993) "Investigation into methods for determining depth from focus," *IEEE Trans. Pattern Analysis & Machine Intelligence*, **15**(2), 97–108.

[611] Panerai, F., and Sandini, G. (1998) "Oculo-motor stabilisation reflexes: Integration of inertial and visual information," *Neural Networks*, **11**, 1191–1204.

[612] Panerai, F., Metta, G., and Sandini, G. (2000) "Visuo-inertial stabilisation in space-invariant binocular systems," *Robotics & Autonomous Systems*, **30**, 195–214.

[613] Panerai, F., Metta, G., and Sandini, G. (2000) "Adaptive image stabilisation: A need for vision-based active robotic agents," in J-A. Meyer et al. (Eds.), *From Animals to Animats: Proceedings 6th International Conference on Simulation of Adaptive Behaviour*, MIT Press, Cambridge, MA.

[614] Terzopoulos, D., and Rabie, T. (1995) "Animat vision: Active vision in artificial animals," *Proceedings IEEE International Conference Robotics and Automation*, pp. 801–808.

[615] Hosseini, H., Neal, M., and Labrosse, F. (2008) "Visual stabilization of an intelligent kite aerial photography platform," *Towards Autonomous Robotic Systems (TAROS)*, preprint.

[616] Papanikolopoulos, N., and Khosla, P. (1993) "Adaptive robotic visual tracking: Theory and experiment," *IEEE Trans. Automatic Control*, **38**(3), 429–445.

[617] Brown, C. (1990) "Gaze control with interactions and delays," *IEEE Trans. Systems Man & Cybernetics*, **20**(1), 518–527.

[618] Coombs, D., and Brown, C. (1991) "Cooperative gaze holding in binocular vision," *IEEE Control Systems*, June, 24–33.

[619] Shibata, T., Vijayakumar, S., Conradt, J., and Schaal, S. (2001) "Biomimetic oculomotor control," *Adaptive Behaviour*, **9**(3/4), 189–207.

[620] Dias, J., Paredes, C., Fonseca, I., Araujo, H., Batista, J., and Almeida, A. (1998) "Simulating pursuit with machine experiments with robots and artificial vision," *IEEE Trans. Robotics & Automation*, **14**(1), 1–18.

[621] Fukushima, K. (2003) "Frontal cortical control of smooth pursuit," *Current Opinion in Neurobiology*, **13**, 647–654.

[622] Das, S. (1997) "Biologically motivated neural network architecture for visuomotor control," *Information Sciences*, **96**, 27–45.

[623] Fukushima, K. (2003) "Frontal cortical control of smooth pursuit," *Current Opinion in Neurobiology*, **13**, 647–654.

[624] Shibata, T., and Schaal, S. (2001) "Biomimetic gaze stabilization based on feedback error learning with nonparametric regression networks," *Neural Networks*, **14**, 201–216.

[625] Ross, J., and Ellery, A. (2014) "Use of feedforward controllers for panoramic camera deployment on planetary rovers," submitted to *IEEE Trans. Robotics*.

[626] Shibata, T., Vijayakumar, S., Conradt, J., and Schaal, S. (2001) "Biomimetic oculomotor control," *Adaptive Behaviour*, **9**(3/4), 189–207.

[627] Peixto, P., Batista, J., and Araujo, H. (2000) "Integration of information from several vision systems for a common task of surveillance," *Robotics & Autonomous Systems*, **31**, 99–108.

[628] Grasso, E., and Tistarelli, M. (1995) "Active/dynamic stereo vision," *IEEE Trans. Pattern Analysis & Machine Intelligence*, **17**(9), 868–879.

[629] Fox, D., Burgand, W., and Thrun, S. (1998) "Active Markov localisation for mobile robots," *Robotics & Autonomous Systems*, **25**, 195–207.

[630] Davidson, A., and Murray, D. (2002) "Simultaneous localisation and map-building using active vision," *IEEE Trans. Pattern Analysis & Machine Intelligence*, **24**(7), 865–880.

[631] Whaite, P., and Ferrie, F. (1997) "Autonomous exploration: Driven by uncertainty," *IEEE Trans. Pattern Analysis & Machine Intelligence*, **19**(3), 193–205.

[632] Fitts, P. (1954) "Information capacity of the human motor system in controlling the amplitude of movement," *J. Exp. Psychol.*, **47**, 381–391.

[633] Bullock, D., and Grossberg, S. (1988) "Neural dynamics of planned arm movements: Emergent invariants and speed–accuracy during trajectory formation," *Psychological Review*, **95**(1), 49–80.

[634] Radix, C., Robinson, P., and Nurse, P. (1999) "Extension of Fitts' Law to modelling motion perormance in man–machine interfaces," *IEEE Trans. Systems Man & Cyber A: Systems & Humans*, **29**(2), 205–209.
Johannesen, L., and Woods, D. (1991) "Human interaction with intelligent systems: Trends, problems and new directions," *Proc. IEEE Int. Conf. Systems Man & Cybernetics*, 1337–1341.

[635] Greafe, V., and Bischoff, R. (1997) "Human interface for an intelligent mobile robot," *RoMan 97, Human Interfaces for Intelligent Mobile Robots*.

[636] Winfield, A., and Holland, O. (2000) "Applications of wireless local area network technology to the control of mobile robots," *Microprocessors & Microsystems*, **23**, 1065–1075.

[637] Hu, H. et al. (2001) "Internet-based robotic systems for teleoperation," *Assembly Automation*, **21**(2), 143–151.

[638] Penin, L. (2000) "Teleoperation with time delay: A survey and its use in space robotics," *Proceedings Sixth Advanced Space Technologies for Robotics and Automation (ASTRA 2000)*.

[639] Conway, L., Volz, R., and Walker, M. (1990) "Teleautonomous systems: Projecting and coordinating intelligent action at a distance," *IEEE Trans. Robotics & Automation*, **6**(2), 146–158.

[640] Bejczy, A., and Kim, W, (1990) "Predictive displays and shared compliance control for time-delayed telemanipulation," *Proceedings IEEE International Workshop on Intelligent Robots & Systems (IROS)*, pp. 407–412.

[641]

[642]

[643] Ambrose, R., Aldridge, H., Askew, R., Burridge, R., Bluethmann, W., Diftler, M., Lovchik, C., Magruder, D., and Rehnmark, F. (2000) "Robonaut: NASA's space humanoid," *IEEE Intelligent Systems*, July/August, 57–62.

[644] Rochlis, J., Clarke, J.-P., and Goza, S. (2001) *Space Station Telerobotics: Designing a Human–Robot Interface* (AIAA 2001-5110), American Institute of Aeronautics and Astronautics, Washington, D.C.

[645] Burdea, G. (1999) "Synergy between virtual reality and robotics," *IEEE Trans. Robotics & Automation*, **15**(3), 400–410.

[646] Brooker, J., Sharkey, P., Wann, J., and Plooy, A. (1999) "Helmet mounted display system with active gaze control for visual telepresence," *Mechatronics*, **9**, 703–716.
Poli, R., Cinel, C., Matran-Fernandez, A., Sepulveda, F., and Stoica, A. (2013) "Towards cooperative brain–computer interfaces for space navigation," *Proc. Int. Conf. Intelligent User Interfaces*, 149–160.

[647] Sheridan, T. (1993) "Space teleoperation through time delay: Review and prognosis," *IEEE Trans. Robotics & Automation*, **9**(5), 592–606.

[648] Arcaca, P., and Melchiorri, C. (2002) "Control schemes for teleoperation with time delay: A comparative study," *Robotics & Autonomous Systems*, **38**, 49–64.

[649]

[650]

[651] Flanagan, J. (1994) "Technologies for multimedia communication," *Proc. IEEE*, **82**(4), 590–603.

[652] Hirzinger, G. et al. (1994) "Multisensory shared autonomy and tele-sensor programming: Key issues in space robotics," *Robotics & Autonomous Systems*, **11**, 141–162.

[653] Hirzinger, G. et al. (1998) "Preparing a new generation of space robots: A survey of research at DLR," *Robotics & Autonomous Systems*, **23**, 99–106.

[654] Backes, P. (1994) "Prototype ground–remote telerobot control system," *Robotica*, **12**, 481–490.

[655] Fontaine, B. (2000) *Payload Support for Planetary Exploration, PSPE/ROBUST: Operations Scenario* (Space Application Services Tech. Rep. ROBUST-SAS-OS, Iss 1), Space Application Services, Leuvensesteenweg, Belgium.

[656] Fontaine, B. et al. (2000) "Autonomous operations of a micro-rover for geoscience on Mars," *Sixth ESA Workshop on Advanced Space Technologies for Robotics and Automation Conference (ASTRA), December 2000*, ESA/ESTEC, Noordwijk, The Netherlands.

[657] Van Winnendael, M., and Joudrier, L. (2004) *Control of ExoMars Rover Surface Operations: Potentially Relevant Results of R&D Activities* (Information Paper ESA-ESTEC TEC-MMA), ESA/ESTEC, Noordwijk, The Netherlands.

[658] Joudrier, L. (2004) *Summary of the DREAMS Concept, Its Main Functionalities and Its Applicability for Ground Control and Onboard Autonomy of the ExoMars Rover* (Information Paper ESA-ESTEC TEC-MMA), ESA/ESTEC, Noordwijk, The Netherlands.

[659] Visentin, G., Putz, P., and Columbina, G. (1997) "Towards a common European controller for space robots," *ESA Preparing for the Future*, **7**(2), 23–24.

[660] Bormann, G., Joudrier, L., and Kapellos, K. (2004) "FORMID: A formal specification and verification environment for DREAMS," *Eighthth ESA Workshop on Advanced Technologies for Robotics and Automation (ASTRA 2000), November 2000* (ESA WP-236), ESA/ESTEC, Noordwijk, The Netherlands.

[661] Sifakis, J. (1997) *Formal Methods and Their Evaluation*, FEMSYS, Munich, Germany.

[662] Kapellos, K. (2004) *A-DREAMS: A Synthetic View* (Trasys Space Document 08501.500-TI-70-TELEMAN = TRA-ADREAMS), Trasys, Hoeilaart, Belgium.

[663] Galardini, D., Kapellos, K., and Didot, F. (2002) "Distributed robot and automation environment and monitoring supervision utilisation in EUROPA," *Seventh ESA Workshop on Advanced Technologies for Robotics and Automation (ASTRA 2000), May 2000* (ESA WPP-179), ESA/ESTEC, Noordwijk, The Netherlands.

[664] Maurette, H. (1997) "Control and operation of planetary rover vehicles," *Preparing for the Future*, **7**(2), 10–11.

[665] Ziegler, M., Falkenhagen, L., ter Horst, R., and Kalivias, D. (1998) "Evolution of stereoscopic and three dimensional video," *Signal Processing: Image Communication*, **14**, 173–194.

[666] Gitelson, J., Bartsev, S., Mezhevikin, ., and Okhonin, V. (2003) "Alternative approach to solar system exploration providing safety of human mission to Mars," *Advances in Space Research*, **31**(1), 17–24.

[667] Ellery, A. (2001) "Robotics perspective on human spaceflight," *Earth Moon & Planets*, **87**, 173–190.

[668] Ellery, A. (2003) "Humans versus robots for space exploration and development," *Space Policy*, **19**, 87–91.

[669] Mendell, W. (2004) "Roles of humans and robots in exploring the solar system," *Acta Astronautica*, **55**, 149–155.

[670] Santos-Victor, J., and Sentiero, J. (2000) "Visual control of teleoperated cellular robots," *Sixth ESA Workshop on Advanced Space Technologies for Robotics and Automation (ASTRA), December 2000*, ESA/ESTEC, Noordwijk, The Netherlands.

[671] Oleson, S., Landis, G., McGuire, M., and Schmidt, G. (2011) "HERRO mission to Mars using telerobotic surface exploration from orbit," *J. British Interplanetary Society*, **64**, 304–313.
Drucker, J. (2011) "Humanities approach to interface theory," *Culture Machine*, **12**, ISSN 1465-4165.

[672] Fong, T., Nourbakhsh, I., and Dautenhahn, K. (2003) "Survey of socially interactive robots," *Robotics & Autonomous Systems*, **42**, 143–166.

[673] Chellappa, R., Wilson, C., and Sirohey, S. (1995) "Human and machine recognition of faces: A survey," *Proc. IEEE*, **83**(5), 705–740.

[674] Kumagai, J. (2006) "Halfway to Mars," *IEEE Spectrum*, March, 33–37.

[675] Li, R. et al. (2005) "Initial results of rover localisation and topographic mapping for the 2003 Mars Exploration Rover mission," *Photogrammetric Engineering & Remote Sensing*, **71**(10), 1129–1142.

[676] Biesiadecki, J., Leger, C., and Maimone, M. (2005) "Tradeoffs between directed and autonomous driving on the Mars Exploration Rovers," reprint.

[677] Erickson, J., Adler, M., Crisp, J., Mishkin, A., and Welch, R. (2002) "Mars exploration rover surface operations," *53rd International Astronautics Congress (WSC), Houston* (IAC-02-Q.3.1.03).

[678] McFarland, D., and Spier, E. (1997) "Basic cycles, utility and opportunism in self-sufficient robots," *Robotics & Autonomous Systems*, **20**, 179–190.

[679] Gat, E., Desai, R., Ivlev, R., Loch, J., and Miller, D. (1994) "Behaviour control for robotic exploration of planetary surfaces," *IEEE Trans. Robotics & Automation*, **10**(4), 490–503.

[680] Morrison, J., and Nguyen, T. (1998) *Onboard Software for the Mars Pathfinder Microrover* (IAA-L-0504P), International Academy of Astronautics, Stockholm, Sweden.

[681] Miller, D. (1993) "Mass of massive rover software," *Proceedings International Symposium on Missions, Technologies and Design of Planetary Mobile Vehicles, Toulouse, September 1992*, CNES/Cipaduhs-Iditions, Toulouse, France (ISBN 2854283317).

[682] Brooks, R. (1991) "New approaches to robotics," *Science*, **253**, 1227–1232.

[683] Mataric, M. (1998) "Behaviour-based robotics as a tool for synthesis of artificial behaviour and analysis of natural behaviour," *Trends in Cognitive Science*, **2**, 82–87.

[684] Braitenburg, V. (1984) *Vehicles: Experiments in Synthetic Psychology*, MIT Press, Cambridge, MA.

[685] Nolfi, S. (2002) "Power and limits of reactive agents," *Neurocomputing*, **42**, 119–145.

[686] Pfeifer, R. (1996) "Building 'fungus eaters': Design principles of autonomous agents," *Proceedings Fourth International Conference Simulation of Adaptive Behaviour*, pp. 3–12.

[687] Dean, J. (1998) "Animats and what they can tell us," *Trends in Cognitive Science*, **2**(2), 60–67.

[688] Hallam, J., and Malcolm, C. (1994) "Behaviour: Perception, action and intelligence—the view from situated robotics," *Phil. Trans. Royal Society*, **A349**(1689), 29–42.

[689] Moravec, H. (1983) "Stanford cart and the CMU rover," *Proc. IEEE*, **71**(7), 872–884.

[690] Brooks, R. (1991) *Intelligence without Reason* (MIT AI Memo 1293), Massachusetts Institute of Technology AI Laboratory, Boston.

[691] Malcolm, C., Smithers, T., and Hallam, J. (1989) "An emerging paradigm in robot architecture," *Proceedings International Conference on Intelligent Autonomous Systems 2, Amsterdam*.

[692] Brooks, R. (1991) "Intelligence without representation," *Artificial Intelligence*, **47**, 139–159.

[693] Brooks, R. (1990) "Elephants don't play chess," *Robotics & Autonomous Systems*, **6**, 3–15.

[694] Brooks, R. (1997) "From earwigs to humans," *Robotics & Autonomous Systems*, **20**, 191–304.

[695] Brooks, R., and Stein, L. (1994) "Building brains for bodies," *Autonomous Robots*, **1**, 7–25.

[696] Anderson, T., and Donath, M. (1990) "Animal behaviour as a paradigm for developing robot autonomy," *Robotics & Autonomous Systems*, **6**, 145–168.

[697] Columbetti, M., Dorigo, M., and Borghi, G. (1996) "Behaviour analysis and training: A methodology for behaviour engineering," *IEEE Trans. Systems Man & Cybernetics: Part B—Cybernetics*, **26**(3), 365–372.

[698] Brooks, R. (1983) "Robust layered control system for a mobile robot," *IEEE J. Robotics & Automation*, **2**(1), 14–23.

[699] Brooks, R. (1985) *Robust Layered Control System for a Mobile Robot* (MIT AI Memo 864), Massachusetts Institute of Technology AI Laboratory, Boston.

[700] Amir, E., and Maynard-Zhang, P. (2004) "Logic-based subsumption architecture," *Artificial Intelligence*, **153**, 167–237.

[701] Brooks, R. (1989) "A robot that walks: Emergent behaviours from a carefully evolved network," *Neural Computing*, **1**, 253–262.

[702] Connell, J. (1989) "Behaviour-based arm controller," *IEEE Trans. Robotics & Automation*, **5**(6), 784–791.

[703] Rosenblatt, J., Williams, S., and Durrant-Whyte, H. (2002) "Behaviour-based architecture for autonomous underwater exploration," *Information Sciences*, **145**, 69–87.

[704] Rosenblatt, J., and Payton, D. (1989) "Fine-grained alternative to the subsumption architecture for mobile robot control," *Proceedings IEEE/INNS Joint Conference Neural Networks*, pp. 317–324.

[705] Hu, H., and Brady, M. (1994) "Bayesian approach to real-time obstacle avoidance for a mobile robot," *Autonomous Robots*, **1**, 69–92.

[706] Barnes, D., Ghanea-Hercock, R., Aylett, R., and Coddington, A. (1997) "Many hands make light work? An investigation into behaviourally controlled co-operant autonomous mobile robots," *Proceedings First International ACM Conference on Autonomous Agents, Marina del Rey, CA*, pp. 413–420; Aylett, R., and Barnes, D. (1998) "Multi-robot architecture for planetary rovers," preprint.

[707] Kording, K., and Wolpert, D. (2006) "Bayesian decision theory in sensorimotor control," *Trends in Cognitive Sciences*, **10**(7), 319–328.

[708] Rosenblatt, J. (1997) "Utility fusion: Map-based planning in a behaviour-based system," *Field & Service Robotics* reprint.

[709] Miller, D. (1993) "Mass of massive rover software," *Proceedings International Symposium Missions, Technologies and Design of Planetary Mobile Vehicles, Toulouse, September 1992*, CNES/Cipaduhs-Iditions, Toulouse, France (ISBN 2854283317).

[710] Miller, D., and Varsi, G. (1993) "Microtechnology for planetary exploration," *Acta Astronautica*, **29**(7), 561–567.

[711] Miller, D. (1990) "Mini-rovers for Mars exploration," *Proc. Vision-21 Workshop, NASA Lewes Research Center.*
Miller, D. et al. (1989) "Autonomous navigation through rough terrain: Experimental results," *IFAC on Automatic Control, Tsukuba, Japan*, pp. 111–114.

[712] Miller, D. et al. (1993) "Experiments with a small behaviour-controlled planetary rover," *Proceedings International Symposium on Missions, Technologies and Design of Planetary Mobile Vehicles, Toulouse, September 1992*, CNES/Cipaduhs-Iditions, Toulouse, France (ISBN 2854283317).

[713] Gat, E., Desai, R., Ivlev, R., Loch, J., and Miller, D. (1994) "Behaviour control for robotic exploration of planetary surfaces," *IEEE Trans. Robotics & Automation*, **10**(4), 490–503.

[714] Miller, D., Desai, R., Gat, E., Ivlev, R., and Loch, J. (1992) "Reactive navigation through rough terrain: Experimental results," *Proceedings 10th National Conference Artificial Intelligence (AAAI-92), San Jose, CA.*

[715] Volpe, R. et al. (1997) "Rocky 7: A next generation Mars rover prototype," *J. Advanced Robotics*, **11**(4), 341–358.

[716] Brooks, R. (1991) *Intelligence without Reason* (MIT Artificial Intelligence Laboratory AI Memo No. 1293), MIT Press, Cambridge, MA.

[717] Mali, A. (1998) "Tradeoffs in making the behavior-based robotic systems goal-directed," *Proceedings IEEE International Conference Robotics and Automation*, pp. 1128–1133.

[718] Mali, A. (2002) "On the behaviour-based architectures of autonomous agency," *IEEE Trans. Systems Man & Cybernetics, C: Applications & Reviews*, **32**(3), 231–242.

[719]

[720] Gershenson, C. (2004) "Cognitive paradigms: Which one is best?" *Cognitive Systems Research*, **5**, 135–156.

[721] Tsoksos, J. (1995) "Behaviourist intelligence and the scaling problem," *Artificial Intelligence*, **75**, 135–160.

[722] Clark, A. (1997) "The dynamical challenge," *Cognitive Science*, **21**(4), 461–481.

[723] Antsaklis, P., Passino, K., and Wang, J. (1991) "Introduction to autonomous control systems," *IEEE Control Syst. Mag.*, June, 5–13.

[724] Tigli, J., Fayek, R., Liscano, R., and Thomas, M. (1994) "Methodology and computing model for a reactive mobile robot controller," *IEEE Int. Conf. Systems Man & Cybernetics*, **2**, 317–322.

[725] McManus, J., and Bynum, W. (1996) "Design and analysis techniques for concurrent blackboard systems," *IEEE Trans. Systems Man & Cybernetics A: Systems & Humans*, **26**(6), 669–680.

[726] Liscano, R., Manx, A., Stuck, E., Fayek, R., and Tigli, J.-Y. (1995) "Using a blackboard to integrate multiple activities and achieve strategic reasoning for mobile robot navigation," *IEEE Expert*, April, pp. 24–36.

[727] Maes, P. (1990) "Situated agents can have goals," *Robotics & Autonomous Systems*, **6**, 49–70.

[728] Mataric, M. (1992) "Integration of representation into goal-driven behaviour-based robots," *IEEE Trans. Robotics & Automation*, **8**(3), 304–312.

[729] Prescott, T., Redgrave, P., and Gurney, K. (1999) "Layered control architectures in robots and vertebrates," *Adaptive Behaviour*, **7**(1), 99–127.

[730] Hawes, N. (2000) "Real-time goal-oriented behaviour for computer game agents," *Proceedings First International Conference Intelligent Games and Simulation*, pp. 71–75.

[731] Chatila, R. (1995) "Deliberation and reactivity in autonomous mobile robots," *Robotics & Autonomous Systems*, **16**, 197–211.

[732] Kaelbling, L. (1990) "Action and planning in embedded agents," *Robotics & Autonomous Systems*, **6**, 35–48.

[733] Murphy, R., Hughes, K., Marzilli, A., and Noll, E. (1999) "Integrating explicit path planning with reactive control of mobile robots using Trulla," *Robotics & Autonomous Systems*, **27**, 225–245.

[734] Wilkins, D., Myers, K., Lowrance, J., and Wesley, L. (1995) "Planning and reacting in uncertain and dynamic environments," *J Experimental Theoretical Artificial Intelligence*, **7**, 121–152.

[735] Payton, D. et al. (1990) "Plan-guided reaction," *IEEE Trans. Systems Man & Cybernetics*, **20**(6), 1370–1382.

[736] Gat, E., Slack, M., Miller, D., and Firby, R. (1990) "Path planning and execution monitoring for a planetary rover," *IEEE International Conference on Robotics and Automation*, pp. 20–25.

[737] Courtois, P. (1975) "Decomposability, instabilities and saturation in multiprogramming systems," *Communications ACM*, **18**(7), 371–390.

[738] Courtois, P. (1985) "On time and space decomposition of complex structures," *Communications ACM*, **28**(6), 590–603.

[739] Conant, R. (1972) "Detecting subsystems of a complex system," *IEEE Trans. Systems Man & Cybernetics*, **2**(4), 550–553.

[740] Conant, R. (1974) "Information flows in hierarchical systems," *Int. J. General Systems*, **2**, 9–18.

[741] Conant, R. (1976) "Laws of information governing systems," *IEEE Trans. Systems Man & Cybernetics*, **6**(4), 240–255.

[742] Albus, J. (1999) "Engineering of mind," *Information Sciences*, **117**, 1–18.

[743] Albus, J., Lumia, R., and McCain, H. (1988) "Hierarchical control of intelligent machines applied to space station telerobotics," *IEEE Trans. Aero. & Elect. Syst.*, **24**(5), 535–541.

[744] Albus, J., McCain, H., and Lumia, R. (1987) *NASA/NBS Standard Reference Model for Telerobotic Control Systems Architecture* (NASA TN 1235), NASA, Washington, D.C.

[745] Albus, J. (1991) "Outline for a theory of intelligence," *IEEE Trans. Systems Man & Cybernetics*, **21**(3), 473–509.

[746] Noreils, F., and Chatila, R. (1995) "Plan execution monitoring and control architecture for mobile robots," *IEEE Trans. Robotics & Automation*, **11**(2), 255–266.

[747] Alami, R., Chatila, R., Fleury, S., Ghallab, M., and Ingrand, F. (1998) "Architecture for autonomy," *Int. J. Robotics Research*, **17**, 315–337.

[748] Saridis, G. (1983) "Intelligent robotic control," *IEEE Trans. Automatic Control*, **28**(5), 547–557.

[749] Chatila, R. (1995) "Deliberation and reactivity in autonomous mobile robots," *Robotics & Autonomous Systems*, **16**, 197–211.

[750] Schenker, P., Pirjanian, P., Balaram, B., Ali, K., Trebi-Ollennu, A., Huntsberger, T., Aghazarian, H., Kennedy, B., Baumgartner, E., Iagnemma, K. et al. (2000) "Reconfigurable robots for all terrain exploration," in G. McKee and P. Schenker (Eds.), *Proceedings SPIE Conference Sensor Fusion and Decentralised Control in Robotic Systems III*, Vol. 4196, pp. 454–467.

[751] Pirjanian, P., Huntsberger, T., Trebi-Ollennu, A., Aghazarian, H., Das, H., Joshi, S., and Schenker, P. (2000) "CAMPOUT: A control architecture for multi-robot planetary outposts," in G. McKee and P. Schenker (Eds.), *Proceedings SPIE on Sensor Fusion and Decentralised Control in Robotic Systems III*, Vol. 4196, pp. 221–229.

[752] Huntsberger, T., Pirjanian, P., Trebi-Ollennu, A., Nayar, H., Aghazarian, H., Ganino, A., Garrett, M., Joshi, S., and Schenker, P. (2003) "CAMPOUT: A control architecture for tightly coupled coordination of multirobot systems for planetary surface exploration," *IEEE Trans. Systems Man & Cybernetics A: Systems & Humans*, **33**(5), 550–559.

[753] Pirjanian, P., Christensen, H., and Fayman, J. (1998) "Application of voting to fusion of purposive modules: An experimental investigation," *Robotics & Autonomous Systems*, **23**, 253–266.

[754]

[755] Huntsberger, T. (2001) "Biologically inspired autonomous rover control," *Autonomous Robots*, **11**, 341–346.

[756] Huntsberger, T., Aghazarian, H., Baumgartner, E., and Schenker, P. (2000) "Behaviour-based control systems for planetary autonomous robot operations," *Proceedings IEEE International Conference Robotics and Automation*.

[757] Huntsberger, T., Mataric, M., and Pirjanian, P. (1999) "Action selection within the context of a robotic colony," *Proceedings Sensor Fusion and Deceentralised Control in Robotic Systems II, Boston* (SPIE 3839), Society of Photo-Optical Instrumentation Engineers, Bellingham, WA.

[758] Estlin, T., Rabideau, G., Mutz, D., and Chien, S. (2000) *Using Continuous Planning Techniques to Coordinate Multiple Rovers* (JPL Report), NASA Jet Propulsion Laboratory, Pasadena, CA.

[759] Bresina, J., Dorais, G., and Golden, K. (1999) "Autonomous rovers for human exploration of Mars," reprint.

[760] Vera, A., and Simon, H. (1993) "Situated action: A symbolic interpretation," *Cognitive Science*, **17**, 7–48.

[761] Durrant-Whyte, H. (2001) *Critical Review of the State-of-the-Art in Autonomous Land Vehicle Systems and Technology* (SNL Report SAND2001-3685), Sandia National Laboratories, Albuquerque, NM.

[762] Duckett, T., and Nehmzow, U. (1998) "Mobile robot self-localisation and measurement of performance in middle-scale environments," *Robotics & Autonomous Systems*, **24**, 57–69.

[763] Nehmzow, U. (1995) "Animal and robot navigation," *Robotics & Autonomous Systems*, **15**, 71–81.

[764] Lemon, O., and Nehmzow, U. (1998) "The scientific status of mobile robotics: Multi-resolution mapbuilding as a case study," *Robotics & Autonomous Systems*, **24**, 5–15.

[765] Tomatis, N., Nourbakhsh, I., and Siegwart, R. (2003) "Hybrid simultaneous localisation and mapbuilding: A natural integration of topological and metric," *Robotics & Autonomous Systems*, **44**, 3–14.

[766] Kuipers, B., and Byun, Y.-T. (1981) "Robot exploration and mapping strategy based on a semantic hierarchy of spatial representations," *Robotics & Autonomous Systems*, **8**, 47–63.

[767] Filliat, D., and Meyer, J.-A. (2003) "Map-based navigation in mobile robots: I. A review of localisation strategies," *Cognitive Systems Research*, **4**, 243–282.

[768] Thrun, S. (1998) "Learning metric–topological maps for indoor mobile robot navigation," *Artificial Intelligence*, **99**, 21–71.

[769] Nehmzow, U. (1985) "Animal and robot navigation," *Robotics & Autonomous Systems*, **15**, 71–82.

[770] Blanco, J.-L., Fernández-Madrigal, J.-A., and González, J. (2008) "Towards a unified Bayesian approach to hybrid metric–topological SLAM," *IEEE Trans. Robotics*, **24**(2), 259–270.

[771] Milford, M., and Wyeth, G. (2009) "Persistent navigation and mapping using a biologically inspired SLAM system," *Int. J. Robotics Research*, **29**(9), 1131–1153.

[772] Lambrinos, D., Moller, R., Labhart, T., Pfeifer, R., and Wehner, R. (2000) "Mobile robot employing insect strategies for navigation," *Robotics & Autonomous Systems*, **30**, 39–64.

[773] Nehmzow, U. (1993) "Animal and robot navigation," in L. Steels (Ed.), *The Biology and Technology of Intelligent Autonomous Agents* (NATO ASI series 920908), North Atlantic Treaty Organization, Brussels, Belgium.

[774] Marsland, U., Nehmzow, U., and Duckett, T. (2001) "Learning to select distinctive landmarks for mobile robot navigation," *J. Robotics & Autonomous Systems*, **37**(4), 241–260.

[775] Olson, C. (2000) "Probabilistic self-localisation for mobile robots," *IEEE Trans. Robotics & Automation*, **16**(1), 55–66.

[776] Li, R. et al. (2005) "Initial results of rover localisation and topographic mapping for the 2003 Mars Exploration Rover mission," *Photogrammetric Engineering & Remote Sensing*, **71**(10), 1129–1142.

[777] Hoffman, R., and Krotkov, E. (1992) "Terrain mapping for long-duration autonomous walking," *Proceedings IEEE/RSJ International Conference Intelligent Robots and Systems*, pp. 563–568.

[778] Morlans, R., and Liegois, A. (1993) "DTM-based path planning method for planetary rovers," *Proceedings of the International Symposium on Missions, Technologies and Design of Planetary Mobile Vehicles, Toulouse, September 1992*, CNES/Cipaduhs-Iditions, Toulouse, France (ISBN 2854283317).

[779] Maurette, H. (1997) "Control and operation of planetary rover vehicles," *Preparing for the Future*, **7**(2), 10–11.

[780] Hu, H., and Brady, M. (1994) "A Bayesian approach to real-time obstacle avoidance for a mobile robot," *Autonomous Robots*, **1**, 68–92.

[781] Gancet, J., and Lacroix, S. (2003) "PG2P: A perception-guided path planning approach for long range autonomous navigation in unknown natural environments," *Proceedings IEEE/RSJ International Conference Intelligent Robots and Systems*, pp. 2992–2997.

[782] Tarokh, M., Shiller, Z., and Hayati, S. (1999) "Comparison of two traversability based path planners for planetary rovers," *Proceedings Fifth International Symposium Artificial Intelligence Robotics and Automation in Space* (ESA SP-440), pp. 151–157, ESA, Noordwijk, The Netherlands.

[783] Saffiotti, A. (1997) "Uses of fuzzy logic in autonomous navigation: A catalogue raisonné," *Soft Computing*, **1**(4), 180–197.

[784] Seraji, H. (1999) "Traversability index: A new concept for planetary rovers," *Proceedings Fifth International Symposium Artificial Intelligence Robotics and Automation in Space* (ESA SP-440), pp. 159–164, ESA, Noordwijk, The Netherlands.

[785] Collins, A., and Loftus, E. (1974) "Spreading activation theory of semantic networks," *Psychological Review*, **82**, 407–428.

[786] Gat, E. et al. (1994) "Behaviour control for robotic exploration of planetary surfaces," *IEEE Trans. Robotics & Automation*, **10**(4), 490–503.

[787] Lacroix, S., Mallet, A.D., Brauzil, G., Fleury, A., Herrb, M., and Chatila, R. (2002) "Autonomous rover navigation on unknown terrains functions and integration," *Int. J. Robotics Research*, **21**(10/11), 917–942.

[788] Olson, C. (2000) "Probabilistic self-localisation for mobile robots," *IEEE Trans. Robotics & Autonomous Systems*, **16**(1), 55–66.

[789] Olson, C., Matthies, L., Schoppers, M., and Maimone, M. (2001) "Stereo ego-motion improvements for robust rover navigation," *Proceedings IEEE International Conference Robotics and Automation*, pp. 1099–1104.

[790] Goldberg, S., Maimone, M., and Matthies, L. (2002) "Stereovision and rover navigation software for planetary exploration," *Proceedings IEEE Aerospace Conference, Big Sky, MT*, pp. 2025–2036.

[791] Lacroix, S., Mallet, A.D., Brauzil, G., Fleury, A., Herrb, M., and Chatila, R. (2002) "Autonomous rover navigation on unknown terrains functions and integration," *Int. J. Robotics Research*, **21**(10/11), 917–942.

[792] Maurette, M. (2003) *CNES Autonomous Navigation: Basic Description and Preliminary Requirements* (CNES Direction des Techniques Spatiale Report Ref DTS/AE/SEA/ER/2003-009), Centre National d'Études Spatiales, Toulouse, France.

[793] Chatila, R. et al. (1999) *Motion Control for a Planetary Rover* (LAAS/CNRS Report 99311), Laboratory for Analysis and Architecture of Systems, Toulouse, France.

[794] Lacroix, S., Mallet, A., and Chatila, R. (1998) *Rover Self-localisation in Planetary-like Environments* (LAAS/CNRS Report 99234), Laboratory for Analysis and Architecture of Systems, Toulouse, France.

[795] Lacroix, S., Mallet, A., Chatila, R., and Gallo, L. (1999) "Rover self-localisation in planetary-like environments," *Proceedings Fifth International Symposium Artificial Intelligence Robotics and Automation in Space*, ESA/ESTEC, Noordwijk, The Netherlands.

[796] Lacroix, S. et al. (2000) "Autonomous long range rover navigation in planetary-like environments," *Sixth ESA Workshop on Advanced Space Technologies for Robotics and Automation (ASTRA), December 2000*, ESA/ESTEC, Noordwijk, The Netherlands.

[797] Mallet, A., Lacroix, S., and Gallo, L. (2000) "Position estimation in outdoor environments using pixel tracking and streovision," *IEEE International Conference Robotics and Automation*, pp. 3519–3524.

[798] Devy, M., Chatila, R., Fillatreau, P., Lacroix, S., and Nashashibi, F. (1995) "On autonomous navigation in a natural environment," *Robotics & Autonomous Systems*, **16**, 5–16.

[799] Chatila, R., and Laumond, J.-P. (1985) "Position referencing and consistent world modelling for mobile robots," *IEEE International Conference Robotics & Automation*, pp. 138–145.

[800] Maurette, M., and Baumgartner, E. (2001) "Autonomous navigation ability: FIDO test results," *Proceedings Advanced Space Technologies in Robotics & Automation (ASTRA)*, ESTEC, Noordwijk, The Netherlands.

[801] Baumgartner, E., Aghazarian, H., and Trebi-Ollennu, A. (2001) "Rover localisation results for the FIDO rover," in G. McKee and P. Schenker (Eds.), *Proceedings SPIE Sensor Fusion & Decentralised Control in Robotic Systems IV*, Vol. 4571.

[802] Cherif, M. (1999) "Motion planning for all-terrain vehicles: A physical modeling approach for coping with dynamic and contact interaction constraints," *IEEE Trans. Robotics & Automation*, **15**(2), 202–218.

[803] Brooks, R. (1991) "Challenges for complete creature architectures," *Proceedings First International Conference on Simulation of Adaptive Behaviour (from Animals to Animats)*, pp. 434–443.

[804] Brady, M., and Hu, H. (1994) "Mind of a robot," *Phil. Trans. Royal Society*, **349**, 15–28.

[805] Durrant-Whyte, H. (2001) *Critical Review of the State-of-the-Art in Autonomous Land Vehicle Systems and Technology* (SNL Report SAND2001-3685), Sandia National Laboratories, Albuquerque, NM.

[806] Berthoz, A., and Viaud-Delmon, I. (1999) "Multisensory integration in spatial orientation," *Current Opinion in Neurobiology*, **9**, 708–712.

[807] Murphy, K. (1998) *Inferene and Learning in Hybrid Bayesian Networks* (University of California Computer Science Division Report No. UCB/CSD-98-990), University of Caliornia Berkeley.

[808] Kalman, R. (1960) "New approach to linear filtering and prediction problems," *ASME J. Basic Engineering*, **82**, 35–45.

[809] Dasarathy, B. (1997) "Sensor fusion potential exploitation: Innovative architectures and illustrative applications," *Proc. IEEE*, **85**(1), 24–38.

[810] Kam, M., Zhu, X., and Kalata, P. (1997) "Sensor fusion for mobile robot navigation," *Proc. IEEE*, **85**, 108–119.

[811] Hostetler, L., and Andreas, R. (1983) "Nonlinear Kalman filtering techniques for terrain-aided navigation," *IEEE Trans. Automatic Control*, **28**(3), 315–323.

[812] Sorenson, H. (1970) "Least squares estimation: From Gauss to Kalman," *IEEE Spectrum*, **7**(7), 63–68.

[813] Luo, R., Yih, C-C., and Su, K. (2002) "Multisensor fusion and integration: Approaches, applications, and future directions," *IEEE Sensors J*, **2**(2), 107–119.

[814] Kalman, R. (1960) "New approach to linear filtering and prediction problems," *Trans. ASME J. Basic Engineering*, March, 35–45.

[815] Baumgartner, E., Aghazarian, H., and Trebi-Ollennu, A. (2001) "Rover localisation results for the FIDO rover," in G. McKee and P. Schenker (Eds.), *Proceedings SPIE Sensor Fusion & Decentralised Control in Robotic Systems IV*, Vol. 4571.

[816] Oriolo, G., Ulivi, G., and Vendittelli, M. (1998) "Real-time map building and navigation for autonomous robots in unknown environments," *IEEE Trans. Systems Man & Cyber B: Cybernetics*, **28**(3), 316–333.

[817] Chatila, R., and Laumond, J.-P. (1985) "Position referencing and consistent world modelling for mobile robots," *IEEE International Conference Robotics & Automation*, pp. 138–145.

[818] Singh, S., Simmons, R., Stentz, A., Verma, V., Yahja, A., and Schwehr, K. (2000) "Recent progress in local and global traversability for planetary rovers," *Proceedings IEEE International Conference Robotics and Autom, San Francisco*.

[819] Pai, D., and Reissell, L.-M. (1998) "Multiresolution rough terrain motion planning," *IEEE Trans. Robotics & Automation*, **14**(1), 19–33.

[820] Zhang, Z., and Faugeras, O. (1992) "3D world model builder with a mobile robot," *Int. J. Robotics Research*, **11**(4), 269–285.

[821] Barsham, B., and Durrant-Whyte, H. (1995) "Inertial navigation systems for mobile robots," *IEEE Trans. Robotics & Automation*, **11**(3), 328–342.

[822] Guivant, J., and Nebot, E. (2001) "Optimisation of the simultaneous localisation and map-building algorithm for real-time implementation," *IEEE Trans. Robotics & Automation*, **17**(3), 242–257.

[823] Guivant, J., Masson, F., and Nebot, E. (2002) "Simultaneous localisation and map building using natural features and absolute information," *Robotics & Autonomous Systems*, **40**, 79–90.

[824] Dissanayake, M., Newman, P., Clark, S., Durrant-Whyte, H., and Csorba, M. (2001) "Solution to the simultaneous localisation and map building (SLAM) problem," *IEEE Trans. Robotics & Automation*, **17**(3), 229–241.

[825] Madhavan, R., and Durrant-Whyte, H. (2004) "Natural landmark-based autonomous vehicle navigation," *Robotics & Autonomous Systems*, **46**, 79–95.

[826] Barshan, B., and Durrant-Whyte, H. (1995) "Inertial navigation systems for mobile robots," *IEEE Trans. Robotics & Automation*, **11**(3), 328–342.

[827]

[828] Arras, K., Tomatis, N., Jensen, B., and Siegwart, R. (2001) "Multisensor on-the-fly localisations: Precision and reliability for applications," *Robotics & Autonomous Systems*, **34**, 131–143.

[829] Pulford, G. (2005) "Taxonomy of multiple target tracking methods," *IEE Proc. Radar Sonar Navigation*, **152**(5), 291–297.

[830] Durrant-Whyte, H., and Bailey, T. (2006) "Simultaneous localisation and mapping (SLAM): The essential algorithms," *IEEE Robotics & Automation Magazine*, **13**(2), 99–110.

[831] Bakambu, J., Langley, C., and Mukherji, R. (2008) "Visual motion estimation: Localization performance evaluation tool for planetary rovers," *Proceedings*

International Symposium Artificial Intelligence & Robotics in Space, ESTEC, Noordwijk, The Netherlands.

[832] Masreliez, C., and Martin, R. (1977) "Robust Bayesian estimation for the linear model and robustifying the Kalman filter," *IEEE Trans. Automatic Control*, **22**(3), 361–371.

[833] Celebi, M. (1996) "Robust locally optimal filters: Kalman and Bayesian estimation theory," *Informatics & Computer Science*, **92**, 1–32.

[834] Deng, Z-L., Zhang, H-S., Liu, S-J., and Zhou, L. (1996) "Optimal and self-tuning white noise estimators with applications to deconvolution and filtering problems," *Automatica*, **32**(2), 199–216.

[835] Castellanos, J., Martinez-Cantin, R., Tardos, J., and Neira, J. (2007) "Robocentric map-joining: Improving the consistency of EKF-SLAM," *Robotics & Autonomous Systems*, **55**, 21–29.

[836] Bezdek, J. (1993) "Review of probabilistic, fuzzy and neural models for pattern recognition," *J. Intelligent & Fuzzy Systems*, **1**(1), 1–25.

[837] Chaer, W., Bishop, R., and Ghosh, J. (1997) "Mixture of experts framework for adaptive Kalman filtering," *IEEE Trans. Systems Man & Cybernetics*, **27**(3), 452–464.

[838] Sasiadek, J. (2002) "Sensor fusion," *Annual Reviews in Control*, **26**, 203–228.

[839] Lee, C., and Salcic, Z. (1997) "High performance FPGA-based implementation of Kalman filter," *Microprocessors & Microsystems*, **21**, 257–265.

[840] Palis, M., abd Krecker, D. (1990) *Parallel Kalman Filtering on the Connection Machine* (MS-CIS-90-81 LINC LAB 186 Report), University of Pennsylvania, Philadelphia, PA.

[841] Stubberud, S., Lobbia, R., and Owen, M. (1995) "Adaptive extended Kalman filter using artificial neural networks," *Proceedings 34th Conference Decision and Control*, pp. 1852–1856.

[842] Swarz, M., Ellery, A., and Marshall, J. (2008) "Towards adaptive localisation for rover navigation using multilayer feedforward neural networks," *ASTRO2008, Montreal, Canada*, Paper No. 74.

[843] Choi, M., Sakthivel, R., and Chung, W. (2007) "Neural network-aided extended Kalman filter for SLAM problem," *Proceedings IEEE International Conference Robotics & Automation*, pp. 1686–1696.

[844] Julier, S., Uhlmann, J., and Durrant-Whyte, H. (1995) "New approach for filtering nonlinear systems," *Proceedings American Control Conference, Seattle, WA*, pp. 1628–1632.

[845] Julier, S., Uhlmann, J., and Durrant-Whyte, H. (2000) "New method for the nonlinear transformation of means and covariances in filters and estimators," *IEEE Trans. Automatic Control*, **45**(3), 477–482.

[846] Julier, S., and Uhlmann, J. (2004) "Unscented filtering and nonlinear estimation," *Proc. IEEE*, **92**(3), 401–422.

[847] Huag, A. (2005) *Tutorial on Bayesian Estimation and Tracking Techniques Applicable to Nonlinear and Non-Gaussian Processes* (MITRE Technical Report MTR05W00000004).

[848] Daum, F. (2005) "Nonlinear filters: Beyond the Kalman filter," *IEEE Aerospace & Electronic Systems Magazine*, **20**(8), 57–69.

[849] Arasaratnam, I., and Haykin, S. (2009) "Cubature Kalman filters," *IEEE Trans. Automatic Control*, **54**(6), 1254–1269.

[850] Duckett, T., and Nehmzow, U. (1998) "Mobile robot self-localisation and measurement of performance in middle-scale environments," *Robotics & Autonomous Systems*, **24**, 57–69.

[851] Thrun, S., Burgard, W., and Fox, D. (1998) "Probabilistic approach to concurrent mapping and localisation for mobile robots," *Machine Learning*, **31**, 29–53.

[852] Pederson, L. (2001) "Autonomous characterisation of unknown environments," *Proceedings IEEE International Conference Robotics and Automation*, pp. 277–284.

[853] Thrun, S. (2002) "Probabilistic robotics," *Communications ACM*, **45**(3), 52–57.

[854] Zhou, H., and Sakane, S. (2007) "Mobile robot localization using active sensing based on Bayesian network inference," *Robotics & Autonomous Systems*, **55**, 292–305.

[855] Fox, D., Thrun, S., Burgard, W., and Dellaert, F. (2001) "Particle filters for mobile robot localisation," in A. Doucet, N. de Freitas, and N. Gordon (Eds.), *Sequential Monte Carlo Systems in Practice*, Springer-Verlag, New York.

[856] Thrun, S. (2000) *Probabilistic Algorithms in Robotics* (CMU Technical Report CMU-CS-00-126), Carnegie Mellon University, Pittsburgh, PA.

[857] Thrun, S. (1998) "Bayesian landmark learning for mobile robot localisation," *J. Machine Learning*, **33**(1), 41–76.

[858] Lazkano, E., Sierra, B., Astigarraga, A., and Martinez-Otzeta, J. (2007) "On the use of Bayesian networks to develop behaviours for mobile robots," *Robotics & Autonomous Systems*, **55**, 253–265.

[859] Pearl, J. (1986) "Fusion, propagation and structuring in belief networks," *Artificial Intelligence*, **29**(3), 241–288.

[860] Han, K., and Veloso, M. (2000) "Automated robot behaviour recognition," *Proceedings Ninth International Symposium Robotics Research*, pp. 199–204.

[861] Thrun, S., Fox, D., Burgard, W., and Dellaert, F. (2001) "Robust Monte Carlo localisation for mobile robots," *Artificial Intelligence*, **128**, 99–141.

[862] Raeside, D. (1976) "Monte Carlo principles and applications," *Physics Medicine Biology*, **21**(2), 181–197.

[863] Fitzgerald, W. (2001) "Markov chain Monte Carlo methods with applications to signal processing," *Signal Processing*, **81**, 3–18.

[864] Dellaert, F., Fox, D., Burgard, W., and Thrun, S. (1999) "Monte Carlo localisation for mobile robots," *Proceedings IEEE International Conference Robotics & Automation*, pp. 1322–1328.

[865] Arulampalam, M., Maskell, S., Gordon, N., and Clapp, T. (2002) "Tutorial on particle filters for online nonlinear/non-Gaussian Bayesian tracking," *IEEE Trans. Signal Processing*, **50**(2), 174–188.

[866] Yuen, D., and MacDonald, B. (2002) "Comparison between extended Kalman filtering and sequential Monte Carlo techniques for simultaneous localisation and map-building," *Proceedings Australasian Conference on Robotics and Automation, Auckland*, pp. 111–116.

[867] Carpenter, J., Clifford, P., and Fearnhead, P. (1999) "Improved particle filter for nonlinear problems," *Proc. IEE Radar Sonar Navigation*, **146**(1), 2.

[868] Fox, D. (2003) "Adapting the sample size in particle filters through KLD sampling," *Int. J. Robotics Research*, **22**(12), 985–1003.

[869] Aydogmus, O., and Talu, M. (2012) "Comparison of extended Kalman and particle filter based sensorless speed control," *IEEE Trans. Instrumentation & Measurement*, **61**(2), 402–410.

[870] Charniak, E. (1991) "Bayesian networks without tears," *AI Magazine*, Winter, 50–63.

[871] Gutmann, J.-S., Burgard, W., Fox, D., and Konolige, K. (1998) "Experimental comparison of localisation methods," *Proceedings IEEE/RSJ International Conference Intelligent Robots and Systems*, pp. 736–743.

[872] Fox, D., Burgard, W., and Thrun, S. (1998) "Active Markov localisation for mobile robots," *Robotics & Autonomous Systems*, **25**, 195–207.

[873] Thrun, S., Montermerlo, M., Koller, D., Wegbreit, B., Nieto, J., and Nebot, E. (2004) "FastSLAM: An efficient solution to the simultaneous localisation and mapping problem with unknown data association," *J. Machine Learning Research*, reprint.

[874] Montemerlo, M., Thrun, S., Koller, D., and Wegbreit, B. (2002) "FastSLAM: A factored solution to the simultaneous localisation and mapping problem," *Proceedings AAAI National Conference on Artificial Intelligence*.

[875] Montemerlo, M., and Thrun, S. (2003) "Simultaneous localisation and mapping with unknown data association using FastSLAM," *IEEE International Conference Robotics and Automation*, pp. 1985–1991.

[876] Thrun, S. (2002) "Probabilistic robotics," *Communications ACM*, **45**(3), 52–57.

[877] Montemerlo, M., Thrun, S., Koller, D., and Wegbreit, B. (2003) "FastSLAM 2.0: An improved particle filtering algorithm for simultaneous localisation and mapping that converges," *Proceedings International Joint Conference Artificial Intelligence*.

[878] Kim, C., Sakthivel, R., and Chung, W. (2008) "Unscented fastSLAM: A robust and efficient solution to the SLAM problem," *IEEE Trans. Robotics*, **24**(4), 808–820.

[879] Folkesson, J., and Christensen, H. (2004) "Graphical SLAM: Self-correcting map," *Proceedings IEEE International Conference Robotics and Automation*, pp. 383–390.

[880] Huang, G., Mourukis, A., and Roumeliotis, S. (2009) "On the complexity and consistency of UKF-based SLAM," *IEEE International Conference Robotics and Automation*, pp. 4401–4408.

[881] Gustafsson, F., Gunnarsson, F., Bergman, N., Forsell, U., Jansson, J., Karlsson, R., and Nordlund, P.-J. (2002) "Particle filters for positioning, navigation and tracking," *IEEE Trans. Signal Processing*, **50**(2), 425–437.

[882] Cummins, M., and Newman, P. (2008) "FAB-MAP: Probabilistic localization and mapping in the space of appearance," *Int. J. Robotics Research*, **27**(6), 647–665.

[883] Hewitt, R., de Ruiter, A., and Ellery, A. (2010) "Artificial neural networks and Kalman filters: A review of their hybridisation and potential uses," *15th CASI ASTRO Conference 2010, Toronto, Canada*.

[884] Hewitt, R., Ellery, A., and de Ruiter, A. (2012) "Efficient navigation and mapping techniques for the Kapvik analogue micro-rover," *Proceedings Global Space Exploration Conference (GLEX) 2012, Washington D.C.*

[885]

[886] Hewitt, R., Ellery, A., and de Ruiter, A. (2014) "Training a terrain traversability classifier for a planetary rover through simulation," submitted to *Autonomous Robots*.

[887] Hewitt, R., de Ruiter, A., and Ellery, A. (2010) "Hybridizing neural networks and Kalman filters for robotic exploration," *Proceedings 15th CASI Conference ASTRO, Toronto, Canada*.

[888] Hewitt, R., Ellery, A., and de Ruiter, A. (2012) "FastSLAM on a planetary micro-rover prototype," *16th CASI ASTRO Conference 2012, Quebec City, Canada*.

[889] Laubach, S., and Burdick, J. (1999) "Autonomous sensor-based path-planner for planetary microrovers," *Proc. IEEE Int. Conf. Robotics & Automation*, **1**, 347–354.

[890] Laubach, S., and Burdick, J. (1998) "A practical autonomous path-planner for turn-of-the-century planetary micro-rovers," *Mobile Robots III, SPIE Symposium Intelligent Systems and Advanced Manufacturing, Boston, MA, November*.

[891] Laubach, S., and Burdick, J. (1999) "Autonomous sensor-based path-planner for planetary environments," *IEEE Trans. Robotics & Automation*, **1**, 347–354.

[892] Laubach, S., and Burdick, J. (1999) "RoverBug: Long-range navigation for Mars rovers," *International Symposium Experimental Robotics (ISER99), Sydney, Australia.*

[893] Laubach, S. et al. (1999) "Long-range navigation for Mars rovers using sensor-based path-planning and visual locations," *Fifth International Symposium AI, Robotics and Automation in Space (iSAIRAS99).*

[894]

[895]

[896]

[897]

[898] Rowe, N., and Richburg, R. (1990) "Efficient Snell's law method for optimal path planning across multiple 2D irregular homogeneous-cost regions," *Int. J. Robotics Research,* **9**(6), 48–66.

[899] Meyer, J.-A., and Filliat, D. (2003) "Map-based navigation in mobile robots: II. A review of map-learning and path-planning strategies," *Cognitive Systems Research,* **4**, 283–317.

[900] Morlans, R., and Liegois, A. (1993) "DTM-based path planning method for planetary rovers," *Proceedings of the International Symposium on Missions, Technologies and Design of Planetary Mobile Vehicles, Toulouse, September 1992,* CNES/Cipaduhs-Iditions, Toulouse, France (ISBN 2854283317).

[901] Schiller, Z. (1999) "Motion planning for Mars rovers," *Proceedings First Workshop Robot Motion and Control,* pp. 28–29.

[902] Al-Hasan, S., and Vachtsevanos, G. (2002) "Intelligent route planning for fast autonomous vehicles operating in a large natural terrain," *Robotics & Autonomous Systems,* **40**, 1–24.

[903] Stentz, A. (1995) "Optimal and efficient path planning for unknown and dynamic environments," *Int. J. Robotics & Autonomous Systems,* **10**(3), 89–100.

[904] Stentz, A., and Hebert, M. (1995) "Complete navigation system for goal acquisition in unknown environments," *Autonomous Robots,* **2**(2), 1–28.

[905] Stentz, A. (1994) "Optimal and efficient path planning for partially-known environments," *Proceedings IEEE International Conference Robotics and Automation.*

[906] Stentz, A. (1995) "Focussed D* algorithm for real-time replanning," *Proc. Int. J. Conf. Artificial Intelligence,* **2**, 1652–1659.

[907] Tompkins, P., Stentz, A., and Wettergreen, D. (2006) "Mission-level path planning and re-planning for rover exploration," *Robotics & Autonomous Systems,* **54**, 174–183.

[908] Nash, A., Daniel, K., Koenig, S., and Aelner, A. (2007) "Theta*: Any angle path planning on grids," *Proceedings AAAI Conference on Artificial Intelligence,* pp. 1177–1183.

[909] Hwang, Y., and Ahua, N. (1992) "Gross motion planning: A survey," *ACM Computing Surveys,* **24**(3), 230–291.

[910] Horswill, I. (1997) "Visual architecture and cognitive architecture," *J. Experimental & Theoretical Artificial Intelligence,* **9**, 277–292.

[911] Shiller, Z. (1991) "Dynamic motion planning of autonomous vehicles," *IEEE Trans. Robotics & Automation,* **7**(2), 241–249.

[912] Khatib, O. (1985) "Real time obstacle avoidance for manipulators and mobile robots," *Proceedings IEEE International Conference Robotics and Automation,* pp. 500–505.

[913] Giszter, S.F. (2002) "Biomechanical primitives and heterarchical control of limb motion in tetrapods," in J. Ayers, J. Davis, and A. Rudolph (Eds.), *Neurotechnology for Biomimetic Robots,* MIT Press, Cambridge, MA, pp. 223–240.

[914] Kweon, I. et al. (1992) "Behaviour-based mobile robot using active sensor fusion," *Proceedings IEEE International Conference Robotics and Automation*, pp. 1675–1682.

[915] Veelaert, P., and Bogaerts, W. (1999) "Ultrasonic potential field sensor for obstacle avoidance," *IEEE Trans. Robotics & Automation*, **15**(4), 774–779.

[916] Payton, D., Rosenblatt, J., and Keirsey, D. (1990) "Plan guided reaction," *IEEE Trans. Systems Man & Cybernetics*, **20**(6), 1370–1382.

[917] Olin, K., and Tseng, D. (1991) "Autonomous cross-country navigation: An integrated perception and planning system," *IEEE Expert*, August, 16–29.

[918] Khatib, O. (1985) "Real time obstacle avoidance for manipulators and mobile robots," *Proceedings IEEE International Conference Robotics and Automation*, pp. 500–505.

[919] Lynch, B., Ellery, A., and Nitzsche, F. (2008) "Two-dimensional robotic vehicle path planning based on artificial potential fields," *CSME Forum (CCToMM), Ottawa University, Canada.*

[920] Shadmehr, R., Mussa-Ivaldi, F., and Bizzi, E. (1993) "Postural force fields of the human arm and their role in generating multijoint movements," *J. Neuroscience*, **13**(1), 45–62.

[921] Howard, A., Mataric, M., and Sukhatme, G. (2002) "Mobile sensor network deployment using potential fields: A distributed, scalable solution to the area coverage problem," *Proceedings Sixth International Symposium Distributed Autonomous Robotics Systems (DARS02), Fukuoka, Japan.*

[922] Herman, H., and Schempf, H. (1992) *Serpentine Manipulator Planning and Control for NASA Space Shuttle Payload Servicing* (CMU-RI-TR-92-10), Carnegie Mellon University, Pittsburgh, PA.

[923] Sawaragi, T., Shiose, T., and Akashi, G. (2000) "Foundations for designing an ecological interface for mobile robot teleoperation," *Robotics & Autonomous Systems*, **31**, 193–207.

[924] Balkenius, C. (1994) *Biological Learning and Artificial Intelligence* (LU Cognitive Studies 30, ISSN 1101-8453), Lund University, Lund, Sweden.

[925] Valavanis, K., Hebert, T., and Kolluru, R. (2000) "Mobile robot navigation in 2D dynamic environments using an electrostatic potential field," *IEEE Trans. Systems Man & Cybernetics A: Systems & Humans*, **30**(2), 187–196.

[926] Tsourveloudis, N., Valavanis, K., and Hebert, T. (2001) "Autonomous vehicle navigation utilising electrostatic potential fields and fuzzy logic," *IEEE Trans. Robotics & Automation*, **17**(4), 490–497.

[927] Tarassenko, L., and Blake, A. (1991) "Analogue computation of collision-free paths," *Proceedings IEEE International Conference Robotics and Automation*, pp. 540–545.

[928] Mussa-Ivaldi, F. (1992) "From basis functions to basis fields: Vector field approximation from sparse data," *Biological Cybernetics*, **67**, 479–489.

[929] Rimon, E., and Koditschek, D. (1992) "Exact robot navigation using artificial potential functions," *IEEE Trans. Robotics & Automation*, **8**(5), 501–518.

[930] Ge, S., and Cui, Y. (2000) "New potential functions for mobile robot path planning," *IEEE Trans. Robotics & Automation*, **16**(5), 615–620.

[931] Ge, S., and Cui, Y. (2002) "Dynamic motion planning for mobile robots using potential field method," *Autonomous Robots*, **13**, 207–222.

[932] Ota, J., Arai, T., Yoshida, E., Kuruabayashi, D., and Sasaki, J. (1996) "Motion skills in multiple robot system," *Robotics & Autonomous Systems*, **19**, 57–65.

[933] Haddad, H., Khatib, M., Lacroix, S., and Chatila, R. (1998) "Reactive navigation in outdoor environments using potential fields," *Proceedings IEEE International Conference Robotics and Automation*, pp. 1232–1237.

[934] Khosla, P., and Volpe, R. (1988) "Superquadratic artificial potentials for obstacle avoidance and approach," *Proceedings IEEE Conference Robotics and Automation, Philadelphia.*

[935] Chella, A. (1997) "Cognitive architecture for artificial vision," *Artificial Intelligence,* **89**, 73–111.

[936]

[937] Chella, A., Frixione, M., and Gaglio, S. (2000) "Understanding dynamic scenes," *Artificial Intelligence,* **123**, 89–132.

[938] Chella, A., Gaglio, S., and Pirrone, R. (2001) "Conceptual representations of actions for autonomous robots," *Robotics & Autonomous Systems,* **89**(9), 1–15.

[939] Warren, C. (1989) "Global path planning using artificial potential fields," *Proceedings IEEE International Conference Robotics and Automation,* pp. 316–321.

[940] Borenstein, J., and Koren, Y. (1989) "Real-time obstacle avoidance for fast mobile robots," *IEEE Trans. Systems Man & Cybernetics,* **19**(5), 1179–1187.

[941] Koren, Y., and Borenstein, J. (1991) "Potential field methods and their inherent limitations for mobile robot navigation," *Proceedings IEEE Conference Robotics and Automation,* pp. 1398–1404.

[942] Borenstein, J., and Koren, Y. (1991) "Vector field histogram: Fast obstacle avoidance for mobile robots," *IEEE Trans. Robotics & Automation,* **7**(3), 278–288.

[943] Labrosse, F. (2007) "Short and long range visual navigation using warped panoramic images," *Robotics & Autonomous Systems,* **55**(9), 675–684.

[944] Hong, J., Tan, X., Pinette, B., Weiss, R., and Riseman, E. (1992) "Image-based homing," *IEEE Control Systems,* February, 38–45.

[945] Schoner, G., and Dose, M. (1992) "Dynamical systems approach to task-level system integration used to plan and control autonomous vehicle motion," *Robotics & Autonomous Systems,* **10**, 253–267.

[946] Bicho, E., and Schoner, G. (1997) "Dynamical approach to autonomous robotics demonstrated on a low-level vehicle platform," *Robotics & Autonomous Systems,* **21**, 23–35.

[947] Huang, W., Fajen, B., Fink, J., and Warren, W. (2006) "Visual navigation and obstacle avoidance using a steering potential function," *Robotics & Autonomous Systems,* **54**, 288–299.

[948] Fajen, B., and Warren, W. (2003) "Behavioural dynamics of steering, obstacle avoidance and route selection," *J. Experimental Psychology: Human Perception & Performance,* **29**(2), 343–362.

[949] Mack, A., and Ellery, A. (2010) "Potential steering function and its application to planetary exploration rovers," *CASI Astronautics Conference, Toronto, Canada.*

[950] Mack, A., and Ellery, A. (2015) "Application of the potential steering function to planetary exploration rovers," submitted to *Robotica.*

[951] Schmajuk, N., and Blair, H. (1993) "Place learning and the dynamics of spatial navigation: A neural network approach," *Adaptive Behaviour,* **1**(3), 353–385.

[952] Gelenbe, E., Schmajuk, N., Stadden, J., and Reif, J. (1997) "Autonomous search by robots and animals: A survey," *Robotics & Autonomous Systems,* **22**, 23–34.

[953] Collett, T. (2000) "Animal navigation: Birds as geometers?" *Current Biology,* **10**(19), R718–R721.

[954] Frazier, C., Lynch, B., Ellery, A., and Baddour, N. (2015) "Re-active virtual equilibrium: A method of autonomous planetary rover navigation using potential fields," submitted to *Robotica.*

[955] Kim, J-O., and Khosla, P. (1992) "Real-time obstacle avoidance using harmonic potential functions," *IEEE Trans. Robotics & Automation*, **8**(3), 338–349.

[956] McInnes, C. (1995) *Potential Function Methods for Autonomous Spacecraft Guidance and Control*, American Astronomical Society, Pasadena, CA, pp. 95–447.

[957] Roger, A., and McInnes, C. (2000) "Safety constrained free-flyer path planning at the International Space Station," *J. Guidance Control & Dynamics*, **23**(6), 971–979.

[958] Cariani, P. (2002) "Extradimensional bypass," *BioSystems*, **64**, 47–53.

[959] Rutenbar, R. (1989) "Simulated annealing algorithms: An overview," *IEEE Circuits & Devices Mag.*, January, 19–26.

[960] Kirkpatrick, S., Gellatt, C., and Vecchi, M. (1983) "Optimisation by simulated annealing," *Science*, **320**, 671–680.

[961] Bandyopadhyay, S., Pal, S., and Murthy, C. (1998) "Simulated annealing based pattern classification," *J. Information Sciences*, **109**, 165–184.

[962] Park, M., Jeon, J., and Lee, M. (2001) "Obstacle avoidance for mobile robots using artificial potential field approach with simulated annealing," *Proceedings ISIE, Busan, South Korea*, pp. 1530–1535.

[963] Sherwood, D., and Lee, T. (2003) "Schema theory: Critical review and implications for the role of cognition in a new theory of motor learning," *Research Quarterly for Exercise & Sport*, **74**(4), 376–382.

[964] Minsky, M. (1975) "Framework for representing knowledge," in P. Winston (Ed.), *Psychology of Computer Vision*, McGraw-Hill, pp. 211–277.

[965] Anderson, T., and Donath, M. (1990) "Animal behaviour as a paradigm for developing robot autonomy," *Robotics & Autonomous Systems*, **6**, 145–168.

[966] Saffiotti, A., Konolige, K., and Ruspini, E. (1995) "Multivalued logic approach to integrating planning and control," *Artificial Intelligence*, **76**, 481–526.

[967] Arkin, R. (1987) "Motor schema based mobile robot navigation," *Int. J. Robotics Research*, **8**(4), 92–112.

[968] Arkin, R. (1990) "Integrating behavioural, perceptual, and world knowledge in reactive navigation," *Robotics & Autonomous Systems*, **6**, 105–122.

[969] Arkin, R. (1990) "Impact of cybernetics on the design of a mobile robot system: A case study," *IEEE Trans. Systems Man & Cybernetics*, **20**(6), 1245–1257.

[970] Arkin, R., and Murphy, R. (1990) "Autonomous navigation in a manufacturing environment," *IEEE Trans. Robotics & Automation*, **6**(4), 445–454.

[971] Laubach, S., Olson, C., Burdick, J., and Hayati, S. (1999) "Long-range navigation for Mars rovers using sensor-based path planning and visual localisation," *Proceedings Fifth International Conference Artificial Intelligence Robotics and Automation* (ESA SP-440), ESA, Noordwijk, The Netherlands, pp. 455–460.

[972] Arbib, M., and Liaw, J.-S. (1995) "Sensorimotor transformations in the world of frogs and robots," *Artificial Intelligence*, **72**, 53–79.

[973] Arkin, R., Ali, K., Wetzenfeld, A., and Cervantes-Perez, F. (2000) "Behavioural models of the praying mantis as a basis for robotic behaviours," *Robotics & Autonomous Systems*, **32**, 39–60.

[974] Beom, H., and Cho, H. (1995) "Sensor-based navigation for a mobile robot using fuzzy logic and reinforcement learning," *IEEE Trans. Systems Man & Cybernetics*, **25**(3), 464–477.

[975] Xu, H., and Van Brussel, H. (1997) "Behaviour-based blackboard architecture for reactive and efficient task execution of an autonomous robot," *Robotics & Autonomous Systems*, **22**, 115–132.

[976] Weitzenfeld, A., Arkin, R., Cervantes, F., Olivares, R., and Corbacho, F. (1998) "Neural schema architecture for autonomous robots," *Proceedings International Conference Robotics and Automation.*

[977] Lyons, D., and Hendricks, A. (1995) "Planning as incremental adaptation of a reactive system," *Robotics & Autonomous Systems*, **14**, 255–288.

[978] Balch, T., and Arkin, R. (1993) "Avoiding the past: A simple but effective strategy for reactive navigation," *Proceedings IEEE International Conference Robotics and Automation*, pp. 678–685.

[979] Moorman, K., and Ram, A. (1992) "Case-based approach to reactive control for autonomous robots," *Proceedings AAAI Fall Symposium on AI for Real-World Autonomous Mobile Robots*, Georgia Institute of Technology, Atlanta, GA.

[980] Arkin, R., and Balch, T. (1997) "AuRA: Principles and practice in review," *J. Experimental Theoretical Artificial Intelligence*, **9**, 175–189.

[981] Reif, J., and Wang, H. (1999) "Social potential fields: A distributed behavioural control for autonomous robots," *Robotics & Autonomous Systems*, **27**, 171–194.

[982] Brooks, R. (1985) *A Robust Layered Control System for a Mobile Robot* (MIT AI Memo 864), Massachusetts Institute of Technology, Cambridge, MA.

[983] Beer, R. (1995) "Dynamical systems perspective on agent–environment interaction," *Artificial Intelligence*, **72**, 173–215.

[984] Beer, R. (1997) "The dynamics of adaptive behaviour: A research program," *Robotics & Autonomous Systems*, **20**, 257–289.

[985] Van Rooij, I., Bongers, R., and Haselager, W. (2002) "Non-representational approach to imagined action," *Cognitive Science*, **26**, 345–375.

[986] Payton, D., Rosenblatt, J., and Keirey, D. (1990) "Plan guided reaction," *IEEE Trans Systems Man & Cybernetics*, **20**(6), 1370–1882.

[987] Nikolos, I., Valavanis, K., Tsourveloudis, N., and Kosturas, A. (2003) "Evolutionary algorithm based offline/online path planner for UAV navigation," *IEEE Trans. Systems Man & Cybernetics B—Cybernetics*, **33**(6), 898–912.

[988] Saffiotti, A., Konolige, K., and Ruspini, E. (1995) "Multivalued logic approach to integrating planning and control," *Artificial Intelligence*, **76**, 481–526.

[989] Malone, T., and Crowstone, K. (1994) "Interdisciplinary study of coordination," *ACM Computing Surveys*, **26**(1), 81–199.

[990] Kleinrock, L. (1985) "Distributed systems," *Communications ACM*, **28**(11), 1200–1213.

[991] Vamos, T. (1983) "Cooperative systems," *IEEE Control Syst Mag.*, August, 9–13.

[992] Balch, T. (2000) "Hierarchic social entropy: An information theoretic measure of robot group diversity," *Autonomous Robots*, **8**, 209–237.

[993] Pacala, S., Gordon, D., and Godfray, H. (1996) "Effects of social group size on information transfer and task allocation," *Evolutionary Ecology*, **10**, 127–165.

[994] Lerman, K., Galstyan, A., Martinolli, A., and Ijspeert, A. (2001) "Macroscopic analytical model of collaboration in distributed robotic systems," *Artificial Life*, **7**(4), 375–393.

[995] Asada, M., Kitano, H., Noda, I., and Veloso, M. (1999) "RoboCup: Today and tomorrow: What we have learned," *Artificial Intelligence*, **110**, 193–214.

[996] Van Dyke Parunak, H. (1997) "Go to the ant: Engineering principles from natural multi-agent systems," *Annals of Operations Research*, **75**, 69–101.

[997] Simmons, R., Apfelbaum, D., Burgard, W., Fox, D., Moors, M., Thrun, S., and Younes, H. (2000) "Coordination for multi-robot exploration and mapping," *Proceedings 17th AAAI National Conference Artificial Intelligence*, pp. 852–858.

[998] Pereira, G., Das, A., Kumar, V., and Campos, M. (2003) "Formation control with configuration space constraints," *Proceedings IEEE/RJS International Conference Intelligent Robots and Systems, Las Vegas, NV.*

[999] Roumeliotis, S., and Bekey, G. (2002) "Distributed multirobot localization," *IEEE Trans. Robotics & Automation*, **18**(5), 781–795.

[1000] Feddema, J., Lewis, C., and Schoenwald, D. (2002) "Decentralised control of cooperative robotic vehicles: Theory and application," *IEEE Trans. Robotics & Automation*, **18**(5), 852–864.

[1001] Dudek, G., Jenkin, M., Milios, E., and Wilkes, D. (1995) "Experiments in sensing and communication for robot convoy navigation," *Proceedings IEEE International Conference Intelligent Robots & Systems*, pp. 268–273.

[1002] Stone, P., and Veloso, M. (1999) "Task decomposition, dynamic role assignment and low bandwidth communication for real-time strategic teamwork," *Artificial Intelligence*, **110**(2), 241–273.

[1003] Kube, C., and Zhang, H. (1994) "Collective robotics: From social insects to robots," *Adaptive Behaviour*, **2**(2), 189–218.

[1004] Cao, Y., Fukunaga, A., and Kahng, A. (1997) "Cooperative mobile robotics: Antecedents and directions," *Autonomous Robots*, **4**, 1–23.

[1005] Castelfranchi, C. (1998) "Modelling social action for AI agents," *Artificial Intelligence*, **103**, 157–182.

[1006] Werger, B. (1999) "Cooperation without deliberation: A minimal behaviour-based approach to multi-robot teams," *Artificial Intelligence*, **110**, 293–320.

[1007] Reynolds, C. (1987) "Flocks, herds and schools: A distributed behavioural model," *Computer Graphics*, **21**(4), 25–34.

[1008] Mataric, M. (1994) "Issues and approaches in the design of collective autonomous agents," *Robotics & Autonomous Systems*, **16**, 321–331.

[1009] Mataric, M. (1994) "Learning to behave socially," in D. Cliff et al. (Eds.), *From Animals to Animats 3: Proceedings Third International Conference on Simulation of Adaptive Behaviour*, MIT Press, Cambridge, MA, pp. 453–462.

[1010] Mataric, M. (1997) "Learning social behaviour," *Robotics & Autonomous Systems*, **20**, 191–204.

[1011] Mataric, M. (1996) "Designing and understanding adaptive group behaviour," *Adaptive Behaviour*, **4**(1), 51–80.

[1012] Axelsson, H., Muhammed, A., and Egersted, M. (2003) "Autonomous formation switching for mobile robots," *IFAC Conference Analysis and Design Hybrid Systems, Brittany, France.*

[1013] Egerstedt, H., and Hu, X. (2001) "Formation constrained multi-agent control," *IEEE Trans. Robotics and Automation*, **17**(6), 947–951.

[1014] Wang, P. (1989) "Navigation strategies for multiple autonomous mobile robots," *IEEE/RSJ International Workshop Intelligent Robots and Systems*, 486–493.

[1015] Pirolli, P., and Card, S. (1997) *Evolutionary Ecology of Information Foraging* (Technical Report UIR-R97-01), Office of Naval Research, Arlington, VA.

[1016] Smith, E. (1983) "Anthropological applications of optimal foraging theory: A critical review," *Current Anthropology*, **24**(5), 625–651.

[1017] Molnar, P., and Starke, J. (2001) "Control of distributed autonomous robotic systems using principles of pattern formation in nature and pedestrian behaviour," *IEEE Trans. Systems Man & Cybernetics B: Cybernetics*, **31**(3), 433–436.

[1018] Molnar, P. (2000) *Self-organised Navigation Control for Manned and Unmanned Vehicles in Space Colonies* (USRA Grant 07600-044 Final Report), NASA Institute for Advanced Concepts, Atlanta, GA.

[1019] Arkin, R., and Balch, T. (1998) "Cooperative multiagent robotic systems," *Artificial Intelligence and Mobile Robots*, MIT Press, Cambridge, MA, pp. 277–296.

[1020] Balch, T., and Arkin, R. (1998) "Behaviour-based formation control for multirobot teams," *IEEE Trans. Robotics & Automation*, **14**(6), 926–939.

[1021] Lee, S-H., Pak, H., and Chon, T-S. (2006) "Dynamics of prey–flock escaping behaviour in response to predator's attack," *J. Theoretical Biology*, **240**, 250–259.

[1022] Kraus, S. (1997) "Negotiation and cooperation in multi-agent environment," *Artificial Intelligence*, **94**, 78–97.

[1023] Oubbati, M., and Palm, G. (2007) "Neural fields for controlling formation of multiple robots," *Proceedings IEEE International Symposium Computational Intelligence Robotics and Automation*, pp. 90–94.

[1024] Schoner, G., Dose, M., and Engels, C. (1995) "Dynamics of behaviour: Theory and applications for autonomous robot architectures," *Robotics & Autonomous Systems*, **16**, 213–245.

[1025] Quoy, M., Moga, S., and Gaussier, P. (2003) "Dynamical neural networks for planning and low-level robot control," *IEEE Trans. Systems Man & Cybernetics A: Systems & Humans*, **33**(4), 523–532.

[1026] Cole, B. (1991) "Is animal behaviour chaotic? Evidence from the activity of ants," *Proc. Royal Society London*, **B244**, 253–259.

[1027] Martinoli, A., Yamamoto, M., and Mondana, F. (1997) "On the modelling of bio-inspired collective experiments with real robots," *Proceedings Third International Symposium Distributed Autonomous Robotic Systems*, pp. 25–27.

[1028] Franks, N. (1989) "Army ants: A collective intelligence," *American Scientist*, **77**, 139–145.

[1029] Poggio, M., and Poggio, T. (1994) *Cooperative Physics of Fly Swarms: An Emergent Behaviour* (AI Memo 1512/CBCL paper 103), MIT AI Laboratory/Centre for Biological & Computation Learning.

[1030] Parunak, H. (1997) "Go to the ant: Engineering principles from natural multiagent systems" *Annals Operations Research*, **75**, 69–101.

[1031] Kube, C., and Zhang, H. (1994) "Collective robotic intelligence," *Proceedings Second International Conference Simulation Adaptive Behaviour*, pp. 460–468.

[1032] Kube, C., and Zhang, H. (1994) "Collective robotics: From social insects to robots," *Adaptive Behaviour*, **2**(8), 189–218.

[1033] Daniels, R., Vanderleyden, J., and Michiels, J. (2004) "Quorum sensing and swarming migration in bacteria," *FEMS Microbiology Reviews*, **28**, 261–289.

[1034] Crespi, B. (2001) "Evolution of social behaviour in microorganisms," *Trends in Ecology & Evolution*, **16**(1), 178–183.

[1035] Payton, D., Estkowski, R., and Howard, M. (2001) "Compound behaviours in pheromone robotics," *Robotics & Autonomous Systems*, **44**, 229–240.

[1036]

[1037] Bonabeau, E., Theraulaz, G., Deneubourg, J.-L., Franks, N., Rafelsberger, O., Joly, J.-L., and Blanco, S. (1998) "Model for the emergence of pillars, walls and royal chambers in termite nests," *Philosophical Trans. Royal Society London B*, **353**, 1561–1576.

[1038] Stickland, T., Britton, N., and Franks, N. (1995) "Complex trails and simple algorithms in ant foraging," *Proc. Royal Society B*, **260**, 53–58.

[1039] Sahin, E., and Franks, N. (2002) "Measurement of space: From ants to robots," *EPSRC/BBRC Proceedings International Workshop Biologically Inspired Robotics: Legacy of W. Grey Walter, Bristol, U.K.*, pp. 241–247.

[1040]

[1041] Bonabeau, E., Theraulaz, G., Deneubourg, J.-L., Aron, S., and Camazine, S. (1997) "Self-organisation in social insects," *Trends in Ecology & Evolution*, **12**(5), 188–193.

[1042] Theraulaz, G., Gautrais, J., Camazine, S., and Deneubourg, J.-L. (2003) "Formation of spatial patterns in social insects: From simple behaviours to complex structures," *Phil. Trans. Royal Society London*, **A361**, 1263–1282.

[1043] Franks, N., and Deneubourg, J.-L. (1997) "Self-organising nest construction in ants: Individual worker behaviour and the nest's dynamics," *Animal Behaviour*, **54**, 779–796.

[1044] Mason, Z. (2002) "Programming with stigmergy: Using swarms for construction," in R. Standish, H. Abbas, and M. Bedau (Eds.), *Artificial Life VIII*, MIT Press, Cambridge, MA, 371-374.

[1045] Dorigo, M., and Gambardella, L. (1997) "Ant colony system: A cooperative learning approach to the travelling salesman problem," *IEEE Trans. Evolutionary Computation*, **1**(1), 53–66.

[1046] Verbeeck, K., and Nowe, A. (2002) "Colonies of learning automata," *IEEE Trans. Systems Man & Cybernetics—B: Cybernetics*, **32**(6), 772–780.

[1047] Stone, P., and Veloso, M. (2000) "Multiagent systems: A survey from a machine learning perspective," *Autonomous Robots*, **8**(3), 345–383.

[1048] Sloman, A., and Croucher, M. (1981) "Why robots will have emotions," *Proceedings Seventh International Joint Conference Artificial Intelligence*, pp. 197–202.

[1049] Murphy, R., Lisetti, C., Tardif, R., Irish, L., and Gage, A. (2002) "Emotion-based control of cooperating heterogeneous mobile robots," *IEEE Trans. Robotics & Automation*, **18**(5), 744–757.

[1050] Vaughan, R., Stoy, K., Sukhatme, G., and Mataric, M. (2000) "Go ahead, make my day: Robot conflict resolution by aggressive competition," *Proceedings Sixth International Conference Simulation Adaptive Behaviour*, pp. 491–500.

[1051] Kube, C., and Zhang, H. (1993) "Collective robotics: From social insects to robots," *Adaptive Behaviour*, **2**(2), 189–219.

[1052] Mataric, M., Nilsson, M., and Simsarian, K. (1995) "Cooperative multi-robot box-pushing," *Proceedings IEEE/RJS International Conference Intelligent Robots and Systems*, pp. 556–561.

[1053] Brown, R., and Jennings, J. (1995) "Pusher/steerer model for strongly cooperative mobile robot cooperation," *Proceedings IEEE/RSJ International Conference Intelligent Robots and Systems*, pp. 562–568.

[1054] Kube, C., and Bonabeau, E. (1998) "Cooperative transport by ants and robots," *Robotics & Autonomous Systems*, **30**, 85–101.

[1055] Kube, C., and Zhang, H. (1992) "Collective robotic intelligence," *Proceedings Second International Conference Simulation Adaptive Behaviour*, pp. 460–468.

[1056] Kube, C., and Zhang, H. (1992) *Collective Task Achieving Group Behaviour by Multiple Robots* (Technical Report TR93-06), University of Alberta, Edmonton, Alberta.

[1057] Deneubourg, J., and Goss, S. (1989) "Collective patterns and decision-making," *Ethology, Ecology & Evolution*, **1**, 295–311.

[1058] Schoonderwoerd, R., Holland, O., Bruten, J., and Rothkrantz, L. (1996) *Ant-based Load Balancing in Telecommunications Networks* (HP Labs Technical Report HPL-96-76), HP Labs, Palo Alto, CA.

[1059] Pereira, G., Pimentel, B., Chaimowicz, L., and Campos, M. (2002) "Coordination of multiple mobile robots in an object carrying task using implicit communication," *Proceedings 2002 IEEE International Conference Robotics and Automation*, pp. 281–286.

[1060] Chaimowicz, L., Sugar, T., Kumar, V., and Campos, M. (2001) "An architecture for tightly-coupled multi-robot cooperation," *Proceedings IEEE International Conference Robotics and Automation*, pp. 2292–2297.

[1061] Donald, B., Gariepy, L., and Rus, D. (2000) "Distributed manipulation of multiple objects using ropes," *Proceedings IEEE International Conference Robotics and Automation*, pp. 450–457.

[1062] Bowyer, A. (2000) *Automated Construction Using Co-operating Biomimetic Robots* (University of Bath Department of Mechanical Engineering Technical Report 11/ 00), University of Bath, Bath, U.K.

[1063] Parker, C., Zhang, H., and Kube, C. (2003) "Blind bulldozing: Multiple robot nest construction," *Proceedings IEEE/RSJ International Conference Intelligent Robots & Systems, Las Vegas, Nevada*, pp. 2010–2015.

[1064] Werfel, J., Petersen, K., and Nagpal, R. (2014) "Designing collective behaviour in a termite-inspired robot construction team," *Science*, **343**, 754–758.

[1065] Robinson, G. (1992) "Regulation of division of labour in insect societies," *Ann. Rev. Entomol.*, **37**, 637–702.

[1066] Wahl, L. (2002) "Evolving the division of labour: Generalists, specialists and task allocation," *J. Theoretical Biology*, **219**, 371–388.

[1067] Bongard, J. (2000) "Reducing collective behavioural complexity through heterogeneity," *Artificial Life VII: Proceedings Seventh International Conference*, pp. 327–336.

[1068] Chaimowicz, L., Sugar, T., Kumar, V., and Campos, M. (2001) "Architecture for tightly coupled multi-robot cooperation," *Proceedings IEEE International Conference Robotics and Automation*, pp. 2992–2997.

[1069] Mataric, M., and Sukhatme, G. (2001) "Task-allocation and coordination of multiple robots for planetary exploration," *Proceedings 10th International Conference Advanced Robotics*, pp. 61–70.

[1070] Mataric, M., Sukhatme, G., and Ostergaard, E. (2003) "Multi-robot task allocation in uncertain environments," *Autonomous Robots*, **14**, 255–263.

[1071] Zlot, R., Stentz, A., Dias, M., and Thayer, S. (2002) "Multi-robot exploration controlled by a market economy," *Proceedings IEEE International Conference Robotics and Automation*.

[1072] Smith, R. (1980) "Contract net protocol: High level communication and control in a distributed problem solver," *IEEE Trans. Computing*, **29**(2), 1104–1113.

[1073] Ephrati, E., and Rosenschein, J. (1996) "Deriving consensus in multiagent systems," *Artificial Intelligence*, **87**, 21–74.

[1074] Yamashita, A., Arai, T., Ota, J., and Asama, H. (2003) "Motion planning of multiple mobile robots for cooperative manipulation and transportation," *IEEE Trans. Robotics & Automation*, **19**(2), 1–15.

[1075] Mueller, R. (2006) *Surface Support Systems for Co-operative and Integrated Human/ Robotic Lunar Exploration* (IAC-06-A5.2.09), International Astronautical Federation, Valencia, Spain.

[1076] Cockell, C., and Ellery, A. (2003) "Human exploration of the Martian poles, Part 1: From early expeditions to a permanent station," *J. British Interplanetary Society*, **56**, 33–42.

[1077] Parkinson, B., and Wright, P. (2006) "Systems modelling and systems trade for pole station," *Project Boreas: A Station for the Martian Geographic North Pole*, British Interplanetary Society, London, pp. 24–31.

[1078] Greene, M. (2006) "Base design for pole station," *Project Boreas: A Station for the Martian Geographic North Pole*, British Interplanetary Society, London, pp. 32–48.

[1079] Ellery, A., and Cockell, C. (2003) "Human exploration of the Martian pole, Part 2: Support technologies," *J. British Interplanetary Society*, **56**, 43–55.

[1080] Benroya, H., Bernold, L., and Chua, K. (2002) "Engineering, design and construction of lunar bases," *J. Aerospace Engineering*, April, 33–46.

[1081] Satish, H., Radziszewski, P., and Ouellet, J. (2005) "Design issues and challenges in lunar/Martian mining applications," *Mining Technology*, **114**(June), A107–A117.

[1082] Isard, W. (1952) "General location principle of optimum space economy," *Econometrica*, **20**, 406–430.

[1083] Smith, T., Simmons, R., Singh, S., and Hershberger, D. (2001) "Future directions in multi-robot autonomy and planetary surface construction," *Proc. Space Studies Institute Conf.*, CMU-RJ-TR-00-02.

[1084] Ha, Q., Santos, M., Nguyen, Q., Rye, D., and Durrant-Whyte, H. (2002) "Robotic excavation in construction automation," *IEEE Robotics & Automation Magazine*, March, 20–28.

[1085] Nakaruma, A., Ota, J., and Arai, T. (2002) "Human-supervised multiple mobile robot system," *IEEE Trans. Robotics & Automation*, **18**(5), 728–743.

[1086] Huntsberger, T., Rodriguez, G., and Schenker, P. (2001) "Robotics challenges for robotic and human Mars exploration," *ASCE Proceedings Space & Robotics*, pp. 340–346.

[1087] Huntsberger, T., Pirjanian, P., and Schenker, P. (2001) "Robotic outposts as precursors to a manned Mars habitat," *Space Technology & Applications International Forum—AIP Conf. Proc.*, **552**, 46–51.

[1088] Cockell, C., and Ellery, A. (2003) "Human exploration of the Martian pole, Part 1: Four phases from early exploration to a permanent station," *J. British Interplanetary Society*, **56**(1/2), 33–42.

[1089] Ellery, A., and Cockell, C. (2003) "Human exploration of the Martian pole, Part 2: Support technologies," *J. British Interplanetary Society*, **56**(1/2), 43–55.

[1090] Estlin, T., Rabideau, G., Mutz, D., and Chien, S. (1999) "Using continuous planning techniques to coordinate multiple rovers," *International Joint Conference Artificial Intelligence Workshop on Scheduling & Planning*, pp. 4–45.

[1091] Huntsberger, T., Mataric, M., and Pirjanian, P. (1999) "Action selection within the context of a robotic colony," *Proc. SPIE Sensor Fusion & Decentralised Control in Robotic Systems II*, **3839**.

[1092] Parker, L. (1998) "Alliance: An architecture for fault tolerant multi-robot cooperation," *IEEE Trans Robotics & Automation*, **14**(2), 220–240.

[1093] Parker, L., Guo, Y., and Jung, D. (2001) "Cooperative robot teams applied to the site preparation task," *Proceedings 10th International Conference Advanced Robotics*, pp. 71–77.

[1094] Parker, L. (1996) "On the design of behaviour-based multi-robot systems," *Advanced Robotics*, **10**(6), 547–578.

[1095] Parker, L. (2002) "Distributed algorithms for multi-robot observation of multiple moving targets," *Autonomous Robots*, **12**(3), 231–255.

[1096] Murphy, R., Lisetti, C., Tardif, R., Irish, L., and Gage, A. (2002) "Emotion-based control of cooperating heterogeneous mobile robots," *IEEE Trans. Robotics & Automation*, **18**, 744–757.

[1097] Sukhatme, G., Montgomery, J., and Mataric, M. (1999) "Design and implementation of a mechanically heterogeneous robot group," *Proceedings SPIE Conference Sensor Fusion and Decentralised Control Robotic Systems II*, pp. 122–133.

[1098] Brooks, R., and Flynn, A. (1989) "Fast, cheap and out of control: A robot invasion of the solar system," *J. British Interplanetary Society*, **42**, 478–485.

[1099] Carrier, W. et al. (1991) "Physical properties of the lunar surface," in G. Heiken et al. (Eds.), *Lunar Sourcebook*, Cambridge University Press, Cambridge, U.K., pp. 522–530.

[1100] Lammer, H., Lichtenegger, H., Kolb, C., Ribas, I., Guinan, E., Abart, R., and Bauer, S. (2003) "Loss of water from Mars: Implications for the oxidation of the soil," *Icarus*. **165**, 9–25.

[1101] Zent, A. (1998) "On the thickness of the oxidised layer of the Martian regolith," *Journal of Geophysical Research*, **103**(E13), 31491–31498.

[1102] Kolb, C., Lammer, H., Ellery, A., Edwards, H., Cockell, C., and Patel, M. (2002a) "The Martian oxygen surface sink and its implications for the oxidant extinction depth," *Proceedings Second European Workshop on Exo/Astrobiology, September 2002, Graz, Austria* (ESA SP-518), ESA, Noordwijk, The Netherlands, pp. 181–184.

[1103] Zent, A., and McKay, C. (1994) "Chemical reactivity of the Martian soil and implications for future missions," *Icarus*, **108**, 146–157.

[1104] Stoker, C., and Bullock, M. (1997) "Organic degradation under simulated Martian conditions," *J. Geophysical Research—Planets*, **102**, 10881–10888.

[1105] Ellery, A., Ball, A., Cockell, C., Dickensheets, D., Edwards, H., Kolb, C., Lammer, H., Patel, M., and Richter, L. (2004) "Vanguard: A European robotic astrobiology-focussed Mars sub-surface mission proposal," *Acta Astronautica*, **56**(3), 397–407.

[1106] Ellery, A., Kolb, C., Lammer, H., Parnell, J., Edwards, H., Richter, L., Patel, M., Romstedt, J., Dickensheets, D., Steel, A. et al. (2004) "Astrobiological instrumentation on Mars: The only way is down," *Int. J. Astrobiology*, **1**(4), 365–380.

[1107] Ellery, A. (2000) *An Introduction to Space Robotics*, Springer/Praxis, Heidelberg, Germany/Chichester, U.K.

[1108] Das, H. et al. (1999) "Robot manipulator technologies for planetary exploration," *Proceedings Sixth Annual International Symposium Smart Structures and Materials, CA* (No. 3668-17).

[1109] Volpe, R. et al. (1998) "A prototype manipulation system for Mars rover science operations," *Space Tech. J.*, **17**(3/4), 219–222.

[1110] Zeng, X., Burnoski, L., Agui, J., and Wilkinson, A. (2007) "Calculation of excavation force for ISRU on lunar surface," *Proceedings 45th AIAA Aerospace Sciences Meeting, Reno, NV*.

[1111] Reece, A. (1965) "Fundamental equation of earthmoving mechanics," *Symposium Earthmoving Machinery, Proc. Inst. Mech. Eng. E*, **179**(3F), 16–22.

[1112] Shmulevich, I., Asaf, Z., and Rubinstein, D. (2007) "Interaction between soil and a wide cutting blade using the discrete element method," *Soil & Tillage Research*, **97**, 37–50.

[1113] Tong, J., and Moayad, B. (2006) "Effects of rake angle of chisel plough on soil cutting factors and power requirements: A computer simulation," *Soil & Tillage Research*, **88**, 55–64.

[1114] Johnson, L., and King, R. (2010) "Measurement of force to excavate extraterrestrial regolith with a small bucket-wheel excavator," *J. Terramechanics*, **47**, 87–95.

[1115] Wilkinson, A., and DeGennaro, A. (2007) "Digging and pushing lunar regolith: Classical soil mechanics and the forces needed for excavation and traction," *J. Terramechanics*, **44**, 133–152.

[1116] Baumgartner, E., Bonitz, R., Melko, J., Shiraishi, L., and Leger, P. (2005) "Mars Exploration Rover instrument positioning system," *Proceedings IEEE Aerospace Conference, Big Sky, MT.*

[1117] Pillinger, C., Sims, M., and Clemmet, J. (2003) *The Guide to Beagle 2*, Open University, Milton Keynes, UK.

[1118] Clemmet, J. (2001) *Beagle 2: Engineering an Integrated Lander for Mars* (ESA SP-468), ESTEC, Noordwijk, The Netherlands.

[1119] Phillips, N. (2000) "Robotic arm for the Beagle 2 Mars," *Proceedings Sixth ESA Workshop on Advanced Space Technologies for Robotics & Automation (ASTRA), December, 2000,* ESA/ESTEC, Noordwijk, The Netherlands.

[1120] Bonitz, R., Shiraishi, L., Robinson, M., Carsten, J., Volpe, R., Trebi-Ollennu, A., Arvidson, R., Chu, P., Wilson, J., and Davis, K. (2008) *Phoenix Mars Lander Robotic Arm* (IEEEAC Paper 1695), Institute of Electrical and Electronic Engineers, Piscataway, NJ.

[1121]

[1122] Sanger, T., and Kumar, V. (2002) "Control of cooperating mobile manipulators," *IEEE Trans. Robotics & Automation*, **18**(1), 94–103.

[1123] Mailah, M., Pitowarno, E., and Jamaluddin, H. (2005) "Robust motion control for mobile manipulator using resolved acceleration and proportional integral active force control," *Int. J. Advanced Robotic Systems*, **2**(2), 125–134.

[1124] Yamamoto, Y., and Yun, X. (1996) "Effects of the dynamic interaction on coordinated control of mobile manipulators," *IEEE Trans. Robotics & Automation*, **12**(5), 816–824.

[1125] Peters, R., Bishay, M., Cambron, M., and Negishi, K. (1996) "Visual servoing for a service robot," *Robotics & Autonomous Systems*, **18**, 213–224.

[1126] Espiau, B., Chaumette, F., and Rives, P. (1992) "New approach to visual servoing in robotics," *IEEE Trans. Robotics & Automation*, **8**(3), 313–324.

[1127] Hutchinson, S., Hager, G., and Corke, P. (1996) "Tutorial on visual servo control," *IEEE Trans. Robotics & Automation*, **12**(5), 651–670.

[1128] Kelly, R., Carelli, R., Nasisi, O., Kuchen, B., and Reyes, F. (2000) "Stable visual servoing of camera-in-hand robotic systems," *IEEE/ASME Trans. Mechatronics*, **5**(1), 39–48.

[1129] Perry, R., and Sephton, M. (2006) "Desert varnish: An environmental recorder for Mars," *Astronomy & Geophysics*, **47**, August, 4.34–4.35.

[1130] Ng, T. et al. (2000) "Hong Kong micro-end effectors and rind grinders," *Proceedings Sixth ESA Workshop on Advanced Space Technologies for Robotics and Automation (ASTRA), December 2000,* ESA/ESTEC, Noordwijk, The Netherlands.

[1131] Bar-Cohen, Y. et al. (2001) "Ultrasonic/sonic drilling/coring (USDC) for planetary application," *Proceedings SPIE Eighth Annual Symposium on Smart Structures and Materials, CA,* pp. 4327–4355.

[1132] Bar-Cohen, Y. et al. (2000) "Utrasonic/sonic drilling/coring (USDC) for in-situ planetary applications," *SPIE Smart Structures 2000, March 2000, Newport Beach, CA,* Paper No 33992-101.

[1133] Potthast, C., Tweifel, J., and Wallaschek, J. (2007) "Modelling approaches for an ultrasonic percussion drill," *J. Sound & Vibration*, **308**, 405–417.

[1134] Das, H. et al. (1999) "Robot manipulator technologies for planetary exploration," *Proceedings Sixth Annual International Symposium Smart Structures and Materials, CA* (No. 3668-17).

[1135] Furutani, K., Kamishi, H., Murase, Y., Kubota, T., Ohtake, M., Saiki, K., Okada, T., Otake, T., Honda, C., Kurosaki, H. et al. (2012) "Prototype of percussive rock surface crusher using solenoid for lunar and planetary exploration," *Proceedings International Symposium Artificial Intelligence, Robots & Automation in Space, Turin, Italy.*

[1136] Moore, H. et al. (1997) "Surface materials of the Viking landing sites," *Journal of Geophysical Research*, **82**(28), 4497–4523.

[1137] Talbot, J. (1999) "Digging on Mars—preliminary report," preprint.

[1138] Beaty, D., Clifford, S., Briggs, G., and Blacic, J. (2000) *Strategic Framework for the Exploration of the Martian Subsurface* (White Paper), NASA Jet Propulsion Laboratory, Pasadena, CA.

[1139] Blacic, J., Dreesen, D., and Mockler, T. (2000) *Report on Conceptual Systems Analysis of Drilling Systems for 200 m Depth Penetration and Sampling of the Martian Subsurface* (Report LAUR-4742), Los Alamos National Laboratory, Los Alamos, NM.

[1140] Clifford, S., and Mars 07 Drilling Feasibility Team (2001) *Science Rationale and Priorities for Subsurface Drilling in '07* (Final Report Rev. 8), LPS, University of Arizona, Tucson, AZ.

[1141] Zacny, K., and Cooper, G. (2006) "Considerations, constraints and strategies for drilling on Mars," *Planetary & Space Science*, **54**, 345–356.

[1142] Zacny, K., and Cooper, G. (2007) "Coring basalt under Mars low pressure conditions," *Mars*, **3**, 1–11.

[1143] Mellor, M. (1989) "Introduction to drilling technology," *Proceedings International Workshop on Physics and Mechanics of Cometary Materials, Munster, Germany* (ESA SP-302), ESA, Noordwijk, The Netherlands, pp. 95–114.

[1144] Blacic, J., Dreesen, D., and Mockler, T. (2000) *Report on Conceptual Systems Analysis of Drilling Systems for 200 m Depth Penetration and Sampling of the Martian Subsurface* (Report LAUR00-4742), Los Alamos National Laboratory, Los Alamos, NM.

[1145]

[1146] Ellery, A., Ball, A., Cockell, C., Coste, P., Dickensheets, D., Edwards, H., Hu, H., Kolb, C., Lammer, H., Lorenz, R. et al. (2002) "Robotic astrobiology: The need for sub-surface penetration of Mars," *Proceedings Second European Workshop on Exo/Astro-Biology, Graz, Austria* (ESA SP-518), ESA, Noordwijk, The Netherlands, pp. 313–317.

[1147] Zacny, K., and Cooper, G. (2007) "Methods for cuttings removal from holes drilled on Mars," *Mars*, **3**, 42–56.

[1148] Kawashima, N. et al. (1993) "Development/drilling of auger boring machines onboard Mars rovers for Mars exploration," *Proceedings of the International Symposium on Missions, Technologies and Design of Planetary Mobile Vehicles, Toulouse, September 1992*, CNES/Cipaduhs-Iditions, Toulouse, France (ISBN 2854283317).

[1149] Peeters, M., and Kovats, J. (2000) "Drilling and logging in space: An oil well perspective," *Space Resources Roundtable II*, Abstract 7025.

[1150] Prensky, S. (1994) "Survey of recent developments and emerging tecnologies in well-logging and rock characteristics," *Log Analyst*, **35**(2), 15–45; **35**(5), 78–84.

[1151] Susmela, J. et al. (2000) "A robotic deep driller for Mars exploration," *Proceedings Sixth ESA Workshop on Advanced Space Technologies for Robotics and Automation (ASTRA)* (Paper No. 3.5a-4), ESA/ESTEC, Noordwijk, The Netherlands.

[1152] Rafeek, S. et al. (2000) "Sample acquisition systems for sampling the surface down to 10m below the surface for Mars exploration," *Concepts and Approaches for Mars Exploration*, Abstract 6239.

[1153] Pozzi, E., and Mugnuolo, R. (1998) "Robotics for ROSETTA cometary landing mission," *Robotics & Autonomous Systems*, **23**, 73–77.

[1154] Di Pippo, S. (1997) "Automation and robotics: The key tool for space exploration," *Acta Astronautica*, **41**(4/10), 247–254.

[1155] Magnani, P., Re, E., Senese, S., Cherubini, G., and Olivieri, A. (2006) "Different drill tool concepts," *Acta Astronautica*, **59**, 1014–1019.

[1156] Magnani, P., Re, E., Ylikorpi, T., Cherubini, G., and Olivieri, A. (2004) "Deep drill (DeeDri) for Mars application," *Planetary & Space Science*, **52**, 79–82.

[1157] Hill III, J., Shenhar, J., and Lombardo, M. (2003) "Tethered down-hole-motor drilling system: A benefit to Mars exploration," *Advances in Space Research*, **31**(11), 2431–2426.

[1158] Vincent, J., and King, M. (1995) "Mechanism of drilling by wood wasp ovipositors," *Biomimetics*, **3**(4), 187–201.

[1159] Menon, C., Lan, N., Ellery, A., Zangani, D., Manning, C., Vincent, J., Bilhaut, L., Gao, Y., Carosio, S., and Jaddou, M. (2006) "Bio-inspired micro-drills for future planetary exploration," *Proceedings CANEUS, Toulouse* (Paper No. 11022).

[1160] Gao, Y., Ellery, A., Jaddou, M., and Vincent, J. (2006) "Bio-inspired drill for planetary subsurface sampling: Literature survey, conceptual design and feasibility study," *Proceedings Adaptation in Intelligent Systems and Biology (AISB) Conference 2, University of Bristol*, pp. 71–77.

[1161] Gao, Y., Ellery, A., Vincent, J., Eckersley, S., and Jaddou, M. (2007) "Planetary micro-penetrator concept study with biomimetic drill and sampler design," *IEEE Trans. Aerospace & Electronic Systems*, **43**(3), 875–885.

[1162] Gao, Y., Ellery, A., Sweeting, M., and Vincent, J. (2007) "Bio-inspired drill for planetary sampling: Literature survey, conceptual design and feasibility study," *J. Spacecraft & Rockets*, **44**(3), 703–709.

[1163] Gao, Y., Ellery, A., Jaddou, M., Vincent, J., and Eckersley, S. (2005) "Novel penetration system for in-situ astrobiological studies," *Int. J. Advanced Robotic Systems*, **2**(4), 281–286.

[1164] Gao, Y., Ellery, A., Vincent, J., Eckersley, S., and Jaddou, M. (2007) "Planetary micro-penetrator concept study with biomimetic drill and sampler design," *IEEE Trans. Aerospace & Electronic Systems*, **43**(3), 875–885.

[1165] Gouache, T., Gao, T., Gourinat, Y., and Coste, P. (2010) "Wood wasp inspired planetary and Earth drill," in A. Mukherjee (Ed.), *Biomimetics Learning from Nature*, InTech, Rijeka, Croatia.

[1166] Hopkins, T., and Ellery, A. (2008) "Biomimetic drill design for in-situ astrobiological studies," *Planetary and Terrestrial Mining Sciences Symposium*, NORCAT, Sudbury, Canada.

[1167] Meehan, R. et al. (1998) "Drill-bit seismic technology," *Oil & Gas J. (Technology)*, **96**, November, 19.

[1168] Cannon, H., Stoker, C., Dunagan, S., Davis, K., Gomez-Elvira, J., Glass, B., Lemke, L., Miller, D., Bonaccorsi, R., Branson, M. et al. (2007) "MARTE: Technology development and lessons learned from a Mars drilling mission simulation," *J. Field Robotics*, **24**(10), 877–905.

[1169] Jerby, E., Dikhtyar, V., and Aktushev, O. (2003) "Microwave drill for ceramics," *Ceramic Bulletin*, **82**, 35–42.

[1170] Jerby, E., and Dikhtyar, V. (2001) "Drilling into hard non-conductive materials by localised microwave radiation," *Proceedings Eighth Ampere Conference, Bayreuth, Germany*.

[1171] Jerby, E., Dikhtyar, V., Aktushev, O., and Grosglick, U. (2002) "Microwave drill," *Science*, **298**, 587–589.

[1172] Grosglik, U., Dikhtyar, V., and Jerby, E. (2002) "Coupled thermal-electromagnetic model for microwave drilling," *Proceedings European Symposium Numerical Methods in Electromagnetics*, pp. 146–151.

[1173] Meir, Y., and Jerby, E. (2011) "Transistor-based miniature microwave drill applicator," *IEEE International Conference Microwaves, Communications, Antennas & Electronic Systems*, pp. 1–4.

[1174] Meir, Y., and Jerby, E. (2012) "Localised rapid heating by low power solid state microwave drill," *IEEE Trans. Microwave Theory & Techniques*, **60**(8), 2665–2672.

[1175] Eustes, A. et al. (2000) "Percussive force magnitude in permafrost," *Space Resources Roundtable II*, Abstract 7028.

[1176] Eustes, A. et al. (2000) "Summary of issues regarding the Martian subsurface explorer," *Space Resources Roundtable II*, Abstract 7029.

[1177] Gromov, V. et al. (1997) "Mobile penetrometer, a mole for sub-surface soil investigation," *Proceedings Seventh European Space Mechanism and Tribology Symposium* (ESA SP-410), ESA, Noordwijk, The Netherlands, pp. 151–156.

[1178] Richter, L. et al. (2001) "Development of the 'Planetary Underground Tool' subsurface soil sampler for the Mars Express Beagle 2 lander," *Advances in Space Research*, **28**(8), 1225–1230.

[1179] Richter, L. et al. (2000) "Development of the mole with sampling mechanism sub-surface sampler," *Proceedings Advanced Space Technologies for Robotics and Automation (ASTRA 2000)* (Paper No. 3.5a-3), ESA/ESTEC, Noordwijk, The Netherlands.

[1180] Kochan, H., Hamacher, H., Richter, L., Hirschmann, L., Assanelli, S., Nadalani, R., Pinna, S., Gromov, V., Matrossov, S., Yudkin, N. et al. (2001) "Mobile penetrometer (mole): A tool for planetary subsurface investigations," in N. Komle, G. Kargl, A. Ball, and R. Lorenz (Eds.), *Proceedings International Workshop on Penetrometry in the Solar System*.

[1181] Gromov, V. et al. (1997) "Mobile penetrometer, a mole for subsurface soil investigation," *Proceedings Seventh European Space Mechanisms and Tribology Symposium* (ESA SP-410), ESA/ESTEC, Noordwijk, The Netherlands, pp. 151–156.

[1182] Richter, L., Dickensheets, D., and Wynn-Williams, D.D., Edwards, H., and Sims, M. (2001) "An instrumented mole for Mars exobiological research," *First European Exo/ Astrobiology Workshop, Frascati, May 2001* (Poster Paper).

[1183] Wilcox, B. (2000) "Nanorovers and subsurface explorers for Mars," *Concepts & Approaches for Mars Exploration*, Abstract 6002.

[1184] Goraven, S., Kong, K., Myrock, T., Bartlett, P., Singh, S., Stroescu, S., and Rafeek, S. (2000) "Inchworm deep drilling system for kilometer scale subsurface exploration of Mars (IDDS)," *Concepts & Approaches for Mars Exploration*, Abstract 6239.

[1185] Gagnon, H., Abou-khalil, E., Azrak, O., Morozov, A., Jones, H., and Ravindran, G. (2007) "Space Worm: Borehole anchoring mechanism for micro-g planetary exploration drill," *Proc. Canadian Engineering Education Association*, reprint.

[1186]

[1187] Mancinelli, R. (2000) "Accessing the martian deep subsurface to search for life," *Planetary & Space Science*, **48**, 1035–1042.

[1188] Wilcox, B. (2000) "Nanorovers and subsurface explorers for Mars," *Concepts & Approaches for Mars Exploration*, Abstract 6002.

[1189] Schunnesson, H. (1996) "RQD predictions based on drill performance parameters," *Tunnelling & Underground Space Technology*, **11**(3), 345–351.

[1190] Kahnaman, S., Bilgin, N., and Feridunoglu, C. (2003) "Dominant rock properties affecting the penetration rate of percussive drills," *Int. J. Rock Mechanics & Mining Sciences*, **40**, 711–723.

[1191] Rabia, H. (1985) "Unified prediction model for percussive and rotary drilling," *Mining Science & Technology*, **2**, 207–216.

[1192] Wilkinson, A., and DeGennaro, A. (2007) "Digging and pushing lunar regolith: Classical soil mechanics and the forces needed for excavation and traction," *J. Terramechanics*, **44**, 133–152.

[1193] Nishimatsu, Y. (1972) "Mechanics of rock cutting," *Int. J. Rock Mechanics & Mining Science*, **9**, 261–270.

[1194] Finzi, A., Lavagna, M., and Rocchitelli, G. (2004) "Drill–soil system modelisation for future Mars exploration," *Planetary & Space Science*, **52**, 83–89.

[1195] Wijk, G. (1991) "Rotary drilling prediction," *Int. J. Rock Mechanics, Mining Science & Geomechanics Abstracts*, **28**(1), 35–42.

[1196] Kahraman, S., Bilgin, N., and Feridunoglu, C. (2003) "Dominant rock properties affecting the penetration rate of percussive drills," *Int. J. Rock Mechanics & Mining Sciences*, **40**, 711–723.

[1197] Kahraman, S. (1999) "Rotary and percussive drilling using regression analysis," *Int. J. Rock Mechanics & Mining Sciences*, **36**, 981–989.

[1198] Hopkins, T., and Ellery, A. (2008) "Drilling model and applications of space drilling," *ASTRO2008* (Paper No. 79), Canadian Aeronautics & Space Institute (CASI) Montreal.

[1199] Hopkins, T., and Ellery, A. (2008) "Rotary and percussive drilling penetration rate prediction model verification," *CSME Forum, Ottawa University, Canada*.

[1200] Rabia, H. (1985) "Unified prediction model for percussive and rotary drilling," *Mining Science & Technology*, **2**, 207–216.

[1201] Zacny, K., and Cooper, G. (2007) "Methods for cuttings removal from holes drilled on Mars," *Mars*, **3**, 42–56.

[1202] Zacny, K., and Cooper, G. (2006) "Considerations, constraints and strategies for drilling on Mars," *Planetary & Space Science*, **54**, 345–356.

[1203] Venkataraman, S., Gulati, S., Brahen, J., and Toomarian, N. (1993) "Neural network based identification of environments models for compliant control of space robots," *IEEE Trans. Robotics & Automation*, **9**(5), 685–697.

[1204] Cannon, H., Stoker, C., Dunagan, S., Davis, K., Gomez-Elvira, J., Glass, B., Lemke, L., Miller, D., Bonaccorsi, R., Branson, M. et al. (2007) "MARTE: Technology development and lessons learned from a Mars drilling mission simulation," *J. Field Robotics*, **24**(10), 877–905.

[1205] Parnell, J., Mazzini, A., and Hingham, C. (2002) "Fluid inclusion studies of chemosynthetic carbonates: Strategy for seeking life on Mars," *Astrobiology*, **2**(1), 43–57.

[1206] Beaty, D., Miller, S., Zimmerman, W., Bada, J., Conrad, P., Dupuis, E., Huntsberger, T., Ivlev, R., Kim, S., Lee, B. et al. (2004) "Planning for a Mars in situ sample preparation and distribution (SPAD) system," *Planetary & Space Science*, **52**, 55–66.

[1207] Clancey, P. et al. (2000) "A & R needs for a multi-user facility for exo-biology research," *Proceedings Sixth ESA Workshop on Advanced Space Technologies for Robotics and Automation (ASTRA), December 2000*, ESA/ESTEC, Noordwijk, The Netherlands.

[1208] Hansen, C., Paige, D., Bearman, G., Fuerstenau, S., Horn, J., Mahoney, C., Patrick, S., Peters, G., Scherbenski, J., Shiraishi, L. et al. (2007) "SPADE: A rock-crushing and sample-handling system developed for Mars missions," *Journal of Geophysical Research*, **112**, E06008.

[1209] Elfving, A. (1993) "Automation technology for remote sample acquisition," *Proceedings International Symposium on Missions, Technologies and Design of Planetary Mobile Vehicles, Toulouse, September 1992*, CNES/Cipaduhs-Iditions, Toulouse, France (ISBN 2854283317).

[1210] Sunshine, D. (2010) "Mars Science Laboratory CHIMRA: A device for processing powdered Martian samples," *Proceedings 40th Aerospace Mechanisms Symposium* (NASA/CP-2010-216272), NASA, Washington, D.C., pp. 249–262.

[1211] Cecil, J., Powell, D., and Vasquez, D. (2007) "Assembly and manipulation of microdevices: A state of the art survey," *Robotics & Computer-Integrated Manufacturing*, **23**, 580–588.

[1212] King, R. (2011) "Rise of the robo scientists," *Scientific American*, January, 73–77.

[1213] Backes, P., Diaz-Calderon, A., Robinson, M., Bajracharya, M., and Helmick, D. (2005) "Automated rover positioning and instrument placement," *IEEE Proceedings Aerospace Conference*, pp. 60–71.

[1214] Maimon, M., Nesnas, I., and Das, H. (1999) "Autonomous rock tracking and acquisition from a Mars rover," *Proceedings Fifth International Symposium Artificial Intelligence Robotics and Automation in Space* (ESA SP-440), ESA, Noordwijk, The Netherlands, pp. 329–334.

[1215] Estlin, T., Gaines, D., Chouinard, C., Castano, R., Bornstein, B., Judd, M., Nesnas, I., and Anderson, R. (2007) "Increased Mars rover autonomy using AI planning, scheduling and execution," *IEEE International Conference Robotics and Automation*, pp. 4911–4918.

[1216] Castano, R., Estlin, T., Gaines, D., Castano, A., Chouinard, C., Bornstein, B., Anderson, R., Chien, S., Fukunega, A., and Judd, M. (2006) "Opportunistic rover science: Finding and reacting to rocks, clouds and dust devils," *IEEE Proceedings Aerospace Conference, Big Sky, Montana*.

[1217] Castano, R., Estlin, T., Anderson, R., Gaines, D., Castano, A., Bormstein, B., Chouinard, C., and Judd, M. (2007) "OASIS: Onboard autonomous science investigation system for opportunistic rover science," *J. Field Robotics*, **24**(5), 379–397.
Cousins, C., Gunn, M., Prosser, B., Barnes, D., Crawford, I., Griffiths, A., Davis, L., and Coates, A. (2012) "Selecting the geology filter wavelengths for the ExoMars Panorama Camera instrument," *Planetary & Space Science*, **71**, 80–100.
Harris, J., Cousins, C., Gunn, M., Grindrod, P., Barnes, D., Crawford, I., Cross, R., and Coates, A. (2015) "Remote detection of past habitability at Mars—analogue

hydrothermal alteration terrains using an ExoMars Panoramic Camera emulator," *Icarus*, **252**, 284–300.

[1218] Thompson, D., Smith, T., and Wettergreen, D. (2005) "Data mining during rover traverse: From images to geological signatures," *Proceedings Eighth International Symposium on Artificial Intelligence Robotics and Automation in Space, Munich, Germany*.

[1219] Gulick, V., Morris, R., Ruzon, M., and Roush, T. (2001) "Autonomous image analyses during the 1999 Marsokhod rover field test," *Journal of Geophysical Research*, **106**(4), 7745–7763.

[1220] Burl, M., and Lucchetti, D. (2000) *Autonomous Visual Discovery*, SPIE AeroSense DMKD, Orlando, FL.

[1221] Mack, A., and Ellery, A. (2012) "Methods of vision-based autonomous science for planetary exploration," *Proceedings CASI ASTRO 2012, Quebec City, Canada*.

[1222] Sharif, H., Ralchenko, M., Samson, C., and Ellery, A. (2014) "Autonomous rock classification using Bayesian image analysis for rover-based planetary exploration," accepted by *Computers & Geosciences*.

[1223] Haralick, R., Shanmugam, K., and Dinstein, I. (1973) "Textural features for image classification," *IEEE Trans. Systems Man & Cybernetics*, **3**(6), 610–621.

[1224] McGuire, P., Ormo, J., Martinez, E., Rodriguez, J., Elvira, J., Ritter, H., Oesker, M., and Ontrup J (2004) "Cyborg astrobiologist: First field experience," *Int. J. Astrobiology*, **3**, 189–207.

[1225] McGuire, P., Gross, C., Wendt, L., Bonnici, A., Souza-Egipsy, V., Ormo, J., Diaz-Martinez, E., Foing, B., Bose, R., Walter, S. et al. (2010) "Cyborg astrobiologist: Testing a novelty detection algorithm on two mobile exploration systems at Rivas Vaciamadrid in Spain and at the Mars Desert Research Station in Utah," *Int. J. Astrobiology*, **9**(1), 11–27.

[1226] Gilmore, M., Castano, R., Mann, T., Anderson, R., Mjolsness, E., Manduchi, R., and Saunders, S. (2000) "Strategies for autonomous rovers at Mars," *Journal of Geophysical Research*, **105**(E12), 29223–29237.

[1227] Griffiths, A., Ellery, A., and the Camera Team (2006) "Context for the ExoMars rover: The panoramic camera (pancam) instrument," *Int. J. Astrobiology*, **5**(3), 269–275.

[1228] Griffiths, A., Coates, A., Josset, J.-L., Paar, G., Hofmann, B., Pullan, D., Ruffer, P., Sims, M., and Pillinger, C. (2005) "Beagle 2 stereocamera system," *Planetary & Space Science*, **53**, 1466–1482.

[1229] Maki, J., Bell III, J., Herkenhoff, K., Squyres, S., Kiely, A., Klimesh, M., Schwochert, M., Litwin, T., Willson, R., Johnson, A. et al. (2003) "Mars Exploration Rover engineering cameras," *Journal of Geophysical Research*, **108**(E12), 12-1–12-24.

[1230] Gazis, P., and Roush, T. (2001) "Autonomous identification of carbonates using near-IR reflectance spectra during the February 1999 Marsokhod field tests," *Journal of Geophysical Research*, **106**(E4), 7765–7773.

Ruckebusch, G. (1983) "Kalman filtering approach to natural gamma ray spectroscopy in well logging," *IEEE Trans. Automatic Control*, **28**(3), 372–380.

Anderson, R., and Bell III, J. (2013) "Correlating multispectral imaging and ccmpositional data from the Mars Exploration Rovers and implications for Mars Science Laboratory," *Icarus*, **233**, 157–180.

[1231] Chou, C-H.,Su,M-C., and Lai, E. (2004) "New cluster validity measure and its application to image compression," *Pattern Analysis Applications*, **7**, 205–220.

[1232] Sung, K-K., and Poggio, T. (1998) "Example-based learning for view-based human face detection," *IEEE Trans. Pattern Analysis & Machine Intelligence*, **20**(1), 39–51.

[1233] Filippone, M., Camastra, F., Masulli, F., and Rovetta, S. (2008) "Survey of kernel and spectral methods of clustering," *Pattern Recgnition*, **41**, 176–190.

[1234] Miller, K.-R., Mika, S., Ratsch, G., Tsuda, K., and Scholkopf, B. (2001) "Introduction to kernel-based learning algorithms," *IEEE Trans. Neural Networks*, **12**(2), 181–201.

[1235] Jain, A., Duin, R., and Mao, J. (2000) "Statistical pattern recognition: A review," *IEEE Trans. Pattern Analysis & Machine Intelligence*, **22**(1), 4–37.

[1236] Markou, M., and Singh, S. (2003) "Novelty detection: A review, Part 1: Statistical approaches," *Signal Processing*, **83**, 2481–2497.

[1237] Wagner, M., Apostopoulos, D., Shillcutt, K., Shamah, B., Simmons, R., and Whittaker, W. (2001) "Science autonomy system of the Nomad robot," *Proceedings IEEE International Conference Robotics and Automation*, pp. 1742–1749.

[1238] Apostoulopoulos, D., Wagner, M., Shamah, B., Pedersen, L., Shillicut, K., and Whittaker, W. (2000) "Technology and field demonstration for robotic search for Antarctic meteorites," *Int. J. Robotics Research*, **19**(11), 1015–1032.

[1239] Basye, K., Dean, T., Kirman, J., and Lejter, M. (1992) "Decision-theoretic approach to planning, perception and control," *IEEE Expert*, August, pp. 58–65.

[1240] Buntine, W. (1996) "Guide to the literature on learning probabilistic networks from data," *IEEE Trans. Knowledge & Data Engineering*, **8**(2), 195–210.

[1241] Pearl, J. (1986) "Fusion, propagation and structuring in belief networks," *Artificial Intelligence*, **29**, 241–288.

[1242] Heckerman, D., Geiger, D., and Chickering, D. (1995) "Learning Bayesian networks: The combination of knowledge and statistical data," *Machine Learning*, 293–301.

[1243] Lauritzen, S., and Spiegelhalter, D. (1988) "Local computations with probabilities on graphical structures and their application to expert systems," *J. Royal Statistical Society B*, **50**, 157–224.

[1244] Glymour, C. (2003) "Learning, prediction and causal Bayes nets," *Trends in Cognitive Sciences*, **7**(1), 43–48.

[1245] Heckerman, D. (1996) *Tutorial on Learning with Bayesian Networks* (Microsoft Research Technical Report MSR-TR-95-06), Microsoft, Redmond, WA.

[1246] Charniak, E. (1991) "Bayesian networks without tears," *AI Magazine*, Winter, 50–63.

[1247] Thompson, D., Niekum, S., Smith, T., and Wettergreen, D. (2006) *Automatic Detection and Classification of Features of Geological Interest* (IEEEAC Paper 1251), Institute of Electrical and Electronic Engineers, Piscataway, NJ.

[1248] Cooper, G., and Herskovits, E. (1992) "Bayesian method for the induction of probabilistic networks from data," *Machine Learning*, **9**, 309–347.

[1249] Friedman, N., Geiger, D., and Goldszmidt, M. (1998) "Bayesian network classifier," *Machine Learning*, **29**, 131–163.

[1250] Grossman, D., and Domingos, P. (2004) "Learning Bayesian network classifiers by maximizing conditional likelihood," *Proceedings 21st International Conference Machine Learning, Banff, Canada*.

[1251] Burge, J., and Lane, B. (2005) "Learning class-discriminative dynamic Bayesian networks," *Proceedings 22nd International Conference Machine Learning, Bonn, Germany*.

[1252] Frey, B., Lawrence, N., and Bishop, C. (1998) "Markovian inference in belief networks," *NIP 98 Algorithms andArchitectures*.

[1253] Heckerman, D., Geiger, D., and Chickering, D. (1995) "Learning Bayesian networks: The combination of knowledge and statistical data," *Machine Learning*, **20**, 197–243.

[1254] Rejimin, T., and Bhanja, S. (2005) "Scalable probabilistic computing using Bayesian networks," *Proceedings IEEE International Conference Robotics and Automation*, pp. 712–715.

[1255] Geiger, D., and Heckerman, D. (1996) "Knowledge representation and inference in similarity networks and Bayesian multinets," *Artificial Intelligence*, **82**, 45–74.

[1256] Sharif, H., Ellery, A., and Samson, C. (2012) "Autonomous rock identification based on image processing techniques," *CASI ASTRO Conference 2012, Quebec City, Canada*.

[1257] Sharif, H., Ellery, A., and Samson, C. (2012) "Strategies for sampling of planetary materials based on images," *Proceedings Global Space Exploration Conference (GLEX) 2012, Washington, D.C.*, 2012.03.P.6x12403.

[1258] Sharif, H., Ralchenko, M., Samson, C., and Ellery, A. (2014) "Autonomous rock classification using Bayesian image analysis for rover-based planetary exploration," accepted by *Computers & Geosciences*.

[1259] Gallant, M., Ellery, A., and Marshall, J. (2011) "Science-influenced mobile robot guidance using Bayesian networks," *Proceedings 24th Canadian Conference on Electrical and Computer Engineering, Canada*.

[1260]

[1261]

[1262] Gallant, M., Ellery, A., and Marshall, J. (2013) "Rover-based autonomous science by probabilistic identification and evaluation," *J. Intelligent & Robotic Systems*, **72**(3), 591–613.

[1263] Gallant, M., Ellery, A., and Marshall, J. (2010) "Exploring salience as an approach to rover-based planetary exploration," *Proceedings of ASTRO Conference, Toronto, Canada*.

[1264] Ellery, A., Wynn-Williams, D., Parnell, J., Edwards, H., and Dickensheets, D. (2004) "The role of Raman spectroscopy as an astrobiological tool," *J. Raman Spectroscopy*, **35**, 441–457.

[1265] Ellery, A., and Wynn-Williams, D. (2003) "Why Raman spectroscopy on Mars? A case of the right tool for the right job," *Astrobiology*, **3**(3), 565–579.

[1266] Hanlon, E., Manoharan, R., Koo, T.-W., Shafer, K., Motz, J., Fitzmaurice, M., Kramer, J., Itzkan, I., Dasari, R., and Feld, M. (2000) "Prospects for in-vivo Raman spectroscopy," *Physiology & Medical Biology*, **45**, R1–R59.

[1267] Ye, Z. (2005) "Artificial intelligence approach for biomedical sample characterization using Raman spectroscopy," *IEEE Trans. Automation Science & Engineering*, **2**(1), 67–73.

[1268] Markou, M., and Singh, S. (2003) "Novelty detection: A review, Part 2: Neural network based approaches," *Signal Processing*, **83**, 2499–2521.

[1269]

[1270] Smith, T., Niekum, S., Thompson, D., and Wettergreen, D. (2005) *Concepts for Science Autonomy during Robotic Traverse and Survey* (IEEEAC Paper 1249), Institute of Electrical and Electronic Engineers, Piscataway, NJ.

[1271] Desimone, R., and Duncan, J. (1995) "Neural mechanisms of selective visual attention," *Annual Reviews Neuroscience*, **18**, 193–222.

[1272] Tarabanis, K., and Allen, P. (1995) "Survey of sensor planning in computer vision," *IEEE Trans. Robotics & Automation*, **11**(1), 86–104.

[1273] Abbott, A. (1992) "Survey of selective fixation control for machine vision," *IEEE Control Systems*, August, 25–31.

[1274] Itti, L., Koch, C., and Niebur, E. (1996) "Model of saliency-based visual attention for rapid scene analysis," *IEEE Trans. Pattern Analysis & Machine Intelligence*, **20**(11), 1254–1259.

[1275] Itti, L. (2004) "Automatic foveation for video compression using a neurobiological model of visual attention," *IEEE Trans Image Processing*, **13**(10), 1304–1318.

[1276] Lee, D., Itti, L., Koch, C., and Braun, J. (1999) "Attention activates winner-takes-all competition among visual filters," *Nature*, **2**(4), 375–381.

[1277] Singer, W., and Gray, C. (1995) "Visual feature integration and the temporal correlation hypothesis" *Annual Reviews Neuroscience*, **18**, 555–586.

[1278] Henderson, J. (2003) "Human gaze control during real world scene perception," *Trends in Cognitive Sciences*, **7**(11), 498–504.

[1279] Fecteau, J., and Munoz, D. (2006) "Salience, relevance and firing: A priority map for target selection," *Trends in Cognitive Sciences*, **10**(8), 382–389.

[1280] Sela, G., and Levine, M. (1997) "Real-time attention for robotic vision," *Real-Time Imaging*, **3**, 173–194.

[1281] Paletta, L., and Pinz, A. (2000) "Active object recognition by view integration and reinforcement learning," *Robotics & Autonomous Systems*, **31**, 71–86.

[1282] Buxton, H., and Gong, S. (1995) "Visual surveillance in a dynamic and uncertain world," *Artificial Intelligence*, **78**, 431–459.

[1283] de Croon, G., Postma, E., and van den Herik, H. (2005) "Situated model for sensory-motor coordination in gaze control," *Pattern Recognition Letters*, **27**(11), July, 1181–1190.

[1284] Torralba, A., Oliva, A., Castehano, M., and Henderson, J. (2007) "Contextual guidance of eye movements and attention in real-world scenes: The role of global features on object search," *Psychological Review*, **113**(4), 766–786.

[1285] Ballard, D., and Brown, C. (1992) "Principles of animate vision," *CVGIP: Image Understanding*, **56**(1), 3–21.

[1286] Frintrop, S., Jensfelt, P., and Christensen, H. (2006) "Attentional landmark selection for visual SLAM," *Proceedings IEEE/RSJ International Conference Intelligent Robots and Systems*, pp. 2582–2587.

[1287] Sebe, N., and Lew, M. (2003) "Comparing salient point detectors," *Pattern Recognition Letters*, **24**, 89–96.

[1288] Sebe, N., Tian, Q., Loupias, E., Lew, M., and Huang, T. (2003) "Evaluation of salient point techniques," *Image & Vision Computing*, **21**, 1087–1095.

[1289] Treisman, A., and Gelade, G. (1980) "Feature integration theory of attention," *Cognitive Psychology*, **12**, 97–136.

[1290] Itti, L., Koch, C., and Niebur, E. (1998) "Model of saliency-based visual attention for rapid scene analysis," *IEEE Trans. Pattern Analysis & Machine Intelligence*, **20**(11), 1254–1259.

[1291] Schiavone, G., Izzo, D., Simoes, L., and de Croon, G. (2012) "Autonomous spacecraft landing through human pre-attentive vision," *Bioinspiration & Biomimetics*, **7**, 025007.

[1292] Gross, A., Smith, B., Muscettola, N., Cannon, H., Barrett, A., Mjolssness, E., Clancy, D., and Dorais, G. (2002) "Advances in autonomous systems for space exploration missions," *World Space Congress, Houston* (IAC-02-U.5.03).

[1293] Michalski, R. (1989) "Pattern recognition as rule-guided inductive inference," *IEEE Trans. Pattern Analysis & Machine Intelligence*, **2**(4), 349–361.

[1294] Chien, S. et al. (1999) "Integrated planning and execution for autonomous spacecraft," *Proceedings IEEE Aerospace Conference*, Vol. 1, pp. 263–271.

[1295] Chien, S. et al. (1999) "Using iterative repair to increase the responsiveness of planning and scheduling for autonomous spacecraft," *International Joint Conference Artificial Intelligence 99 (Workshop on Scheduling and Planning meet Real-time Monitoring in a Dynamic and Uncertain World), Sweden, August.*

[1296] Woods, M., Shaw, A., Barnes, D., Price, D., Long, D., and Pullan, D. (2009) "Autonomous science for an ExoMars rover-like mission," *J. Field Robotics*, **26**(4), 358–390.

[1297] Barnes, D., Pugh, S., and Tyler, L. (2009) "Autonomous science target identification and acquisition (ASTIA) for planetary exploration," *International Conference Intelligent Robots and Systems*, pp. 3329–3335.

[1298] Norris, J., Powell, M., Vona, M., Backes, P., and Wick, J. (2005) "Mars exploration rover operations with the science activity planner," *Proceedings IEEE International Conference on Robotics and Automation*, pp. 4629–4634.

[1299]

[1300] Duda, R., and Shortliffe, E. (1983) "Expert systems research," *Science*, **220**, 261–268.

[1301] Ruckebusch, G. (1983) "Kalman filtering approach to natural gamma ray spectroscopy in well logging," *IEEE Trans. Automatic Control*, **28**(3), 372–380.

[1302] Smith, T., Thompson, D., Wettergreen, D., Cabrol, N., Warren-Rhodes, A., and Weinstein, S. (2007) "Life in the Atacama: Science autonomy for improving data quality," *Journal of Geophysical Research*, **112**, G04S03.

[1303] Kumagai, J. (2006) "Halfway to Mars," *IEEE Spectrum*, March, pp. 33–37.

[1304] Wettergreen, D., Cabrol, N., Baskaran, V., Calderon, F., Heys, S., Jonak, D., Luders, A., Pane, D., Smith, T., Teza, J. et al. (2005) "Second experiments in the robotic investigation of life in the Atacama desert of Chile," *Proceedings International Symposium Artificial Intelligence Robotics and Automation in Space (iSAIRAS), Munich, Germany* (ESA SP-603), ESA, Noordwijk, The Netherlands.

[1305] Warren-Rhodes, K., Weinstein, S., Piatek, J., Dohm, J., Hock, A., Minkley, E., Pane, D., Ernst, A., Fisher, G., Emani, S. et al. (2007) "Robotic ecological mapping: Habitats and the search for life in the Atacama Desert," *Journal of Geophysical Research*, **112**, G04S06.

[1306] Cabrol, N., Wettergreen, D., Warren-Rhodes, K., Grin, E., Moersch, J., Diaz, G., Cockell, C., Coppin, P., Demergasso, C., Dohm, J. et al. (2007) "Life in the Atacama: Searching for life with rovers (science overview)," *Journal of Geophysical Research*, **112**, G04S02, 1–25.

[1307] Cabrol, N., Bettis, E., Glenisyter, B., Ching, G., Herrera, C., Jensen, A., Pereira, M., Stoker, C., Grin, E., Landheim, R. et al. (2001) "Nomad rover field experiment, Atacama Desert, Chile, 2: Identification of palaeolife evidence using a robotic vehicle: Lessons and recommendations for a Mars sample return mission," *Journal of Geophysical Research*, **106**(E4), 7807–7815.

[1308] Arvidson, R., Squyres, S., Baumgartner, E., Schenker, P., Neibur, C., Larsen, K., Seelos, F., Snider, N., and Jolliff, B. (2002) "FIDO prototype Mars rover field trials, Black Rock Summit, Nevada, as a test of the ability of robotic mobility systems to conduct field science," *Journal of Geophysical Research*, **107**(E11), 2-1–2-16.

[1309] Mumma, M., Villanueva, G., Novak, R., Hewagama, T., Boney, B., DiSanti, M., Mandell, A., and Smith, M. (2009) "Strong release of methane on Mars in northern summer 2003," *Science*, **323**, 1040–1041.

[1310] Formisano, V., Atreya, S., Encrenaz, T., Ignatiev, N., and Giuranna, M. (2004) "Detection of methane in the atmosphere of Mars," *Science*, **306**, 158–161.

[1311] Zahnle, K., Freedman, R., and Catling, D. (2011) "Is there methane on Mars?" *Icarus*, **212**(2), 483–503.

[1312] Cloutis, E., Whyte, L., Qadi, A., Ellery, A. et al. (2012) "Mars methane analogue mission (M3): Results of the 2011 field deployment," *Lunar and Planetary Sciences Conference (LPSC)*.

[1313] Cloutis, E., Qadi, A., Ellery, A. et al. (2013) "Mars methane analogue mission (M3): Results of the 2012 field deployment," *Lunar & Planetary Science Conference 44*, abstract 1579.

[1314] Ellery, A., Nicol, C., and Cloutis, E. (2012) *Scent of Science: Model Creation for Odour Based Control of Robotic Vehicles* (ESA Advanced Concepts Team Report 11-6301), European Space Agency, Noordwijk, The Netherlands.

[1315] Krasnopolsky, V., Maillard, J., and Owen, T. (2004) "Detection of methane in the Martian atmosphere: Evidence for life?" *Icarus*, **172**, 537–547.

[1316] Christensen, P., and the THEMIS Team (2003) "Morphology and composition of the surface of Mars: Mars Odyssey THEMIS results," *Science*, **300**, 2056–2061.

[1317] Chastain, B., and Chevrier, V. (2007) "Methane clathrate hydrates as a potential source for martian atmospheric methane," *Planetary & Space Science*, **55**, 1246–1256.

[1318] Parnell, J., Boyce, A., and Blamey, N. (2010) "Follow the methane: The search for a deep biosphere and the case for sampling serpentinites on Mars," *Int. J. Astrobiology*, **9**(4), 193–200.

[1319] Murlis, J., Elkington, J., and Carde, R. (1992) "Odour plumes and how insects use them," *Annual Reviews in Entomology*, **37**, 505–532.

[1320] Kowadlo, G., and Russell, R. (2008) "Robot odour localization: A taxonomy and survey," *Int. J. Robotics Research*, **27**(8), 869–894.

[1321] Balkovsky, E., and Shraiman, B. (2002) "Olfactory search at high Reynolds number," *Proc. National Academy Sciences*, **99**(20), 12589–12593.

[1322] Russell, R., Bab-Hadiashar, A., Shepherd, R., and Wallace, G. (2003) "Comparison of reactive robot chemotaxis algorithms," *Robotics & Autonomous Systems*, **45**, 83–97.

[1323] Passino, K. (2002) "Biomimicry of bacterial foraging for distributed optimization and control," *IEEE Control Systems Magazine*, June, 52–67.

[1324] Shraiman, B., and Siggia, E. (2000) "Scalar turbulence," *Nature*, **405**, 639–646.

[1325] Kowadlo, G., and Russell, R. (2006) "Using naïve physics for odour localization in a cluttered indoor environment," *Autonomous Robotics*, **20**, 215–230.

[1326] Vergassola, M., Villermaux, E., and Shraiman, B. (2007) "Infotaxis as a strategy for searching without gradients," *Nature*, **445**, 406–409.

[1327] Kanzaki, R. (1996) "Behavioural and neural basis of instinctive behaviour in insects: Odour-source searching strategies without memory and learning," *Robotics & Autonomous Systems*, **18**, 33–43.

[1328] Farrell, J., Pang, S., and Li, W. (2005) "Chemical plume tracing via an autonomous underwater vehicle," *IEEE J. Oceanic Engineering*, **30**(2), 428–442.

[1329] Martinez, D., Rochel, O., and Hugues, E. (2006) "Biomimetic robot for tracking specific odours in turbulent plumes," *Autonomous Robots*, **20**, 185–195.

[1330] De Croon, G., O'Connor, L., Nicol, C., and Izzo, D. (2013) "Evolutionary robotics approach to odour source localisation," *Neurocomputing*, **121**, 481–487.

[1331] Vergassola, M., Villermaux, E., and Sharaiman, B. (2007) "Infotaxis as a strategy for searching without gradients," *Nature*, **445**, 406–409.

[1332] Moraud, E., and Martinez, D. (2010) "Effectiveness and robustness of robot infotaxis for searching in dilute conditions," *Frontiers in Neurorobotics*, **4**, March, 1–8.

[1333] Ellery, A., Nicol, C., and Cloutis, E. (2014) "Chasing methane plumes on Mars," submitted to *J. British Interplanetary Society*.

[1334] Farrell, J., Pang, S., and Li, W. (2003) "Plume mapping via hidden Markov methods," *IEEE Trans. on Systems Man and Cybernetics—B: Cybernetics*, **33**(6), 850–863.

[1335]

[1336]

[1337] Pang, S., and Farrell, J. (2006) "Chemical plume source localization," *IEEE Trans. on Systems Man & Cybernetics—B: Cybernetics*, **36**(5), 1068–1080.

[1338] Ferri, G., Jakuba, M., Caselli, E., Mattoli, V., Mazzolai, B., Yoerger, D., and Dario, P. (2007) "Localizing multiple gas/odour sources in an indoor environment using Bayesian occupancy grid mapping," *Proceedings IEEE/RSJ International Conference on Intelligent Robots and Systems*, pp. 566–571.

[1339] Qadi, A., Cloutis, E., Whyte, L., Ellery, A., Bell, J., Berard, G., Boivin, A., Haddad, E., Jamroz, W., Kruzelecky, R. et al. (2012) "Mars methane analogue mission: Rover operations at Jeffrey mine deployment," *CASI ASTRO Conference 2012, Quebec City*, Paper No. 189.

[1340] Qadi, A., Cloutis, E., Whyte, L., Ellery, A., Bell, J., Berard, G., Boivin, A., Haddad, E., Jamroz, W., Kruzelecky, R. et al. (2012) "Mars methane analogue mission: Mission simulation and rover operations at Jeffery Mine deployment," *Advances in Space Research*, **55**(10), 2414–2426.

[1341]

[1342] Nicol, C., Ellery, A., Cloutis, E., and de Croon, G. (2012) "Sniffing as a strategy for detecting life on Mars," *Proceedings Global Space Exploration Conference (GLEX) 2012, Washington, D.C.*, 2012.08.1.8x12332.

[1343] Horvath, J., Carsey, F., Cutts, J., Jones, J., Johnson, E., Landry, B., Lane, L., Lynch, G., Jezek, K., Chela-Flores, J. et al. (1997) "Searching for ice and ocean biogenic activity on Europa and Earth," *Proceedings SPIE 3111, Instruments, Methods and Missions for the Investigation of Extraterrestrial Microorganisms*, pp. 490–500.

[1344] Stillwagen, F. (2002) "Exobiological exploration of Europa (E3): Europa lander," *Proceedings 53rd Internatonal Astronautical Congress, Houston, TX* (Paper IAC-02-Q.2.04).

[1345] Di Pippo, S., Mugnuolo, R., Vielmo, P., and Prendin, W. (1999) "Exploitation of Europa ice and water basins: An assessment on required technological developments, on system design approaches and on relevant expected benefits to space and Earth based activities," *Planetary & Space Science*, **47**, 921–933.

[1346] Chyba, C. (2002) "Searching for life on Europa from a spacecraft lander," in *Signs of Life: A Report Based on the April 2000 Workshop on Life Detection Techniques*, National Research Council, National Academies Press, Washington, D.C., pp. 86–90.

[1347] Zimmerman, W. (1999) "Europa: Extreme communication technologies for extreme conditions," *JPL Deep Space Communication and Navigation Symposium*.

[1348] Powell, J., Maise, G., and Paniagua, J. (2001) "Self-sustaining Mars colonies utilizing the north polar cap and the Martian atmosphere," *Acta Astronautica*, **48**(5/12), 737–765.

[1349] Powell, J., Maise, G., and Paniagua, J. (2005) "NEMO: A mission to search for and return to Earth possible life forms on Europa," *Acta Astronautica*, **57**, 579–593.

[1350] Atkinson, D.. (1999) "Autonomy technology challenges of Europa and Titan exploration missions," *Proceedings Fifth International Symposium Artificial Intelligence Robotics and Automation in Space* (ESA SP-440), ESA, Noordwijk, The Netherlands, pp. 175–181.

[1351] Carsey, F., Chen, G-S., Cutts, J., French, L., Kern, R., Lane, A., Stolorz, P., and Zimmerman, W. (2000) "Exploring Europa's ocean: A challenge for marine technology of this century," *Marine Technology Society J.*, **33**(4), 5–12.

[1352] Bruhn, F., Carsey, F., Kohler, J., Mowlem, M., German, C., and Behar, A. (2005) "MEMS enablement and analysis of the miniature autonomous submersible explorer," *IEEE J Oceanic Engineering*, **30**(1), 165–178.

[1353] Priede, I., Bagley, P., Armstrong, J., Smith, K., and Merrett, N. (1991) "Direct measurement of active dispersal of food-falls by deep-sea demersal fishes," *Nature*, **351**, 647–649.

[1354] Bagley, P., Bradley, S., Priede, I., and Gray, P. (1999) "Measurement of fish movements at depths to 6000 m using a deep-ocean lander incorporating a short baseline sonar utilizing miniature code-activated transponder technology," *Measurement Science & Technology*, **10**, 1214–1221.

[1355] Bailey, D., Jamieson, A., Bagley, P., Collins, M., and Priede, I. (2002) "Measurement of in situ oxygen consumption of deep-sea fish using an autonomous lander vehicle" *Deep Sea Research I*, **49**, 1519–1529.

[1356] Leonard, N. (1995) ""Control synthesis and adaptation of an underactuated autonomous underwater vehicle," *IEEE J. Oceanic Engineering*, **20**(3), 211–220.

[1357] Ballard, R. (1993) "Medea/Jason remotely operated vehicle system," *Deep Sea Research 1*, **40**, 1673–1687.

[1358] Schubak, G., and Scott, D. (1995) "Techno-economic comparison of power systems for autonomous underwater vehicles," *IEEE J. Oceanic Engineering*, **20**(1), 94–100.

[1359] Eriksen, C., Osse, T., Light, R., Wen, T., Lehman, T., Sabin, P., Ballard, J., and Chiodi, A. (2001) "Seaglider: A long-range autonomous underwater vehicle for oceanographic research," *IEEE J. Oceanic Engineering*, **26**(4), 424–436.

[1360] Vaneck, T., Scire-Scappuzzo, F., Hunter, A., and Joshi, P. (2000) *System of Mesoscale Biomimetic Roboswimmers for Underwater Exploration and Search for Life on Europa* (NIAC Final Report NAS5-98051/PSI-1337/TR-1723), NASA Institute for Advanced Concepts, Atlanta, GA.

[1361] Perry Slingsby Systems (2002) *Triton MRV#5 ROV System Technical Specification* (A865-000-004), Perry Slingsby Systems Ltd., York, U.K.

[1362] Brighenti, A. (1990) "Parametric analysis of the configuration of autonomous underwater vehicles," *IEEE J. Oceanic Engineering*, **15**(3), 179–188.

[1363] Gad-El-Hak, M (1996) "Compliant coatings: A decade of progress," *Applied Mechanics Review*, **49**(10/2), S147–S157.

[1364] Webb, P. (2004) "Maneuverability: General issues," *IEEE J. Oceanic Engineering*, **29**(3), 547–555.

[1365] Bandyopadhyay, P. (2005) "Trends in biorobotic autonomous undersea vehicles," *IEEE J. Oceanic Engineering*, **30**(1), 109–139.

[1366] Sfakiotakis, M., Lane, D., and Davies, J. (1999) "Review of fish swimming modes for aquatic locomotion," *IEEE J. Oceanic Engineering*, **24**(2), 237–252.

[1367] Colgate, J., and Lynch, K. (2004) "Mechanics and control of swimming: A review," *IEEE J. Oceanic Engineering*, **29**(3), 660–673.

[1368] Yu, J., Tan, M., Wang, S., and Chen, E. (2004) "Development of a biomimetic robot fish and its control algorithm," *IEEE Trans. Systems Man & Cybernetics B: Cybernetics*, **34**(4), 1798–1810.

[1369] Sayers, C., Paul, R., Whitcomb, L., and Yoerger, D. (1998) "Teleprogramming for subsea teleoperation using acoustic communication," *IEEE J. Oceanic Engineering*, **23**(1), 60–71.

[1370] Kilfoyle, D., and Baggeroer, A. (2000) "State of the art in underwater acoustic telemetry," *IEEE J. Oceanic Engineering*, **25**(1), 4–27.

[1371] Stojanovic, M. (1996) "Recent advances in high speed underwater acoustic communications," *IEEE J Oceanic Engineering*, **21**(2), 125–136.

[1372] Cox, R., and Wei, S. (1995) "Advances in the state of the art for AUV inertial sensors and navigation systems," *IEEE J. Oceanic Engineering*, **20**(4), 361–366.

[1373] Whitcomb, L., Yoerger, D., Singh, H., and Howland, J. (1999) "Advances in underwater robot vehicles for deep ocean exploration: Navigation, control and survey operations," *Proceedings Ninth International Symposium Robotics Research*.

[1374] Helweg, D., Moore, P., Martin, S., and Dankiewicz, L. (2006) "Using a binaural biomimetic array to identify bottom objects ensonified by echolocating dolphins," *Bioinspiration & Biomimetics*, **1**, 41–51.

[1375] Jonsson, J., Edquist, E., Kratz, H., Almqvist, M., and Thornell, G. (2010) "Simulation, manufacturing and evaluation of a sonar for a miniaturized submersible explorer," *IEEE Trans. Ultrasonics, Ferroelectrics & Frequency Control*, **57**(2), 490–495.

[1376] Feng, T-J., Li, X., Ji, G-R., Zheng, B., Zhang, H-Y., Wang, G-Y., and Zheng, G-X. (1996) "New laser scanning sensing technique for underwater engineering inspection," *Artificial Intelligence in Engineering*, **10**, 363–368.

[1377] Webb, P. (2004) "Maneuverability: General issues," *IEEE J. Oceanic Engineering*, **29**(3), 547–555.

[1378]

[1379] Dumlu, D., and Istefanopulos, Y. (1995) "Design of an adaptive controller for submersibles via multimodel gain scheduling," *Ocean Engineering*, **22**(6), 593–614.

[1380] Antonelli, G., Caccavale, F., Chiaverini, S., and Villani, L. (2000) "Tracking control for underwater vehicle–manipulator systems with velocity estimation," *IEEE J. Oceanic Engineering*, **25**(3), 399–413.

[1381] Caccia, M., Bono, R., Bruzzone, G., and Veruggio, G. (1999) "Variable-configuration UUVs for marine science applications," *IEEE Robotics & Automation Magazine*, June, 22–32.

[1382] Yoerger, D., Cooke, J., and Slotine, J.-J. (1990) "Influence of thruster dynamics on underwater vehicle behaviour and their incorporation into control system design," *IEEE J. Oceanic Engineering*, **15**(3), 167–178.

[1383] Caccia, M., and Veruggio, G. (2000) "Guidance and control of a reconfigurable unmanned underwater vehicle," *Control Engineering Practice*, **8**, 21–37.

[1384] Yoerger, D., and Slotine, J.-J. (1985) "Robust trajectory control of underwater vehicles," *IEEE J. Oceanic Engineering*, **10**(4), 462–470.

[1385] Yuh, J., and Gonugunta, K. (1993) "Learning control of underwater robotic vehicles," *IEEE International Conference Robotics and Automation*, pp. 106–111.

[1386] Yuh, J. (1990) "Neural net controller for underwater robotic vehicles," *IEEE J. Oceanic Engineering*, **15**(3), 161–166.

[1387] Fjellstad, O.-E., and Fossen, T. (1994) "Position and attitude tracking of AUVs: A quaternion feedback approach," *IEEE J. Oceanic Engineering*, **19**(4), 512–518.

[1388] Yoerger, D., and Slotine, J. (1985) "Robust trajectory control of underwater vehicles" *IEEE J. Oceanic Engineering*, **10**(4), 462–470.

[1389] Yuh, J., and Lakshmi, R. (1993) "Intelligent control system for remotely operated vehicles," *IEEE J. Oceanic Engineering*, **18**(1), 55–62.

[1390] Van de Ven, P., Flanagan, C., and Toal, D. (2005) "Neural network control of underwater vehicles," *Engineering Applications of Artificial Intelligence*, **18**, 533–547.

[1391] Whitcomb, L., and Yoerger, D. (1999) "Preliminary experiments in model-based thruster control for underwater vehicle positioning," *IEEE J. Oceanic Engineering*, **24**(4), 495–506.

[1392] Healey, A., Rock, S., Cody, S., Miles, D., and Brown, J. (1995) "Towards an improved understanding of thruster dynamics for underwater vehicles," *IEEE J. Oceanic Engineering*, **20**(4), 354–361.

[1393] Bachmayer, R., Whitcomb, L., and Grosenbaugh, M. (2000) "Accurate four-quadrant nonlinear dynamical model for marine thrusters: Theory and experimental validation," *IEEE J. Oceanic Engineering*, **25**(1), 146–159.

[1394] Fossen, T., and Blanke, M. (2000) "Nonlinear output feedback control of underwater vehicle propellers using feedback from estimated axial flow velocity," *IEEE J. Oceanic Engineering*, **25**(2), 241–255.

[1395] Driscoll, F., Lueek, R., and Nahon, M. (2000) "Development and validation of a lumped-mass dynamics model of a deep-sea ROV system," *Applied Ocean Research*, **22**, 169–182.

[1396] Hsu, L., Costa, R., Lizarralde, F., and da Cunha, J. (2000) "Dynamic positioning of remotely operated underwater vehicles," *IEEE Robotics & Automation Magazine*, September, 21–31.

[1397] Antonelli, G., Chiaverini, S., and Sarkar, N. (2001) "External force control for underwater vehicle-manipulator systems," *IEEE Trans. Robotics & Automation*, **17**(6), 931–938.

[1397] Ellery, A. (2004) "Robotic on-orbit servicers—the need for control moment gyroscopes for attitude control," *Aeronautical J.*, **108**(2), 207–214.

[1398] Yuh, J. (1990) "Modelling and control of underwater robotic vehicles," *IEEE Trans. Systems Man & Cybernetics*, **20**(6), 1475–1483.

[1399] Ellery, A. (2004) "Robotic on-orbit servicers—the need for control moment gyroscopes for attitude control," *Aeronautical J.*, **108**(2), 207–214.

[1399] McMillan, S., Orin, D., and McGhee, R. (1994) "Efficient dynamic simulation of an unmanned underwater vehicle with a manipulator," *Proceedings IEEE International Conference Robotics & Automation*, pp. 1133–1140.

[1400] Dunnigan, M., and Russell, G. (1998) "Evaluation and reduction of the dynamic coupling between a manipulator and an underwater vehicle," *IEEE J. Oceanic Engineering*, **23**(3), 260–273.

[1401] McIsaac, K., and Ostrowski, J. (1999) "Geometric approach to anguilliform locomotion: Modeling of an underwater eel robot," *Proceedings IEEE International Conference Robotics & Automation*, pp. 2843–2848.

[1402] Lynch, B., and Ellery, A. (2014) "Efficient control of an AUV vehicle–manipulator system: An application for the exploration of Europa," *IEEE J. Oceanic Engineering*, **39**(3), 552–570.

[1403] Ellery, A. (2004) "An engineering approach to the dynamic control of space robotic on-orbit servicers," *Proc. Inst. Mech. Engineers, Part G: Aerospace Engineering*, **218**, 79–98.

[1404] Canudas de Wit, C., Diaz, E., and Perrier, M. (2000) "Nonlinear control of an underwater vehicle/manipulator with composite dynamics," *IEEE Trans. Control Systems Technology*, **8**(6), 948–960.

[1405] Ryu, J-H., Kwon, D-S., and Lee, P-M. (2001) "Control of underwater manipulators mounted on an ROV using base force information," *Proceedings IEEE International Conference Robotics and Automation*, pp. 3238–3243.

[1406] Sarkar, N., and Podder, T. (2001) "Coordinated motion planning and control of autonomous underwater vehicle–manipulator systems subject to drag optimization," *IEEE J. Oceanic Engineering*, **26**(2), 228–239.

[1407] Antonelli, G., Caccavale, F., and Chiaverini, S. (2004) "Adaptive tracking control of underwater vehicle–manipulator systems based on the virtual decomposition approach," *IEEE Trans. Robotics & Automation*, **20**(3), 594–602.

[1408] Elliott, J., and Waite, J. (2010) "In-situ missions for the exploration of Titan's lakes," *J. British Interplanetary Society*, **63**, 376–383.

[1409] Morrow, M., Woolsey, C., and Hagerman, G. (2006) "Exploring Titan with buoyancy-driven gliders," *J. British Interplanetary Society*, **59**, 27–34.

[1410] Stevenson, D. (2003) "Mission to Earth's core: A modest proposal," *Nature*, **423**, 239–240.

[1411] Ellis-Evans, J.C. (2001) "Sub-glacial Lake Vostok: A possible analogue for a Europan ocean?" *J. British Interplanetary Society*, **54**, 159–168.

[1412] Studinger, M., Karner, G., Bell, R., Levin, V., Raymond, C., and Tikku, A. (2003) "Geophysical models for the tectonic framework of the Lake Vostok region, East Antarctica," *Earth & Planetary Sciences*, **216**, 663–677.

[1413] Siegert, M. (2000) "Antarctic subglacial lakes," *Earth-Science Reviews*, **50**, 29–50.

[1414] Siegert, M., Carter, S., Tabacco, I., Popov, S., and Blankenship, D. (2005) "Revised inventory of Antarctic subglacial lakes," *Antarctic Science*, **17**(3), 453–460.

[1415] Siegert, M. (2000) "Antarctica's Lake Vostok," *American Scientist*, **87**, November/December, 511–517.

[1416] Souchez, R., Petit, J., Tison, J-L., Jouzel, J., and Verbeke, V. (2000) "Ice formation in subglacial Lake Vostok, Central Antarctica," *Earth & Planetary Science Letters*, **181**, 529–538.

[1417] Abyzov, S., Miskevich, I., Poglazona, M., Barkov, N., Lipenkov, V., Bobin, N., Koudryashov, B., Pashkevich, V., and Ivanov, M. (2001) "Microflora in the basal strata at Antarctica ice core above the Vostok lake," *Advances in Space Research*, **28**(4), 701–706.

[1418] Priscu, J., Fritsen, C., Adams, E., Giovannoni, S., Paerl, H., McKay,C., Doran, P., Gordon, D., Lanoil, B., and Pinckney, J. (1998) "Perennial Antarctic lake ice: An oasis for life in a polar desert," *Science*, **280**, 2095–2098.

[1419] Price, B. (2000) "Habitat for psychrophiles in deep Antarctic ice," *Proc. National Academy Sciences*, **97**(3), 1247–1251.

[1420] Siegert, M., Ellis-Evans, J., Tranter, M., Mayer, C., Petit, J-R., Salamatini, A., and Priscu, J. (2001) "Physical, chemical and biological processes in Lake Vostok and other Antarctic subglacial lakes," *Nature*, **414**, 603–609.

[1421] Siegert, M., Hindmarsh, R., Corr, H., Smith, A., Woodward, J., King, E., Payne, A., and Joughin, I. (2004) "Subglacial Lake Ellsworth: A candidate for in situ exploration in West Antarctica," *Geophysical Research Letters*, **31**, L23404.

[1422] Siegert, M., Ellery, A., and the Lake Ellsworth Consortium (2006) "Exploration of Ellsworth subglacial lake: A concept for the development, organization and execution of an experiment to explore, measure and sample the environment of a West Antarctic subglacial lake," *Reviews in Environmental Science & Biotechnology*, **6**(1/3), 161–179.

[1423] Behar, A. et al. (2001) "Antarctic ice borehole probe," *IEEE Proc. Aerospace Conf.*, **1**, 325–330.

[1424] Paulson, L. (2004) "Biomimetic robots," *IEEE Computer*, September, 48–53.

[1425] Ellery, A. et al. (2004) *Bionics and Space Systems, Design 1: Overview of Biomimetics Technology* (ESA-ESTEC Technical Note 1, ESA Contract No. 18203/04/NL/PA), ESA/ESTEC, Noordwijk, The Netherlands.

[1426] Ellery, A. et al. (2005) *Bionics and Space Systems, Design 3: Application of Biomimetics to Space Technology* (ESA-ESTEC Technical Note 3, ESA Contract No. 18203/04/NL/ PA), ESA/ESTEC, Noordwijk, The Netherlands.

[1427] Menon, C., Ayre, M., and Ellery, A. (2006) "Biomimetics: A new approach to space systems design," *ESA Bulletin*, **125**, February, 21–26.

[1428] Bar-Cohen, Y. (2006) "Biomimetics: Using nature to inspire human innovation," *Bioinspiration & Biomimetics*, **1**, P1–P12.

[1429] Webb. B. (2002) "Robots in invertebrate neuroscience," *Nature*, **417**, 359–363.

[1430] Thakoor, S. (2000) "Bio-inspired engineering of exploration systems," *J. Space Mission Architecture*, Fall, 49–79.

[1431] Beer, R., Chiel, H., Quinn, R., and Ritzmann, R. (1998) "Biorobotic approaches to the study of motor systems," *Current Opinion in Neurobiology*, **8**, 777–782.

[1432] Menon, C., Broschart, M., and Lan, N. (2007) "Biomimetics and robotics for space applications: Challenges and emerging technologies," *Proceedings IEEE International Conference Robotics & Automation: Workshop on Biomimetic Robotics*.

[1433] Ellery, A., and Cockell, C. (2006) "Bio-inspired microrobots for support of exploration from a Mars polar station" in C. Cockell (Ed.), *Project Boreas: A Station for the Martian Geographic Pole*, British Interplanetary Society, London.

[1434] McGeer, T. (1990) "Passive dynamic walking," *Int. J. Robotics Research*, **9**(2), 62–82.

[1435] Scott, G., and Ellery, A. (2005) "Using Bekker theory as the primary performance metric for measuring the benefits of legged locomotion over traditional wheeled vehicles for planetary robotic explorers," *Proceedings Space 2005, Long Beach, California* (AIAA-2005-6736), American Institute of Aeronautics and Astronautics, Washington, D.C.

[1436] Graham, D. (1985) "Pattern and control of walking in insects," *Advances in Insect Physiology*, **18**, 31–140.

[1437] Celaya, E., and Porta, J. (1998) "Control structure for the locomition of a legged robot on difficult terrain," *IEEE Robotics & Automation Magazine*, June, 43–51.

[1438] Waldron, K., Volinout, V., Peng, A., and McGhee, R. (1984) "Configuration design of the Adaptive Suspension Vehicle," *Int. J. Robotics Research*, **3**(2), 37–48.

[1439] Song, S-M., and Waldron, K. (1987) "Analytical approach for gait study and its applications on wave gaits," *Int. J. Robotics Research*, **6**(2), 60–71.

[1440] Bay, J., and Hemami, H. (1987) "Modelling of a neural pattern generator with coupled nonlinear oscillators," *IEEE Trans. Biomedical Engineering*, **34**(4), 297–306.

[1441] Venkataraman, S. (1997) "Simple legged locomotion gait model," *Robotics & Autonomous Systems*, **22**, 75–85.

[1442] Celaya, E., and Porta, J. (1998) "Control structure for the locomotion of a legged robot on difficult terrain," *IEEE Robotics & Automation Magazine*, June, 43–51.

[1443] Galvez, J., Estremera, J., and de Santos, P. (2003) "New legged-robot configuration for research in force distribution," *Mechatronics*, **13**, 907–932.

[1444] Full, R. (2000) "Biological inspiration: Lessons from many-legged locomotors," *International Symposium Robotics Research*, pp. 337–342.

[1445] Juinenez, M., and Gonzalez de Santos, P. (1998) "Attitude and position control method for realistic legged vehicles," *Robotics & Autonomous Systems*, **18**, 345–354.

[1446] Klein, C., Olson, K., and Pugh, D. (1983) "Use of force and attitude sensors for locomotion of a legged vehicle rover over irregular terrain," *Int. J. Robotics Research*, **2**(3), 3–17.

[1447] Zielinka, T., and Heny, J. (2002) "Development of a walking machine: Mechanical design and control problems," *Mechatronics*, **12**, 737–754.

[1448] Pfeiffer, F., Eltze, J., and Weidenmaron, H. (1995) "Six legged technical walking considering biological principles," *Robotics & Autonomous Systems*, **16**, 223–232.

[1449] Klassen, B., Linneman, R., Spenneberg, D., and Kirchner, F. (2001) "Biomimetic walking robot scorpion: Control and modelling," *Proceedings ASME Design Engineering Technical Conference*, pp. 1105–1112.

[1450] Waldron, K. (1995) "Terrain adaptive vehicles," *Trans. ASME*, Spec 50th Anniversary Design Issue, **117**, June, 107–112.

[1451] Waldron, K. (1985) "Mechanics of mobile robots," *Proceedings International Conference Advanced Robotics (ICAR) 85, Tokyo, Japan, September 1985*, pp. 533–544.

[1452] Krotkov, E. et al. (1995) "Ambler: Performance of a six-legged planetary rover," *Acta Astronautica*, **35**(1), 75–81.

[1453] Bares, J., and Wethergreen, D. (1999) "Dante II: Technical description, results and lessons learned," *Int. J. Robotics Research*, **18**(7), 621–649.

[1454] Guccione, S., and Muscato, G. (2003) "Wheeleg robot," *IEEE Robotics & Automation Magazine*, December, 33-43.

[1455] Lacagnina, M., Muscato, G., and Sinatra, R. (2003) "Kinematics, dynamics and control of a hybrid robot Wheeleg," *Robotics & Autonomous Systems*, **45**, 161–180.

[1456] Hemle, A. et al. (2000) "Hybrid locomotion of wheel-legged machine," *Proceedings Third International Conference on Climbing and Walking Robots, Madrid*, pp. 167–173.

[1457] Hemle, A. et al. (1999) "Development of Work Partner robot: Design of actuating and motion control system," *Climbing and Walking Robot Conference (CLAWAR 99), September, Portsmouth, U.K.*

[1458] Lepannen, I. et al. (1998) "Work Partner: HUT Automation's new hybrid walking machine," *Climbing and Walking Robots Conference (CLAWAR 98), Brussels, Belgium.*

[1459]

[1460]

[1461]

[1462] Bogatchev, A., Gromov, V., Koutcherenko, V., Matrossov, S., Petriga, V., Solomnikov, V., and Fedoseev, S. (2000) "Walking and wheel-walking robots," *Climbing and Walking Robots Conference (CLAWAR 2000)*.

[1463] Genta, G., and Amati, N. (2004) "Mobility on planetary surfaces: May walking machines be a viable alternative," *Planetary & Space Sciences*, **52**, 31–40.

[1464] Amati, N., Genta, G., and Reyner, L. (2002) "Three-rigid-frames walking planetary rovers: A new concept," *Acta Astronautica*, **50**(12), 729–736.

[1465] Brooks, R., and Flynn, A. (1989) "Fast, cheap and out of control: A robot invasion of the solar system," *J. British Interplanetary Society*, **42**, 478–485.

[1466] Hill, W., Mausli, P.-A., Estier, T., Huber, R., and van Winnendael, M. (2000) "Using microtechnologies to build micro-robot systems," *Advanced Space Technologies for Robotics and Automation (ASTRA 2000)*.

[1467] Bekker, M. (1985) "Development of a moon rover," *J. British Interplanetary Society*, **38**, 537–543.

[1468] Hirose, S., and Morishima, A. (1990 "Design and control of a mobile robot with an articulated body," *Int. J. Robotics Research*, **9**(2), 99–114.

[1469] Hirose, S. et al. (1995) "Fundamental considerations for the design of a planetary rover," *IEEE International Conference Robotics and Automation*, pp. 1939–1944.

[1470] Chirikjian, G. (1995) "Kinematics of hyper-redundant robot locomotion," *IEEE Trans. Robotics & Automation*, **11**(6), 781–793.

[1471] Chirikjian, G., and Burdick, J. (1991) "Kinematics of hyper-redundant robot locomotion with applications to grasping," *Proceedings IEEE International Conference Robotics and Automation*, pp. 720–725.

[1472] Klaasen, B., and Paap, K. (1999) "GMD-SNAKE2: A snake-like robot driven by wheels and a method for motion," *Proceedings IEEE International Conference Robotics & Automation*, pp. 3014–3019.

[1473] Rincon, D., and Sotelo, J. (2003) "Ver-Vite: Dynamic and experimental analysis for inchworm-like biomimetic robots," *IEEE Robotics & Automation Magazine*, December, 53–57.

[1474] Kim, B., Lee, M., Lee, Y., Kim, Y., and Lee, G. (2006) "Earthworm-like micro-robot using shape memory alloy actuator," *Sensors & Actuators A*, **125**, 429–437.

[1475] Kim, J., Kim, J., and Choi, S. (2002) "A hybrid inchworm linear motor," *Mechatronics*, **12**, 525–542.

[1476] Lee, S., and Esashi, M. (1995) "Design of the electrostatic linear microactuator based on the inchworm motion," *Mechatronics*, **5**(8), 963–972.

[1477] Chu, B., Jung, K., Han, C-S., and Hong, D. (2010) "Survey of climbing robots: Locomotion and adhesion," *Int. J. Precision Engineering & Manufacturing*, **11**(4), 633–647.

[1478] Soper, N. (2005) "Biomimetic approach to the design of a climbing robot for planetary exploration," *Second Mars Expedition Planning Workshop*, Abstract.

[1479] Sitti, M., and Fearing, R. (2003) "Synthetic gecko foot-hair micro/nano-structures for future wall-climbing robots," *IEEE International Conference Robots & Automation*, pp. 1164–1170.

[1480] Bretl, T., Miller, T., Rock, S., and Latombe, J-C. (2003) "Climbing robots in natural terrain," *Proceedings International Symposium Artificial Intelligence & Robotics in Space, Nara, Japan*.

[1481] Linder, S., Wei, E., and Clay, A. (2005) "Robotic rock climbing using computer vision and force feedback," *Proceedings IEEE International Conference Robotics & Automation*, pp. 4685–4689.

[1482] Behar, A., Matthews, J., Carsey, F., and Jones, J. (2004) "NASA/JPL Tumbleweed polar rover," *IEEE Aerospace Conference, Big Sky, MT*, 0-7803-8155-6/4.

[1483] Antol, J., Calhoun, P., Flick, J., Hajos, G., Kolacinski, R., Minton, D., Owens, R., and Parker, J. (2003) *Low Cost Mars Surface Exploration: The Mars Tumbleweed* (NASA TM-2003-212411), NASA, Washington, D.C.

[1484] Yoshimitsu, T., Kubota, T., Nakatani, I., Adachi, T., and Saito, H. (2003) "Micro-hopping robot for asteroid exploration," *Acta Astronautica*, **52**, 441–446.

[1485] Richter, L. (1998) "Principles for robotic mobility on minor solar system bodies," *Robotics & Autonomous Systems*, **23**, 117–124.

[1486] Shafirovitch, E., Saloman, M., and Gokalp, I. (2006) "Mars hopper versus Mars rover," *Acta Astronautica*, **59**, 710–716.

[1487] Armour, R., Paskins, K., Bowyer, A., Vincent, J., and Megill, W. (2007) "Jumping robots: A biomimetic solution to locomotion across rough terrain," *Bioinspiration & Biomimetics*, **2**, S65–S82.

[1488] Hale, E., Schara, N., Burdick, J., and Fiorini, P. (2000) "Minimally actuated hopping rover for exploration of celestial bodies," *IEEE International Conference Robotics & Automation*, pp. 420–427.

[1489] Harbick, K., and Sukhatme, G. (2001) *Height Control for a One-legged Hopping Robot Using a One-dimensional Model* (IRIS Technical Report IRIS-01-405), Institute for Robotics and Intelligent Systems, University of Southern California, Los Angeles, CA.

[1490] Scholz, M., Bobbert, M., and van Soest, A. (2006) "Scaling and jumping: Gravity loses grip on small jumpers," *J. Theoretical Biology*, **240**, 554–561.

[1491] Confente, M., Cosma, C., Fiorini, P., and Burdick, J. (2002) "Planetary exploration using hopping robots," *Proceedings Seventh ESA Workshop on Advanced Space Technologies for Robotics and Automation (ASTRA), Noordwijk, Netherlands*.

[1492] Martinez-Cantin, R. (2004) "Bio-inspired multi-robot behaviour for exploration in low-gravity environments," *Proceedings 55th International Astronautical Congress, Vancouver*.

[1493] Wong, J. (2001) *Theory of Ground Vehicles*, John Wiley & Sons, New York.

[1494] Elfes, A., Bueno, S., Berrgerman, M., De Paiva, E., Ramos, J., and Azinheira, J. (2003) "Robotic airships for exploration of planetary bodies with an atmosphere: Autonomy challenges," *Autonomous Robots*, **14**, 147–164.

[1495] Barnes, D. et al. (2000) "Investigation into aerobot technologies for planetary exploration," *Proceedings Sixth ESA Workshop on Advanced Space Technologies for Robotics and Automation (ASTRA), December 2000*, ESA/ESTEC, Noordwijk, The Netherlands, 3.6-5.

[1496] Cutts, J., Nock, K., Jones, J., Rodriguez, G., and Balaram, J. (1995) "Planetary exploration by robotic aerovehicles," *Autonomous Robots*, **2**, 261–282.

[1497] Blamont, J., and Jones, J. (2002) "New method for landing on Mars," *Acta Astronautica*, **51**(10), 723–726.

[1498] Joshi, S., Nagai, G., and Sullivan, C. (2004) "Planetary navigation architecture for cooperative autonomous ground-air-space robots," *Proceedings First AIAA Intelligent Systems Technical Conference, Chicago* (AIAA 2004-6256), American Institute of Aeronautics and Astronautics, Washington, D.C.

[1499] Pankine, A., Aaron, K., Heun, M., Nock, K., Schlaifer, R., Wyszkowski, C., Ingersoll, A., and Lorenz, R. (2004) "Directed aerial robotic explorers for planetary exploration," *Advances in Space Research*, **33**, 1825–1830.

[1500] Jones, J., Saunders, S., Blamont, J., and Yavrouvian, A. (1999) "Balloons for controlled roving/landing on Mars," *Acta Astronautica*, **45**(4/9), 293–300.

[1501] Galeev, A. et al. (1995) "Russian programs of planetary exploration: Mars-94/98 missions," *Acta Astronautica*, **35**, 9–33.

[1502] Lorenz, R. (2008) "Review of balloon concepts for Titan," *J. British Interplanetary Society*, **61**, 2–13.

[1503] Hall, J., Kerzhanovich, V., Yavrouian, A., Jones, J., White, C., Dudik, B., Plett, G., Mannella, J., and Elfes, A. (2006) "Aerobot for global in situ exploration of Titan," *Advances in Space Research*, **37**, 2108–2119.

[1504] Jones, J. (2000) "Meeting the challenge to balloon science," *Advances in Space Research*, **26**(9), 1303–1311.

[1505] Rand, J., and Crenshaw, J. (1996) "Superpressure balloon design," *Advances in Space Research*, **17**(9), 5–13.

[1506] Raque, S., and Simpson, J. (1998) "Automated pressure control in pressurized balloon systems," *Advances in Space Research*, **21**(7), 971–974.

[1507] Winker, J. (2002) "Pumpkins and onions and balloon design," *Advances in Space Research*, **30**(5), 1199–1204.

[1508] Yajima, N. (2002) "Survey of balloon design problems and prospects for large super-pressure balloons in the next century," *Advances in Space Research*, **30**(5), 1183–1192.

[1509] Jones, W. (1990) "Recent developments and near-term expectations for the NASA balloon programme," *AIAA J. Spacecraft*, **27**(3), 306–311.

[1510] Kerzhanovich, V., Cutts, J., Cooper, H., Hall, J., MacDonald, B., Pauken, M., White, C., Yavrouian, A., Castano, A., Cathey, H. et al. (2004) "Breakthrough in Mars balloon technology," *Advances in Space Research*, **33**, 1836–1841.

[1511]

[1512]

[1513] Wu, H., Sun, D., and Zhou, Z. (2004) "Micro air vehicle: Configuration, analysis, fabrication and test," *IEEE/ASME Trans. Mechatronics*, **9**(1), 108–117.

[1514] Gursul, I., Taylor, G., and Wooding, C. (2002) *Vortex Flows over Fixed Wing Micro-air Vehicles* (AIAA 2002-0698), American Institute of Aeronautics and Astronautics, Washington, D.C.

[1515] Goraj, Z. (2000) "Dynamics of a high altitude long duration UAV," *ICAS 2000 Congress*, pp. 362.1–362.10.

[1516] Goraj, Z. (2000) "Ultralight wing structure for high altitude long endurance UAV," *ICAS 2000 Congress*, pp. 476.1–476.10.

[1517] Hill, W. et al. (2000) "Using microtechnologies to build micro-robot systems," *Proceedings ESA Workshop on Advanced Space Technologies for Robotics and Automation 2000*, ESA/ESTEC, Noordwijk, The Netherlands.

[1518] Fielding, J., and Underwood, C. (2004) "MASSIVA: Mars surface sampling and imaging VTOL aircraft," *J. British Interplanetary Society*, **57**, 306–312.

[1519] Underwood, C., Sweeting, M., and Song, H. (2008) "Design of an autonomous Mars VTOL aerobot," *International Astronautics Congress Conference*, IAC-08.A3.3.B5.

[1520] Kroo, I., Prinz, F., Shantz, M., Kunz, P., Fay, G., Cheng, S., Fabian, T., and Partridge, C. (2001) *The Mesicopter: A Miniaturized Rotorcraft Concept, Phase I/II* (NIAC Final Report), NASA Institute for Advanced Concepts, Atlanta, GA.

[1521] Colozza, A. (2002) *Solid State Aircraft, Phase 1 Project* (NIAC NAS5-98051 Final Report), NASA Institute for Advanced Concepts, Atlanta, GA.

[1522]

[1523] Agrawal, A., and Agrawal, S. (2009) "Design of bio-inspired flexible wings for flapping wing micro-sized air vehicle applications," *Advanced Robotics*, **23**, 979–1002.

[1524] Shyy, W., Berg, M., and Ljungqvist, D. (1999) "Flapping and flexible wings for biological and micro-air vehicles," *Progress in Aerospace Sciences*, **35**, 455–505.

[1525] Michelson, R., and Navqi, M. (2003) "Beyond biologically-inspired insect flight," *Low Re Aerodynamics on Aircraft Including Applications in Emerging UAV Technology* (RTO AVT von Kármán Institute for Fluid Dynamics Lecture Series), von Kármán Institute for Fluid Dynamics, Rode, Belgium.

[1526] Jones, K., and Platzer, M. (2002) "On the design of efficient micro-air vehicles," *Proceedings First International Conference Design & Nature: Comparing Design in Nature with Science & Engineering, Udine, Italy*.

[1527] Jones, K., Duggan, S., and Platzer, M. (2001) "Flapping wing propulsion for a micro-air vehicle," *Proceedings 39th Aerospace Sciences Meeting & Exhibition, Reno, Nevada*, AIAA-2001-0126.

[1528] Jones, K., and Platzer, M. (2002) "On the design of efficient micro-air vehicles," *Proceedings First International Conference Design & Nature, Udine, Italy*.

[1529] Jones, K., Bradshaw, C., Papadopoulos, J., and Platzer, M. (2003) "Development and flight testing of flapping wing propelled micro air vehicles," *Proceedings Second AIAA Unmanned Unlimited Systems Technologies and Operations Conference & Workshop, San Diego, California*, AIAA-2003-6549.

[1530] Jones, K., Bradshaw, C., Papadopoulos, J., and Platzer, M. (2004) "Improved performance and control of flapping wing propelled micro air vehicles," *Proceedings 42nd Aerospace Sciences Meeting & Exhibition, Reno, Nevada*, AIAA-2004-0399.

[1531] Michelson, R. (2002) "Entomopter," in J. Ayers, J. Davis, and A. Rudolph (Eds.), *Neurotechnology for Biomimetic Robots*, Bradford Books, MIT Press, Cambridge, MA, pp. 481–510.

[1532] Michelson, R.C. (2002) "The Entomopter," in J. Ayers, J. Davis, and A. Rudolph (Eds.), *Neurotechnology for Biomimetic Robots*, Bradford Books, MIT Press, Cambridge, MA, pp. 481–510.

[1533] De Croon, G., Groen, M., de Wagter, C., Remes, B., Ruijsink, R., and van Oudheusden, B. (2012) "Design, aerodynamics and autonomy of the DelFly," *Biomimetics & Bioinspiration*, **7**(2), 025003.

[1534] De Croon, G., de Clercq, K., Ruijsink, R., Remes, B., and de Wagter, C. (2009) "Design, aerodynamics and vision-based control of the DelFly," *Int. J. Micro-Air Vehicles*, **1**(2), 71–97.

[1535] Kaelbling, L., Littman, M., and Cassandra, A. (1998) "Planning and acting in partially observable stochastic domains," *Artificial Intelligence*, **101**, 99–134.

[1536] Roweis, S., and Ghahramani, Z. (1999) "Unifying review of linear Gaussian models," *Neural Computation*, **11**(2), 1–33.

[1537] Eddy, S. (1996) "Hidden Markov models," *Current Opinion in Structural Biology*, **6**, 361–365.

[1538] Bahl, L., Jelinek, F., and Mercer, R. (1983) "Maximum likelihood approach to continuous speech recognition," *IEEE Trans. Pattern Analysis & Machine Intelligence*, **5**(2), 179–190.

[1539] Rabiner, L., and Juang, B. (1986) "Introduction to hidden Markov models," *IEEE ASSP Magazine*, January, 4–16.

[1540] Fine, S., Singer, Y., and Tishby, N. (1998) "Hierarchical hidden Markov model: Analysis and applications," *Machine Learning*, **32**, 41–62.

[1541] Kaelbling, L., Littman, M., and Cassandra, A. (1988) "Planning and acting in partially observable stochastic domains," *Artificial Intelligence*, **101**, 99–134.

[1542] Wang, P. (2004) "Limitation of Bayesianism," *Artificial Intelligence*, **158**, 97–106.

[1543] Horvitz, E., Breese, J., and Henrion, M. (1988) "Decision theory in expert systems and artificial intelligence," *Int. J. Approximate Reasoning*, **2**, 247–302.

[1544] Grove, A., Halpern, J., and Koller, D. (1994) "Random worlds and maximum entropy," *J. Artificial Intelligence Research*, **2**, 33–88.

[1545] Shannon, C. (1948) "Mathematical theory of communications," *Bell Systems Technical J.*, **27**, 379–423, 623–656.

[1546] Kolmogorov, A. (1968) "Logical basis for information theory and probability theory," *IEEE Trans. Information Theory*, **14**(5), 662–664.

[1547] Jaynes, E. (1982) "On the rationale of maximum entropy methods," *Proc. IEEE*, **70**(9), 939–952.

[1548] Jaynes, E. (1968) "Prior probabilities," *IEEE Systems Science Cybernetics*, **4**(3), 227–244.

[1549] Hummel, R., and Landy, M. (1988) "Statistical viewpoint on the theory of evidence," *IEEE Trans. Pattern Analysis & Machine Intelligence*, **10**(2), 235–247.

[1550] Liau, C-J., and Lin, B. (1996) "Possibilistic reasoning: A mini-survey and uniform semantics," *Artificial Intelligence*, **88**, 163–193.

[1551] Chandra, K., Gu, D-W., and Postlethwaite, I. (2013) "Square root cubature information filter," *IEEE Sensors J.*, **13**(2), 750–758.

Index

Multibeam Doppler sonar 526–7
multimodal probability distributions
367–74
multiple cues, stereoscopic imaging systems
261–2
multiple microrover units 15, 283–4,
310–29, 406–18
multirover teams 15, 283–4, 310–29,
406–18, 486–95, 499–500, 546–61
concepts 406–18, 499–500
divisions of labor 414–18
path planning 406–18
multisensor fusion 185–97, 315–29,
348–57, 360–7, 383–418, 486–95,
541–3
see also sensors
MUROCO 309–29
muscles 541–2, 547–9
MUSES-C Hayabusa near-Earth asteroid
mission 9–10, 152–4, 172–4
mutually exclusive and exhaustive aspects,
hypotheses 362–7
Mylar 415–16, 455–6, 518–19, 554–6

name origins, rovers 2
Nanokhod tethered tracked rover 6, 8–9,
45–6, 89–90, 107, 127, 147–54,
443–54, 466–7
concepts 6, 8–9, 89–90, 107, 127, 443,
466–7
samples 443, 466–7
nanorovers 6–8, 9–10, 79–83, 89–90,
120–32, 466–7, 492–5, 546–61
see also rovers; SpaceCat...
definition 7–9
NASA Ames Mars general circulation
model 54
NASA/JPL Field Integrated Design and
Operations testbed 63–4, 67–9, 83,
87–9, 115, 121, 158, 166–7, 201–3,
208–9, 265, 307, 326–7, 335, 338, 339,
423, 426, 433, 468, 472, 515, 550
NASREM 325–9
NATO Reference Mobility Model
(NRMM) 39, 95–8
NavCams 10, 66–9, 205–62, 343–5,
472–512
navigation systems 3, 6, 8–10, 62–9,
71–132, 146–66, 167–71, 199–262,

266–300, 301–29, 331–74, 375–418,
472–512, 513–39, 546–61
see also locomotion; path planning; self-
localization...; sensors; steering...;
telerobotics; vehicle control...
algorithms 308–29, 337–74, 375–83,
494–5
architecture levels 325–9
architectures 301–29, 331–74
autonomous navigation 301–29, 331–74,
472–512
behavior-based autonomy 314–29,
332–7, 383–418, 499–500
behaviors 301–29, 332–7, 383–418,
499–500
CNES/LAAS autonomous navigation
system 340–5
concepts 146–66, 167–71, 199–213,
266–7, 269–300, 301–29, 331–74,
472–512, 526–39, 546–61
definition 331–2
distributed architecture for mobile
navigation 318–22, 327–9, 379–83,
417–18, 473–4
Europa case study 514–39
intelligent rover architectures 322–9,
331–3, 417–18, 471–2, 474–84
MERs 307–9, 313–14, 337–45
optic flow–based navigation 269–300
safety issues 326–8
the Sun 169–71
surface operations 313–14, 337–40
NdFeB magnets 137–54
near-Earth objects 449–50
see also comets
near-infrared spectrometers (NIR) 9–10,
42–3, 47–8, 65–9, 165–6, 171, 173–4,
252–3, 420–1, 464–5, 466–7, 475–6,
480–512
near-surface water, Mars 50–1, 52–7
nearest-neighbor tracking, multirover teams
408–18
NEMO Europa melt probe 517–18
neodynium magnets 135–54
Nepean wheeled vehicles 119–20
Nereus 520
nested arc approaches 340
net present values (NPVs) 26–7
NetLander 6